EMERGING MODEL ORGANISMS

A LABORATORY MANUAL

VOLUME 1

ALSO FROM COLD SPRING HARBOR LABORATORY PRESS

RELATED LABORATORY MANUALS

Basic Methods in Microscopy
Gene Transfer: A Laboratory Manual: Delivery and Expression of DNA and RNA
Imaging in Neuroscience and Development: A Laboratory Manual
Live Cell Imaging: A Laboratory Manual
Manipulating the Mouse Embryo: A Laboratory Manual
Molecular Cloning: A Laboratory Manual, Third Edition
The Condensed Protocols from Molecular Cloning: A Laboratory Manual

OTHER RELATED TITLES

At the Bench: A Laboratory Navigator, Updated Edition
Binding and Kinetics for Molecular Biologists
C. elegans *Atlas*
C. elegans *II*
Drosophila: *A Laboratory Handbook,* Second Edition
Drosophila *Protocols*
Early Development of Xenopus laevis: *A Laboratory Manual*
Evolution
Experimental Design for Biologists
Fly Pushing, The Theory and Practice of Drosophila *Genetics,* Second Edition
Lab Dynamics: Management Skills for Scientists
Lab Math: A Handbook of Measurements, Calculations, and Other Quantitative Skills for Use at the Bench
Lab Ref, Volume 1: A Handbook of Recipes, Reagents, and Other Reference Tools for Use at the Bench
Lab Ref, Volume 2: A Handbook of Recipes, Reagents, and Other Reference Tools for Use at the Bench
Methods in Yeast Genetics: A Cold Spring Harbor Laboratory Course Manual, 2005 Edition
The Nematode Caenorhabditis elegans

WEBSITE

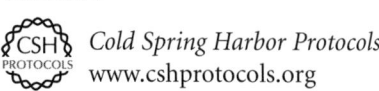 *Cold Spring Harbor Protocols*
www.cshprotocols.org

EMERGING MODEL ORGANISMS

A LABORATORY MANUAL

VOLUME 1

Also published in

Cold Spring Harbor Protocols
www.cshprotocols.org/emo

COLD SPRING HARBOR LABORATORY PRESS
Cold Spring Harbor, New York • www.cshlpress.com

EMERGING MODEL ORGANISMS
A Laboratory Manual
Volume 1

All rights reserved.
© 2009 by Cold Spring Harbor Laboratory Press, Cold Spring Harbor, New York
Printed in the United States of America

Publisher	John Inglis
Acquisition Editors	Alexander Gann and David Crotty
Director of Development, Marketing, & Sales	Jan Argentine
Managing Editors	Kaaren Janssen and Maria Smit
Developmental Editors	Heather Cerne, Nick Oswald, Irene Pech, Catriona Simpson, Martin Winer
Project Coordinator	Inez Sialiano
Production Editor	Rena Steuer
Desktop Editor	Susan Schaefer
Production Manager	Denise Weiss
Book Marketing Manager	Ingrid Benirschke
Sales Account Manager	Elizabeth Powers
Cover Designer	Michael Albano

Cover artwork: The front cover depicts images of a representative collection of the emerging model organisms described in this book. Shown clockwise from the top are the comb jelly (Ctenophora), the sea lamprey (*Petromyzon marinus*), an example of Darwin's finches (the Galápagos finches), the nematode (*Pristionchus pacificus*), the leech (*Helobdella robusta*), the garden snapdragon (*Antirrhinum majus*), surface (*above*) and cave (*below*) individuals of the Mexican Cave Fish (*Astyanax mexicanus*), and the American wandering spider (*Cupiennius salei*). The back cover displays the image of a African butterfly (*Bicyclus anynana*).

Material from this book was originally published in *Cold Spring Harbor Protocols*, 2008, Volume 3, Issues 10–12 (http://www.cshprotocols.org).

Library of Congress Cataloging-in-Publication Data

Emerging model organisms : a laboratory manual
      p. cm.
  Includes bibliographical references and index.
  ISBN 978-0-87969-826-3 (hardcover : alk. paper) -- ISBN 978-0-87969-872-0
(pbk. : alk. paper)
  1. Biological models--Laboratory manuals. I. Title.

QH324.8.E44 2009
570.1'1--dc22

           2008034154

10 9 8 7 6 5 4 3 2 1

Students and researchers using the procedures in this manual do so at their own risk. Cold Spring Harbor Laboratory makes no representations or warranties with respect to the material set forth in this manual and has no liability in connection with the use of these materials. All registered trademarks, trade names, and brand names mentioned in this book are the property of the respective owners. Readers should please consult individual manufacturers and other resources for current and specific product information.

All suppliers mentioned in this manual can be found on the BioSupplyNet Website at http://www.biosupplynet.com.

All World Wide Web addresses are accurate to the best of our knowledge at the time of printing.

Procedures for the humane treatment of animals must be observed at all times. Check with the local animal facility for guidelines.

Certain experimental procedures in this manual may be the subject of national or local legislation or agency restrictions. Users of this manual are responsible for obtaining the relevant permissions, certificates, or licenses in these cases. Neither the authors of this manual nor Cold Spring Harbor Laboratory assume any responsibility for failure of a user to do so.

The materials and methods in this manual may infringe the patent and proprietary rights of other individuals, companies, or organizations. Users of this manual are responsible for obtaining any licenses necessary to use such materials and to practice such methods. COLD SPRING HARBOR LABORATORY MAKES NO WARRANTEE OR REPRESENTATION THAT USE OF THE INFORMATION IN THIS MANUAL WILL NOT INFRINGE ANY PATENT OR OTHER PROPRIETARY RIGHT.

Authorization to photocopy items for internal or personal use, or the internal or personal use of specific clients, is granted by Cold Spring Harbor Laboratory Press, provided that the appropriate fee is paid directly to the Copyright Clearance Center (CCC). Write or call CCC at 222 Rosewood Drive, Danvers, MA 01923 (978-750-8400) for information about fees and regulations. Prior to photocopying items for educational classroom use, contact CCC at the above address. Additional information on CCC can be obtained at CCC Online at http://www.copyright.com/.

All Cold Spring Harbor Laboratory Press publications may be ordered directly from Cold Spring Harbor Laboratory Press, 500 Sunnyside Blvd., Woodbury, New York 11797-2924. Phone: 1-800-843-4388 in the Continental U.S. and Canada. All other locations: (516) 422-4100. FAX: (516) 422-4097. E-mail: cshpress@cshl.edu. For a complete catalog of all Cold Spring Harbor Laboratory Press publications, visit our World Wide Web site http://www.cshlpress.com/.

Contents

A Note from the Publisher, ix

1 The Choanoflagellates, 1
Heterotrophic Nanoflagellates and Sister Group of the Metazoa
N. King, S.L. Young, M. Abedin, M. Carr, and B.S.C. Leadbeater

2 *Dictyostelium discoideum*, 29
The Social Ameba
P. Gaudet, P. Fey, and R. Chisholm

3 The Moss *Physcomitrella patens*, 69
A Novel Model System for Plant Development and Genomic Studies
D.J. Cove, P.-F. Perroud, A.J. Charron, S.F. McDaniel, A. Khandelwal, and R.S. Quatrano

4 The Genus *Antirrhinum* (Snapdragon), 105
A Flowering Plant Model for Evolution and Development
A. Hudson, J. Critchley, and Y. Erasmus

5 Tomato (*Solanum lycopersicum*), 119
A Model Fruit-bearing Crop
S. Kimura and N. Sinha

6 The Demosponge *Amphimedon queenslandica*, 139
Reconstructing the Ancestral Metazoan Genome and Deciphering the Origin of Animal Multicellularity
B.M. Degnan, M. Adamska, A. Craigie, S.M. Degnan, B. Fahey, M. Gauthier, J.N.A. Hooper, C. Larroux, S.P. Leys, E. Lovas, and G.S. Richards

7 Comb Jellies (Ctenophora), 167
A Model for Basal Metazoan Evolution and Development
K. Pang and M.Q. Martindale

8 Planarians, 195
A Versatile and Powerful Model System for Molecular Studies of Regeneration, Adult Stem Cell Regulation, Aging, and Behavior
N.J. Oviedo, C.L. Nicolas, D.S. Adams, and M. Levin

9 **The Snail *Ilyanassa*, 219**
   A Reemerging Model for Studies in Development
   M. Gharbiah, J. Cooley, E.M. Leise, A. Nakamoto, J.S. Rabinowitz, J.D. Lambert, and L.M. Nagy

10 ***Helobdella* (Leech), 245**
   A Model for Developmental Studies
   D.A. Weisblat and D.-H. Kuo

11 ***Pristionchus pacificus*, 275**
   A Genetic Model System for the Study of Evolutionary Developmental Biology and the Evolution of Complex Life-history Traits
   R. Rae, B. Schlager, and R.J. Sommer

12 **The African Butterfly *Bicyclus anynana*, 291**
   A Model for Evolutionary Genetics and Evolutionary Developmental Biology
   P.M. Brakefield, P. Beldade, and B.J. Zwaan

13 **The Two-spotted Cricket *Gryllus bimaculatus*, 331**
   An Emerging Model for Developmental and Regeneration Studies
   T. Mito and S. Noji

14 **The American Wandering Spider *Cupiennius salei*, 347**
   A Model for Behavioral, Evolutionary, and Developmental Studies
   N.-M. Prpic, M. Schoppmeier, and W.G.M. Damen

15 **The Crustacean *Parhyale hawaiensis*, 373**
   A New Model for Arthropod Development
   E.J. Rehm, R.L. Hannibal, R. Crystal Chaw, M.A. Vargas-Vila, and N.H. Patel

16 **The Sea Lamprey *Petromyzon marinus*, 405**
   A Model for Evolutionary and Developmental Biology
   N. Nikitina, M. Bronner-Fraser, and T. Sauka-Spengler

17 **The Dogfish *Scyliorhinus canicula*, 431**
   A Reference in Jawed Vertebrates
   M. Coolen, A. Menuet, D. Chassoux, C. Compagnucci, S. Henry, L. Lévèque, C. Da Silva, F. Gavory, S. Samain, P. Wincker, C. Thermes, Y. D'Aubenton-Carafa, I. Rodriguez-Moldes, G. Naylor, M. Depew, P. Sourdaine, and S. Mazan

18 **The Genus *Polypterus* (Bichirs), 447**
   A Fish Group Diverged at the Stem of Ray-finned Fishes (Actinopterygii)
   M. Takeuchi, M. Okabe, and S. Aizawa

19 ***Astyanax mexicanus*, The Blind Mexican Cave Fish, 469**
   A Model for Studies in Development and Morphology
   R. Borowsky

20 **Darwin's Finches, 481**
   Analysis of Beak Morphological Changes During Evolution
   A. Abzhanov

21   Japanese Quail, 501
     An Efficient Animal Model for the Production of Transgenic Avians
     G. Poynter, D. Huss, and R. Lansford

22   The Short-tailed Fruit Bat *Carollia perspicillata*, 519
     A Model for Studies in Reproduction and Development
     J.J. Rasweiler IV, C.J. Cretekos, and R.R. Behringer

23   Opossum *(Monodelphis domestica)*, 557
     A Marsupial Developmental Model
     A.L. Keyte and K.K. Smith

General Cautions Appendix, 577

Index, 587

A Note from the Publisher

Much of 20th century biological research has focused on a limited number of model organisms, such as A*rabidopsis, C. elegans*, mouse, *Drosophila*, and *E. coli*. These classical model species, chosen because they are amenable to laboratory research and suitable for studying a range of biological problems, have served to elucidate many biological processes that can be generalized across a wider array of organisms. It is only a slight exaggeration to say that the basic workings of the cell were elucidated mostly from experiments on a few single-celled organisms—primarily *E. coli* and yeast. Our understanding of animal development was largely based on the genetics of fruit fly and worm and on the manipulation of a handful of amphibians and mouse; most of what we learned about the molecular and developmental biology of plants came from examining *Arabidopsis* and just a few other species. But biology was not always done this way.

As an example, early efforts to understand gastrulation were informed by careful analysis of a wide variety of organisms (Brauckmann and Gilbert 2004). Ernst Haeckel's Gastrea theory (Haeckel 1874), as he openly acknowledged, relied heavily on the exhaustive work of Alexander Onufreevich Kowalevsky, who described early embryonic development in *Amphioxus, Phallusia, Ascidia, Phoronix, Echinus, Ophiura, Limnaeus*, the frog, comb jellies, sturgeons, and *Petromyzon*. Modern biologists, in contrast, have only rarely strayed from the relatively small handful of established model organisms. Studies using these models have resulted in a great depth of knowledge, because the concentration of many on a limited number of systems has proven to be a valuable approach. But things are changing again; with the evolution of new technologies, new questions and approaches are now possible.

The time it takes to sequence genomes, as well as the cost, continues to drop. Techniques for selectively altering the expression patterns of genes have become more generally applicable across species. And more and more biologists are expanding their interests from the purely mechanistic to embrace evolutionary considerations. Because of these factors, we are now seeing a great expansion in the variety of organisms regularly studied. Researchers are now introducing new species to the laboratory, opening new avenues of research, and allowing comparison and refinement of our understanding of already-established models.

This is the first volume of *Emerging Model Organisms*, a new laboratory manual series. The goal of the series is to introduce researchers to this new generation of model organisms and to provide a diverse catalog of potential species useful for extending research in new directions. Each chapter presents a new organism (or group of related organisms) and supplies detailed explanations of why they are useful for laboratory research, along with information on husbandry, genetics and genomics, pointers toward further resources, and a set of basic laboratory protocols for working with that organism. The material in this book is also available online and in full color, from *Cold Spring Harbor Protocols* (http://www.cshprotocols.org/emo). We hope that *Emerging Model Organisms* will serve as a practical guide for finding just the right species for addressing your research needs and as a primer for introducing new systems to your laboratory.

This series would not have been possible without the efforts of our editorial advisors, Richard Behringer, Sandy Johnson, Robb Krumlauf, Mike Levine, Nipam Patel, and Neelima Sinha. They

1 The Choanoflagellates
Heterotrophic Nanoflagellates and Sister Group of the Metazoa

Nicole King,[1-3] Susan L. Young,[1] Monika Abedin,[1] Martin Carr,[4] and Barry S.C. Leadbeater[5]

[1]Department of Molecular and Cell Biology, University of California, Berkeley 94704; [2]Department of Integrative Biology, University of California, Berkeley 94704; [3]Canadian Institute for Advanced Research, Toronto, Ontario M5G 1Z8 Canada; [4]Department of Biology, University of York, York YO10 5YW, United Kingdom; [5]School of Biosciences, University of Birmingham, Birmingham B15 2TT, United Kingdom

ABSTRACT

Choanoflagellates are the closest living relatives of the metazoa, and the study of their cell biology and genomes promises to provide new insights into metazoan ancestry and origins. These heterotrophic flagellates are a cosmopolitan group of small, colorless protozoa that are present in marine and freshwater environments as well as in hydrated soils. *Monosiga brevicollis*, a marine species, has emerged as a representative of the group because it is easily grown and manipulated in the laboratory and was the subject of a recently completed genome project. Because of the similarity in morphology and ecology shown across the order Choanoflagellida, the protocols presented here should be transferable to most choanoflagellate species.

PROTOCOLS

1. Isolation of Single Choanoflagellate Cells from Field Samples and Establishment of Clonal Cultures, 5
2. Starting and Maintaining *M. brevicollis* Cultures, 8
3. Long-term Frozen Storage of Choanoflagellate Cultures, 10
4. Visualizing the Subcellular Localization of Actin, β-tubulin, and DNA in *M. brevicollis*, 11
5. Preparation of Total RNA from *M. brevicollis* and Other Choanoflagellates, 14
6. Rapid Preparation of Genomic DNA from *M. brevicollis* and Other Choanoflagellates, 16
7. Preparation of High-molecular-weight Genomic DNA from *M. brevicollis* and Other Choanoflagellates, 18
8. Separation of Choanoflagellate and Bacterial Genomic DNA, 21

BACKGROUND INFORMATION

Choanoflagellates consist of a spherical to ovoid cell with a single anterior flagellum surrounded by a funnel-shaped collar composed of 30 or more actin-based tentacles or microvilli (Fig. 1). Choanoflagellate species were first described by Henry James-Clark (1866, 1868) and subsequently recognized by William Saville-Kent (1880–1882) as a formal group. A close evolutionary link with the metazoa was swiftly proposed due to the morphological similarities between choanoflagellate cells and the choanocytes (feeding cells) of sponges. Molecular phylogenies have subsequently

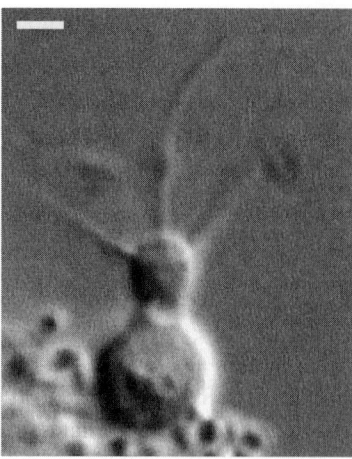

FIGURE 1. *Choanoeca perplexa* cell with a single anterior flagellum and a widely angled collar. Bar, 2 μm.

confirmed the monophyly of the choanoflagellates and their sister-grouping to metazoa (Fig. 2) (Steenkamp et al. 2006; Ruiz-Trillo et al. 2008).

Currently, choanoflagellates are grouped into three families according to the structure and composition of their coverings. All cells are surrounded by a fine fibrillar glycocalyx. Members of the Codosigidae lack any other form of covering. In contrast, members of the Salpingoecidae possess a continuous organic covering that prevents lateral division, so cells become amoeboid and emerge from the theca to divide. Members of the Acanthoecidae possess a siliceous basket-like lorica. The first two families include marine and freshwater representatives; the third family is exclusively marine.

Here, we focus on the choanoflagellate *Monosiga brevicollis*, which was the focus of a recently completed genome project (King et al. 2008). *M. brevicollis* is a member of Codosigidae and has emerged as a representative of the group because it is easily grown in the laboratory, can be stored under liquid nitrogen, and is readily available from the American Type Culture Collection (ATCC).

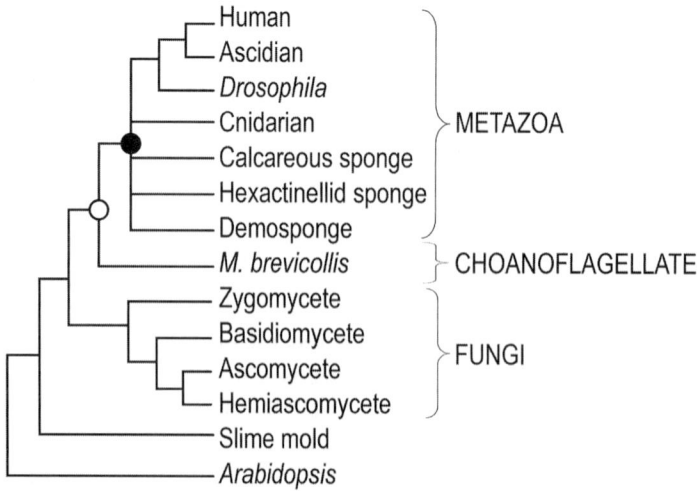

FIGURE 2. Choanoflagellates represent the closest known relatives of metazoans. Their study offers insights into the biology of the last common ancestor of metazoans (*filled circle*) and the biology of the last common ancestor of choanoflagellates and metazoans (*open circle*). (Modified from King et al. 2008.)

SOURCES AND HUSBANDRY

M. brevicollis, as well as many other choanoflagellate species, is available from the ATCC (Table 1). This species has been deposited under ATCC number 50154. *M. brevicollis* is a widespread marine choanoflagellate, so if new or multiple populations of *M. brevicollis* are required, they may be easily isolated from seawater (see Protocol 1). Protocol 2 describes how to maintain *M. brevicollis* in the laboratory. Both protocols can be applied to a wide variety of choanoflagellate species. Most choanoflagellate species, including *M. brevicollis*, are sedentary organisms and attach themselves to the walls of media flasks. This will occur within a few days of establishing a fresh culture. As a result, when harvesting cells, the sides of flasks must carefully be scraped in order to maximize yield.

Most, if not all, choanoflagellates are heterotrophs that prey on bacteria. Therefore, the basis of all choanoflagellate media is to provide nutrition for bacteria, which are in turn consumed by the choanoflagellates. A variety of media can be used to grow choanoflagellates; media for both marine and freshwater species are provided beginning on page 25. If a culture is axenic or has low bacterial content, it can be supplemented with a specific bacterial strain that has been grown separately. On the other hand, if a culture already contains a well-established flora of bacteria, it can be co-cultured with the choanoflagellate isolate. In either case, it is common to enrich the medium with an organic extract, such as liver extract, cereal grass infusion, or a mixture of proteose peptone and yeast extract. Refer to the American Type Culture Collection (http://www.atcc.org) and Culture Collection of Algae and Protozoa (http://www.ccap.ac.uk) websites for further information.

RELATED SPECIES

Much of the current molecular biology involving choanoflagellates has centered on *M. brevicollis*. Although this species has rapidly assumed the status of a model organism, a number of other choanoflagellate species are available from culture collections. In addition to *M. brevicollis*, the ATCC has 17 species of choanoflagellates in its collection (Table 1). The media and culture conditions required for each of these species are provided by the ATCC.

TABLE 1. Choanoflagellate species available from the ATCC (www.atcc.org)

Species	ATCC number
Acanthoeca spectabilis	PRA-103
Choanoeca perplexa[1]	50453
Codosiga gracilis	50454
Diaphanoeca grandis	50111
Helgoeca nana[2]	50073
Monosiga brevicollis[3]	50154
Monosiga gracilis	50964
Monosiga ovata	50635
Proterospongia sp.	50818
Salpingoeca amphoridium	50788
Stephanoeca diplocostata	50456
Salpingoeca gracilis	50959
Salpingoeca infusionum	50559
Salpingoeca minuta	50938
Salpingoeca napiformis	50153
Salpingoeca pyxidium	50929
Salpingoeca sp.	50931
Salpingoeca urceolata	50566

[1] Also deposited as *Proterospongia choanojuncta* (ATTC 50455).
[2] Deposited under the name *Acanthoecopsis unguiculata*.
[3] The clonal culture *Monosiga brevicollis* MX1 is also available (ATCC PRA-258).

USES OF THE CHOANOFLAGELLATE MODEL SYSTEM

Choanoflagellates are biologically important and have been studied extensively for three reasons:

1. **Cell biology.** The unique siliceous lorica is a "model" structure for understanding the production and assembly of biomineralized exoskeletons.

2. **Ecology.** Choanoflagellates are a ubiquitous and important group of aquatic free-living heterotrophic nanoflagellates.

3. **Molecular phylogeny and genomics.** Choanoflagellates are a sister group of the metazoa and are therefore of great importance in understanding the origins of metazoan multicellularity and the animal lineage.

The last decade has seen considerable progress in understanding the molecular biology of choanoflagellates. This work has concentrated mainly on determining the exact taxonomic position of the choanoflagellates (Wainright et al. 1993; Snell et al. 2001; Steenkamp et al. 2006) as well as on studying genetic pathways shared between choanoflagellates and metazoans (King and Carroll 2001; Abedin and King 2008; King et al. 2008; Ruiz-Trillo et al. 2008).

Loricate choanoflagellates have received the most attention in ultrastructural and physiological studies because of their distinctive, easily recognizable loricae. However, several nonloricate species, such as *M. brevicollis*, have become the focus of intensive molecular study.

GENOMICS RESOURCES

In recent molecular biology studies, considerable amounts of DNA sequence data have accumulated for the choanoflagellates. The entire genome of *M. brevicollis* (ATCC 50154) has been sequenced by the Joint Genome Institute (JGI) with eightfold coverage (King et al. 2008). Version 1.0 of the genome was released in July 2006 and is publicly available from the JGI website (http://genome.jgi-psf.org/Monbr1/Monbr1.home.html). The sequenced genome is approximately 41 Mb (megabases) in length and contains more than 9000 predicted gene models. Two other choanoflagellates, the colony-forming *Proterospongia* sp. (ATCC 50818) and the single-celled choanoflagellate *M. ovata*, are also due to be sequenced within the next few years by the Broad Institute (MIT/Harvard) and the RIKEN Genomic Sciences Center (Japan), respectively.

In addition to the genome sequences, there are in excess of 43,000 *M. brevicollis* nucleotide sequences, including 29,000 expressed sequence tags (ESTs), in the GenBank database. EST libraries have been publicly released for two additional species, *M. ovata* (strain 2A4) and *Proterospongia* sp. (ATCC 50818). There are approximately 76,500 EST sequences for the former in GenBank and more than 1000 sequences for the latter in the Taxonomically Broad EST Database (TBestDB; http://amoebidia.bcm.umontreal.ca/pepdb/searches/welcome.php). TBestDB also contains thousands of EST sequences from both *M. brevicollis* and *M. ovata*.

Currently, nucleotide sequences from a total of 19 named choanoflagellate species (as well as a number of unnamed environmental samples) have been deposited into GenBank, and this number is likely to increase rapidly in the near future. Deposited sequences are currently a mixture of ribosomal RNA and protein-coding genes. Ribosomal RNA sequences (mainly small-subunit ribosomal RNA sequences) have proven to be reliable for the phylogenetic reconstruction of the choanoflagellates (Leadbeater et al. 2008).

TECHNICAL APPROACHES

The following eight protocols present fundamental techniques for manipulating choanoflagellates in the laboratory and they can be applied to a wide range of studies.

Protocol 1

Isolation of Single Choanoflagellate Cells from Field Samples and Establishment of Clonal Cultures

This protocol describes how to enrich freshly collected field samples and isolate single choanoflagellate cells. Two methods are commonly used: isolation by micropipetting single cells and isolation by dilution. The two methods are complementary and each has its advantages and disadvantages. The micropipette technique requires considerable experience. However, provided a single cell has been isolated, any culture arising from it will be clonal. The dilution technique does not require the same level of skill but single-cell isolation is not ensured. For the latter technique, repeating the isolation procedure using higher dilutions is advised to ensure that only one cell is the basis of an established culture.

MATERIALS

The recipes for items marked with <R> begin on page 25.

Reagents

Culture medium, sterile:
  Pratt's freshwater medium <R>, with organic enrichment <R> (for freshwater species only)
  Seawater <R>, with organic enrichment <R> (for marine species only)
Natural field medium, sterile (for Step 15) <R>

Equipment

Bunsen burner with small flame (for isolation by micropipette only; see Steps 6–13)
Capillary tubes or hematocrit tubes (approximately 1 mm diameter × 100 mm long) (for isolation by micropipette only; see Steps 6–13)
Conical flasks (250 ml) with plugs (sterile)
Culture plates with flat-bottomed wells (96-well, 0.4 ml/well)
Distilled H_2O, hot (for isolation by micropipette only; see Step 6–12)
Field collecting bottles, sterile or ethanol-cleaned
  Use Van Dorn bottles to collect samples at a specific depth.
Hemocytometer (for isolation by dilution only; see Steps 14–20)
Incubator, set at 15°C or other temperature as appropriate (see Step 4)
Inverted microscope with magnification of 200×–400×
  Phase-contrast optics are an advantage.
Portable box, insulated with cold ice packs
Silicone tubing (for isolation by micropipette only; see Steps 6–13)
  The tubing should be approximately 300–400 mm long and have a diameter sufficient to fit snugly over the end of the capillary tube.
Slides, sterile (for isolation by micropipette only; see Steps 6–13)
Whatman No. 1 filter paper, cut into 20-mm-long triangular slivers and sterilized

METHOD

Water Bottle Collection and Enrichment of Field Samples

1. Immerse a field collecting bottle (or a Van Dorn bottle for samples taken at a specified depth) into the water to be sampled and, when positioned correctly, close the entrance aperture.
2. Transfer the water sample from the collecting bottle to a sterile bottle, place the bottle containing the sample into a cool box, and return it to the laboratory.
3. Place 100-ml aliquots of field water into 250-ml sterile flasks, each containing approximately ten slivers of sterile filter paper.
   The filter paper enhances the development of a microbial biofilm and slowly releases minute amounts of organic enrichment into the culture.
4. Incubate the flasks for 2–3 weeks in the dark at 15°C.
   This temperature is optimal for most field samples; however, other temperatures may be selected for cold- or warm-water samples. Static conditions without illumination encourage the growth of heterotrophic nanoflagellates. Initially, rapidly swimming bodonids and chrysophytes are apparent, but later, after about 10 days, choanoflagellates predominate. See Troubleshooting.
5. When choanoflagellates are present in abundance, use the micropipette (Steps 6–13) or dilution (Steps 14–20) method described below to isolate single cells.

Isolation of Single Cells and Establishment of Clonal Cultures

The Micropipette Method

6. With a fine flame from a Bunsen burner, heat and draw out the capillary tube (holding it at both ends) to form two micropipettes.
   The narrow end of each micropipette should be about twice the diameter of the cell to be isolated.
7. Attach the wider end of the micropipette to silicone tubing and insert the other end of the tubing into your mouth.
8. Dilute a small volume of enriched sample (from Step 5) with sterile culture medium so that the cells are moderately dispersed. Transfer a drop of the mixture to a sterilized slide and view with an inverted light microscope.
9. With your mouth, suck up and blow out a small amount of the hot distilled H_2O.
   This sterilizes the micropipette.
10. Locate a choanoflagellate cell. While observing the cell, gently suck it up into the micropipette.
11. Transfer the cell to 0.3 ml of culture medium in one well of a multiwell plate.
12. Repeat Steps 9–11 for each well of the multiwell plate.
13. Place the multiwell plate in a darkened incubator at 15°C. On a regular basis, use an inverted microscope to regularly observe the wells.
    If appropriate, use another temperature (see Step 4). Choanoflagellate cultures will grow after several days. Cells will be sedentary except for the occasional motile cell in nonloricate and nudiform loricate species.

The Dilution Method

14. Place a drop of enriched field culture (from Step 5) on a hemocytometer. Count the cells and determine their concentration (per microliter).

15. Add sterile natural field medium to adjust the dilution to yield approximately one cell per 5 µl.
16. Pipette 0.2 ml of sterile culture medium into each well of a 96-well plate.
17. Pipette 1.0 µl of diluted culture (from Step 15) into each well.
18. Place a lid on the plate and place the plate in an incubator at 15°C. Leave the plate undisturbed for 4–5 days.

    If appropriate, use another temperature (see Step 4).

19. After 5 days, examine the plate at daily intervals with an inverted microscope.

    Wells containing choanoflagellate cells alone can usually be distinguished by the shape of the cell and its lack of swimming activity. See Troubleshooting.

20. If necessary, repeat Steps 14–19 until single-species cultures have been produced.

TROUBLESHOOTING

Problem (Step 4): The culture contains an overwhelming number of motile flagellates.
Solution: Dilute the culture with sterile, membrane-filtered natural field medium or culture medium.

Problem (Step 19): The culture is contaminated with many species, particularly flagellates.
Solution: Dilute the inoculum; it is important that the calculated concentration be approximately 5× or more dilute than necessary to give, on average, one cell per well.

Protocol 2

Starting and Maintaining *M. brevicollis* Cultures

This protocol describes how to start *M. brevicollis* cultures from frozen stock and, starting with a culture of *M. brevicollis* at maximum density (10^6–10^7 cells/ml), how to maintain and expand cultures in preparation for DNA or RNA isolation or cell biological assays. *M. brevicollis* is cocultured with bacteria as a food source.

MATERIALS

The recipes for items marked with <R> begin on page 25.

Reagents

Cereal grass medium <R>
M. brevicollis stock, frozen
> Stocks are available from the American Type Culture Collection (ATCC 50154; http://www.atcc.org).

Rice grains, sterile
> Rice can be sterilized by autoclaving in a capped test tube.

Equipment

Cell lifter
Incubator, set at 25°C (optional; see Steps 4 and 9)
Inverted microscope with magnification of 200x–400x
Pipette (15 ml), sterile glass or disposable
Pipette tip (1 ml), with filter
Pipetter (1 ml)
Tissue culture dish (100 x 20 mm), polystyrene
Tissue culture hood, sterile

METHOD

To avoid contamination, perform all steps in a sterile tissue culture hood if possible.

Starting Cultures from Frozen Stock

1. Place one grain of sterile rice into a standard tissue culture dish. Using a sterile 15-ml pipette under a tissue culture hood, add 14 ml of cereal grass medium.

2. Thaw a vial of frozen *M. brevicollis* cells in hand or at room temperature.

3. When thawed, immediately pipette 1 ml of cells into the dish from Step 1.

4. Incubate the culture in a 25°C incubator or at room temperature.
   > *The culture may initially look overgrown with bacteria, but within 2–3 days, the choanoflagellates should out-compete their bacterial food.*

Subculturing and Maintaining Cultures

5. Start with a culture of *M. brevicollis* at maximum density (10^6–10^7 cells/ml), typically seeded from a frozen stock (as in Steps 1–4).
6. Using a cell lifter, scrape the bottom of the dish to resuspend any adherent cells. Then, use a 1-ml pipette with a filtered pipette tip to gently pipette up and down to disrupt clumps of cells.
7. Place one grain of sterile rice into a standard tissue culture dish and add 14 ml of cereal grass medium.
8. Add 1 ml of cells from Step 6 to the dish from Step 7.
9. Incubate the culture in a 25°C incubator or at room temperature.
10. Monitor growth and viability of cells on an inverted microscope at 200×–400× magnification. Every 2–3 days, split the culture (as in Steps 6–8) to fresh cereal grass medium.
    See Troubleshooting.

TROUBLESHOOTING

Problem (Step 10): The culture contains an overgrowth of cocultured bacteria.

Solution: Bacterial growth is a fact of life in choanoflagellate cultures. If bacteria begin to dominate the culture and interfere with choanoflagellate proliferation, start over with a freshly thawed stock of cells. *M. brevicollis* has an approximately 6-hour doubling time during log phase and should reach a maximal density of 10^6–10^7 cells/ml.

Protocol 3

Long-term Frozen Storage of Choanoflagellate Cultures

Choanoflagellates may be stored under stable conditions by freezing them with liquid nitrogen. To prevent cross-contamination among vials stored in the liquid phase of a liquid nitrogen freezer, we encourage the use of CryoFlex™ tubing to fully seal the vials.

MATERIALS

CAUTION: See the Cautions Appendix for appropriate handling of materials marked with <!>.

Reagents

Culture of *M. brevicollis* (10^6–10^7 cells/ml; see Protocol 2) or other choanoflagellate species grown to maximal density
DMSO <!>
Isopropyl alcohol <!>

Equipment

Bunsen burner
CryoFlex™ tubing (Nunc 343958)
CryoTube™ vials (1 ml, internally threaded, with screw cap; Nunc 366656)
Dewar with liquid nitrogen <!>
Freezing container, rate-controlled (e.g., "Mr. Frosty" Freezing Container; Nalgene 5100-0001)
Mechanical freezer (–80°C)
Pliers, needle-nose
Scissors

METHOD

1. Precut the CryoFlex™ tubing into 6.5-cm lengths with a scissors.

2. Add 900 µl of *M. brevicollis* cells and 100 µl of DMSO to a CryoTube™ vial. Repeatedly invert the tube to mix the cells and the DMSO.

3. Slide the CryoTube™ vial into one precut CryoFlex™ tube (from Step 1). Quickly pass the covered CryoTube™ vial through the flame of a Bunsen burner so that the tube shrinks around the vial. Crimp the ends with a pair of needle-nose pliers.

4. Freeze the covered CryoTube™ vial overnight at –80°C in a rate-controlled freezer filled with isopropyl alcohol.

5. The next day, transfer the frozen vial to a liquid nitrogen dewar for long-term storage.

    Choanoflagellates cannot be stably stored at –80°C and may not recover from freezing if stored at this temperature. Make sure to transfer frozen cells to liquid nitrogen after 24 hours. To resuscitate frozen cells, see Protocol 2.

Protocol 4

Visualizing the Subcellular Localization of Actin, β-tubulin, and DNA in *M. brevicollis*

The microvillar collar and apical flagellum of *M. brevicollis* can be visualized by staining for actin and β-tubulin (Abedin and King 2008). A simple DNA stain highlights the nucleus. In this protocol, cells are fixed in dilute formaldehyde, attached to coverslips, and then exposed to antibodies and stains for β-tubulin, actin, and DNA. The central challenge of this protocol is to treat the cells gently so that they retain their ultrastructure and remain attached to the coverslips during processing.

MATERIALS

The recipes for items marked with <R> begin on page 25.

CAUTION: See the Cautions Appendix for appropriate handling of materials marked with <!>.

Reagents

Antibody, primary (E7 anti-β-tubulin; Developmental Studies Hybridoma Bank, product form: supernatant)
Antibody, secondary (Alexa Fluor® 488 goat antimouse IgG [H+L]; Invitrogen)
Blocking solution (2.5 ml/coverslip, freshly prepared) <R>
Culture of *M. brevicollis* (10^6–10^7 cells/ml; see Protocol 2)
Formaldehyde (37%) <!>
H_2O (Milli-Q)
PEM (2.5 ml/coverslip) <R>
Phalloidin solution <R> <!>
Poly-L-lysine (0.1%, w/v, prepared in H_2O)
ProLong® Gold antifade reagent with DAPI <!> (Invitrogen)
  DAPI is a simple DNA stain. Bring the reagent to room temperature before performing Step 32).

Equipment

Coverslips (18 x 18 mm)
Fluorescence microscope
Microcentrifuge
Microcentrifuge tubes (1.5 ml)
Nail polish, clear
Parafilm
Pasteur pipette (optional)
Petri dishes (150 x 25 mm)
Slides (3 inch x 1 inch x 1 mm), glass
Vacuum aspirator
Whatman filter paper

METHOD

Preparation of Poly-L-Lysine–Coated Coverslips

1. Prepare a hydration chamber as follows:
    i. Cut a piece of Whatman filter paper to cover the bottom of a 150 × 25 mm Petri dish.
    ii. Wet the filter paper, shake off excess H_2O, and place it inside the Petri dish.
2. Cut a square of Parafilm such that the four corners of the square reach the edges of the dish.
3. Place the square of Parafilm on top of the filter paper. Make sure that the Parafilm lies flat.
4. Place clean, dry coverslips on the Parafilm approximately 4 cm apart.
    No more than four coverslips should be placed in a single dish. Do not place the coverslips too close to the edge of the dish or they will be difficult to manipulate.
5. Apply 200 µl of poly-L-lysine solution to the center of each coverslip.
6. Incubate the coverslips for 10 minutes at room temperature with the lid on the Petri dish.
7. Remove and save the poly-L-lysine.
    Poly-L-lysine can be stored at room temperature until turbidity or bacterial growth develops.
8. Wash the coverslips by applying 200 µl of Milli-Q H_2O to each coverslip and immediately aspirating it.
9. Repeat Step 8 nine times more.
10. Allow the coverslips to air-dry for at least 10 minutes. Make sure to leave the lid of the Petri dish off for this step.
    Coverslips can be stored in a sealed container, protected from dust, for approximately one year at room temperature.

Cell Fixation

Unless otherwise specified, all incubations from this point forward should be performed with the Petri dish lid on.

11. For each sample of *M. brevicollis* cells to be stained, transfer 1 ml of culture to each of two 1.5-ml microcentrifuge tubes.
12. Add 125 µl of 37% formaldehyde to each 1 ml of culture.
13. Incubate the cells with the formaldehyde for 15 minutes at room temperature.
    The formaldehyde acts as a fixative.
14. Gently centrifuge the cells in a microcentrifuge at 500g for 5 minutes at room temperature.
15. Using a micropipette, gently remove and discard the supernatant.
16. Add 100 µl of PEM to each tube and very gently pipette up and down three to five times to resuspend the cells.
17. Pool the two samples.
    One tube should now contain 200 µl of resuspended cells.
18. Apply all 200 µl of cells to a poly-L-lysine–coated coverslip (from Step 10) and incubate on the benchtop for at least 30 minutes to allow the cells to attach to the coverslip.
19. Using a micropipette or a Pasteur pipette, gently aspirate the remaining fluid.

20. Wash the cells with PEM as follows:

    i. Apply 100 µl of PEM to the cells.

    ii. Gently aspirate the PEM with a micropipette or a Pasteur pipette.

    iii. Repeat Steps 20.i and 20.ii for a total of four washes and then proceed immediately to Step 21. Do not allow the cells to dry out.

Staining

21. Apply 100 µl of blocking solution to the cells and incubate for at least 30 minutes at room temperature.

22. Dilute 1 µl of E7 β-tubulin primary antibody in 99 µl of blocking solution.

23. Gently aspirate the blocking solution and apply 100 µl of diluted E7 β-tubulin primary antibody. Incubate on the benchtop for 1 hour at room temperature.

24. Aspirate the remaining fluid and then wash the cells with blocking solution as follows:

    i. Apply 100 µl of blocking solution to the cells.

    ii. Gently aspirate the blocking solution with a micropipette or a Pasteur pipette.

    iii. Repeat Steps 24.i and 24.ii for a total of four washes. Do not allow the cells to dry out.

25. Dilute 1 µl of secondary antibody in 999 µl of blocking solution.

26. After gently aspirating the blocking solution from the final wash in Step 24, apply 100 µl of diluted secondary antibody to the cells. Incubate for 1 hour at room temperature in a dark drawer.

    The absence of light preserves the fluorescence of the secondary antibody.

27. Gently aspirate the secondary antibody and wash the cells four times with blocking solution as in Step 24.

28. After gently aspirating off the blocking solution from the final wash in Step 27, apply 100 µl of phalloidin solution to the cells.

29. Incubate the cells for 15 minutes in a dark drawer.

30. Gently aspirate the phalloidin solution and wash twice with PEM as described in Step 20.

31. Gently aspirate the PEM. With the Petri dish lid off in a dark drawer, allow the coverslip to dry almost completely (about 5 or 10 minutes).

32. Place 10 µl of room-temperature ProLong® Gold in a drop on a clean, dry slide.

33. Place the cell-coated coverslip from Step 31 face down on the slide from Step 32. Make sure to apply the coverslip slowly and gently.

34. Allow the ProLong® Gold to equilibrate (about 30 minutes at room temperature or overnight at 4°C) and seal the edges of the coverslip with clear nail polish.

35. Visualize staining on a microscope equipped for fluorescence microscopy.

    See Troubleshooting.

TROUBLESHOOTING

Problem (Step 35): Poor collar integrity (i.e., no collars, short collars, or tangled collars).
Solution: Decrease vacuum strength during aspiration.

Protocol 5

Preparation of Total RNA from *M. brevicollis* and Other Choanoflagellates

The preparation of RNA is an important prerequisite for characterizing the gene expression profile of choanoflagellates under different conditions. Although most standard protocols (e.g., those using TRIzol) are sufficient for many purposes, the approach described here minimizes RNA degradation and provides a high-quality template for downstream protocols, including real-time polymerase chain reaction (PCR).

MATERIALS

CAUTION: See the Cautions Appendix for appropriate handling of materials marked with <!>.

Reagents

Culture of *M. brevicollis* (10^6–10^7 cells/ml; see Protocol 2)
> *Harvest 10^7–10^8 cells per RNeasy midi column.*

Ethanol (70%, prepared with DEPC-treated H_2O) <!>

RNase-free DNase Set (QIAGEN)
> *This kit contains RNase-free DNase I and Buffer RDD.*

RNeasy Midi Kit (QIAGEN)
> *This kit contains Buffer RLT, Buffer RPE, Buffer RW1, and RNase-free H_2O as well as midi columns and collection tubes. Mix 4 ml of Buffer RLT with 40 µl of β-mercaptoethanol <!>.*

Equipment

Cell lifter
Centrifugal filter unit (Microcon YM-100; Millipore) (optional; see Step 21)
Clinical centrifuge (refrigerated, with swinging bucket rotors)
Conical centrifuge tubes (50 ml, sterile)
Needle (18 gauge, attached to a 5-ml syringe)
Spectrophotometer
Vortex
Water bath, set at 37°C (optional; see Step 5)

METHOD

1. Using a cell lifter, scrape the cells off of the bottom of the culture plate and pipette the equivalent of 10^7–10^8 cells into a 50-ml conical tube.
   > *Overloading the column can lead to a lower RNA yield; do not use more than 10^8 cells per midi column.*

2. Pellet the cells by centrifugation at 3220g in a swinging bucket rotor for 10 minutes at 4°C.

3. Discard the supernatant. Add to the cell pellet the 4 ml of Buffer RLT that contains β-mercaptoethanol.

4. Vortex to loosen the cells and resuspend the pellet.

5. Homogenize the suspension by drawing it through an 18-gauge needle ten times.
   At this stage, the homogenate can be stored at –80°C. To continue, thaw the homogenate for 15 minutes in a 37°C water bath.

6. Centrifuge the sample at 3220*g* for 10 minutes at room temperature to pellet any unlysed cells and debris.

7. Transfer the supernatant to a new conical tube.
   The supernatant contains the RNA.

8. Add 4 ml of 70% ethanol and mix vigorously.

9. Apply 4 ml of lysate to a midi column, centrifuge at 3220*g* for 5 minutes at room temperature, and dispose of the flow-through.

10. Apply the remaining lysate to the column, centrifuge at 3220*g* for 5 minutes at room temperature, and dispose of the flow-through. Repeat this step until the entire sample has been applied to the column.

11. Apply 2 ml of Buffer RW1 to the column. Centrifuge at 3220*g* for 5 minutes at room temperature. Dispose of the flow-through.

12. Combine 20 µl of DNase I stock solution and 140 µl of Buffer RDD.

13. Apply the DNase mix from Step 12 to the column and incubate it for 15 minutes on the benchtop.

14. Apply 2 ml of Buffer RW1 to the column and incubate it for 5 minutes on the benchtop.

15. Centrifuge the column at 3220*g* for 5 minutes at room temperature and dispose of the flow-through.

16. Add 2.5 ml of Buffer RPE to the column, centrifuge at 3220*g* for 2 minutes at room temperature, and dispose of the flow-through.

17. Apply another 2.5 ml of Buffer RPE to the column and centrifuge at 3220*g* for 5 minutes at room temperature. Transfer the column to a new conical tube (supplied with the kit).

18. To elute the RNA, proceed as follows:

    i. Pipette 250 µl of RNase-free H_2O onto the column.

    ii. Incubate the column for 1 minute at room temperature.

    iii. Centrifuge the column at 3220*g* for 3 minutes at room temperature.

19. Repeat Steps 18.i–18.iii.

20. Using a spectrophotometer, determine the concentration of RNA.
    Starting with 10^7 cells, expect RNA yields of 30–80 ng/µl.

21. If the concentration of RNA is lower than needed for downstream applications, concentrate the RNA using a Microcon YM-100 tube (follow the manufacturer's instructions).

22. Store the RNA at –80°C.

Protocol 6

Rapid Preparation of Genomic DNA from *M. brevicollis* and Other Choanoflagellates

This protocol describes a straightforward but crude genomic DNA isolation strategy that has been effective on all tested choanoflagellate species (Leadbeater et al. 2008). The resulting genomic DNA provides a suitable template for a variety of PCR techniques, but is too impure for procedures that require very high-quality DNA such as Southern blotting (see Protocol 7 for the isolation of high-molecular-weight DNA). The following protocol has been successfully used on 15 species of choanoflagellates (Leadbeater et al. 2008).

MATERIALS

The recipes for items marked with <R> begin on page 25.

CAUTION: See the Cautions Appendix for appropriate handling of materials marked with <!>.

Reagents

Cereal grass growth medium
> *Prepare 38 ml of sterile cereal grass growth medium by combining 2 ml of cereal grass medium <R> with 36 ml of seawater <R>.*

Culture of *M. brevicollis* MX1 or any other clonal choanoflagellate culture
> *Stocks of M. brevicollis MX1 are available from the American Type Culture Collection (ATCC PRA-258; http://www.atcc.org).*

Ethanol (70% and 100%, ice-cold) <!>
NaCl (5 M)
Proteinase K (10 mg/ml) <!>
TNES solution <R>

Equipment

Cell scraper
Centrifuge, prechilled to 4°C
Centrifuge tubes, 50 ml
Incubator, preset to 15°C for Step 1 and to 37°C for Step 5
Media flask (250 ml), sterile
Microcentrifuge
Microcentrifuge tubes (1.5 ml)

METHOD

1. Add 2 ml of *M. brevicollis* MX1 to 38 ml of cereal grass growth medium in a 250-ml flask. Incubate the culture for 14 days at 15°C.

2. Using a cell scraper, scrape the bottom and sides of the culture flask. Pour the culture medium into a 50-ml centrifuge tube.
3. Balance the tube in a prechilled (4°C) centrifuge, and centrifuge the cells for 40 minutes at 2700g.
4. Carefully remove and discard the supernatant. Resuspend the cells in 300 µl of TNES and transfer them to a 1.5-ml microcentrifuge tube.
5. Add 10 µl of proteinase K to the tube and incubate it overnight at 37°C.
6. Add 85 µl of 5 M NaCl to the tube and shake it vigorously for 30 seconds.
7. Centrifuge the tube at maximum speed in a microcentrifuge for 5 minutes.
8. Transfer the supernatant to a fresh 1.5-ml microcentrifuge tube and shake it vigorously for 30 seconds.
9. Centrifuge the tube at maximum speed in a microcentrifuge for 5 minutes.
10. Transfer the supernatant to a fresh 1.5-ml microcentrifuge tube, add 600 µl of ice-cold 100% ethanol, and let stand for 30 minutes at −20°C.
11. Centrifuge the sample at maximum speed in a microcentrifuge for 5 minutes at 4°C.
12. Carefully remove and discard the supernatant.
13. Add 400 µl of ice-cold 70% ethanol and flick the centrifuge tube until the DNA pellet becomes loose.
14. Centrifuge the sample at maximum speed in a microcentrifuge for 5 minutes at 4°C.
15. Decant the supernatant and allow the DNA pellet to air-dry.
16. Resuspend the DNA pellet in 75 µl of H_2O.
    The DNA yield should be sufficient to provide 1 µl for each of 75 PCR reactions.
17. Store the DNA at 4°C.

Protocol 7

Preparation of High-molecular-weight Genomic DNA from *M. brevicollis* and Other Choanoflagellates

This protocol describes a simple genomic DNA isolation strategy that should be effective for almost all choanoflagellate species and will provide genomic templates suitable for PCR and Southern analysis (King et al. 2008). The preparation of genomic DNA from cultures of *M. brevicollis* (and, indeed, from all choanoflagellates) can be complicated by the presence of abundant bacterial DNA. In our experience, bacterial DNA can represent as much as 90% of the total yield from a standard genomic DNA preparation. See Protocol 8 for a procedure that significantly reduces contamination from bacterial DNA.

MATERIALS

The recipes for items marked with <R> begin on page 25.

CAUTION: See the Cautions Appendix for appropriate handling of materials marked with <!>.

Reagents

Choanoflagellate growth medium, sterile <R>
Culture of *M. brevicollis* MX1 or any other clonal choanoflagellate culture

> Stocks of M. brevicollis *MX1 are available from the American Type Culture Collection (ATCC PRA-258; http://www.atcc.org). The culture should be started from frozen stock and grown for 3 days in cereal grass medium <R> as described in Protocol 2.*

Ethanol (70%) <!>
Gentra Puregene Cell Kit (QIAGEN)

> *The kit contains DNA Hydration Solution, Cell Lysis Solution, Protein Precipitation Solution, and RNase A Solution.*

Isopropanol <!>

Equipment

Benchtop clinical centrifuge (refrigerated, with swinging bucket rotors)
Cell lifter, sterile
Cell scraper, sterile
Centrifuge bottles (500 ml)
Conical centrifuge tubes (50 ml), sterile
Culture flasks (T75), sterile

> *Erlenmeyer flasks (and a shaking incubator) may be substituted for T75 flasks (see Step 4).*

Floor centrifuge with fixed-angle rotor for 500-ml bottles (e.g., Beckman centrifuge J2-MI with rotor JA-10)
Ice
Kimwipes
Liquid nitrogen <!>
Microcentrifuge tubes
Mortar and pestle, sterilized by autoclaving and then prechilled to –80°C

Pipette tips (1 ml, wide-bore)
  These may be prepared by cutting off the ends of 1-ml pipette tips using a single-edged razor.
Vortex
Water baths (set at 37°C for Step 17 and at 65°C for Step 26)

METHOD

Expanding Choanoflagellate Cultures for Batch Growth

1. Using a sterile cell scraper, scrape the bottom of the culture flask containing the choanoflagellates.

2. In each of three separate T75 flasks, add 5 ml of the choanoflagellate culture to 45 ml of choanoflagellate growth medium. Incubate them for 3 days at room temperature.

3. Scrape the bottoms of the flasks with a cell scraper and pool the cultures.
   The total volume should be about 150 ml.

4. Add the pooled culture to 3 liters of sterile choanoflagellate growth medium. Then, aliquot 50 ml of freshly diluted choanoflagellate culture to each of 60 sterile T75 flasks. Work quickly and agitate the mixture frequently to prevent the choanoflagellates from adhering to the flask wall.
   Using T75 flasks is recommended to prevent contamination, but it is also possible to grow cultures in Erlenmeyer flasks under gentle agitation (e.g., 100 rpm).

5. Incubate the flasks for approximately 3 days at room temperature, until the cultures have reached a density of about 10^7 choanoflagellate cells/ml.

Harvesting Cells

6. When the cultures have reached a density of nearly 10^7 cells/ml, harvest the cells by scraping the bottoms of the flasks with a cell scraper. To each of three centrifuge bottles, add 400 ml of culture from each of ten flasks. Prepare a fourth bottle with H_2O for balance during centrifugation.
   There may be unused culture left over.

7. Centrifuge the cells in a floor centrifuge at 17,700g for 30 minutes at 4°C.

8. Gently pour off the supernatant and overlay the pellets with the remainder of the cell culture (approximately 375–400 ml per bottle).

9. Rebalance the tubes in the centrifuge and centrifuge again at 17,700g for 30 minutes at 4°C.

10. Gently pour off the supernatant, freeze the pellets overnight at –80°C, and proceed to Step 11.

Isolating DNA

Proceed with DNA isolation from a single pellet and reserve the remaining two pellets for future use.

11. Using a sterile cell lifter, chip a frozen pellet into a sterile –80°C frozen mortar and pestle. Apply liquid nitrogen to keep the pellet frozen.

12. Grind the pellet to a powder, keeping the powder frozen by frequently applying liquid nitrogen.

13. Transfer 25 ml of Cell Lysis Solution to a 50-ml conical tube, add the frozen cell powder, and then add an additional 25 ml of Cell Lysis Solution. Lyse the cells by gently pipetting and by inverting the tube.

14. Split the lysed cells into two 50-ml conical tubes, each containing 25 ml.
15. Centrifuge the tubes in a benchtop clinical centrifuge at 800g for 2 minutes at room temperature to pellet unlysed cells and precipitates.
16. Transfer the supernatant from each tube to fresh 50-ml conical tubes.
17. Add 125 µl of RNase A Solution to each tube and invert the tubes 25 times to mix. Incubate for 35 minutes at 37°C.
18. Transfer the tubes to ice and incubate for 5 minutes.
19. Add 8.3 ml of Protein Precipitation Solution to each tube. Vortex the tubes for 20 seconds to mix thoroughly. Then, centrifuge the tubes in a benchtop clinical centrifuge at 3220g for 20 minutes at 16°C.
20. Transfer the supernatants to four 50-ml conical tubes, each containing approximately 16.7 ml of cell lysate.
21. Add 12.5 ml of isopropanol (at room temperature) to each of the four tubes and invert the tubes 50 times to mix. Incubate overnight at room temperature to precipitate the DNA.
22. Centrifuge the tubes in a benchtop clinical centrifuge at 2000g for 3 minutes at 16°C.
23. Gently pour off the supernatant. Overlay each pellet with 12.5 ml of 70% ethanol at room temperature. Centrifuge the tubes at 2000g for 3 minutes at 16°C.
24. Pour off the supernatant. Invert the tubes and drain residual ethanol by dabbing the edge of each tube with a Kimwipe.
25. Air-dry the pellets for 10–15 minutes. Do not allow the pellets to overdry because the DNA will not go into solution.
26. Add 1.5 ml of DNA Hydration Solution to each pellet and incubate for 1 hour at 65°C.
27. Flick the tubes to loosen the pellets and touch-spin in a benchtop clinical centrifuge.
28. To further dissolve the DNA, gently pipette the DNA solution using a wide-bore pipette tip.
29. Incubate the DNA solution for 2 hours (to overnight) at room temperature to allow complete rehydration.
30. Using a wide-bore pipette tip to prevent shearing of the DNA, transfer the DNA to microcentrifuge tubes. Store the DNA at 4°C.

    Starting with 800 ml of culture, you can expect a yield of 2 mg total genomic DNA from this protocol.

Protocol 8

Separation of Choanoflagellate and Bacterial Genomic DNA

This protocol describes the use of a CsCl gradient to separate choanoflagellate genomic DNA from the DNA of its bacterial prey. This strategy works only when the G+C content of the choanoflagellate genomic DNA is significantly different from that of its bacterial food. A culture of *M. brevicollis* containing a single bacterial prey source, the A+T-rich *Flavobacterium* sp., is available as *M. brevicollis* MX1 from the ATCC (PRA-258). The final DNA sample from this protocol is expected to be 90–95% enriched for choanoflagellate genomic DNA and to provide adequate template for genome sequencing projects.

MATERIALS

CAUTION: See the Cautions Appendix for appropriate handling of materials marked with <!>.

Reagents

n-butanol, saturated with NaCl <!>
> Mix equal volumes of 0.2 M NaCl and n-butanol and allow the phases to separate. Use the upper organic phase for extractions (see Step 21).

CsCl <!>
Ethanol (70% and 100%) <!>
Genomic DNA (2 mg), prepared from *M. brevicollis* MX1 as described in Protocol 7
Pellet Paint® Coprecipitant (Novagen)
Sodium acetate (3 M, pH 5.2)
Tris-EDTA (TE) (10 mM Tris-Cl, 1 mM EDTA, pH 8.0)
TE/Hoechst 33258 solution
> Dilute a 10-mg/ml stock of Hoechst 33258 1:10 into TE (pH 8.0) and protect it from light.

Equipment

Aluminum foil
Centrifuge, benchtop
Centrifuge tubes (5.1 ml, 13 × 51 mm Quick-Seal polyallomer tubes; Beckman)
Conical centrifuge tubes (15 ml)
Dialysis membrane, rinsed in sterile H_2O (Spectra/Por 7, MWCO 8000 D, 20.4 mm diameter; Spectrum)
Microbalance
Microcentrifuge tubes (2 ml)
Needles (18 and 27 gauge)
Nutator
Pasteur pipettes, glass
Pipette tips (1 ml, wide-bore)
> These may be prepared by cutting off the ends of 1-ml pipette tips using a single-edged razor.

Ring stand with clamp in a dark room
Stir plate, set at 4°C
Syringe (5 ml)

Tube sealer for heat-sealing the Quick-Seal centrifuge tubes
Ultracentrifuge with fixed-angle rotor and accessories (e.g., Beckman centrifuge L8-70M centrifuge with rotor VTI-80, as well as a torque wrench, metal caps, and hex nuts)
UV lamp, hand held <!>

METHOD

Preparing the CsCl Gradient

You will need six tubes of CsCl solution (density 1.69 g/ml) to separate a sufficient quantity of genomic DNA for most downstream purposes. A seventh solution of CsCl (density 1.69 g/ml) is necessary to top off and balance each of the sample tubes before ultracentrifugation.

1. Place 4.73 g of CsCl in each of seven 15-ml conical tubes.

    Note that preparation of a CsCl solution of density 1.69 g/ml requires the dissolution of 1.24 g of CsCl in 1 ml of H_2O. The final volume of this solution at 25°C is 1.31 ml (see Flamm et al. 1972). So, for a final volume of 5 ml (in Step 7), use 4.73 g of CsCl.

2. Add 2.8 ml of TE to each of the six sample tubes. To the seventh tube, which will be used for topping off each of the sample tubes, add 3.5 ml of TE.

    With the addition of TE, these mixtures will initially become cold to the touch.

3. Gently agitate the tubes on a nutator at room temperature until the CsCl has fully dissolved and the solution has returned to room temperature.

4. Add 200 µl of the TE/Hoechst 33258 solution to each of the seven sample tubes to reach a final concentration of 40 µg/ml Hoechst 33258. Protect the tubes from light by wrapping them with foil.

5. Gently agitate the samples on a nutator for 10 minutes at room temperature. Leave the seventh ("topping-off") tube nutating at room temperature as you proceed through Steps 6 and 7.

6. Add 1 ml of genomic DNA (~400 µg) to each sample tube. Gently agitate the mixture on a nutator for 10 minutes at room temperature. Confirm that the DNA is fully in solution before proceeding.

7. Check the volume of the CsCl/DNA mixture in each sample tube with a pipette. If the volume is less than 5 ml, add sufficient TE to bring the volume to 5 ml.

8. Centrifuge all seven tubes at 1800g for 10 minutes at room temperature to pellet the precipitates. While the tubes are spinning, preheat the tube sealer.

9. Use a glass Pasteur pipette to load each CsCl/DNA mixture into separate Quick-Seal tubes. Be careful to avoid introducing bubbles.

    To avoid bubbles, submerge the tip of the Pasteur pipette below the surface when the tube is close to full.

10. Top off each tube with CsCl/TE prepared without DNA (from the seventh tube) so that each tube is filled to just above the start of the neck.

11. Seal the tubes with the tube sealer. Gently squeeze each tube to check for leaks.

12. Weigh each tube and cap/gasket set to identify pairs within 0.02 g of one another.

13. Load the sealed, balanced tubes into the ultracentrifuge rotor and immediately proceed to Step 14.

Separating Choanoflagellate and Bacterial DNA

14. Carefully place the rotor in an ultracentrifuge and centrifuge at 30,500g for 40 hours at 25°C. Set the deceleration to 0.

15. After centrifugation, turn off the vacuum to open the lid of the ultracentrifuge, carefully move the rotor to a solid benchtop, and loosen the gaskets on the rotor stand. Handle the rotor carefully to avoid mixing the bands of bacterial and choanoflagellate genomic DNA.

16. In a dark room, mount one tube on a ring stand, using a small clamp. Expose the tube briefly to UV light to visualize the DNA.

    For MX1 cultures (M. brevicollis cultured with Flavobacterium *sp.) you should observe two distinct bands. The brighter, upper band is bacterial DNA, and the fainter, lower band is choanoflagellate DNA. See Troubleshooting.*

17. To release the pressure, pierce the top of the tube, near the neck, with a 27-gauge needle.

18. With an 18-gauge needle attached to a 5-ml syringe, pierce the side of tube just below the band of choanoflagellate DNA and pull out the entire band slowly to avoid mixing with bacterial DNA.

19. Repeat Steps 16–18 for each of the remaining samples.

20. Pool all bands of choanoflagellate DNA together and proceed to Step 21.

Cleaning, Precipitating, and Rehydrating Choanoflagellate Genomic DNA

21. To extract the Hoechst dye from the DNA, mix equal volumes of DNA and NaCl-saturated *n*-butanol. Invert the tube 15 times.

22. Centrifuge the tube at 500*g* for 5 minutes at room temperature.

23. After centrifugation, transfer the lower phase to a fresh 15-ml conical tube. Discard the upper phase appropriately. Check for the presence of dye by exposing the tube to UV light.

24. Repeat Steps 21–23 until no dye remains in the upper phase. Then, perform Steps 21–23 one final time.

    Usually, Steps 21–23 are performed a total of five times.

25. Prepare the dialysis tubing. Cut lengths of tubing that will allow for doubling of the sample volume.

26. Dialyze the DNA sample from Step 24 in 1 liter of TE for 2 hours at 4°C with slow stirring (i.e., with a stir bar).

27. Change the TE. Dialyze 48 more hours at 4°C, changing the TE every 12 hours.

28. Remove the dialyzed DNA from the tubing with the wide-bore pipette tip.

29. To precipitate the DNA, mix 600 µl of dialyzed DNA, 60 µl of 3 M sodium acetate (pH 5.2), and 2 µl of Pellet Paint Coprecipitant in a 2-ml microcentrifuge tube. Gently invert the tubes to mix.

30. Add 1.2 ml of 100% ethanol to each tube and freeze overnight at −20°C.

31. Centrifuge the samples at 16,000*g* for 30 minutes at 4°C.

    The pellet should be pink.

32. Wash each pellet with 500 µl of 70% ethanol.

33. Centrifuge the samples at 16,000*g* for 5 minutes at room temperature.

34. Remove the supernatant with a pipette.

35. Wash each pellet with 50 µl of 100% ethanol.

36. Centrifuge the samples at 16,000*g* for 30 seconds at room temperature.

37. Remove the supernatant with a pipette. Let the pellets air-dry for 15 minutes.

38. Resuspend each pellet in 5 µl of TE and then combine all samples.

    Starting with 2 mg total genomic DNA, you can expect 20 µg of choanoflagellate-enriched genomic DNA.

TROUBLESHOOTING

Problem (Step 16): Chunks or strands of genomic DNA fail to separate.
Solution: Make sure that the DNA loaded on the CsCl gradient is fully in solution before centrifugation; otherwise, it will not separate properly.

Recipes

The recipes for items marked with <R> are also listed here.

CAUTION: See the Cautions Appendix for appropriate handling of materials marked with <!>.

BLOCKING SOLUTION

Reagent	Quantity (for 5 ml)	Final concentration
BSA	50 mg	1%
Triton X-100 <!>	15 µl	0.3%
PEM <R>	to 5 ml	

CEREAL GRASS MEDIUM

Reagent	Quantity (for 1 liter)	Final concentration
Cereal grass powder (WARD'S 944 V 8602)	2.5 g	2.5 mg/ml
H_2O, distilled	to 1 liter	

Prepare this medium under sterile conditions to facilitate downstream DNA and RNA isolation protocols. Boil for 15 minutes. Allow the solution to cool and filter it through Whatman No. 1 paper and then through a 0.22-µm membrane filter. The final product should be a transparent, straw-colored liquid.

CHOANOFLAGELLATE GROWTH MEDIUM

Reagent	Quantity (for 2 liters)	Final concentration
Cereal grass powder (WARD'S 944 V 8602)	10 g	5 g/liter
Seawater, natural	2 liters	

Autoclave the natural seawater in a 4-liter Erlenmeyer flask for 20 minutes until boiling. Add the cereal grass powder to the boiling seawater. Allow the mixture to steep until cool (~3–5 hours or overnight). To remove macroscopic cereal grass particles, filter the mixture through Whatman No. 1 paper in a Buchner funnel under vacuum (the funnel should be attached to the top of a 4-liter vacuum flask with a rubber stopper). Then, sterilize by filtering through a 22-µm membrane filter. Filtered media may be stored at room temperature, but it is prone to contamination; use aseptic technique when opening and closing the bottle.

NATURAL FIELD MEDIUM

Using Whatman No. 1 filter paper, filter 1 liter of natural field water and allow it to stand for 24 hours. Then, sterilize the medium by autoclaving. After the medium is cool, filter it a second time using a 0.22-µm membrane filter to remove fine particulates.

ORGANIC ENRICHMENT

Reagent	Quantity (for 100 ml)	Final concentration
Proteose peptone	0.4 g	0.4%
Yeast extract	0.08 g	0.08%
H_2O	100 ml	

Dispense aliquots into 2-ml Bijou bottles, autoclave, and store in a refrigerator. For standard growth medium, add 40 µl of organic enrichment to 10 ml of seawater.

PEM

Reagent	Quantity (for 5 ml)	Final concentration
EGTA (10 mM)	0.5 ml	1 mM
$MgSO_4$ (10 mM) <!>	50 µl	0.1 mM
PIPES (0.5 M, pH 6.9)	1 ml	100 mM
H_2O, Milli-Q	to 5 ml	

When preparing the stock solution of 0.5 M PIPES (pH 6.9), use 10 N NaOH <!> to adjust the pH. Readjust the pH as necessary.

PHALLOIDIN SOLUTION

Reagent	Quantity (for 100 µl)	Final concentration
Rhodamine phalloidin (0.2 units/µl) <!>	3 µl	0.006 units/µl
PEM <R>	97 µl	

PRATT'S FRESHWATER MEDIUM

Reagent	Quantity (for 1 liter)	Final concentration
KNO_3 <!>	0.1 g	1 mM
$MgSO_4 \cdot 7H_2O$ <!>	0.01 g	40 µM
$K_2HPO_4 \cdot 3H_2O$	0.1 g	400 µM
$FeCl_3 \cdot 6H_2O$ <!>	0.001 g	3.6 µM
H_2O, distilled	to 1 liter	

Adjust pH to 7.0. Autoclave and store at 5°C. From Pratt (1984).

SEAWATER

Autoclave and membrane-filter (pore size 0.2 µm) 950 ml of natural seawater. (Natural seawater can be replaced with high-quality synthetic seawater made with Milli-Q H_2O.) When cool, combine seawater with 50 ml of sterile distilled H_2O.

TNES SOLUTION

Reagent	Quantity (for 100 ml)	Final concentration
EDTA	0.75 g	25 mM
NaCl	2.34 g	400 mM
SDS <!>	0.5 g	0.5%
Tris <!>	0.61 g	50 mM
H_2O, sterile distilled	to 100 ml	

Store at 4°C. Preheat to 37°C before use.

REFERENCES

Abedin, M.A. and King, N. 2008. The premetazoan ancestry of cadherins. *Science* **319:** 946–948.

Flamm, W.G., Birnstiel, M.L., and Walker, P.M.B. 1972. Isopycnic centrifugation of DNA methods and applications. In *Subcellular components: Preparation and fractionation* (ed. G.D. Birnie), pp. 279–310. Butterworth, London and University Park Press, Baltimore.

James-Clark, H. 1866. Note on the infusoria flagellata and the spongiae ciliate. *Am. J. Sci.* **1:** 113–114.

James-Clark, H. 1868. On the spongiae ciliatae as infusoria flagellata; or observations on the structure, animality and relationship of *Leucosolenia botryoides*, Bowerbank. *Ann. Mag. Nat. Hist.* **1:** 133–142.

King, N. and Carroll, S.B. 2001. A receptor tyrosine kinase from choanoflagellates: Molecular insights into early animal evolution. *Proc. Natl. Acad. Sci.* **98:** 15032–15037.

King, N., Westbrook, M.J., Young, S.L., Kuo, A., Abedin, M., Chapman, J., Fairclough, S., Hellsten, U., Isogai, Y., Letunic, I., et al. 2008. The genome of the choanoflagellate *Monosiga brevicollis* and the origin of the metazoans. *Nature* **451:** 783–788.

Leadbeater, B.S.C., Hassan, R., Nelson, M., Carr, M., and Baldauf, S.L. 2008. A new genus, *Helgoeca* gen. nov., for a nudiform choanoflagellate. *Eur. J. Protistol.* **44:** 227–237.

Pratt, J.M. 1984. Coupled transcription-translation in prokaryotic cell-free systems. In *Transcription and translation: A practical approach* (eds. B.D. Hames and S.J. Higgins), pp. 179–209. IRL, Oxford and New York.

Ruiz-Trillo, I., Roger, A.J., Burger, G., Gray, M.W., and Lang, B.F. 2008. A phylogenomic investigation into the origin of metazoa. *Mol. Biol. Evol.* **25:** 664–672.

Saville-Kent, W. 1880–1882. *A manual of the infusoria*, vol. 5, pp. 1–3. D. Bogue, London.

Snell, E.A., Furlong, R.F., and Holland, P.W. 2001. Hsp70 sequences indicate that choanoflagellates are closely related to animals. *Curr. Biol.* **11:** 967–970.

Steenkamp, E.T., Wright, J., and Baldauf, S.L. 2006. The protistan origins of animals and fungi. *Mol. Biol. Evol.* **23:** 93–106.

Wainright, P.O., Hinkle, G., Sogin, M.L., and Stickel, S.K. 1993. Monophyletic origins of the metazoa: An evolutionary link with fungi. *Science* **260:** 340–342.

WWW RESOURCES

http://amoebidia.bcm.umontreal.ca/pepdb/searches/welcome.php Taxonomically Broad EST Database (TBestDB). This is a repository for EST data generated by the labs of the Protist EST Program.

http://genome.jgi-psf.org/Monbr1/Monbr1.home.html JGI *Monosiga brevicollis* (v1.0) Genome Portal. The complete genome sequence of *M. brevicollis* is publicly available from this site.

http://www.atcc.org The American Type Culture Collection (ATCC). The ATCC collects, catalogs, and distributes cell lines and cultures from a broad range of species.

http://www.ccap.ac.uk The Culture Collection of Algae and Protozoa (CCAP). The CCAP collects and supplies cultures of algae and protozoa in the U.K. and is linked to other service collections worldwide.

2 | *Dictyostelium discoideum*
The Social Ameba

Pascale Gaudet,[1] Petra Fey,[2] and Rex Chisholm[1]

[1]*Center for Genetic Medicine and Department of Cell and Molecular Biology, Northwestern University Medical School, Chicago, Illinois 60611;* [2]*dictyBase, Northwestern University, Center for Genetic Medicine, Chicago, Illinois 60611*

ABSTRACT

Dictyostelium discoideum is a unicellular eukaryote often referred to as a social ameba because it can form a multicellular structure when nutrient conditions are limiting. *D. discoideum* and related organisms, known as the Dictyostelia, have been studied for almost 150 years. The multicellular part of their life cycle has interested developmental biologists ever since these organisms were first described. The cellular and molecular aspects of their multicellular lifestyle have been studied in detail, and general principles for cell-to-cell communication, intracellular signaling, and cytoskeleton organization during cell motility have been derived from this work and have been found to be conserved across all eukaryotes. The bacteriovore lifestyle of the unicellular stage provides an excellent model in which to study phagocytosis and the mechanisms of bacterial virulence. *D. discoideum* has been used successfully to explore the molecular basis of various human diseases, as well as the mechanisms of drug action and the pathways that lead to resistance to certain therapeutic agents. The presence of a complete genome sequence is further widening the scope of studies using *D. discoideum*. A large potential for secondary metabolism has become apparent, which opens the door to the discovery of new compounds with potential medical applications. Numerous putative orthologs of genes responsible for diseases in humans, but whose molecular functions are still uncharacterized, are present in the *D. discoideum* genome. Finally, the availability of community resources, including the genome database dictyBase and the Dicty Stock Center, make *D. discoideum* an easily accessible and powerful model organism to study.

PROTOCOLS
1. Growth and Maintenance of *Dictyostelium* Cells, 39
2. Multicellular Development of *Dictyostelium*, 43
3. Making Permanent Stocks of *Dictyostelium*, 46
4. Transformation of *Dictyostelium* with Plasmid DNA: Calcium Phosphate Precipitation, 49
5. Electroporation, 52
6. Selection of Transformants, 54
7. DNA Extraction, 56
8. RNA Extraction, 59

BACKGROUND INFORMATION

Natural History and Distribution

D. discoideum is a unicellular eukaryotic soil ameba that possesses the distinctive ability to form a multicellular structure. When nutrients are limited, cells in close proximity attract one another to an aggregate center by chemotaxis in response to cAMP signaling. When a certain size has been reached (under standard laboratory conditions, approximately 100,000 cells), the cells form a coherent tissue that undergoes morphological changes to form a fruiting body (Fig. 1). The fruiting body consists of two main cell types: spores and stalk cells. The role of the stalk cells is to support and protect the spores, and they undergo a primitive form of programmed cell death during the developmental process. These unusual features have attracted the attention of biologists for more than 100 years. However, *D. discoideum* most often exists as a unicellular haploid ameba, also known as a myxameba. During the vegetative phase of its life cycle, it engulfs bacteria by phagocytosis as a nutrient source and is the prey of nematodes and other soil microorganisms (Kessin 2001).

D. discoideum and related species, known as the Dictyostelia, have been isolated and subsequently cultivated from numerous sites worldwide and from various types of environments, including forests, prairie, desert, and marshland (Raper 1984). Fruiting bodies are almost exclusively encountered on dung, presumably because the bacterial density is high enough to favor multiplication and fruiting body formation (Gilbert et al. 2007).

Organism

Dictyostelium discoideum is the scientific name of the most widely studied member of the Dictyostelia family. Although there are several species, in the literature, the genus name "*Dictyostelium*" is often used to refer specifically to *D. discoideum*. Usual common names are "social ameba," which refers to the multicellular phase of its life cycle that occurs through the aggregation of cells in the same location, and (becoming rarer) "cellular slime mold," which describes the trail of proteins and carbohydrates that is left on the substratum when the migratory slug travels from one site to another. The name is occasionally abbreviated to Dicty or Dictyo.

Taxonomic Information

Phylogenetic analysis places *Dictyostelium* in one of the earliest branches to emerge after the divergence of plants and animals (Fig. 2). *Dictyostelium* has retained more of the diversity of the ancestral

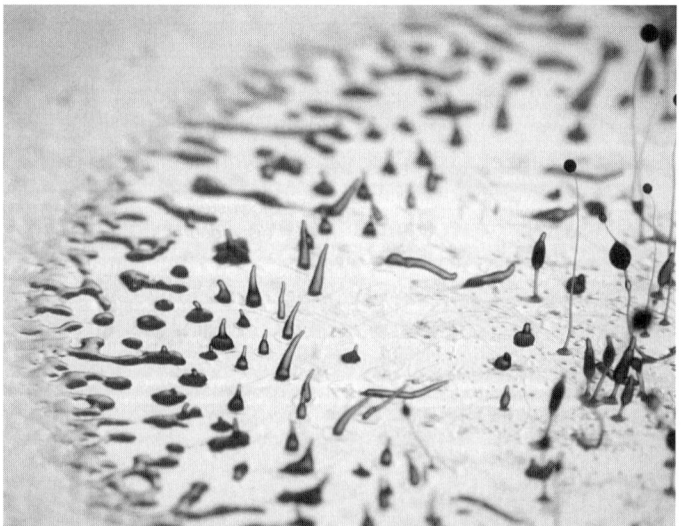

FIGURE 1. Development of *D. discoideum*. Different stages are visible, with the cells in the earliest stage of development (aggregate) visible on the left. Tipped mounds can be seen; these are rather compact with a tip emerging from the cell mass. As the tip elongates, it forms a finger-shaped structure (also known as a standing slug). The finger has the ability to migrate by bending into a horizontal position. On the right are culminants, the stage in which the stalk is formed, the spore head elevates, and terminal cell differentiation takes place. (Figure from Dormann and Weijer 2006, reproduced, with permission, from the *Nature* publishing group.)

eukaryotic genome than have plants, animals, or fungi (Eichinger et al. 2005). It is the best studied organism in that central phylogenetic position. *D. discoideum* belongs to the supergroup Amoebazoa, division Mycetozoa (Baldauf et al. 2000). The taxonomy of Dictyostelia has been difficult to establish. The group was once considered to be a fungus because it is a nonphotosynthetic organism that obtains food by absorptive nutrition. However, with more molecular data available, it has become clear that the Dictyostelia and the fungi are only distantly related. Even within the Mycetozoa, the establishment of relationships among the different known species has been misled by the fact that the morphological traits used to characterize each of these species failed to be supported by the molecular evidence that later became available. The Dictyostelia were once classified into several families and genera, including *Dictyostelium* and *Polysphondylium*, which are distinguished by the presence of unbranched and branched fruiting bodies, respectively, and *Acytostelium*, in which the mass of spores is supported by an acellular stalk. Other parameters that were taken into account to classify species were aggregation patterns, the presence of a migratory multicellular stage, spore morphology, as well as the presence of alternative life cycles such as microcysts and macrocysts (Raper 1984). It is easy to imagine that those limited morphological characteristics could be encoded by minor gene modifications or differential regulation of gene expression. A recent study by Schaap et al. (2006) analyzing small-subunit rRNA gene sequences from more than 100 species of Dictyostelia has led to a reorganization of the family into four taxa that do not correspond to the previously used classification. The groups are (1) Parvisporids, which have small (parvi) spores; (2) Heterostelids, which exhibit varied morphological characteristics; (3) Rhizostelids, several species of which possess root-like support structures for their fruiting bodies; and (4) Dictyostelia, the group that includes *D. discoideum*, the most intensely studied member of the family.

When Established as an Organism of Study and by Whom?

D. discoideum was first described by Kenneth Raper in 1935. At that time, a few related species were known, including *D. mucoroides*, *D. purpureum*, and *Polysphondylium pallidum*, and their morphological characteristics, life cycle, and phagocytic lifestyle had been described by Brefeld (1896), Olive (1902), and Harper (1926). *D. discoideum* was found in a forest in North Carolina in 1933, and the name of the strains isolated at that site start with NC, including NC-4, which is the most commonly used nonaxenic laboratory strain (for review, see Kessin 2001). One distinguishing feature of this organism compared with other Dictyostelia is the presence of a stalkless migra-

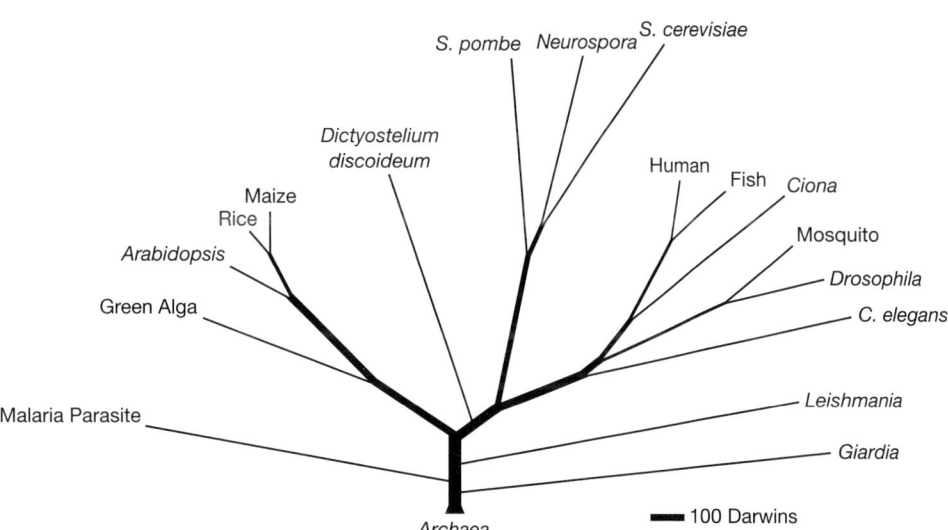

FIGURE 2. Phylogenetic tree of eukaryotes. This tree was built using 5279 orthologous protein clusters from 17 eukaryotic species. It illustrates the central position of *Dictyostelium* among eukaryotes. (Figure from Eichinger et al. 2005, reproduced, with permission, from the *Nature* publishing group.)

tory stage, which has been described in detail by Raper as well as by John Bonner (1944). In the late 1960s, Sussmann, Watts, and Ashworth generated axenic strains that can grow in defined or semidefined media (Sussman and Sussman 1967; Watts and Ashworth 1970). Those have been invaluable for developing molecular and biochemical techniques using this organism.

What Is This Organism Used to Study? Why?

Dictyostelium provides several features that make it an attractive model system for research and educational purposes. A small eukaryote, it nevertheless has a large number of genes (~12,500), many of which are absent in other simple eukaryotes, and accordingly, it exhibits more complex behavior. A wide variety of molecular genetics techniques is available to investigate diverse biological problems. The remarkable synchrony of its multicellular stage formation makes it an excellent system for studying enzyme regulation and cellular signaling during development. Another major advantage is the independence of the growth and developmental stages in *Dictyostelium*, which allows the study of genes that are essential in other organisms. *Dictyostelium* is used to study many important and varied cellular processes, including development, pattern formation, cytoskeleton function, the molecular bases of human diseases, and intraspecies and interspecies interactions, as described in the following sections.

SOURCES AND HUSBANDRY

Wild-type strains of *Dictyostelium* are obligate bacteriovores, but axenic mutants that can grow free of bacteria were created in the late 1960s. The use of axenic strains is widespread because they are easier to maintain in the laboratory and they facilitate biochemical and molecular biology experiments by avoiding bacterial contamination of samples. Moreover, the protocols for transforming *Dictyostelium* work best with axenic strains, although small modifications to the protocol allow efficient transformation of nonaxenic strains. Details for growing and transforming the cells have been published recently by Fey et al. (2007) and Gaudet et al. (2007). Strains grown in axenic media double two to three times more slowly than cells grown on bacteria (8–12 hours vs. 3–4 hours). Two isolates of axenic strains are used for research: AX2 and AX3. AX2 is more widely used in Europe, whereas AX3 and its derivative AX4 are the predominant strains used in the United States. AX4 has provided the starting material for the genome sequencing project. A number of genetic differences exist between these strains, the most important being a 750-kb duplication on chromosome 2 that is present in AX3 and AX4, but not in AX2.

There is a centralized strain, plasmid, and clone repository for the *Dictyostelium* community, the Dicty Stock Center (http://dictybase.org/StockCenter/StockCenter.html), where all common wild-type strains and numerous mutants can be obtained. During the past 6 years, the Stock Center has assembled an impressive collection of more than 1200 strains, representing more than 80 different species. The collection contains the parental strains used by different laboratories around the world, more than 300 null mutants provided by researchers, and strains marked with green fluorescent protein (GFP) or other genes useful for tracking gene expression and regulated by various constitutive or cell-type-specific promoters. Other strains include tester strains required to make use of parasexual genetics, chemical and physical mutants, and personal strain collections. In addition, the Stock Center holds close to 400 plasmids. There is a set of parent vectors that are useful for making gene knockouts, overexpression, or fusions with different marker and reporter genes.

Any academic researcher can order strains from the Dicty Stock Center through a simple Amazon-type ordering system on the dictyBase website that allows ordering as many strains as needed in a single ordering form (http://dictybase.org/StockCenter/StockCenter.html). The user only pays the shipping fees. Strains can be found by using the dictyBase search engine located at the top-right-hand corner of each page. In addition, more strain characteristics can be queried using the Stock Center Search Tool (http://dictybase.org/db/cgi-bin/dictyBase/SC/searchform.pl), for example, the mutagenesis method or depositor. In addition, each strain that has a mutation(s) in a known gene(s) is linked to the appropriate gene(s). On each gene page, a section called "Strains and Phenotypes" lists all strains known to have mutations in the gene; strains with mutations in multiple genes are listed on all the appropriate gene pages (some of which are not available in the Stock Center but have been described in the literature). Clicking on a strain name opens a new page entitled "Phenotype and Strain Details" that lists phenotypes and shows the complete strain record, including the genotype, parental strain, mutagenesis method, and reference describing the strain. If the strain is available, a button appears at the bottom of the page from which strains can be added to the shopping cart. At any point, the contents of the order can be viewed by clicking on the shopping cart icon at the top right of each strain record page.

RELATED SPECIES

More than 100 species of Dictyostelia have been described and new ones are discovered regularly (by Vadell and Cavender in 2007, for instance). Some examples are illustrated in Figure 3. Although *D. discoideum* is by far the most widely used species, a number of studies make use of *P. pallidum* and *D. mucoroides*, in which some molecular genetic techniques such as DNA-mediated transformation are available. All species of Dictyostelia can be grown in conjunction with bacteria using the same conditions as those used for *D. discoideum* (Raper 1984; Schaap et al. 2006). Outside of *D. discoideum*, axenic strains are not available. It should be noted that the known species of Dictyostelia are quite distantly related to one another: There is about as much evolutionary distance within the Dictyostelia as there is among the various species of animals (Schaap et al. 2006). Other, more distantly related, members of the amoebozoan include *Physarum* and *Acanthamoeba* (Baldauf et al. 2000).

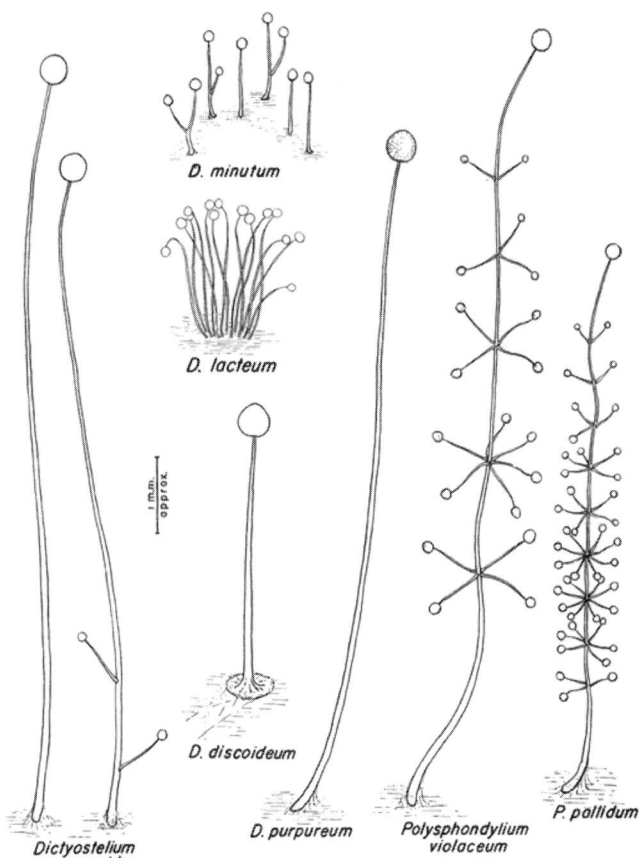

FIGURE 3. Other Dictyostelid species. Shown are fascinating multicellular patterns of some representatives of the Dictyostelia. (Illustration by Raper 1951.)

Recently, the sequencing of a number of species related to *Dictyostelium* has been undertaken. The genome sequence of the amebae *Entamoebae histolytica* is already available (Song et al. 2005) and the sequencing of the highly related *D. purpureum* as well as that of *D. citrinum* and *P. violaceum* is also under way (A. Kuspa, pers. comm.). These new sequences open the door to more powerful comparative genomic analyses. dictyBase will be providing genome databases for these additional genome sequences.

USES OF THE *DICTYOSTELIUM* MODEL SYSTEM

Development and Pattern Formation

Dictyostelium has classically been used as a model system for the study of many fundamental principles of developmental biology, including pattern formation, cell differentiation, and tip signaling. The same general principles govern these events in *Dictyostelium* as in other systems: Both intrinsic and extrinsic signals lead to asymmetry in the multicellular organism. Extrinsic factors that regulate development in *Dictyostelium* include cAMP and the differentiation-inducing factor DIF. Some cells have a higher propensity to secrete cAMP upon starvation, and these are more likely to end up in the center of the cell mass. Other cells are very responsive to DIF, the stalk cell differentiation-inducing factor. The intrinsic cellular mechanisms that are responsible for these initial differences among individual cells remain unclear, but they have been correlated with a number of factors such as cell cycle position at starvation, intracellular pH, and calcium ion concentration (for review, see Strmecki et al. 2005). The generation of these stimuli and the response to them are followed by differential gene expression, which leads to the differentiation of the aggregate cells into either prespore cells (which will form spores) or prestalk cells (which will form the stalk, basal disc, lower cup, or apical disc) (for review, see Gaudet et al. 2008). Although at the organism level the observations made using *Dictyostelium* cannot generally be applied to organisms outside the group of Dictyostelia, it turns out that the molecular mechanisms regulating these processes are highly conserved, including the involvement of G-protein-coupled receptors, adenylate cyclase, protein kinase A, glycogen synthase kinase 3, and mitogen-activated protein kinase (MAPK) (for review, see Chisholm and Firtel 2004).

Cytoskeleton Function

The cytoskeleton has an essential role in many cellular processes such as cytokinesis, cell migration, phagocytosis, intracellular vesicle transport, and cellular morphogenesis. Unlike comparable microorganisms, *Dictyostelium* exhibits all of these behaviors and has proven to be a most useful organism to study these events. The force required for producing cellular movements is provided by motor proteins. Myosins are actin-based molecular motors that control the rearrangement of the cortical cytoskeleton during cytokinesis, phagocytosis, and chemotaxis. Conventional myosin from *Dictyostelium*, myosin II, has been extensively studied at both the cellular and molecular levels. *Dictyostelium* provided the first myosin II knockout in a motile cell. The mutant has a defect in cytokinesis that can be overcome by traction-mediated cytofission when grown on substratum; myosin-null mutants also exhibit poor chemotaxis, reduced capping of surface receptors, and developmental arrest at the aggregation stage (Spudich 1989). Physical properties, molecular mechanism of action, and regulation of myosin by phosphorylation have been studied extensively (for review, see Ma et al. 2001; De la Roche et al. 2002). This research has been critical to the understanding of myosin function.

Chemotaxis is the movement of a cell toward or away from a stimulus. This directed movement requires sensing the stimulus, determining the direction in which to move, and reorganizing the cytoskeleton to generate the movement. The initial developmental stage in the *Dictyostelium* life cycle involves aggregation of single cells to form a multicellular mass, a process that is mediated by the positive attraction toward cAMP (Konijn et al. 1967). The similarities between chemotaxis during early development in *Dictyostelium* and directional sensing in other multicellular organisms during development, immunity, and metastasis formation have led to a rapid increase in our understanding of these processes in higher organisms. Any cell can start secreting cAMP upon starvation, so when responding, each cell must detect the direction in which the highest concentration of chemoattractant is located to orient itself toward the closest group of cells. Accordingly, cells can sense differences in ligand concentration that are as low as 1% between opposite sides of the cell. cAMP is secreted by early aggregating cells and binds to a G-protein-coupled receptor on the surface of neighboring cells. Upon binding of the ligand to cAR1 (the main cAMP receptor during aggregation), multiple downstream signaling pathways are activated. Among other effects, this binding leads to the production of second messengers cAMP, cGMP, and phosphatidylinositol (3,4,5)-trisphosphate (PIP3); activation of small GTPases; and release of calcium from intracellular compartments. These signals lead to different events at the front and the back of the cell. At the front of the cell, the response leads to actin polymerization through the action of the conserved Scar/WAVE complex and extension of a pseudopodium in the direction of the highest concentration of chemoattractant. Conversely, at the sides and back of the cell, myosin filaments are formed that result in the contraction of the back of the cell. The extension of the front of the cell and the retraction of the back result in an overall movement toward the chemoattractant (Chisholm and Firtel 2004). Many of the proteins and signaling molecules found to have a role in *Dictyostelium* chemotaxis, including G-protein-coupled receptors, small GTPases, and phosphoinositides, have been implicated in that same process in other systems such as neutrophils and fibroblasts (Devreotes and Janetopoulos 2003; Parent 2004; von Philipsborn and Bastmeyer 2007).

Phagocytosis—the process through which cells engulf other cells or particles and digest their component parts—has various roles in all multicellular organisms. During development and in tissue remodeling, cells targeted for removal are eliminated by neighboring cells through phagocytosis. When insects and animals are infected by bacteria, part of the immune response is to eliminate the infectious agent through phagocytosis. *Dictyostelium* obtains nutrients through the engulfment and subsequent lysosomal digestion of other small microorganisms such as bacteria and yeasts. This behavior makes it an excellent system in which to study phagocytosis (Desjardins et al. 2005). At the cellular level, phagocytosis can be divided into distinct events: particle recognition by extracellular receptors, internalization of the particle mediated by reorganization of the cytoskeleton at the site of engulfment, fusion of the phagosome with lysosomes that result in the digestion of the particle, and membrane recycling. *Dictyostelium* research has provided many insights into the molecular mechanisms of phagocytosis (for review, see Maniak 2003).

The Molecular Basis of Human Diseases

The *Dictyostelium* genome contains orthologs of several disease-causing genes and its ease of manipulation makes it an attractive system in which to investigate genes with poorly characterized function.

One notable case in which research in *Dictyostelium* has led to a treatment in human patients is in the type-1b congenital disorder of glycosylation that manifests itself by chronic gastrointestinal problems in children. This defect is due to a mutation in phosphomannose isomerase (PMI), which is responsible for the conversion of fructose-6-phosphate to mannose-6-phosphate. The growth defect in *Dictyostelium* mutant cells that cannot synthesize mannose-6-phosphate can be rescued by the addition of mannose to the medium, and similarly, this condition can be treated very effectively by simply feeding children mannose (Alper 2001).

Another important group of diseases for which little molecular data are available is the lysosomal storage diseases, such as type-2 Hermansky-Pudlak syndrome and Chediak-Higashi syndrome, which are caused by conserved genes required for proper lysosomal function. Work using *Dictyostelium* is providing much insight into lysosomal function, including the roles of G-protein-coupled receptors, phosphatidylinositol (3,4,5)-trisphosphate, protein kinase B (Akt), syntaxin 7, SNAP, NSF, rab7, clathrin, clathrin adaptor proteins, and BEACH family proteins (De Lozanne 2003; Maniak 2003; Haas 2007).

The mechanism of action of several drugs has been elucidated using *Dictyostelium*. Studies with *Dictyostelium* helped to establish that the mood-stabilizing drugs lithium and valproic acid act via inositol-3-phosphate depletion (Williams et al. 2006). *Dictyostelium* research was also instrumental in identifying the target of bisphosphonates, drugs that inhibit bone resorption used in the treatment of osteoporosis. The molecular target is farnesyl diphosphate synthase in the mevalonate pathway (Rogers 2004).

Resistance to chemotherapeutic agents is one major difficultly for treatment of cancer. Cisplatin, one of the most widely used drugs in chemotherapy, acts by causing DNA damage that results in apoptosis of the cells and, when successful, removal of the tumor. *Dictyostelium* was used to investigate the mechanisms that lead to resistance to cisplatin. The results implied for the first time that the sphingolipid signaling pathway is involved in the sensitivity to cisplatin. These results have been translated to human cells where a similar mechanism operates (for review, see Williams et al. 2006).

Several groups have recently taken advantage of the ability of *Dictyostelium* to engulf various species of bacteria to study the molecular mechanisms that enhance or reduce the ability of pathogenic bacteria to infect eukaryotic cells. Mutants that affect uptake or replication of intracellular pathogens have been isolated, and they contain alterations in conserved signaling molecules such as G proteins and small GTPases, as well as in cytoskeleton components such as coronin, profilin, and the unconventional myosin I (Hilbi et al. 2007). A novel mechanism of bacterial toxicity has also been elucidated using *Dictyostelium*: Bacteria infecting eukaryotic cells secrete virulence factors through a previously uncharacterized pathway, termed type-VI secretion system, that does not depend on known signal peptides (Pukatzki et al. 2006).

Intraspecies and Interspecies Interactions

As a soil dweller, *Dictyostelium* must cope with toxins secreted by bacteria and fungi and escape predation from nematodes and other microorganisms. The sequencing of the genome has revealed a surprisingly high number of genes that could function in producing secondary metabolites, which may be a source of novel compounds such as antibiotics and antiproliferative molecules that could be useful for cancer treatment. The genome contains 40 polyketide synthase genes, which encode highly modular proteins that polymerize two-carbon fragments into an amazing variety of biologically active compounds. Also present are numerous chlorinating and dechlorinating enzymes as well as O-methyl transferases, which could increase the diversity of produced metabolites (Eichinger et al. 2005).

The primitive form of communal life exhibited by *Dictyostelium* offers a new way to consider cooperation, altruism, and cheating that relate ecology and evolution. Because the multicellular organism is formed through the aggregation of cells rather than from a single zygote, like in most organisms, it is possible (and even likely) that not all cells comprising the multicellular organism are genetically identical. The fact that the stalk cells (which make up a quarter of the cells of the mature organism) die to allow sporulation makes it theoretically possible for "cheaters" to emerge in a population that would ensure that their own genetic material gets propagated by avoiding the stalk cell differentiation pathway. In practice, however, this rarely happens (Shaulsky and Kessin 2007). One possible reason for the fact that cheater mutants do not seem to have a selective advantage in the wild may be because individual *Dictyostelium* cells can recognize cells that are most highly related to themselves, and they specifically form fruiting bodies with those cells. This behavior could provide important new insights into the molecular mechanisms of self/nonself recognition.

GENETICS

A large number of mutant strains bearing useful markers for genetic analysis was generated in the 1970s and 1980s by the laboratories of Maurice Sussman, Peter Newell, Barrie Couckell, Reg Deering, and Keith Williams. The mutated genes affect easily tractable traits such as drug resistance, color, and auxotrophy. These strains, in conjunction with chemical mutants affecting various interesting biological processes such as development, chemotaxis, and DNA-damage response, have been used successfully to create libraries of mutants and to organize these mutants into complementation groups. Several linkage and physical maps describe these data (Newell 1978; Loomis et al. 1995). To address the fact that classical genetic complementation experiments had not yet been successful for cloning the genes affected in these mutant strains, Kuspa and Loomis (1992) adapted REMI (restriction-enzyme-mediated integration), a method for random mutagenesis that was first used in yeast, for *Dictyostelium*. In REMI, cells are transformed with a plasmid bearing a drug resistance marker (the most efficient in *Dictyostelium* being the *bsr* gene that confers resistance to blasticidin) together with a restriction enzyme, such as BamHI or DpnII. Mutants thus produced have an insertion in a gene or its flanking sequence, and the gene affected can be isolated from genomic DNA and the mutant recapitulated. This method has been very successful and is widely used to generate nontargeted mutations. More recently, the adaptation of the Cre-*lox* system for use in *Dictyostelium* has proven to be an easy and reliable method for producing multiple gene disruptions (Faix et al. 2004). Using the Cre-*lox* system, the number of gene disruptions is, in theory, unlimited with the use of a single selection marker.

Have Gene Nomenclature Guidelines Been Established for Your Species?

In the mid 1990s, the *Dictyostelium* community established nomenclature guidelines for genes, proteins, strains, alleles, and genotypes. The genome database dictyBase now acts as the clearing house for *Dictyostelium* gene names. The guidelines can be found on the dictyBase website under Research Tools or directly at http://dictybase.org/Dicty_Info/nomenclature_guidelines.html.

Gene Nomenclature Guidelines

Gene names are italicized and lowercase, except for the last letter if it distinguishes different genes sharing the same prefix. These two main rules are followed when attributing a new gene name:

1. When there is an accepted naming system for the gene or gene family across other organisms, that name is adopted to facilitate cross-species comparisons. Recently, as more genomes have been sequenced and their annotation has been of higher quality, an increasing number of genes have been given names that reflect gene nomenclature in other organisms.

2. For uncharacterized genes that do not belong to families with established nomenclature, the naming follows the rules established by Demerec for bacterial gene nomenclature: a description consisting of three lowercase italicized letters, followed by an uppercase italicized letter to distinguish genes with the same descriptor that are related in a significant way. In certain cases (such as large gene families), numbers replace letters as the discriminating character of the gene name.

Strain Nomenclature Guidelines

Each strain generated by a laboratory should be given a systematic name consisting of two or three letters (usually based on the researcher's initials) followed by a unique serial number. Laboratory strain prefixes should be registered with dictyBase (http://dictybase.org/Dicty_Info/nomenclature_guidelines.html#appen1) to avoid name conflicts. This is to ensure that similar strains, for example, those made in the same parental background but in different laboratories, can be distinguished. When researchers do not provide a systematic strain name, dictyBase assigns one that starts with DBS (dictyBase Strain) and is followed by seven digits. In addition, a "strain descriptor" has been implemented by dictyBase that provides a quick overview of the key genetic modifications that produced the strain, including the gene name, promoter, mutations, and tags or reporter genes. For more information, see the dictyBase nomenclature page.

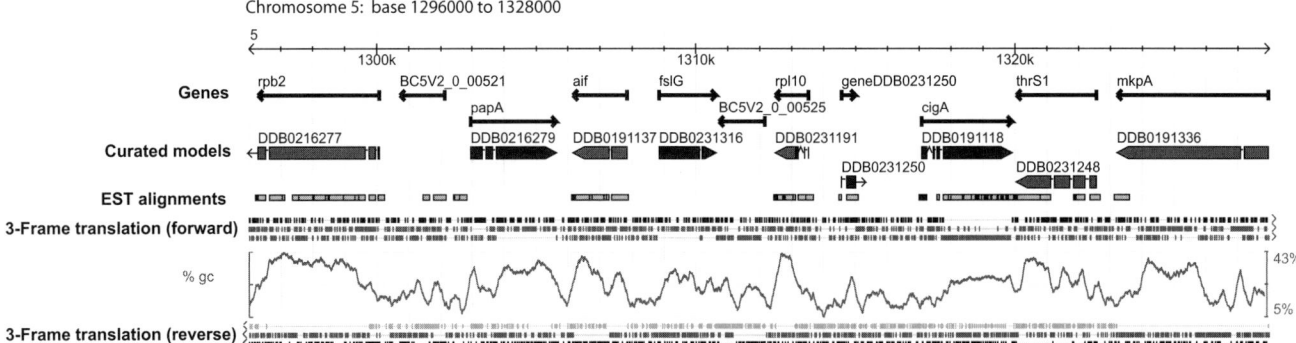

FIGURE 4. The Genome Browser of dictyBase. All sequence information in dictyBase is displayed graphically by the Genome Browser. The tool can be configured to show different genomic information. Shown here are genes, curated gene models, EST alignments, translations of genomic sequence in all six frames, and GC content from a 32-kb region of chromosome 5. Clicking on the "Gene" track leads to the gene page that contains annotations for that gene. Clicking on any of the sequences (gene models or ESTs) leads to the sequence of that feature.

GENOMICS RESOURCES

The *Dictyostelium* genome has been completely sequenced. The nuclear genome consists of 34 Mbp distributed on six chromosomes and is predicted to encode about 12,500 proteins (Eichinger et al. 2005). In addition, mitochondria bear a 55-kb genome that contains 41 genes (Ogawa et al. 2000). Researchers interested in *Dictyostelium* benefit from the presence of a manually curated model organism database, dictyBase (www.dictybase.org) (Fig. 4). The database contains the entire genome sequence and gene predictions associated with the sequence. Curators manually review each gene prediction, and by using supporting data such as published sequences, expressed sequence tags (ESTs), and similarity to other genes, they can correct the gene models if necessary. About 15% of the gene models need to be adjusted. Manually annotated gene models are shown as "curated models." As of April 2008, more than 5000 gene models had been manually curated. The 155,000 ESTs from the *Dictyostelium* cDNA project (Morio et al. 1998) are also in the database, as are all of the GenBank records submitted by individual researchers. This sequence information can be viewed on the corresponding gene pages, downloaded in bulk, or browsed through the Genome Browser.

More than 7000 genes in the database have functional annotations such as gene products (including >5600 manually annotated) and a short description detailing the function of the gene product or other notable features. Genes are also annotated with terms from the Gene Ontology Project (in which close to 30,000 annotations have been made to 7000 genes) with greater than 60% of these being manual annotations. Mutant phenotypes are also being captured; nearly 700 genes now have phenotypic annotations. Annotation coverage is constantly increasing: In 2007, the number of genes with annotations in each of the curated fields increased by 15% on average. A smaller number of genes have a summary paragraph giving a more detailed overview of the knowledge about the gene. Publications referring to *Dictyostelium* genes are also part of dictyBase and can be found on individual gene pages. Topics such as "chemotaxis/motility," "cell-type localization," and "signal transduction" are assigned to each paper to facilitate navigation through long lists of articles.

TECHNICAL APPROACHES

We describe here methods for the growth, maintenance, and development of stocks of *Dictyostelium* strains. Further protocols include various approaches for delivery of DNA, selection of transformants, and the preparation of DNA and RNA.

Protocol 1

Growth and Maintenance of *Dictyostelium* Cells

Most *Dictyostelium* strains used in the laboratory can be grown either with bacteria or in axenic medium (Fey et al. 2007). When grown in the presence of bacteria, cells double approximately every 4 hours, whereas axenically grown cells double more slowly, every 8–12 hours. The cells can be grown in a standard microbiology incubator or on the laboratory bench provided the room temperature is consistently approximately 22°C. Because development is triggered by limited nutrients, it is highly recommended that the culture does not reach a density that allows entry into development ($\geq 4 \times 10^6$ cells/ml). If cells have overgrown, a new culture should be started. Moreover, to ensure genetically uniform strains, it is imperative to start fresh cultures every 2–4 weeks from frozen or silica stocks.

MATERIALS

The recipes for the items marked with <R> begin on page 62.

Reagents

Antibiotics: Ampicillin (20 mg/ml) and streptomycin (20 mg/ml)
Bacterial strain: *Aerobacter aerogenes* or *Escherichia coli* B/R
DB (phosphate buffer/development buffer) <R>
Dictyostelium strain
FM (defined medium) <R>
HL5 (rich axenic medium) <R>
SM broth <R>
SM plates

Add 20 g/liter of bacto agar to SM broth and autoclave. Pour thick plates (35 ml per 100-mm Petri dish). A solution of 1 liter yields about 30 plates. For best results, allow the plates to dry overnight at room temperature and then store them at 4°C. If the plates are too dry, the bacteria will not grow well.

Equipment

Autoclave
Culture flasks (sterile)
Glass spatula
Hemocytometer
Incubator (set at 21–23°C)

This may be either a shaking incubator for growth in suspension or a stationary incubator for growth in plastic dishes. Alternatively, cultures can be maintained at room temperature, provided the temperature in the laboratory is constant and below 25°C.

Light microscope
Micropipettes (1000 µl and 200 µl)
Petri dishes (sterile, 100 mm)

METHODS

Cell densities must be determined using a hemocytometer or a Coulter counter because they are too heavy for accurate reading with light diffraction. Ideal densities for maintaining *Dictyostelium* cell cultures range from about 5×10^4 to 4×10^6 cells/ml. The maximal cell density that axenic cultures reach is 2×10^7 cells/ml.

I. Axenic Growth in Suspension

1. Inoculate cells at a density of about 5×10^4 cells/ml in a flask containing HL5 or FM medium.
   To achieve sufficient aeration, make sure that the volume of medium does not exceed 20% of the volume of the flask.

2. Grow the cells at 21–23°C with shaking at 180 rpm.
   Never allow the cells to reach a density higher than 4×10^6 cells/ml. Stock cultures must be diluted every 2–3 days.

II. Axenic Growth in Plastic Dishes

To observe cells during growth or to produce smaller quantities of cells, grow them axenically in regular bacteriological Petri dishes. Coated tissue culture dishes are not required for Dictyostelium *amebae to grow and attach.*

1. Add 10 ml of HL5 or FM medium and 2×10^4 cells/ml to a 100-mm Petri dish. Gently mix the cells with the medium to obtain an even distribution.
   Cells attach to the plastic surface and can be observed with a standard tissue culture microscope at 30x–100x magnification.

2. Grow the cells at 21–23°C. Dilute the cells when they are subconfluent or reach a density of 2×10^6 to 4×10^6 cells/ml.

III. Growth with Bacteria on SM Plates

1. The most commonly used bacterial species on which to grow *Dictyostelium* are *A. aerogenes* (also known as *Klebsiella aerogenes*) and *E. coli B/r* (Raper 1937). Use 0.2 ml of an overnight culture of bacteria per 100-mm plate and mix with about 8×10^4 *Dictyostelium* cells. Spread the mixture evenly on an SM agar plate using a sterilized glass spatula. The bacterial culture is prepared according to standard protocols by inoculating a colony from a plate or some cells from a frozen tube into 2–3 ml of LB broth in a test tube overnight at 37°C with shaking (Sambrook and Russell 2001).
   If spores are used, take 2–3 spore heads (i.e., 2×10^5 to 4×10^5 cells) mixed with bacteria to inoculate each SM plate. Pipette up and down to mix the spores well with the bacteria. The inoculum is higher when spores are used because the germination process provides time for the bacteria to grow.

2. Incubate at 21–23°C. Invert the plates so that condensation, if formed, does not drip into the culture.

3. Grow the culture until the appearance of the plate changes from opaque (because of the bacterial lawn) to more translucent.
   The bacteria initially produce a confluent lawn because they have a faster growth rate than the amebae, but the latter subsequently "clear" the plate of the bacteria. At this point, the smell of the plate changes from bacterial to a smell similar to forest soil. The expected yield is about 1×10^9 cells.

IV. Growth in Suspension in SM Broth in Association with Heat-killed Bacteria

1. Grow the bacteria in a sterile culture flask containing SM broth until they reach stationary phase.
2. Autoclave the culture to kill the bacteria.
   The heat-killed bacteria serve as the main nutrient for growing Dictyostelium *cells.*
3. Inoculate *Dictyostelium* to give a density of between 3×10^3 and 1×10^4 cells/ml. Incubate with shaking at 180 rpm at 22°C.
   Doubling time is approximately 4 hours and the maximum yield is 1×10^7 to 2×10^7 cells/ml.

V. Growth in Suspension in Phosphate Buffer in Association with Live Bacteria

1. Prepare a 1-liter culture of bacteria by inoculating 1 ml of an overnight culture (as described in Method III) in 1 liter of LB medium.
2. Pellet the bacteria by spinning at 3000g for 15 minutes at 4°C.
3. Resuspend the pellet in 200 ml of DB and spin at 3000g for 15 minutes at 4°C. Repeat the washing step.
4. After the second wash, resuspend the bacteria in DB at an OD_{600} of 8.0. The approximate volume of DB required to resuspend the bacteria is 0.5 liter.
5. Inoculate *Dictyostelium* cells to give a density of between 3×10^3 and 1×10^4 cells/ml.
   The bacteria are still alive but cannot grow in phosphate buffer. They serve as the food source for Dictyostelium.
6. Incubate with shaking at 180 rpm at 22°C.
   Doubling time is approximately 4 hours and the maximum yield is 1×10^7 to 2×10^7 cells/ml.

VI. Initiating an Axenic Culture from Bacterially Grown Cells

Dictyostelium *cells can be easily transferred from a bacterial plate to axenic growth in suspension.*

1. Use a sterile loop to remove amebae from a single colony on a bacterial plate and suspend in 0.2 ml of fresh bacterial overnight culture.
   Grow single colonies as described in Method III above, but use vegetative cells from the feeding front at the perimeter of the clone where cells are feeding and dividing.
2. Spread the bacteria–*Dictyostelium* mixture onto an SM agar plate.
3. Harvest the cells when the plate has been cleared of bacteria but before fruiting bodies have formed (2–3 days) by flooding the plate with 3–4 ml of DB. Recover the cells by gently scraping with a sterile spatula and transfer them to a 50-ml conical plastic tube. Fill the tube with 20 ml of DB, vortex briefly, and spin at 500g for 4 minutes at room temperature.
4. Discard the cloudy supernatant containing the bacteria and resuspend the cells in 25 ml of DB four more times to remove as many bacterial cells as possible.
5. Resuspend the final pellet in 25 ml of HL5 containing streptomycin sulfate (300 µg/ml) and ampicillin (100 µg/ml) and add to a 125-ml flask. Incubate with shaking at 180 rpm for 2–3 days at 22°C.
6. Transfer a small volume (1–2 ml) into a 250-ml flask containing 50 ml of fresh HL5 (antibiotics are not essential).
   Be sure to retain the bacterial plates with clones (from Step 1) until it is clear that the axenic cultures are growing well and are not contaminated.
7. Maintain the cells axenically, as described in Method I.

TROUBLESHOOTING

Problem (all Methods): Cells are not growing or are growing much more slowly than expected.
Solutions:

- Some batches of peptone seem to be lacking certain nutrients, resulting in an unusually long doubling time. This problem can be successfully overcome by adding vitamins from the FM medium recipe. It is also important to use high-quality water such as Milli-Q and glassware free of soap residues. Alternatively, HL5 medium is readily available from Formedium™ (HLG0101; http://www.formedium.com).
- Check the number of cells inoculated. If the cell density is below 2×10^4 cells/ml, the cells might go through a lag phase and have difficulty growing in suspension.
- Check the water quality.
- Check the pH of the medium before and after autoclaving.

Protocol 2

Multicellular Development of *Dictyostelium*

The multicellular phase of *Dictyostelium* is induced when nutrients are depleted from the immediate environment of the cells. In the laboratory, this is accomplished simply by replacing the growth medium with a buffer solution (Fey et al. 2007). For best results, it is important to harvest the cells while they are still in exponential growth (1×10^6 to 4×10^6 cells/ml). At high cell density, many of the cells in a culture will have initiated development, thus yielding asynchronous development. *Dictyostelium* cells can be developed on solid media, either on filter paper (for best synchronicity and if developmental stages need to be harvested for further procedures; see Method I below) or on KK2 plates (for optical examination of morphology or to allow agar blocks to be cut out and reoriented for better visualization; see Method II). If only the early stages of development are important, for example, to study chemotaxis, cells may be developed in suspension (see Method III). Under these conditions, cells will only progress through the first 6–8 hours of development. Addition of cAMP pulses to the starved suspension culture will allow development to progress up to the 12-hour stage, corresponding to the beginning of culmination.

MATERIALS

The recipes for the items marked with <R> begin on page 62.

Reagents

Agar plates (2%, non-nutrient)
> Add 15 g/liter agar to water and autoclave. Pour about 20 ml per 100-mm plate. Make sure that the agar plates are leveled when poured, otherwise the cells will be unevenly distributed, which may result in asynchronous development.

cAMP solution (1 mM) <R>
DB (phosphate buffer/development buffer) <R>
Dictyostelium strain
10x KK2 buffer <R>
KK2 plates
> Add 15 g/liter agar to 1x KK2 buffer. Pour about 20 ml per 100-mm plate.

Equipment

Forceps (two pairs)
Glass pipette (5 ml or 10 ml)
Glass spatula
Peristaltic pump
Petri dishes (150 mm)
Plastic wrap
Whatman no. 50 filter and Whatman no. 3 filters (150 mm)

METHODS

I. Development of *Dictyostelium* Cells on Filter Paper

1. Collect 5×10^8 cells.
   If cells are at a concentration of 2×10^6 cells/ml, use 250 ml.
2. Spin at 500g for 4 minutes at room temperature. Pour off the supernatant.
3. Wash the cells by resuspending the pellet in 0.5 volumes (125 ml) of cold, sterile DB and spin again.
4. Repeat the washes twice for a total of three washes.
5. Resuspend the final pellet in 5 ml of DB.
6. Prepare a 150-mm Petri dish with one Whatman no. 50 filter resting on four Whatman no. 3 filters soaked in DB.
7. Remove any air bubbles trapped between the filters by moving a sterile glass spatula over the surface.
8. Remove any excess liquid.
9. Distribute the cells over the surface of the filter.
   This is best accomplished by taking the entire 5 ml of cells up into a 5- or 10-ml pipette. Starting at the center of the plate, allow the cell suspension to flow onto the filter by moving the pipette tip in an outward spiral.
10. Allow the filters to sit for about 5 minutes or until all of the liquid has soaked in.
11. Carefully aspirate excess fluid.
12. Place a wet paper towel in the center of a piece of plastic wrap about 30 cm in length.
13. Wrap the plastic around the plate to ensure adequate humidity for development.
    Alternatively, the plates and wet paper towels can also be placed into a sealed plastic box that has one small opening for aeration. Do not invert the plates, because filters may not stick to the Petri dish.
14. Incubate at 21–23°C and allow the cells to develop for the desired time period.
    Under optimal conditions, development takes 24 hours to complete.
15. Harvest *Dictyostelium* cells from the filter by lifting the top (Whatman no. 50) filter from the Petri dish. Hold the filter with two forceps and tear it in half.
16. Roll one-half filter carrying cells into a cylinder and place it into a 50-ml conical plastic centrifuge tube.
17. Add 20–25 ml of ice-cold DB to the tube and vortex vigorously to remove the cells from the filter.
18. Open the tube, remove the filter, and place the other half of the filter in the tube and vortex again.
19. Harvest the cells by centrifuging at 500g for 4 minutes at 4°C and use as desired.

II. Development of *Dictyostelium* Cells on KK2 Agar Plates

1. Collect and wash cells as described in Method I, Steps 1–4.
2. After the final wash, resuspend the cells in DB at a density of 1×10^9/ml.
3. Use 200 µl of the cell suspension (2×10^8 cells) per 100-mm KK2 plate and use a sterile glass spatula to spread the cells evenly over the surface of the plate.

4. Wrap the plates with plastic wrap including a wet paper towel as described in Method I, Steps 12–13.

5. Invert the plates and incubate the cells at 22°C.

   Under optimal conditions, development takes 24 hours to complete. Aggregation is complete in about 12 hours, slugs can be observed approximately 16 hours after initiation of development, and culmination begins after about 18 hours.

III. Development of *Dictyostelium* Cells in Suspension

1. Collect and wash the cells as described in Method I, Steps 1–4.
2. After the final wash, resuspend the cells in DB at a density of 1×10^7 cells/ml.
3. Place the cell suspension in a sterile flask and shake at 22°C.

Optional: cAMP Pulsing

4. Start cAMP pulsing 60–120 minutes after the initiation of starvation.

   cAMP pulses are added using a timer to control a peristaltic pump that delivers cAMP to the cell suspension through tubing that goes into the flask.

5. Turn on the pump every 6 minutes for 5 seconds.

   During this 5-second period, cAMP is added to a final concentration of 50 nM in the cell suspension. Use a volume of added cAMP corresponding to less than 0.01% of the cell suspension. For example, for a 1-liter cell suspension, 50 µl of a 1 mM cAMP solution may be added at each pulse.

6. Pulse the cells for approximately 5 hours. The cells are now aggregation-competent and can be used for migration studies.

TROUBLESHOOTING

Problem: Cells are not developing uniformly.
Solutions:

- Use sterile techniques when setting up the cells for development. Because development is induced by starvation and *Dictyostelium* cells feed on bacteria, contamination of the buffer with bacteria or leftover medium interferes with development.
- If developing cells on filters, make sure to remove all air bubbles. The filter must be flat or the cells will collect in puddles in low spots. In addition, ensure that the filters are evenly humid but not wet with puddles. The filters must be able to absorb the liquid from the cells.
- Make sure to use fresh KK2 plates. For good, uniform development, plates should be neither too dry nor wet when inverted and incubated.
- Make sure that the culture is fresh (<1 month old) and has never reached densities higher than 4×10^6 cells/ml. If that is not the case, start fresh cultures.

5. To start cultures from silica stocks, first spread 200 µl of bacteria from a fresh overnight culture on an SM plate.
6. Sprinkle a few silica crystals onto the surface of the plate and incubate at 22°C.
   Spores will germinate and cell growth on the bacterial lawn should be visible after 2–3 days. See Troubleshooting.

TROUBLESHOOTING

Problem: Low cell recovery.
Solutions:
- Make sure that the cells are not exposed to high concentrations of DMSO. When thawing the cells, change the medium rapidly to remove traces of DMSO.
- Do not warm the cells too much upon thawing. Instead of incubating the cells at 37°C, warm up the tube in your hands and transfer to the HL5 plate as soon as enough has thawed to be able to transfer the contents of the tube.

Protocol 4

Transformation of *Dictyostelium* with Plasmid DNA: Calcium Phosphate Precipitation

Dictyostelium is amenable to genetic manipulations that require the introduction of DNA into cells, such as gene knockout, overexpression, antisense RNA expression, RNAi-mediated gene knockdown, and restriction-enzyme-mediated mutagenesis. Two commonly used methods for DNA-mediated transformation in *Dictyostelium* are calcium phosphate precipitation and electroporation (described in Protocol 5) (Gaudet et al. 2007). Calcium phosphate precipitation produces high-copy-number transformants and is often used for overexpression experiments in conjunction with the G418 resistance gene, which needs to be present at high levels to produce efficient selection.

MATERIALS

The recipes for the items marked with <R> begin on page 62.

CAUTION: See the Cautions Appendix for appropriate handling of materials marked with <!>.

Reagents

Bis-Tris HL5 <R>
$CaCl_2$ (1.25 M) <!>
    Dissolve 1.84 g of $CaCl_2$ in 10 ml of distilled H_2O. Filter-sterilize and store at room temperature.
Dictyostelium strain
DNA, for transformation (1 µg/ml)
Glycerol (18%)
    To prepare 4 ml, combine 1.44 ml of autoclaved 50% glycerol in H_2O with 0.56 ml of sterile distilled H_2O and 2.0 ml of 2x HBS. Prepare quantities of this solution according to the number of transformations. It is essential to make this solution fresh for every transformation, because the H_2O may evaporate during storage, which would increase the glycerol concentration.
2x HBS <R>
HL5 medium <R>
10x KK2 buffer <R>
QIAprep Spin Midiprep Kit (QIAGEN 12143; http://www1.qiagen.com)

Equipment

Incubator, set at 22°C and at 37°C
Micropipettes
Sterile hood
Sterile pipettes (10 ml)

METHOD

Preparation of DNA

1. Prepare plasmid DNA according to standard protocols or use a commercially available plasmid preparation kit such as QIAGEN's QIAprep Spin Midiprep Kit.

Transformation

2. Grow *Dictyostelium* cells in HL5 medium at 21–23°C until they reach a density of 1×10^6 to 2×10^6 cells/ml.

    Cultures used for transformations should be fresh; ideally, maintained for less than 2 weeks. If the culture is older, start a new one from silica or frozen stocks. In addition, do not let cultures reach high densities ($>4 \times 10^6$ cells/ml). If this happens, start a new culture. Healthy cells produce better transformation efficiencies.

3. For each transformation, place 10 ml of log-phase cells in a 10-cm Petri dish and allow the cells to attach for 15–30 minutes.

4. Replace the HL5 medium with 12.5 ml of Bis-Tris HL5. Leave the cells in Bis-Tris HL5 for 30 minutes.

5. During this incubation, prepare the DNA solution in a plastic test tube by combining the components shown in the table below, in order, to a final volume of 540 µl.

Component	Amount per sample	Final amount/concentration
Sterile water	230 µl	n/a
DNA (1 µg/µl)	10 µl	10 µg
2x HBS	300 µl	1x

    It is good practice to perform transformations in duplicate to have a backup in case of contamination or other problems.

    See Troubleshooting.

6. Add 60 µl of 1.25 M $CaCl_2$ dropwise to the DNA/HBS mixture while vortexing (the final volume is now 600 µl). Incubate the DNA solution for 15–30 minutes at room temperature.

    Remember to transform a sample lacking DNA (negative control) as well as an empty vector (positive control).

7. Remove the medium from the Petri dish and gradually add the DNA solution to the cells. Add the DNA to the center of the plate to ensure that the small volume used (600 µl) will cover the entire plate. It is important to add the solution slowly to avoid disturbing the cells. Swirl the plate gently to distribute the DNA solution across the plate. Cover the plate and let it stand for 30 minutes while the cells take up the vector.

    Do not incubate longer than 30 minutes, because there is a very small volume of liquid and the cells can dry if the liquid is allowed to evaporate.

8. Without removing the DNA, add 12.5 ml of Bis-Tris HL5 and allow the plate to stand for a minimum of 4 hours to allow the cells to take up the DNA.

Glycerol Shock

9. Carefully aspirate the medium and gently add 4 ml of 18% glycerol in HBS by allowing it to run down the side of the dish.

10. Let the plate stand for exactly 5 minutes.

    The glycerol concentration and the length of the osmotic shock are both highly critical for transformation success. If the concentration is too strong or the incubation is prolonged, the glycerol shock will lyse and kill the cells. If transformation efficiencies are unacceptably low, it may be helpful to titrate the glycerol concentration between 10% and 20%, with increases between 1% and 2%.

11. Gently aspirate the glycerol solution and replace it with 12.5 ml of HL5 medium.

    This step must be done rapidly because until the new medium is added, the cells are still undergoing osmotic shock. To ensure that the duration of the osmotic shock does not exceed 5 minutes, remove the glycerol a few seconds before the 5 minutes are up and add the medium at the end of the last (5th) minute. The efficiency of the glycerol shock can be assessed the day after transformation (before adding the selection drug) by looking at the appearance of the cells under an inverted microscope. Ideally, between 30% and 50% of the cells should look sick (i.e., rounded up). If all of the cells appear to be sick, the glycerol shock was too harsh. It is still possible to obtain transformants, but the efficiency will be low.

12. Incubate overnight at 22°C to allow for expression of the selection marker.

    Selective pressure (drugs or the transfer to FM [defined medium] without uracil and/or thymidine) should be done between 12 and 24 hours following the transformation. For more details, see Protocol 6, which describes the selection of transformants.

    See Troubleshooting.

TROUBLESHOOTING

Problem (Step 6): The DNA/calcium phosphate precipitate is too cloudy or too clear.

Solution: Use MOPS-phosphate (2x stock solution contains 40 mM MOPS and 1.4 mM sodium phosphate) instead of HBS as a buffer to precipitate the DNA. After the addition of the $CaCl_2$, if the DNA does not seem to precipitate within 5 minutes, add 0.5 µl of 500 mM Tris base and mix again. Repeat as necessary, stopping as soon as the precipitate forms. The pH and the temperature affect the quality of the precipitate.

Problem (Step 12): Few or no transformants.
Solutions:

- Make sure that the cells are healthy. The culture must be fresh (<2 weeks old) and must not have reached saturation.

- Titrate the amount of DNA. Most researchers use 5–10 µg of DNA. Increase the amount of DNA by 5 µg up to 35 µg.

- Titrate the glycerol concentration for the glycerol shock. Vary the glycerol concentration by 1–2% to give a final concentration of between 10% and 20%.

Protocol 5

Electroporation

MATERIALS

The recipes for the items marked with <R> begin on page 62.

Reagents

$CaCl_2$ and $MgCl_2$ solution (100 mM each), optional (see Step 8)
Dictyostelium strain
DNA for electroporation (see Step 1)
Electroporation buffer (E buffer) <R>
H-50 buffer <R>

Equipment

Electroporation cuvettes
Electroporator
Incubator, set at 37°C
QIAprep Spin Midiprep Kit (QIAGEN 12143; http://www1.qiagen.com)
Sterile hood
Water bath, set at 80°C

METHOD

Preparation of DNA

1. Prepare plasmid DNA according to standard protocols or use a commercially available plasmid preparation kit such as QIAGEN's QIAprep Spin Midiprep Kit.

Transformation

2. Grow *Dictyostelium* cells in HL5 medium at 21–23°C until they reach a density of 1×10^6 to 2×10^6 cells/ml.

    Cultures used for transformations should be fresh; ideally, maintained for less than 2 weeks. If the culture is older, start a new one from silica or frozen stocks. In addition, do not let cultures reach high densities ($>4 \times 10^6$ cells/ml). If this happens, start a new culture. Healthy cells produce better transformation efficiencies.

3. For ten transformations, pellet approximately 4×10^7 cells by spinning 20 ml of a 2×10^6 cell/ml culture at 500g for 5 minutes at 4°C.

    It is very important to keep everything cold. Prechill sterile H-50 buffer, cuvettes, and sterile tubes containing the DNA on ice. Note that the H-50 buffer may be substituted by E buffer.

4. Wash the cells twice with a half volume (10 ml) of ice-cold H-50 buffer. Pellet the cells by spinning at 500g for 5 minutes at 4°C.

   The cells must be treated gently: Resuspend them with a sterile pipette rather than vortexing.

5. Resuspend the pellet in 1 ml of H-50 buffer to give a density of 4×10^7 cells/ml.

6. Transfer 100 µl of the cell suspension to a cold 0.1-cm electroporation cuvette containing 10 µg of DNA in a volume of approximately 10 µl. Pipette up and down to mix the DNA and the cells.

   The volume of DNA should be less than 10% of the total volume. However, if the volume of DNA is too small, the mixing might not be effective. Vectors to be used for integration by homologous recombination must be linearized before transformation. The restriction enzyme should be inactivated either by heat if that is sufficient, ethanol precipitation, or gel purification of the DNA to avoid undesired random insertional mutagenesis events. It is good practice to do transformations in duplicate to have a backup in case of contamination or other problems. Remember to transform a sample lacking DNA (negative control) as well as an empty vector (positive control).

7. Electroporate the cells twice in the cold at settings of 0.85 kV and 25 µF, waiting about 5 seconds between pulses.

   The time constant should be about 0.6 msec. There is no need for a resistor.

8. Incubate the cuvette on ice for 5 minutes.

   Optional: Some researchers add 1/100th of a volume of a solution of 100 mM $CaCl_2$ and 100 mM $MgCl_2$ to a final concentration of 1 mM each to allow the cells to heal before incubation on ice.

9. Transfer the cells out of the cuvette by adding a few hundred microliters of HL5 from the dish to the cuvette. Pipette up and down to mix and then withdraw the medium and cells and add to a Petri dish containing 12.5 ml of HL5. Repeat once to ensure that all cells are collected. Swirl the dish gently to distribute the cells evenly.

   Note that cell death might be expected upon electroporation. However, the cell survival rate should be above 25%.

10. Incubate overnight at 22°C to allow for expression of the selection marker.

    Selective pressure (drugs or the transfer to FM [defined medium] without uracil and/or thymidine) should be done between 12 and 24 hours following the transformation. For more details, see Protocol 6, which describes the selection of transformants.

    See Troubleshooting.

TROUBLESHOOTING

Problem (Step 10): There are few or no transformants.
Solutions:

- Try different conditions for electroporation: (1) capacitance 3 µF, field strength 1.0–1.5 kV; (2) capacitance 1 µF, field strength 1.0 kV, 0.4-cm cuvette, pulse twice at 5-second intervals; (3) capacitance 3 µF, field strength 1.0–1.1 kV, resistance 5 Ω, 0.4-cm cuvette.
- Try using more cells, for example, 0.8 ml of cells in 0.4-cm cuvettes or 0.4 ml of cells in 0.2-cm cuvettes.
- Use oscillating electroporation instead of the more standard exponential decay electroporation (Alibaud et al. 2003).

Protocol 6

Selection of Transformants

Transformants may be selected in liquid media or on bacterial plates. The latter method reduces the chances of contamination because the cells are grown in buffered agar containing live or dead bacteria, rather than in a rich broth. This method also facilitates the isolation of clones from transformations because each transformant produces a single colony on the plate instead of the pools of transformants obtained in liquid culture. For gene ablation experiments, it is important to obtain a minimum of two independent clones with the same phenotype to exclude the possibility that the phenotype is due to a nonspecific mutation.

MATERIALS

The recipes for the items marked with <R> begin on page 62.

CAUTION: See the Cautions Appendix for appropriate handling of materials marked with <!>.

Reagents

Antibiotics: Ampicillin (20 mg/ml) and streptomycin (20 mg/ml)
Bacterial strain: A. aerogenes or E. coli B/r
Blasticidin S hydrochloride (10 mg/ml in H_2O; use at 10 µg/ml final concentration) <!>
Bleomycin sulfate (20 mg/ml in normal saline) <!>
> Use at 50 µg/ml for the first 5 days and at 15 µg/ml thereafter. The solution is more stable when stored in a glass container in normal saline. Solutions made in distilled H_2O are only stable for a few days.

Dictyostelium strain, transformed from Protocol 4 or 5
G100 plates
> Add 100 µg/ml of G418 (final concentration) to KK2 agar plates (1.5% agar in 1x KK2 and autoclaved) after the agar has cooled to 60°C. Use 20 ml of agar per 10-cm plate. The accuracy is important because it determines the stringency of the selection, so it is recommended to use a pipette to prepare G100 plates.

G418 sulfate (10 mg/ml in H_2O; use at 10 µg/ml final concentration) <!>
Heat-killed bacteria stock (HKB) (for preparation, see Method I)
HL5 medium <R>
Hygromycin B (25 mg/ml in 10 mM HEPES; use at 25 µg/ml final concentration) <!>
10x KK2 buffer <R>
Phleomycin (15 mg/ml; use at 15 µg/ml final concentration; store at –20°C) <!>
SM agar plates (see Protocol 1)

Equipment

Incubator, set at 22°C and at 37°C
Water bath, set at 80°C

METHODS

I. Preparation of Heat-killed Bacteria Stock

1. Grow bacteria on an SM plate at 37°C until a thick lawn is formed (2–3 days).

2. Collect the bacteria by adding about 2 ml of 1x KK2 buffer to the bacterial plate and scraping the cells loose, taking care to avoid breaking the agar surface. Transfer the cells to a 15-ml conical plastic tube. Repeat twice to collect all of the bacteria.

3. Wash the cells twice in 1x KK2 by spinning at 3000g for 10 minutes.
4. Resuspend the cells in 2.5 ml of 1x KK2.
5. Heat-kill the cells by incubating them in a water bath for 20 minutes at 80°C.
6. Freeze in aliquots of 100–200 µl and store at –80°C.

II. Selection of Transformants in Liquid Media

1. Aspirate the medium from the transformed *Dictyostelium* cells and replace it with 12.5 ml of fresh HL5 containing the selection drug. Add 300 µg/ml streptomycin sulfate and 100 µg/ml ampicillin <!> to avoid bacterial contamination.

2. To select stable transformants, replace the medium after 12–24 hours with HL5 plus the selection drug. If desired, add 20 µl of HKB per plate.

3. *Optional:* To isolate clones, pipette the medium up and down to remove cells from the plate and distribute into microtiter well plates. Save the original plate until the clones are stably growing as a backup in case of contamination. Keep replacing the medium every 2–3 days as for other plates.
   The cells must be diluted to yield one cell per well. For a normal transformation, expect between 100 and 1000 transformants. Thus, use 100–1000 wells if the entire cell population is to be plated.

4. Replace the medium two to three times per week until transformants are visible. If desired, add 20 µl of HKB per plate.
   See Troubleshooting.

III. Selection of Transformants on Solid Bacterial Plates

Note that this procedure is not compatible with the use of auxotrophic markers because the bacteria provide nucleotides that must be absent for the selection to be effective.

1. The day after transformation, collect all of the *Dictyostelium* cells from the transformation plate in 5–10 ml of 1x KK2.

2. Wash the cells with 5 ml of 1x KK2.

3. Plate approximately 3×10^5 *Dictyostelium* cells per G100 plate with 0.5 ml of HKB. Add 300 µg/ml streptomycin sulfate and 100 µg/ml ampicillin to avoid bacterial contamination.
   Live bacteria can be used rather than HKB. Avoid using nutrient plates such as SM, because these require a lot more G418 than do KK2 plates, and it is thus harder to block the growth of untransformed cells.

4. Incubate in a moist chamber at 22°C until transformants appear. Separate plaques visible on the bacterial lawn are clones of single transformed cells.
   See Troubleshooting.

TROUBLESHOOTING

Problem (Methods II and III, Step 4): Untransformed cells are present, but there are very few or no transformed cells.
Solution: Titrate the concentration of selection drug using wild-type cells. Use the concentration at which all cells that do not harbor the drug resistance gene die. This may need to be done for every cell line, new batches of media, and new stocks of antibiotics, in particular when the drug comes from a different lot. It is possible that the optimal drug concentration will be outside the range suggested in the Reagents section by up to tenfold.

Problem (Methods II and III, Step 4): Transformants grow in the control without DNA.
Solution: Verify that the drug is active. Titrate the drug concentration until the growth of wild-type cells is completely suppressed.

Protocol 7

DNA Extraction

Extraction of genomic DNA is used to clone gene fragments and for analysis of mutants to determine the site of vector integration. Because *Dictyostelium* cells contain relatively high levels of carbohydrate and nucleases, commercially available DNA preparation kits are not very successful. The DNA isolated according to the following protocol is suitable for digestion by restriction enzymes, amplification by PCR, and Southern blotting (Pilcher et al. 2007).

MATERIALS

The recipes for the items marked with <R> begin on page 62.

CAUTION: See the Cautions Appendix for appropriate handling of materials marked with <!>.

Reagents

Chloroform <!>
Dictyostelium strain
Ethanol (70% and 100%) <!>
NaOAc (3 M)
Nuclei buffer <R>
Phenol (pH 8.0) <!>

> Add 10 ml of 50 mM Tris <!> (pH 8.0) to 35 ml of liquefied phenol. Shake vigorously and spin at 2500g. Remove the upper layer (Tris) and repeat by adding 10 ml of 50 mM Tris (pH 8.0), shaking, spinning, and removing the Tris layer. Top off the phenol with 5 ml of 50 mM Tris (pH 8.0) and store at 4°C in the dark. (Cover with aluminum foil if an opaque container is not available.) The pH of phenol is very important. An acidic pH degrades DNA and thus partitions it to the organic phase. If the pH is below 7.0, the DNA may be lost due to improper phase separation. In addition, phenol oxidizes by a free radical process indicated by a pinkish color. Do not use oxidized phenol because it can result in DNA damage.

Phenol:chloroform (1:1, pH 8.0) <!>

> Mix an equal volume of phenol (pH 8.0) with chloroform. Store the phenol:chloroform mixture at 4°C in the dark. (Cover with aluminum foil if an opaque container is not available.) Work in a chemical fume hood while using phenol:chloroform.

Proteinase K buffer <R>
RNase A (10 mg/ml; Sigma-Aldrich R6513)
TE buffer (pH 7.4) <R>
Triton X-100 (20%) <!>

Equipment

Centrifuge
Microfuge
Spectrophotometer
Vortex
Water bath, set at 65°C

METHOD

1. Grow *Dictyostelium* cells as described in Protocol 1.

   Use vegetative cells or cells developed in suspension for a short period to avoid cell aggregation, which makes lysis less effective and reduces the yield. Development for a short period reduces the amount of protein, making for a cleaner, larger yield.

2. Pellet 2×10^7 to 3×10^7 *Dictyostelium* cells in a 50-ml polypropylene tube by spinning at 500g for 4 minutes at room temperature. Pour off the supernatant. Keeping the tube inverted, carefully drain the remaining supernatant onto a paper tissue.

3. Resuspend the cells by gentle pipetting in 1 ml of nuclei buffer. The pellet may be loosened by scraping the tube over an uneven surface such as a microfuge tube rack. Transfer to a clean 1.5-ml microfuge tube.

4. Add 200 µl of 20% Triton X-100 and incubate on ice for 5 minutes.

5. Spin at 12,000g (highest setting for a typical benchtop microfuge) for 5 minutes. Carefully remove the supernatant by pipetting.

6. Loosen the pellet by vortexing briefly and resuspend the cells in 300 µl of proteinase K buffer.

7. Incubate for 30 minutes at 65°C.

8. Extract the nucleic acids by adding an equal volume (300 µl) of phenol:chloroform (1:1). Invert the tubes or vortex gently.

   Be careful not to vortex vigorously because this can lead to shearing of genomic DNA.

9. Spin at 12,000g for 10 minutes.

   Following centrifugation, two distinct phases should be seen. The DNA will be contained in the aqueous (upper) layer, whereas the phenol:chloroform will make up the organic (bottom) layer. A distinct interphase containing the degraded proteins will appear between the layers.

10. Carefully pipette the aqueous (upper) layer into a fresh 1.5-ml tube.

    Do not disturb the interphase while removing the aqueous layer. Failure to do so will result in a DNA sample containing proteins and other contaminants, which can have an inhibitory effect on subsequent enzymatic reactions.

11. Extract the aqueous layer by adding 1 volume (300 µl) of chloroform. Invert the tubes or vortex very gently.

    This final extraction with chloroform is intended to remove any remaining phenol because it is inhibitory for downstream applications.

12. Spin at 12,000g for 10 minutes. Carefully pipette the aqueous layer into a fresh 1.5-ml tube.

13. Precipitate the aqueous layer with 2.5 volumes (750 µl) of ice-cold 100% ethanol. Allow the tube to sit for 5 minutes at room temperature or longer at –20°C.

    Salt does not need to be added to this precipitation step. The residual MgOAc from the nuclei buffer is sufficient. However, the subsequent precipitation (Step 16) does require the addition of salt.

14. Pellet the DNA by spinning at 12,000g for 15 minutes at 4°C. Remove the supernatant with a pipette without disturbing the pellet.

    The DNA will appear as a tight, whitish pellet at the bottom of the tube. Be sure to look for it before discarding the ethanol.

15. Resuspend the pellet in 100 µl of TE buffer (pH 7.4) containing 10 µg/ml RNase A and incubate for 15 minutes at room temperature.

16. Precipitate again by adding 1/10 volume (10 µl) of 3 M NaOAc and 2.5 volumes (250 µl) of ice-cold 100% ethanol. Allow the tube to sit for 15 minutes on ice or longer at –20°C.

17. Spin at 12,000g for 15 minutes at 4°C and then remove the supernatant with a pipette without disturbing the pellet.
18. Wash the pellet with two volumes (200 µl) of 70% ethanol and spin at 12,000g for 2 minutes.
19. Carefully remove the supernatant with a pipette and leave the pellet to dry at room temperature.

    Do not allow the pellet to dry for too long. An overdry pellet will be difficult to dissolve in TE.

20. Resuspend the DNA pellet in 50 µl of TE buffer (pH 7.4).
21. Find the DNA concentration using the optical density (OD). Dilute the DNA 1:5 in TE buffer (pH 7.4) and determine the OD_{260} using a spectrophotometer. Calculate the DNA concentration with the following formula: OD_{260} reading × dilution factor × ([50 µg/ml]/OD_{260} unit) = DNA concentration (µg/ml). To calculate the DNA concentration in µg/µl, divide by 1000. To determine the quality of the resulting DNA, take an OD_{280} reading as well. Pure DNA has an OD_{260}/OD_{280} ratio of in the range of 1.8 to 2.0. OD_{260}/OD_{280} ratios below 1.8 indicate contamination with proteins and/or phenol.
22. Store the genomic DNA sample at 4°C.

    See Troubleshooting.

TROUBLESHOOTING

Problem (Step 22): The DNA will not be digested by restriction enzymes or yields poor PCR results.

Solution: The DNA may be contaminated by phenol. Be careful when pipetting the aqueous layer during the phenol:chloroform extraction. Leave a small amount (10–20%) near the interphase.

Protocol 8

RNA Extraction

Northern blots, reverse transcriptase–polymerase chain reaction (RT-PCR), and microarrays are common techniques for the analysis of gene expression. The extraction of RNA from *Dictyostelium* is relatively easy because RNA levels are very high in comparison to DNA levels; they are estimated to be approximately 40 times those of DNA. The RNA isolation protocol presented here is adapted from Franke et al (1977). Certain commercially available kits, such as Trizol (Invitrogen) and RNeasy (QIAGEN) have been used successfully, although lysis conditions need to be adjusted. RNA samples are stable for several years at –80°C in diethylpyrocarbonate (DEPC)-treated H_2O. For longer-term storage, the RNA pellet can be stored in 100% ethanol at –80°C.

MATERIALS

The recipes for the items marked with <R> begin on page 62.

CAUTION: See the Cautions Appendix for appropriate handling of materials marked with <!>.

Reagents

Chloroform <!>

DEPC-treated H_2O

    Add 0.1% diethylpyrocarbonate (DEPC) <!> to distilled H_2O. Shake vigorously, leave at room temperature for 3 hours to overnight, and autoclave.

Dictyostelium strain

Ethanol (70% and 100%) <!>

    Make 70% ethanol solution with DEPC-treated H_2O. To minimize RNase contamination, always use fresh aliquots of ethanol or keep ethanol used for RNA preparations separate from reagents used for other experiments.

GSEM buffer <R>

10x KK2 <R>

NaOAc (4 M)

    Adjust pH to 6.0 with acetic acid <!>. Use DEPC-treated H_2O to make the solution in a baked bottle and autoclave.

Phenol (pH 4.7) <!>

    Add 10 ml of 50 mM Tris (pH 4.7) to 35 ml of liquefied phenol. Shake vigorously and spin at 2500g. Remove the upper layer (Tris) and repeat by adding 10 ml of 50 mM Tris (pH 4.7), shaking, spinning, and removing the Tris layer. Top off the phenol with 5 ml of 50 mM Tris (pH 4.7) and store at 4°C in the dark. (Cover with aluminum foil if an opaque container is not available.) The pH of phenol is very important. An acidic pH degrades DNA and thus partitions it to the organic phase. A pH below 7.0 allows the RNA to separate to the aqueous layer for subsequent extraction. Use RNA grade phenol to minimize possible exposure to RNases.

Phenol:chloroform (1:1, pH 4.7) <!>

    Mix an equal volume of phenol (pH 4.7) with chloroform. Store the phenol:chloroform mixture at 4°C in the dark. (Cover with aluminum foil if an opaque container is not available.) Work in a chemical fume hood while using phenol:chloroform.

Equipment

> Centrifuge
> Vortex

METHOD

1. Grow *Dictyostelium* cells as described in Protocol 1.

2. Pellet 1×10^8 *Dictyostelium* cells in a 50-ml polypropylene tube by spinning at 500*g* for 4 minutes at 4°C. Pour off the supernatant.

   It is important to wash the cells at a cold temperature to make sure they do not change their gene expression pattern before lysis. Changes in gene expression may occur very rapidly once nutrients are removed; thus, when working with vegetative cells, this is particularly important.

3. Wash the cells twice by adding 0.5 volume (25 ml) of KK2 buffer. Resuspend the cells by vortexing and spin at 500*g* for 4 minutes at 4°C. Pour off the supernatant. Remove excess buffer with a pipette.

4. Resuspend the cells in 1 ml of KK2 buffer. Transfer to a 1.5-ml microfuge tube and pellet the cells by spinning in a microfuge at 12,000*g* (highest setting for a typical benchtop microfuge) for 1 minute at room temperature. Remove the supernatant using a 1000-μl micropipettor.

   It is essential to perform the remaining steps quickly and on ice to minimize degradation of the RNA.

5. Dissolve the pellets in 200 μl of GSEM buffer with vigorous vortexing for 1–2 minutes.

   It is important to lyse the cells thoroughly for optimal efficiency of RNA extraction, which might be difficult for cells of later developmental stages. Vortex time can be increased to 5 minutes for postaggregative cells to ensure proper lysis.

6. Add 200 μl of phenol:chloroform 1:1 (1 volume) to separate the nucleic acids from the proteins. Vortex for 30 seconds and then spin at 12,000*g* for 5 minutes at room temperature.

   Following centrifugation, two distinct phases should be seen: The RNA will be contained in the aqueous (upper) layer, whereas the phenol and chloroform will make up the organic (bottom) layer. An interphase containing the denatured proteins should be visible between the aqueous and organic layers.

7. Remove the aqueous (upper) phase with a pipette and transfer to a clean 1.5-ml microfuge tube.

8. Reextract the organic (bottom) phase by adding 100 μl of DEPC-treated H_2O, vortexing, and spinning as detailed in Step 5. This maximizes the yield.

9. Remove the upper aqueous layer and combine it with that from Step 6.

10. Extract twice with 300 μl (1 volume) of phenol:chloroform 1:1, as in Steps 6 and 7.

11. Extract twice with 300 μl (1 volume) of chloroform, as in Steps 6 and 7 (omitting phenol).

    These final extractions with chloroform are intended to remove any remaining phenol because it is inhibitory for downstream applications.

12. Add 15 μl of 4 M NaOAc and 600 μl (2 volumes) of ice-cold 100% ethanol to precipitate the RNA. Leave on ice for at least 10 minutes or at –20°C.

13. Spin in a microfuge at 12,000*g* for 10 minutes at room temperature. Discard the supernatant (ethanol) with a pipette.

14. Rinse the RNA pellet twice by adding 70% ice-cold ethanol and spinning for 1–2 minutes at 4°C. Pipette off the supernatant.

As in the previous step, care should be taken to avoid disturbing the RNA pellet when removing the ethanol.

15. Allow the pellet to dry for approximately 15 minutes at room temperature and dissolve in 20 µl of DEPC-treated H_2O. Note that RNA does not dissolve easily. Leave the samples on ice for 15–30 minutes and then resuspend by pipetting up and down.

16. Determine the RNA concentration using the optical density (OD). Dilute the RNA 1:1000 in DEPC-treated H_2O and determine the OD_{260} using a spectrophotometer. Calculate the RNA concentration with the following formula: OD_{260} reading × dilution factor × ([40 µg/ml]/OD_{260} unit) = RNA concentration (µg/ml). To calculate the RNA concentration in µg/µl, divide by 1000. To determine the quality of the resulting RNA, take an OD_{280} reading as well. Pure RNA has an OD_{260}/OD_{280} ratio of in the range of 1.8 to 2.0. OD_{260}/OD_{280} ratios below 1.8 indicate contamination with proteins and/or phenol.

    See Troubleshooting.

17. Store RNA samples at –80°C; they will be stable for several years.

TROUBLESHOOTING

Problem (Step 16): Low yield of RNA.
Solutions:

- The RNA may be degraded. Make fresh solutions and be sure to use DEPC-treated H_2O. Ensure that all equipment is RNA grade (tips, tubes, etc.).
- The RNA may not be fully resuspended. Leave the RNA on ice for an additional 30 minutes, pipette up and down to resuspend, and repeat quantification.

Recipes

CAUTION: See the Cautions Appendix for appropriate handling of materials marked with <!>.

4x AMINO ACIDS (in 8 mM NaOH)

Reagent	Quantity (for 5 liters)	Final concentration
NaOH <!>	1.6 g	8 mM
Arginine · HCl <!>	14 g	1.3 mM
Asparagine	6 g	9 mM
Cysteine <!>	4 g	6.6 mM
Glutamic acid	10 g	13.6 mM
Glycine	18 g	48 mM
Histidine · HCl · H_2O	6 g	5.7 mM
Isoleucine	12 g	18.3 mM
Leucine	18 g	27.4 mM
Lysine · HCl	18 g	19.7 mM
Methionine	6 g	8 mM
Phenylalanine	10 g	12 mM
Proline	16 g	28 mM
Threonine	10 g	16.8 mM
Tryptophan <!>	4 g	4 mM
Valine <!>	14 g	24 mM

Dissolve in 5 liters of distilled H_2O. The pH should be about 4.2. Store at –20°C in 200-ml aliquots. Use the amino acid forms indicated, otherwise solubility problems will arise.

BIS-TRIS HL5

Reagent	Quantity (for 1 liter)	Final concentration
Bis-Tris	2.1 g	10 mM
Proteose peptone (Oxoid)	10 g	1%
Yeast extract	5 g	0.5%
D-Glucose	10 g	1%

Dissolve in distilled H_2O; adjust the volume to 1 liter. Autoclave to fully solubilize. Readjust pH to 7.1 (it will have gone down to about pH 6.8). Filter-sterilize. Store at room temperature.

cAMP SOLUTION (1 mM)

Dissolve 330 mg of cAMP (Sigma A9501) in 100 ml of DB. Adjust the pH to 6.5 with NaOH <!> and filter-sterilize. Store in 5-ml aliquots at –20°C.

DB (PHOSPHATE BUFFER/DEVELOPMENT BUFFER)

Reagent	Final concentration
Na_2HPO_4 (1 M)	5 mM
KH_2PO_4 (1 M)	5 mM
$CaCl_2$ (10 mM) <!>	1 mM
$MgCl_2$ (20 mM) <!>	2 mM

Prepare the phosphate solution as 25 mM (5x), adjust the pH to 6.5, and autoclave. Make 10x $CaCl_2$ (10 mM) and $MgCl_2$ (20 mM) solutions separately and autoclave. To make 1 liter of DB, mix 600 ml of distilled, autoclaved H_2O with 200 ml of 5x phosphate solution and 100 ml each of 10x $CaCl_2$ and 10x $MgCl_2$.

ELECTROPORATION BUFFER (E BUFFER)

Reagent	Quantity (10 ml)	Final concentration
Sodium phosphate (1 M, pH 6.1)	0.1 ml	10 mM
Sucrose (5 M)	0.1 ml	50 mM

Autoclave. E buffer is an alternative to H-50 buffer below.

FM (DEFINED MEDIUM)

FM is a defined medium useful for maintaining auxotrophic cell lines. It is made up from four stock solutions (amino acids, vitamins, salts, and trace metals), which can be stored indefinitely at –20°C. Recipes for the stock solutions are given below. The recipes for items marked with <R> are also listed here.

Reagent	Quantity (for 1 liter)	Final concentration
4x Amino acids (in 8 mM NaOH) <R>	250 ml	1x
20x Vitamins <R>	50 ml	1x
50x Salts <R>	20 ml	1x
Trace metals (10,000x) <R>	0.1 ml	1x
Glucose	10 g	56 mM
K_2HPO_4	0.87 g	5 mM
Dihydrostreptomycin sulfate	50 mg	35 mM

Check the pH and adjust to 6.5 (if necessary) with HCl or NaOH <!>. Adjust the volume to 1 liter with distilled H_2O. Autoclave or filter-sterilize and store in the dark at 4°C. Useful tip: Adjust the volume to 90% of the final volume to leave room for supplements. FM medium is available from Formedium (http://www.formedium.com; FMM0101).

GSEM BUFFER

Reagent	Quantity (1 ml)	Final concentration
Guanidine thiocyanate <!>	5 g	50%
Sarkosyl (Na salt) <!>	0.05 g	0.5%
EDTA (0.5 M, pH 8.0)	0.5 ml	25 mM

Adjust the pH to 7.0. Do not autoclave. Store at 4°C. Add 0.1% β-mercaptoethanol <!> just before use.

H-50 BUFFER

Reagent	Quantity (for 1 liter)	Final concentration
HEPES	4.76 g	20 mM
KCl <!>	3.73 g	50 mM
NaCl	0.58 g	10 mM
$MgSO_4$ <!>	0.12 g	1 mM
$NaHCO_3$	0.42 g	5 mM
$NaH_2PO_4 \cdot 2H_2O$	0.156 g	1 mM

Dissolve in distilled H_2O to give a final volume of 1 liter. Adjust the pH to 7.0 with HCl <!> or NaOH <!> as appropriate. Autoclave and store cold or frozen. H-50 buffer is an alternative to E buffer on page 63.

2x HBS

Reagent	Quantity (for 1 liter)	Final concentration
NaCl	4.0 g	270 mM
KCl <!>	0.18 g	10 mM
Na_2HPO_4	0.05 g	1.5 mM
HEPES	2.5 g	40 mM
D-Glucose	0.5 g	10 mM

Dissolve in 250 ml of distilled H_2O. Adjust pH to 7.1 with NaOH <!>. Filter-sterilize and store at –20°C.

HL5 (RICH AXENIC MEDIUM)

Reagent	Quantity (for 1 liter)	Final concentration
Proteose peptone 2 (Difco 212120)	5 g	0.5%
Thiotone E peptone (BBL 212302)	5 g	0.5%
Glucose	10 g	56 mM
Yeast extract (Oxoid LP0021)	5 g	0.5%
$Na_2HPO_4 \cdot 7H_2O$	0.35 g	1.3 mM
KH_2PO_4	0.35 g	2.6 mM

Dissolve the reagents in just under 1 liter of distilled H_2O, adjust the pH with HCl to 6.4–6.7 (if necessary), and then bring the volume to 1 liter and autoclave. HL5 medium must be stored at 4°C, but to ensure consistent growth rates, warm to 22°C before use. It is possible to keep a working solution at room temperature for several days. Antibiotics such as ampicillin <!> (100 μg/ml) and/or streptomycin sulfate (300 μg/ml) may be added to reduce bacterial contamination.

10x KK2 BUFFER

Reagent	Quantity (for 1 liter)	Final concentration
KH_2PO_4 (monobasic)	22 g	1.6 mM
K_2HPO_4 (dibasic)	7 g	40 mM

Dissolve in 1 liter of distilled H_2O. Autoclave and dilute 1:10 to use at a 1x concentration.

NUCLEI BUFFER

Reagent	Quantity (100 ml)	Final concentration
Tris-HCl (1 M, pH 7.4)	2 ml	20 mM
MgOAc (1 M)	0.5 ml	5 mM
EDTA (0.5 M, pH 8.0)	0.1 ml	0.5 mM
Sucrose	5 g	5%

PROTEINASE K BUFFER

Reagent	Quantity (10 ml)	Final concentration
Tris-HCl (1 M, pH 7.4)	1 ml	100 mM
EDTA (0.5 M, pH 8.0)	0.1 ml	5 mM
Proteinase K <!>	1 mg	0.1 mg/ml
SDS <!>	0.1 g	1%

50x SALTS

Reagent	Quantity (for 1 liter)	Final concentration
1 M NH_4Cl <!>	50 ml	50 mM
1 M $CaCl_2$ <!>	1 ml	1 mM
0.1 M $FeCl_3$ <!>	50 ml	5 mM
1 M $MgCl_2$ <!>	20 ml	20 mM

Adjust the volume to 1 liter with distilled H_2O. Filter-sterilize and store in 40-ml aliquots at –20°C.

SM BROTH

Reagent	Quantity (for 1 liter)	Final concentration
Glucose	10 g	56 mM
Proteose peptone 2	10 g	1%
Yeast extract	1 g	0.1%
NH_4Cl <!>	1 g	20 mM
KH_2PO_4	1.9 g	14 mM
K_2HPO_4	0.6 g	3.4 mM

Dissolve in 1 liter of distilled H_2O. Adjust the pH to 6.0–6.4 with KOH <!>.

TE BUFFER (pH 7.4)

Reagent	Quantity (w/v)	Final concentration
Tris-HCl (1 M, pH 7.4)	10 ml	10 mM
EDTA (0.5 M)	2 ml	1 mM

10,000x TRACE METALS

Reagent	Quantity (for 100 ml)	Final concentration
Na_2-EDTA · $2H_2O$	4.84 g	130 mM
$ZnSO_4$ · $7H_2O$	2.30 g	80 mM
H_3BO_3	1.11 g	18 mM
$MnCl_2$ · $4H_2O$	0.51 g	26 mM
$CoCl_2$ · $6H_2O$	0.17 g	70 mM
$CuSO_4$ · $5H_2O$	0.15 g	60 mM
$(NH_4)_6Mo_7O_{24}$ · $4H_2O$	0.10 g	8 mM

Dissolve in under 100 ml of distilled H_2O. Adjust the pH to 6.5 and bring the volume to 100 ml. Freeze small aliquots (≤1 ml) at –20°C.

20x VITAMINS

Reagent	Quantity (for 1 liter)
Biotin (4 mg in 1 liter of 10 mM $NaHCO_3$)	100 ml
Cyanocobalamin (5 mg in 1 liter of H_2O)	20 ml
Folic acid (40 mg in 1 liter of 10 mM $NaHCO_3$)	100 ml
Lipoic acid (80 mg in 1 liter of 10 mM $NaHCO_3$)	100 ml
Riboflavin (100 mg in 1 liter of 10 mM $NaHCO_3$)	100 ml
Thiamine · HCl (120 mg in 1 liter of H_2O)	100 ml

Adjust the volume to 1 liter with distilled H_2O and store in 100-ml aliquots at –20°C.

REFERENCES

Alibaud, L., Cosson, P., and Benghezal, M. 2003. *Dictyostelium discoideum* transformation by oscillating electric field electroporation. *Biotechniques* **35:** 78–80, 82–83.

Alper, J. 2001. Saving lives with sugar. *Science* **291:** 2339.

Baldauf, S.L., Rogern, A.J., Wenk-Siefertn, I., and Doolittlen, W.F. 2000. A kingdom-level phylogeny of eukaryotes based on combined protein data. *Science* **290:** 972–977.

Bloomfield, G., Tanaka, Y., Skelton, J., Ivens, A., and Kay, R.R. 2008. Widespread duplications in the genomes of laboratory stocks of *Dictyostelium discoideum*. *Genome Biol.* **9:** R75.

Bonner, J.T. 1944. A descriptive study of the development of the slime mold *Dictyostelium discoideum*. *Am. J. Bot.* **31:** 175–182.

Brefeld, O. 1896. *Dictyostelium mucoroides*. Ein neuer organismus aus der verwandtschaft der myxomyceten. *Adh. Senck. Natur. Ges. Frankfurt* **7:** 85–107.

Chisholm, R.L. and Firtel, R.A. 2004. Insights into morphogenesis from a simple developmental system. *Nat. Rev. Mol. Cell Biol.* **5:** 531–541.

De Lozanne, A. 2003. The role of BEACH proteins in *Dictyostelium*. *Traffic* **4:** 6–12.

De la Roche, M.A., Smith, J.L., Betapudi, V., Egelhoff, T.T., and Cote, G.P. 2002. Signaling pathways regulating *Dictyostelium* myosin II. *J. Muscle Res. Cell Motil.* **23:** 703–718.

Desjardins, M., Houde, M., and Gagnon, E. 2005. Phagocytosis: The convoluted way from nutrition to adaptive immunity. *Immunol. Rev.* **207:** 158–165.

Devreotes, P. and Janetopoulos, C. 2003. Eukaryotic chemotaxis: Distinctions between directional sensing and polarization. *J. Biol. Chem.* **278:** 20445–20448.

Dormann, D. and Weijer, C.J. 2006. Imaging of cell migration. *EMBO J.* **25:** 3480–3493.

Eichinger, L., Pachebat, J.A., Glockner, G., Rajandream, M.A., Sucgang, R., Berriman, M., Song, J., Olsen, R., Szafranski, K., Xu, Q., et al. 2005. The genome of the social amoeba *Dictyostelium discoideum*. *Nature* **435:** 43–57.

Faix, J., Kreppel, L., Shaulsky, G., Schleicher, M., and Kimmel, A.R. 2004. A rapid and efficient method to generate multiple gene disruptions in *Dictyostelium discoideum* using a single selectable marker and the Cre-*loxP* system. *Nucleic Acids Res.* **32:** e143.

Fey, P., Kowal, A.S., Gaudet, P., Pilcher, K.E., and Chisholm, R.L. 2007. Protocols for growth and development of *Dictyostelium discoideum*. *Nat. Protoc.* **2:** 1307–1316.

Franke, J. and Kessin, R. 1977. A defined minimal medium for axenic strains of *Dictyostelium discoideum*. *Proc. Natl. Acad. Sci.* **74:** 2157–2161.

Gaudet, P., Pilcher, K.E., Fey, P., and Chisholm, R.L. 2007. Transformation of *Dictyostelium discoideum* with plasmid DNA. *Nat. Protoc.* **2:** 1317–1324.

Gaudet, P., Williams, J.G., Fey, P., and Chisholm, R.L. 2008. An anatomy ontology to represent biological knowledge in *Dictyostelium discoideum*. *BMC Genomics* **9:** 130.

Gilbert, O.M., Foster, K.R., Mehdiabadi, N.J., Strassmann, J.E., and

Queller, D.C. 2007. High relatedness maintains multicellular cooperation in a social amoeba by controlling cheater mutants. *Proc. Natl. Acad. Sci.* **104:** 8913–8917.

Harper, R.A. 1926. Morphogenesis in *Dictyostelium*. *Bull. Torrey Bot. Club* **53:** 229–268.

Hass, A. 2007. The phagosome: Compartment with a license to kill. *Traffic* **8:** 311–330.

Hilbi, H., Weber, S.S., Ragaz, C., Nyfeler, Y., and Urwyler, S. 2007. Environmental predators as models for bacterial pathogenesis. *Environ. Microbiol.* **9:** 563–575.

Hughes, P., Marshall, D., Reid, Y., Parkes, H., and Gelber, C. 2007. The costs of using unauthenticated, over-passaged cell lines: How much more data do we need? *Biotechniques* **5:** 575–586.

Kessin, R. 2001. Dictyostelium: *Evolution, cell biology, and the development of multicellularity*. Cambridge University Press, United Kingdom.

Konijn, T.M., Van De Meene, J.G., Bonner, J.T., and Barkley, D.S. 1967. The acrasin activity of adenosine-3,5-cyclic phosphate. *Proc. Natl. Acad. Sci.* **58:** 1152–1154.

Kuspa, A. and Loomis, W.F. 1992. Tagging developmental genes in *Dictyostelium* by restriction enzyme-mediated integration of plasmid DNA. *Proc. Natl. Acad. Sci.* **89:** 8803–8807.

Loomis, W.F., Welker, D., Hughes, J., Maghakian, D., and Kuspa, A. 1995. Integrated maps of the chromosomes in *Dictyostelium discoideum*. *Genetics* **141:** 147–157.

Ma, S., Fey, P., and Chisholm, R.L. 2001. Molecular motors and membrane traffic in *Dictyostelium*. *Biochim. Biophys. Acta* **1525:** 234–244.

Maniak, M. 2003. Fusion and fission events in the endocytic pathway of *Dictyostelium*. *Traffic* **4:** 1–5.

Morio, T., Urushihara, H., Saito, T., Ugawa, Y., Mizuno, H., Yoshida, M., Yoshino, R., Mitra, B.N., Pi, M., Sato, T., et al. 1998. The *Dictyostelium* developmental cDNA project: Generation and analysis of expressed sequence tags from the first-finger stage of development. *DNA Res.* **5:** 335–340.

Newell, P.C. 1978. Genetics of the cellular slime molds. *Annu. Rev. Genet.* **12:** 69–93.

Ogawa, S., Yoshino, R., Angata, K., Iwamoto, M., Pi, M., Kuroe, K., Matsuo, K., Morio, T., Urushihara, H., Yanagisawa, K., and Tanaka, Y. 2000. The mitochondrial DNA of *Dictyostelium discoideum*: Complete sequence, gene content and genome organization. *Mol. Gen. Genet.* **263:** 514–519.

Olive, F.W. 1902. Monograph of the Acrasieae. *Proc. Boston Soc. Natur. Hist.* **30:** 451–513.

Parent, C.A. 2004. Making all the right moves: Chemotaxis in neutrophils and *Dictyostelium*. *Curr. Opin. Cell Biol.* **16:** 4–13.

Pilcher, K.E., Fey, P., Gaudet, P., Kowal, A.S., and Chisholm, R.L. 2007. A reliable general purpose method for extracting genomic DNA from *Dictyostelium* cells. *Nat. Protoc.* **2:** 1325–1328.

Pukatzki, S., Ma, A.T., Sturtevant, D., Krastins, B., Sarracino, D., Nelson, W.C., Heidelberg, J.F., and Mekalanos, J.J. 2006. Identification of a conserved bacterial protein secretion system in *Vibrio cholerae* using the *Dictyostelium* host model system. *Proc. Natl. Acad. Sci.* **103:** 1528–1533.

Raper, K.B. 1935. *Dictyostelium discoideum*, a new species of slime mold from decaying forest leaves. *J. Agr. Res.* **50:** 135–147.

Raper, K.B. 1937. Growth and development of *Dictyostelium discoideum* with different bacterial associates. *J. Agric. Res.* **55:** 289–316.

Raper, K.B. 1951. Isolation, cultivation, and conservation of simple slime molds. *Q. Rev. Biol.* **26:** 169–190.

Raper, K.B. 1984. *The Dictyostelia*. Princeton University Press, Princeton, New Jersey.

Rogers, M.J. 2004. From molds and macrophages to mevalonate: A decade of progress in understanding the molecular mode of action of bisphosphonates. *Calcif. Tissue Int.* **75:** 451–461.

Sambrook, J. and Russell, D. 2001. *Molecular cloning: A laboratory manual*. Cold Spring Harbor Laboratory, Cold Spring Harbor, New York.

Schaap, P., Winckler, T., Nelson, M., Alvarez-Curto, E., Elgie, B., Hagiwara, H., Cavender, J., Milano-Curto, A., Rozen, D.E., Dingermann, T., Mutzel, R., and Baldauf, S.L. 2006. Molecular phylogeny and evolution of morphology in the social amoebas. *Science* **314:** 661–663.

Shaulsky, G. and Kessin, R.H. 2007. The cold war of the social amoebae. *Curr. Biol.* **17:** R684–R692.

Song, J., Xu, Q., Olsen, R., Loomis, W.F., Shaulsky, G., Kuspa, A., and Sucgang, R. 2005. Comparing the *Dictyostelium* and *Entamoeba* genomes reveals an ancient split in the Conosa lineage. *PLoS Comput. Biol.* **7:** e71.

Spudich, J.A. 1989 In pursuit of myosin function. *Cell Regul.* **1:** 1–11.

Strmecki, L., Greene, D.M., and Pears, C.J. 2005. Developmental decisions in *Dictyostelium discoideum*. *Dev. Biol.* **284:** 25–36.

Sussman, R. and Sussman, M. 1967. Cultivation of *Dictyostelium discoideum* in axenic medium. *Biochem. Biophys. Res. Commun.* **29:** 53–55.

Vadell, E.M. and Cavender, J.C. 2007. Dictyostelids living in the soils of the Atlantic forest, Iguazu region, Misiones, Argentina: Description of new species. *Mycologia* **99:** 112–124.

von Philipsborn, A. and Bastmeyer, M. 2007. Mechanisms of gradient detection: A comparison of axon pathfinding with eukaryotic cell migration. *Int. Rev. Cytol.* **263:** 1–62.

Watts, D.J. and Ashworth, J.M. 1970. Growth of myxameobae of the cellular slime mould *Dictyostelium discoideum* in axenic culture. *Biochem. J.* **119:** 171–174.

Williams, R.S., Boeckeler, K., Graf, R., Muller-Taubenberger, A., Li, Z., Isberg, R.R., Wessels, D., Soll, D.R., Alexander, H., and Alexander, S. 2006. Towards a molecular understanding of human diseases using *Dictyostelium discoideum*. *Trends Mol. Med.* **12:** 415–424.

3 The Moss *Physcomitrella patens*
A Novel Model System for Plant Development and Genomic Studies

David J. Cove, Pierre-François Perroud, Audra J. Charron, Stuart F. McDaniel, Abha Khandelwal, and Ralph S. Quatrano
Department of Biology, Washington University, St. Louis, Missouri 63130

ABSTRACT

The moss *Physcomitrella patens* has been used as an experimental organism for more than 80 years. Within the last 15 years, its use as a model to explore plant functions has increased enormously. The ability to use gene targeting and RNA interference methods to study gene function, the availability of many tools for comparative and functional genomics (including a sequenced and assembled genome, physical and genetic maps, and >250,000 expressed sequence tags [ESTs]), and a dominant haploid phase that allows direct forward genetic analysis have all led to a surge of new activity. *P. patens* can be easily cultured and spends the majority of its life cycle in the haploid state, allowing the application of experimental techniques similar to those used in microbes and yeast. Its development is relatively simple, and it generates only a few tissues that contain a limited number of cell types. Although mosses lack vascular tissue, true roots/stems/leaves, and flowers and seeds, many signaling pathways found in angiosperms are intact in moss. For example, the phytohormones auxin, cytokinin, and abscisic acid, as well as the photomorphogenic pigments phytochrome and cryptochrome, are all interwoven into distinct but overlapping pathways and linked to clear developmental phenotypes. In addition, about one quarter of the moss genome contains genes with no known function based on sequence motifs, raising the likelihood of successful discovery efforts to identify new and novel gene functions. The methods outlined in this chapter will enhance the use of the *P. patens* model system in many laboratories throughout the world.

> **PROTOCOLS**
> 1 Culturing the Moss *Physcomitrella patens*, 75
> 2 Isolation and Regeneration of Protoplasts, 80
> 3 Somatic Hybridization in *P. patens* Using PEG-induced Protoplast Fusion, 82
> 4 Chemical and UV Mutagenesis of Spores and Protonemal Tissue, 84
> 5 Transformation Using Direct DNA Uptake by Protoplasts, 87
> 6 Transformation Using T-DNA Mutagenesis, 89
> 7 Transformation of Gametophytes Using a Biolistic Projectile Delivery System, 91
> 8 Isolation of DNA, RNA, and Protein from *P. patens* Gametophytes, 93

This chapter, with full-color images, can be found online at www.cshprotocols.org/emo.

BACKGROUND INFORMATION

The moss *P. patens* (Hedw.) Bruch & Schimp was first established as a laboratory experimental system in the 1920s by Fritz von Wettstein (1924), who studied the effects of ploidy variation and inheritance patterns in interspecific and intergeneric crosses within the moss family Funariaceae. The modern era of *Physcomitrella* research dates to the work of Paulinus Engel (1968), who generated the first biochemical and morphological mutants in the species.

Like all land plants, the moss life cycle consists of a multicellular haploid gametophyte generation that alternates with a morphologically distinct diploid sporophyte generation. But unlike vascular plants, the gametophyte (Fig. 1C) is the dominant portion of the moss life cycle. Haploid spores germinate to produce a filamentous protonemal stage (Fig. 1D). Protonemata are initially

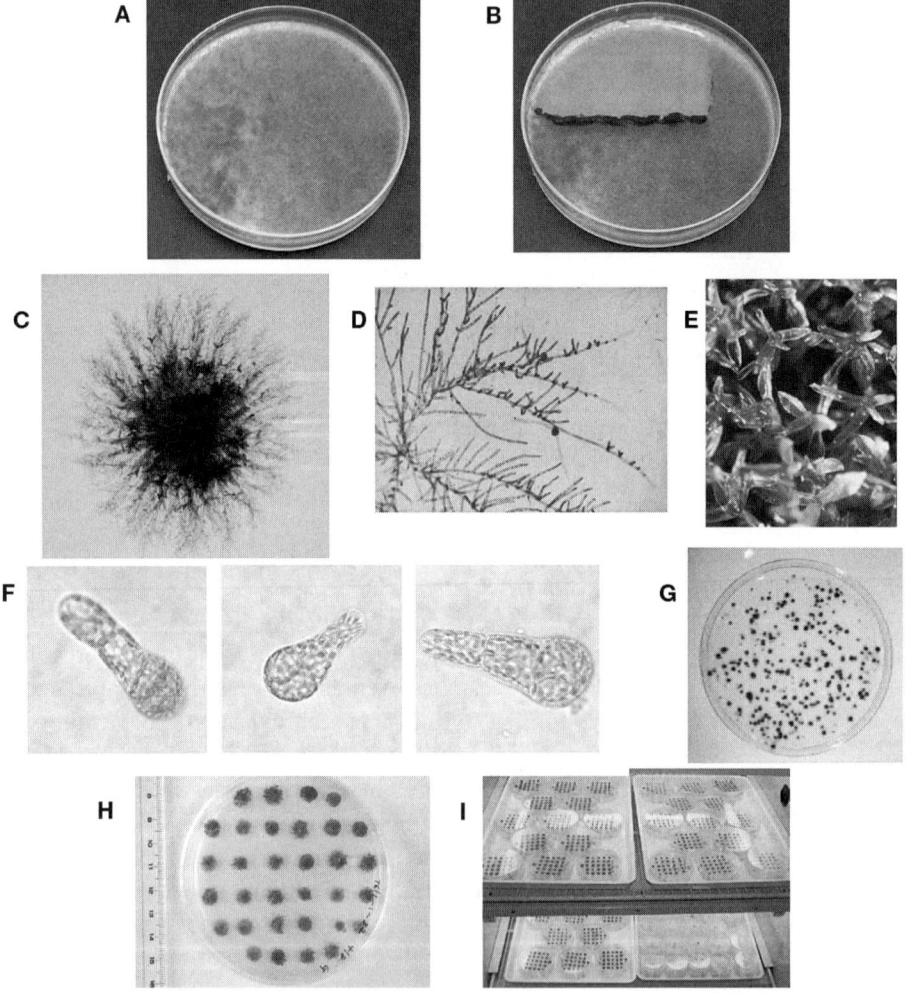

FIGURE 1. *P. patens* cultures. (A) Six-day-old *P. patens* protonemata grown on cellophane over solid BCD medium supplemented with diammonium tartrate. (B) Easy harvesting, with a spatula, of protonemata grown on solid media overlaid with cellophane. (C) Four-week-old inoculum grown on BCD supplemented with diammonium tartrate. Note the presence of the two major tissues that are characteristic of the haploid growth phase of *P. patens* development: the filamentous protonemata and the leafy gametophore. (D) *P. patens* protonemata displaying the characteristic branching pattern. (E) Six-week-old *P. patens* gametophores. (F) Three 3-day-old filaments regenerating from protoplast on PRMB medium. (G) Transformation plate after 2 weeks on antibiotic selection. Transformants surviving selection are easily identified as individual growing plants. (H) Individual spot-inoculums of *P. patens* strains on a 9-cm Petri dish after 2 weeks' growth. Up to 32 independent isolates can be grown and stored this way. (I) Multiple plates can be easily stored after growth in an incubator with 2 hours of light per day at 10°C.

composed of chloronemal cells that are full of large chloroplasts. Chloronemal cells extend by serial division of the apical cell, and subapical cells branch to form new apices. Some apical chloronemal cells develop into a second cell type, caulonemata. Caulonemal filaments contain fewer and less-well-developed chloroplasts. But they extend more rapidly than chloronema; the division times of the apical cells of caulonema and chloronema are about 6 and 24 hours, respectively. The subapical cells of caulonemal filaments branch to form more filaments and leafy stems, called gametophores (Fig. 1E), on which gametes are produced. Moss is monoecious: Both male and female gametes are produced on the same gametophore. Although self-fertilization is common, cross-fertilization can occur when two strains are grown adjacent to each other. Fertilized zygotes develop into sporophytes that remain attached to the gametophore. Within the sporophyte, spore mother cells give rise to spores meiotically.

P. patens is small, and in nature, the gametophores seldom reach more than 5 mm in height. It is mostly found on wet soil and, in particular, on sites that are exposed to seasonal flooding, such as the banks of lakes, ponds, rivers, and drainage ditches (Crum and Anderson 1981). Although it is distributed widely in the northern hemisphere, it is uncommon throughout its range. Natural populations produce spores from September to March, depending on the locality. Although *P. patens* itself is restricted to North America and Europe, other morphologically similar species are found in Africa, Asia, South America, and Australia. Many variants of *P. patens* have been given species or subspecies rank, although the degree to which the morphological features that distinguish these taxa have a genetic basis has not been established experimentally. Natural hybrids between closely related species in the family Funariaceae, similar to those produced in culture by von Wettstein, have been documented in several localities (Pettet 1964).

SOURCES AND HUSBANDRY

The strain of *P. patens* used by Engel was generated from a single spore that was obtained in 1962 from a plant in Gransden Wood, Huntingdonshire, England by Dr. H.W.K. Whitehouse. This strain has since been used by many laboratories, but it is routinely taken through its sexual cycle about every year, during which time cultures are reestablished from individual spores. Therefore, Gransden strains will often be identified by the laboratory and the year in which a spore was used to start a new culture (e.g., the material used to sequence the *P. patens* genome came from the Gransden St. Louis 2004 strain). Because of gametophytic haploidy, all such strains differ only as a result of mutation or epigenetic variation. More recently, additional collections of *P. patens* have been made from Europe and North America, and these are now being genetically and morphologically characterized (von Stackelberg et al. 2006). This new collection is curated and distributed from the University of Freiburg, Germany (see http://www.cosmoss.org).

P. patens can be grown on either solid (agar-based) or liquid media (for details, see Protocol 1 and Fig. 1). Temperatures between 24°C and 26°C are used for routine culture, although little difference in growth rate is observed from 20°C to 26°C. Growth is slower but still satisfactory at 15°C, and this has been used as the permissive temperature when temperature-sensitive mutants are sought. For routine culture, continuous light from fluorescent tubes at an intensity of between 5 and 20 Wm^{-2} is generally satisfactory, although the exact quality of light is not critical. Many laboratories use intermittent light, most commonly a 16-hour light/8-hour dark cycle. This regime entrains the cell cycle of chloronemata. Development is slower under intermittent light regimes; developmental landmarks are achieved in response to the total hours of illumination experienced.

RELATED SPECIES

Two additional moss species are currently used for experimental research: *Ceratodon purpureus* and *Tortula ruralis*. *C. purpureus* is one of the most common mosses in exposed rock and soil in temperate regions of the northern and southern hemispheres. In the spring, it is easily recognized by the purple seta that elevates the diploid sporophyte. This species also has a long history in experimental biology; the term "heterochromatin" was coined for the dark-staining sex chromosomes of this and other moss species (Heitz 1928). A genetic map of *C. purpureus* has been constructed, and several natural isolates have been extensively characterized (McDaniel et al. 2008). Cultures of *C. purpureus*, isolated by E. Hartmann in Germany and by D.J. Cove in Austria, are used by several labs worldwide, principally to study phototropism and gravitropism. *C. purpureus* is so abundant that the isolation of additional cultures is fairly straightforward; isolates currently in use are available from D.J. Cove (Washington University, St. Louis). All of the experimental procedures that are described for *P. patens* are also used for *C. purpureus* with few modifications.

T. ruralis has been used principally to study water stress because it is able to withstand complete desiccation, similar to seeds. Although a modest collection of ESTs is available for *T. ruralis* (Oliver et al. 2004), it is less amenable to growth in culture than either *C. purpureus* or *P. patens* and has not been shown to be easily transformable or to undergo efficient gene targeting.

USES OF THE *P. PATENS* MODEL SYSTEM

The common ancestor of mosses, such as *P. patens*, and seed plants, such as *Arabidopsis thaliana* and pines, lived approximately 480 million years ago (Mya). Comparative studies including members of both of these lineages allow us to infer biological properties of this common ancestor, giving us a richer understanding of the diversity of plant life. This may have practical value to the extent that understanding diverse plant systems yields novel solutions to problems in crop breeding, for example.

During the last several years, *P. patens* has been used as a model to study various components of cell, developmental, and evolutionary plant biology. Its development is relatively simple, and it generates only a few tissues that contain a limited number of cell types. Although mosses lack vascular tissue, true roots/stems/leaves, and flowers and seeds, many signaling pathways found in angiosperms are intact in moss. For example, the phytohormones auxin, cytokinin, and abscisic acid, as well as the photomorphogenic pigments phytochrome and cryptochrome, are all interwoven into distinct but overlapping pathways and linked to clear developmental phenotypes (Quatrano et al. 2007; Rensing et al. 2008).

RNA interference (RNAi) methods (Bezanilla et al. 2005) have been used to analyze the role of ARPC1 (Harries et al. 2005), a member of the Arp2/3 complex, and profilin (Vidali et al. 2007) in tip growth. Khandelwal et al. (2007) studied the role of the *P. patens* presenilin protein using RNAi. Presenilin possesses γ-secretase activity, is involved in Alzheimer's disease (AD), and is an intermediate in the NOTCH signaling pathway of animal cells; however, unlike animal cells, *P. patens* and other plants do not possess this pathway although the protein is present. The observed mutant phenotype indicated a possible role for presenilin that is independent of γ-secretase activity and the NOTCH pathway, thus raising the possibility of using *P. patens* as a novel system for studying the off-target effects of AD therapy and drug discovery.

Targeted gene deletion and replacement methods have been used to study the role of another member of the Arp2/3 complex, ARPC4 (Perroud and Quatrano 2006), and BRICK1, a member of the Scar/Wave family (Perroud and Quatrano 2008). Like the other papers referenced above, the excellent cell biology of *P. patens* was used to localize ARPC4 and BRICK1 in growing tip filaments. Transcriptome (Nishiyama et al. 2003; Cuming et al. 2007) and metabolic (Thelander et al. 2005; Kaewsuwan et al. 2006; Schulte et al. 2006) studies, as well as detailed analyses of microRNAs (Axtell et al. 2006, 2007), have now appeared using *P. patens*. Finally, comparative genomic studies have elucidated the role of the transcriptional regulators LEAFY (Maizel et al. 2005) and ABI3 (Marella et al. 2006) and transcription factors involved in rooting function (Menand et al. 2007) in *P. patens* as well as in *A. thaliana*.

GENETICS, GENOMICS, AND ASSOCIATED RESOURCES

P. patens is amenable to classical genetics studies, with the haploidy of the gametophyte allowing straightforward analysis (Cove 2005). Efforts to identify polymorphisms among isolates of *P. patens* are proceeding and will enable map-based cloning of both ethylmethanesulfonate (EMS)- and UV-generated mutants (for a method to generate such mutants, see Protocol 4), as well as quantitative trait locus (QTL) mapping of natural variants (von Stackelberg et al. 2006). A number of mutant strains are available from different laboratories (http://www.cosmoss.org; http://biology4.wustl.edu/moss), including those having requirements for the vitamins p-aminobenzoic acid, nicotinic acid, and thiamine. Vitamin requirements have been exploited to increase the frequency of cross-fertilization. When two complementary p-aminobenzoic acid or nicotinic acid auxotrophs are grown together on a medium with only a limited level of supplementation, the sporophytes produced are the result of cross-fertilization (Courtice and Cove 1978).

No universally accepted system of gene nomenclature has been adopted, but there has been general agreement that annotated genes will be identified by numbers. Trivial names can then be added as synonyms. Originally, the system used for trivial names was similar to that used for many bacteria and fungi: Each symbol was composed of a three-letter lowercase code to designate the mutant gene family, an uppercase letter to designate the family member, and a number to designate the allele (e.g., *pab*A4). More recently, some laboratories have adopted the system used by the yeast and *A. thaliana* communities, designating the family member by a number rather than by an uppercase letter (e.g., *pab*1-4). But no general agreement has yet been reached as to which system should be adopted.

The assembled *P. patens* genome (~487 Mb), representing eightfold coverage, has been released by the Joint Genome Institute (http://shake.jgi-psf.org/Phypa1/Phypa1.home.html; Rensing et al. 2008). In parallel, sequences of full-length cDNAs, additional ESTs, and bacterial artificial chromosome (BAC) ends are being developed, and updates can be accessed through the *Physcomitrella* Genome Consortium website (http://www.mossgenome.org). Various libraries and vectors are available (see links at http://biology4.wustl.edu/moss/links.html), as is an Agilent microarray (MO gene; http://www.mogene.com), which contains 41,382 features (~28,000 gene models) based on all of the open reading frames (ORFs) in the draft genome (Fig. 2).

Several tools are available for the functional analysis of genes in *P. patens*. For example, the dexamethasone- (Chakhparonian 2001), heat-shock- (Saidi et al. 2005), and homoserine-lactone- (You et al. 2006) inducible promoter systems have all been successfully used in this system. Forward genetics can be used to dissect gene function using a shuttle-mutagenesis library (Nishiyama et al. 2000; Hayashida et al. 2005). A targeted deletion library that was created using ESTs (Schween et al. 2005) has also been used for functional analysis (Schulte et al. 2006). Transformation can be performed via polyethylene glycol (PEG)–mediated DNA uptake by isolated protoplasts (see Protocol 5), via *Agrobacterium* (see Protocol 6), or via a gene gun (see Protocol 7), and somatic hybridization has been used to analyze mutants genetically (see Protocol 3) (Cove and Quatrano 2006). Reverse genetics using gene targeting is a tool of choice for manipulating individual genes in *P. patens*, and RNAi allows the down-regulation of gene families. An RNAi system has been developed in *P. patens* that silences the nucleus-localized green fluorescent protein::β-glucuronidase (GFP::GUS) fusion protein at the same time that it silences the gene(s) of interest (Bezanilla et al. 2005).

TECHNICAL APPROACHES

The following eight protocols outline techniques for manipulating *P. patens* in the laboratory and include three transformation methods (Protocols 5–7). *P. patens* (and *C. purpureus*) have a high frequency of gene targeting (Kamisugi et al. 2005, 2006); when a transforming construct contains

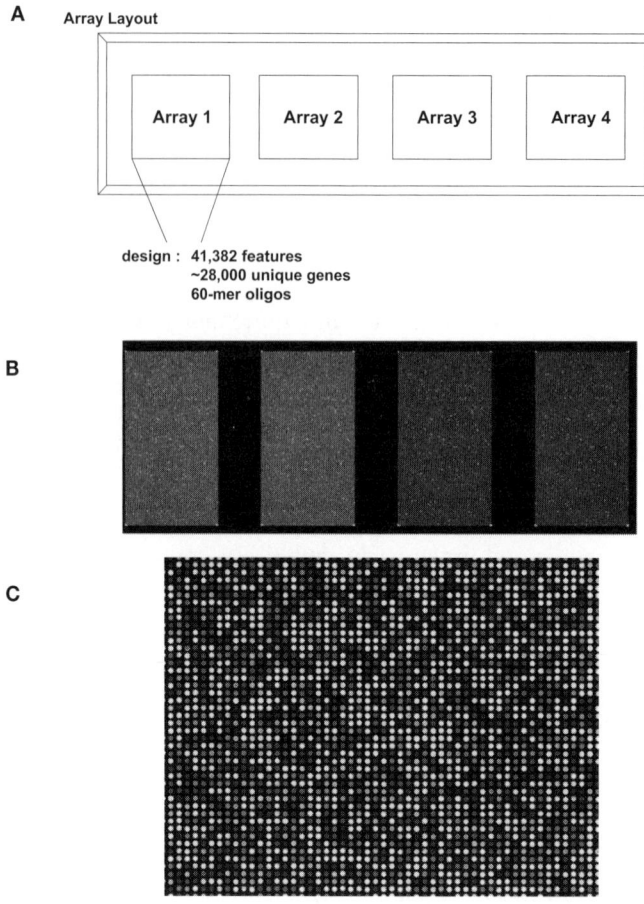

FIGURE 2. *P. patens* microarray. (*A*) Layout of the *P. patens* 4 x 44K Agilent microarray. (*B*) Photograph of a scanned microarray slide diagrammed in *A*. (*C*) Enlarged image of a single array showing differential gene expression.

a genomic sequence, the construct is targeted to the corresponding sequence in the genome. This can be exploited to knock out or modify a gene.

For deletion or disruption, aim to replace the coding sequence with a selection cassette and to border this on each side by about 1000 bp of genomic sequence. Linear DNA fragments generated by polymerase chain reaction (PCR) give the highest rates of targeting. It is convenient to perform a number of transformations at the same time (ten is not difficult). For each experiment, make sure to include a minus DNA control to assess protoplast viability.

Figure 1G shows a plate of transformants that have been growing for 2 weeks on selective media. Following transformation, the regenerants are of three types:

- *Transient:* These do not retain resistance upon subculture.
- *Unstable:* These exhibit slow growth on selective medium. Resistance is probably not transmitted through meiosis and is rapidly lost when selection is relaxed.
- *Stable:* These grow on selective medium almost as fast as on nonselective medium. Resistance is transmitted regularly through meiosis and is retained even when selection is absent.

Protocol 1

Culturing the Moss *Physcomitrella patens*

Here, we present a series of methods for culturing the moss *P. patens* at all stages of its life cycle. Gametophytes are axenically cultured on solid agar-based media (Methods IA and IB, adapted from Grimsley et al. 1977) and in shaken liquid cultures (Method IC). Growth rates in shaken liquid cultures are not as great as those obtained on solid media or in bioreactors (Boyd et al. 1988), especially if these are supplied with CO_2. However, to obtain large quantities of tissue for biochemical studies, to accumulate secreted products from tissue for characterization, or for tracer studies, liquid culture may be preferred. For long-term storage of gametophytes, cultures are maintained on solid medium at 10°C in a very short day (see Method IIA), but cryopreservation (Method IIB, adapted from Grimsley and Withers 1983) may also be used. Finally, sporophytes are generated by self-fertilization and sexual crossing (Method III, adapted from Ashton and Cove 1977).

MATERIALS

The recipes for items marked with <R> begin on page 97.

CAUTION: See the Cautions Appendix for appropriate handling of materials marked with <!>.

Reagents

BCD medium (liquid and solid) <R> containing common moss media supplements <R> as necessary
  A common supplement is diammonium tartrate, which is added to BCD medium <R> at a final concentration of 5 mM. For cryopreservation, also include mannitol (500 mM) in the appropriately supplemented liquid BCD medium (see Step 1 of Method IIB).

Dimethylsulfoxide (DMSO) <!>–glucose solution <R>

Ethanol (70%) <!>

Liquid nitrogen <!>

Nitrogen-free medium <R> containing 400 µM KNO_3 <!>

Somatic tissue or spores from *P. patens*
  To establish a new culture, a fragment of tissue 1–2 mm in diameter is sufficient. For routine subculture, it is best to use protonemal tissue from a vigorously growing culture that is no more than 20 days old, which will be composed mostly of chloronemal tissue. Chloronemal tissue, the growth of which is enhanced when ammonium is provided as nitrogen source, is easiest to subculture. Tissue other than chloronemata (e.g., leaf cells) may take a long time to regenerate. For sexual crosses, it is best that at least one of the two strains to be crossed is self-sterile. Strains containing vitamin auxotrophies (e.g., for thiamine, p-aminobenzoic acid, or nicotinic acid) are normally self-sterile but cross-fertile (Courtice and Cove 1978).

Equipment

Blender (e.g., Fisher Scientific PowerGen Model 125 Homogenizer)
  Different tissue blenders are used in different laboratories, and some have been made specifically for blending moss tissue. Provided the blending assembly can be sterilized, it appears that there is little difference among blenders in the end result—a tissue inoculum that will grow rapidly.

Cellophane discs (type 325P), sterile (AA Packaging Ltd.)
  These are most conveniently sterilized dry, by autoclaving interleaved with disks of filter paper.

Dissecting microscope (optional; see Step 1 of Method IA)

Erlenmeyer flasks (1 liter, sterile)

Ethanol bath (controlled, low temperature) <!>

Forceps (fine)
Incubator with temperature control and white light at intensities between 5 and 20 Wm^{-2} (e.g., Percival Scientific Model CU-36L5)

Because of the radiant heat from light sources, it may be necessary to keep the air temperature below the desired culture temperature. In this protocol, the temperatures given are those of the medium in which the cultures are grown.

Liquid nitrogen storage facility
Magenta jars (optional; see Method III)
Microcentrifuge (optional; see Step 16 of Method III)
Microcentrifuge tubes (1.5 ml)
Micropore surgical tape (3M, 1530-0)
Parafilm (optional; see Step 4 of Method IIA)
Petri dishes (sterile)

Most work uses 90-mm-diameter, presterilized disposable plastic Petri dishes, but glass Petri dishes are also suitable. Ideally, the dishes should be vent-free to slow evaporation and limit contamination.

Shaker (platform)
Spatula (sterile)
Test tubes (25 × 150 mm, sterile) with caps
Tubes (2 ml, plastic) containing 1.5 ml of agar medium (see Step 1 of Method IIA)

If space is not at a premium, 15-ml tubes containing 10 ml of medium provide a better resource for long-term storage.

Vials (2 ml, plastic)
Water bath, set at 30°C

METHODS

I. Growth of Gametophytes

A. Using Petri Dishes Containing Solid Medium

1. Inoculate a Petri dish containing appropriately supplemented solid BCD medium as follows:

   - For spores, use standard microbiological procedures to spread them on the agar. If individual plants are required, add about 100–200 spores per 90-mm-diameter Petri dish.

   - For somatic tissue, place a clump of gametophyte tissue (usually 1–2 mm in diameter) on the agar. The size of the clump is not critical. However, for growth tests, the clumps should be as uniform as possible, and uniformity is best achieved by picking the tissue while looking through a microscope.

2. Seal the Petri dishes with Micropore surgical tape.

   This reduces the risk of contamination without affecting the growth or development of the cultures. Do not seal cultures with Parafilm; this slows growth and prevents regeneration.

3. Incubate the cultures in an incubator at 25°C with constant white light at intensities between 5 and 20 Wm^{-2}.

   Depending on the purpose of the procedure, growth under these conditions may range from a few days (e.g., to score antibiotic resistance) to several weeks (e.g., to score the requirements of some vitamins).

B. Using Petri Dishes Containing Solid Medium Overlaid with Cellophane

1. Overlay a Petri dish containing appropriately supplemented solid BCD medium with a sterilized cellophane disc. Allow the dish to stand for at least 10 minutes to allow the cellophane to hydrate and then, if necessary, straighten the disc while maintaining sterility.

2. Obtain one dish of protonemal tissue from plants that have developed for about 10 days from tissue-clump inocula growing on appropriately supplemented BCD medium (see Method IA). Harvest the tissue with a spatula as shown in Figure 1B.

3. Add the tissue to 10 ml of H_2O, and blend it for about 2 minutes.

   The exact procedure will depend on the type of blender used. Blending should result in an easily pipettable suspension, but it should still consist of tissue clumps containing 20–50 cells. Overblending leads to poor regeneration.

4. Pipette 1–2 ml of the protonemal suspension from Step 3 onto each Petri dish from Step 1. Spread the suspension evenly.

   There should be sufficient tissue to inoculate 5–10 Petri dishes.

5. Incubate the culture for 7 days at 25°C under a 16-hour light/8-hour dark cycle (with white light at intensities between 5 and 20 Wm^{-2}).

   For an image of 6-day-old protonemata grown under these conditions, see Figure 1A. For a wild-type culture, each 90-mm-diameter Petri dish will yield about 200 mg (fresh weight) of vigorously growing protonemal tissue, consisting mainly of chloronemata.

6. Harvest the tissue by scraping it from the cellophane using a sterile spatula.

   Once tissue is growing on cellophane, it is convenient to use this tissue (instead of the tissue described in Step 2) for further inocula by repeating Steps 3–6. However, do not repeat this cycle more than three times.

C. Using Liquid Medium

1. Add tissue from one cellophane-overlay plate (see Method IB) to 10 ml of H_2O and blend as described in Step 3 of Method IB.

2. Inoculate 200 ml of appropriately supplemented liquid BCD medium (in a 1-liter Erlenmeyer flask) with 2 ml of the tissue suspension from Step 1.

   Other quantities of the liquid BCD medium and/or the tissue suspension may be used instead.

3. Shake the cultures on a platform shaker at 25°C under continuous white light.

   Vigorous agitation is not necessary for growth. In these conditions, tissue weight doubles in 3 to 4 days.

II. Long-term Storage of Gametophyte Tissue

A. Storage on Solid Medium

1. Place a clump of gametophyte tissue (usually 1–2 mm in diameter) in each plastic tube containing agar medium (usually, appropriately supplemented solid BCD medium).

2. Make sure that the lids are not tightly sealed. For screw-cap lids, tighten the cap and then release it about one quarter of a turn.

3. Grow the cultures for about 3 weeks in an incubator at 25°C with continuous white light at intensities between 5 and 20 Wm^{-2}.

4. After 3 weeks, tightly seal each tube. If the tube has an air-tight seal, this is sufficient; if it does not, seal the tube with Parafilm.

5. Transfer the tubes to an incubator at 10°C with a 2-hour light/22-hour dark cycle (with white light at intensities between 5 and 20 Wm^{-2}) for long-term storage.

   Cultures can be kept in a healthy state for a considerable period of time (at least 3 years) under these conditions.

B. Cryopreservation

1. For each strain to be preserved, grow gametophyte tissue as described in Steps 1–5 of Method IB. After 7 days of incubation, transfer 100 mg of tissue onto a Petri dish overlaid with fresh cellophane. Pipette 1 ml of appropriately supplemented liquid BCD containing 500 mM of mannitol onto the surface of the tissue.
2. Incubate the culture for an additional 7 days at 25°C under a 16-hour light/8-hour dark cycle (with white light at intensities between 5 and 20 Wm^{-2}).
3. To each of ten 2-ml sterile plastic vials, add 1.5 ml of DMSO–glucose solution. Then add one tenth of the tissue on the plate to each vial. Incubate the vials for 1 hour at 20°C.
4. Freeze the vials at a rate of 1°C per minute to –35°C using a controlled low-temperature ethanol bath.
5. Place the vials in liquid nitrogen for storage.

    It is advisable to freeze multiple aliquots of tissue. Recovery from cryopreservation is usually good for vigorous strains, but some mutant strains recover poorly. To thaw the cultures, proceed with Steps 6–8. Successful recovery has been achieved after 10 years of storage; longer storage periods may be possible.

6. Retrieve the vials from the liquid nitrogen and place them in a water bath at 30°C until thawed.
7. After thawing, add tissue from one vial to 10 ml of H_2O. Allow the mixture to stand for 30 minutes at room temperature.
8. Inoculate the tissue suspension from Step 7 onto appropriate solid medium as described in Method IA (use 1–2 ml of the tissue suspension per plate).

III. Production of Sporophytes and Isolation of Spores

1. Prepare test tubes (25 x 150 mm) with between 15 and 20 ml of solid nitrogen-free medium containing 400 µM KNO_3.

    Alternatively, fill Magenta jars two-thirds full with the same medium.

2. Inoculate the medium with protonemata by placing a clump of tissue (usually 1–2 mm in diameter) on the agar.

    - If sporophytes resulting from self-fertilization are required, place a single inoculum into a test tube. Alternatively, place four inocula into a Magenta jar.
    - To establish a sexual cross, place one or more inocula of each strain into a tube (or a Magenta jar).

        Harvest and test the sporophytes produced from sexual crosses individually because they may result from either self- or cross-fertilization. Strains that have been kept for prolonged periods in vegetative culture may loose fertility.

3. Place the cap on the tube, but do not completely tighten it.
4. Culture for about 4 weeks at 25°C in continuous white light at intensities between 5 and 20 Wm^{-2}.

    Continuous white light accelerates sporophyte production somewhat, but sporophyte yield may be reduced.

5. Transfer the culture to 15°C for an additional 3 weeks under an 8-hour light/16-hour dark cycle.
6. Irrigate the culture with H_2O to facilitate fertilization. Make sure to thoroughly wet but not submerge the culture. Allow it to stand for 24 hours.

7. After 24 hours, decant excess H₂O. Place the culture for 1 week at 15°C under an 8-hour light/16-hour dark cycle.

8. Repeat the irrigation procedure as described in Step 6.

9. After 24 hours, decant excess H₂O. Place the culture for 2–5 weeks at 15°C under an 8-hour light/16-hour dark cycle.

   After this incubation period, sporophytes will have developed.

10. Use fine forceps to harvest pale brown sporophytes by separating them from the gametophytic material. Do not take the green or yellow sporophytes (these are immature) or the dark-brown sporophytes (these burst easily).

11. Pinch the base of the seta, which can be identified by a zone of pigmentation, to release the spore capsule.

    Each capsule of a mature sporophyte contains 1×10^3 to 4×10^3 viable, haploid, uninucleate spores.

12. Place one or more sporophytes in a sterile 1.5-ml microcentrifuge tube. Sterilize the sporophytes by adding 1 ml of 70% ethanol and incubating for 4 minutes at room temperature.

13. Remove the ethanol. Gently rinse the sporophytes three times with 1 ml of H₂O at room temperature.

14. Add 1 ml of H₂O and place the tube for 7 days in the dark at 4°C.

    This step increases spore germination, but some spores will germinate if this step is omitted.

15. Crush the sporophyte capsules and mix to produce a spore suspension.

    Any residual sporophytic tissue will not regenerate and can be ignored. The spore suspension may be kept for several weeks at 4°C.

16. If desired, dry the spores by removing as much water as possible after centrifugation at $100g$ for 4 minutes and then allowing the excess water to evaporate at room temperature under sterile conditions.

    Dry spores may be stored for several years.

Protocol 2

Isolation and Regeneration of Protoplasts

This protocol describes how to isolate individual protoplasts from young gametophyte tissue (Method I) and how to regenerate them into plants (Method II) if desired. It is adapted from a protocol described by Grimsley et al. (1977).

MATERIALS

The recipes for items marked with <R> begin on page 97.

Reagents

Driselase solution (0.5%, sterile) <R>
D-Mannitol solution (8.5%)
Protonemal tissue from *P. patens*
  This tissue is grown on cellophane-overlay plates as described in Method IB of Protocol 1.
Protoplast regeneration medium for the bottom layer (PRMB) <R>
  Prepare Petri dishes (50 mm or 90 mm) containing PRMB and overlay with sterile cellophane.
Protoplast regeneration medium for the top layer (PRMT) <R>
  This medium should be melted and kept at 45°C in a water bath.
Protoplast wash (PW) solution <R>

Equipment

Centrifuge
Filters (pore sizes of 50 μm and 100 μm, sterile)
  One filter of each pore size is are needed for each protoplast isolation. Filters can be made of stainless steel or nylon.
Hemocytometer
Incubator with temperature control and white light at intensities between 5 and 20 Wm^{-2} (e.g., Percival Scientific Model CU-36L5)

METHODS

I. Protoplast Isolation

1. After culturing protonemal tissue for 7 days (Method IB, Steps 1–5), harvest the tissue as described in Step 6 of Method IB. Weigh the tissue and add 1 ml of sterile 0.5% driselase solution for every 40 mg of tissue (fresh weight) (i.e., use 5 ml of driselase solution per 90-mm-diameter plate of tissue).

   It is difficult to obtain protoplasts from tissue other than gametophytes growing on cellophane-overlay plates. Use young protonemal tissue; older tissue may leave clumps of undigested tissue, which can act like glue and stick protoplasts together.

2. Incubate the tissue-driselase mixture for 30–60 minutes at room temperature with occasional gentle shaking. Monitor the breakdown of the tissue until abundant, round, free-floating protoplasts are observed.

For some strains, the incubation time may need to be longer. Protoplast viability is not greatly affected by exposure to driselase for up to 2 hours, but it will decrease as the incubation time increases.

3. To isolate protoplasts from the digested tissue, sterilely filter the tissue-driselase mixture through mesh with a pore size of 100 µm.

4. Sterilely refilter the filtrate from Step 3 through mesh with a pore size of 50 µm.

5. Sediment the protoplasts by centrifuging the filtrate from Step 4 at 100–200g for 4 minutes at room temperature with no braking.

6. Discard the supernatant.

7. Resuspend the protoplast pellet in PW. Use about the same volume as the volume of driselase used in Step 1.

8. Repeat Steps 5–7.

   This should yield about 10^6 protoplasts per 90-mm-diameter plate of tissue cultured for 7 days as described in Method IB. Protoplasts can be used in a number of techniques (e.g., Protocols 3, 5, and 6). If protoplasts are to be plated without further manipulation, proceed to Method II.

II. Protoplast Regeneration

1. Estimate the density of protoplasts using a hemocytometer. Adjust the volume of the protoplast suspension with 8.5% D-mannitol solution so that the density of protoplasts per milliliter is approximately 1.5×10^4 ml.

2. Determine the quantity of molten PRMT needed (use 1 volume of protoplast suspension for every 2 volumes of molten PRMT). Gently add the protoplast suspension from Step 1 to the appropriate volume of molten PRMT.

3. Pipette 1 ml of the protoplast-PRMT mixture gently but quickly onto a 90-mm PRMB plate overlaid with cellophane.

   Alternatively, use 400 µl of the protoplast-PRMT mixture to cover a 50-mm dish.

4. Incubate the plate for 3 or 4 days at 25°C with strong, constant light (intensities >5 Wm^{-2}).

   For images of 3-day-old filaments regenerating from protoplasts, see Figure 1F.

5. After about 4 days, transfer the cellophane overlaid with protoplasts to appropriately supplemented, fresh solid BCD medium.

Protocol 3

Somatic Hybridization in *P. patens* Using PEG-induced Protoplast Fusion

As an alternative to sexual crossing (see Protocol 1), protoplasts from two strains of moss (*P. patens*) can be hybridized using polyethylene glycol (PEG). This protocol for PEG-induced protoplast fusion was adapted from the protocol described by Grimsley et al. (1977). Although the efficiency is low, it requires no sophisticated apparatus. Hybrids are readily obtained using complementary auxotrophic mutants or strains with transgenic antibiotic resistance markers. It is now routine to obtain hybrids using transgenic strains that are hygromycin- or G418-resistant by selecting hybrids on medium containing both antibiotics.

MATERIALS

The recipes for items marked with <R> begin on page 97.

Reagents

BCD medium (solid) <R> containing common moss media supplements <R> and antibiotics <R> as necessary in 90-mm Petri dishes
PEG solution for protoplast fusion (PEG/F) <R>
   For each hybridization, 750 µl of PEG/F is required.
Protoplast regeneration medium for the bottom layer (PRMB) <R>
   For each hybridization, prepare eight 90-mm Petri plates containing PRMB and overlay with sterile cellophane.
Protoplast regeneration medium for the top layer (PRMT) <R>
   This medium should be melted and kept at 45°C in a water bath.
Protoplast wash (PW) solution <R>
   For each hybridization, prepare 30 ml of PW solution.
Protoplasts from two different *P. patens* strains
   These protoplasts are isolated as described in Method I of Protocol 2.

Equipment

Centrifuge
Hemocytometer (optional; see Step 2)
Incubator with temperature control and white light at intensities between 5 and 20 Wm^{-2} (e.g., Percival Scientific Model CU-36L5)

METHOD

1. For each strain to be hybridized, resuspend 10^6 protoplasts (approximately the number produced on one 90-mm plate in Method I of Protocol 2) in 2.5 ml of PW.

2. Combine 2.5 ml of each of the two strains to be hybridized.

Judge the density of protoplasts by eye (if preferred, density can be determined using a hemocytometer). If the yield of protoplasts from one strain is poor, the yield of hybrids involving that strain can be maximized by mixing with an excess of the other strain.

3. Sediment the mixed protoplasts by centrifugation at 100–200g for 4 minutes at room temperature with no braking.

4. Discard the supernatant. Resuspend the protoplasts in 250 µl of PW.

5. Hybridize the two strains of protoplasts by performing the following steps on schedule:

    i. At 0 minutes, add 750 µl of PEG/F and mix gently.

    ii. At 40 minutes, add 1.5 ml of PW and mix gently.

    iii. At 50 minutes, add 10 ml of PW and mix gently.

    iv. At 60 minutes, add 10 ml of PW and mix gently.

    v. At 70 minutes, sediment the protoplasts by centrifugation at 100–200g for 4 minutes at room temperature with no braking. Discard the supernatant and resuspend the pellet in 1 ml of PW.

    Multiple hybridizations can be performed in parallel. It should be possible to do these at 30-second intervals.

6. Add the protoplast suspension to 7 ml of molten PRMT and plate 2 ml onto each of four plates of PRMB overlaid with cellophane. In addition, to estimate the survival of the individual component strains, add 50 µl of the protoplast suspension to 8 ml of molten PRMT and plate 2 ml onto each of four plates of PRMB overlaid with cellophane. Incubate all plates at 25°C with strong, constant light (intensities >5 Wm^{-2}).

7. After 5–6 days (when protoplasts have regenerated, but growth is still not great), transfer the protoplasts from Step 6 to Petri dishes containing the appropriate BCD selective medium to select for hybrids. In addition, to estimate the survival of the individual component strains, transfer the cellophane overlays from Step 6 onto the appropriate BCD media to select for one or the other component strain. Incubate all plates under the conditions described in Step 6.

    Selection of hybrids using vitamin auxotrophies usually takes about 3 weeks; using transgenic antibiotic resistances, hybrids can usually be identified after only 7 days.

Protocol 4

Chemical and UV Mutagenesis of Spores and Protonemal Tissue

This protocol describes how to mutagenize spores and protonemal tissue from moss (*P. patens*) using chemicals or UV light. Spores are mutagenized using the alkylating agents *N*-methyl-*N'*-nitro-*N*-nitrosoguanidine (NTG) and ethylmethanesulfonate (EMS) (Methods IA and IB, respectively), and protonemal tissue is mutagenized with NTG and UV light (Methods IIA and IIB, respectively). The chemical mutagenesis protocols were adapted from those described by Ashton and Cove (1977) and Boyd et al. (1988). Compared to alkylating agents, UV is less effective as a mutagen, but it may be advantageous because it is less hazardous and may not lead to clustered lesions.

MATERIALS

The recipes for items marked with <R> begin on page 97.

CAUTION: See the Cautions Appendix for appropriate handling of materials marked with <!>.

Reagents

BCD medium (solid) <R> containing common moss media supplements <R> as necessary in 90-mm Petri dishes
Ethylmethanesulfonate (EMS) <!> (for Method IB only)
N-Methyl-*N'*-nitro-*N*-nitrosoguanidine (NTG) <!> (for Methods IA and IIA only)
Moss (*P. patens*) spores (see Method III of Protocol 1) or protonemal tissue on a cellophane-overlay plate (see Method IB of Protocol 1)
Orthophosphoric acid (H_3PO_4; 100 mM, pH 7.0; sterile) <!> (for Method IB only)
Sodium thiosulfate ($Na_2S_2O_3$; 400 mM, sterile) (for Method IB only)
Tris-maleate buffer (sterile) <R> (for Methods IA and IIA only)

Equipment

Centrifuge (for Method I only)
Centrifuge tubes (for Method I only)
Filter (pore size 100 µm, sterile) (for Method IIA only)
UV light source <!> (for Method IIB only)
  Any standard mutagenic UV source is likely to be effective, but this will need to be calibrated.

METHODS

Perform all chemical mutagenesis procedures in safe handling facilities for mutagens.

I. Mutagenesis of Spores

A. Using NTG

1. Prepare 10 ml of a spore suspension containing about 10^6 spores in sterile Tris-maleate buffer. Incubate the mixture for 30 minutes at 25°C.

2. Observing strict safety procedures, carefully add 1.2 mg of NTG to a separate tube containing 10 ml of Tris-maleate buffer. Incubate the tube for 30 minutes at 25°C.

3. Mix the spore suspension with the NTG solution and shake it gently for an additional 30 minutes at 25°C.

4. Terminate the treatment by sedimenting the spores by centrifugation (100g for 4 minutes at room temperature) and dispose of the supernatant safely.

    Usually about 10% of the originally viable spores survive this treatment.

5. Wash the spores three times as follows:

    i. Add 10 ml of H$_2$O to the tube.

    ii. Centrifuge the tube at 100g for 4 minutes at room temperature.

    iii. Discard the supernatant.

6. After the final wash, resuspend the spores in sterile distilled H$_2$O to give about 10^4 spores/ml. Spread 1 ml of spore solution onto the surface of appropriately supplemented BCD medium in 90-mm Petri dishes.

7. Incubate the plate under the desired conditions. For standard cultivating procedures and conditions, see Protocol 1.

    The mutagenic treatment results in an initial delay in development, but once germinated, the sporelings grow as usual. This should result in about 100 developing sporelings/dish.

B. Using EMS

1. Prepare 9.5 ml of a spore suspension containing about 10^5 spores in sterile 100 mM orthophosphoric acid (pH 7.0).

2. Observing strict safety procedures, add 500 µl of EMS.

3. Incubate the mixture for 45 minutes at 25°C.

4. Terminate the treatment by adding 10 ml of sterile 400 mM sodium thiosulfate.

5. Sediment the spores by centrifugation (100g for 4 minutes at room temperature) and dispose of the supernatant safely.

    Usually 35–40% of the originally viable spores survive this treatment.

6. Wash the spores three times as follows:

    i. Add 10 ml of H$_2$O to the tube.

    ii. Centrifuge the tube at 100g for 4 minutes at room temperature.

    iii. Discard the supernatant.

7. After the final wash, resuspend the spores in sterile distilled H$_2$O to give about 3×10^3 spores/ml. Spread 1 ml of spore solution onto the surface of appropriately supplemented BCD medium in 90-mm Petri dishes.

8. Incubate the plate under the desired conditions. For standard cultivating procedures and conditions, see Protocol 1.

    The mutagenic treatment results in an initial delay in development, but once germinated, the sporelings grow as usual. This should result in about 100 developing sporelings/dish. The mutagenized spore suspension may be stored for at least 1 year at 4°C.

II. Mutagenesis of Protonemal Tissue

A. Using NTG

1. Add 1 g (fresh weight) of young protonemal tissue from a cellophane-overlay plate to 10 ml of sterile Tris-maleate buffer. Incubate the mixture for 30 minutes at 25°C.

2. Observing strict safety procedures, carefully add 1.2 mg of NTG to a separate tube containing 10 ml of Tris-maleate buffer. Incubate the tube for 30 minutes at 25°C.

3. Mix the tissue with the NTG solution and shake it gently for an additional 30 minutes at 25°C.

4. Terminate the treatment by filtering the mixture through a sterile filter.

   The filter retains the protonemal tissue. This procedure results in a cell survival rate of about 10%.

5. Wash the tissue thoroughly, with at least 100 ml of H_2O.

   If individual cells are required, these may be obtained by protoplasting the tissue as described in Method I of Protocol 2.

B. Using UV Light

1. Remove the lid from a cellophane-overlay plate and irradiate the tissue with UV light. The kill level is difficult to determine, but a dose of 2000 Jm^{-2} should give about 10% survival. It is best to perform a number of experiments to calibrate the UV source in order to get a feel for the correct exposure. Petri dishes are opaque to UV light, so make sure to remove the lid.

2. Immediately after irradiation, transfer the plate to darkness for at least 24 hours to prevent photoreactivation.

   Following incubation in the dark, the tissue can be protoplasted (see Protocol 2, Method I), but very few protoplasts regenerate following UV irradiation. Yields are better if the tissue is allowed to grow following incubation (see Method IB of Protocol 1), but this may yield mutant clones.

Protocol 5

Transformation Using Direct DNA Uptake by Protoplasts

This protocol describes how to transform moss (*P. patens*) protoplasts using PEG-mediated DNA uptake. The transformation rates for direct uptake by protoplasts of DNA with and without genomic sequence (a targeting construct) are typically 10^{-5} and 10^{-3}, respectively (these are the frequencies of stable transformants among regenerants surviving the transformation procedure).

MATERIALS

The recipes for items marked with <R> begin on page 97.

Reagents

BCD medium (solid) <R> containing common moss media supplements <R> and antibiotics <R> as necessary

DNA to be used in transformation

Use 15–30 μg of DNA in a volume no greater than 30 μl.

D-Mannitol (8.5%, w/v)

MMM solution <R>

Prepare 10 ml of MMM solution; this is sufficient for 10–15 transformations.

PEG solution for transformation (PEG/T)

To prepare sufficient PEG/T for 15 transformations, melt 2 g of PEG (MW 6000) in a microwave on the day of transformation. Add 5 ml of MCT solution <R>, mix well, and allow the mixture to stand for 2–3 hours at room temperature before use.

Protoplast regeneration medium for the bottom layer (PRMB) <R>

For each transformation, prepare four 90-mm Petri plates containing PRMB and overlay with sterile cellophane.

Protoplast regeneration medium for the top layer (PRMT) <R>

This medium should be melted and kept at 45°C in a water bath.

Protoplasts, isolated as described in Method I of Protocol 2

Equipment

Centrifuge

Hemocytometer

Incubator with temperature control and white light at intensities between 5 and 20 Wm^{-2} (e.g., Percival Scientific Model CU-36L5)

Tube (10 ml, sterile)

Water bath, set at 45°C

METHOD

Except where stated otherwise, perform all procedures at room temperature, providing this is between 20°C and 25°C. Otherwise, use a water bath. The ten-step dilution described in Steps 8–11 maximizes the recovery rate; however, the dilution can be made in fewer steps if maximum recovery is not required.

1. Resuspend the protoplasts in 8.5% (w/v) D-mannitol solution. Use 2.5 ml for each plate of tissue digested.
2. Use a hemocytometer to estimate protoplast density.
3. Centrifuge at 300g for 5 minutes at room temperature and discard the supernatant. Resuspend the protoplasts in MMM solution at a concentration of 1.6×10^6 protoplasts/ml.
4. Prepare the DNA to be used in transformation by dispensing 15–30 µg of DNA into a sterile 10-ml tube. Centrifuge briefly to bring the DNA to the bottom of the tube.
5. Add 300 µl of protoplast suspension from Step 3 and 300 µl of PEG/T solution to the DNA from Step 4. Mix by tapping the tube gently.
6. Heat the mixture in a water bath for 5 minutes at 45°C.
7. Transfer the tube to room temperature (20°C) and leave it for 5 minutes.
8. Add 300 µl of 8.5% D-mannitol solution. Gently invert the tube to mix. Wait at least 1 minute.
9. Repeat Step 8 four times.
10. Add 1 ml of 8.5% D-mannitol solution. Invert gently to mix. Wait at least 1 minute.
11. Repeat Step 10 four times.
12. Centrifuge at 100–200g for 5 minutes at room temperature.
13. Discard the supernatant and resuspend the pellet in 500 µl of 8.5% D-mannitol solution.
14. Add 2.5 ml of molten PRMT. Dispense 1 ml of the mixture to each of three 90-mm plates containing PRMB and overlaid with cellophane.
15. Incubate the plates in continuous white light for 5 days at 25°C.
16. Transfer cellophane and regenerating protoplasts to BCD medium, containing the appropriate antibiotic, to select for transformants.

Protocol 6

Transformation Using T-DNA Mutagenesis

In this protocol, the transformation of moss (*P. patens*) protoplasts is performed via *Agrobacterium*-mediated transfer of T-DNA. The transformation rate for this protocol is typically 10^{-4} (expressed as the frequency of stable transformants among regenerants surviving the transformation procedure).

MATERIALS

The recipes for items marked with <R> beging on page 97.

CAUTION: See the Cautions Appendix for appropriate handling of materials marked with <!>.

Reagents

Acetoseringone (3′5′-dimethoxy-4′-hydroxyacetophenone) <!> (100 mM, prepared in ethanol <!>)
Agrobacterium tumefaciens culture (strain Agl-1)
  About 1 ml of a mid-log-phase culture (OD_{600} = 0.4–0.6) is needed.
BCD medium (modified)
  Prepare Petri dishes with solid BCD medium <R> containing 5 mM diammonium tartrate (920 mg/liter of BCD) and 0.05 mg/ml vancomycin <!> (1 ml of 1000x stock per liter of BCD).
Protoplast regeneration medium for the bottom layer (PRMB, modified)
  After autoclaving 1 liter of PRMB <R>, add 1 ml of 1000x cefotaxime <!> and 1 ml of 1000x vancomycin <!> (for final concentrations of 0.2 mg/ml and 0.05 mg/ml, respectively). Pour the solution into 90-mm Petri dishes and overlay each with a sterile cellophane disc. Four dishes are required per transformation.
Protoplast regeneration medium for the top layer (PRMT, modified)
  Melt 2.5 ml of PRMT <R> for each transformation and keep it at 45°C in a water bath. Add 1 µl of 1000x cefotaxime <!> and 1 µl of 1000x vancomycin <!> per milliliter of PRMT (for final concentrations of 0.2 mg/ml and 0.05 mg/ml, respectively).
Protoplast regeneration medium, liquid (PRML) <R>
Protoplast wash solution for *Agrobacterium* transformation (APW), maintained at 25°C <R>
  Prepare 35 ml of APW for each transformation.
Protoplasts, isolated as described in Method I of Protocol 2
Selection medium
  For the selection medium, prepare 90-mm Petri dishes with solid BCD medium <R> containing 5 mM diammonium tartrate (920 mg/liter of BCD), 0.2 mg/ml cefotaxime (1 ml of 1000x stock per liter of BCD), 0.05 mg/ml vancomycin (1 ml of 1000x stock per liter of BCD), and an appropriate antibiotic <R> used to select for T-DNA insertion.

Equipment

Centrifuge
Hemocytometer
Incubator with temperature control and white light at intensities between 5 and 20 Wm^{-2} (e.g., Percival Scientific Model CU-36L5)

Micropore surgical tape (3M, 1530-0)
Petri dishes (90-mm diameter, sterile)

METHOD

1. Use a hemocytometer to estimate the density of protoplasts.
2. Centrifuge the protoplasts at 100–200g for 4 minutes at room temperature with no braking and discard the supernatant. Resuspend the protoplasts in PRML at a concentration of 5×10^5 protoplasts/ml.
3. For each milliliter of protoplast suspension, prepare a Petri dish containing 9 ml of PRML.
4. Add 1 ml of protoplast suspension to each prepared Petri dish.
5. Add 200 µl of an *Agrobacterium* culture with an OD_{600} of 0.4–0.6 to each Petri dish.
6. Add 20 µl of 100 mM acetoseringone to each Petri dish.
7. Wrap each dish with Micropore surgical tape and incubate for 48 hours at 25°C under a 16-hour light/8-hour dark cycle.

   The protoplasts will tend to adhere to the Petri dish surface during incubation.

8. Use a pipette to carefully resuspend the protoplasts.
9. Centrifuge the protoplasts at 100–200g for 4 minutes at room temperature with no braking and discard the supernatant. Resuspend the pellet in 10 ml of APW.
10. Repeat Step 9 twice, but after the last centrifugation step, resuspend the pellet in 2 ml of APW and add 2.4 ml of modified PRMT.
11. Dispense 1.1 ml of the protoplast suspension into each of four Petri dishes containing modified PRMB and overlaid with sterile cellophane discs. Incubate the dishes for 7 days at 25°C under continuous white light.
12. Transfer the cellophane overlays onto plates containing selection medium and incubate for 7 days at 25°C under a 16-hour light/8-hour dark cycle.
13. Transfer the cellophane overlays to plates containing modified BCD medium and incubate for 7 days at 25°C under continuous white light.
14. Transfer the cellophane overlays to plates containing selection medium and incubate for 7–10 days at 25°C under a 16-hour light/8-hour dark cycle.
15. Pick small clumps of tissue from each vigorously growing plant and place them on plates containing modified BCD medium. Incubate the plates for 7 days at 25°C under continuous white light.

    These should be stable transformants. However, to confirm that these are transformants, test them once more for resistance by growing small clumps of tissue on selection medium for 7–10 days under continuous white light.

Protocol 7

Transformation of Gametophytes Using a Biolistic Projectile Delivery System

This method is especially suitable for transient gene expression studies, but it can be used to obtain stable transformants.

MATERIALS

CAUTION: See the Cautions Appendix for appropriate handling of materials marked with <!>.

Reagents

$CaCl_2$ (2.5 M) <!>
Ethanol (70% and 100%) <!>
Gold beads (1 µM)
  Prepare in 100% ethanol and resuspend by vigorous mixing for 5 minutes.
Macrocarriers
  Wash in 100% ethanol and allow to dry on sterile filter paper.
Plasmid DNA
  Use 2.5 µg of plasmid DNA contained in less than 4 µl.
Protonemal tissue from *P. patens*
  This tissue is grown as described in Method IB in Protocol 1.
Spermidine (0.1 M) <!>

Equipment

Biolistic particle delivery system (Bio-Rad, PDS-1000/He)
  Before use, clean the chamber with H_2O followed by 100% ethanol and allow it to dry. The disruption disks (900 psi) should be immersed in 100% isopropanol <!> until bombardment.
Centrifuge
Incubator with temperature control and white light at intensities between 5 and 20 Wm^{-2} (e.g., Percival Scientific Model CU-36L5)

METHOD

1. While mixing vigorously, add the following solutions to 2.5 µg of plasmid DNA:

Gold beads (1 µM)	25 µl
$CaCl_2$ (2.5 M)	25 µl
Spermidine (0.1 M)	10 µl

2. Mix for an additional minute. Allow the beads to settle and then centrifuge the tube at 16,000g for 2 minutes at room temperature.

3. Discard the supernatant. Wash the pellet with 70% ethanol by adding 200 µl of 70% ethanol and centrifuging at 16,000g for 2 minutes at room temperature.

   Do not resuspend the gold particles during this step.

4. Discard the supernatant. Resuspend the gold particles in 24 µl of 100% ethanol and apply 6 µl to each macrocarrier. Allow the ethanol to evaporate.

5. Place the protonemal tissue so that it is about 150 mm below the stopper plate of the Bio-Rad PDS-1000/He biolistic particle delivery system. Make two discharges into each dish, moving the position of the dish between discharges.

6. Following bombardment, incubate the tissue for 48 hours at 25°C under continuous white light at intensities between 5 and 20 Wm^{-2}.

7. After the 48-hour incubation, harvest the tissue and blend and plate it onto selective medium as described in Method IB of Protocol 1.

   Alternatively, if clones derived from single cells are required, isolate protoplasts from the tissue as described in Method I of Protocol 2.

Protocol 8

Isolation of DNA, RNA, and Protein from *P. patens* Gametophytes

This protocol describes a series of procedures for isolating nucleic acids (DNA and RNA) and proteins from moss (*P. patens*) tissue. Method IA, a rapid small-scale procedure for isolating DNA, results in genomic DNA that is suitable for PCR only. However, larger amounts of genomic DNA suitable for Southern analysis may be obtained using Method IB (adapted from Luo and Wing 2003) together with the Nucleon PhytoPure Genomic DNA Extraction Kit (GE Healthcare). The RNA isolation procedure (Method II) uses the Plant RNA Isolation Mini Kit (Agilent) with some modifications. Proteins can be obtained from moss gametophytes using Method III.

MATERIALS

The recipes for items marked with <R> begin on page 97.

CAUTION: See the Cautions Appendix for appropriate handling of materials marked with <!>.

Reagents

Isopropanol <!> (for Method IA only)
Liquid nitrogen <!>
Nuclei isolation buffer containing β-mercaptoethanol (NIBM) <R> <!> (for Method IB only)
Nuclei isolation buffer containing Triton X-100 (NIBT) <R> <!> (for Method IB only)
Protein extraction solution, prechilled to 4°C <R> (for Method III only)
Protein wash solution, prechilled to 4°C <R> (for Method III only)
5× Shorty buffer <R> (diluted to 1× before use) (for Method IA only)
Tissue of interest from *P. patens* (see Protocol 1)

> *For DNA isolation (Method I), use protonemal tissue grown for 6–7 days on cellophane-overlay plates as described in Method IB of Protocol 1.*

Tris-Cl (10 mM) containing EDTA (1 mM) (TE buffer, pH 7.5) (for Method IA only)

Equipment

Centrifuge, at 4°C (for Method IB only)
Centrifuge tubes (50 ml) (for Method IB only)
Cheesecloth (pore size of 1.4 mm × 2.4 mm) (for Method IB only)
Erlenmeyer flask (1 liter) (for Method IB only)
Filter paper
Incubator, set at 65°C (optional; see Step 9 of Method IA)
Microcentrifuge (for Methods IA, II, and III only)
Microcentrifuge tube pestles (prechilled; Pellet Pestles, Kontes) (for Methods IA and II only)
Microcentrifuge tubes (prechilled)

> *Methods II and III require 1.5-ml tubes; Method IA requires 2-ml tubes.*

Miracloth (pore size of 25 µm; Calbiochem) (for Method IB only)
Mortar and pestle (prechilled) (for Methods IB and III only)
Nucleon PhytoPure Genomic DNA Extraction Kit (GE Healthcare) (for Method IB only)

> *The pellet that results from Step 15 of Method IB is the starting material for the large sample of the kit (RPN8511).*

Paintbrush (small) (optional; see Steps 11 and 13 of Method IB)
Paper towels (for Method IA only)
Plant RNA Isolation Mini Kit (Agilent) (for Method II only)
> *Before performing the procedure, prepare the Extraction Solution from the kit by adding 10 µl of β-mercaptoethanol <!> per milliliter of Extraction Buffer and store at 4°C.*

Rotator, at 4°C (for Method IB only)
Spatula (prechilled)

METHODS

I. DNA Isolation

A. Rapid Small-scale Preparation of DNA

1. Blot the tissue of moisture by sandwiching the tissue between sheets of filter paper and firmly compressing the tissue with the thumb.
2. Weigh 100 mg of tissue and freeze it in liquid nitrogen. Then, using a prechilled spatula, transfer the frozen tissue to a prechilled 2-ml microcentrifuge tube.
3. Grind the frozen tissue in liquid nitrogen using a prechilled microcentrifuge tube pestle. Make sure the tissue remains frozen.
4. Add 400 µl of 1× Shorty buffer to the tube and mix well.
5. Centrifuge the sample at 16,000g for 5 minutes at room temperature.
6. Transfer 300 µl of the supernatant to a fresh 2-ml microcentrifuge tube containing 300 µl of isopropanol. Mix by inverting the tube.
7. Centrifuge the sample at 16,000g for 10 minutes at room temperature.
8. Discard the supernatant and dry the pellet by inverting the tube on a paper towel.
   > *The pellet may not be obvious, but DNA is there—don't worry!*
9. Once the tube is dry, add 300 µl of TE. Resuspend the pellet by incubating for 30 minutes at 65°C or overnight at 4°C.
10. Store the DNA at 4°C for up to 1 month.
    > *For PCR, use 1 µl of DNA per reaction.*

B. Isolation of High-molecular-weight DNA

1. Blot the tissue of moisture by sandwiching the tissue between sheets of filter paper and firmly compressing the tissue with the thumb.
   > *A typical extraction uses 100 Petri dishes of 7-day-old protonemal tissue (Method IB of Protocol 1).*
2. Weigh the blotted tissue and record the mass in grams.
3. Freeze the tissue in liquid nitrogen. Then, using a liquid-nitrogen-chilled mortar and pestle, grind the frozen tissue to a coarse powder. Transfer the ground tissue to a 1-liter Erlenmeyer flask.
4. To the flask, add 13.3 ml of cold NIBM per gram of tissue.
5. Gently shake the flask by slowly rotating it for 15 minutes at 4°C.
6. Filter the ground tissue through four layers of cheesecloth plus one layer of Miracloth.
   > *The filtrate will contain nuclei.*
7. Squeeze the pellet to allow maximum recovery of nucleus-containing solution.

8. Filter the filtrate again through one layer of Miracloth.
9. To the filtrate, add 1 ml of NIBT for every 30 ml of filtrate. Gently shake the mixture by slowly rotating it for 15 minutes at 4°C.
10. Transfer the mixture to a 50-ml centrifuge tube and centrifuge at 2400g for 15 minutes at 4°C.
11. Discard the supernatant and gently resuspend the pellets in the residual buffer by tapping the tubes or by mixing with a small paintbrush.
12. Dilute the nuclei suspensions with NIBM and combine them into two 50-ml centrifuge tubes. Adjust the total volume in each tube to 50 ml with NIBM and centrifuge at 2400g for 15 minutes at 4°C.
13. Discard the supernatant and gently resuspend the pellets in the residual buffer by tapping the tubes or by mixing with a small paintbrush.
14. Dilute the nuclei suspensions with NIBM and combine them into one 50-ml centrifuge tube. Adjust the total volume to 50 ml with NIBM and centrifuge at 2400g for 15 minutes at 4°C.
15. Discard the supernatant and resuspended the pellet in the residual buffer (1 ml). Use the resuspended pellet, which is enriched in nuclei, as the starting material for genomic DNA extraction with the Nucleon PhytoPure Genomic DNA Extraction Kit (GE Healthcare).

II. RNA Isolation

1. Blot the tissue of moisture by sandwiching the tissue between sheets of filter paper and firmly compressing the tissue with the thumb.
2. Weigh 100 mg of the tissue and, without delay, freeze the tissue in liquid nitrogen. Use a prechilled spatula to transfer the frozen tissue to a prechilled 1.5-ml microcentrifuge tube.
3. Grind the frozen tissue in liquid nitrogen using a prechilled microcentrifuge tube pestle. Make sure that the tissue remains frozen.
4. Add 500 µl of Extraction Buffer (containing β-mercaptoethanol) from the Plant RNA Isolation Mini Kit (Agilent) and grind continuously for at least 7 minutes to maximize yield.
   This and all subsequent steps can be performed at room temperature.
5. Centrifuge the sample at 12,000g for 1.5 minutes and transfer the supernatant to the prefiltration column that is supplied with the Plant RNA Isolation Mini Kit (Agilent). Proceed per the kit protocol. Store all RNA samples at –80°C.

III. Protein Isolation

1. Blot the tissue of moisture by sandwiching the tissue between sheets of filter paper and firmly compressing the tissue with the thumb.
2. Weigh about 700 mg of the blotted tissue and freeze it in liquid nitrogen.
3. Using a liquid-nitrogen-chilled mortar and pestle, grind the frozen tissue to a fine powder.
4. Divide the tissue powder into 100-µl aliquots in 1.5-ml microcentrifuge tube and store them at –80°C until needed for protein extraction.
   Approximately 6 aliquots may be obtained from 700 mg of tissue.
5. When required, suspend about 100 µl of powder in 1 ml of prechilled protein extraction solution. Incubate the mixture for 2 hours at –20°C.
6. Centrifuge the mixture at 20,000g for 30 minutes at 4°C and discard the supernatant.
7. To remove pigments, lipids, and nucleic acids, add 1 ml of prechilled protein wash solution, mix, and incubate for 1 hour at –20°C.

8. Centrifuge the mixture at 20,000*g* for 30 minutes at 4°C and discard the supernatant.
9. Repeat Steps 7 and 8 until the supernatant is completely clear and uncolored.
   The number of washes depends on the tissue type and its age. Gametophore tissue requires more washes than does protonemal tissue.
10. Dry the pellet for approximately 2 hours by leaving the centrifuge tubes open at room temperature. Store the pellets at –80°C.

Recipes

The recipes for items marked with <R> are also listed here.

CAUTION: See the Cautions Appendix for appropriate handling of materials marked with <!>.

BCD MEDIUM, LIQUID OR SOLID

Reagent	Quantity (for 1 liter)	Final concentration
Agar (Sigma-Aldrich A9799)	7 g	0.7% (w/v)
$CaCl_2$ <!>	111 mg	1 mM
$FeSO_4 \cdot 7H_2O$ <!>	12.5 mg	45 µM
Solution B <R>	10 ml	1 mM $MgSO_4$
Solution C <R>	10 ml	1.84 mM KH_2PO_4
Solution D <R>	10 ml	10 mM KNO_3
TES <R>	1 ml	trace
H_2O	to 1 liter	

Do not include $CaCl_2$ if the medium is to be used to prepare protoplast regeneration media (PRMB, PRMT, or PRML). If using $CaCl_2$, add after autoclaving, immediately before use. Include agar for solid media only. Most high-grade agars can be used, but some may affect the pH of the medium. If an agar other than Sigma-Aldrich A9799 is used, the quantity (7 g) may need to be altered. Sterilize the medium by autoclaving for 40 minutes at 121°C. Add common moss media supplements <R> as necessary. Store for 2 months at room temperature.

COMMON ANTIBIOTICS INCLUDED WITH MOSS MEDIA

Supplement	Stock solution (1000x)	Concentration in medium (1x)
G418 <!>	50 mg/ml	50 µg/ml
Hygromycin <!>	30 mg/ml	30 µg/ml
Sulfadiazine <!>	150 mg/ml	150 µg/ml
Zeocin	50 mg/ml	50 µg/ml

Prepare each stock solution in H_2O. Filter-sterilize Zeocin.

COMMON MOSS MEDIA SUPPLEMENTS

Supplement	Quantity (per liter of H_2O) for stock solution	Concentration in stock solution	Concentration in medium (1x)
p-Amino-benzoic acid <!>	247 mg	1.8 mM (1000x)	1.8 µM
Diammonium tartrate	92 g	500 mM (100x)	5 mM
Nicotinic acid	1 g	8 mM (1000x)	8 µM
D-Sucrose	500 g	1.5 M (100x)	15 mM
Thiamine-HCl	500 mg	1.5 mM (1000x)	1.5 µM

Stock solutions may be either autoclaved or filter-sterilized and added to growth media as required to give the concentrations listed above. For crossing thiA1 strains, add thiamine-HCl to a final concentration of 15 nM (0.01x), not 1.5 µM (1x). Store for up to 1 year at 4°C in the dark.

DMSO–GLUCOSE SOLUTION

Reagent	Quantity (for 100 ml)	Final concentration
Dimethylsulfoxide (DMSO) <!>	5 ml	5% (v/v)
Glucose	10 g	10% (w/v)
H_2O	to 100 ml	

Prepare fresh.

DRISELASE SOLUTION

Reagent	Quantity (for 100 ml)	Final concentration
Driselase	2 g	2%
D-Mannitol solution (8.5% w/v)	to 100 ml	

Stir to mix (but do not shake vigorously) for at least 15 minutes. Centrifuge at 2500g for 5 minutes and filter-sterilize the clear supernatant. Dilute 1 part 2% driselase with 3 parts 8.5% D-mannitol solution to give a final concentration of 0.5%. Use on the same day. There may be variation among batches of driselase, so adjustment to the driselase quantity (2 g) may be necessary.

HOAGLAND'S A-Z TRACE ELEMENT SOLUTION (TES)

Reagent	Quantity (for 1 liter)	Final concentration
$Al_2(SO_4)_3 \cdot K_2SO_4 \cdot 24H_2O$ <!>	55 mg	0.006% (w/v)
$CoCl_2 \cdot 6H_2O$ <!>	55 mg	0.006% (w/v)
$CuSO_4 \cdot 5H_2O$ <!>	55 mg	0.006% (w/v)
H_3BO_3 <!>	614 mg	0.061% (w/v)
KBr <!>	28 mg	0.003% (w/v)
KI <!>	28 mg	0.003% (w/v)
LiCl <!>	28 mg	0.003% (w/v)
$MnCl_2 \cdot 4H_2O$ <!>	389 mg	0.039% (w/v)
$SnCl_2 \cdot 2H_2O$ <!>	28 mg	0.003% (w/v)
$ZnSO_4 \cdot 7H_2O$ <!>	55 mg	0.006% (w/v)
H_2O	to 1 liter	

The exact composition of this solution is probably not important.

MCT SOLUTION

Reagent	Quantity (for 10 ml)	Final concentration
$Ca(NO_3)_2$ (1 M) <!>	1 ml	100 mM
D-Mannitol (8.5% w/v)	9 ml	7.65% (w/v)
Tris-Cl (1 M, pH 8.0)	100 µl	10 mM

Prepare fresh on day of use and filter-sterilize.

MMM SOLUTION

Reagent	Quantity (for 10 ml)	Final concentration
D-Mannitol	910 mg	9.1%
2-[N-morpholino]ethanesulfonic acid (MES) (1% w/v, pH 5.6) <!>	1 ml	10%
$MgCl_2$ (1 M)	150 μl	15 mM
H_2O	8.85 ml	

Dissolve the D-mannitol in the H_2O, sterilize by autoclaving, and store at room temperature. On the day of use, add the MES and $MgCl_2$ to the D-mannitol solution and filter-sterilize.

NITROGEN-FREE MEDIUM

Reagent	Quantity (for 1 liter)	Final concentration
Agar (Sigma-Aldrich A9799)	7 g	0.7% (w/v)
$CaCl_2$ <!>	111 mg	1 mM
$FeSO_4 \cdot 7H_2O$ <!>	12.5 mg	45 μM
Solution B <R>	10 ml	1 mM $MgSO_4$
Solution C <R>	10 ml	1.84 mM KH_2PO_4
TES <R>	1 ml	trace
H_2O	to 1 liter	

Most high-grade agars can be used, but some may affect the pH of the medium. If an agar other than Sigma-Aldrich A9799 is used, the quantity (7 g) may need to be altered. Add $CaCl_2$ after autoclaving, immediately before use.

NUCLEI ISOLATION BUFFER CONTAINING β-MERCAPTOETHANOL (NIBM)

Reagent	Quantity (for 1 liter)	Final concentration
EDTA (500 mM)	20 ml	10 mM
KCl (1 M)	100 ml	100 mM
β-Mercaptoethanol <!>	20 μl	2% (v/v)
Spermidine (1 M) <!>	4 ml	4 mM
Spermine (1 M) <!>	1 ml	1 mM
D-Sucrose	171 g	17.1% (w/v)
H_2O	to 1 liter	

Mix all ingredients except for the β-mercaptoethanol and filter-sterilize. Store for up to 7 days at 4°C. Add the β-mercaptoethanol immediately before use.

NUCLEI ISOLATION BUFFER CONTAINING TRITON X-100 (NIBT)

Reagent	Quantity (for 1 liter)	Final concentration
EDTA (500 mM)	20 ml	10 mM
KCl (1 M)	100 ml	100 mM
Spermidine (1 M) <!>	4 ml	4 mM
Spermine (1 M) <!>	1 ml	1 mM
D-Sucrose	171 g	17.1% (w/v)
Triton X-100 (20%) <!>	100 ml	2%
H_2O	to 1 liter	

Filter-sterilize. Store for up to 7 days at 4°C.

PEG SOLUTION FOR PROTOPLAST FUSION (PEG/F)

Reagent	Quantity
$CaCl_2 \cdot 6H_2O$ <!>	109 mg
H_2O	10 ml
Polyethylene glycol (PEG) (MW 6000)	5 g

Melt the PEG in a microwave oven or water bath. Dissolve the $CaCl_2 \cdot 6H_2O$ in H_2O and then mix the solution with the melted PEG. Dispense into appropriate aliquots (750 µl is needed for each fusion). Store for up to 7 days at 4°C.

PROTEIN EXTRACTION SOLUTION

Reagent	Quantity (for 100 ml)	Final concentration
β-Mercaptoethanol <!>	70 µl	0.07% (v/v)
Trichloroacetic acid (TCA) (10% w/v) <!>	10 ml	1% (w/v)
Ethanol <!>	to 100 ml	

Filter-sterilize. Store for up to 2 months at 4°C.

PROTEIN WASH SOLUTION

Reagent	Quantity (for 100 ml)	Final concentration
EDTA (500 mM)	400 µl	2 mM
β-Mercaptoethanol <!>	70 µl	0.07% (v/v)
Phenylmethylsulfonyl fluoride (PMSF) (1 M) <!>	100 µl	100 mM
Acetone <!>	to 100 ml	

Filter-sterilize. Store for up to 1 year at –20°C.

PROTOPLAST REGENERATION MEDIUM FOR THE BOTTOM LAYER (PRMB)

Reagent	Quantity (for 1 liter)	Final concentration
Agar (Sigma-Aldrich A9799)	7 g	0.7% (w/v)
$CaCl_2$ <!>	1.1 g	10 mM
Diammonium tartrate	920 mg	5 mM
D-Mannitol	60 g	6% (w/v)
BCD medium, liquid <R>	to 1 liter	

Most high-grade agars can be used, but some may affect the pH of the medium. If an agar other than Sigma-Aldrich A9799 is used, the quantity (7 g) may need to be altered. Sterilize the medium by autoclaving for 40 minutes at 121°C. Store for up to 2 months at room temperature. Add $CaCl_2$ after autoclaving, immediately before use.

PROTOPLAST REGENERATION MEDIUM FOR THE TOP LAYER (PRMT)

Reagent	Quantity (for 1 liter)	Final concentration
Agar (Sigma-Aldrich A9799)	4 g	0.4% (w/v)
$CaCl_2$ <!>	1.1 g	10 mM
Diammonium tartrate	920 mg	5 mM
D-Mannitol	80 g	8% (w/v)
BCD medium, liquid <R>	to 1 liter	

Most high-grade agars can be used, but some may affect the pH of the medium. If an agar other than Sigma-Aldrich A9799 is used, the quantity (4 g) may need to be altered. Sterilize the medium by autoclaving for 40 minutes at 121°C. Store for up to 2 months at room temperature. Add $CaCl_2$ after autoclaving, immediately before use.

PROTOPLAST REGENERATION MEDIUM, LIQUID (PRML)

Reagent	Quantity (for 1 liter)	Final concentration
$CaCl_2$ <!>	1.1 g	10 mM
Diammonium tartrate	920 mg	5 mM
D-Mannitol	85 g	8.5% (w/v)
BCD medium, liquid <R>	to 1 liter	

Sterilize the medium by autoclaving for 40 minutes at 121°C. Store for up to 2 months at room temperature. Add $CaCl_2$ after autoclaving, immediately before use.

PROTOPLAST WASH (PW) SOLUTION

Reagent	Quantity (for 1 liter)	Final concentration
$CaCl_2 \cdot 6H_2O$ <!>	2.19 g	10 mM
D-Mannitol	85 g	8.5% (w/v)
H_2O	to 1 liter	

Sterilize by autoclaving for 40 minutes at 121°C. Store for up to 2 months at room temperature. This recipe contains $CaCl_2$ at 10 mM, but different laboratories use concentrations between 0 mM and 50 mM with no apparent differences in protoplast viability.

PROTOPLAST WASH SOLUTION FOR *AGROBACTERIUM* TRANSFORMATION (APW)

Reagent	Quantity (for 100 ml)	Final concentration
$CaCl_2 \cdot 6H_2O$ <!>	219 mg	10 mM
1000x Vancomycin stock solution (50 mg/ml) <!>	100 μl	1x
1000x Cefotaxime stock solution (200 mg/ml) <!>	100 μl	1x
D-Mannitol solution (8.5% w/v)	100 ml	

Prepare fresh.

5x SHORTY BUFFER STOCK SOLUTION

Reagent	Quantity (for 500 ml)	Final concentration (5x)
EDTA (500 mM)	25 ml	25 mM
LiCl (2 M) <!>	100 ml	400 mM
Sodium dodecyl sulfate (SDS) (10% w/v) <!>	50 ml	1% (w/v)
Tris-Cl (1 M, pH 9.0)	100 ml	200 mM
H_2O	225 ml	

Store for up to 1 year at room temperature. Dilute fivefold with H_2O as required.

SOLUTION B

Reagent	Quantity (for 1 liter)	Final concentration
$MgSO_4 \cdot 7H_2O$ <!>	25 g	0.1 M
H_2O	to 1 liter	

Sterilize by autoclaving for 20 minutes at 120°C and store for up to 2 months at room temperature.

SOLUTION C

Reagent	Quantity (for 1 liter)	Final concentration
KH_2PO_4	25 g	184 mM
H_2O	to 1 liter	

Dissolve the KH_2PO_4 in 500 ml of H_2O and adjust the pH to 6.5 with a minimal volume of 4 M KOH <!>. Bring the final volume to 1 liter with additional H_2O. Sterilize by autoclaving for 20 minutes at 120°C and store for up to 2 months at room temperature.

SOLUTION D

Reagent	Quantity (for 1 liter)	Final concentration
KNO_3 <!>	101 g	1 M
H_2O	to 1 liter	

Sterilize by autoclaving for 20 minutes at 120°C and store for up to 2 months at room temperature.

TRIS-MALEATE BUFFER

Reagent	Quantity (for 1 liter)	Final concentration
Maleic acid <!>	6 g	50 mM
Tris-(hydroxymethyl)amino-methane <!>	6 g	50 mM
H_2O	to 1 liter	

Adjust the pH to 6.0 with 10 M NaOH <!> or KOH <!>. Store for up to 1 year at room temperature.

REFERENCES

Ashton, N.W. and Cove, D.J. 1977. The isolation and preliminary characterization of auxotrophic and analogue resistant mutants in the moss *Physcomitrella patens*. *Mol. Gen. Genet.* **154:** 87–95.

Axtell, M.J., Snyder, J.A., and Bartel, D.P. 2007. Common functions for diverse small RNAs of land plants. *Plant Cell* **19:** 1750–1769.

Axtell, M., Jan, C., Rajagopalan, R., and Bartel, D. 2006. A two-hit trigger for siRNA biogenesis in plants. *Cell* **127:** 565–577.

Bezanilla, M., Perroud, P.-F., Pan. A., Klueh, P., and Quatrano, R.S. 2005. An RNAi system in *Physcomitrella patens* with an internal marker for silencing allows for rapid identification of loss of function phenotypes. *Plant Biol.* **7:** 251–257.

Boyd, P.J., Hall, J., and Cove, D.J. 1988. An airlift fermenter for the culture of the moss *Physcomitrella patens*. In *Methods in bryology*. Proc. Bryological Methods Workshop, Mainz, Germany (ed. J.M. Glime), pp. 41–45. Hattori Botany Laboratory, Japan.

Chakhparonian, M. 2001. "Développement d'outils de la mutagenèse ciblée par recombinaison homologue chez *Physcomitrella patens*." Ph.D. Thesis, Université de Lausanne, Lausanne, Switzerland. http://www.unil.ch/lpg/docs/theses/MCthesis.pdf

Cuming, A.C., Cho, S.H., Kamisugi, Y., Graham, H., and Quatrano, R.S. 2007. Microarray analysis of transcriptional responses to abscisic acid and osmotic, salt, and drought stress in the moss, *Physcomitrella patens*. *New Phytol.* **176:** 275–287.

Courtice, G.R.M. and Cove, D.J. 1978. Evidence for the restricted passage of metabolites into the sporophyte of the moss, *Physcomitrella patens*. *J. Bryol.* **10:** 191–198.

Cove, D.J. 2005. The moss *Physcomitrella patens*. *Annu. Rev. Genet.* **39:** 339–358.

Cove, D.J. and Quatrano, R.S. 2006. Agravitropic mutants of the moss *Ceratodon purpureus* do not complement mutants having a reversed gravitropic response. *Plant Cell Environ.* **29:** 1379–1387.

Crum, H.A. and Anderson, L.E. 1981. *Mosses of eastern North America*, vols. 1 and 2. Columbia University Press, New York.

Engel, P.P. 1968. The induction of biochemical and morphological mutants in the moss, *Physcomitrella patens*. *Am. J. Bot.* **55:** 438–446.

Grimsley, N.H. and Withers, L.A. 1983. Cryopreservation in the moss, *Physcomitrella patens*. *CryoLetters* **4:** 251–258.

Grimsley, N.H., Ashton, N.W., and Cove, D.J. 1977. Complementation analysis of auxotrophic mutants of the moss, *Physcomitrella patens*, using protoplast fusion. *Mol. Gen. Genet.* **155:** 103–107.

Harries, P., Pan, A., and Quatrano, R.S. 2005. Actin-related protein 2/3 complex component ARPC1 is required for proper cell morphogenesis and polarized cell growth in *Physcomitrella patens*. *Plant Cell* **17:** 2327–2339.

Hayashida, A., Takechi, K., Sugiyama, M., Kubo, M., Itoh, R.D., Takio, S., Fujita, T., Hiwatashi, Y., Hasebe, M., and Takano, H. 2005. Isolation of mutant lines with decreased numbers of chloroplasts per cell from a tagged mutant library of the moss *Physcomitrella patens*. *Plant Biol.* **7:** 300–306.

Heitz, E. 1928. Das heterochromatin der moose. I. *Jahrb. Wiss. Bot.* **69:** 762–818.

Kaewsuwan, S., Cahoon, E.B., Perroud, P.-F., Wiwat, C., Panvisavas, N., Quatrano, R.S., Cove, D.J., and Bunyapraphatsara, N. 2006. Identification and functional characterization of the moss *Physcomitrella patens* Δ^5-desaturase gene involved in arachidonic and eicosapentaenoic acids biosynthesis. *J. Biol. Chem.* **281:** 21988–21997.

Kamisugi, Y., Cuming, A.C., and Cove, D.J. 2005. Parameters determining the efficiency of gene targeting in the moss *Physcomitrella patens*. *Nucleic Acids Res.* **33:** e173.

Kamisugi, Y., Schlink, K., Rensing, S.A., Schween, G., von Stackelberg, M., Cuming, A.C., Reski, R., and Cove, D.J. 2006. The mechanism of gene targeting in *Physcomitrella patens*: Homologous recombination, concatenation and multiple integration. *Nucleic Acid Res.* **34:** 6205–6214.

Khandelwal, A., Chandu, D., Roe, C.M., Kopan, R., and Quatrano, R.S. 2007. Moonlighting activity of presenilin in plants is independent of γ-secretase and evolutionarily conserved. *Proc. Natl. Acad. Sci.* **104:** 13337–13342.

Luo, M. and Wing, R.A. 2003. An improved method for plant BAC library construction. *Methods Mol. Biol.* **236:** 3–20.

Maizel, A., Busch, M.A., Tanahashi, T., Perkovic, J., Kato, M., Hasebe, M., and Weigel, D. 2005. The floral regulator LEAFY evolves by substitutions in the DNA binding domain. *Science* **308:** 260–263.

Marella, H.H., Sakata, Y., and Quatrano, R.S. 2006. Characterization and functional analysis of *ABSCISIC ACID INSENSITIVE3*-like genes from *Physcomitrella patens*. *Plant J.* **46:** 1032–1044.

McDaniel, S.F., Willis, J.H., and Shaw, A.J. 2008. The genetic basis of abnormal development in interpopulation hybrids of the moss *Ceratodon purpureus*. *Genetics* **179:** 1425–1435.

Menand, B., Yi, K.K., Jouannic, S., Hoffmann, L., Ryan, E., Linstead, P., Schaefer, D.G., and Dolan, L. 2007. An ancient mechanism controls the development of cells with a rooting function in land plants. *Science* **316:** 1477–1480.

Nishiyama, T., Hiwatashi, Y., Sakakibara, I., Kato, M., and Hasebe, M. 2000. Tagged mutagenesis and gene-trap in the moss, *Physcomitrella patens*, by shuttle mutagenesis. *DNA Res.* **7:** 9–17.

Nishiyama, T., Fujita, T., Shin-I, T., Seki, M., Nishide, H., Uchiyama, I., Kamiya, A., Carninci, P., Hayashizaki, Y., Shinozaki, K., et al. 2003. Comparative genomics of *Physcomitrella patens* gametophytic transcriptome and *Arabidopsis thaliana*: Implication for land plant evolution. *Proc. Natl. Acad. Sci.* **100:** 8007–8012.

Oliver, M.J., Velten, J., and Mishler, B.D. 2004. Desiccation tolerance in bryophytes: A reflection of the primitive strategy for plant survival in dehydrating habitats? *Integr. Comp. Biol.* **45:** 788–799.

Pettet, A. 1964. Hybrid sporophytes in the Funariaceae. *Trans. Br. Bryol. Soc.* **4:** 642–648.

Perroud, P.-F. and Quatrano, R.S. 2006. The role of ARPC4 in tip growth and alignment of the polar axis in filaments of *Physcomitrella patens*. *Cell Motil. Cytoskelet.* **63:** 162–171.

Perroud, P.-F. and Quatrano, R.S. 2008. BRICK1 is required for apical cell growth in filaments of the moss *Physcomitrella patens* but not for gametophore morphology. *Plant Cell* **20:** 411–422.

Quatrano, R.S., McDaniel, S.F., Khandelwal, A., Perroud, P.-F., and Cove, D.J. 2007. *Physcomitrella patens*: Mosses enter the genomic age. *Curr. Opin. Plant Biol.* **10:** 182–189.

Rensing, S.A., Lang, D., Zimmer, A., Terry, A., Salamov, A., Shapiro, H., Nishiyama, T., Perroud, P.-F., Lindquist, E.A., Kamisugi, Y., et al. 2008. The *Physcomitrella* genome reveals insights into the conquest of land by plants. *Science* **319:** 64–69.

Saidi, Y., Finka, A., Chakhporanian, M., Zryd, J.P., Schaefer, D.G., and Goloubinoff, P. 2005. Controlled expression of recombinant proteins in *Physcomitrella patens* by a conditional heat-shock promoter: A tool for plant research and biotechnology. *Plant Mol. Biol.* **59:** 697–711.

Schween, G., Egener, T., Fritzowsky, D., Granado, J., Guitton, M.C., Hartmann, N., Hohe, A., Holtorf, H., Lang, D., Lucht, J.M., et al. 2005. Large-scale analysis of 73,329 *Physcomitrella* plants transformed with different gene disruption libraries: Production parameters and mutant phenotypes. *Plant Biol.* **7:** 228–237.

Schulte, J., Erxleben, A., Schween, G., and Reski, R. 2006. High throughput metabolic screen of *Physcomitrella* transformants. *Bryologist* **109:** 247–256.

Thelander, M., Olsson, T., and Ronne, H. 2005. Effect of the energy supply on filamentous growth and development in *Physcomitrella patens*. *J. Exp. Bot.* **56:** 653–662.

You, Y.S., Marella, H.H., Zentella, R., Zhou, Y., Ulmasov, T., Ho, T.-H.D., and Quatrano, R.S. 2006. Use of bacterial quorum sensing components to regulate gene expression in plants. *Plant Physiol.* **140:** 1205–1212.

Vidali, L., Augustine, R.C., Kleinman, K.P., and Bezanilla, M. 2007. Profiling is essential for tip growth in the moss *Physcomitrella patens*. *Plant Cell* **19:** 3705–3722.

von Stackelberg, M., Rensing, S., and Reski, R. 2006. Identification of genic moss SSR markers and a comparative analysis of twenty-four algal and plant gene indices reveal species-specific rather than group-specific characteristics of microsatellites. *BMC Plant Biol.* **6:** 9.

von Wettstein, F. 1924. Morphologie und physiologie des formwechsels der moose auf genetischler grundlage I. *Z. Indukt. Abstammungs–Vererbungesl.* **33:** 1–236.

4 The Genus *Antirrhinum* (Snapdragon)
A Flowering Plant Model for Evolution and Development

Andrew Hudson, Joanna Critchley, and Yvette Erasmus
University of Edinburgh, Institute of Molecular Plant Sciences, Edinburgh EH9 3JH, United Kingdom

ABSTRACT

The *Antirrhinum* species group comprises approximately 20 morphologically diverse members that are able to form fertile hybrids. It includes the cultivated snapdragon *Antirrhinum majus*, which has been used as a model for biochemical and developmental genetics for more than 75 years. The research infrastructure for *A. majus*, together with the interfertility of the species group, allows *Antirrhinum* to be used to examine the genetic basis for plant diversity.

PROTOCOLS
1 Cultivating *Antirrhinum*, 111
2 Propagating *Antirrhinum*, 114

BACKGROUND INFORMATION

The garden snapdragon *A. majus* has several centuries' history of cultivation as a flowering ornamental. It emerged as a model organism during early studies of inheritance and mutation (e.g., Darwin 1868) because of its diploid inheritance, ease of cultivation, and variation in morphology and flower color. Laboratory lines of *A. majus* were produced from cultivars, and a substantial collection of mutants had amassed during the course of the 20th century. This collection included lines with unstable mutations in pigment genes, which produced variegated flowers (Fig. 1a).

Transposons responsible for flower variegation were identified in the 1980s at the John Innes Centre in Norwich, United Kingdom and the Max-Planck-Institut in Cologne, Germany, allowing genes involved in flower and leaf development and in pigmentation to be isolated by transposon tagging. *A. majus* subsequently provided the first insights into the regulation of many developmental processes that are conserved in flowering plants, including the specification of flower and floral organ identity, leaf and flower asymmetry, and the pollen component of gametophytic self-incompatibility. Because *A. majus* diverged from the more commonly used eudicot model *Arabidopsis thaliana* early in the history of flowering plants, it has proven to be useful in comparative developmental studies.

The *Antirrhinum* species group also has a history of use in studies of natural variation. The close relatives of *A. majus* form a monophyletic group of about 20 species native to the Mediterranean region, particularly southwestern Europe and northern Africa. The species vary

This chapter, with full-color images, can be found online at www.cshprotocols.org/emo.

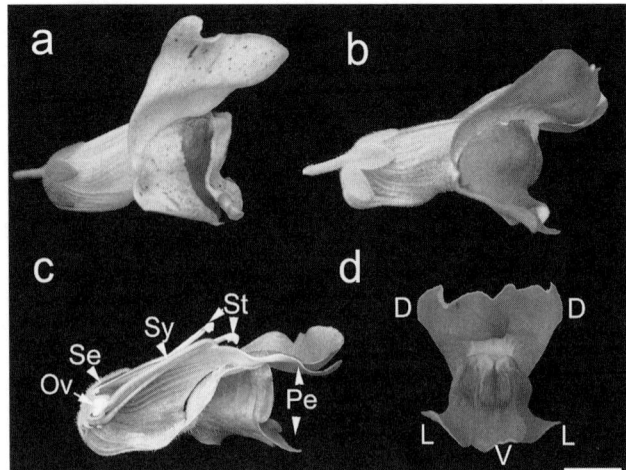

FIGURE 1. *Antirrhinum majus.* (a) The variegated flower caused by a transposon-induced mutation in a gene required for synthesis of anthocyanin pigments. Excision of the transposon can give rise to clones of wild-type cells able to produce magenta anthocyanins. (b) Wild-type *A. majus* flower. (c) Section of an *A. majus* flower, showing the four types of organs. (Ov) Ovary; (Se) sepal; (Sy) style; (St) stamen; (Pe) petal. (d) Dorsoventral asymmetry of the *Antirrhinum* corolla. (D) dorsal, (L) lateral, and (V) ventral petals. Bar, 10 mm.

widely in morphology and ecology and are adapted to different—often extreme—habitats. However, all are able to form fertile hybrids with one another and with *A. majus*, allowing the identification of genes that underlie their differences (see, e.g., Langlade et al. 2005). Population genetic studies that have been applied to *Antirrhinum* species show different population sizes, geographic distributions, and breeding systems, from self-fertility to obligate out-crossing (see, e.g., Jiménez et al. 2002; Mateu-Andres and de Paco 2006).

SOURCES AND HUSBANDRY

Seeds from a large collection of *A. majus* mutants, their wild-type progenitors (Fig. 1b), and a limited number of other *Antirrhinum* species can be obtained from the Leibniz Institute of Plant Genetics and Crop Plant Research (IPK) in Gatersleben, Germany (http://gbis.ipk-gatersleben.de). Protocols for *Antirrhinum* husbandry can be found in most gardening books, and the necessary materials are readily available. Although tolerant of both frost and high temperatures, *Antirrhinum* species grow best at daytime temperatures of 17–25°C. In temperate regions, they are usually treated as half-hardy annuals and are transferred from a glasshouse to the open in the spring or are grown entirely in a glasshouse. All species grow well from seeds, most flowering within 3–4 months of sowing, and are readily propagated clonally from cuttings. Detailed methods for *Antirrhinum* culture and propagation are provided in Protocols 1 and 2.

RELATIVES OF *A. MAJUS*

Antirrhinum is a member of the asterid clade of flowering plants. The more commonly used model species *Arabidopsis* is a member of the second major clade of broad-leaved plants—the rosids—from which asterids diverged an estimated 120 million years ago. Within the asterids, *Antirrhinum* belongs to the order Lamiales, a close relative of the order Solanales, which includes other model species such as petunia and tomato. *Antirrhinum* was recently placed in the family Plantaginaceae (synonymous with Veronicaceae) following a revision of the classical family Scrophulariaceae based on DNA sequence variation (Olmstead et al. 2001).

Other aspects of *Antirrhinum* taxonomy remain controversial. The generic epithet *Antirrhinum* is now usually reserved for the monophyletic group of Old World perennials with a diploid chromosome number of 16. However, it is still applied to a broader monophyletic group that includes species with different chromosome numbers, such as the New World *Sairocarpus* and the annual *Misopates*, with which *Antirrhinum* species are unable to form fertile hybrids (see, e.g., Oyama and Baum 2004). Within *Antirrhinum sensu stricto*, a variable number of different species have been proposed and relationships between taxa are currently unresolved. These taxonomic problems largely reflect the young age of the genus (<5 million years; Gübitz et al. 2003) and the effects of hybridization (see, e.g., Whibley et al. 2006); thus, attempts to reconstruct phylogenies based on nuclear or chloroplast DNA sequence variation have so far been unsuccessful (see, e.g., Jiménez et al. 2005). In the absence of a taxonomic revision based on a resolved phylogeny, the descriptions of approximately 20 species and their likely hybrids in *Flora Europaea* (Webb 1972) provide a realistic working guide.

The genus has traditionally been divided into three subsections or morphological groups (Rothmaler 1956) that have received support from studies of isozyme and DNA variation. The subsection *Antirrhinum* includes the close relatives of *A. majus* and consists of species with similar upright growth, large organs, and pink or yellow flowers (Fig. 2a). These tend to be geographically widespread and grow in a variety of habitats. *A. majus* was probably domesticated in northeastern Spain or southwestern France from *A. pseudomajus* (also known as *A. majus* subspecies *pseudomajus*), from which it differs by having more darkly pigmented flowers, although traits such as flower color variation might have been introduced by introgression from other species. Members of subsection *Kickxiella*, in contrast, are usually restricted to rock faces and walls. They are also smaller,

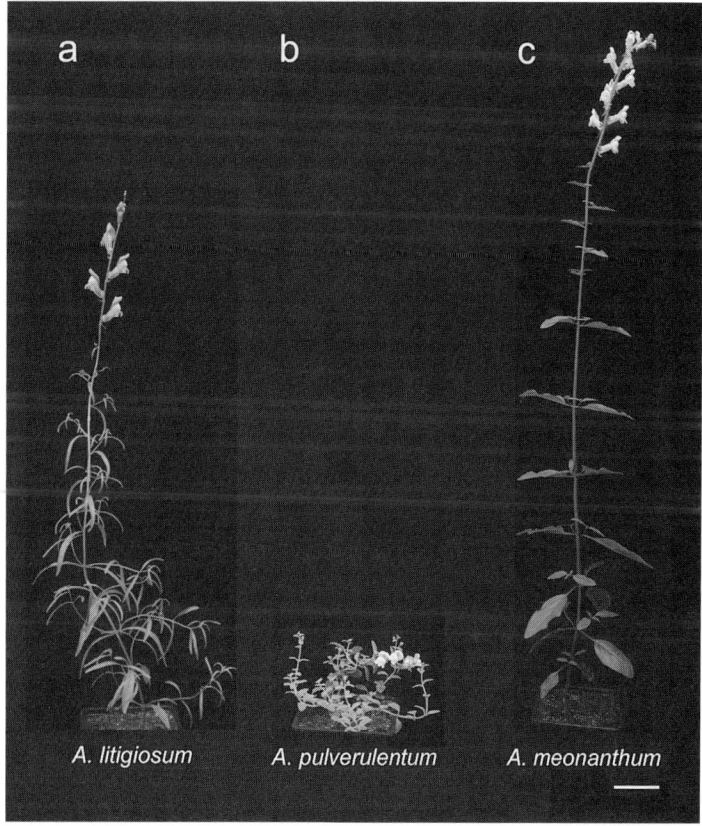

FIGURE 2. Representatives of the three traditional subsections of the genus *Antirrhinum*. (a) *A. litigiosum* (also known as *A. majus* subspecies *litigiosum*), a member of subsection *Antirrhinum*; (b) *A. pulverulentum*, a member of subsection *Kickxiella*; (c) *A. meonanthum*, a member of subsection *Streptosepalum*. Bar, 50 mm.

typically prostrate in habit, and have small white or pale pink flowers (Fig. 2b). They are geographically more restricted and many are endemic to particular mountain regions. The two members of subsection *Streptosepalum* are pale- or yellow-flowered, of upright habit, and grow in hedges and rocky outcrops in northern Spain and Portugal (Fig. 2c). Distribution maps for all *Antirrhinum* species were produced by Rothmaler (1956) and can also be found in Stubbe (1966).

USES OF THE *A. MAJUS* MODEL SYSTEM

Biochemistry

Two aspects of *Antirrhinum* biochemistry relevant to their attraction of pollinating bees have been studied in detail. First, genes encoding the enzymes involved in the production of floral scents—phenylpropanoids and isoprenoids—have been identified from *A. majus* and used to study the regulation of scent production and the effects of variation in scent composition on pollinator attraction (see, e.g., Wright et al. 2005). Second, the basis for flower color variation in *Antirrhinum* has a long history of study, including the pioneering biochemical genetics of Muriel Wheldale in the early 20th century. Structural genes encoding most enzymes involved in the biosynthesis of magenta anthocyanins and the structurally related yellow aurones have been identified and isolated through a combination of genetics and biochemistry. Several regulatory genes encoding MYB or basic helix-loop-helix (bHLH) transcription factors are known to affect the intensity or pattern of pigmentation (Schwinn et al. 2006). In parallel, many of the genes underlying natural variation in *Antirrhinum* flower pigmentation were mapped or shown to be allelic to loci identified from mutations in *A. majus*. Three of the loci involved in natural variation have been isolated: Two encode the MYB transcription factors ROSEA and VENOSA and the third is the structural gene INCOLORATA. Two additional genes—SULFUREA (SULF), which controls aurone pigmentation, and ELUTA (EL), which is responsible for variation in the pigment pattern within the flower—have been mapped but not yet isolated. Flower color variation in *Antirrhinum* has been shown to affect pollinator behavior, and selection at loci including SULF and EL is likely to maintain the distinction between yellow-flowered *A. striatum* and magenta-flowered *A. pseudomajus* in the face of hybridization (Whibley et al. 2006).

Development

The use of *A. majus* in parallel with other species, notably *Arabidopsis*, has lead to an understanding of how the identity of flowers and floral organs is specified (for recent review, see Davies et al. 2006). These processes are broadly conserved between the asterid *Antirrhinum* and the rosid *Arabidopsis*, which have similar inflorescence and floral structures (see Fig. 1c,d for the structure of the *Antirrhinum* flower). Both species, for example, contributed to the "ABC" model of floral organ specification, in which the combination of genes expressed in each of the whorls determines the identity of the floral organs (A genes specify sepals, A + B genes specify petals, B + C genes specify stamens, and C genes specify the carpel). However, the comparative use of *A. majus* further revealed subtle evolutionary differences in the way that the development of similar flowers can be regulated. *Antirrhinum* and *Arabidopsis*, for instance, use different genes to exclude the expression of C genes from the outermost parts of the developing flower—a role originally attributed to A-function genes—leading to a reevaluation of the A-function genes in both species (Keck et al. 2003; Davies et al. 2006). Similarly, the ancestral C function appears to have been transferred to different duplicated genes after the divergence of *Antirrhinum* from *Arabidopsis* (Causier et al. 2005).

One notable aspect of the *Antirrhinum* flower that differs from *Arabidopsis* is its marked dorsoventral asymmetry (zygomorphy). Zygomorphy, considered to have coevolved with insect pollination, is apparent in the different morphologies of the dorsal, lateral, and ventral petals of *Antirrhinum* (Fig. 1d) and in stamen development. The dorsal and lateral petals are specified by the paralogous TCP transcription factor genes *CYCLOIDEA* (*CYC*) and *DICHOTOMA* (*DICH*), which act partly by activating dorsal expression of *RADIALIS* (*RAD*). RAD is a protein with a single MYB transcription factor repeat that is thought to compete with the two-repeat MYB protein encoded by the *DIVARICATA* (*DIV*) gene to antagonize its ventralizing effect (for review, see Almeida and Galego 2005). *CYC*-like genes have also been implicated in the evolution of floral asymmetry in other lineages. For example, ectopic *CYC* expression has been proposed to account for the evolution of the derived radially symmetrical flower of *Mohavea*, which is within the tribe Antirrhineae (Hileman et al. 2003). *CYC*-like genes are expressed asymmetrically in both rosids and asterids, including species that have radially symmetrical flowers (for review, see Cubas 2004). This indicates that ancestral, asymmetrically expressed *CYC*-like genes might have been recruited independently to produce zygomorphic flowers in different flowering plant lineages, a view recently supported by the effects of *cyc*-like mutants in the asterid legume *Lotus* (Feng et al. 2006).

Aided by a large collection of leaf-shape mutants, *A. majus* has also been useful in studies of leaf development. These studies have included the identification of genes involved in specifying leaf identity and promoting leaf growth (see, e.g., Golz et al. 2004) and in coordinating growth in the leaf blade to ensure the development of a flat organ (see, e.g., Nath et al. 2003).

Ecology and Population Genetics

Antirrhinum species have been the subject of population genetic studies, including those aimed at assessing genetic diversity in order to inform conservation strategies for rare endemic species (see, e.g., Mateu-Andres 2004). As with other taxa, genetic diversity has generally been found to be lower in smaller populations, and the distribution of genetic variants within and between populations has been correlated with their level of self-incompatibility. In addition, several genetically well-characterized aspects of *Antirrhinum* development and physiology are relevant to their reproductive ecology, including petal cell morphology, genetic self-incompatibility, flower color, and scent production.

Whereas cultivated *A. majus* and some wild species (e.g., *A. valentinum*, *A. subbaeticum*, and *A. siculum*) are self-fertile, the majority of *Antirrhinum* species show gametophytic self-incompatibility that is determined by a single, complex *S* locus. Individuals of self-incompatible species reject pollen carrying an *S* allele that corresponds to one of their own alleles and are therefore obligate outbreeders. Through studies of relatives in the family Solanaceae, rejection of pollen was known to involve an *S*-encoded RNase that was expressed in the pistil (McClure et al. 1989), although the pollen-expressed component had remained elusive. Mapping and sequence analysis of the active *S* locus of *A. hispanicum*, followed by expression and functional studies, identified the likely pollen component to be an F-box protein involved in targeting the RNase for degradation (Lai et al. 2002).

Cells of the petal epidermis of *Antirrhinum*, like those of many flowering plants, have a conical shape that is dependent on activity of the MYB transcription factor encoded by *MIXTA*. These conical cells intensify flower color by reducing reflection, as revealed by *MIXTA* mutants with flat epidermal cells, and are important in attracting pollinating bees (Noda et al. 1994). In addition to having a role in the reproductive ecology of *Antirrhinum*, the misexpression of *MIXTA* or related genes can give rise to the formation of epidermal hairs (trichomes), indicating a common regulation of these two cell types and a role for other *MIXTA*-like genes in regulating multicellular trichome development in *Antirrhinum*. This latter function of *MIXTA*-like genes does not appear to be conserved in *Arabidopsis*, which has unicellular trichomes (for review, see Serna and Martin 2006).

GENETICS, GENOMICS, AND ASSOCIATED RESOURCES

A. majus is amenable to classical genetics. It has a relatively short generation time of about 4 months, is diploid ($2n = 16$), and is easily self- and cross-pollinated (see Protocol 2). A collection of classical mutants and corresponding wild-type lines originating from the work of Erwin Baur, Hans Stubbe, and their colleagues is maintained at the IPK. The phenotypes of most of the mutants in this collection have been described by Stubbe (1966) and catalogued by Hammer et al. (1990). Additional mutants and wild-type lines, generated at the John Innes Centre, are described at http://www.jic.ac.uk/STAFF/enrico-coen/Rosemary/start.htm.

A molecular recombination map, currently comprising more than 250 loci with an average distance between loci of approximately 2 cM, has been produced for *Antirrhinum* using hybrids between *A. majus* and wild species, and the map is maintained at http://www.antirrhinum.net. It has been aligned with a map of classical mutants and with the eight *Antirrhinum* chromosomes using fluorescence in situ hybridization (FISH; Zhang et al. 2005). Maps constructed using hybrids between different *Antirrhinum* species are largely colinear, suggesting an absence of extensive chromosomal rearrangements, although distorted transmission of some genomic regions in interspecies hybrids can hinder genetic mapping (Schwarz-Sommer et al. 2003). Recombinant inbred lines (RILs) and near-isogenic lines (NILs) have been produced from hybrids between different *Antirrhinum* species, allowing those genes that underlie differences between species to be identified.

Native *Antirrhinum* transposons have provided the basis for a number of genetic resources. Most spontaneous mutations in *A. majus* have been found to carry transposons belonging to one of two families: the *Tam3* family, which is similar to *Ac* in maize, and the *CACTA* family, which is homologous to *Spm/En* in maize. Mutagenic retroposons and miniature inverted repeat transposable elements (MITEs) have also been found. Transposition of the *CACTA* and *Tam3* families occurs by excision from the donor site and reintegration elsewhere in the genome and is promoted by low temperature. A number of *A. majus* lines were selected for high transposon activity based on flower variegation and have been used for forward genetic screens and for inactivating known genes to study their functions (see, e.g., Keck et al. 2003).

Transposon excision from pigment genes, which can be controlled with temperature, has been used to mark cells for fate and growth analysis (e.g., Rolland-Lagan et al. 2003), and excision from developmental genes has been used to examine the cell autonomy of gene action (see, e.g., Golz et al. 2004). In many cases, it has been possible to establish stable periclinal chimeras in which layers of the shoot apical meristem comprise genetically different cells as the result of transposon excision (see, e.g., Carpenter and Coen 1995). Such chimeras can be maintained by vegetative propagation through cuttings.

Although stable transgenic *Antirrhinum* can be produced using *Agrobacterium tumefaciens* (Cui et al. 2003), transformation efficiency is low. Particle bombardment and *Agrobacterium* infiltration of developing petals have been used successfully to obtain transient expression of pigment genes and suppression by RNA interference (RNAi; Schwinn et al. 2006; Shang et al. 2007).

Bacterial artificial chromosome (BAC) libraries have been produced from *A. majus* (Causier et al. 2005) and from an *A. majus* x *A. hispanicum* hybrid (Lai et al. 2002). Although the genome size of *A. majus* has been estimated at 430 Mb (Bennett and Leitch 1995), the recovery of clones from genomic libraries indicates that it might be at least twofold larger; therefore, the depth of coverage of existing BAC libraries is uncertain. Currently, only short BAC contigs have been assembled to allow positional cloning in specific target regions (Cartolano et al. 2007). FISH has been used to locate BAC clones on *Antirrhinum* chromosomes (Zhang et al. 2005; Yang et al. 2007). An expressed sequence tag (EST) collection of approximately 12,000 unigenes from *A. majus* is available for searching at http://www.antirrhinum.net. Other DNA libraries include cDNA clones in yeast one-hybrid and two-hybrid vectors (see, e.g., Egea-Cortines et al. 1999).

TECHNICAL APPROACHES

Antirrhinum species are amenable to molecular genetic techniques. Most protocols that have been developed for *Arabidopsis*, including those for nucleic acid, protein, and organelle purification and for detection of RNA and proteins in situ, can be applied directly to *Antirrhinum*. Here, we present protocols for cultivating (Protocol 1) and propagating (Protocol 2) *Antirrhinum* plants.

Protocol 1

Cultivating *Antirrhinum*

In this protocol, we describe methods for cultivating *Antirrhinum* species. These plants are easily grown, provided that they have sufficient light and are not overwatered. In good conditions, most species will flower and produce seeds within 3–4 months. Strongly growing plants should suffer from few pests or diseases, but we also prescribe methods for handling microbes and insects that commonly damage *Antirrhinum*.

MATERIALS

The recipe for the item marked with <R> is on page 117.

Reagents

Agar (0.1%, w/v) (optional; see Step 1)
Compost <R>

As an alternative to the peat-based compost recipe provided, Antirrhinum *can be grown in loam-based compost (e.g., John Innes No. 1) with additional feeding as required.* Antirrhinum *will grow well in a range of soils and commercial composts, but avoid those containing ammonium-based fertilizers, which encourage fungal wilting diseases.*

Gibberellin (10 µM) (optional; see Step 3)
Seeds

Seeds from a large collection of A. majus *mutants and from a limited number of other* Antirrhinum *species can be obtained from the IPK (http://gbis.ipk-gatersleben.de). Other seeds are available from members of the* Antirrhinum *research community.*

Equipment

Canes and wire ties (see Step 9)
Capillary matting (optional; see Step 7)
Covers for pots, clear plastic
Glasshouse (17–23°C, <70% humidity), with adequate ventilation and fans (see Step 8) and supplemental lighting in winter (see Step 5)
Pots (5 cm)

Larger (e.g., 10 cm) pots may also be needed (see Step 7).

METHOD

1. Fill each 5-cm pot with lightly firmed compost. In each pot, sow multiple seeds on the surface of the compost.

   Alternatively, seeds can be sown singly in 5-cm pots, which eliminates the need to transplant the seedlings in Step 4. To facilitate sowing, the seeds can be suspended in 0.1% agar and pipetted onto the soil.

2. Cover the pots with a clear plastic cover to keep the soil moist.

3. Keep the plants at approximately 17°C.

    A. majus laboratory strains and cultivars will germinate within 7–10 days, whereas seeds of wild accessions may show dormancy. Germination can be encouraged by soaking seeds in 10 µM gibberellin solution at 4°C for 3–5 days before sowing.

4. After germination, transplant the seedlings individually to 5-cm pots for flowering in the glasshouse.

    Alternatively, transplant the plants to soil outside, once the weather is warm enough.

5. Keep plants under high light (>150 µM $m^{-2}s^{-1}$).

    Flowering is promoted by long days; therefore, strong supplemental lighting is usually needed in glasshouses during the winter to provide a long day (~16 hours of light) and sufficient intensity for growth.

6. Maintain the plants at a temperature of 17–23°C.

    A nighttime drop in temperature to 15–17°C increases apical meristem size and encourages robust stem growth.

7. Water the plants as necessary.

    Compost should not be allowed to dry to the point where plants wilt, and during the summer, it may be necessary to water the plants twice daily in 5-cm pots. Increasing the size of the pot (e.g., to 10 cm) and placing pots on capillary matting can reduce the need for watering. Avoid wetting the foliage when watering to discourage fungal infections. All Antirrhinum species are intolerant of waterlogged soil. Avoid standing pots in water.

8. Use adequate ventilation and fans, if necessary, to maintain the relative humidity below 70%.

9. Using canes and wire ties, stake plants as necessary.

    A. majus grown under high light and in lower temperatures may not require support. Several species in the subsection Kickxiella are naturally procumbent and can be allowed to trail over the sides of the pot.

10. Monitor the plants for pests and disease.

    Strongly growing plants should suffer from few pests or diseases, but if they do, see Troubleshooting.

TROUBLESHOOTING

Problem (Step 10): Plants are wilted.

Solution: Wilting can be caused by either infection with *Pythium* species or lack of water. If the compost is moist, the wilting is caused by *Pythium*. In this case, drench the compost with a solution of commercial fosetylaluminium fungicide (e.g. Aliette from Bayer CropScience), following manufacturer instructions. Alternatively, if wilting is due to drought, water the compost well or soak it by submerging the pot in water for several minutes. The shoots of plants that have wilted due to mild *Pythium* infection or short-term drought should recover. If not, cut them off and allow new shoots to grow.

Problem (Step 10): Plants are infected with other fungal and oomycete diseases (mildews, *Botrytis*, and rust). (Mildews are apparent as grey or white powdery growths on leaves; *Botrytis* causes rapid localized death of shoots, leaves, and flowers, with dead tissues appearing dry and pale brown. Rust is first apparent on stems and on the undersides of leaves as lighter green circles that subsequently erupt into brown, spore-bearing pustules.)

Solution: Spray the shoots with suspended sulphur or other proprietary fungicide, using manufacturer recommendations. Remove dead and/or badly infected tissues.

Problem (Step 10): Shoots and flower buds show small lesions of dead tissue, and pollen is missing from anthers.

Solution: Plants may be infested with the western flower thrip *Frankliniella occidentalis*. Adult thrips are small (~1 mm long) buff-colored insects that crawl rapidly. They can usually be seen within newly opened flowers. Even strongly growing *Antirrhinum* are susceptible to infestation. This insect is both difficult to exclude from glasshouses and to treat with chemical insecticides, although biological control by the predatory mite *Neoseiulus* (formerly *Amblyseius*) *cucumeris* can keep infestation to a manageable level. The mite is available from biological control specialists.

Protocol 2

Propagating *Antirrhinum*

This protocol describes general strategies for propagating *Antirrhinum* species: self- and cross-pollination, cuttings, and grafting. *A. majus* cultivars and some wild species are self-fertile, but they require self-pollination for high seed yields. Although self-fertile, *A. majus* shows unilateral incompatibility and can only be crossed to other self-incompatible species as the female parent. All *Antirrhinum* species can be propagated clonally from cuttings. *Antirrhinum* also readily forms grafts within and between species.

MATERIALS

The recipe for the item marked with <R> is on page 117.

CAUTION: See the Cautions Appendix for appropriate handling of materials marked with <!>.

Reagents

Auxin-based rooting compound (for cuttings only; see Steps 7–10)
> *These compounds are often sold as "rooting hormone" and typically contain approximately 0.5% synthetic auxin. Gel-based solutions are easier and safer to handle than powders.*

Compost <R> or washed sand (for cuttings only; see Steps 7–10)
Ethanol (70%) <!> (only if using forceps for pollination; see Steps 2 and 3.ii)
Plants, mature (see Protocol 1)

Equipment

Bags, mesh or glassine paper (for self- or cross-pollination only; see Steps 1–6)
Boxes, plastic, with silica desiccant (optional; see Step 6)
Covers, clear plastic, or mist propagation unit (for cuttings only; see Steps 7–10)
Forceps or toothpicks (for self- or cross-pollination only; see Steps 1–6)
Jewelry tags (for cross-pollination only; see Step 3.iii)
Lab film or transparent silicone rubber tubing (2–5-mm internal diameter) (for grafting only; see Steps 11–17)
Scalpel or razor blade (for grafting only; see Steps 11–17)

METHOD

Three separate procedures are presented here. For self- or cross-pollination, follow Steps 1–6; for cuttings, follow Steps 7–10; and for grafts, follow Steps 11–17.

Self- and Cross-pollination

1. Collect pollen from newly dehisced anthers by scraping the pollen from the anthers with forceps or a toothpick. For self-pollination, proceed to Step 2; for cross-pollination, proceed to Step 3.
   > *The pollen should appear bright yellow and stick together in clumps. Older pollen that is lighter yellow and powdery is unlikely to be viable. To store pollen, freeze newly dehisced anthers at –80°C.*

2. For self-pollination, use forceps or a toothpick to transfer pollen (see Step 1) to the stigma after the petals have opened. Between pollinations, remove pollen from the forceps with 70% ethanol. Discard used toothpicks or autoclave for reuse. Proceed to Step 4.

   For rapid self-pollination, the lip of the lower petals can be used to smear pollen from anthers to stigma.

3. For cross-pollination, proceed as follows:

   i. Emasculate a flower by removing the anthers from an unopened bud. Fold back the flower petals of the bud using forceps to expose the stamens. Remove the stamens by pulling the anther filaments away with forceps.

      Most wild Antirrhinum species are self-incompatible and can only be pollinated by a plant carrying at least one different allele at the self-incompatibility locus. These species, therefore, do not require emasculation before cross-pollination.

   ii. Using forceps or a toothpick, transfer pollen (Step 1) to the stigma of the emasculated flower (Step 3.i). Between pollinations, remove pollen from the forceps with 70% ethanol. Discard used toothpicks or autoclave for reuse.

   iii. Label the crosses with jewelry tags looped around the peduncle.

4. Cover the inflorescences with mesh or glassine paper bags to prevent contamination with pollen from visiting bees.

   Alternatively, exclude pollinating bees from the glasshouse or remove the lower three flower petals to prevent their use as a landing platform by bees.

5. Between 2 and 5 days after pollination, monitor the plants for abscission of the flower petals and stamens, which indicates that fertilization was successful. Remove any unfertilized flowers by cutting the peduncle close to the stem to prevent *Botrytis* infection of the inflorescence.

6. Once the pores in the fruit have dehisced (3–5 weeks after pollination), collect the seeds by shaking them out of the detached fruit.

   The seeds will remain viable for several years under ambient conditions. For longer storage, seal the seeds in plastic boxes with silica desiccant at 4°C. Under these conditions (low temperature and low humidity), the seeds will remain viable for at least 10 years.

Cuttings

7. Cut a vegetative shoot 20–50 mm in length immediately below a leaf node. Cut off the leaves from the bottom two thirds of the shoot.

8. Dip the cut end in an auxin-based rooting compound and then insert it into compost or washed sand.

9. Grow the cutting at 17–20°C. Maintain high (~100%) humidity with a clear plastic cover or a mist propagation unit.

10. When new shoot growth is apparent (after 2–3 weeks), transplant to a larger pot of compost (see Protocol 1).

Grafts

Grafts of Young Plants

11. Graft young stems of similar diameter. Make slanting cuts in stock (recipient) and scion (donor) with a scalpel or razor blade. Cut off any large leaves from the scion to reduce water loss.

12. Hold the stock and scion together by binding them with lab film or by inserting them into opposite ends of a short length of transparent silicone rubber tubing.
13. Once the scion resumes growth, indicating a successful graft union, carefully cut away the lab film or tubing with a scalpel.

Grafts to Woody Stock

14. To graft to a large woody stock, make a T-shaped cut in the surface of the stem and ease back the soft green tissue with a scalpel blade.
15. Remove a vegetative axillary bud from the donor plant, together with a short length of leaf petiole, by cutting vertically through the soft stem tissue behind the bud.
16. Insert the donor bud (Step 15), in the correct orientation, into the T-shaped cut (in the stem from Step 14). Bind around the bud with lab film.
17. After 2–3 weeks, remove the film and cut the stock above the graft to promote bud growth.

Recipe

CAUTION: Horticultural chemicals and fertilizers may have potential health and environmental effects. Always follow manufacturer-recommended safety precautions and guidelines.

COMPOST

Ingredient	Quantity (per m^3)
Dolomitic limestone, powdered, to increase the pH to 5.5–6.5	4.0 kg
Fertilizer granules, slow-release, with an N:P:K ratio of ~20:10:20	1.0 kg
Fertilizer with micronutrients and an N:P:K ratio of ~15:10:20	0.75 kg
Horticultural wetting agent	according to supplier instructions
Peat or peat substitute	to 1 m^3
Perlite or vermiculite	150 liters

The fertilizer granules may be omitted, and instead, the established plants can be watered weekly with a soluble fertilizer. All *Antirrhinum* species are intolerant of waterlogged soil, particularly members of subsection *Kickxiella*, and an additional 20% (v/v) of washed sand or grit can be added to aid drainage. Avoid ammonium-based fertilizers, which encourage fungal wilting diseases.

REFERENCES

Almeida, J. and Galego, L. 2005. Flower symmetry and shape in *Antirrhinum*. *Int. J. Dev. Biol.* **49:** 527–537.

Bennett, M.D. and Leitch, I.J. 1995. Nuclear DNA amounts in angiosperms. *Ann. Bot.* **76:** 113–176.

Carpenter, R. and Coen, E.S. 1995. Transposon induced chimeras show that *floricaula*, a meristem identity gene, acts non-autonomously between cell layers. *Development* **121:** 19–26.

Cartolano, M., Castillo, R., Efremova, N., Kuckenberg, M., Zethof, J., Gerats, T., Schwarz-Sommer, Z., and Vandenbussche, M. 2007. A conserved microRNA module exerts homeotic control over *Petunia hybrida* and *Antirrhinum majus* floral organ identity. *Nat. Genet.* **39:** 901–905.

Causier, B., Castillo, R., Zhou, J., Ingram, R., Xue, Y., Schwarz-Sommer, Z., and Davies, B. 2005. Evolution in action: Following function in duplicated floral homeotic genes. *Curr. Biol.* **15:** 1508–1512.

Cubas, P. 2004. Floral zygomorphy, the recurring evolution of a successful trait. *Bioessays* **26:** 1175–1184.

Cui, M.L., Handa, T., and Ezura, H. 2003. An improved protocol for *Agrobacterium*-mediated transformation of *Antirrhinum majus* L. *Mol. Genet. Genomics* **270:** 296–302.

Darwin, C.R. 1868. *Variation of animals and plants under domestication.* John Murray, London.

Davies, B., Cartolano, M., and Schwarz-Sommer, Z. 2006. Flower development: The *Antirrhinum* perspective. *Adv. Bot. Res.* **44:** 279–321.

Egea-Cortines, M., Saedler, H., and Sommer, H. 1999. Ternary complex formation between the MADS-box proteins SQUAMOSA, DEFICIENS and GLOBOSA is involved in the control of floral architecture in *Antirrhinum majus*. *EMBO J.* **18:** 5370–5379.

Feng, X., Zhao, Z., Tian, Z., Xu, S., Luo, Y., Cai, Z., Wang, Y., Yang, J., Wang, Z., Weng, L., et al. 2006. Control of petal shape and floral zygomorphy in *Lotus japonicus*. *Proc. Natl. Acad. Sci.* **103:** 4970–4975.

Golz, J.F., Roccaro, M., Kuzoff, R., and Hudson, A. 2004. *GRAMINIFOLIA* promotes growth and polarity of *Antirrhinum* leaves. *Development* **131:** 3661–3670.

Gübitz, T., Caldwell, A., and Hudson, A. 2003. Rapid molecular evolution of *CYCLOIDEA*-like genes in *Antirrhinum* and its relatives. *Mol. Biol. Evol.* **20:** 1537–1544.

Hammer, K., Knüpffer, H., and Knüpffer, S. 1990. Das Gaterlebener Antirrhinum-sortiment. *Kulturpflanze* **38:** 91–117.

Hileman, L.C., Kramer, E.M., and Baum, D.A. 2003. Differential regulation of symmetry genes and the evolution of floral morphologies. *Proc. Natl. Acad. Sci.* **100:** 12814–12819.

Jiménez, J.F., Sánchez-Gómez, P., Güemes, J., Werner, O., and Rosselló, J.A. 2002. Genetic variability in a narrow endemic snapdragon (*Antirrhinum subbaeticum*, Scrophulariaceae) using RAPD markers. *Heredity* **89:** 387–393.

Jiménez, J.F., Sánchez-Gómez, P., Güemes, J., and Rosselló, J.A. 2005. Phylogeny of snapdragon species (*Antirrhinum*; Scrophulariaceae) using non-coding cpDNA sequences. *Isr. J. Plant Sci.* **53:** 47–53.

Keck, E., McSteen, P., Carpenter, R., and Coen, E. 2003. Separation of genetic functions controlling organ identity in flowers. *EMBO J.* **22:** 1058–1066.

Lai, Z., Ma, W., Han, B., Liang, L., Zhang, Y., Hong, G., and Xue, Y. 2002. An F-box gene linked to the self-incompatibility (*S*) locus of *Antirrhinum* is expressed specifically in pollen and tapetum. *Plant Mol. Biol.* **50:** 29–42.

Langlade, N.B., Feng, X., Dransfield, T., Copsey, L., Hanna, A.I.,

Thebaud, C., Bangham, A., Hudson, A., and Coen, E. 2005. Evolution through genetically controlled allometry space. *Proc. Natl. Acad. Sci.* **102:** 10221–10226.

Mateu-Andres, I. 2004. Low levels of allozyme variability in the threatened species *Antirrhinum subbaeticum* and *A. pertegasii* (Scrophulariaceae): Implications for conservation of the species. *Ann. Bot.* **94:** 797–804.

Mateu-Andres, I. and de Paco, L. 2006. Genetic diversity and the reproductive system in related species of *Antirrhinum*. *Ann. Bot.* **98:** 1053–1060.

McClure, B.A., Haring, V., Ebert, P.R., Anderson, M.A., Simpson, R.J., Sakijama, F., and Clarke, A.E. 1989. Style self-incompatibility gene products of *Nicotiana alata* are ribonucleases. *Nature* **342:** 955–957.

Nath, U., Crawford, B.C., Carpenter, R., and Coen, E. 2003. Genetic control of surface curvature. *Science* **299:** 1404–1407.

Noda, K., Glover, B.J., Linstead, P., and Martin, C. 1994. Flower colour intensity depends on specialized cell shape controlled by a Myb-related transcription factor. *Nature* **369:** 661–664.

Olmstead, R.G., dePamphilis, C.W., Wolfe, A.D., Young, N.D., Elisons, W.J., and Reeves, P.A. 2001. Disintegration of the Scrophulariaceae. *Am. J. Bot.* **88:** 348–361.

Oyama, R.K. and Baum, D.A. 2004. Phylogenetic relationships of North American *Antirrhinum* (Veronicacae). *Am. J. Bot.* **91:** 918–925.

Rolland-Lagan, A.G., Bangham, J.A., and Coen, E. 2003. Growth dynamics underlying petal shape and asymmetry. *Nature* **422:** 161–163.

Rothmaler, W. 1956. *Taxonomische monographie der gattung* Antirrhinum. Academie Verlag, Berlin.

Schwarz-Sommer, Z., de Andrade, S.E., Berndtgen, R., Lonnig, W.E., Muller, A., Nindl, I., Stuber, K., Wunder, J., Saedler, H., Gubitz, T., et al. 2003. A linkage map of an F2 hybrid population of *Antirrhinum majus* and *A. molle*. *Genetics* **163:** 699–710.

Schwinn, K., Venail, J., Shang, Y., Mackay, S., Alm, V., Butelli, E., Oyama, R., Bailey, P., Davies, K., and Martin, C. 2006. A small family of MYB-regulatory genes controls floral pigmentation intensity and patterning in the genus *Antirrhinum*. *Plant Cell* **18:** 831–851.

Serna, L. and Martin, C. 2006. Trichomes: Different regulatory networks lead to convergent structures. *Trends Plant Sci.* **11:** 274–280.

Shang, Y., Schwinn, K.E., Bennett, M.J., Hunter, D.A., Waugh, T.L., Pathirana, N.N., Brummell, D.A., Jameson, P.E., and Davies, K.M. 2007. Methods for transient assay of gene function in floral tissues. *Plant Methods* **3:** 1.

Stubbe, H. 1966. *Genetik und zytologie von* Antirrhinum *L. sect.* Antirrhinum. Gustav Fischer, Jena.

Webb, D.A. 1972. *Antirrhinum* L. In *Flora europaea* (ed. T.G. Tutin, et al.), pp. 221–224. Cambridge University Press, New York.

Whibley, A.C., Langlade, N.B., Andalo, C., Hanna, A.I., Bangham, A., Thebaud, C., and Coen, E. 2006. Evolutionary paths underlying flower color variation in *Antirrhinum*. *Science* **313:** 963–966.

Wright, G.A., Lutmerding, A., Dudareva, N., and Smith, B.H. 2005. Intensity and the ratios of compounds in the scent of snapdragon flowers affect scent discrimination by honeybees (*Apis mellifera*). *J. Comp. Physiol. A Neuroethol. Sens. Neural Behav. Physiol.* **191:** 105–114.

Yang, Q., Zhang, D., Li, Q., Cheng, Z., and Xue, Y. 2007. Heterochromatic and genetic features are consistent with recombination suppression of the self-incompatibility locus in *Antirrhinum*. *Plant J.* **51:** 140–151.

Zhang, D., Yang, Q., Bao, W., Zhang, Y., Han, B., Xue, Y., and Cheng, Z. 2005. Molecular cytogenetic characterization of the *Antirrhinum majus* genome. *Genetics* **169:** 325–335.

FURTHER READING

Schwarz-Sommer, Z., Davies, B., and Hudson, A. 2003. An everlasting pioneer: The story of *Antirrhinum* research. *Nat. Rev. Genet.* **4:** 657–666.
Reviews the history of *Antirrhinum* genetics.

Davies, B., Cartolano, M., and Schwarz-Sommer, Z. 2006. Flower development: The *Antirrhinum* perspective. *Adv. Bot. Res.* **44:** 279–321.
Provides a detailed review of the control of flower development in *Antirrhinum*, in comparison to that of other species.

WWW RESOURCES

http://www.antirrhinum.net DragonDB homepage. The *Antirrhinum majus* (snapdragon) genetic and genomic database. The contents of this searchable database in AceDB schema include *Antirrhinum* sequences, mutants, maps, and publications.

http://www.jic.ac.uk/STAFF/enrico-coen/Rosemary/start.htm *Antirrhinum* stock collection, The John Innes Centre.

5 Tomato (*Solanum lycopersicum*)
A Model Fruit-bearing Crop

Seisuke Kimura and Neelima Sinha
Department of Plant Biology, University of California, Davis, California 95616

ABSTRACT

Tomato (*Solanum lycopersicum*) is one of the most important vegetable plants in the world. It originated in western South America, and domestication is thought to have occurred in Central America. Because of its importance as food, tomato has been bred to improve productivity, fruit quality, and resistance to biotic and abiotic stresses. Tomato has been widely used not only as food, but also as research material. The tomato plant has many interesting features such as fleshy fruit, a sympodial shoot, and compound leaves, which other model plants (e.g., rice and *Arabidopsis*) do not have. Most of these traits are agronomically important and cannot be studied using other model plant systems. There are 13 recognized wild tomato species that display a great variety of phenotypes and can be crossed with the cultivated tomato. These wild tomatoes are important for breeding, as sources of desirable traits, and for evolutionary studies. Current progress on the tomato genome sequencing project has generated useful information to help in the study of tomato. In addition, the tomato belongs to the extremely large family Solanaceae and is closely related to many commercially important plants such as potato, eggplant, peppers, tobacco, and petunias. Knowledge obtained from studies conducted on tomato can be easily applied to these plants, which makes tomato important research material. Because of these facts, tomato serves as a model organism for the family Solanaceae and, specifically, for fleshy-fruited plants.

> **PROTOCOLS**
> 1 How to Grow Tomatoes, 128
> 2 Crossing Tomato Plants, 130
> 3 Grafting Tomato Plants, 131
> 4 Tomato Transformation, 132

BACKGROUND INFORMATION

The tomato (*Solanum lycopersicum*) (Fig. 1) and its wild relatives (genus *Solanum*, section *Lycopersicon*) originated in western South America (Ecuador, Peru, and Chile). Wild tomatoes can still be found along the western coast of South America, in the Andes, and on the Galapagos Islands. Although the center of diversity of wild tomatoes is in Peru (Rick 1991), genetic analysis of primitive cultivars has shown that the center of diversity of cultivated tomatoes is in Mexico. This indicates that tomato domestication may have occurred in Central America (Rick 1991). When the conquistadors from Europe came to the Americas, there was widespread cultivation of tomato. It is likely that Europeans distributed the tomato from the Americas to Europe and European colonies

This chapter, with full-color images, can be found online at www.cshprotocols.org/emo.

FIGURE 2. Many economically important plants belong to the Solanaceae family: (*Counterclockwise from top, right*) Tomato, potato, eggplant, red and green bell peppers, chili peppers, tomatillo, and tobacco.

belladonna], mandrake [*Mandragora officinarum*], and jimson weed [*Datura stramonium*]), and as ornamentals (e.g., petunia [*Petunia hybrida*]). The presence of many economically important plants in the Solanaceae family makes tomato pivotal as a model plant species.

USES OF THE TOMATO MODEL SYSTEM

Because of its importance as a food source, tomato has been bred and studied for a long time. The tomato has many interesting features that other model plants, such as rice and *Arabidopsis*, do not have. For example, tomato plants produce fleshy fruits that are important for the human diet (Fig. 1E); botanically, tomato fruits are berries. The tomato has sympodial shoots, and it is the only model plant with compound leaves (Fig. 1C) that has extensive genome sequence data available. Most of these traits are agronomically important and have been the major targets for domestication and breeding.

Phenotypes of most of these traits are determined by the collective effects of many loci (quantitative trait loci, or QTL). The tomato is an ideal plant for QTL analysis because of the phenotypic diversity of these traits in cultivated and wild tomatoes. In addition, a large number of mutants have been generated, which facilitates the study of these important traits. In this section, we describe current research on the genes and the genetic and molecular mechanisms controlling several of the traits that are well studied in tomatoes.

Fruit Characteristics

Fruit Size

Fruit size is one of the most important traits in agriculture and was a major selection target during domestication and breeding. The presumed ancestor of cultivated tomato, *S. pimpinellifolium*, has tiny fruits that weigh only a few grams each. The fruit of the cherry tomato plant (*S. lycopersicum* subspecies *cerasiforme*), which is thought to be the direct ancestor of cultivated tomatoes, is only slightly larger than that of *S. pimpinellifolium*. Domestication and breeding resulted in modern cultivated tomatoes, with fruits that are more than 10 cm in diameter and weighing more than 500 grams.

Fruit size in tomato is a quantitative trait, and studies have identified many QTL controlling variation in fruit size (Tanksley 2004). So far, only one of the QTL, *fw2.2* (the second fruit weight [fw] QTL on chromosome 2), has been cloned (Frary et al. 2000). This QTL explains as much as 30% of the difference in fruit size between wild and cultivated tomatoes. *fw2.2* encodes a protein that shares homology with the human RAS oncogene and acts as negative regulator of cell division during fruit development. Differences in fruit size between wild and cultivated tomatoes are thought to be caused by differences in the transcriptional regulation of this gene.

Fruit Shape

Fruit shape is also an important agricultural trait and several loci controlling fruit shape have been identified and characterized (Tanksley 2004). One of the QTL, *SUN*, accounts for 58% of the phenotypic variation in a cross between *S. lycopersicum* variety Sun1642 and wild tomato *S. pimpinellifolium* (Xiao et al. 2008). *S. pimpinellifolium* has round fruit, whereas Sun1642 has elongated fruits. Map-based cloning revealed that the *SUN* locus includes the *IDQ12* gene; the corresponding IDQ12 protein is a member of the IQ67 domain-containing family. Interestingly, a retrotransposon (*Rider*)-mediated gene duplication resulted in the fusion of the *DEFL1* gene promoter region with the *IDQ12* gene, which in turn resulted in the overexpression of *IDQ12*. This *IDQ12* overexpression is thought to cause the elongated fruit shape by altering plant hormone and/or secondary metabolite levels (Xiao et al. 2008).

The *OVATE* gene, also identified by QTL analysis, is involved in making pear-shaped fruits and was found in the *S. lycopersicum* subspecies *cerasiforme* variety "Yellow Pear" (Liu et al. 2002). Overexpression of the *OVATE* gene converts pear-shaped tomatoes to round-shape tomatoes but does not affect fruit size. *OVATE* is a novel regulatory gene with a putative nuclear localization signal and homology with human von Willebrand factor genes.

Fruit Color

Tomato fruit colors vary from green and yellow to orange and red. In most cases, the color of the fruit is determined by the quantities of carotenoids, such as lycopene and β-carotene, and chlorophylls. Red and orange colors result from the accumulation of lycopene and β-carotene, respectively. Green fruits contain chlorophylls, which usually degrade during the ripening process. Multiple genetic loci control tomato fruit color, and most of the loci encode enzymes of the carotenoid biosynthetic pathway or their regulators.

Solanum cheesmaniae (previously known as *Lycopersicon cheesmanii* f. *major*) and *Solanum galapagense* (previously known as *Lycopersicon cheesmanii* f. *minor*), wild tomatoes endemic to the Galapagos Islands, have orange fruits that contain fivefold to tenfold more β-carotene than red-colored fruits (Ronen et al. 2000). This color change is regulated by a dominant allele (*Beta*) of the *B* gene. The *B* gene encodes a chromoplast-specific lycopene β-cyclase (CYC-B), an enzyme that converts lycopene to β-carotene. Expression of the *B* gene in ripening fruit is dramatically higher in the *Beta* mutant than in wild-type fruit and results in an accumulation of β-carotene in the fruit. The recessive *old-gold* (*og*) mutation results in deep-red-colored fruits. Molecular analysis has shown that *og* is a null allele of the *B* gene (Ronen et al. 2000). The complete absence of β-carotene in *og* fruits (due to the lack of CYC-B) causes their deep-red color.

The *DELTA* locus encodes lycopene ε-cyclase, which converts lycopene to δ-carotene (Ronen et al. 1999). The dominant allele *Delta* exhibits increased gene expression, which results in an accumulation of δ-carotene and orange-colored fruit. The recessive *tangerine* mutant also has orange fruit. The *TANGERINE* gene encodes a carotenoid isomerase (CRTISO) that is required during carotenoid desaturation. The expression of CRTISO is up-regulated during fruit ripening, but its expression is abolished in the *tangerine* mutant. In the *tangerine* mutant, prolycopene (instead of all-*trans*-lycopene) accumulates, thereby making the fruit color orange (Isaacson et al. 2002).

The yellow-fruited tomato mutant *yellow flesh* has a mutation in the *PHYTOENE SYNTHASE 1* (*PSY1*) gene (Fray and Grierson 1993), which results in the production of a truncated protein. The yellow color is thought to be caused by lack of carotenoids because the truncated protein is unable to convert geranylgeranyl diphosphate to phytoene.

Total Soluble Solids

One of the major objectives for tomato breeders is to increase the contents of total soluble solids (Brix, measured by a refractometer in Brix units) in the fruit (Eshed and Zamir 1995). Brix is a measure of sugars and acids in fruit, is an important fruit taste trait, and is associated with commercial tomato yields. Brix values in the wild tomato *S. pennellii* are significantly higher than those in cultivated tomatoes, and QTL analyses have identified more than 20 QTL that increase Brix (Eshed and Zamir 1995). One of the QTL identified in *S. pennellii*, *Brix9-2-5*, improves Brix by more than 20% and was cloned by map-based cloning (Fridman et al. 2004). The *Brix9-2-5* locus encompasses a gene for a cell wall invertase (*LIN5*). The functional polymorphism of the *Brix9-2-5* locus causes an amino acid substitution near the catalytic site of the invertase, affecting enzyme kinetics and resulting in fruits with high sugar contents.

Floral System and Plant Architecture

The shoot systems of flowering plants display great variation in their architecture and growth habits. The model plant *Arabidopsis thaliana* has monopodial shoot architecture; the apical meristem is indeterminant and active throughout the plant life cycle (Pnueli et al. 1998). In contrast, tomato has sympodial shoot development; the primary vegetative shoot terminates in a flower after the development of 8–12 leaves. Subsequently, new vegetative shoots arise from the axillary bud just below the terminating inflorescence. This new shoot, in turn, terminates again after making three leaves, and the next shoot arises from the newly formed axillary bud. This cycle is repeated continuously to form sympodial shoots. By definition, tomato shoots are considered to be "determinate" because each shoot terminates in a flower. However, the wild-type growth habit of tomatoes is classified as "indeterminate" because they continuously produce sympodial units.

The *SELF-PRUNING* (*SP*) gene regulates the switch from vegetative to reproductive growth. *sp* mutants show a "determinate" shoot habit (a limited number of sympodial shoots), resulting in limited shoot growth, a reduction in the number of leaves between inflorescences (which in turn results in closely set fruit), and bushy compact plants with nearly homogeneous fruit settings. In contrast, the overexpression of *SP* results in an extended vegetative phase of sympodial shoot development. The tomato *SP* gene is the ortholog of the *A. thaliana TERMINAL FLOWER 1* gene, which maintains the indeterminate state of the inflorescence meristem. *SP* is thought to have a role in preventing early flowering in each of the developing shoot meristems (Pnueli et al. 1998). The efficiency of mechanical fruit harvesting in field-grown tomatoes has been increased by introducing the *sp* mutation into cultivated tomatoes (Atherton and Harris 1986).

Compound Leaf Development

In general, plant leaves are characterized by a striking diversity in shape. The most conspicuous characteristic of leaf shape is the degree to which the leaf is subdivided into smaller segments. Leaves lacking subdivision are termed "simple," whereas divided leaves are termed "compound." Leaf complexity is known to directly affect several processes critical to plant survival, including efficiency of photosynthesis, rate of gas exchange, and resistance to herbivore damage. This indicates that the diversity in leaf form is the result of adaptation to growth in varied environments.

The tomato has unipinnately compound leaves (Fig. 1C) and is the only model plant with compound leaves that has extensive genomic and genetic data available. Leaf shape varies among wild and cultivated tomatoes, which makes QTL analyses possible (Holtan and Hake 2003; Frary et al. 2004). A large number of leaf mutants are also known. These facts make tomato ideally suited to study the molecular mechanisms of compound leaf development.

Tomato leaf mutants have varying degrees of leaf developmental defects. Some mutations change compound leaves into simple or nearly simple leaves. *Lanceolate* (*La*) and *entire* (*e*) mutants belong to this category. The *LA* gene encodes a TCP family transcription factor that contains a microRNA (*miR319*)-binding site (Ori et al. 2007). Mutations are in the *miR319*-binding site, so the mutated transcripts cannot be recognized by *miR319*, resulting in a partial resistance of the *La* transcript to miRNA-directed inhibition. Elevated *La* transcripts in young leaf primordia are thought to promote the precocious differentiation of leaf margins, leading to the *La* phenotype (Ori et al. 2007). The *entire* mutant also has nearly simple leaves. An alteration in the coding region of the tomato *IAA9* gene, a member of the *Aux/IAA* gene family, was shown to result in the *entire* mutant phenotype (Zhang et al. 2007). The function of *IAA9* in compound leaf development is still unclear, but results indicate the involvement of the auxin pathway in compound leaf development.

Other mutations, such as *Mouse ear* (*Me*), *Peteroselinum* (*Pts*), *Curl* (*Cu*), and *bipinnata* (*bip*), increase leaf complexity. *Mouse ear* (*Me*) is a dominant mutation that arose spontaneously in the tomato cultivar Rutgers and shows excessively proliferated leaves, with three or four orders of leaflets (Chen et al. 1997; Parnis et al. 1997). The *Me* phenotype resembles the phenotype of transgenic plants that overexpress the tomato gene *LeT6*, a member of the *KNOX* gene family (*L. esculentum* [*S. lycopersicum*] T6). Molecular analysis showed that the *Me* phenotype was caused by a spontaneous fusion of the *PFP* gene, which encodes a metabolic enzyme (β-subunit of PPi-dependent phophofructokinase), and *LeT6*, resulting in the altered expression pattern and elevated expression levels of *LeT6* (Chen et al. 1997; Parnis et al. 1997). The *Cu* is a dominant mutant with ramified compound leaves that have wrinkled and curled blades (Parnis et al. 1997). The *Me* and *Cu* loci are tightly linked to each other. *LeT6* is overexpressed in *Cu* (as it is in *Me*), indicating that *Cu* is also caused by the overexpression of *LeT6* (Parnis et al. 1997).

The most dramatic example of leaf natural variation was found between the closely related wild tomato species *S. cheesmaniae* and *S. galapagense* (Darwin et al. 2003). *S. cheesmaniae* leaves are similar to leaves of cultivated tomatoes, whereas the leaves of *S. galapagense* are more complex. The introgression of the leaf complexity trait from *S. galapagense* into cultivated tomato demonstrated that this phenotype is conferred by a single semidominant allele named *Petroselinum* (*Pts*) (Rick 1980). Map-based cloning of *Pts* showed that this locus encompasses a novel *KNOX* gene that lacks a homeodomain (Kimura et al. 2008). A mutation in the promoter of this gene up-regulates the gene product in leaves and is responsible for increased leaf complexity. The classic tomato mutation *bipinnata* (*bip*) closely resembles the *Pts* phenotype and has been shown to be a loss-of-function mutant of the *BEL-LIKE HOMEODOMAIN* (*BELL*) gene (Kimura et al. 2008).

GENETICS AND GENOMICS RESOURCES

Tomato is one of the most well-studied organisms for genetics research and, because of its agricultural importance, has a long history of breeding. The cultivated tomato is diploid and self-pollinating and has a relatively short generation time. Wild tomatoes are a valuable resource for desirable traits because they show great phenotypic variation (e.g., morphology and resistance against diseases and abiotic stresses) and can be crossed with cultivated tomato. These features make tomato ideal material for genetic research.

Classical genetics research has created more than 1000 tomato mutants that include spontaneous and induced (via radiation and chemicals) mutations. Loci responsible for many of these mutants have been genetically mapped (Stevens and Rick 1986). Recently, an isogenic tomato mutant library was developed on the genetic background of the inbred variety M82 (Menda et al. 2004). Currently, 3417 mutations have been cataloged in this mutant library. These collections of mutants are used to investigate the functions of genes in tomato.

Nomenclature guidelines for tomato genetics were published in 1955 (Barton et al. 1955). Names and symbols of mutant genes are designated by letters and are italicized. The mutant name should indicate the main character of the phenotype. The initial letter of a dominant allele is uppercased, whereas a recessive allele is written with all letters in lowercase. The normal (wild-type) allele is indicated by adding the superscript "+" after the name of the gene, and additional alleles at the same locus, except for the first allele, are designated by numbered superscripts (e.g., bip^+, bip, bip^2, and bip^3). Mutations at different loci whose effects are indistinguishable by phenotype are called mimics. Mimics are preferably designated by different names, but a numbered series can be applied, too. In the latter case, mimics are designated by the same name, followed by a hyphen and a number (e.g., *wiry-1* and *wiry-3*).

The first-generation linkage map of the tomato contained more than 200 morphological and isozyme loci (Stevens and Rick 1986). Subsequently, high-density molecular linkage maps were constructed by mapping restriction fragment length polymorphism (RFLP) and amplified fragment length polymorphism (AFLP) markers to the tomato genome using *S. lycopersicum* × *S. pennellii* F_2 populations (Haanstra et al. 1999). Some of the morphological and isozyme markers were mapped to the RFLP map to integrate the linkage and molecular maps. Linkage maps have also been generated using *S. lycopersicum* × *S. pimpinellifolium* (Doganlar et al. 2002) and *S. lycopersicum* × *S. habrochaites* (Bernacchi and Tanksley 1997). These linkage maps have made map-based cloning, QTL analysis, marker-based breeding, and genome sequence assembly feasible in tomato. Information about these maps is available on the website for the International Tomato Genome Sequencing Project (http://www.sgn.cornell.edu/about/tomato_sequencing.pl).

The ongoing International Tomato Genome Sequencing Project is the cornerstone of a larger project, the International Solanaceae Genome Project (SOL), that aims to develop the family Solanaceae as a model for understanding plant adaptation and diversification through systems biology (http://sgn.cornell.edu/solanaceae-project). Macrosynteny and microsynteny conservation among the genomes of tomato, potato, pepper, and eggplant is very high because there were no large-scale duplication events such as polyploidization early in the radiation of the Solanaceae family. This allows one to easily apply the knowledge obtained from tomato genome sequence data to other Solananceae species.

The tomato genome contains 950 Mb of DNA ($n = 12$) (Arumuganathan and Earle 1991), and of that, only 220 Mb (25%) is gene-rich euchromatin (Wang et al. 2006). The remaining 730 Mb (75%) is heterochromatin and largely devoid of genes. In tomato, heterochromatin is concentrated around centromeres, and euchromatin is located in the distal portion of each chromosome. In contrast, in the chromosomes of maize and rice, heterochromatin and euchromatin are interspersed. This interesting feature of the tomato genome allowed the International Tomato Genome Sequencing Project to focus on sequencing only the euchromatic regions. Bacterial artificial chromosome (BAC) clones that mapped to euchromatic regions were selected and used as "seed BACs" to identify BAC clones that covered entire euchromatic regions. These BACs are currently being sequenced by shotgun sequencing. This strategy is expected to give us more useful data than would be obtained from the whole-genome shotgun approach. Currently, about 25% of euchromatic DNA has been sequenced and the sequencing project is expected to be finished in 2008. The sequence data, other useful information (annotations, maps, markers, etc.), and tools are available on the website for the International Tomato Genome Sequencing Project (http://www.sgn.cornell.edu/index.pl).

Expressed sequence tag (EST), unigene, and BAC information is also available on the website for the International Tomato Genome Sequencing Project. The entire tomato BAC library, BAC filters, or individual BAC clones can be purchased from the Clemson University Genomics Institute (CUGI; http://www.genome.clemson.edu/capabilities/bacCenter.shtml). Tomato EST clones can also be ordered online (http://ted.bti.cornell.edu/order/index.html). Information on full-length cDNAs from Micro-Tom is available from the Kazusa Full-length Tomato cDNA Dababase (KaFTom; http://www.pgb.kazusa.or.jp/kaftom). Microarray expression data are available from the Tomato Expression Database (http://ted.bti.cornell.edu). A database for metabolites detected in tomato fruit is available from KOMICS (http://webs2.kazusa.or.jp/komics).

TECHNICAL APPROACHES

Tomato is excellent research material. In this section, some of the protocols that are useful for tomato research are described. Most basic techniques, such as genomic DNA isolation, RNA isolation, and protein extraction, are easy, and protocols for the model plant *Arabidopsis* can be applied to tomato. Tomatoes can be stably transformed using tissue culture methods (see Protocol 4), so investigators can subsequently perform overexpression, RNA interference (RNAi), promoter-GUS analysis, and other assays to functionally characterize a gene of interest.

Protocol 1

How to Grow Tomatoes

Tomatoes can be easily grown in a field, in a greenhouse, or in a growth cabinet. They need acidic soil (pH 6.0–6.8), a lot of light, and water. The optimum temperature for growing tomato plants and fruit is 18–24°C. This protocol describes how to germinate tomato seeds, cultivate adult plants, and harvest seeds from fruit.

MATERIALS

CAUTION: See the Cautions Appendix for appropriate handling of materials marked with <!>.

Reagents

Bleach (sodium hypochlorite) (2.7%) <!> (optional; see Step 1)
Tomato seeds

Equipment

Container for seed collection (see Step 9)
Flats (optional; see Step 1)
Gardening soil (pH 6.0–6.8)
Greenhouse or growth cabinet, set at 18–24°C (optional; see Steps 2 and 4)
Knife
Paper towels
Petri dishes (100 mm diameter x 15 mm high) or other transparent container lined with wet filter paper
> *Alternatively, use flats with pastic covers (see Steps 1–2).*

Plastic cover
Pots (15 cm diameter x 15 cm high)
Scissors (optional; see Step 6)
Sieve (optional; see Step 10)
Stakes or tomato cages (optional; see Step 5)

METHOD

Germination

1. Sow the seeds in flats containing soil (0.5–2 cm from the surface of the soil) or on wet filter paper in a Petri dish. Keep the seeds 2.5–5 cm apart.
   > *Seeds of some tomato accessions, such as wild tomato and introgression lines, are resistant to germination. To improve the germination rate, completely immerse the seeds in 2.7% bleach (sodium hypochlorite) for 30 minutes at room temperature before sowing.*

2. After sowing the seeds directly in the soil, place the flats in a greenhouse or growth cabinet. Cover them with a plastic cover to keep the soil moist while the seeds are germinating. If using wet filter paper, keep the seeds in the dark at room temperature until germination.
   > *The seeds will germinate in 7–14 days.*

Cultivation

3. After germination, transplant the seedlings from the flats or Petri dishes to a field (30–60 cm apart) or into pots (15 cm diameter × 15 cm high).
4. Keep the soil moist by watering the plants once a day. If maintaining the plants in a greenhouse or growth cabinet, keep the temperature between 18°C and 24°C and the light intensity relatively strong (50,000–60,000 lx).
5. If necessary, use stakes or a tomato cage to support the tomato vine after it grows in height.
6. If necessary, prune the plants using a scissors.
   Determinate tomatoes need no pruning. For indeterminate tomatoes, pruning "suckers" (lateral shoots that form between the main stem and the leaf) makes the main stem stronger.
7. Once flowering begins, shake the tomato plants gently once or twice each week to promote pollination.
   This increases fruit production. Usually, fruit ripens within 90–120 days after germination.

Seed Collection

8. After ripening, harvest the fruit.
9. Cut the fruit in half with a knife and squeeze the seeds into a container.
10. Add tap water to the container so that it is approximately 70–80% full. Keep it for about 3 days at room temperature.
    This ferments the seeds, which allows the removal of the gelatinous coating that covers them. Once fermented, mold will form on the surface of the water. The coating can also be mechanically removed without fermentation by smashing the seeds in a sieve with a lot of water.
11. Repeatedly wash the seeds with tap water at room temperature (~5–10 washes, for 5–10 seconds each wash) to remove the coating.
12. Dry the seeds by leaving them on a paper towel overnight at room temperature.

Protocol 2

Crossing Tomato Plants

This protocol describes how to cross tomato plants. Crossing is important for the genetic analysis and breeding of tomatoes. Tomatoes are self-pollinating plants; thus, emasculation (removal of the anthers from the female parent) is essential. All wild tomato species can be crossed with cultivated tomatoes (although it may be difficult); this is useful because wild tomatoes are a great source of desirable traits. Most commercial tomatoes are F_1 hybrids, and the seeds for them were produced by crossing two parent tomatoes.

MATERIALS

Reagents

Tomato plants, adult (see Protocol 1)

Equipment

Forceps
Microcentrifuge tubes (1.5 ml)
   Alternatively, use empty gelatin capsules (available at most drug stores).
Paint brush, small (optional; see Step 3)
Pencil
Razor blade, sharp
Tags for labeling crosses

METHOD

1. Select a female parent. Choose flowers that have not yet opened but are about to turn yellow in color. Emasculate the flowers by carefully removing the sepals, petals, and anther cone with forceps to expose the stigma and style.

2. Select a male parent (pollen source). Choose flowers that are opened or partially opened. Use a sharp razor blade to cut a slit in the anther cone to facilitate the release of pollen. Collect the pollen in a 1.5-ml microcentrifuge tube by holding the anther with a forceps and tapping the forceps with a pencil. Do not tap the anther directly.

3. Transfer pollen to the stigma by one of the following methods:

   i. Directly dip the stigma into the collected pollen.

   ii. Use a small paint brush or finger to transfer the pollen to the stigma.
       It is important to ensure that the stigma gets plenty of pollen but suffers no damage.
       Leftover pollen can be stored at 0°C and in dry conditions for up to a few months.

4. Mark the crossed flowers with tags.

Protocol 3

Grafting Tomato Plants

Grafting is agronomically important because one can combine desirable aboveground characteristics (such as fruit size) and underground characteristics (such as resistance to soil-borne diseases). The top portion, which contains the shoot, is called the "scion," and the bottom part, which contains the roots, is called the "stock." This protocol describes the simplest way of grafting tomato plants using "top wedge grafting" or "cleft grafting." Potatoes, eggplants, and tobacco plants are closely related to tomatoes, and they can be grafted onto each other as well. Although the grafting of vegetable crops is still rare, this technique has been useful in reducing infections caused by pathogens, increasing resistance to drought, and enhancing nutrient uptake. Grafting techniques have been used to investigate the long-distance movement of signals through vascular tissue (Xoconostle-Cazares et al. 1999).

MATERIALS

Reagents

Tomato plants, adult (see Protocol 1)
It is best to use scion and stock stems with the same (or similar) diameter to increase the success of the grafting.

Equipment

Pruning shears
Razor blades, sharp
Surgical tape (Micropore™, 3M), plastic tube (~5 mm in diameter), or grafting clip (see Step 4)

METHOD

1. Use pruning shears to cut off the stock about 100 mm above the soil. Use a sharp razor blade to make a 5- to 15-mm-long slit in the center of the stem. Leave a few leaves on the stock.

2. Cut off the scion from the plant with pruning shears. Use a sharp razor blade to cut the lower end into a V shape (wedge).

3. Insert the scion into the slit of the stock.

4. Tie the junction with tape or use a tiny plastic tube or grafting clip to hold the scion and stock together and keep moisture in until they heal.
   A graft junction takes about 1 week to heal.

Protocol 4

Tomato Transformation

Transformation is an essential technique to analyze the function of genes. Tomato can be stably transformed using *Agrobacterium tumefaciens*–mediated transfer of T-DNA. The first report of tomato transformation was by McCormick et al. (1986). This protocol describes an *Agrobacterium*-mediated transformation method for tomato and is a modified version of the protocol by McCormick et al. (1986).

MATERIALS

The recipes for items marked with <R> begin on page 135.

CAUTION: See the Cautions Appendix for appropriate handling of materials marked with <!>.

Reagents

Agrobacterium (containing gene of interest)
Bleach (50%) <!>
Ethanol (70%) <!>
Germination medium (in Magenta boxes) <R>
H_2O (sterile distilled)
LB medium (solid [in Petri dishes] and liquid) <R>
MSO liquid medium <R>
Rooting medium (in Magenta boxes) <R>
Selection medium (in Petri dishes) <R>
Temporary medium (in Petri dishes) <R>
Tomato seeds

Equipment

Centrifuge
Culture tubes
Forceps (sterilized)
Greenhouse
Growth chamber, set at 26°C (with 16-hour photoperiod)
Incubator (shaking, preset to 28°C)
Pots, soil (moist), and clear plastic covers
Razor blade (single edged, sterilized)
Scissors (sterilized)
Spectrophotometer
Sterile hood
Surgical tape (Micropore™, 3M)
Transfer pipette

METHOD

Preparing the Plant Material

1. Surface-sterilize approximately 30 tomato seeds by immersing them in 70% ethanol and swirling them for 2 minutes at room temperature.

2. Immerse the seeds in 50% bleach for 15–20 minutes at room temperature with gentle swirling.

3. Wash off the bleach completely by rinsing the seeds 5–10 times with sterile distilled H_2O under a sterile hood.

4. Add 5 ml of sterile distilled H_2O to the seeds. Pour the seeds into a Magenta box containing germination medium. Place the cover on the Magenta box and put it in a growth chamber (26°C, 16-hour photoperiod).

5. Use sterilized forceps and scissors to harvest a cotyledon from an 8–10-day-old plant and place it in a Petri dish. Keep the cotyledons moist by adding about 20 ml of MSO liquid medium to the dish.

6. Cut the tip and base of each cotyledon with a razor blade (see Fig. 3). Wound them with one to three shallow transverse cuts across the main vein on the adaxial side to facilitate *Agrobacterium* infection. Place the explants onto a Petri dish containing 20 ml of temporary medium; they are ready for cocultivation (Step 10).

Preparing the *Agrobacterium* Suspension

7. Streak the *Agrobacterium* onto an LB plate and incubate the plate at 28°C until colonies form.
   It typically takes 2 to 3 days for colonies to form.

8. Inoculate 10 ml of LB liquid medium with a single *Agrobacterium* colony in a culture tube. Incubate the culture with shaking overnight at 28°C and then measure the OD_{600} of the culture using a spectrophotometer.
   The incubation time may be increased to get enough Agrobacterium. *The optimum* OD_{600} *is 0.6–0.7.*

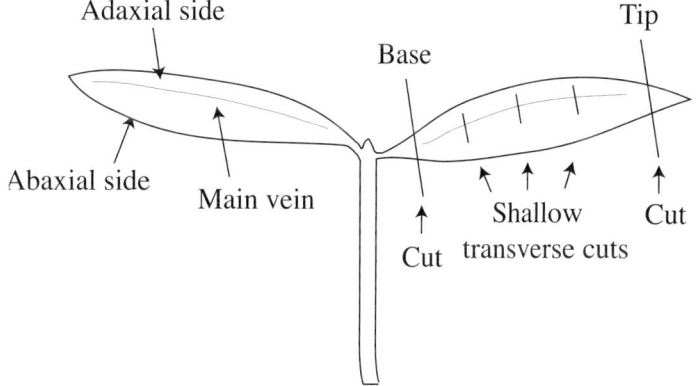

FIGURE 3. Tomato cotyledon. The cotyledon is the primary leaf of the embryo of seed plants and emerges first after germination. The tomato has two cotyledons. After germination, the size of each cotyledon is about 3 cm. For transformation, cut the tip and base of each cotyledon and wound them with up to three shallow transverse cuts on the adaxial side.

9. Harvest the *Agrobacterium* by centrifugation at 3000g for 15 minutes at room temperature and resuspend the cells with an appropriate amount of MSO liquid medium to make the OD_{600} of the suspension around 0.5; this suspension is now ready for cocultivation (Step 10).

Cocultivation

10. Pour about 5 ml of the *Agrobacterium* suspension (from Step 9) onto the temporary medium with the explants (from Step 6).
11. Incubate the explants on the temporary medium for 2 hours at room temperature.
12. Remove excess *Agrobacterium* suspension with a transfer pipette. Seal the plate with surgical tape.
13. Cocultivate explants in the growth chamber (26°C, 16-hour photoperiod) for 48 hours.

Selection of Transformants

14. Transfer the explants onto a Petri dish containing selection medium. (Keep the adaxial side up.) Seal the plate with surgical tape.
15. Keep the explants in the growth chamber (26°C, 16-hour photoperiod) until a callus forms. Transfer the explants to new selection medium every 2 weeks or when *Agrobacterium* is growing in the medium.
    After a few weeks, the first shoot should form.

Rooting

16. When shoots get to about 2–4 cm, excise them from the explants and transfer them to a Magenta box that contains rooting medium.
17. Keep the Magenta boxes in a growth chamber (26°C, 16-hour photoperiod).
    Roots should form 2–3 weeks after transferring to the rooting medium.

Transplanting

18. When the shoots are about 5 cm and the roots are established, take the transformants out of the Magenta box and gently wash off the agar.
19. Transplant the transformants to a pot with wet soil.
20. Place a clear plastic cover on the pot to keep moisture in and put them in the growth chamber (26°C, 16-hour photoperiod) for about 1 week.
21. After 1 week, transfer the pot to the greenhouse and let them grow.
    For general tomato cultivation practices, see Protocol 1.

Recipes

CAUTION: See the Cautions Appendix for appropriate handling of materials marked with <!>.

GERMINATION MEDIUM (pH 5.5)

Reagent	Quantity (for 1 liter)	Final concentration
Agar	8 g	0.8%
500x Gamborg's B5 vitamins stock solution	2 ml	1x
MS (Murashige and Skoog) basal salt mixture	4.3 g	0.43%
Sucrose	30 g	3%

Adjust the pH to 5.5 using KOH <!>, autoclave the solution, and pour 50 ml into each Magenta box (6 cm long x 6 cm wide x 10 cm high). When stored at 4°C, the medium will keep for up to 1 month.

LB MEDIUM (pH 7.0)

Reagent	Quantity (for 1 liter)	Final concentration
NaCl	10 g	1%
Tryptone	10 g	1%
Yeast extract	5 g	0.5%

For LB plates, include 15 g/liter (1.5%) agar. Include the appropriate types and amounts of antibiotics for selection and for the prevention of *Agrobacterium* growth. Adjust the pH to 7.0 using NaOH <!> and then autoclave the solution. For LB plates, pour 20 ml into each Petri dish (100 mm diameter x 20 mm high). Store the plates or liquid medium for up to 1 month at 4°C.

MSO LIQUID MEDIUM (pH 5.5)

Reagent	Quantity (for 1 liter)	Final concentration
500x Gamborg's B5 vitamins stock solution	2 ml	1x
MS (Murashige and Skoog) basal salt mixture	4.3 g	0.43%
Sucrose	30 g	3%

Adjust the pH to 5.5 using KOH <!> and then autoclave the solution. Store it for up to 1 month at 4°C.

ROOTING MEDIUM (pH 5.5)

Reagent	Quantity (for 1 liter)	Final concentration
Agar	8 g	0.8%
500x Gamborg's B5 vitamins stock solution	2 ml	1x
MS (Murashige and Skoog) basal salt mixture	4.3 g	0.43%
Sucrose	30 g	3%

Include the appropriate types and amounts of antibiotics for selection and for the prevention of *Agrobacterium* growth. Adjust the pH to 5.5 using KOH <!>, autoclave the solution, and pour 30 ml into each Magenta box (6 cm long x 6 cm wide x 10 cm high). When stored at 4°C, the medium will keep for up to 1 month.

SELECTION MEDIUM (pH 5.5)

Reagent	Quantity (for 1 liter)	Final concentration
Agar	8 g	0.8%
500x Gamborg's B5 vitamins stock solution	2 ml	1x
MS (Murashige and Skoog) basal salt mixture	4.3 g	0.43%
Sucrose	30 g	3%
Zeatin	0.1 mg	0.00001%

Include the appropriate types and amounts of antibiotics for selection and for the prevention of *Agrobacterium* growth. Adjust the pH to 5.5 using KOH <!>, autoclave the solution, and pour 30 ml into each Petri dish (100 mm diameter × 20 mm high). Store the plates for up to 1 month at 4°C.

TEMPORARY MEDIUM (pH 5.5)

Reagent	Quantity (for 1 liter)	Final concentration
Agar	8 g	0.8%
500x Gamborg's B5 vitamins stock solution	2 ml	1x
Glucose	30 g	3%
MS (Murashige and Skoog) basal salt mixture	4.3 g	0.43%

Adjust the pH to 5.5 using KOH <!>, autoclave the solution, and pour 30 ml into each Petri dish (100 mm diameter × 20 mm high). Store the plates for up to 1 month at 4°C.

REFERENCES

Arumuganathan, K. and Earle, E. 1991. Estimation of nuclear DNA content of plants by flow cytometry. *Plant Mol. Biol. Rep.* **9**: 208–218.

Atherton, J.G. and Harris, G.P. 1986. Flowering. In *The tomato crop: A scientific basis for improvement* (ed. J.G. Atherton and J. Rudich), pp. 167–200. Chapman and Hall, New York.

Barton, D.W., Butler, L., Jenkins, J.A., Rick, C.M., and Young, P.A. 1955. Rules for nomenclature in tomato genetics. *J. Hered.* **46**: 22–26.

Bernacchi, D. and Tanksley, S.D. 1997. An interspecific backcross of *Lycopersicon esculentum* × *L. hirsutum*: Linkage analysis and a QTL study of sexual compatibility factors and floral traits. *Genetics* **147**: 861–877.

Bohs, L. and Olmstead, R.G. 1997. Phylogenetic relationships in *Solanum* (Solanaceae) based on *ndhF* sequences. *Syst. Bot.* **22**: 5–17.

Chen, J.J., Janssen, B.J., Williams, A., and Sinha, N. 1997. A gene fusion at a homeobox locus: Alterations in leaf shape and implications for morphological evolution. *Plant Cell* **9**: 1289–1304.

Child, A. 1990. A synopsis of *Solanum* subgenus *Potatoe* (G. Don) (D'Arcy) (*Tuberarium* (Dun.) Bitter (s.l)). *Feddes Repert.* **101**: 209–235.

Darwin, S., Knapp, S., and Peralta, I. 2003. Taxonomy of tomatoes in the Galapagos Islands: Native and introduced species of *Solanum* section *Lycopersicon* (Solanaceae). *Syst. Biodivers.* **1**: 29–53.

Doganlar, S., Frary, A., Ku, H.M., and Tanksley, S.D. 2002. Mapping quantitative trait loci in inbred backcross lines of *Lycopersicon pimpinellifolium* (LA1589). *Genome* **45**: 1189–1202.

Eshed, Y. and Zamir, D. 1995. An introgression line population of *Lycopersicon pennellii* in the cultivated tomato enables the identification and fine mapping of yield-associated QTL. *Genetics* **141**: 1147–1162.

Frary, A., Nesbitt, T.C., Grandillo, S., Knaap, E., Cong, B., Liu, J., Meller, J., Elber, R., Alpert, K.B., and Tanksley, S.D. 2000. *fw2.2*: A quantitative trait locus key to the evolution of tomato fruit size. *Science* **289**: 85–88.

Frary, A., Fritz, L.A., and Tanksley, S.D. 2004. A comparative study of the genetic bases of natural variation in tomato leaf, sepal, and petal morphology. *Theor. Appl. Genet.* **109**: 523–533.

Fray, R.G. and Grierson, D. 1993. Identification and genetic analysis of normal and mutant phytoene synthase genes of tomato by sequencing, complementation and co-suppression. *Plant Mol. Biol.* **22**: 589–602.

Fridman, E., Carrari, F., Liu, Y.S., Fernie, A.R., and Zamir, D. 2004. Zooming in on a quantitative trait for tomato yield using interspecific introgressions. *Science* **305**: 1786–1789.

Haanstra, J.P.W., Wye, C., Verbakel, H., Meijer-Dekens, F., van den Berg, P., Odinot, P., van Heusden, A.W., Tanksley, S.D., Lindhout, P., and Peleman, J. 1999. An integrated high-density RFLP-AFLP map of tomato based on two *Lycopersicon esculentum* × *L. pennellii* populations. *Theor. Appl. Genet.* **99**: 254–271.

Holtan, H.E. and Hake, S. 2003. Quantitative trait locus analysis of leaf dissection in tomato using *Lycopersicon pennellii* segmental introgression lines. *Genetics* **165**: 1541–1550.

Isaacson, T., Ronen, G., Zamir, D., and Hirschberg, J. 2002. Cloning of *tangerine* from tomato reveals a carotenoid isomerase essential for the production of β-carotene and xanthophylls in plants. *Plant Cell* **14**: 333–342.

Kimura, S., Koenig, D., Kang, J., Yoong, F.Y., and Sinha, N. 2008. Natural variation of leaf morphology results from mutation of a novel *KNOX* gene. *Curr. Biol.* **18**: 672–677.

Linnaeus, C. 1753. *Species plantarum*. Stockholm.

Liu, J., Van Eck, J., Cong, B., and Tanksley, S.D. 2002. A new class of regulatory genes underlying the cause of pear-shaped tomato fruit. *Proc. Natl. Acad. Sci.* **99:** 13302–13306.

McCormick, S., Niedermeyer, J., Fry, J., Barnason, A., Horsch, R., and Fraley, R. 1986. Leaf disc transformation of cultivated tomato (*L. esculentum*) using *Agrobacterium tumefaciens*. *Plant Cell Reports* **5:** 81–84.

Meissner, R., Jacobson, Y., Melamed, S., Levyatuv, S., Shalev, G., Ashri, A., Elkind, Y., and Levy, A. 1997. A new model system for tomato genetics. *Plant J.* **12:** 1465–1472.

Menda, N., Semel, Y., Peled, D., Eshed, Y., and Zamir, D. 2004. In silico screening of a saturated mutation library of tomato. *Plant J.* **38:** 861–872.

Miller, P. 1754. *The gardener's dictionary*. London.

Ori, N., Cohen, A.R., Etzioni, A., Brand, A., Yanai, O., Shleizer, S., Menda, N., Amsellem, Z., Efroni, I., Pekker, I., et al. 2007. Regulation of *LANCEOLATE* by *miR319* is required for compound-leaf development in tomato. *Nat. Genet.* **39:** 787–791.

Parnis, A., Cohen, O., Gutfinger, T., Hareven, D., Zamir, D., and Lifschitz, E. 1997. The dominant developmental mutants of tomato, *Mouse-ear* and *Curl*, are associated with distinct modes of abnormal transcriptional regulation of a *Knotted* gene. *Plant Cell* **9:** 2143–2158.

Peralta, I., Knapp, S., and Spooner, D. 2005. New species of wild tomatoes (*Solanum* section *Lycopersicon*: Solanaceae) from northern Peru. *Syst. Bot.* **30:** 424–434.

Pnueli, L., Carmel-Goren, L., Hareven, D., Gutfinger, T., Alvarez, J., Ganal, M., Zamir, D., and Lifschitz, E. 1998. The *SELF-PRUNING* gene of tomato regulates vegetative to reproductive switching of sympodial meristems and is the ortholog of *CEN* and *TFL1*. *Development* **125:** 1979–1989.

Rick, C. 1980. *Petroselinum* (*Pts*), a new marker for chromosome 6. *Tomato Genet. Coop. Rep.* **20:** 32.

Rick, C.M. 1991. *Tomato*. Longman Scientific and Technical, Essex, England.

Ronen, G., Cohen, M., Zamir, D., and Hirschberg, J. 1999. Regulation of carotenoid biosynthesis during tomato fruit development: Expression of the gene for lycopene epsilon-cyclase is down-regulated during ripening and is elevated in the mutant *Delta*. *Plant J.* **17:** 341–351.

Ronen, G., Carmel-Goren, L., Zamir, D., and Hirschberg, J. 2000. An alternative pathway to β-carotene formation in plant chromoplasts discovered by map-based cloning of *Beta* and *old-gold* color mutations in tomato. *Proc. Natl. Acad. Sci.* **97:** 11102–11107.

Spooner, D.M., Anderson, G.J., and Jansen, R.K. 1993. Chloroplast DNA evidence for the interrelationships of tomatoes, potatoes and pepinos (Solanaceae). *Am. J. Bot.* **80:** 676–688.

Spooner, D.M., Peralta, I.E., and Knapp, S. 2005. Comparison of AFLPs with other markers for phylogenetic inference in wild tomatoes [*Solanum* L. section *Lycopersicon* (Mill.) Wettst.]. *TAXON* **54:** 43–61.

Stevens, M.A. and Rick, C.M. 1986. Genetics and breeding. In *The tomato crop* (ed. J.G. Atherton and J. Rudich), pp. 35–109. Chapman and Hall, New York/London.

Tanksley, S.D. 2004. The genetic, developmental, and molecular bases of fruit size and shape variation in tomato. *Plant Cell* (suppl.) **16:** S181–S189.

Wang, Y., Tang, X., Cheng, Z., Mueller, L., Giovannoni, J., and Tanksley, S.D. 2006. Euchromatin and pericentromeric heterochromatin: Comparative composition in the tomato genome. *Genetics* **172:** 2529–2540.

Xiao, H., Jiang, N., Schaffner, E., Stockinger, E.J., and van der Knaap, E. 2008. A retrotransposon-mediated gene duplication underlies morphological variation of tomato fruit. *Science* **319:** 1527–1530.

Xoconostle-Cázares, B., Xiang, Y., Ruiz-Medrano, R., Wang, H.L., Monzer, J., Yoo, B.C., McFarland, K.C., Franceschi, V.R., and Lucas, W.J. 1999. Plant paralog to viral movement protein that potentiates transport of mRNA into the phloem. *Science* **283:** 94–98.

Zhang, J., Chen, R., Xiao, J., Qian, C., Wang, T., Li, H., Ouyang, B., and Ye, Z. 2007. A single-base deletion mutation in *SlIAA9* gene causes tomato (*Solanum lycopersicum*) entire mutant. *J. Plant Res.* **120:** 671–678.

FURTHER READING

Doebley, J.F., Gaut, B.S., and Smith, B.D. 2006. The molecular genetics of crop domestication. *Cell* **127:** 1309–1321.
Discusses the genes and genetic changes involved in the domestication of crop plants, including tomato.

Knapp, S., Bohs, L., Nee, M., and Spooner, D.M. 2005. Solanaceae—A model for linking genomics with biodiversity. *Comp. Funct. Genomics* **5:** 285–291.
Outlines an effort to integrate genomic and taxonomic information for the members of the diverse Solanaceae plant family.

Lippman, Z.B., Semel, Y., and Zamir, D. 2007. An integrated view of quantitative trait variation using tomato interspecific introgression lines. *Curr. Opin. Genet. Dev.* **17:** 545–552.
Reviews QTL analyses in tomato.

Mueller, L.A., Solow, T.H., Taylor, N., Skwarecki, B., Buels, R., Binns, J., Lin, C., Wright, M.H., Ahrens, R., Wang, Y., et al. 2005. The SOL Genomics Network: A comparative resource for Solanaceae biology and beyond. *Plant Physiol.* **138:** 1310–1317.
Provides an overview of the SOL Genomics Network (SGN).

Paran, I. and van der Knaap, E. 2007. Genetic and molecular regulation of fruit and plant domestication traits in tomato and pepper. *J. Exp. Bot.* **58:** 3841–3852.
Reviews the key traits involved in tomato (and pepper) domestication, as well as efforts to identify the genes responsible for those traits.

Rick, C.M. 1991. Tomato paste: A concentrated review of genetic highlights from the beginnings to the advent of molecular genetics. *Genetics* **128:** 1–5.
Provides a history of tomato genetics.

Schmitz, G. and Theres, K. 1999. Genetic control of branching in *Arabidopsis* and tomato. *Curr. Opin. Plant Biol.* **2:** 51–55.
Reviews the genetic control and development of shoot branching in tomato (and *Arabidopsis*).

Sinha, N. 1999. Leaf development in angiosperms. *Annu. Rev. Plant Physiol. Plant Mol. Biol.* **50:** 419–446.
Reviews the genetic control and development of compound leaves in tomato (and other plant species).

WWW RESOURCES

http://ted.bti.cornell.edu Tomato Expression Database. This site includes the tomato microarray data warehouse, tomato microarray expression data, and tomato digital expression data.

http://ted.bti.cornell.edu/order/index.html Tomato cDNA Clone Order Information. Tomato EST clones are available for purchase through this website.

http://tgrc.ucdavis.edu The C.M. Rick Tomato Genetics Resource Center (TGRC). A collection of wild relatives, monogenic mutants, and miscellaneous genetic stocks of tomato.

http://webs2.kazusa.or.jp/komics KOMICS (Kazusa OMICS). A database for metabolites.

http://www.genome.clemson.edu/capabilities/bacCenter.shtml BAC/EST Resource Center at the Clemson University Genomics Institute. Tomato BAC libraries and clones are available for purchase from this institute.

http://www.pgb.kazusa.or.jp/kaftom Kazusa Full-length Tomato cDNA Database. This site provides information about full-length cDNA clones derived from the miniature cultivar Micro-Tom.

http://www.sgn.cornell.edu/index.pl The International Tomato Sequencing Project. This website tracks the progress of the tomato genome sequencing project and includes links to additional tomato- and Solanaceae-related resources. It is part of the International Solanaceae Genome Project (SOL) Genomics Network (SGN).

http://zamir.sgn.cornell.edu/mutants The Genes That Make Tomatoes. This site provides information about the saturated mutation library of tomato.

6 | The Demosponge *Amphimedon queenslandica*
Reconstructing the Ancestral Metazoan Genome and Deciphering the Origin of Animal Multicellularity

Bernard M. Degnan,[1] Maja Adamska,[1,2] Alina Craigie,[1]
Sandie M. Degnan,[1] Bryony Fahey,[1] Marie Gauthier,[1]
John N.A. Hooper,[3] Claire Larroux,[1] Sally P. Leys,[4]
Erica Lovas,[1] and Gemma S. Richards[1]

[1]*School of Integrative Biology, University of Queensland, Brisbane QLD 4072 Australia;* [2]*Sars International Centre for Marine Molecular Biology, N-5008 Bergen, Norway;* [3]*Queensland Museum, Queensland, Australia;* [4]*Department of Biological Sciences CW 405, University of Alberta, Edmonton, Alberta T6G 2E9, Canada*

ABSTRACT

Sponges are one of the earliest branching metazoans. In addition to undergoing complex development and differentiation, they can regenerate via stem cells and can discern self from nonself (allorecognition), making them a useful comparative model for a range of metazoan-specific processes. Molecular analyses of these processes have the potential to reveal ancient homologies shared among all living animals and critical genomic innovations that underpin metazoan multicellularity. *Amphimedon queenslandica* (Porifera, Demospongiae, Haplosclerida, Niphatidae) is the first poriferan representative to have its genome sequenced, assembled, and annotated. *Amphimedon* exemplifies many sessile and sedentary marine invertebrates (e.g., corals, ascidians, bryozoans): They disperse during a planktonic larval phase, settle in the vicinity of conspecifics, ward off potential competitors (including incompatible genotypes), and ensure that brooded eggs are fertilized by conspecific sperm. Using genomic and expressed sequence tag (EST) resources from *Amphimedon*, functional genomic approaches can be applied to a wide range of ecological and population genetic processes, including fertilization, dispersal, and colonization dynamics, host–symbiont interactions, and secondary metabolite production. Unlike most other sponges, *Amphimedon* produce hundreds of asynchronously developing embryos and larvae year-round in distinct, easily accessible brood chambers. Embryogenesis gives rise to larvae with at least a dozen cell types that are segregated into three layers and patterned along the body axis. In this chapter, we describe some of the methods currently available for studying *A. queenslandica*, focusing on the analysis of embryos, larvae, and postlarvae.

> **PROTOCOLS**
>
> 1 Isolation of *Amphimedon* Developmental Material, 144
> 2 Whole-mount In Situ Hybridization, 148
> 3 Analysis of Cell Movement by Injection of Fluorescent Tracers, 155
> 4 Genotyping Individual Embryos, Larvae, and Adults, 158

This chapter, with full-color images, can be found online at www.cshprotocols.org/emo.

BACKGROUND INFORMATION

Sponges (phylum Porifera) are sessile, aquatic (largely marine) animals with external pores connected to a flowthrough system of canals and chambers through which water is pumped to extract food. Passage of water through the body is driven by a single layer of specialized flagellated cells called choanocytes. Unlike most other metazoans, sponges do not construct true tissues, lack a centralized gut, and do not possess conventional nerves and muscle. Nonetheless, they have a range of other cell types, including a population of totipotent stem cells and skeletogenic cells that form siliceous or calcitic spicules (Hooper and Van Soest 2002). Although many have not yet been described or named, there are an estimated 15,000 sponge species alive today. Their evolution and ecology is associated tightly with a range of microbial symbionts and the ability of these sponge–microbe communities to synthesize and deploy unique bioactive compounds with a variety of ecological roles (Taylor et al. 2007).

A. queenslandica is named for Queensland and was described originally from individuals found at Heron and One Tree Island Reefs in the Capricorn-Bunker Group, southern Great Barrier Reef (GBR) (Hooper and Van Soest 2006). Its discovery off Magnetic Island on the central/northern GBR suggests a pan-GBR distribution. Recent analysis of an *Amphimedon* species collected from Dahab, Egypt revealed identical external characteristics and brood chambers with identically colored and sized embryos and larvae, suggesting a possibly wider distribution, with Red Sea specimens having 93% sequence identity with the *A. queenslandica* 28S rRNA gene. Anecdotal evidence from southern Japan suggests that a species similar to *A. queenslandica*, as well as closely related sister species, live there, and thus it might be distributed beyond the South Pacific Ocean.

A. queenslandica growths range from thin to thick encrustings over coral or other substrata. The latter form has massive lobate or digitate bulbs arising from the base, measuring no more than several centimeters in diameter. Live sponges are gray-blue to green, with a lighter shade of gray around the rim of the oscules. They have large brood chambers (up to 1 cm in diameter) containing as many as 200 embryos at any time (Leys and Degnan 2001). Hooper and Van Soest (2006) provide a full description of the species and comparison with similar species in the Indo-West Pacific.

A. queenslandica's taxonomic history is checkered by misidentification and misallocation. This is the result of the state of poriferan taxonomy (partly remedied by a phylum synopsis by an international consortium of taxonomists; Hooper and Van Soest 2002), the difficulty in identifying haplosclerid sponges in general and Niphatidae in particular (see, e.g., Desqueyroux-Faúndez and Valentine 2002), and the limited knowledge of Australian sponge fauna. Indeed, many species that initially appear to be new have been described previously in the largely ancient Australian sponge literature (see, e.g., Hooper and Wiedenmayer 1994; http://www.environment.gov.au/cgi-bin/abrs/fauna/tree.pl?pstrVol=PORIFERA&pintMode=1). *A. queenslandica*, for example, was initially identified as a species of *Reniera* (Haplosclerida, Chalinidae) based on a fragment of a single specimen (Leys and Degnan 2001). Its eventual allocation as a new species of *Amphimedon* in the Niphatidae family was confirmed by collection of several more specimens from different habitats within Heron and One Tree Islands.

A. queenslandica was discovered in 1998 on Heron Island Reef by S.P. Leys of B.M. Degnan's group during a survey to identify a sponge species with which to study development. Unlike most sponges, *A. queenslandica* broods embryos in large chambers year-round. Larvae are relatively large (~0.6 mm) and overtly negatively phototactic, moving quickly to the darker side of any collecting vessel. These attributes led to detailed morphological studies of its embryogenesis (Leys and Degnan 2002) and behavioral studies of its photoresponsiveness (Leys and Degnan 2001; Leys et al. 2002). These studies established nondestructive methods in which adult animals are maintained while attached to boulders in corals on the reef flat (i.e., "ranching"), allowing larvae to be harvested from the same adults over time. Methods developed subsequently in the Degnan laboratory include in vitro cultivation of individual late-stage embryos and postlarvae and cultivation of transplanted embryos in situ.

As with most marine invertebrates, *A. queenslandica* has a biphasic life cycle, including a planktonic larval phase and benthic juvenile and adult phases (Degnan and Degnan 2006). Eggs are fertilized internally and embryos brood until they hatch as parenchymella larvae. For the first 4 hours after emergence, larvae appear to be unable to settle and they undergo metamorphosis. During this period (up to 24 hours postemergence), larvae are conspicuously negatively phototactic, although this diminishes in older larvae (Leys and Degnan 2002). Negative phototaxis apparently results from the sum of responses to changing light intensity by individual ciliated cells in the posterior ring. Rotating larvae expose some ciliated cells to light, but shade others. Shaded cilia cease beating, causing the animal to swim into darker crevices among the coral boulders on the reef flat. Similar phototaxis has been documented in other parenchymella larvae (for review, see Maldonado 2006), confirming the widespread use of this mechanism.

SOURCES AND HUSBANDRY

A. queenslandica is reasonably common on coral reefs of the southern GBR. They or sibling species have been found as far away as the Red Sea, suggesting that these populations as model organisms are widely available as sources throughout the Indo-Pacific and possibly beyond. They live on the shallow subtidal reef flat and crest, mainly under boulders, in crevices, among coral rubble, in sand patches, and sometimes on hard algal pavement, partially exposed during low tide. Distribution on a given reef tends toward the patchy, often with adults found in areas of low current. In addition to easy accessibility, it is a robust sponge that tolerates being pried off the substratum. *Amphimedon* can be collected from the field and maintained in local aquaria under ambient conditions (i.e., in flowthrough systems). Alternatively, collected *Amphimedon* can be ranched in a local embayment where they can be readily accessed. During collection, care must be taken not to touch the sponge directly nor expose it to air at any time. Ranched *Amphimedon* can live for more than 1 year.

Although adult *Amphimedon* can be transported in small volumes of seawater (for a maximum transport time of up to 10 hours) and maintained in closed, seawater aquaria (24–27°C) for months, there is no evidence that they flourish in captivity. They can be maintained by being fed liquid foods developed for filter-feeding marine invertebrates twice per week. However, the number of viable embryos in captivity appears to diminish with time.

Ideally, *Amphimedon* are best studied in a marine research laboratory near the point of collection, with access to high-quality flowing seawater. This ensures that adults, embryos, larvae, and experimentally manipulated cultures are maintained under ambient conditions. Most studies on the southern GBR populations are based at the University of Queensland Heron Island Research Station (http://www.marine.uq.edu.au/index.html?page=54940) and the University of Sydney One Tree Island Research Station (http://www.bio.usyd.edu.au/OTI/).

RELATED SPECIES

Sponge systematics are undergoing major revision, based on recent morphological data (Hooper and Van Soest 2002) and a range of studies using molecular data (for review, see Wörheide et al. 2005). Traditionally, sponges were divided into three classes: Demospongiae, Hexactinellida, and Calcarea (Hooper and Van Soest 2002). However, there is debate about the monophyly of phylum Porifera and class Demospongiae (see, e.g., Borchiellini et al. 2004). The order of branching of "basal" metazoan phyla—Porifera, Placozoa, Ctenophora, and Cnidaria—and whether sponges are monophyletic or paraphyletic directly impacts the reconstruction of the ancestor from which all living metazoans stem. Undertaking comparative molecular, cellular, and developmental analyses within the Porifera will yield significant insights into the early metazoan genome and its role into the evolution of the first multicellular animals.

Most modern analyses of sponge genomics and development focus on demosponges, with the *A. queenslandica* genome being the only one sequenced to date. Many developmental and structural genes have been isolated and characterized in a number demosponges, including *Ephydatia*, *Geodia*, *Spongilla*, and *Suberites* (Segawa et al. 2006; Wiens et al. 2007). Emerging models include the homoscleromorph *Oscarella* (Nichols et al. 2006), the calcareous sponges *Sycon* and *Leucetta* (Manuel and LeParco 2000), and the hexactinellid *Oopsacas* (Leys et al. 2006). Detailed studies of these species will allow for a more comprehensive view of sponge genomics, evolution, and development, and thus a more mature view of the earliest metazoans.

USE OF THE *A. QUEENSLANDICA* MODEL SYSTEM

Amphimedon is a demosponge and, as such, represents one of the most (if not *the* most) ancient phyla of multicellular animals alive today. Sponges lack many fundamental metazoan attributes, including tissues, yet comprehensive molecular phylogenies place them firmly within the Metazoa. They are usually considered to be an example of the earliest extant branching metazoan lineage, although two recent hypotheses have suggested that placozoans or ctenophores could be more basal (Dellaporta et al. 2006; Dunn et al. 2008).

Knowledge of the formation and patterning of cell types via embryogenesis, skeletogenesis, or the cellular and molecular basis of sponge behavior is rudimentary at best. The ancestor from which all extant metazoans stemmed was likely more sophisticated than is widely appreciated, and the morphogenetic tools used by all modern animals evolved well before the Cambrian explosion. A detailed understanding of these features in sponges will contribute to the reconstruction the ancestral metazoan genome and elucidate its role in directing the construction and maintenance of the first multicellular animals (Larroux et al. 2006, 2007, 2008; Fahey et al. 2008).

Amphimedon larva morphogenesis provides a framework within which to address gastrulation and tissue formation in early animals. Embryogenic studies in poriferans have been largely descriptive, but many parallels can be drawn with other sponge larvae, of both parenchymella and other types, from studies on *Amphimedon*. The expression of "polarity" genes during *Amphimedon* larval morphogenesis (Adamska et al. 2007a) suggests that similar mechanisms function more broadly within the Porifera. Expression of a suite of genes involved in postsynaptic signaling (Sakarya et al. 2007) similarly hints at complex signaling systems in sponges.

Amphimedon embryos are large yolky spheres numbering several hundred to a single chamber. Oogenesis has not been studied, but oocytes presumably arise from amoeboid cells within the maternal tissue, because feeding chambers with choanocytes (an alternative source of gametes in other sponges) are not always adjacent to chambers. Spermatocysts are encountered in only one of every 50 sponges examined in early surveys (Leys and Degnan 2001). Cleavage is irregular because of the yolky cytoplasm; early in development, unequal cleavages begin and small and large cells (micromeres and macromeres) are produced throughout the embryo. During cellular differentiation, micromeres produce cilia and other features characteristic of differentiated cells (e.g., pigmentation in cells fated to form the pigment ring), whereas macromeres retain lipid and yolk reserves. The two cell types gradually sort, with micromeres localizing at the embryo periphery. Sclerocytes (i.e., skeletogenic cells) differentiate and start to form incipient spicules early in development. At the same time, a subpopulation of cells with dark pigment granules begins to migrate toward the future posterior pole. *Wnt* gene expression in this region preempts pigment cell migration (Adamska et al. 2007a). Cellular differentiation and sorting continue as the pigment cells form first a dense spot at the pole and subsequently tighten into a discrete ring of pigment that designates the larval posterior pole. Sclerocytes, with fully differentiated spicules secreted intracellularly, aggregate toward the posterior pole with spicules largely oriented in an anterior-posterior direction. The fully differentiated larva has distinct anterior-posterior polarity with a ciliated columnar epithelium, but it also has bare anterior and posterior poles, spicules in sclerocytes at the posterior pole, and a ring of pigment adjacent and posterior to a ring of cells with very long cilia. Interspersed among the columnar epithelial-like cells are globular cells (formerly called mucous cells) and flask cells. Current evidence suggests that globular cells express genes found in eumetazoan neurons and they have a sensory role (Richards et al. 2008).

Because *Amphimedon* adults produce many embryos year-round, parenchymella larvae can be harvested continually. In addition, the larvae are large and easy to study, both for morphology and behavior. *Amphimedon* development and developmental gene structure and expression have been studied extensively (Leys and Degnan 2001, 2002; Degnan et al. 2005; Larroux et al. 2006, 2007, 2008; Adamska et al. 2007a,b; Sakarya et al. 2007; Fahey et al. 2008; Gauthier and Degnan 2008a,b). Excellent recent reviews on sponge embryos and larvae are also available (Leys and Ereskovsky 2006; Maldonado 2006).

GENETICS, GENOMICS, AND ASSOCIATED RESOURCES

A. queenslandica is not amenable to classical genetic studies and currently has no available genetic markers, maps, or mutants. Given the inability to culture this species, it is unlikely that such resources will come available in the near future. All genes isolated and characterized in *A. queenslandica* are given a name with prefix "*Amq*" followed by a descriptor based on its metazoan/eukaryotic gene family. Ideally, names are based on orthology to known eumetazoan genes, e.g., *AmqNF-κB* (Gauthier and Degnan 2008a) or *AmqSix1/2* (Larroux et al. 2008). Sequence divergence often makes such precise naming difficult. In such cases, a more generic name is applied (e.g., *AmqbHLH1*; Simionato et al. 2007; Richards et al. 2008). The *Amphimedon* genome also contains genes with novel domain architectures. Names are created for these genes that reflect their relation to known eumetazoan genes. For example, *Amq-hedgling* encodes a large transmembrane protein with the Hedgehog amino-terminal signaling domain (Adamska et al. 2007b). Novel genes are named based on contig number and location; these may be revised in the future if homologs are identified.

The *A. queenslandica* genome has been sequenced by the United States Department of Energy Joint Genome Institute (http://www.jgi.doe.gov/). Genomic resources currently include genomic traces (~14-fold coverage) and more than 75,000 ESTs. These sequences are available in the National Center for Biotechnology Information Trace archive (http://www.ncbi.nlm.nih.gov/Traces/trace.cgi?) and can be retrieved by searching with species_code = "RENIERA SP. JGI-2005" (reflecting the incorrect original designation of *Amphimedon* as *Reniera* sp.). Release of the annotated draft assembly by the Joint Genome Institute is scheduled for late 2008.

Biological material (fixed specimens, RNA, cDNA, genomic DNA, etc.) can generally be obtained by direct collection from field sites or by contacting the corresponding investigator. Natural populations are managed sustainably and collection of adults might be restricted periodically. Permits are required to collect animals from the GBR.

TECHNICAL APPROACHES

The following protocols describe methods to obtain *Amphimedon* tissues at various stages of development (with particular emphasis on embryonic, larval, and juvenile samples; see Protocol 1) and process them for studies such as in situ hybridization (Protocol 2) and genotyping (Protocol 4). Methods are also described for microinjection of individual embryos to study cell migration during development (Protocol 3).

Protocol 1

Isolation of *Amphimedon* Developmental Material

Fertilization occurs internally in *Amphimedon* and embryos are brooded in multiple chambers throughout the adult. Each chamber contains a mixture of developmental stages, from egg to late ring stages (i.e., prehatch late embryos). At the end of embryogenesis, swimming parenchymella larvae emerge from the adult. After several hours in the water column, the larvae settle and metamorphose into juvenile sponges (Fig. 1). This protocol details how to obtain *Amphimedon* larvae and postlarvae/juveniles (Method I) as well as embryos (Method II). Once isolated, these biological stages can be used for a variety of molecular and cellular analyses.

MATERIALS

Reagents

Agarose
Seawater (filtered [FSW], 0.22 µm)

Equipment

Coverslips (round, plastic, or glass; 12 mm)
Field collection equipment (hammer, chisel, snorkeling gear, collection bucket/bag, 0.5–1-liter screw-cap jars, large black tub [e.g., 32-liter Nally bins])
Forceps (fine; e.g., Dumont #5)
Light source (cold)
Microscope (dissecting)
Needle (teasing, curved)
Needle holder (equipped with screw chuck)

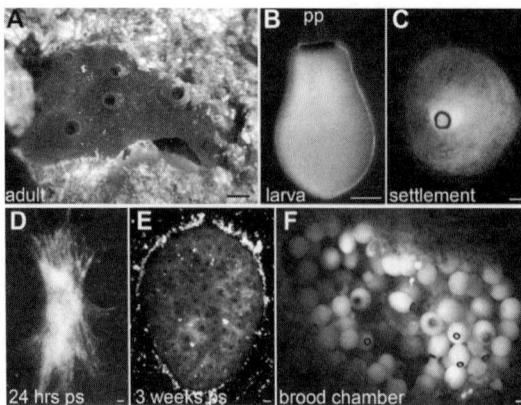

FIGURE 1. Stages of the *A. queenslandica* life cycle. (*A*) Adult animal; (*B*) swimming larva ([*pp*] posterior pole); (*C*) larva undergoing settlement with anterior part flattened on substrate; (*D*) postlarva 24 hours postsettlement (*ps*); (*E*) 3-week-old juvenile; (*F*) sliced brood chamber showing developing embryos at different stages. Scale bar, 1 cm (*A*) or 100 µm (*B*–*F*). (Modified from Adamska et al. 2007a).

Pasteur pipettes
> *Use a Bunsen burner to reduce the diameter of the Pasteur pipette until it approximates the size of an embryo.*

Petri dishes (plastic, including deep Petri dishes 90 x 25 mm; e.g., Labserv, LBS61014)
Pins (stainless steel)
Pipette (mouth)
Plates (cell culture, 6 and 24 well)
Scalpel and blades
Sharpening stone
Sucker cup
> *Drill two holes in the lid of a 250-ml plastic screw-top jar. Insert two lengths of 5-mm plastic tubing through the holes.*

Thermometer

METHODS

I. Larvae and Juveniles

1. Snorkel at low tide to locate adult sponges. Chisel away the rock around the sponge to collect adults on substrate.
   > *Avoid squeezing or touching adult tissue or exposing it to air. Although* Amphimedon *is reproductive year-round, developmental materials are more readily available in the summer. After a strong tropical storm, brood chambers can be empty for a few days.*

2. Place sponges in a black tub filled with seawater. Leave the tub in full sunlight, uncovered, and protected from the wind.
   > *Transfer adult sponges gently and rapidly to minimize contact with air.*

3. Monitor the temperature of the seawater. Do not allow it to increase more than 3–4°C above the ambient water temperature.
   > *This treatment induces larval release. The white larvae, clearly visible against the black background, will initially swim upward and tend to cluster in the corners of the tub. See Troubleshooting.*

4. Check for larvae every 30 minutes (up to a maximum of 3–4 hours).

5. Use a sucker cup to collect larvae:
   i. Place one tube into your mouth and the other into the seawater.
   ii. Use short bursts of suction to collect larvae into the jar.
   iii. Transfer the seawater with the larvae to screw-cap jars.
      > *Store these in a cool, shaded location.*

6. Gradually lower the temperature in the black tub to acclimatize the adult sponges back to ambient water temperature.

7. Return adult sponges to where they were found. Secure to rocks with ties if necessary.

8. Use a Pasteur pipette to transfer larvae (with as little seawater as possible) into a beaker containing fresh filtered seawater (FSW).
   > *These can then be maintained in the laboratory for days with regular changes of FSW (e.g., every 6–10 hours).*

9. Transfer larvae into a beaker containing fresh FSW.
   > *This ensures that the seawater carried over from the initial larval collection is diluted, minimizing the risks of contamination and infection.*

10. Transfer approximately 40 larvae into a deep plastic Petri dish with 100 ml of FSW. Allow the larvae to settle undisturbed in the dark.

    It is important not to overcrowd the dishes because this increases the rate of fusion between settling larvae. Settlement and metamorphosis in FSW occur 4–72 hours after release. See Troubleshooting.

11. Within 6 hours after settlement, use a curved teasing needle to gently detach newly settled larvae. Transfer to small round coverslips placed in wells or dishes submerged in FSW. Use 24-well plates if subsequently conducting whole-mount in situ hybridization (see Protocol 2).

    Postlarvae reattach rapidly to the coverslips and proceed with development.

12. Raise postlarvae/juveniles to the age of interest. Change the FSW regularly and scan the dishes for protist outbreaks.

    Osculae and choanocyte tracts become apparent 3 days after metamorphosis. See Troubleshooting.

II. Embryos

Keep the sponge, slices, and embryos submerged in FSW throughout the procedure.

1. Snorkel at low tide to locate adult sponges. Chisel away the rock around the sponge to collect adults on substrate. Carefully "peel" an adult sponge away from its substrate. Gently remove debris from the sponge.

    Avoid squeezing or touching adult tissue or exposing it to air. Brood chambers are often located at the base of the sponge, and squeezing can damage embryos. Although Amphimedon is reproductive year-round, developmental materials are more readily available in the summer. After a strong tropical storm, brood chambers can be empty for a few days.

2. Use a scalpel to slice the sponge into 3-mm sections. Inspect each section for the presence of brood chambers (Fig. 2).

    Glass spicules embedded in the adult can readily dull scalpel blades. Replace these regularly because cutting with dull blades crushes the embryos. See Troubleshooting.

3. Transfer a slice with chambers to a deep Petri dish containing a thin base of 2% agarose in FSW. Use pins to secure the slice to the agarose.

4. Under a dissecting microscope, carefully use fine forceps and/or sharpened pins to release embryos from the white maternal tissue. Use a mouth or Pasteur pipette to gently blow on and/or extract embryos.

5. Transfer isolated embryos to a separate dish of FSW. Ensure that there is space around each embryo to prevent them from fusing together.

    If embryos detach from the chamber in a clump, fix the clumps within 1 hour to avoid fusion.

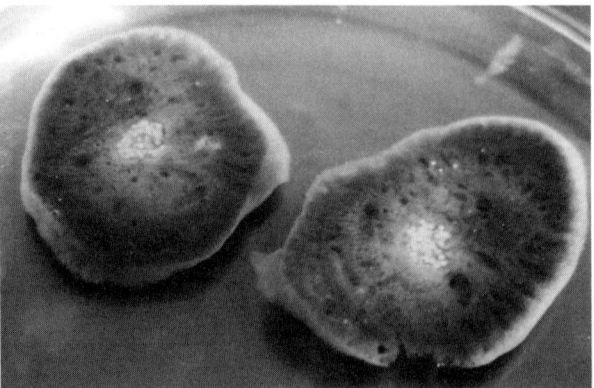

FIGURE 2. Detecting a brood chamber (white embryos) by making thin slices through an adult.

TROUBLESHOOTING

Problem: Embryos, larvae, or postlarvae stick to the inside of Pasteur pipettes or plastic pipette tips.
Solution: Precoat tips with seawater or FSW every time a fresh pipette or tip is used. To dislodge an embryo from the pipette/tip, draw liquid and air bubbles up past the embryo or fill the pipette/tip with liquid and tap the outside.

Problem (Method I, Step 3): Weather conditions are not appropriate to significantly raise the temperature of the seawater in the tub.
Solution: Use an aquarium heater (100 W) to heat the water.

Problem (Method I, Step 10): Larval settlement rate is low.
Solution: Settlement occurs at a higher rate on certain types of plastic. Use new deep Petri dishes or test various plastic dishes for settlement rate. If desired, add small shards of coral rubble collected from the natural habitat of *Amphimedon* to induce settlement. These shards are typically a decaying calcium carbonate substratum consisting of crustose coralline algae, the boring sponge *Cliona* sp., and undescribed biofilm biota. Alternatively, Petri dishes "aged" in running seawater for 1–4 days develop a natural biofilm that also can induce settlement. Note that whereas induction increases settlement rates, it can also lower the chances of postlarval/juvenile survival because of the higher risk of bacterial and protozoan infections.

Problem (Method I, Step 12): Settling larvae or settled postlarvae arrest development and appear to degenerate.
Solution: Avoid contaminating the FSW with organisms introduced from seawater. Larvae intended for settlement should be transferred to fresh FSW several times with a Pasteur pipette to dilute seawater carried over from the initial collection. Check cultures regularly under a dissecting microscope for the presence of contaminants. If possible, avoid using coral rubble or biofilm to induce metamorphosis.

Problem (Method II, Step 2): Brood chambers only contain late embryonic stages.
Solution: This can occur when sponges are kept in an open system aquarium for more than 1 week. Collect new sponges from the field or keep sponges in a closed system.

Protocol 2

Whole-mount In Situ Hybridization

Developmental gene expression is analyzed predominantly via whole-mount in situ hybridization using digoxigenin-labeled RNA probes. The following protocol (Larroux et al. 2006), adapted from Hinman and Degnan (2000) and Finnerty et al. (2003), describes how to perform this procedure in *Amphimedon*, including fixation, hybridization, and sectioning of embryonic, larval, and postlarval juvenile stages.

MATERIALS

Recipes for the items marked with <R> begin on page 161.

CAUTION: See the Cautions Appendix for appropriate handling of materials marked with <!>.

Reagents

Acetic anhydride <!>
Alkaline phosphatase (AP) buffer <R>
Amphimedon developmental specimens
> Collect and process Amphimedon *embryos, larvae, and juveniles as described in Protocol 1.*

Antibody (antidigoxigenin, alkaline phosphatase–conjugated, Fab fragments; Roche)
AP buffer (Mg-free) <R>
Benzyl alcohol <!>
Benzyl benzoate <!>
Blocking Reagent (10% [Roche], prepared in MAB) <R>
> *Aliquot and store at –20°C. Dilute to 2% with MAB <R> before use. (The 2% blocking solution can be stored for up to 1 month at 4°C.)*

Color reagent 1 <R>
Color reagent 2 <R>
DIG (digoxigenin) RNA–labeling mix (Roche) <!>
DNA template (e.g., polymerase chain reaction [PCR] product; for gene of interest)
Epon (ProSci Tech) <!>
Ethanol (50, 70, 90, and 100%) <!>
Fixative <R>
Formamide (deionized) <!>
Glycine
Hybridization buffer (HB) <R>
Maleic acid buffer (MAB) <R>
NaCl (5 M)
Observation reagent <R>
PBT <R>
Proteinase K (10 mg/ml) <!> (prepared in PBT <R> and stored as 5-μl aliquots at –80°C)
Seawater (filtered [FSW], 0.22 μm)
Shur/mount (Triangle Biomedical Sciences)

20x SSC (pH 7) <R>
> Add 0.1% Tween 20 when diluting to 4x SSC or 2x SSC.

Triethanolamine (TEA; 0.1 M, pH 8, stored as 10-ml aliquots at –20°C) <!>
Tween 20 (20% [Sigma], prepared in H$_2$O)
> Dilute to 0.05% before procedure.

Equipment

Biowave (Pelco)
Camera (digital, equipped with microscope mounts)
Coverslips (22 mm x 22 mm)
Dishes (tissue culture, 35 mm x 10 mm; Nunclon)
Embedding molds (flat)
Heating blocks, preset to 51°C and 80°C
Histological knife (diamond; Diatome)
Hybridization oven, preset to 51°C
Incubator, preset to 37°C
Microscope (compound, equipped with differential interference contrast [DIC] optics)
Microscope (dissecting)
Microtome (e.g., Leica Ultracut UCT)
Needle (25 gauge; Terumo)
Nutator
Oven, preset to 60°C
Pasteur pipettes
Plates (cell culture; 6, 12, and 24 well)
Rotator
Shaker
Slides (microscope; concave and flat)
Tissues (e.g., Kimwipes)
Tubes (microcentrifuge, 1.5 ml)
Watch glasses
Water bath (equipped with shaker), preset to 51°C

METHOD

Use RNase-free equipment and solutions for all steps from postfixation through the addition of the probe. Always keep embryos and juveniles in solution. Because embryos and larvae stick to plastic, use full tubes for washes and, when transferring samples, use a moistened glass Pasteur pipette or P1000 tip with its end cut off. All washes and incubations are for 5 minutes at room temperature unless stated otherwise.

Probe Preparation

Hybridizations have been successful using probes from 0.2 to 1.7 kb, but success is more readily obtained with probes larger than 1 kb.

1. Use 500 ng of template and the DIG RNA–labeling mix, according to manufacturer instructions, to synthesize probes. Include the ethanol precipitation but omit the DNase treatment.
2. Dilute the precipitated probe in 100 µl of HB.
   > *Synthesis generally yields 10–20 µg of probe (100–200 ng/µl), which can be assessed with a dot blot.*
3. Dilute aliquots of the stock 1:3 and 1:10 in HB. Store dilutions and undiluted stock at –80°C.

Fixation

4. Fix the tissues as appropriate for their developmental stage.

To Fix Embryos

Fixing brood chamber slices whole results in uneven fixation and inconsistent staining during whole-mount in situ hybridization.

   i. Collect individual embryos in a 6-well plate (<100 embryos/well). Rinse several times in FSW.
   ii. Tilt dish so that embryos are in a bottom corner. Remove as much FSW as possible without exposing the embryos to air.
   iii. Add 1 ml of fixative. Allow the embryos to settle. Immediately replace with 3–5 ml of fresh fixative.

To Fix Larvae

   i. Transfer larvae into a 6-well plate (<100 larvae/well) in a minimal amount of FSW.
   ii. Add several drops of fixative to immobilize larvae. Quickly remove as much solution as possible and replace with 3–5 ml of fixative.

To Fix Juveniles

   i. Process juveniles in a 24-well plate, generally one juvenile per well. Remove FSW from each well, leaving just enough liquid so that the osculum is not exposed to air. Add 300 µl of fixative.
   ii. Gently swirl the plate for 5 seconds. Immediately replace with 300 µl of fresh fixative.

5. Fix samples on a slow shaker for 1 hour at room temperature.
   During the first 30 minutes, use a moistened Pasteur pipette to periodically mix embryos to prevent clumping.

6. After fixation, transfer embryos and larvae to 1.5-ml tubes. Leave juveniles in 24-well plates.
   Perform all subsequent washes on a nutator at room temperature. See Troubleshooting.

7. Exchange half of the fixative with 70% ethanol. Wash for 5 minutes. Repeat once.
8. Wash twice in 70% ethanol for 5 minutes.
9. Wash twice in 70% ethanol for 10 minutes.
   At this point, samples can be stored at –20°C.

Rehydration

For embryos and larvae, start at Step 10. For juveniles, skip Steps 10–11 and proceed to Step 12.

10. Under a dissecting microscope, sort fixed embryos and larvae in a dish of 70% ethanol. Separate developmental stages and remove damaged specimens.
11. Transfer an appropriate number of samples into a 6-well plate.
    Samples should be organized into wells based on subsequent proteinase K treatments (see Step 14).
12. Wash all developmental stages through a graded series of decreasing ethanol concentrations into 100% PBT.
13. Wash four times in PBT.

Proteinase K Digestion

All stages are very fragile in proteinase K. Change solutions gently and mix very gently by hand at the start of each wash. Follow incubation times closely for proteinase K digestion.

14. Incubate samples in freshly made proteinase K solution for 9–11 minutes at 37°C.

    Each new batch of proteinase K must be tested: Generally, a small proportion of embryos should be damaged after digestion.

    i. For cleavage embryos, larvae, and juveniles, use a final proteinase K concentration of 5 µg/ml.

    ii. For other embryonic stages, use a final proteinase K concentration of 10 µg/ml.

15. Wash gently twice in fresh PBT containing 2 mg/ml of glycine.

Acetylation

All stages are very fragile in TEA. Change solutions gently and mix very gently by hand at the start of the wash. Follow incubation times closely for acetylation.

16. Wash twice in TEA.
17. Wash in freshly made TEA containing 1.5 µl/ml of acetic anhydride.
18. Wash in freshly made TEA containing 3 µl/ml of acetic anhydride.
19. Gently wash twice in PBT.

Postfixation

20. Postfix samples for 1 hour in fixative.
21. Wash five times in PBT.

Prehybridization

22. Transfer embryos and larvae from 6-well plates to 1.5-ml tubes. (Continue to process juveniles in 24-well plates).

    If desired, separate different stages into different tubes (corresponding to different probe concentrations).

23. Incubate in HB for 10 minutes.
24. Replace with fresh HB. Prehybridize in a gently shaking water bath (for embryos and larvae) or a hybridization oven (for juveniles) overnight at 51°C.

Hybridization

All stages are very fragile in HB. Change solutions gently and mix very gently by hand at the start of each wash.

25. Dilute probe in HB. Prepare approximately 50 µl/sample for embryos and larvae and 150 µl/sample for juveniles.

    When first testing a probe, use three different concentrations: 1 µl of undiluted stock in a 100 µl hybridization volume (1–2 ng/µl), 1:3 dilution (0.33–0.66 ng/µl), and 1:10 dilution (0.1–0.2 ng/µl). Generally, a signal develops for all three concentrations, but the signal-to-noise ratio will vary. Further probe concentration optimization may be necessary.

26. Remove the majority of HB from the samples, leaving about 50 µl in each tube/well.

    Do not pipette embryos and larvae when hot.

27. Denature the probe in a heating block for 10 minutes at 80°C.
28. Add probe in HB to samples (50 µl for embryos and larvae and 150 µl for juveniles).
29. Hybridize in a gently shaking water bath (for embryos and larvae) or a hybridization oven (for juveniles) for 24–72 hours at 51°C.

Probe Rinse

Timing is crucial for these washes. Embryos and larvae are fragile in formamide and sink slowly; periodically mix gently during incubation and allow mixture to settle before the next change.

30. Incubate in 50% formamide:4x SSC three times for 15 minutes at room temperature.
31. Incubate in 50% formamide:4x SSC three times for 10 minutes at 51°C.
32. Incubate in 50% formamide:2x SSC twice for 10 minutes at 51°C.
    To reduce stringency, only wash once with formamide:2x SSC. To increase stringency, perform an additional incubation in 50% formamide:1x SSC.
33. Wash in 2x SSC three times for 20 minutes.

Anti-DIG Antibody Incubation

34. Wash in 2x SSC:MAB for 10 minutes.
35. Wash twice in MAB.
36. Block tissues in 2% blocking solution for 1–2 hours on a nutator.
37. Incubate samples in AP-conjugated anti-DIG antibody (freshly diluted 1:5000 in 2% blocking buffer) on a nutator overnight at 4°C.

Color Development

Amphimedon *embryos and larvae are opaque. When staining is internal, expression patterns are not readily observed when developing and will only become apparent when the embryos are cleared and mounted for observation under a compound microscope. Embryos and larvae must be stained longer if sectioning is intended. See Troubleshooting.*

38. Wash four times in MAB over 2 hours.
39. Transfer embryos and larvae into a 12-well plate. (Continue to process juveniles in 24-well plates).
40. Wash twice in Mg-free AP buffer on a shaker table.
41. Wash twice in freshly made AP buffer for 10 minutes on a shaker table.
    Samples are fragile and sticky in AP buffer. Monitor time carefully.
42. Mix equal volumes of color reagents 1 and 2 (color solution).
    Prepare and filter (0.45 µm) just before use. Keep in the dark.
43. Replace the AP buffer with freshly prepared color solution. Develop in the dark at 4°C on a slow shaker table.
    See Troubleshooting.
44. Monitor color development under a dissecting microscope after 15 minutes.
    If long periods of observations are necessary, to avoid excess exposure of embryos and larvae to light and room temperature, mix equal volumes of color reagent 1 and observation reagent just before use. Transfer samples temporarily to a small dish of this solution for observation and then return them to the color solution.

i. Monitor color development hourly for up to 6 hours.

ii. Monitor color development daily for up to 1 week.
Replace the color solution daily.

45. Remove samples when stained as desired. Wash twice in MAB on a wheel to stop reaction.
46. Store in PBT for up to 1 week at 4°C.

Postfixation

47. Postfix in fixative for 1 hour.
48. Wash five times in PBT.
Samples can be stored for several months at 4°C.

Microscopy

49. Observe and photograph juveniles from Step 48 under a dissecting microscope.
In certain cases, photography of whole-mount embryos and larvae in watch glasses under a dissecting microscope also can be informative.

50. Observe whole-mount specimens from Step 48 with DIC optics or embed and section the specimens for light microscopy.

Whole-mount Observation with DIC Optics

i. Wash five times in H$_2$O containing 0.05% Tween.

ii. Wash samples through a graded series of increasing ethanol concentrations (50, 70, 90, and 100% ethanol).

iii. Wash four times in 100% ethanol.

iv. Transfer samples in 1 ml of ethanol to a clean, dry watch glass.
Avoid storing in ethanol.

v. Clear samples in freshly made benzyl benzoate:benzyl alcohol (2:1). Monitor clearing under a dissecting microscope; it will take approximately 3–5 minutes.
Do not dehydrate or clear juveniles in benzyl benzoate:benzyl alcohol because they will lose their stain; clearing in glycerol can be useful for juveniles. Avoid excess exposure to light after clearing.

vi. Once cleared, mount and photograph samples within 1–2 hours.
- Mount cleavage embryos in 200 µl on a concave slide with an elevated coverslip.
- Mount larvae and other embryological stages in 100 µl on a concave slide or on a slide with an elevated coverslip. Remove excess benzyl benzoate:benzyl alcohol with a tissue.
- Mount juveniles in 200 µl on a slide with an elevated coverslip.

vii. Photograph under a compound microscope with DIC optics.
Samples can be rolled by gently pushing the coverslip.

Embedding and Sectioning for Light Microscopy

Use the Biowave (Pelco), according to manufacturer recommendations, to perform embedding and infiltration.

i. Transfer samples into 6-well plates containing enough 50% ethanol to completely cover.

ii. Place the plate into the Biowave for 40 seconds at 250 watts with no vacuum.

iii. Remove as much ethanol as possible, replacing with 70% ethanol.
This must be performed quickly to ensure that samples do not dry out.

iv. Repeat the previous two steps, once with 90% ethanol and twice with 100% ethanol.

v. Remove as much ethanol as possible. Replace with Epon/ethanol (1:1). Place in the Biowave for 3 minutes at 250 watts with vacuum.

vi. Remove as much Epon/ethanol as possible. Replace with 100% Epon. Place in the Biowave for 3 minutes at 250 watts with vacuum.

vii. Repeat the previous step with fresh 100% Epon.

viii. Use Pasteur pipettes to transfer samples into flat embedding molds. Fill molds completely with Epon. Use a 25-gauge needle to orient samples under a dissecting microscope.

ix. Allow samples to polymerize for 3 days in a 60°C oven.

x. Trim the faces of the Epon blocks into trapezoids, ensuring that the top and bottom of the trapezoids are parallel.

xi. Use a microtome to cut 5-µm-thick sections from the blocks.

xii. Transfer groups of floating sections onto drops of H_2O on a slide. Place the slide on hot plate until the H_2O evaporates.

xiii. Use Shur/mount to mount slides with a coverslip. Allow slides to dry completely (in a flat position) before viewing by light microscopy.

TROUBLESHOOTING

Problem: Embryos and larvae stick to insides of Pasteur pipettes or plastic pipette tips or become damaged during pipetting.

Solution: Precoat fresh pipette or tips with relevant solution before pipetting embryos and larvae. Cut the ends of plastic pipettes or tips to ensure that the opening is wide enough to cause minimal damage to embryos and larvae during transfer. To dislodge an embryo from the pipette/tip, draw liquid and air bubbles up past the embryo or fill the pipette/tip with liquid and tap the outside.

Problem (Step 6): Larval morphology is not preserved correctly after fixation. Larvae appear shriveled (but still intact) or burst at the anterior pole.

Solution: Preservation of larval morphology appears to be particularly sensitive to alterations in salt concentrations and the overall strength of the fixative. Ensure fixative and fixative buffers are prepared carefully and that all fixative stock solutions are freshly made. Damage to larvae can also occur mechanically during fixation and wash steps. Make sure that enough liquid is present in the tube to allow for only gentle movements of the larvae within.

Problem (Step 43): Color solution is too viscous to allow for easy pipetting or turning of embryos and larvae.

Solution: Allow plate or dish to come to room temperature before attempting to manipulate the embryos.

Problem (Steps 43–50): Crystals form in solutions and adhere to embryos and larvae during color development and subsequent processing steps.

Solution: Replace the relevant solution with a freshly made 0.45-µm filtered solution. Filter color reagent 2 before adding color reagent 1.

Protocol 3

Analysis of Cell Movement by Injection of Fluorescent Tracers

Although attempts to culture prepigmentation stage embryos (i.e., blastulas and early gastrulas) outside of brood chambers have so far been unsuccessful in *Amphimedon*, it is possible to manipulate embryos within the brood chamber and follow their development under laboratory conditions. This protocol describes microinjection of lipophilic tracers such as DiI into embryos embedded in their native brood chamber (Figs. 1 and 3). DiI does not appear to perturb embryonic development and is relatively resistant to photobleaching. As long as care is taken not to damage the fragile embryos during observation and photography, the same embryo can be photographed multiple times, permitting its development to be tracked (up to 4 weeks) from early cleavage stages to hatching of free-swimming parenchymella larvae. The embryos or larvae also can be fixed (as described in Protocol 2) without loss of fluorescence. This method also can be used to deliver other types of solutions to embryos or individual cells of early embryos.

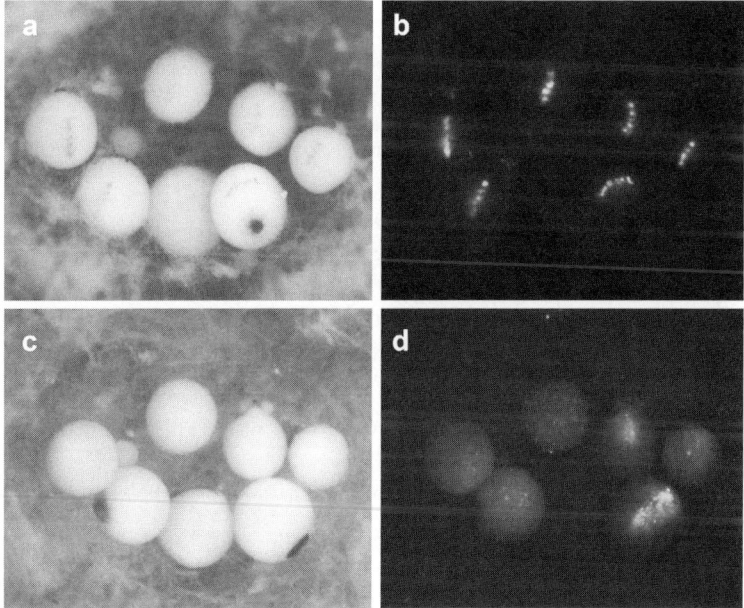

FIGURE 3. Labeling cells on the surface of the *Amphimedon* embryo with DiI. (*A,C*) light micrographs; (*B,D*) corresponding epifluorescence micrographs. (*A,B*) Initial "tattooing" of the surface of embryos at different stages with DiI. (*C,D*) Migration of labeled cells after 48 hours.

MATERIALS

CAUTION: See the Cautions Appendix for appropriate handling of materials marked with <!>.

Reagents

Agarose

Amphimedon brood chamber slices (2–3-mm thick)

Collect and prepare from adult Amphimedon as described in Protocol 1. Take care not to expose the slices to air and to avoid squeezing the slices while cutting. Firm, uniform slices with large margins of healthy sponge tissue surrounding the exposed brood chamber have a better chance of survival in culture.

CellTracker CM-DiI (Molecular Probes)

Dimethyl sulfoxide (DMSO) <!>

Seawater (filtered [FSW], 0.22 µm)

Equipment

Bunsen burner
Fishing line (thin)
Forceps (fine; e.g., Dumont #5)
Injector (equipped with pressurized nitrogen cylinder and regulator; e.g., Harvard Apparatus)
Light source (cold)
Magnetic stand (GJ-1)
Microcentrifuge (tabletop)
Micromanipulator and joystick (e.g., Narishige, MN-151)
Micromanipulator (oil hydraulic, single axis; e.g., Narishige, MMO-220A)
Needle (teasing, curved)
Parafilm
Pasteur pipettes
Petri dishes (plastic, including deep Petri dishes 90 x 25 mm; e.g., Labserv, LBS61014)
Pins (stainless steel)
Pipette grinder (e.g., Narishige, EG-44)
Pipette puller (e.g., Narishige, PC-10)
Stereomicroscope (fluorescence, equipped with digital camera; e.g., Zeiss Lumar)
Tubes (microcentrifuge, 1.5 ml)
Tubing (capillary, borosilicate glass, with inner filament; e.g., Harvard Apparatus, 30-0038)

METHOD

Preparation of Injection Capillaries and Solutions

The quality of capillaries is critical for successful injection, and the optimal parameters for pulling and sharpening must be established by trial and error. The capillary should be thin enough to avoid damaging the embryo, with straight shanks, especially for injections deeper within the embryos.

1. Pull injection capillaries from thin walled borosilicate glass tubing with inner filament.

2. To round the edges, briefly flame the back end of each capillary over a Bunsen burner.

3. Use a pipette grinder to open and sharpen the injection tip of the capillary at an approximately 30° angle.

4. Inspect each capillary under 400x magnification. The tip should be sharp, with an external diameter of approximately 5 μm.

   The tip should be sharpened for easy piercing of the follicle membrane, with the opening small enough to avoid uncontrolled flow of the dye solution, yet large enough to avoid clogging.

5. Store pulled capillaries in a dust-free environment until ready to use.

6. Prepare 2 mg/ml CellTRacker CM-DiI in DMSO. Mix well and store in the dark at −20°C.

   The solution is stable for at least several weeks.

Injection of Embryos

7. Affix a 2–3-mm brood chamber slice to a deep Petri dish containing a thin base of 2% agarose in FSW. Secure the slice to the agarose with sharpened entomological pins.

8. To remove damaged embryos from the slice, gently blow FSW through a Pasteur pipette with a rounded tip.

9. Briefly centrifuge the DiI solution to remove larger particles.

   Use only the top portion of the solution for capillary loading.

10. Load the DiI solution into capillaries by either backfilling or positioning a drop of solution on Parafilm and using the vacuum function of the injector to draw it into the capillary.

    The dye solution is intensely pink and visible in bright light if concentrated. To be able to see small volumes or a diluted sample, use a stereomicroscope with a rhodamine epifluorescence filter set.

11. Position the dish with the brood chamber slice under a stereomicroscope and make sure that the pins are not in the way of the capillary. Lower the capillary into water.

    Correct the balance pressure if the dye solution is free-flowing or if water appears to be entering the capillary. Ensure that the injection pressure and time are sufficient; try injecting dye into the water with very short bursts so that the capillary is not clogged. Correct the pressure or injection time if necessary or use the "clean" function if the capillary appears to be clogged.

12. Use a three-axis manipulator to maneuver the capillary into the vicinity of the selected embryo. Use the fine hydraulic manipulator to insert the capillary into the embryo.

13. Use short bursts to inject the desired volume. If necessary, use fluorescence to aid visualization.

14. If labeling of a larger area is required, withdraw the capillary after delivering the desired volume and shift and reinsert the capillary in an adjacent area.

    It is easy to "tattoo" stripes and dots on the embryos using this method (see Fig. 3).

15. When all targeted embryos in a given slice are injected, photograph the slice in bright light and under fluorescence.

16. Store the slice in FSW for several hours at room temperature.

17. For longer experiments, place slices in a darkened aquarium with running seawater as follows:

    i. Thread thin fishing line through a natural opening in the slice.

    ii. Affix the slice to a rack or netting in the tank.

       The slice should not lie on the bottom nor should it float on the surface. Within approximately 12 hours, regenerating tissue will begin to cover the exposed embryos in the brood chamber.

18. To observe and photograph the injected embryos, use fine forceps or dissecting needles to gently remove regenerated tissue from the surface of the brood chamber.

    Use extreme care when removing the tissue to avoid damage to the embryos.

Protocol 4

Genotyping Individual Embryos, Larvae, and Adults

The distribution of *A. queenslandica* is patchy on coral reefs in the GBR, with small, localized populations detected in shallow, still water reef-flat environments. *A. queenslandica* is a spermcast spawner, in which fertilization occurs internally. Sperm presumably originate from neighboring reproductive individuals within the population. The ability to genotype individual embryos within a single brood chamber has the potential to shed light on the fertilization biology and generation/maintenance of genetic diversity in this sessile invertebrate. Here, we describe a protocol for rapidly genotyping individuals using polymorphic microsatellite loci. The loci are amplified by PCR using a pair of primers specifically designed for the region of interest with a fluorescent dye attached to the 5' end to enable easy detection of the amplified product. An advantage of this procedure is that fluorescently labeled PCR products can be combined (i.e., multiplexed) to reduce time and cost when using the genotyping machine. The dye label and size of the product must be taken into consideration when multiplexing. For example, three differently labeled PCR products can be multiplexed or PCR products with the same label can be multiplexed as long as the allelic size ranges do not overlap. The amount of each cleaned, labeled PCR product added to the multiplex must be optimized depending on the dye and the PCR efficiency.

MATERIALS

The recipes for items marked with <R> begin on page 161.

CAUTION: See the Cautions Appendix for appropriate handling of materials marked with <!>.

Reagents

Agarose gel and electrophoresis buffer (e.g., 1x TAE or 0.5x TBE; see Step 11)
Ammonium acetate (7.5 M)
Amphimedon adults
Bovine serum albumin (BSA) (optional; see Step 5)
DNeasy Tissue Kit (QIAGEN)
dNTPs (10 mM)
Dry ice <!>
Ethanol (70% and 100%) <!>
H_2O (RNase- and DNase-free)
Lysis buffer <R> (prewarmed to 55°C)

> Add proteinase K to lysis buffer just before use (i.e., for each sample, add 90 µl of lysis buffer and 3.6 µl of 10 mg/ml proteinase K).

10x PCR buffer
Primers (microsatellite specific)

> Label the forward primer during synthesis (e.g., by Sigma Oligos) with a fluorescent dye (e.g., Hex, Fam, Tet) attached to the 5' end.

Proteinase K (10 mg/ml) <!> (prepared in H_2O)
Seawater (filtered [FSW], 0.22 µm)
Taq polymerase

Equipment

Centrifuge and plate
Dissecting instruments
Gel electrophoresis equipment
Genotyping machine and software (e.g., MegaBACE 1000)
Grinder (Eppendorf)
Heating blocks, preset to 99°C
Microcentrifuge
Shaker table, preset to 55°C
Thermal cycler
Tubes (microcentrifuge, 0.6 ml)
Tubes (PCR, 0.2 ml)

METHOD

Sample Collection

1. Collect tissue samples of the developmental stage of interest from adult sponges.

For Embryos and Larvae

   i. Dissect brood chambers as described in Protocol 1.

   ii. Transfer each individual embryo or larva to 10 µl of FSW in a 0.6-ml microcentrifuge tube and freeze at –80°C as soon as possible after removal from the maternal tissue.

For Adult Tissue

   i. Dissect tissue surrounding brood chambers from an adult sponge.

   ii. Transfer to a 1.5-ml microcentrifuge tube and freeze at –80°C as soon as possible after dissection.

2. Place samples on dry ice to transfer them from the field site back to the laboratory. Store at –80°C until extraction.

Quick Total Genomic DNA Extraction

3. Extract DNA from the tissues of interest.

For Individual Embryos and Larvae

   i. Add 90 µl of prewarmed lysis buffer to a microcentrifuge tube containing a single embryo or larva in 10 µl of FSW.

   ii. Pipette up and down several times to immediately lyse the sample. Remove until it completely dissolves.

   iii. Incubate in a heating block for 10 minutes at 99°C to inactivate the proteinase K.

   iv. Centrifuge at 13,000 rpm in a microcentrifuge for 15 minutes at room temperature.

   v. Transfer the supernatant to a fresh tube. Store at –20°C.

For Adult Sponge Tissue

Extract DNA from adult sponge tissue using the "animal tissue protocol" from the DNeasy Tissue Kit (QIAGEN), with the following modifications:

   i. In Step 1 of the manufacturer's protocol, use approximately 25 mg of starting material. Place the tissue in a 1.5-ml microcentrifuge tube with 180 μl of buffer ATL. Use a grinder to squeeze the adult tissue to expose all of the cells.

   ii. In Step 2 of the manufacturer's protocol, add 40 μl of 10 mg/ml proteinase K. Place on a shaker table at 70 rpm for 1 hour at 55°C.

   iii. In Step 3 of the manufacturer's protocol, after shaking, squeeze the adult tissue again with a grinder. Remove the sponge matrix from the tube.

   iv. Perform the remaining steps of the manufacturer's protocol on this sponge tissue lysate.

Microsatellite Amplification

4. Use a thermal cycler to perform PCR by standard methods in 0.2-ml PCR tubes. Each 25-μl reaction should contain fluorescently tagged microsatellite-specific primers, 1 μl of template (from Step 3), and appropriate concentrations of dNTPs, PCR buffer (with $MgCl_2$), and Taq polymerase.

   Optimize PCR conditions for each set of primers. This can include adjusting the annealing temperature (54–58°C), $MgCl_2$ concentration (1–3 mM), additives (e.g., BSA), and the type of Taq polymerase used.

   The size of the amplified loci ranges from 180 to 400 bp.

5. To precipitate the labeled product, add 150 μl of 100% ethanol (i.e., six times the volume of the PCR reaction) and 5 μl of 7.5 M ammonium acetate (0.2x volume).

6. Centrifuge in a plate centrifuge at 4000 rpm for 30 minutes at room temperature.

7. Decant the supernatant. Wash the pellet with 150 μl of 70% ethanol.

8. Centrifuge at 4000 rpm for 5 minutes at room temperature.

9. Centrifuge the plate upside down at 1000 rpm for 1 minute to remove the supernatant entirely.

10. Air-dry the pellet and redissolve it in H_2O.

    Determine the volume empirically, depending on the dye and PCR efficiency. Note that Hex-labeled products should be redissolved in less water because this fluorescent dye has a weaker signal.

11. Visualize PCR products on an agarose gel to determine PCR efficiency.

12. Run the samples on the genotyping machine.

13. Analyze the output with genotyping software, using the adult samples to indicate the maternal alleles.

    To correctly analyze the data, the strength of the signal from each labeled PCR product must be similar.

    When genotyping very young individuals (i.e., white or brown stage embryos), three alleles are sometimes detected. These represent both maternal alleles and a paternally contributed third allele. This is the result of maternal tissue from the brood chamber still being attached to the embryo at the time of collection and genomic DNA extraction.

Recipes

The recipes for items marked with <R> are listed here.

CAUTION: See the Cautions Appendix for appropriate handling of materials marked with <!>.

ALKALINE PHOSPHATASE (AP) BUFFER

Reagent	Quantity (for 100 ml)	Final concentration
5x NaCl-Tris solution <R>	20 ml	1x
$MgCl_2$ (1 M) <!>	5 ml	50 mM
Tween 20 (20%)	500 µl	0.1%
H_2O	to 100 ml	

Filter (0.45 µm) before use. Prepare just before use.

AP BUFFER, Mg-FREE

Reagent	Quantity (for 100 ml)	Final concentration
5x NaCl-Tris solution <R>	20 ml	1x
Tween 20 (20%)	500 µl	0.1%
H_2O	to 100 ml	

Keep for up to 1 week. Filter (0.45 µm) before use.

COLOR REAGENT 1

Reagent	Quantity (for 100 ml)	Final concentration
5x NaCl-Tris solution <R>	20 ml	1x
Polyvinyl alcohol	14 g	14%
H_2O	to 100 ml	

Microwave on low power (with little boiling) and shake until dissolved. Keep for up to 1 month.

COLOR REAGENT 2

Reagent	Quantity (for 10 ml)	Final concentration
5-bromo-4-chloro-3'-indolyl-phosphate, p-toluidine salt (BCIP) (50 mg/ml) <!>	66 µl	330 µg/ml
$MgCl_2$ (1 M) <!>	1 ml	100 mM
5x NaCl-Tris solution <R>	2 ml	1x
Nitro-blue tetrazolium chloride (NBT) (100 mg/ml) <!>	66 µl	660 µg/ml
Tween 20 (20%)	100 µl	0.2%
H_2O	to 10 ml	

Prepare just before use. Combine all ingredients except for BCIP and NBT. Add NBT; mix. Add BCIP; mix. Filter and keep in the dark.

100x DENHARDT'S SOLUTION

Reagent	Quantity (for 50 ml)	Final concentration
Bovine serum albumin (Fraction V)	1 g	2% (w/v)
Ficoll 400	1 g	2% (w/v)
Polyvinylpyrrolidone (PVP)	1 g	2% (w/v)
H_2O	to 50 ml	

Dissolve the components in the H_2O. Filter to sterilize and remove particulate matter. Divide into aliquots and store at –20°C.

FIXATIVE

Reagent	Quantity (for 100 ml)	Final concentration
Glutaraldehyde (25%) <!>	20 µl	0.05%
5x MOPS buffer <R>	20 ml	1x
Paraformaldehyde (20%) <!>	20 ml	4%
H_2O	to 100 ml	

Filter just before use. Store at 4°C.

HYBRIDIZATION BUFFER (HB)

Reagent	Quantity (for 50 ml)	Final concentration
Denhardt's solution (100x) <R>	500 µl	1x
EDTA (0.5 M, pH 8)	500 µl	5 mM
Formamide, deionized <!>	25 ml	50%
Heparin (50 mg/ml) <!>	100 µl	100 µg/ml
20x SSC <R>	12.5 ml	5x
tRNA (10 mg/ml), from yeast	500 µl	100 µg/ml
Tween 20 (20%)	250 µl	0.1%
H_2O	to 50 ml	

Store at –20°C.

LYSIS BUFFER

Reagent	Quantity (for 50 ml)	Final concentration
KCl (0.5 M) <!>	5 ml	50 mM
Tris-Cl (1 M, pH 9.0)	0.5 ml	10 mM
Triton X-100 <!>	100 µl	0.1%
H_2O	to 50 ml	

Store at room temperature.

MALEIC ACID BUFFER (MAB)

Reagent	Quantity (for 50 ml)	Final concentration
Maleic acid (1 M, pH 7.5) <!>	5 ml	0.1 M
NaCl (5 M)	1.25 ml	0.15 M
Tween 20 (20%)	250 µl	0.1%
H_2O	to 50 ml	

5x MOPS BUFFER

Reagent	Quantity (for 100 ml)	Final concentration (5x)
$MgSO_4$ <!>	0.1204 g	10 mM
MOPS <!>	10.463 g	0.5 M
NaCl	14.61 g	2.5 M

Dissolve the reagents listed above in 50 ml of H_2O. Adjust the pH to 7.5 with NaOH and then add H_2O to 100 ml. Filter to sterilize and store in the dark.

5x NaCl-TRIS SOLUTION

Reagent	Quantity (for 50 ml)	Final concentration (5x)
NaCl (5 M)	5 ml	0.5 M
Tris (1 M, pH 9.5) <!>	25 ml	0.5 M
H_2O	to 50 ml	

OBSERVATION REAGENT

Reagent	Quantity (for 10 ml)	Final concentration
5x NaCl-Tris solution <R>	2 ml	1x
$MgCl_2$ (1 M)	1 ml	100 mM
Tween 20 (20%)	100 µl	0.2%
H_2O	to 10 ml	

Prepare just before use.

10x PBS

Reagent	Quantity (for 1 liter)	Final concentration (10x)
NaCl	80 g	1.37 M
KCl <!>	2 g	27 mM
Na_2HPO_4	14.4 g	100 mM
KH_2PO_4	1.3609 g	10 mM

Dissolve the reagents listed above in 800 ml of H_2O. Adjust the pH to 7.4 with HCl <!> and then add H_2O to 1 liter. Dispense the solution into aliquots and sterilize them by autoclaving for 20 minutes at 15 psi (1.05 kg/cm^2) on liquid cycle or filter-sterilizing. Store at room temperature.

PBT

Reagent	Quantity (for 50 ml)	Final concentration
10x PBS <R>	5 ml	1x
Tween 20 (20%)	250 µl	0.1%
H_2O	to 50 ml	

20x SSC (pH 7.0)

Reagent	Quantity (for 1 liter)	Final concentration (20x)
NaCl	175.3 g	3.0 M
Sodium citrate <!>	88.2 g	0.3 M

Dissolve the ingredients in 800 ml of H_2O. Adjust the pH to 7.0 with a few drops of 14 N HCl <!> and then adjust the final volume to 1 liter with H_2O. Dispense into aliquots. Autoclave to sterilize.

ACKNOWLEDGMENT

This research has been supported by grants from the Australian Research Council.

REFERENCES

Adamska, M., Degnan, S.M., Green, K.M., Adamski, M., Craigie, A., Larroux, C., and Degnan, B.M. 2007a. Wnt and TGF-β expression in the sponge *Amphimedon queenslandica* and the origin of metazoan embryonic patterning. *PLoS ONE* **2:** e1031.

Adamska, M., Matus, D.Q., Adamski, M., Green, K.M., Rokhsar, D.S., Martindale, M.Q., and Degnan, B.M. 2007b. The evolutionary origin of hedgehog proteins. *Curr. Biol.* **17:** R836–R837.

Borchiellini, C., Chombard, C., Manuel, M., Alivon, E., Vacelet, J., and Boury-Esnault, N. 2004. Molecular phylogeny of Demospongiae: Implications for classification and scenarios of character evolution. *Mol. Phylogenet. Evol.* **32:** 823–837.

Degnan, S.M. and Degnan, B.M. 2006. The origin of the pelagobenthic metazoan life cycle: What's sex got to do with it? *Integr. Comp. Biol.* **46:** 683–690.

Degnan, B.M., Leys, S.P., and Larroux, C. 2005. Sponge development and antiquity of animal pattern formation. *Integr. Comp. Biol.* **45:** 335–341.

Dellaporta, S.L., Xu, A., Sagasser, S., Jakob, W., Moreno, M.A., Buss, L.W., and Schierwater, B. 2006. Mitochondrial genome of *Trichoplax adhaerens* supports Placozoa as the basal lower metazoan phylum. *Proc. Natl. Acad. Sci.* **103:** 8751–8756.

Desqueyroux-Faúndez, R. and Valentine, C. 2002. Family Niphatidae Van Soest. In *Systema Porifera: A guide to the classification of sponges* (eds. J.N.A. Hooper and R.W.M. Van Soest), vol. 1, pp. 874–890. Kluwer Academic/Plenum Publishers, New York.

Dunn, C.W., Hejnol, A., Matus, D.Q., Pang, K., Browne, W.E., Smith, S.A., Seaver, E., Rouse, G.W., Obst, M., Edgecombe, G.D., et al. 2008. Broad phylogenomic sampling improves resolution of the animal tree of life. *Nature* **452:** 745–749.

Fahey, B., Larroux, C., Woodcroft, B., and Degnan, B.M. 2008. Does the high gene density in the sponge NK homeobox gene cluster reflect limited regulatory capacity? *Biol. Bull.* **214:** 205–217.

Finnerty, J.R., Paulson, D., Burton, P., Pang, K., and Martindale, M.Q. 2003. Early evolution of a homeobox gene: The parahox gene *Gsx* in the Cnidaria and the Bilateria. *Evol. Dev.* **5:** 331–345.

Gauthier, M. and Degnan, B.M. 2008a. The transcription factor NF-κB in the sponge *Amphimedon queenslandica*: Insights into the evolutionary origin of the Rel homology domain. *Dev. Genes Evol.* **218:** 23–32.

Gauthier, M. and Degnan, B.M. 2008b. Partitioning of genetically distinct cell populations in chimeric juveniles of the sponge *Amphimedon queenslandica*. *Dev. Comp. Immunol.* **32:** 1270–1280.

Hinman, V.F. and Degnan, B.M. 2000. Retinoic acid perturbs *Otx* gene expression in the ascidian pharynx. *Dev. Genes Evol.* **210:** 129–139.

Hooper, J.N.A. and Van Soest, R.W.M. 2002. *Systema Porifera: A guide to the classification of sponges*. Kluwer Academic/Plenum Publishers, New York.

Hooper, J.N.A. and Van Soest, R.W.M. 2006. A new species of *Amphimedon* (Porifera, Demospongiae, Haplosclerida, Niphatidae) from the Capricorn-Bunker Group of Islands, Great Barrier Reef, Australia: Target species for the "sponge genome project." *Zootaxa* **1314:** 31–39.

Hooper, J.N.A. and Wiedenmayer, F. 1994. *Zoological Catalogue of Australia: Porifera*. CSIRO Publishing, Collingwood, Australia.

Larroux, C., Fahey, B., Liubicich, D., Hinman, V.F., Gongora, M., Green, K.M., Wörheide, G., Leys, S.P., and Degnan, B.M. 2006. Developmental expression of transcription factor genes in a demosponge: Insights into the origin of metazoan multicellularity. *Evol. Dev.* **8:** 150–173.

Larroux, C., Fahey, B., Degnan, S.M., Adamski, M., Rokhsar, D.S., and Degnan, B.M. 2007. *NK* homeobox gene cluster predates the origin of *Hox* genes. *Curr. Biol.* **17:** 706–710.

Larroux, C., Luke, G.N., Koopman, P., Rokhsar, D.S., Shimeld, S.M., and Degnan, B.M. 2008. Genesis and expansion of metazoan transcription factor gene families and classes. *Mol. Biol. Evol.* **25:** 980–996.

Leys, S.P. and Degnan, B.M. 2001. The cytological basis of photoresponsive behavior in a sponge larva. *Biol. Bull.* **201:** 323–338.

Leys, S.P. and Degnan, B.M. 2002. Embryogenesis and metamorphosis in a haplosclerid demosponge: Gastrulation and transdifferentiation of larval ciliated cells to choanocytes. *Invert. Biol.* **121:** 171–189.

Leys, S.P. and Ereskovsky, A.E. 2006. Embryogenesis and larval differentiation in sponges. *Can. J. Zool.* **84:** 262–287.

Leys, S.P., Cronin, T.W., Degnan, B.M., and Marshall, J.N. 2002. Spectral sensitivity in a sponge larva. *J. Comp. Physiol. A* **188:** 199–202.

Leys, S.P., Cheung, E., and Boury-Esnault, N. 2006. Embryogenesis in the glass sponge *Oopsacas minuta*: Formation of syncytia by fusion of blastomeres. *Integr. Comp. Biol.* **46:** 104–117.

Maldonado, M. 2006. Ecology of the sponge larva. *Can. J. Zool.* **84:** 175–194.

Manuel, M. and Le Parco, Y. 2000. Homeobox gene diversification in the calcareous sponge, *Sycon raphanus*. *Mol. Phylogenet. Evol.* **17:** 97–107.

Nichols, S.A., Dirks, W., Pearse, J.S., and King, N. 2006. Early evolution of animal cell signaling and adhesion genes. *Proc. Natl. Acad. Sci.* **103**: 12451–12456.

Richards, G.S., Simionato, E., Perron, M., Adamska, M., Vervoort, M., and Degnan, B.M. 2008. Sponge genes provide new insight into the evolutionary origin of the neurogenic circuit. *Curr. Biol.* **18**: 1156–1161.

Sakarya, O., Armstrong, K.A., Adamska, M., Adamski, M., Wang, I.-F., Sachdu, D., Tidor, B., Degnan, B.M., Oakley, T., and Kosik, K.S. 2007. A post-synaptic scaffold at the origin of the animal kingdom. *PLoS ONE* **2**: e506.

Segawa, Y., Suga, H., Iwabe, N., Oneyama, C., Akagi, T., Miyata, T., and Okada, M. 2006. Functional development of Src tyrosine kinases during evolution from a unicellular ancestor to multicellular animals. *Proc. Natl. Acad. Sci.* **103**: 12021–12026.

Simionato, E., Ledent, V., Richards, G., Thomas-Chollier, M., Kerner, P., Coornaert, D., Degnan, B.M., and Vervoort, M. 2007. Origin and diversification of the basic helix-loop-helix gene family in metazoans: Insights from comparative genomics. *BMC Evol. Biol.* **7**: 33.

Taylor, M.W., Radax, R., Steger, D., and Wagner, M. 2007. Sponge-associated microorganisms: Evolution, ecology, and biotechnological potential. *Microbiol. Mol. Biol. Rev.* **71**: 295–347.

Wiens, M., Korzhev, M., Perovic-Ottstadt, S., Luthringer, B., Brandt, D., Klein, S., and Müller, W.E. 2007. Toll-like receptors are part of the innate immune defense system of sponges (demospongiae: Porifera). *Mol. Biol. Evol.* **24**: 792–804.

Wörheide, G., Solé-Cava, A.M., and Hooper, J.N.A. 2005. Biodiversity, molecular ecology and phylogeography of marine sponges: Patterns, implications and outlooks. *Integr. Comp. Biol.* **45**: 377–385.

WWW RESOURCES

http://www.bio.usyd.edu.au/OTI/ Homepage for the University of Sydney's One Tree Island Research Station.

http://www.environment.gov.au/cgi-bin/abrs/fauna/tree.pl?pstrVol=PORIFERA&pintMode=1 The Australian Faunal Directory is an online version of the Zoological Catalogue of Australia. The Porifera section is compiled and updated by J.N.A. Hooper and F. Wiedenmayer.

http://www.jgi.doe.gov/ The Joint Genome Institute, administered by the United States Department of Energy, has genomic resources for integrated high-throughput sequencing and computational analysis.

http://www.marine.uq.edu.au/index.html?page=54940 Homepage for the University of Queensland's Centre for Marine Studies Heron Island Research Station.

http://www.ncbi.nlm.nih.gov/Traces/trace.cgi? The Trace Archive at the National Center for Biotechnology Information provides a record of single-pass DNA sequencing reads for a variety of organisms.

7 Comb Jellies (Ctenophora)
A Model for Basal Metazoan Evolution and Development

Kevin Pang and Mark Q. Martindale

Kewalo Marine Laboratory, University of Hawaii, Honolulu, Hawaii 96813

ABSTRACT

Ctenophores, or comb jellies, are a group of marine organisms whose unique biological features and phylogenetic placement make them a key taxon for understanding animal evolution. These gelatinous creatures are clearly distinct from cnidarian medusae (e.g., "jellyfish"). Key features present in the ctenophore body plan include biradial symmetry, an oral–aboral axis that is delimited by a mouth and an apical sensory organ, two tentacles, eight comb rows composed of interconnected cilia, and thick mesoglea. Other morphological features include definitive muscle cells, a nerve net, basal lamina, a sperm acrosome, and light-producing photocytes. Aspects of their development made them attractive to experimental embryologists in as early as the 19th century. In recent years, their role as an invasive species has increased their study in ecology and fisheries-related fields. Although the phylogenetic placement of ctenophores with respect to other animals has proven to be difficult, it is clear that, along with poriferans, placozoans, and cnidarians, ctenophores are one of the earliest diverging extant animal groups. It is becoming clearer that it is important to determine whether some of the complex features of ctenophores are examples of convergence or whether they were lost in other animal branches. Many questions have yet to be answered but, because ctenophores are amenable to modern technical approaches, they could prove to be a highly useful emerging model.

> **PROTOCOLS**
> 1. *M. leidyi* Spawning and Embryo Collection, 177
> 2. Ctenophore Whole-mount Antibody Staining, 179
> 3. Ctenophore Whole-mount In Situ Hybridization, 182
> 4. Ctenophore Tissue Preparation and Extraction of DNA, 186
> 5. Ctenophore Tissue Preparation and Extraction of RNA, 189

BACKGROUND INFORMATION

Ctenophores are a group of animals found in nearly all marine waters, from coastal areas to the deep sea and from the tropics to the poles. They are more commonly known as comb jellies because they have a jelly-like appearance and distinctive rows of comb plates that are used for locomotion. Although they were once grouped with cnidarians as coelenterates (Hyman 1940), ctenophores are not true jellyfish and form a separate monophyletic group. Recent molecular phylogenetic studies have revealed that ctenophores are one of the earliest evolving extant animal phyla (Podar et al. 2001; Wallberg et al. 2004; Dunn et al. 2008). There are an estimated 200 described species of ctenophores;

however, with improved technology and sampling methods, many new species are being discovered in the deep sea (Haddock 2004). Adult morphologies are very diverse in shape and size, from the small and round "sea gooseberry" (*Pleurobrachia*) to the large and vermiform "Venus' girdle" (*Cestum*). However, some characteristics are present in nearly all ctenophores, including biradial symmetry, a unique cleavage program, comb rows composed of ctenes (linked cilia), an apical sensory organ, and two tentacles that bear colloblasts, which are specialized adhesive cells. Almost all ctenophores are capable of bioluminescence, producing flashes of light in photocytes beneath the comb plates. In addition, the refraction of light from the comb plates gives ctenophores their characteristic iridescent appearance. All ctenophores studied thus far have the same stereotyped cleavage program and go through a specific stage of development known as the cydippid larva, after which adult structures develop and diverge greatly among species. Nearly all ctenophores are hermaphroditic, possessing both eggs and sperm in separate gonads beneath the ctene rows. Although most ctenophores are entirely planktonic, members of one order, Platyctenida, possess a benthic adult form; these animals lose their comb plates and move by creeping on the substratum. All ctenophores are omnivorous, feeding on other plankton, except for the beroids: This group, the only ctenophore group that does not form tentacles, preys on other ctenophores.

Mnemiopsis leidyi (A. Agassiz 1865) belongs to the order Lobata, a group of ctenophores named for their conspicuous pair of oral lobes that are used for feeding. Lobate ctenophores are commonly called "sea walnuts." The natural habitat of *M. leidyi* is the Atlantic coast of North America (Mayer 1912) and South America. Within the *Mnemiopsis* genus, additional described species, *M. gardeni* and *M. mccradyi*, are now thought to be different morphs of *M. leidyi* (Seravin 1994). Being holoplanktonic, *M. leidyi* is a voracious omnivore. The larvae feed on ciliates, phytoplankton, and microzooplankton, and the adults feed on all types of zooplankton, even on fish eggs and fish larvae (Sullivan and Gifford 2007). Under optimal conditions, these self-fertile hermaphrodites can release up to 10,000 eggs per day and are capable of reproduction at 2 weeks of age (Baker and Reeve 1974). Consequently, during the summer months, they can appear in coastal waters in large blooms, making it easy to obtain large amounts of material. Cydippid larvae are approximately 0.5 mm at hatching; adults reach more than 12 cm. The main predators of *M. leidyi* include jellyfish medusae, some species of fish such as the butterfish, *Peprilus triacanthus*, as well as the cannibalistic ctenophore *Beroe*. Recently, *M. leidyi* was accidentally introduced into the Black Sea, the Caspian Sea, the North Sea, and the Baltic Sea, where, due to its high fecundity and lack of natural predators, it has wreaked havoc on local fisheries. As a result, in addition to its scientific value, there is great socioeconomic importance for studying *M. leidyi*. Because of its large size, fecundity, abundance in coastal areas, and recent introduction to European waters, *M. leidyi* is the most highly studied ctenophore. It is also popular in the public aquarium industry because of its hardiness and ease of collection compared to other ctenophores.

SOURCES AND HUSBANDRY

M. leidyi can normally be collected in Woods Hole, Massachusetts during the months of June to October, and it can also be found in estuarine areas along the Atlantic coast of North and South America. It tolerates a wide range of salinities. The main factors that influence its abundance are temperature, food availability, and mortality/predation (Kremer 1994; Sullivan et al. 2001). In a survey of U.S. estuaries, peak abundances varied geographically, with the northern areas (Narragansett Bay, Rhode Island and Chesapeake Bay, Maryland) peaking in the summer months and the southern areas (Biscayne Bay, Florida and Nueces Estuary, Texas) peaking in the fall to winter months (Kremer 1994). *M. leidyi* can now be found in European waters, most notably in the Black Sea, eastern Mediterranean, and Caspian Sea. More recently, it has been found in northern European waters, including the North Sea and Baltic Sea (Hansson 2006).

The culturing of *M. leidyi* (*M. mccradyi*) has been well described by Baker and Reeve (1974). Adults can be maintained in large aquaria with gentle aeration as long as they are well fed. The best food source is a variety of zooplankton, including copepods, mollusk veligers, and other crustacean larvae. It should be noted that *Artemia nauplii* alone is not sufficient for maintaining *M. leidyi* for long periods of time. If *M. leidyi* are starved, they will begin to absorb their oral lobes and slowly shrink in size. Overfeeding can also be a problem; this results in regurgitation of undigested food that can foul the water. Under the right conditions, adults can be spawned daily and multiple generations can be raised in the laboratory (see Protocol 1).

RELATED SPECIES

The two most commonly studied ctenophores in addition to *M. leidyi* are *Pleurobrachia pileus* and *Beroe ovata*, mainly because these are some of the more abundant ctenophores in coastal areas. *B. ovata* has the advantage of having extremely large (1-mm diameter) eggs. Another well-studied group of ctenophores is the platyctenes (such as *Coeloplana* and *Vallicula*), which are easily cultured in the lab and have been especially useful for studying regeneration. All of these ctenophores undergo nearly identical development until the cydippid larval stage, but they possess strikingly different adult morphologies, which is beneficial for comparative studies. *Pleurobrachia* species maintain cydippid forms as adults, but the others become radically different: *M. leidyi* greatly reduces its tentacles and forms large oral lobes; *B. ovata* lacks tentacles and forms large oral lips that it uses to engulf or nip at other ctenophores; and platyctenes become compressed in the oral–aboral direction and adopt a sessile or creeping lifestyle, crawling around the substratum rather than locomoting by beating their ctene rows.

USES OF THE CTENOPHORE MODEL SYSTEM

Because ctenophores are amenable to modern technical approaches, they could prove to be a useful emerging model for a variety of studies. Here we review the reproduction, development, morphology, physiology, and ecology of ctenophores. The focus is on *M. leidyi*, the most frequently studied ctenophore.

Reproduction and Development

Due to the ease of embryo collection, optical properties, and rapid development, ctenophores have been very attractive study animals for experimental embryologists. Observations and detailed descriptions of their development first took place in the late 19th century (Chun 1880; Metschnikoff 1885). With clear and relatively large (up to 1 mm in some species) embryos, early researchers were able to not only follow the course of development from egg to larva, but also to perform experimental manipulations on these embryos.

M. leidyi, as well as most ctenophores studied so far, are simultaneous hermaphrodites, whose eggs are approximately 200 µm in size. Separate testes and ovaries are located on opposite sides of the meridional canals below the comb rows. Each canal is associated with both a testis and an ovary (for review, see Dunlap Pianka 1974). Oogenesis has been well described in another lobate ctenophore, *Bolinopsis microptera*, where each oocyte is associated with as many as 100 nurse cells that are connected to the oocyte by cytoplasmic bridges that transport cytoplasm and yolk to the oocyte (Dunlap Pianka 1974). Mature oocytes are then transported from the ovaries through the oviduct and released out of the gonopore. The mature living eggs are described as centrolecithal and are composed of two visible zones. In the center and accounting for most of the egg is the endoplasm, a central core containing dense translucent yolk spheres, formed by the nurse cells, that are mostly composed of proteins and carbohydrates (very few lipids). The second visible layer is the ectoplasm, which is a thin layer of cytoplasm that surrounds the endoplasm. This ectoplasm is yolk-free and contains mostly mitochondria and endoplasmic reticulum, as well as the nucleus. Additionally, in fixed and sectioned eggs of some ctenophores, a third layer, the subcortical plasm, has been observed between the endoplasm and ectoplasm. This layer is composed of cytoplasm that contains dense, small periodic acid–Schiff (PAS)-positive yolk particles.

In *M. leidyi*, eggs are fertilized by acrosome-containing sperm as they are shed, making detailed studies of fertilization difficult. However, in *Beroe ovata*, in which unfertilized eggs can be obtained, there have been more in-depth studies on the mechanism of fertilization. Being able to control fertilization in *B. ovata*, Carre et al. (1991) demonstrated that sperm can enter the egg at any site and that, upon entry, the sperm pronucleus remains near this site. What is interesting and unique is that multiple sperm (up to 20) can enter a single egg. The female pronucleus then migrates from its original site of meiosis to successive sperm pronuclei, often traveling great distances across the egg over the course of a few minutes to several hours. Once the female pronucleus has found a suitable male pronucleus, fusion occurs; the remaining male pronuclei later degenerate. It is not known what mechanisms are involved in determining this subcellular example of female sexual selection; however, this pronuclear site of fusion is very important because it becomes the site of the first cleavage, which sets up subsequent aspects of development.

Ctenophore cleavage is highly stereotyped and unlike that of any other metazoan phylum. The cleavage program (Fig. 1) has been studied in great morphological detail by Chun (1880), Metschnikoff (1885), and others. Later studies involved marking blastomeres with chalk particles (Reverberi and Ortololani 1965) and experimental manipulations (Freeman 1977), culminating in a complete fate map using intracellular tracers (Martindale and Henry 1999). Although most cell fates are autonomously specified, inductive signals are involved later in development. Cleavage is unipolar, with the cleavage furrow beginning at one pole and proceeding to the opposite pole. The first cleavage occurs at the site of fertilization. The first two cleavages are equal; they run through the future oral–aboral axis and give rise to four cells that correspond to the four quadrants of the larval and adult body plan. Ctenophores undergo mosaic development such that if blastomeres are separated at the two-cell stage, each will generate a "half-animal," possessing exactly half of the normal set of adult features. Similarly, blastomeres separated at the four-cell stage will give rise to "quarter-animals." The third cleavage is oblique and unequal, giving rise to four larger cells in the "middle" (M blastomeres) and four smaller cells at the "ends" (E blastomeres). At this point, these macromeres produce three sets of smaller micromeres at the aboral pole. The micromeres are composed mostly of ectoplasm; the endoplasm remains in the macromeres. The micromeres continue to divide to form a cap of cells on top of the macromeres.

Marking experiments have shown that the site of the first cleavage corresponds to the site of gastrulation, which becomes the oral pole (Freeman 1977). At gastrulation, the aboral micromeres undergo epibolic movements as they move toward the oral pole to completely surround the

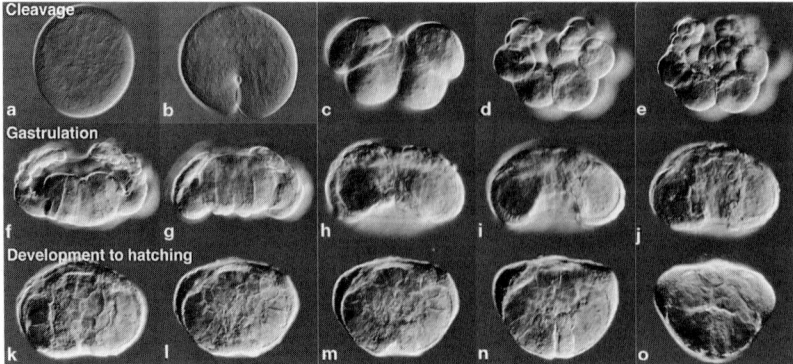

FIGURE 1. Images of *M. leidyi* development from egg to hatching. In all images, the oral pole is at the bottom. An uncleaved egg is shown in *a*, and early cleavage stages are shown (*b*–*e*). (*f*–*g*) Gastrulation begins as the aboral micromeres undergo epiboly to surround the macromeres. Later (*h*–*j*), the macromeres and oral micromeres invaginate. Development continues (*k*–*o*) as the pharynx invaginates and thickenings in the ectoderm give rise to the tentacle apparatus and the apical sensory organ.

macromeres. Before the micromeres completely cover the macromeres, the macromeres buckle inward, thereby becoming more flattened in the oral–aboral axis. The macromeres generate a set of micromeres on the oral side. These oral micromeres enter the blastocoel, proliferate, and migrate to the aboral pole, stopping between the macromeres and the aboral micromeres. They spread out in both the tentacular and sagittal axes to form what has been referred to as a "cruciform mesodermal fundament," which gives rise to the entire musculature and mesenchymal cells (Metschnikoff 1885). Once the aboral micromeres have completely surrounded the macromeres and meet at the oral pole, they invaginate to form the ectodermally derived pharynx. The macromeres, which are now interior, give rise to the gut and endodermal canal system. At the aboral side of the animal, a thickening of the ectoderm gives rise to the apical sensory organ. Additionally, ectodermal thickenings give rise to the tentacular apparati and the eight comb plates begin to form. Therefore, the aboral micromeres give rise to the entire ectoderm of the animal (epidermis, apical organ, pharynx, nerve net, ctene plates, and tentacle sheath); the macromeres give rise to the endoderm, including the anal canals and photocytes; and the oral micromeres give rise to musculature and mesenchyme, which is referred to as the mesoderm.

Another unique feature of ctenophores is their ability to sexually reproduce while they are still larvae, a condition that Chun (1880) referred to as dissogeny. This ability to precociously reproduce has been reported in at least two orders of ctenophores, Lobata (*M. leidyi*) and Cydippida (*Pleurobrachia bachei*), but it is not known how widespread this phenomenon is because it has only been observed in the laboratory. In *M. leidyi*, some cydippid larvae as young as 6 days posthatching and 1.8 mm in size were able to produce viable embryos (Martindale 1987). Only four of the eight gonads in the larvae are mature and capable of producing gametes. After about 7 days of spawning, the larvae stop spawning, grow into lobate forms, and later are again capable of sexual reproduction. It is hypothesized that this ability to reproduce at such a young age reflects the ephemeral ecological conditions experienced by ctenophores; there is often only a narrow window of time when food is available. Predation rates can be very high, so perhaps it is advantageous to produce progeny as early as possible under certain conditions. It would be interesting to see whether different ecological conditions or feeding regimes impact the rate at which ctenophores undergo dissogeny.

Some studies have focused largely on the regenerative capabilities of ctenophores. Much of this work has been on platyctenes, because these benthic ctenophores are not only able to undergo regeneration, but are also capable of asexual fission by budding off pieces of tissue that can grow to become complete adults. In sharp contrast to development, where there is very little regulation and replacement of missing parts, adult animals display remarkable regenerative abilities. All body parts, including the apical organ, tentacles, and comb rows, are capable of regenerating if removed, often taking only a couple days. If an *M. leidyi* embryo is bisected before ctene row coordination, it will generate half-animals, but if it is bisected after ctene row coordination, it will regenerate and form a complete animal (Martindale 1986). Coonfield (1936) performed experiments dissecting adult *M. leidyi* and showed that halves, thirds, and quarters of animals cut along their oral–aboral axes were able to regenerate missing parts, whereas eighths of animals were not.

Only a few groups have examined ctenophore development at the molecular level, in the form of gene expression studies. In *M. leidyi*, these studies have looked at the expression of T-box (Yamada et al. 2007) and homeobox (Pang and Martindale 2008) family members during development. In some cases, gene expression patterns, such as that of *brachyury* in the blastopore, appear to be conserved with bilaterians; however, in many cases, making such comparisons is difficult. It has proven difficult to determine the orthology of some genes, possibly due to the extreme divergence of ctenophore representatives. Of particular interest is the finding that some genes (e.g., *MlTbxE* and *MlPrd1*) are expressed in regions of the ctenophore body that are not morphologically distinct from the adjacent areas (Yamada et al. 2007; Pang and Martindale 2008). Further studies of other gene families should be very informative for gaining a broader knowledge of molecular mechanisms during development. There has also been some work done in *P. pileus*, examining gene expression patterns in the cydippid and adult stages (Derelle and Manuel 2007). At this time, gaining as much information as possible from many different ctenophore species is of great benefit.

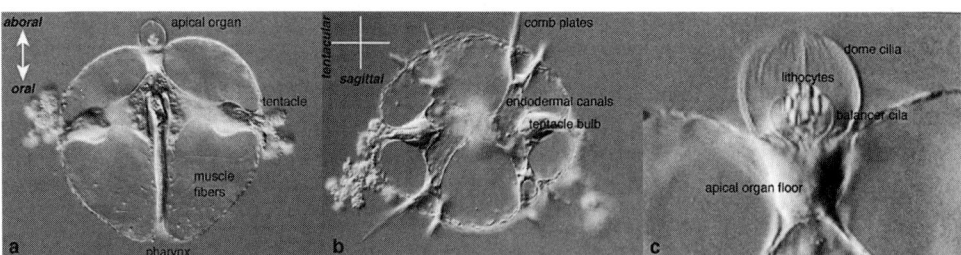

FIGURE 2. The basic cydippid body plan. (a) Lateral view of *M. leidyi* cydippid larvae, revealing the apical organ at the aboral pole and the mouth and pharynx at the oral pole. Also visible are the two tentacles, one on either side of the animal, and numerous muscle fibers located in the mesoglea. (b) View of the same cydippid from the aboral pole, revealing the tentacular and sagittal planes of symmetry. (c) Close-up of the apical sensory organ, showing the apical organ floor, balancer cilia, and lithocytes, which are housed in the dome cilia.

Morphology and Physiology

The main body axis of the ctenophore is the oral–aboral axis (Fig. 2a), with one extremity being the mouth (oral) and the opposite being the apical sensory organ (aboral). There are two major planes of rotational symmetry around the oral–aboral axis. The tentacular plane passes through the two tentacles, whereas the sagittal plane is orthogonal to this and corresponds to the plane of the flattened pharynx (Fig. 2b). The ctenophore body is composed of two epithelial layers: the outer epidermis and the inner gastrodermis. The ectodermal layers, relatively complex, are composed of epithelial cells, gland cells, neural and sensory cells, and muscle cells bound by a basal lamina (Hernandez-Nicaise 1984). However, the bulk of the body is composed of mesoglea, the thick jelly-like layer between the epidermis and gastrodermis. The mesoglea is mostly acellular and composed of extracellular matrix, but it also has some muscle and mesenchymal cells.

Ctenophores are characterized by their many unique forms of cilia. Their main mode of locomotion is via the coordinated beating of their comb plates. These comb plates, or ctenes, are composed of numerous cilia that are attached to basal cushions made up of polster cells (Hernandez-Nicaise 1991). Thus, ctenophores are one of the largest animals to locomote completely via ciliary movements. The total number of cilia in a single comb plate of an adult *M. leidyi* is estimated to be 100,000 or more (Afzelius 1961), and each animal bears several hundred comb plates. In addition to the comb plates, specialized cilia are present in sensory cells, polar fields, tentacle sheaths, auricles, around the mouth, and in the pharynx and tentacular and food grooves (Hernandez-Nicaise 1984; Moss et al. 2001). The aboral pole also has several ciliated structures: balancer cilia, dome cilia, presumed photoreceptors, polar fields, and a ciliated groove.

Feeding

In *M. leidyi*, cydippid larvae hatch after 18–24 hours, bearing two long and highly characteristic branching tentacles that are its primary mode of catching prey. The cydippids wait passively, with all of the tentacle branches, or tentillae, extended like a net. When prey comes into contact with a tentilla, it becomes entangled and stuck via the discharged colloblast cells. The tentacle retracts and prey is transferred to the mouth, where it is consumed. The colloblasts, which are the major component of the tentacle epidermis, are composed of two structures: the axial filament (which appears as a granular dome) and the spiral filament (which has a coiled shape and is inserted into the mesoglea) (Hernandez-Nicaise 1991). On the apical surface of each colloblast are eosinophilic granules that are thought to be responsible for its strong adhesive properties (Franc 1978). When a prey comes into contact with the colloblasts, these granules burst, releasing the adhesive material onto the prey; the spiral filament serves as an anchor or a shock absorber to prevent the prey from breaking free. Tentillae, including colloblasts, are constantly being formed at the base of the tentacle to replace those that are damaged or destroyed. Colloblasts are present in all ctenophores except for beroids, which lack tentacles altogether. Additionally, *Euchlora* does not possess colloblasts but instead has nematocysts that it obtains from the medusae it consumes (Hernandez-Nicaise 1991).

Throughout its life, the feeding mechanisms of *M. leidyi* adjust to changes in morphology (Main 1928; Sullivan and Gifford 2007). As young cydippid larvae (<4 mm), they feed by letting their tentacles trail behind them as they swim. Whenever prey comes into contact, the tentacles retract and the prey is transferred to the mouth. As *M. leidyi* larvae grow larger, the tentacles become reduced and two oral lobes begin to form. In addition to the lobes, four finger-like structures called auricles form between the lobes. These highly ciliated structures generate water currents that bring in food particles. At a certain point, the larvae enter a transition stage where oral lobes are beginning to form but tentacles are still present; therefore, feeding is via a combination of the two mechanisms. Once the lobes have fully formed, the auricles begin to develop and the tentacles become highly reduced. At this stage, feeding is completely through the oral lobes.

Lobate adult animals swim with their lobes spread and catch food via two different mechanisms (Waggett and Costello 1999). One mechanism uses the cilia of the auricles to generate a low-velocity current to direct food toward their lobes and then into their mouths. The flow is so slow that it does not generate a rapid escape response in the prey; therefore, this mechanism is mostly used to capture small prey items. The second mechanism relies on the collision of prey items with the inner surfaces of the lobes themselves, which tends to be more successful for larger plankton that are not entrained by the auricular flow. Thus, *M. leidyi*, as well as other lobate ctenophores, are able to catch both slower-swimming prey and more motile forms.

One physiological feature that enables ctenophores to more efficiently process food is the morphology of the pharynx. Although ctenophores possess only this one major digestive opening, its flat and elongated shape creates different regions with different digestive capabilities (Bumann and Puls 1997) and allows food ingestion and waste egestion to occur simultaneously. When food particles enter the opening of the pharynx, they are first exposed to an acidic environment. The particles then move through two alkaline phases, first along the folds of the pharynx and then near the region where the pharynx meets the infundibulum. It is in the latter region that large densities of cilia break up the food particles. Smaller particles then enter the endodermal system of meridional canals that run subjacent to the comb rows and distribute nutrients to the rest of the body. Larger particles are egested back out of the pharynx. Although most waste is excreted out of the pharynx (Main 1928), some smaller waste particles do exit out of two anal pores that are located in diagonal quadrants next to the apical organ. These pores are connected to the rest of the endoderm via anal canals.

Rates of feeding, growth, and metabolism in *M. leidyi* are reviewed by Purcell et al. (2001). Feeding rates tend to be proportional to prey concentration; lobate forms do not satiate at most natural prey concentrations, although clearing rates do vary depending on prey type. Rates of body carbon and nitrogen turnover range from 6% to 18%. Metabolism can vary greatly depending on the temperature and food availability. When starved, animals will shrink in size and eventually die after a few weeks.

Muscles

The muscle cells of ctenophores can be described as true muscle cells; they lack the epithelial component found in cnidarian epitheliomuscle cells. Ctenophore muscles can be divided into three categories: the parietal muscles, which are located at the base of the integument, the mesogleal muscles, which run through the body and tentacles, and the tentacular muscles (Hernandez-Nicaise 1991). The parietal muscles are ribbon-shaped smooth muscles that reside between the glandular epithelial cells and the basal lamina. The mesogleal muscles include longitudinal fibers that run along the oral–aboral axis, circumferential fibers that circle the pharynx, and radial fibers that run from the outside toward the pharynx. All of these mesogleal fibers are referred to as the "giant smooth muscle" and are often multinucleated. Additionally, there are muscles associated with the tentacles and tentacle apparati. A muscle connects the two tentacle apparati (the *trans*-tentacular muscle); muscles also run from the apparatus of each tentacle to the apical organ (the apical organ muscles). Within the tentacles themselves are mesogleal muscles that are constantly being produced from cells at the base of the tentacle. The only description of striated muscle is in the tentilla of *Euplokamis* (Chun 1880).

Giant smooth muscle fibers located in two sagittal bundles in *M. leidyi*, as well as in *B. ovata*, have been very useful in electrophysiology studies (Hernandez-Nicaise et al. 1984). These large multinucleated bundles, which can reach up to 4 cm in length and 35 μm in diameter in *M. leidyi*, can be isolated by enzymatic digestion of the mesoglea. Resting potentials and acting potentials can be measured in these cells, as well as in isolated muscle cells, in situ (Anderson 1984).

Nerves and Sensory Structures

The nervous system is composed of a subepidermal nerve net, mesogleal nerve fibers, a few subgastrodermal nerves, and an apical sensory organ (Hernandez-Nicaise 1991). In addition, two types of presumed sensory cells are located in the epithelia, named by Hertwig (1880) as "tastborsten" and "taststiften." The tastborsten are characterized by a single thin ciliary projection, whereas the taststiften, later termed "hoplocysts," possess one or several short pointed pegs. The subepidermal nerve net has a polygonal pattern and has been visualized via vital staining (Hernandez-Nicaise 1991), electron microscopy (Hernandez-Nicaise 1973), and cell-lineage studies (Martindale and Henry 1999). The mesogleal nerve fibers cross the basal lamina and run through the mesoglea, where they synapse with the giant smooth muscle fibers and mesenchymal cells (Hernandez-Nicaise 1991). Nerve fibers are also present in the mesoglea of the tentacles. The subgastrodermal nerves are thought to be located in the walls of the meridional canals and innervate the photocytes (Anctil 1985).

Nervous elements can be identified ultrastructurally by the presence of the presynaptic triad. The presynaptic triad is characterized by a row of vesicles lining the synaptic membrane, a sac of endoplasmic reticulum, and one or several mitochondria (Hernandez-Nicaise 1973). Concentrations of nervous elements in the apical sensory organ are associated with the cilia ted grooves and comb rows and are found around the oral pole in beroids. Additionally, a tentacular nerve that links the tentacle to the apical sensory organ has been described in some species. The apical sensory organ is thought to be the main sensory region due to the concentrations of nervous elements surrounding it. It houses a gravity-sensing statolith covered by a dome of nonmotile cilia (Fig. 2c). The thickened epithelium below this is known as the apical organ floor. In the floor are four groups of cells that bear a tuft of a few hundred cilia known as the balancer cilia. The balancer cilia project upward and support a group of mineralized lithocytes. The lithocytes are generated in the floor of the apical organ and are constantly being replaced. The lithocytes control the posture of the animal, and the balancers serve as mechanoreceptors (Tamm 1982). Depending on the stimulus, the statolith exerts pressure on the balancer cilia, which in turn are connected to the comb plates via the ciliated grooves. Running over the apical organ floor is a bridge of axon-like processes that may be an electrical conduction pathway to the balancer cilia (Tamm and Tamm 2002). Also located next to the balancer cilia are four protruding epithelial papillae that are thought to be pressure receptors. In the floor of the apical organ are lamellate bodies whose ciliary structure indicates that they may be involved in photoreception (Horridge 1964). Thus, the apical organ may be able to sense and react to changes in gravity, pressure, and light.

Nearly all ctenophores possess photocytes, or light-producing cells, in the meridional canals adjacent to the male gonads. A photoprotein, called mnemiopsin in *M. leidyi*, emits light in the presence of calcium ions (Ward and Seliger 1974). Any type of mechanical stimulus results in a blue–green flash of light that begins in the canal that was stimulated and runs along it and adjacent canals. Exposure to light inhibits this reaction; thus, only animals that are kept in the dark will display it. Ultrastructurally, photocytes are characterized by numerous mitochondria and rough endoplasmic reticulum, and they are linked by gap junctions (Anctil 1985). Neurites do synapse with photocytes, but the issue of whether bioluminescence is under the control of the nervous system has been debated.

Ecology

Much of the recent research related to the ecology of *M. leidyi* has been due to its role as an invasive species. Because it is such a voracious predator, its presence and abundance in both its natural and introduced habitats have huge ecological impacts. It is thought that *M. leidyi* was first introduced to the Black Sea in the early 1980s via ballast water from North American ships (Kideys 2002). Its strong reproductive and regenerative capabilities resulted in a huge population growth, which, when combined with other factors such as eutrophication and overfishing, resulted in a crash in the anchovy fishing industry in 1989. It is suspected that this was due not only to *M. leidyi* feeding directly on anchovy eggs and larvae, but also to their feeding on the anchovy food supply, the zooplankton. *M. leidyi* then spread to the Sea of Azov, the Sea of Marmara, the Aegean Sea, and the eastern Mediterranean. It invaded the Caspian Sea in 1999 and was observed in the North Sea and the Baltic Sea in 2006. Its effects have varied by location. The Black Sea experienced the most severe impacts, with *M. leidyi* blooms reaching levels of greater than 1 kg per square meter in 1989 (Kideys 2002). In 1997, *B. ovata*, a natural predator of *M. leidyi*, appeared in Black Sea waters, most likely also via ballast water. *B. ovata* greatly reduced the abundance of *M. leidyi*; increases in zooplankton and fish biomass were also observed. Later, *B. ovata* numbers also declined.

In the Caspian Sea, the introduction of *M. leidyi* has greatly influenced the catch of kilka, a planktivorous fish, which in turn has led to declines in both sturgeon and Caspian seal. The low salinity of the Caspian Sea causes the *M. leidyi* to be smaller and less fecund than those in the Black Sea. The main factors determining *M. leidyi* abundance are temperature and food availability. In both native and introduced habitats, *M. leidyi* numbers decrease markedly during winter months, with surviving individuals overwintering in deeper waters. As surface temperatures rise, these survivors repopulate to coastal areas. In some locales, such as the Sea of Azov, low temperatures combined with low salinity prevent *M. leidyi* from surviving the winter, and therefore, populations are reintroduced every year from other sources.

In addition to predation, *M. leidyi* populations are also kept in check in native habitats by the parasitic sea anemone *Edwardsiella lineata*. The planula larvae of this cnidarian enter the ctenophore through the mouth as it feeds. The larvae then bore into the mesoglea and attach themselves to the pharynx or gut of the ctenophore and feed on partially digested material that the ctenophore has consumed. As a result, *E. lineata* competes for the ingested food. *E. lineata* infections can therefore mimic the effects of starvation. Multiple parasites can infect a single animal, and peak infection frequencies have reached 60% (Reitzel et al 2007).

GENETICS, GENOMICS, AND ASSOCIATED RESOURCES

There is little genetic information on *M. leidyi* or on ctenophores in general. In a study of rDNA across 26 ctenophore species (Podar et al. 2001), there was very little difference in sequence length, and sequence divergence was found to be less than 5%, which is the lowest for any metazoan phylum studied so far. This low level of divergence indicates that all extant ctenophores are derived from a relatively recent common ancestor. The reduced variation could also be due to the fact that they are self-fertile hermaphrodites. Whether this level of conservation is the same for other nuclear or mitochondrial markers is not yet known, but a study of the genetic differences among *M. leidyi* populations in the western Atlantic and Europe is currently being undertaken (K. Bayha, pers. comm.).

The genome of *M. leidyi* has not been sequenced; in fact, there are currently no ctenophore genome sequencing projects in the works. *M. leidyi* has a genome size of approximately 300 Mb (Gregory et al. 2006), so it seems likely, with the new sequencing technologies and lower sequencing prices, that its genome will eventually be sequenced. Currently, 15,752 sequences from an *M. ledyi* gastrula-stage cDNA library and 8540 sequences from a *P. pileus* cDNA library are available from the NCBI Expressed Sequence Tag database (dbEST). We have also sequenced 3360 clones from an *M. ledyi* early-cleavage-stage library, and they are available from the NCBI Trace Archive.

TECHNICAL APPROACHES

The following protocols provide some basic guidance for manipulating *M. leidyi* in the laboratory. Protocol 1 outlines a procedure for spawning and collecting embryos from *M. leidyi*. The subsequent methods, antibody staining (Protocol 2), in situ hybridization (Protocol 3), DNA isolation (Protocol 4), and RNA isolation (Protocol 5), focus on handling embryonic and larval tissues, although the procedures can also be applied to adult tissues.

Protocol 1

M. leidyi Spawning and Embryo Collection

This protocol describes how to obtain and collect embryos from *M. leidyi*. Under natural conditions, spawning normally occurs about 8 hours after sunset, such that eggs are released under the cover of darkness. Because spawning is triggered by the onset of darkness, keeping animals under an artificial light regime in the laboratory can alter the time of spawning. This protocol is designed to induce animal spawning at approximately 11:00 am; however, it can be adjusted for other times. The duration from spawning to hatching of cydippid larvae is 18–24 hours.

MATERIALS

CAUTION: See the Cautions Appendix for appropriate handling of materials marked with <!>.

Reagents

Seawater (fresh, filtered, 0.2 µm; FSW)
Seawater (fresh, natural)

Equipment

Culture dishes (4.5 inch, glass; Carolina Biological)
Filter (165 µm) (optional; see Step 7)
> This filter can be constructed by cutting off the bottom of a 50-ml conical tube and then stretching and gluing 165-µm Nytex mesh over the top of the tube.

Forceps
Microscope (dissecting)
Petri dishes (35 mm or 60 mm), coated with 0.2% [w/v] gelatin and 0.074% formaldehyde [w/v] <!> (for devitellinization only; see Step 9)
Plastic bucket (5 gallon) or similar large container
Room with temperature/light control or fluorescent light stick and outlet timer (see Step 3)
Transfer pipettes (plastic) (optional; see Step 7)

METHOD

1. Catch *M. leidyi* animals and carefully transfer them into a 5-gallon plastic bucket filled with natural seawater.

2. Place the bucket of animals in the lab under constant light and at a temperature between 12°C and 18°C.

3. When spawning is desired, place the bucket of animals into a temperature- and light-controlled room. Set the temperature at 12°C and keep them under constant light until 3:00 am.

   If a temperature- and light-controlled room is not available, use a darkened room whose temperature is as close to 12°C as possible. At approximately 12:00 am, place the bucket of animals into this room. Place a fluorescent light stick, attached to an outlet timer, above the bucket. Set the timer to turn off at 3:00 am.

4. The next morning, at approximately 9:30 or 10:00 am, remove the animals from the bucket and place them into individual glass culture dishes filled with FSW. Rinse the animals carefully with FSW to remove debris and mucus.

5. Monitor the animals every 20 minutes to check for signs of spawning.

   Before spawning, the meridional canals below the comb rows may turn cloudy. Sperm are released first; eggs are released soon thereafter and are fertilized upon release. Spawning will be mostly synchronous, although some animals may not spawn.

6. After spawning is complete (20–30 minutes), carefully remove the adult animal from the culture dish by hand and place it back into the bucket.

   The eggs will settle to the bottom of the culture dishes.

7. If desired, concentrate the embryos by pipetting them into a new dish using a transfer pipette or by reverse-filtering through a 165-μm filter.

   Filtering also helps in breaking up the egg jelly and removes debris.

8. Raise the embryos in FSW at 12–18°C to ensure proper development.

   Development should be relatively synchronous, with the first cleavage occurring 1 hour after fertilization and subsequent cleavages every 20 minutes thereafter. Gastrulation occurs at 3–4 hours, comb row formation at 9–10 hours, and hatching at 18–24 hours. If these events do not occur, see Troubleshooting.

9. If manipulations (e.g., injection or ablation) are required, carefully remove the vitelline membrane surrounding an embryo of the desired stage using forceps under a dissecting microscope. Place the denuded embryos in gelatin-coated Petri dishes to prevent them from sticking.

TROUBLESHOOTING

Problem (Step 8): Embryos do not develop properly.

Solution: Make sure all dishes and solutions are free of detergents and harmful chemicals. The embryos are very sensitive, and proper care must be taken to maintain glassware that is safe for the embryos. Additionally, to prevent any abnormalities due to polyspermy, rinse the eggs after spawning and make sure that they are not overcrowded.

Protocol 2

Ctenophore Whole-mount Antibody Staining

This protocol describes how to fix ctenophore embryos and cydippid larvae for antibody staining. Once the samples have been fixed, the tissue is incubated with an antibody to the epitope of interest. A secondary antibody that is conjugated to a fluorescent molecule then reveals the protein expression of the epitope. Finally, fluorescent microscopy is used to visualize and document the signal. The protocol also includes methods for staining or counterstaining with a fluorescent derivative of phalloidin, which reveals F-actin in muscles and cell borders.

MATERIALS

The recipes for items marked with <R> begin on page 191.

CAUTION: See the Cautions Appendix for appropriate handling of materials marked with <!>.

Reagents

Alexa Fluor 488 phalloidin (Molecular Probes) (for phalloidin staining only; see Method IV) <!>
Antibodies (primary and secondary)
Blocking solution for antibody staining <R>
Ctenophore embryos and/or cydippid larvae
    To collect embryos from M. leidyi, *see Protocol 1.*
Fixation buffer 1 for antibody staining (ice cold) <R>
Fixation buffer 2 (ice cold) <R>
Magnesium chloride ($MgCl_2$; 6.5%, prepared in H_2O)
Methanol (for phalloidin staining only; see Method IV) <!>
10× Phosphate-buffered saline (PBS) <R>, diluted to 1× before the procedure
    Also prepare 1× PBS containing 0.2% Triton X-100 <!>.

Equipment

Filter (165 µm)
    This filter can be constructed by cutting off the bottom of a 50-ml conical tube and then stretching and gluing 165-µm Nytex mesh over the top of the tube.
Forceps (optional; see Step 6)
Ice
Microcentrifuge tubes (for phalloidin staining only; see Method IV)
Microscope (fluorescent)
Pipette tip (P-200), cut to size with a razor blade (see Step 6 of Method I)
Tissue culture plates (4 well and 24 well)
Transfer pipettes (plastic)
Tubes (2 ml, with screw caps) (optional; see Step 4 of Method I and Step 5 of Method II)
Tubes (5 ml, with round bottoms)

METHODS

I. Preparation of Embryos

Fixation

1. At the appropriate time in development (see Protocol 1), transfer the embryos to 5-ml round-bottom tubes and allow them to settle. Remove as much liquid as possible.
2. Add ice-cold fixation buffer 1 to fill each tube. Cap the tubes and gently invert them a couple of times. Then, place the tube at 4°C and allow the embryos to settle; this takes about 5 minutes.
3. Remove the fixative and add ice-cold fixation buffer 2 to fill each tube. Gently invert the tubes a couple of times and fix for 15 minutes at 4°C.
4. Remove the fixative. Wash the embryos five times with 1x PBS; mix by inversion and allow the embryos to settle before changing the PBS.

   Use the fixed embryos immediately (start by removing the vitelline membranes as described in Steps 5–7) or transfer them to 2-ml screw-cap tubes and store them in 1x PBS at 4°C (use them within a few weeks).

Membrane Removal

5. Transfer the embryos to 4-well dishes filled with 500 µl of 1x PBS.
6. Gently pipette the embryos with a P-200 tip to remove the membrane.

   The bore should be smaller than the vitelline membrane but larger than the actual embryo; if necessary, cut the tip with a razor blade. Alternatively, use forceps to physically remove the membranes from the individual embryos.
7. After the membranes have been removed, transfer the embryos to a fresh well.

   Proceed to Method III.

II. Preparation of Larvae

1. Concentrate 100–500 cydippid larvae by reverse-filtering using a 165-µm filter. Use a plastic transfer pipette to transfer them to one well of a 24-well plate.
2. Try to remove as much liquid as possible and then add an equal volume of 6.5% $MgCl_2$. Allow the cydippids to sit for 15–20 minutes until they are relaxed (tentacles extended) and begin to settle.
3. Remove as much liquid as possible. Add ice-cold fixation buffer 1 and fix for 3–5 minutes on ice.
4. Remove the fixative. Add ice-cold fixation buffer 2 and fix for 15 minutes on ice.
5. Remove the fixative. Wash the embryos five times with ice-cold 1x PBS; mix by inversion and allow the embryos to settle before changing the PBS.

   Use the fixed larvae immediately (proceed to Method III) or transfer them to 2-ml screw-cap tubes and store them at 4°C in 1x PBS (use them within a few weeks).

III. Processing of Embryos and Larvae

Use 500 µl per wash unless otherwise stated.

Primary Antibody Incubation

1. Place 25–100 embryos (from Step 7 of Method I) or cydippids (from Step 5 of Method II) in one well of a 4-well dish.

2. Wash embryos or cydippids five times with room-temperature 1× PBS containing 0.2% Triton X-100, which permeabilizes the cell membranes. Repeat this step twice.

3. Remove the liquid, and add 500 µl of blocking solution to each well. Incubate for at least 1 hour at room temperature.

4. While the embryos or cydippids are in blocking solution, prepare the primary antibody dilution. Dilute the antibodies to an appropriate concentration using the blocking solution.

5. After the 1-hour incubation, remove the blocking solution and add 200–400 µl of the primary antibody dilution. Incubate overnight at 4°C.

Secondary Antibody Incubation

6. Remove and save the primary antibody dilution.
   The primary antibody dilution can be stored at 4°C and used multiple times.

7. Wash the embryos or cydippids six times with room-temperature 1× PBS containing 0.2% Triton X-100. Incubate for 15–30 minutes each time.

8. While washing, prepare the secondary antibody dilution. Dilute the antibodies to an appropriate concentration using the blocking solution.
   For Alexa Fluor secondary antibodies (Molecular Probes), a 1:250 dilution in blocking solution works well.

9. After the washes are completed, add the secondary antibody dilution. Incubate in the dark overnight at 4°C.

Washing and Detection

10. Remove the secondary antibody dilution.

11. Wash the embryos or cydippids six times with room-temperature 1× PBS containing 0.2% Triton X-100. Incubate for 15–30 minutes each time. While washing, try to keep the embryos or cydippids in the dark as much as possible.

12. Mount the embryos or cydippids in 1× PBS and view them under a fluorescent microscope.

IV. Phalloidin Staining (Optional)

If only doing phalloidin staining, the following steps can be performed after Step 2 of Method III. If using phalloidin as a counterstain to other antibodies, conduct these steps after Step 11 of Method III.

1. Dissolve the Alexa Fluor 488 phalloidin in the appropriate volume of methanol to yield 200 units/ml.

2. In an empty microcentrifuge tube, add 5 µl of the phalloidin solution from Step 1. Allow the methanol to evaporate for a few minutes.

3. Add 500 µl of 1× PBS (i.e., add 100 µl of 1× PBS per 1 µl of phalloidin) to the microcentrifuge tube and mix. Add the mixture to the embryos or cydippids and incubate them in the dark for 1–2 hours at room temperature.

4. Wash the embryos or cydippids with 1× PBS three times at room temperature (5 minutes each wash).

5. Mount the embryos or cydippids in 1× PBS and view them under a fluorescent microscope.

Protocol 3

Ctenophore Whole-mount In Situ Hybridization

This protocol describes how to fix, prepare, and hybridize antisense RNA probes (i.e., complementary to a mRNA of interest) in ctenophore embryos and cydippid larvae, as well as how to detect the probes using an alkaline-phosphatase-conjugated antibody and colorimetric substrates. Using these techniques, it is possible to determine which cells or tissues express the gene of interest.

MATERIALS

The recipes for items marked with <R> begin on page 191.

CAUTION: See the Cautions Appendix for appropriate handling of materials marked with <!>.

Reagents

Acetic anhydride <!>
Alkaline phosphatase (AP) buffer (with and without $MgCl_2$) <R>
Anti-digoxigenin-AP antibody
10x Blocking buffer for in situ hybridization <R> (freshly diluted to 1x with MAB <R>)
Ctenophore embryos and/or cydippid larvae
   To collect embryos from M. leidyi, see Protocol 1.
Developing buffer <R>
Fixation buffer 1 for in situ hybridization (ice cold) <R>
Fixation buffer 2 (ice cold) <R>
Glycerol (70%, prepared in 1x PBS)
Glycine (2 mg/ml, prepared in PTw)
H_2O (sterile, distilled)
Hybridization buffer <R>
Magnesium chloride (6.5%, prepared in H_2O) (for cydippid larvae only; see Step 1)
Methanol (100, 60, and 30%; 60% and 30%, prepared in PTw) <!>
Paraformaldehyde (4%, prepared in PTw) <!>
10x Phosphate-buffered saline (PBS) <R> (diluted to 1x before the procedure)
Proteinase K (0.01 mg/ml, prepared in PTw) <!>
PTw <R>
RNA probe(s) labeled with digoxigenin <!> (diluted to 0.1 ng/µl in hybridization buffer)
Seawater (fresh, filtered, 0.2 µm; FSW)
20x SSC (pH 7.0) <R>
   For a list of buffers to prepare with the 20x SSC stock solution, see Step 22.
Triethanolamine (1% [w/v], prepared in PTw) <!>

Equipment

Forceps (optional; see Step 8)
Heating block (set at 80–90°C)
Humidified chamber
   Place damp paper towels inside a plastic Tupperware tray to construct a humidified chamber.

Hybridization oven, preset to 60°C
Microscope (compound)
Pipette tip (P-200, cut to size with a razor blade; see Step 8)
Tissue culture plates (24 well)
Transfer pipettes (plastic)
Tubes (2 ml, with screw caps)
Tubes (5 ml, with round bottoms)

METHOD

Fixation

1. At the appropriate age of development (see Protocol 1), wash and concentrate live embryos or cydippids in FSW to remove any debris and egg jelly. Transfer them to 5-ml round-bottom tubes and allow them to settle. For cydippids, add an equal volume of 6.5% $MgCl_2$ and allow them to sit for 15–20 minutes until they are relaxed (tentacles extended).

2. Remove as much liquid as possible and add ice-cold fixation buffer 1 to fill the tube. Mix by gently inverting and incubate for 3–5 minutes at 4°C.

3. Remove fixation buffer 1 and add the same volume of ice-cold fixation buffer 2. Mix by gently inverting and incubate for 1 hour at 4°C.

4. Remove fixation buffer 2 and wash the specimens with 1x PBS five times (10 minutes each) at room temperature.

5. Use a plastic transfer pipette to transfer the fixed specimens to 2-ml screw-cap tubes.

6. Wash the specimens twice with room temperature, sterile, distilled H_2O.

7. Wash the specimens twice with room temperature 100% methanol and store them in methanol at –20°C.

    Fixed samples can be stored for years under these conditions.

Pretreatment

Perform all pretreatment steps at room temperature unless otherwise noted.

8. Add 500 µl of 100% methanol to each well of a 24-well plate. Carefully add 50–200 fixed embryos or cydippids to each well. For embryos, break open the vitelline membranes by gently pipetting each embryo up and down with a P-200 tip.

    The bore should be smaller than the vitelline membrane but larger than the actual embryo; if necessary, cut the tip with a razor blade. Alternatively, use forceps to physically remove the membranes from the individual embryos.

    The membranes should float slightly while the denuded embryos should sink. If necessary, wash the embryos with methanol a few times to get rid of the opened membranes.

9. Rehydrate through a methanol series as follows:

    i. Wash with 500–800 µl of 60% methanol for 5 minutes.

    ii. Wash with 500–800 µl of 30% methanol for 5 minutes.

    iii. Wash five times with 500–800 µl of PTw for 5 minutes each.

10. Incubate the specimens with 500 µl of 0.01 mg/ml proteinase K. For embryos, incubate for 10 minutes; for cydippids, incubate for 5 minutes.

11. To stop the proteinase K treatment, wash the specimens twice (5 minutes each) with 500 µl of 2 mg/ml glycine.

12. Wash the specimens twice (5 minutes each) with 500 µl of 1% (w/v) triethanolamine. Leave the specimens in triethanolamine after the second wash.
13. Add 1.5 µl of acetic anhydride. Mix by swirling for 5 minutes.
    The acetic anhydride should appear as a small droplet in the bottom of the wells.
14. After 5 minutes, add another 1.5 µl of acetic anhydride and mix again.
15. Wash twice (5 minutes each) with 500 µl of PTw to wash out the triethanolamine and acetic anhydride.
16. Refix the specimens with 500 µl of the 4% paraformaldehyde solution for 1 hour.
17. Wash the specimens five times (5 minutes each) with 500 µl of PTw. During these rinses, transfer the specimens to a new 24-well plate and divide them into separate wells for different probes.
18. Wash the specimens twice (5 minutes each) with 500 µl of hybridization buffer. Then add 500 µl of fresh hybridization buffer, place the plate in a humidified chamber to prevent evaporation, and place the chamber in a hybridization oven overnight at 60°C.

Hybridization

19. Denature the RNA probe(s) by heating for 10 minutes at 80–90°C in a heating block.
20. Remove the plate from the hybridization oven and remove as much buffer as possible. Add 500 µl of the heated probe to each well and place the plate back in a humidified chamber in the oven. Allow the probes to hybridize for 24–48 hours at 60°C.
21. Remove and save the probe(s). Remember to use a different pipette tip for each probe.
    The probes can be stored at –20°C and reused several times.
22. Wash the specimens in the following series of buffers. Preheat each buffer to 60°C before adding. Each wash is with 500 µl of buffer for 20 minutes in a 60°C hybridization oven (the humidified chamber is no longer necessary).
    i. Wash once in 100% hybridization buffer.
    ii. Wash once in 75% hybridization buffer and 25% 2x SSC (pH 7.0).
    iii. Wash once in 50% hybridization buffer and 50% 2x SSC (pH 7.0).
    iv. Wash once in 25% hybridization buffer and 75% 2x SSC (pH 7.0).
    v. Wash three times in 2x SSC (pH 7.0).
    vi. Wash three times in 0.05x SSC (pH 7.0).
23. Wash five times (5 minutes each) with 500 µl of PTw at room temperature.

Detection

24. Add 500 µl of 1x blocking buffer and incubate for 1 hour at room temperature.
25. While the specimens are incubating in blocking buffer, prepare a working dilution of anti-digoxigenin-AP antibody (1:5000 in 1x blocking buffer). Add 500 µl of diluted antibody to the specimens and incubate overnight at 4°C.
26. The next day, wash the specimens 15 times (10–20 minutes each) with 500 µl of PTw at room temperature to wash out the antibody. Use only 1x PBS (PTw without Tween 20) for the last few washes.
    The duration of these washes is not so important; they can be performed periodically throughout the course of 1 day.

27. Wash the specimens twice (5 minutes each) in 500 µl of AP buffer without $MgCl_2$ at room temperature.

28. Wash the specimens twice (5 minutes each) in 500 µl of AP buffer with $MgCl_2$ at room temperature.

29. Add 500 µl of developing buffer to the specimens and place them in the dark. Monitor the reaction every 30 minutes initially and then later at longer intervals.

    A dark blue/purple precipitate will form in cells that are expressing the gene of interest. Developing can take as little as 30 minutes or as long as a few days, depending on the probe and mRNA expression levels. The developing buffer should start off slightly yellowish in color and turn purple over time. Before it turns purple, replace it with fresh developing buffer.

30. When the specimens have been stained sufficiently, stop the reaction by washing them three to five times (5 minutes each) with 500 µl of 1x PBS. Mount the specimens in 70% glycerol and photograph under a compound microscope.

    See Troubleshooting.

TROUBLESHOOTING

Problem: The embryos fall apart.
Solution: Embryos are very delicate; be gentle during membrane removal (Step 8) and the washes. In addition, it may be helpful to reduce the proteinase K concentration or digestion time in Step 10. Alternatively, the embryos may not have been fixed adequately; if this is the case, a longer fixation (Steps 2 and 3) or a higher concentration of glutaraldehyde in fixation buffer 1 may be required.

Problem (Step 30): There are high levels of nonspecific background staining.
Solution: Increase the hybridization temperature in Step 20, increase the number of washes posthybridization (Step 22) and/or postantibody incubation (Step 26), and make sure that the formamide in the hybridization buffer is fresh.

Problem (Step 30): No signal is detected.
Solution: Reduce the hybridization temperature in Step 20, use a higher probe concentration, or instead of washing with 0.05x SSC (pH 7.0) in Step 22.vi, use 0.2x SSC (pH 7.0).

Protocol 4

Ctenophore Tissue Preparation and Extraction of DNA

This protocol describes how to isolate genomic DNA from ctenophores. The procedure can be applied to adult tissues, but it is best to use embryos and larvae. After washing and concentrating the embryos or larvae, DNA is extracted using DNAzol Reagent (Molecular Research Center), a guanidine-detergent lysing solution. The resulting DNA can be used for polymerase chain reaction (PCR) or other applications.

MATERIALS

CAUTION: See the Cautions Appendix for appropriate handling of materials marked with <!>.

Reagents

Agarose gel (1%) and appropriate electrophoresis buffer (see Step 16)
Ctenophore embryos (and/or cydippid larvae)
  Spawn animals and collect the embryos as described in Protocol 1. It is best to raise the embryos until hatching.
DNAzol Reagent (Molecular Research Center)
Ethanol (100% and 75%) <!>
H_2O (sterile and nuclease-free)
Proteinase K (20 mg/ml) <!>
Seawater (fresh, filtered, 0.2 µm; FSW)

Equipment

Centrifuge (benchtop)
Conical tubes (15 ml)
Culture dishes (4.5 inch, glass; Carolina Biological)
Filters (250 µm and 165 µm)
  These can be constructed by cutting off the bottoms of 50-ml conical tubes and then stretching and gluing 250-µm or 165-µm Nytex mesh over the tops of the tubes.
Gel electrophoresis equipment
Microcentrifuge
Microcentrifuge tubes (1.5 ml)
Microscope (dissecting)
Spectrophotometer
Transfer pipettes (plastic)

METHOD

Perform all steps at room temperature unless otherwise noted.

Tissue Preparation

1. Use a transfer pipette to concentrate the embryos in a dish of FSW. Wash the embryos a few times with FSW to break up the egg jelly and to remove dirt and other debris.

2. Pass the embryos through a 250-μm mesh filter.

    Hatched embryos should pass through undamaged, whereas unhatched embryos and membranes should be trapped in the mesh.

3. To clean the embryos further, reverse-filter the filtrate from Step 2 through a 165-μm mesh filter.

    The hatched embryos will remain in the dish.

4. Collect the embryos by washing them into a dish of FSW. Under a dissecting microscope, check to ensure that the embryos are all right and that there are no contaminants. If necessary, continue to reverse-filter the embryos and wash them with FSW until no contaminants remain.

5. To concentrate the embryos, reverse-filter and wash one final time, using a smaller volume of FSW. Transfer the embryos to 15-ml conical tubes and centrifuge the tubes in a benchtop centrifuge at 1000–2000g for 5 minutes to concentrate the embryos at the bottom of each tube. Then, quickly pipette and discard as much FSW from the top of each tube as possible.

6. Transfer the embryos to 1.5-ml microcentrifuge tubes. Concentrate the embryos at the bottom by centrifuging in a microcentrifuge at 2000g for 5 minutes. Remove as much seawater as possible, so that the final volume in each tube is less than 100 μl.

    There are usually 500–1000 embryos at the bottom of each microcentrifuge tube at this stage.

DNA Extraction

7. Add 1 ml of DNAzol to each tube and mix the contents by inverting. Then, add 5 μl of 20 mg/ml proteinase K and mix again by inverting. Allow the tubes to sit overnight at room temperature.

    After the overnight incubation, the tissue should be completely broken up; however, there may be some insoluble material.

8. Centrifuge the tubes at 10,000g for 10 minutes to pellet any insoluble material.

9. Transfer the supernatants to new tubes without touching the insoluble material.

10. Add 500 μl of 100% ethanol to each tube and mix by inverting. Allow the tubes to sit for at least 5 minutes at room temperature.

11. Centrifuge the tubes at 5000g for 5 minutes and discard the supernatant.

    A small white pellet should be visible at the bottom of the tube.

12. Wash the pellet with a solution of 70% DNAzol and 30% ethanol as follows:

    i. Add 700 μl of DNAzol and 300 μl of 100% ethanol to each tube.

    ii. Centrifuge the tubes at 5000g for 2 minutes.

    iii. Discard the supernatant.

13. Wash the pellet with 75% ethanol as follows:

    i. Add 1 ml of 75% ethanol to each tube.

    ii. Centrifuge the tubes at 5000g for 2 minutes.

    iii. Discard as much of the supernatant as possible.

14. Open the cap on each tube and dry the pellets briefly. Do not overdry the pellets.

15. Resuspend each pellet in an appropriate amount of sterile H_2O. Add as little as 10 μl or as much as 100 μl of sterile H_2O, depending on the size of the pellet. Mix the DNA gently by slowly pipetting up and down.

16. Check the concentration and purity of the DNA by using a spectrophotometer and by running it out on a 1% agarose gel.

    A single spawning of 500–1000 embryos should yield 10–30 µg of genomic DNA. See Troubleshooting.

17. If desired, aliquot the DNA to reduce the number of times that a sample is frozen and thawed. Store the DNA at −20°C.

TROUBLESHOOTING

Problem (Step 16): There is insoluble material in the resuspended DNA.
Solution: Sometimes, insoluble material, such as carbohydrates or proteins, is carried over. This is more common with older cydippids (>24 hours postspawning) and adult tissues because these have more jelly and mucous. Insoluble material does affect the purity of the DNA and could inhibit downstream applications such as PCR. To avoid this, allow the tube to sit overnight at 4°C after Step 15 to completely resuspend the DNA. Centrifuge the sample at maximum speed in a microcentrifuge for 10 minutes. Then, transfer the supernatant, which should contain the DNA, to a new tube; do not touch the pellet, which contains the insoluble material.

Problem (Step 16): DNA yields are low.
Solution: Sometimes too much tissue can inhibit the extraction. If starting with a decent amount of material and still obtaining low yields, try diluting the samples 1:1 or 1:5 with additional DNAzol after Step 7.

Protocol 5

Ctenophore Tissue Preparation and Extraction of RNA

This protocol describes how to isolate total RNA from ctenophore embryos and larvae. After the specimens are sorted, cleaned, and concentrated, they are placed into TRI Reagent (Molecular Research Center), a solution containing phenol and guanidine thiocyanate that allows for the effective isolation of total RNA. The resulting RNA can be used for various applications (e.g., to generate cDNA for reverse transciptase–PCR [RT-PCR]).

MATERIALS

CAUTION: See the Cautions Appendix for appropriate handling of materials marked with <!>.

Reagents

Agarose gel (1%) and appropriate gel electrophoresis buffer (see Step 14)
1-Bromo-3-chloropropane (BCP) <!>
Ctenophore embryos (and/or cydippid larvae)
  Spawn animals and collect the embryos as described in Protocol 1. Raise the embryos until the desired stage.
Ethanol (75%, ice cold) <!>
H_2O (sterile and nuclease-free) or RNA*secure* Resuspension Solution (Ambion) (see Step 13)
Isopropanol <!>
Phase Lock Gel Heavy (2 ml; 5 PRIME)

Seawater (fresh, filtered, 0.2 µm; FSW)
TRI Reagent (Molecular Research Center)

Equipment

Culture dishes (4.5 inch, glass; Carolina Biological)
Gel electrophoresis equipment
Homogenizer (drill and pestle)
Microcentrifuge (at 4°C)
Microcentrifuge tubes (1.5 ml)
Microscope (dissecting)
Spectrophotometer
Transfer pipettes (plastic)
Vortex

METHOD

Tissue Preparation

1. Use a transfer pipette to concentrate the embryos in a dish of FSW. Wash the embryos a few times with FSW to break up the egg jelly and to remove dirt and other debris.
   Embryos can be used with or without their vitelline membranes.
2. Under a dissecting microscope, check to ensure that there are no contaminants.

3. Concentrate 200–500 embryos by pipetting them into a 1.5-ml microcentrifuge tube and allowing them to settle. If necessary, centrifuge the tubes in a microcentrifuge at 2000g for 5 minutes. Remove as much FSW as possible, so that the final volume in each tube is less than 100 µl.

RNA Extraction

4. Add 250 µl of TRI Reagent to each tube of embryos. If the vitelline membranes have been removed, dissolve the embryos by pipetting up and down a few times and by vortexing. If the embryos still have their membranes, attach the pestle to the drill and homogenize for a few minutes until most of the tissue has dissolved.

5. Add 750 µl of TRI Reagent to bring the volume up to approximately 1 ml. Vortex to mix.
   At this stage, the sample can be stored at –80°C or processed immediately.

6. Centrifuge at maximum speed in a microcentrifuge for 10 minutes at 4°C to pellet any insoluble material.

7. Transfer the supernatant to a new tube. Add 100 µl of BCP and vortex for 15 seconds to mix well.

8. Pipette the mixture into a Phase Lock Gel tube. Allow it to sit for 15 minutes at room temperature.

9. Centrifuge at maximum speed in a microcentrifuge for 5 minutes at 4°C.
   The Phase Lock Gel should form a barrier between the aqueous and organic phases.

10. Carefully decant the aqueous phase into a new tube. To precipitate the RNA, add 500 µl of isopropanol and mix by inverting. Allow the tube to sit for 10 minutes at room temperature.
    Alternatively, store the tube overnight at –20°C.

11. Centrifuge at maximum speed in a microcentrifuge for 8 minutes at 4°C to pellet the RNA.
    There should be a small translucent or whitish pellet on the bottom of the tube.

12. Carefully discard the supernatant. Wash the pellet as follows:
    i. Add 1 ml of ice-cold 75% ethanol to the tube.
    ii. Centrifuge the sample at 7500g for 5 minutes at 4°C.
    iii. Discard as much of the supernatant as possible.
    iv. Allow the pellet to air-dry briefly.

13. Resuspend the RNA in an appropriate volume (10–50 µl) of sterile nuclease-free H_2O or RNA*secure* Resuspension Solution. Mix by pipetting gently.

14. Check the concentration and purity of the RNA by using a spectrophotometer and by running it out on a 1% agarose gel. (See Troubleshooting.)
    The typical yield of RNA from one starting tube should be approximately 5–10 µg of total RNA. See Troubleshooting

15. Aliquot the RNA to reduce the number of times that a sample is frozen and thawed. Store the RNA at –80°C.

TROUBLESHOOTING

Problem (Step 14): The RNA is degraded or RNA yields are low.
Solution: Use more tissue in Step 3. In addition, make sure that all solutions and equipment are free of RNases.

Problem (Step 14): There are protein or carbohydrate contaminants in the RNA.
Solution: Perform a postpurification step (after Step 13) using the QIAGEN RNeasy Micro Kit. Passing the RNA sample through one of the columns in this kit will get rid of most contaminants but will also result in a lower yield.

Recipes

The recipes for items marked with <R> are also listed here.

CAUTION: See the Cautions Appendix for appropriate handling of materials marked with <!>.

ALKALINE PHOSPHATASE (AP) BUFFER

Reagent	Quantity (for 10 ml)	Final concentration
$MgCl_2$ (1 M)	0.5 ml	50 mM
NaCl (5 M)	0.2 ml	100 mM
Tris-Cl (1 M, pH 9.5)	1 ml	100 mM
Tween 20 (20%)	0.25 ml	0.5%
H_2O	to 10 ml	

This buffer can be prepared with or without $MgCl_2$. Make sure to use a fresh stock of Tween 20; if it is not fresh, the buffer will turn cloudy. Prepare buffer the same day it is to be used and store at room temperature.

10x BLOCKING BUFFER FOR IN SITU HYBRIDIZATION

Reagent	Quantity (for 10 ml)	Final concentration (10x)
Blocking Reagent (Roche)	1 g	10%
MAB <R>	to 10 ml	

Dissolve the Blocking Reagent in MAB with shaking and heating. Autoclave. Store in aliquots at –20°C.

BLOCKING SOLUTION FOR ANTIBODY STAINING

Reagent	Quantity (for 100 ml)	Final concentration
Goat serum, normal	10 ml	10%
Triton X-100 <!>	0.2 ml	0.2%
1x PBS <!>	to 100 ml	

Store solution at 4°C and it will keep for 1–2 months.

DEVELOPING BUFFER

Reagent	Quantity (for 10 ml)	Final concentration
5-Bromo-4-chloro-3-indolyl-phosphate (BCIP) (50 mg/ml) <!>	33 µl	0.165 mg/ml
4-Nitro blue tetrazolium chloride (NBT) (100 mg/ml) <!>	33 µl	0.33 mg/ml
AP buffer with $MgCl_2$ <R>	10 ml	

Prepare fresh and store at room temperature.

FIXATION BUFFER 1 FOR ANTIBODY STAINING

Reagent	Quantity (for 100 ml)	Final concentration
Glutaraldehyde (25%) <!>	80 μl	0.02%
Paraformaldehyde (16%) <!>	25 ml	4.0%
Seawater (fresh, filtered, 0.2 μm; FSW)	75 ml	

Store buffer at 4°C and it will keep for 1–2 days.

FIXATION BUFFER 1 FOR IN SITU HYBRIDIZATION

Reagent	Quantity (for 100 ml)	Final concentration
Glutaraldehyde (25%) <!>	800 μl	0.2%
Paraformaldehyde (16%) <!>	25 ml	4.0%
Seawater (fresh, filtered, 0.2 μm; FSW)	75 ml	

Store buffer at 4°C and it will keep for 1–2 days.

FIXATION BUFFER 2

Reagent	Quantity (for 100 ml)	Final concentration
Paraformaldehyde (16%) <!>	25 ml	4%
Seawater (fresh, filtered, 0.2 μm; FSW)	75 ml	

Store buffer at 4°C and it will keep for 1–2 days.

HYBRIDIZATION BUFFER

Reagent	Quantity (for 10 ml)	Final concentration
Formamide <!>	5 ml	50%
Heparin (20 mg/ml) <!>	25 μl	50 μg/ml
SDS (20%) <!>	500 μl	1.0%
Salmon sperm DNA (10 mg/ml)	50 μl	100 μg/ml
20x SSC (pH 4.5) <R>	2.5 ml	5x
Tween 20 (20%)	50 μl	0.1%
H_2O	to 10 ml	

Store buffer at −20°C and it will keep for 1–2 months.

MALEIC ACID BUFFER (MAB)

Reagent	Quantity (for 500 ml)	Final concentration
Maleic acid (1 M, pH 7.5) <!>	50 ml	0.1 M
NaCl (5 M)	15 ml	0.15 M
H_2O	to 10 ml	

Adjust the pH to 7.5 with NaOH <!>. Sterilize by autoclaving for 20 minutes at 15 psi and store at room temperature. Buffer will keep for 6–12 months.

10x PBS

Reagent	Quantity (for 10 ml)	Final concentration (10x)
NaCl	80 g	1.37 M
KCl <!>	2 g	27 mM
Na_2HPO_4	14.4 g	100 mM
KH_2PO_4	2.4 g	18 mM

Dissolve the reagents listed above in 800 ml of H_2O. Adjust the pH to 7.4 with HCl <!> and then add H_2O to 1 liter. Dispense the solution into aliquots and sterilize them by autoclaving for 20 minutes at 15 psi (1.05 kg/cm^2) on liquid cycle or by filter-sterilization. Store at room temperature and dilute to 1x with H_2O before use.

PTw

Reagent	Quantity (for 1 liter)	Final concentration
Tween 20 (20%)	5 ml	0.1%
1x PBS	955 ml	

Store PTw at room temperature and it will keep for 6–12 months.

20x SSC

Reagent	Quantity (for 1 liter)	Final concentration
NaCl	175.3 g	3.0 M
Sodium citrate <!>	88.2 g	0.3 M

Dissolve the ingredients in 800 ml of H_2O. Adjust the pH to 4.5 or 7.0, as necessary, with 14 N HCl <!> and then adjust the final volume to 1 liter with H_2O. Dispense into aliquots. Sterilize by autoclaving.

REFERENCES

Afzelius, M.A. 1961. The fine structure of the cilia from ctenophore swimming plates. *J. Cell Biol.* **9:** 383–394.

Anctil, M. 1985. Ultrastructure of the luminescent system of the ctenophore *Mnemiopsis leidyi*. *Cell Tissue Res.* **242:** 333–340.

Anderson, P.A.V. 1984. The electrophysiology of single smooth muscle cells isolated from the ctenophore *Mnemiopsis*. *J. Comp. Physiol. B* **154:** 257–268.

Baker, L.D. and Reeve, M.R. 1974. Laboratory culture of the lobate ctenophore *Mnemiopsis mccradyi* with notes on feeding and fecundity. *Mar. Biol.* **26:** 57–62.

Bumann, D. and Puls, G. 1997. The ctenophore *Mnemiopsis leidyi* has a flow-through system for digestion with three consecutive phases of extracellular digestion. *Physiol. Zool.* **70:** 1–6.

Carre, D., Rouviere, C., and Sardet, C. 1991. *In vitro* fertilization in ctenophores: Sperm entry, mitosis, and the establishment of bilateral symmetry in *Beroe ovata*. *Dev. Biol.* **147:** 381–391.

Chun, C. 1880. Die Ctenophoren des golfes von neapel und der angrenzenden meeres-abschnitte. In *Fauna und flora des golfes neapel*, vol. 1, pp. 1–311. Engelmann, Leipzig.

Coonfield, B.R. 1936. Regeneration in *Mnemiopsis leidyi*, Agassiz. *Biol. Bull.* **71:** 421–428.

Derelle, R. and Manuel, M. 2007. Ancient connection between NKL genes and the mesoderm? Insights from *Tlx* expression in a ctenophore. *Dev. Genes Evol.* **217:** 253–267.

Dunlap Pianka, H. 1974. Ctenophora. In *Reproduction of marine invertebrates* (ed. C. Giese and J.S. Pearse), pp. 201–265. Academic, New York.

Dunn, C.W., Hejnol, A., Matus, D.Q., Pang, K., Browne, W.E., Smith, S.A., Seaver, E., Rouse, G.W., Obst, M., Edgecombe, G.D., et al. 2008. Broad phylogenomic sampling improves resolution of the animal tree of life. *Nature* **452:** 745–749.

Franc, J. 1978. Organization and function of ctenophore colloblasts: An ultrastructural study. *Biol. Bull.* **155:** 527–541.

Freeman, G. 1977. The establishment of the oral-aboral axis in the ctenophore embryo. *J. Embryol. Exp. Morphol.* **42:** 237–260.

Gregory, T.R., Nicol, J.A., Tamm, H., Kullman, B., Kullman, K., Leitch, I., Murray, B.G., Kapraun, D.F., Greilhuber, J., and Bennett, M.D. 2006. Eukaryotic genome size databases. *Nucleic Acids Res.* **35:** D332–D338.

Haddock, S.H.D. 2004. A golden age of gelata: Past and future research on planktonic ctenophores and cnidarians. *Hydrobiologia* **530/531:** 549–556.

Hansson, H.G. 2006. Ctenophores of the Baltic and adjacent seas— The invader *Mnemiopsis* is here! *Aquat. Invasions* **1:** 295–298.

Hernandez-Nicaise, M.-L. 1973. The nervous system of ctenophores. III. Ultrastructure of synapses. *J. Neurocytol.* **2:** 249–263.

Hernandez-Nicaise, M.-L. 1984. The Ctenophora. In *Biology of the integument: Invertebrates* (ed. J. Bereiter-Hahn et al.), vol. 1, pp. 96–111. Springer-Verlag, Berlin/Heidelberg.

Hernandez-Nicaise, M.-L. 1991. Ctenophora. In *Microscopic anatomy of invertebrates: Placozoa, Porifera, Cnidaria and Ctenophora* (ed. F.W. Harrison and J.A. Westfall), vol. 2, pp. 359–418. Wiley, New York.

Hernandez-Nicaise, M.-L., Nicaise, G., and Malaval, L. 1984. Giant smooth muscle fibers of the ctenophore *Mnemiopsis leidyi*: Ultrastructural study of in situ and isolated cells. *Biol. Bull.* **167:** 210–228.

Hertwig, R. 1880. Ueber den bau der Ctenophoren. *Jena Z. Naturw.* **14:** 313–457.

Horridge, G.A. 1964. Presumed photoreceptive cilia in a ctenophore. *Q. J. Microsc. Sci.* **105:** 311–317.

Hyman, L.H. 1940. *The invertebrates: Protozoa through Ctenophora*, pp. 662–695. McGraw, New York.

Kideys, A.E. 2002. Fall and rise of the Black Sea ecosystem. *Science* **297:** 1482–1484.

Kremer, P. 1994. Patterns of abundance for *Mnemiopsis* in US coastal waters: A comparative overview. *ICES J. Mar. Sci.* **51:** 347–354.

Main, R.J. 1928. Observations of the feeding mechanism of a ctenophore, *Mnemiopsis leidyi*. *Biol. Bull.* **55:** 69–78.

Martindale, M.Q. 1986. The ontogeny and maintenance of adult symmetry properties in the ctenophore, *Mnemiopsis leidyi*. *Dev. Biol.* **118:** 556–576.

Martindale, M.Q. 1987. Larval reproduction in the ctenophore *Mnemiopsis mccradyi* (order Lobata). *Mar. Biol.* **94:** 409–414.

Martindale, M.Q. and Henry, J.Q. 1999. Intracellular fate mapping in a basal metazoan, the ctenophore *Mnemiopsis leidyi*, reveals the origins of mesoderm and the existence of indeterminate cell lineages. *Dev. Biol.* **214:** 243–257.

Mayer, A.G. 1912. *Ctenophores of the Atlantic coast of North America*. Carnegie Institute of Washington, Washington, D.C.

Metschnikoff, E. 1885. Vergleichend-embryologische studien. IV. Ueber die gastrulation und mesodermbildung der Ctenophoren. *Z. Wiss. Zool.* **42:** 648–656.

Moss, A.G., Rapoza, R.C., and Muellner, L. 2001. A novel cilia-based feature within the food grooves of *Mnemiopsis mccradyi* Mayer. *Hydrobiologia* **451:** 287–294.

Pang, K. and Martindale, M.Q. 2008. Developmental expression of homeobox genes in the ctenophore *Mnemiopsis leidyi*. *Dev. Genes Evol.* **218:** 307–319.

Podar, M., Haddock, S.H.D., Sogin, M.L., and Harbison, G.R. 2001. A molecular phylogenetic framework for the phylum Ctenophora using 18S rRNA genes. *Mol. Phylogenet. Evol.* **21:** 218–230.

Purcell, J.E., Shiganova, T.A., Decker, M.B., and Houde, E.D. 2001. The ctenophore *Mnemiopsis* in native and exotic habitats: U.S. estuaries versus the Black Sea basin. *Hydrobiologia* **451:** 145–176.

Reitzel, A.M., Sullivan, J.C., Brown, B.K., Chin, D.W., Cira, E.K., Edquist, S.K., Genco, B.M., Joseph, O.C., Kaufman, C.A., Kovitvongsa, K., et al. 2007. Ecological and developmental dynamics of a host-parasite system involving a sea anemone and two ctenophores. *J. Parasitol.* **93:** 1392–1402.

Reverberi, G. and Ortolani, G. 1965. The development of the ctenophores egg. *Riv. Biol.* **58:** 113–137.

Seravin, L.N. 1994. The systematic revision of the genus *Mnemiopsis* (Ctenophora, Lobata). *Zool. Zh.* **73:** 9–18.

Sullivan, B.K., Van Keuren, D., and Clancy, M. 2001. Timing and size of blooms of the ctenophore *Mnemiopsis leidyi* in relation to temperature in Narragansett Bay, RI. *Hydrobiologia* **451:** 113–201.

Sullivan, L.J. and Gifford, D.J. 2007. Growth and feeding rates of the newly hatched larval ctenophore, *Mnemiopsis leidyi* A. Agassiz (Ctenophora, Lobata). *J. Plankton Res.* **29:** 949–965.

Tamm, S.L. 1982. Ctenophora. In *Electrical conduction and behavior in "simple" invertebrates* (ed. G.A.B. Shelton), pp. 266–358. Claredon, Oxford.

Tamm, S.L. and Tamm, S. 2002. Novel bridge of axon-like processes of epithelial cells in the aboral sense organ of ctenophores. *J. Morphol.* **254:** 99–120.

Waggett, R. and Costello, J.H. 1999. Capture mechanisms used by the lobate ctenophore, *Mnemiopsis leidyi*, preying on the copepod *Acartia tonsa*. *J. Plankton Res.* **21:** 2037–2052.

Wallberg, A., Thollesson, M., Farris, J.S., and Jondelius, U. 2004. The phylogenetic position of the comb jellies (Ctenophora) and the importance of taxonomic sampling. *Cladistics* **20:** 558–578.

Ward, W.W. and Seliger, H.H. 1974. Properties of mnemiopsis and berovin, calcium-activated photoproteins from the ctenophores *Mnemiopsis* sp. and *Beroe ovata*. *Biochemistry* **13:** 1500–1510.

Yamada, A., Pang, K., Martindale, M.Q., and Tochinai, S. 2007. Surprisingly complex T-box gene complement in diploblastic metazoans. *Evol. Dev.* **9:** 220–230.

8 Planarians
A Versatile and Powerful Model System for Molecular Studies of Regeneration, Adult Stem Cell Regulation, Aging, and Behavior

Néstor J. Oviedo, Cindy L. Nicolas, Dany S. Adams, and Michael Levin

Center for Regenerative and Developmental Biology, Forsyth Institute and Developmental Biology Department, Harvard School of Dental Medicine, Boston, Massachusetts 02115

ABSTRACT

In recent years, planarians have been increasingly recognized as an emerging model organism amenable to molecular genetic techniques aimed at understanding complex biological tasks commonly observed among metazoans. Growing evidence suggests that this model organism is uniquely poised to inform us about the mechanisms of tissue regeneration, stem cell regulation, tissue turnover, pharmacological action of diverse drugs, cancer, and aging. This chapter provides an overview of the planarian model system with special attention to the species *Schmidtea mediterranea*. Additionally, information is provided about the most popular use of this organism together with modern genomic resources and technical approaches. Furthermore, a detailed protocol for establishing and maintaining planarian colonies for diverse purposes is described. Because RNA interference (RNAi) is increasingly used to study gene function in planarians, details on this technique are also given. Finally, a detailed protocol is included to illustrate why planarians are suitable for the study of complex physiological processes during regeneration and tissue turnover, in real time.

> **PROTOCOLS**
> 1. Establishing and Maintaining a Colony of Planarians, 201
> 2. Gene Knockdown in Planarians Using RNA Interference, 206
> 3. Live Imaging of Planarian Membrane Potential Using $DiBAC_4(3)$, 210

BACKGROUND INFORMATION

Planarians, free-living nonparasitic invertebrates, are one of the most basal triploblastic organisms, with derivatives of all three germ layers (ectoderm, mesoderm, and endoderm). They represent a critical breakthrough in the evolution of the animal body plan, have bilateral symmetry and encephalization, and are capable of detecting environmental stimuli very efficiently. Planarians are perhaps better known for their extraordinary regenerative capacity, which is associated with a large population of adult stem cells (neoblasts).

An extensive amount of literature on planarian research has accumulated during more than 200 years (Reddien and Sánchez Alvarado 2004). In the last century, a number of laboratories around the world have used planarians as a biological model in a surprisingly broad and extensive

This chapter, with full-color images, can be found online at www.cshprotocols.org/emo.

range of applications. Systematic research on planarians was first substantiated by Thomas Hunt Morgan and Harriet Randolph, who studied regeneration in planarians (Randolph 1897; Morgan 1898). The fact that freshwater planarians are commonly found in the wild (e.g., ponds, streams, and rivers) and are easily reared and maintained under laboratory conditions, together with their astonishing regenerative capacity and adult tissue plasticity, were factors that undoubtedly made these organisms a favorite subject for researchers in subsequent years.

Although several hundred planarian species exist, research on freshwater planarians in the last century has been restricted to species either commonly found in the wild or commercially available (mostly members of the Dugesiidae family, including the genera *Dugesia*, *Girardia*, *Neppia*, *Romankenkius*, *Schmidtea*, and *Spathula*). Modern research on freshwater planarians has applied diverse state-of-the-art techniques to understand their cellular biology, genetics, genomics, behavior, and other aspects at the molecular level (Sánchez Alvarado 2006; Oviedo and Levin 2008).

In the last 10 years, attention to planarian molecular and genetic research has been greatly enhanced and has focused on work in two species in which clonal lines have been derived: *Schmidtea mediterranea* and *Dugesia japonica*. Advantages of using *S. mediterranea* over other planarian species for molecular studies have been reviewed elsewhere (Newmark and Sánchez Alvarado 2002; Sánchez Alvarado 2006), and those advantages have been enhanced with the recent introduction of high-throughput molecular tools, large collections of publicly available DNA sequences, automated systems for the analysis of behavior, and genomic resources (Sánchez Alvarado et al. 2002; Zayas et al. 2005; Hicks et al. 2006; Robb et al. 2007). Although most protocols and techniques can be adapted to different planarian species, here we focus on *S. mediterranea*, an increasingly popular planarian species used to study the molecular basis of tissue regeneration and stem cell biology. The taxonomic classification for *S. mediterranea* may follow this order: domain Eukaryota; kingdom Animalia; phylum Platyhelminthes (flatworms); class Turbellaria; order Seriata; suborder Tricladida; family Dugesiidae; genus *Schmidtea*; species *S. mediterranea*. Further taxonomic details for the class Turbellaria can be found elsewhere (http://turbellaria.umaine.edu).

About 40 years ago, in an artificial pond at Montjuïc near Barcelona, Spain, Jaume Baguñà and Rafael Ballester identified the asexual (fissiparous) biotype of *Dugesia lugubris* s.l. (Benazzi et al. 1970, 1972). Karyological and other studies confirmed that this asexual worm matched a diploid ($2n = 8$) sexual form of *D. lugubris* s.l. that was known from the Mediterranean islands of Corsica, Sardinia, and Sicily; these two forms were later classified together as *D. mediterranea* (Benazzi et al. 1975). Further analyses by different investigators led to the renaming of this species as it is currently known: *S. mediterranea* (Baguñà 1999). The primary difference between the asexual and sexual strains of *S. mediterranea* is a chromosomal translocation between chromosomes 1 and 3 that is observed in fissiparous animals only (Baguñà 1999; Newmark and Sánchez Alvarado 2002). The diploid asexual strain of *S. mediterranea* reproduces by transverse fission (i.e., the posterior end attaches to a surface, whereas the anterior part tears away until the animal breaks transversally in the postpharyngeal region), and the missing structures in each fragment are regenerated in about 1 week. The *S. mediterranea* sexual strain is a hermaphrodite but requires cross-fertilization in order to lay cocoons that after few weeks hatch, giving rise to one or several planarians (Sánchez Alvarado 2003). Studies on the geographical distribution of *S. mediterranea* revealed that the natural habitat of the asexual biotype is limited to the Iberian peninsula and Mallorca, whereas the sexual strain is more widely distributed throughout the islands of the western Mediterranean (De Vries 1985).

In the mid 1990s, Alejandro Sánchez Alvarado chose planarians as a convenient model in which to study the molecular basis of tissue regeneration (Sáchez Alvarado 2006). Together with Philip Newmark, Sánchez Alvarado selected *S. mediterranea* because of its extraordinary regenerative properties, karyological features (stable diploid), and relatively small genome size compared to other common planarians (Newmark and Sánchez Alvarado 2002; Sánchez Alvarado 2006). Asexual specimens of *S. mediterranea* were first introduced to the United States by Newmark in 1997; however, those animals did not survive. Then, after an expedition to Montjuïc in 1998, Newmark and Sánchez Alvarado brought to the United States several specimens of fissiparous *S. mediterranea* that, after several rounds of culture optimization, were successfully reared under laboratory conditions. Some of the specimens that thrived in the new laboratory conditions were individually cut and grown separately to establish clonal lines. This led to the establishment of the asexual strain CIW4 (Carnegie Institution of Washington clonal line 4) (Sánchez Alvarado 2006).

Maria Pala (of Italy) provided the Sánchez Alvarado laboratory with wild-type animals of the *S. mediterranea* sexual strain from the Mediterranean island of Corsica (Sánchez Alvarado 2003). Clonal lines were created using a strategy similar to that used to establish CIW4. Laboratory conditions were adjusted to allow sexual reproduction of these clonal worms (Sánchez Alvarado 2003, 2006). This offered a unique opportunity to study embryonic development and functional germ-cell specification in *S. mediterranea* (Sánchez Alvarado 2003). Furthermore, the inbreeding of these animals (performed by Peter W. Reddien) created "more genetically uniform worms" and led to the generation of the S2F2 line, the source of animals used to sequence the *S. mediterranea* genome.

SOURCES AND HUSBANDRY

Protocol 1 describes in detail how to establish and maintain a colony of planarians for behavioral and other experiments.

RELATED SPECIES

Besides *S. mediterranea*, another planarian species commonly used to study regenerative biology is *D. japonica*. Mixoploid and diploid clonal lines (GI, HI, and SSP) of this organism originated in Japan and were subsequently expanded, mainly laboratories of Kenji Watanabe and Kiyokazu Agata. Interestingly, different procedures (e.g., decreasing the temperature of the culture and feeding asexual-state planarians with freeze-thawed sexual-state worms) are commonly used to induce gonad development in asexual worms (Ogawa et al. 1998). Similar to the sexual strain of *S. mediterranea*, germ cells are fully developed in adult *D. japonica*.

Important advances have been made possible by using *D. japonica*; in fact, several protocols (e.g., whole-mount in situ hybridization and neoblast purification with flow cytometry) were initially developed in this species and later optimized and adjusted for *S. mediterranea* (Umesono et al. 1997; Sánchez Alvarado et al. 2002; Hayashi et al. 2006; Sánchez Alvarado 2007). However, there are some important differences between *S. mediterranea* and *D. japonica* (including expression of the *piwi* and *innexin* genes and sensitivity to drugs) (Nogi and Levin 2005; Reddien et al. 2005; Rossi et al. 2006; Oviedo and Levin 2007). Large collections of expressed sequence tags (ESTs) from *D. japonica* can be accessed at Keio University in Japan (http://planarian.bio.keio.ac.jp/go/EST/P/GO0007582/GO0006915.html), and a database with neoblast-associated genes has been recently published (Rossi et al. 2007b).

USES OF THE PLANARIAN MODEL SYSTEM

For the last century, planarians have been a favorite model system in many research areas. Not only are planarians easy to rear under laboratory conditions and relatively inexpensive to maintain, they are amenable to pharmacological, behavioral, physiological, molecular genetic, and classical surgical techniques. Thus, in recent years they have become a popular model for state-of-the-art studies in regenerative biology as well as a low-cost system to produce fast experimental results for classroom science projects and other educational purposes (Oviedo and Levin 2008).

Tissue Regeneration, Cellular Turnover, and Aging

The fact that adult planarians can regenerate large parts of their bodies from tiny fragments (in ~1 week) makes them an outstanding model to study regeneration and morphogenesis in complex tissues (Newmark and Sánchez Alvarado 2002; Agata 2003; Reddien and Sánchez Alvarado 2004). The planarian central nervous system (CNS) is the organ system that has been more extensively studied (Cebrià 2007). The planarian CNS is particularly important because it exhibits evolutionary conservation with higher organisms in many neural receptors (e.g., acetylcholine receptor, fibroblast growth factor receptor, and netrin receptor), neuroactive molecules (e.g., γ-aminobutyric acid and FMRFamide-like peptides), and axon guidance molecules (e.g., netrin and roundabout receptor-robo). Recent functional/behavioral assays suggest that this conservation holds at the structural and physiological levels (for references, see Oviedo and Levin 2008). Moreover, the molecular conservation of the CNS also indicates that planarians are a good model in which to test the effects and mechanisms of drugs using large pharmacological screens (Oviedo and Levin 2008).

Remarkably, the rapid reconstruction of multiple missing parts (e.g., CNS, muscle, digestive, sensory system, and epithelium) in S. mediterranea is accompanied by an adjustment in growth to accommodate a new body size. In other words, small planarian fragments regenerate structures such as the CNS with proportions that are based on the new size of the animal, rather than the size of the original fragment (Newmark and Sánchez Alvarado 2002; Oviedo et al. 2003; Reddien and Sánchez Alvarado 2004). Additionally, intact adult animals have the capacity to regulate their body sizes according to metabolic status (i.e., worms grow when food is available and, if starved, they reduce their size proportionally in a process known as "degrowth"). Interestingly, both growth and degrowth in S. mediterranea involve well-regulated mechanisms for adding and subtracting cells in the body (Baguñà and Romero 1981; Oviedo et al. 2003). Taken together, these observations indicate that planarians possess efficient mechanisms for synchronizing metabolic status and tissue maintenance. The aging process in planarians is fascinating because the neoblasts' progeny regularly replace senescent cells in different organs (see below). Thus, their tissues are continuously renewed. Therefore, S. mediterranea is an exquisite model for studying the regulation of adult differentiated tissue remodeling, tissue maintenance, cellular turnover, and aging (Pellettieri and Sánchez Alvarado 2007).

The anatomy of the reproductive system of S. mediterranea, which consists of paired ovaries, many testicles, and a copulatory apparatus, has been reviewed elsewhere (Newmark and Sánchez Alvarado 2002). Unlike other commonly studied invertebrate models, planarians do not appear to segregate their germ-cell lineage during early embryogenesis; instead, the reproductive system in sexual planarians has been proposed to be determined by epigenetic mechanisms after hatching (postembryonically) (Zayas et al. 2005; Sato et al. 2006; Handberg-Thorsager and Saló 2007; Wang et al. 2007). If a planarian with fully developed gonads is starved for several weeks, the gonads are resorbed, and if the same animal is fed and allowed to reach an appropriate size, the reproductive organs are properly regenerated (Newmark and Sánchez Alvarado 2002). Similarly, if the head of a mature sexual planarian is removed, the remaining gonads in the body are resorbed and will develop again only after the head regenerates completely (Newmark and Sánchez Alvarado 2002; Wang et al. 2007).

Stem Cells and Germ Cells

In planarians, tissue regeneration, germ-cell specification, tissue remodeling, and adult tissue maintenance involve a large population of undifferentiated cells known as neoblasts (planarian stem cells), located throughout the body, which constantly divide. During regeneration, these cells are the source of new tissue. In asexual worms, neoblasts are the only known cell with mitotic activity and are therefore the sole source of new cells. Thus, neoblasts coexist in a microenvironment that is tightly regulated and allows them to respond to signals to self-renew, proliferate, and migrate, giving rise to differentiated progeny that are properly incorporated into demanding tissues. In recent years, an important focus of research in planarians has been centered on understanding, at the genetic and biochemical levels, how neoblasts are regulated in vivo to respond to demanding signals while renewing themselves (Agata 2003; Sánchez Alvarado and Kang 2005; Sánchez Alvarado 2006; Rossi et al. 2007a).

The complexity of the neoblast population is far from being understood (Rossi et al. 2007a). Neoblast populations are mainly recognized by their morphology, spatial distribution, sensitivity to γ-irradiation, ultrastructural composition, and gene expression patterns (Orii et al. 2005; Reddien et al. 2005; Hayashi et al. 2006; Sato et al. 2006; Handberg-Thorsager and Saló 2007; Higuchi et al. 2007; Oviedo and Levin 2007; Rossi et al. 2007a,b; Wang et al. 2007). Study of the molecular conservation of regulatory molecules (e.g., the *piwi* genes) in neoblasts has expanded our knowledge of the evolution of stem cells (Sánchez Alvarado and Kang 2005). Furthermore, the identification and characterization of stem cell regulatory molecules that are conserved between vertebrates and planarians (but absent in other classical invertebrate model organisms such as *Drosophila melanogaster* and *Caenorhabditis elegans*) is now possible in *S. mediterranea* (Oviedo et al. 2008).

The molecular basis of germ-cell specification in adult stages is poorly understood at this time (Zayas et al. 2005). Interestingly, it has been proposed that germ cells in sexual *S. mediterranea* are epigenetically specified at adult stages and that the maintenance of these epigenetic marks can be affected by diverse factors such as metabolic status, regeneration, and seasonal conditions (Zayas et al. 2005). The plasticity of germ-cell specification and maintenance in adult *S. mediterranea* provides a unique opportunity to investigate regulatory cues that modulate this process. Taking advantage of these properties, the Newmark lab at the University of Illinois used clonal animals of the sexual strain of *S. mediterranea* to build an EST database comprising two different developmental stages (Zayas et al. 2005). Thus, this database will be useful for the identification of germ-cell markers as well as signaling pathways involved in germ-cell determination and tissue regeneration.

Memory, Learning, and Behavior

As the simplest animal with a bilaterally symmetrical CNS composed of neurons similar to our own (Cebrià 2007), planarians offer an opportunity to study neural patterning and behavior while working with an easily maintained organism that is highly tractable to pharmacological, surgical, and molecular genetic manipulations. Planarians have developed sensory capabilities for the detection of light, chemical gradients, vibration, electric fields, magnetic fields, and weak γ-radiation. When proper procedures are followed, the data conclusively show that planarians can learn. Thus, planarians are a unique model system in which memory and regeneration can be studied in the same animal. This kind of system provides an ideal context in which to characterize the impact of naïve stem cells on memories stored in an adult brain, which is of relevance to the current approaches in treating degenerative diseases of the human brain (McKay 2004), or to dissect the mechanisms by which existing memories can be imposed on regenerated brain tissue. For an in-depth discussion of memory and learning in planarians, consult Nicolas et al. (2008).

GENETICS AND GENOMICS RESOURCES

At this point in time, genetic tools for gain-of-function and permanent genomic modifications in planarians are not well established. Recently, a proposal involving guidelines for gene and protein nomenclature in the planarian species *S. mediterranea* was made public (Reddien et al. 2008). Applying these nomenclature guidelines will help to standardize naming and facilitate the identification of gene homology and experimental treatments (e.g., RNAi). Briefly, the prefix "*Smed*" should be attached to the beginning of all gene names; the gene name should be selected to match the convention of the homologous gene and should be lowercase and italicized (e.g., *Smed-genex*). When possible, genes names should try to match those of orthologs. In addition, if there exist paralogs for a given gene, the respective planarian gene gets a numerical suffix, not a letter, at the end (e.g., *Smed-genex-1*). The protein name should be uppercase and not italicized (e.g., SMED-

GENEX). A single RNAi should be denoted as follows: *Smed-genex(RNAi)*. If multiple RNAis are used, a semicolon should connect the names: *Smed-genex(RNAi)*; *Smed-geney(RNAi)*. Note that because mutant alleles and transgenes are not yet commonly used in planarian experiments, there is no standard nomenclature for such. For additional details, see Reddien et al. (2008) and http://smedgd.neuro.utah.edu.

The *S. mediterranea* sexual strain S2F2, which originated in the Sánchez Alvarado lab, was used for genomic sequencing at the Washington University Genome Sequencing Center (http://genome.wustl.edu/). The *S. mediterranea* genome database (SmedGD) at http://smedgd.neuro.utah.edu/ includes details about the available DNA sequences, as well as predicted and annotated genes, ESTs, protein homologies, gene expression patterns, and RNAi phenotypes (Robb et al. 2007). Complementary information on a large collection of ESTs from sexual and asexual *S. mediterranea* is publicly available from the laboratories of Newmark and Sánchez Alvarado (Sánchez Alvarado et al. 2002; Zayas et al. 2005).

TECHNICAL APPROACHES

Many published protocols from other well-established invertebrate systems such as *D. melanogaster* and *C. elegans* are quite robust and can be adapted for planarians to produce results in a short time (less than 1 week). Basic molecular approaches include purification of mRNA to evaluate gene expression (both quantitatively and qualitatively), antibody staining to evaluate the spatial distribution of molecular markers and morphogenesis during regeneration, in situ hybridization to evaluate the spatial distribution of messenger RNA, and quantitative real-time–polymerase chain reaction (RT-PCR) to evaluate transcript levels in planarian tissue. One common procedure to evaluate whether a specific transcript is associated with neoblast cells is to treat some worms with γ-irradiation (which eliminates neoblasts) and then compare the patterns and/or levels of gene expression between irradiated and untreated worms.

Methods used successfully for developmental studies include tissue fixation for immunostaining, in situ hybridization, and gene knockdown with RNAi. The latter is most commonly applied by microinjecting double-stranded RNA (dsRNA) as described in Protocol 2 or by feeding the worms with artificial food carrying modified bacteria. Neoblasts are the only known proliferative cell in the worm, so labeling with the thymidine analog bromodeoxyuridine (BrdU) (Newmark and Sánchez Alvarado 2000) can be used to evaluate the cell cycle and the migration and integration of the neoblasts into differentiated tissues. Neoblast subpopulations can be isolated by flow cytometry (fluorescence-activated cell sorting [FACS]) (Reddien et al. 2005; Hayashi et al. 2006) and characterized by gene expression analysis (Reddien et al. 2005; Oviedo and Levin 2007), whereas ultrastructural features can provide more details about different neoblast subpopulations (Hayashi et al. 2006; Higuchi et al. 2007).

The effects of drug compounds on cell behavior can be conveniently tested in planaria because regeneration is a sensitive assay for changes in proliferation, migration, differentiation, and morphogenetic cues. Additionally, diverse physiological processes can be analyzed in the whole organism in real time (e.g., RNAi followed by imaging of physiological parameters; see Protocol 3), providing an opportunity to dissect complex molecular interactions in vivo. Also, because regeneration can reveal subtle changes in cell signaling, planarians are a good system in which to identify and molecularly characterize the effects of weak electromagnetic stimulation (Novikov et al. 2002).

Protocols for running behavioral experiments have been refined over the years since planarians were first used for behavioral work (McConnell 1965). Although these experiments can be performed manually, the future of this field clearly rests in automated systems for training and testing worms, both for high-throughput approaches and for establishing basic quantitative and objective results. The construction and use of such automated systems has most recently been described by Hicks et al. (2006).

Protocol 1

Establishing and Maintaining a Colony of Planarians

To provide sufficient material for experimentation, a laboratory needs to expand and maintain a colony of planarians. It is crucial to keep a stable, healthy population of animals in a consistent environment to avoid interanimal variability and modifier effects that can mask true phenotypes from experimental perturbation. In this protocol, we describe basic procedures for establishing and maintaining healthy colonies of *D. japonica*, *S. mediterranea*, and *G. tigrina* (commonly found in the wild and commercially available in the United States). Although the recommendations are based on our optimization of conditions for *G. tigrina*, many of the procedures (such as food preparation and feeding strategy) can be applied to other species. For best results, the culture water must be carefully monitored and adjusted for each species.

MATERIALS

The recipe for the item marked with <R> is on page 215.

CAUTION: See the Cautions Appendix for appropriate handling of materials marked with <!>.

Reagents

Aquarium buffers, acid and alkaline (e.g., Seachem)
Beef liver (organic), very fresh
H_2O, Poland Spring®

Poland Spring® H_2O gives better results than other types of bottled water. Alternatively, 18 MΩ H_2O containing 1.6 mM NaCl, 1.0 mM $CaCl_2$ <!>, 1.0 mM $MgSO_4$ <!>, 0.1 mM $MgCl_2$, and 1.2 mM KCl can be used.

1× Montjuïch salts <R>

Equipment

Blender
Centrifuge (e.g., Sorvall RC-5B Refrigerated Superspeed Centrifuge, DuPont)
Container, covered and chilled, for collecting processed liver (optional; see Step 3)
Cutting surface, chilled
Dissolved oxygen test kit (e.g., LaMotte)
Incubator or temperature-controlled light cabinet, preset to 17–20°C
Paper towels, unbleached
Petri dishes (35 mm)
Plastic containers, food grade (e.g., Rubbermaid 1.8-liter rectangular boxes)
Scalpel
Strainer
Transfer pipettes, disposable

TABLE 1. Sources of planarians

Laboratories conducting research with planarians	
Kiyokazu Agata	http://mdb.biophys.kyoto-u.ac.jp/index_E.html
Takashi Gojobori	http://www.cib.nig.ac.jp/dda/en/index.html
Brenton Graveley	http://genetics.uchc.edu/Graveley/Welcome.html
Michael Levin	http://www.drmichaellevin.org
Nico Michiels	http://www.uni-tuebingen.de/evoeco
Phillip Newmark	http://www.life.uiuc.edu/newmark
Robert Raffa	http://www.temple.edu/pharmacy/faculty_Raffa_Research.htm
Peter Reddien	http://inside.wi.mit.edu/reddien/pub/PWR_website/Rddn_home.html
Leonardo Rossi	http://www.unipi.it/english/university/index.htm
Emili Saló	http://planarian.bio.ub.es
Alejandro Sánchez Alvarado	http://planaria.neuro.utah.edu
Eva-Maria Schoetz	http://genomics.princeton.edu/schoetzlab/index.html
Ronald Vale	http://valelab.ucsf.edu
Ricardo Zayas	http://www.bio.sdsu.edu/faculty/zayas.html
Commercial sources of planarians	
Carolina Biological Supply Company	http://www.carolina.com
WARD'S Natural Science	http://www.wardsci.com

METHOD

Obtaining Planarians

Planarians may be found in the wild—in ponds and streams (usually under rocks or attached to other surfaces)—or ordered from commercial sources (Table 1). There are no commercial suppliers for either *S. mediterranea* or *D. japonica* at this time; stocks may be obtained from research labs working with the desired species (Table 1). Currently, we know of no commercial supplier that maintains their own stock of *G. tigrina*, so acquiring these may mean obtaining planarians that were recently in the wild.

Planarians may be sent by overnight mail when the temperature is mild enough not to harm the worms; above freezing and below 25°C is recommended. A cold pack and insulation may also help to keep the temperature down in the shipping container, if necessary. For shipping, it is very important that the worms be placed in containers completely filled with water, leaving no air bubbles; this will prevent sloshing, which can kill planarians. The oxygen already dissolved in the water should be sufficient for 1–2 days. On receipt of planarians from a commercial source, immediately replace at least half of the water in which the animals were transported (be careful not to shock them with too great a temperature change) and completely change the water by no later than the next day. Water that arrives with worms is typically quite toxic with ammonia and other metabolic byproducts and is extremely low in oxygen. In our experience, a gradual change of water on arrival renders better results for clonal lines of *S. mediterranea* and *D. japonica* (e.g., reduce to 75, 50, and 25% every 2 days, until worms are in 100% fresh appropriate medium).

Culture Conditions

Media

The aqueous medium in which each species lives differs slightly. In each case, the medium should be freshly prepared (e.g., weekly) because the pH tends to drift and salts may precipitate with time. The conductivity of our water after buffering is approximately 450 µS. Water is typically changed every 2 to 5 days when the planarian containers are cleaned (see Steps 13–20). Planarian cultures should be virtually odorless (if they are not, see Troubleshooting).

For *S. mediterranea*, use the recipe for 1× Montjuïch salts published by Cebrià and Newmark (2005) on page 215. For *G. tigrina*, adapt the recipe for 1× Montjuïch salts by omitting the $NaHCO_3$ and, instead, use aquarium buffers to bring the water to a pH of 7.5. Alternatively, use

Poland Spring® water (which gives better results than several other types of spring water) and adjust the pH in the same way. If Poland Spring® is unavailable, another brand of bottled water may be tried, though some trial and error may be necessary to locate one with adequate characteristics. In either case, use an alkaline buffer:acid buffer ratio of 12.5:1 (approximately 2.5 g of alkaline buffer and 0.2 g of acid buffer in 11 liters of water) to adjust the pH. These numbers are only approximate because water from different sources and even from the same source in different seasons varies slightly in ion composition. The pH will drift upward over time but will not exceed the upper limit of *G. tigrina*'s tolerance, even if the water is not changed for 1 week. A pH range between 7.5 and 9.5 is tolerated by *G. tigrina*. For *D. japonica*, Poland Spring® water may be used alone or the pH may be raised slightly, to as high as 7.5, using the above buffers.

Containers

Worms thrive in plastic containers (e.g., 2000-ml capacity with approximately 1500 ml of culture water). To remove the manufacturing residue from new containers, soak them overnight in pure water (without soap) and then thoroughly wipe them clean. The containers should be covered, but significant air circulation should be ensured. It is important to control the number of animals per container (e.g., keep between 400 and 700 worms in each 2000-ml container). Too many worms can induce stressful situations, leading to infections by opportunistic pathogens (see Troubleshooting).

Temperature

Planarian colonies are kept in an incubator or in a temperature-controlled light cabinet at a temperature of between 17°C and 20°C. However, all three species discussed here can tolerate a variety of temperatures. *G. tigrina* have been found at water temperatures ranging from 3°C to 31°C in the wild (Stokely et al. 1965). Higher temperatures (about 25°C) are acceptable to planarians, but this warmer environment encourages bacterial growth and infections are more likely. Infections are particularly a concern in experimental animals that have been cut; regenerating worms may have higher mortality rates when kept in warmer environments.

Dissolved Oxygen

Dissolved oxygen (DO) is an important factor in worm health, but as long as planarians are kept at an appropriate temperature in boxes that have a broad water surface and reasonable air circulation, sufficient oxygen should be incorporated into the water to avoid the need for manual circulation or aeration. Our DO for *G. tigrina* normally reads between 7.6 and 8.4 ppm using a LaMotte DO test kit; fluctuation of ±0.5 ppm during the course of a day is typical. DO may be checked sporadically or when abnormalities in the colony are observed.

Light/Dark Cycle

Planarians are nocturnal, and *S. mediterranea* and *D. japonica* are best maintained under dark environments (although they are exposed to light during feeding and cleaning).

Worms (of any of the above-mentioned species) used in behavioral experiments are typically kept on a 12-hour light/12-hour dark cycle that helps to synchronize their peak activity time, thereby benefiting many types of training trials. If possible, learning trials on *G. tigrina* should be conducted during the worms' night cycle. The worms can be sluggish during the day, and this creates problems in obtaining consistent responses for behavioral work. We have compared the behavior of *G. tigrina* at the beginning of their night cycle versus midday, and a definite difference was observed (C. Nicolas, unpubl.).

Worms used in behavioral studies have also been kept in dark cabinets, exposed to light only for feeding and cleaning, with no apparent ill effects. These worms have an increased aversion to light, which may be useful in some circumstances.

Food Preparation and Allocation

Preparation of Beef Liver Paste

This method of preparation ensures that the liver will sink to the bottom of the dish where the worms can consume it.

1. Place fresh organic beef liver on a chilled cutting surface.
2. Use a scalpel to cut away all visible veins and connective tissue (including the connective tissue of the capsule) as well as any fatty inclusions.
3. Cut the liver into small pieces (1-inch cubes). If you are cutting a significant amount of liver, make sure to collect the processed liver in a chilled and covered container that is protected from the air while you work.
4. Use a blender to purée the beef liver.
5. Strain the puréed liver to remove any remaining connective tissue. A standard kitchen strainer can be used; only gentle pushing through the strainer should be used to avoid introducing connective tissue into the filtrate.
6. Centrifuge the liver paste at 4000 rpm to remove all air bubbles. Five minutes at 4°C in a Sorvall RC-5B Refrigerated Superspeed Centrifuge (DuPont) is sufficient.
7. Aliquot the liver paste into 35-mm Petri dishes (filling each dish), being careful to avoid introducing new air bubbles.
8. Freeze individual aliquots at –80°C.

Feeding

Feed planarians once a week. Before any experiment, starve them for 7 to 15 days. This fasting period provides a more uniform metabolic status, thereby minimizing variability in data.

9. Thaw an aliquot of prepared liver paste (from Step 8).
10. Drop into the colony box as much liver as the colony can consume in 2 hours (about 2 ml per 2000-ml container with ~400–700 worms). Make sure that liver paste sinks to the bottom; floating food will not be eaten.
11. After 1 or 2 hours, carefully remove any uneaten food with a pipette. Proceed immediately to Steps 12–17.

Cleaning

Planarian containers must be cleaned immediately after the feeding period. In addition, all species must have a second cleaning 2 days after feeding to remove metabolic debris and to prevent the water quality from degrading.

12. Gently agitate the water with a pipette to force any planarians from the surface to the bottom of the container.
13. Pour off all of the old water. If necessary, use the pipette to wash the worms back down to the bottom of the box as you pour off the water.
14. Add a small amount of water to the container (do not add any detergents or chemicals to the water). Use this water to rinse the sides and the bottom of the container and to wash the planarians down to one corner.
15. Pour off the rinse water.
16. Using paper towels, wipe out any mucus and debris attached to the walls and the bottom of the container.
17. Refill the container with fresh medium.

Reproduction

A choice may be made to keep either clonal colonies, where all worms derive from one original worm, or more genetically varied colonies. If clonal colonies are desired, expect one large *G. tigrina* to yield about 40 worms after 6 months of weekly feedings. Both *S. mediterranea* and *D. japonica* reproduce faster, doubling approximately every 2 to 3 weeks. Although happy (and well-fed) worms will spontaneously reproduce by fissioning, cutting may also be used to accelerate the process of populating the colony with genetically similar worms.

TROUBLESHOOTING

Problem: The planarians are lying limply on the bottom of the box, or a gentle tap on the box sends a large number of worms cascading off the sides of the box to land limply at the bottom. They may be scrunched up, looking ruffled around the edges rather than smooth. In advanced cases, they may lie on their sides curled into a "C" shape.

Solution: Under normal conditions, about 50% of our worms will be on the sides of the box at any given time. If the worms appear stressed, consider the following:

- Check the water quality. If the ammonia level of the water is too high, the dissolved oxygen content is too low, or the pH is too high or too low, change the culture medium immediately. Worms should recover from most water-quality problems by the next day or so.
- The container may be overcrowded. Split the colony.

Problem: Animals have white or black lesions on the dorsal side, or they are losing tissue at the anterior end. The lesions may be visible under a dissecting scope. At higher magnifications (~200x), protozoans may be seen clustered around the open wounds of the planarians. There may be a bad odor coming from the water.

Solution: An infestation by protozoa is not uncommon, but it can lead to a secondary bacterial infection that can devastate a colony. The symptoms may vary from worm to worm; some succumb sooner. Worms that are feeling unwell for any reason will not eat until they are recovered from the incident. Because uneaten food can provide sustenance for bacteria, we do not feed worms suffering from either water-quality problems or microorganism attacks until the issue has been resolved. Remove any sick animals with abnormal behavior and then treat the culture for microorganism attacks as described below. Keeping containers with sick worms and materials (e.g., pipettes) away from healthy animals is safe practice.

- The antibiotics metronidazole (3 mg/liter) and gentamicin (50 µg/ml) have been used to treat our *G. tigrina* and are well tolerated, but they were only marginally effective because they target the secondary bacterial infections and not the original problem. If antibiotic treatment is used, it is important to remember to bring the pH of the culture water back up to 7.5 after mixing in the antibiotics. Otherwise, these drugs can drop the pH low enough to harm the planarians.
- The most effective treatment we have found for infections in our *G. tigrina* colonies was not chemical: Simply chill the colonies of worms overnight down to 10°C. Then, after slowly warming them back up to 18°C the next day, clean the containers thoroughly. Daily water changes during the next week are helpful in speeding the recovery of this species. The good news is that lost tissues should be regenerated shortly.

Protocol 2

Gene Knockdown in Planarians Using RNA Interference

This protocol describes how to produce gene knockdown in planarians using RNAi. It is a standard technique to evaluate gene function during regeneration and tissue maintenance in planarians. The procedure involves microinjecting dsRNA synthesized in vitro (Sánchez Alvarado and Newmark 1999). Depending on the gene target, this technique can produce robust phenotypes that can be further evaluated by diverse macroscopic or microscopic procedures.

MATERIALS

Recipes for the items marked with <R> are on page 215.
CAUTION: See the Cautions Appendix for appropriate handling of materials marked with <!>.

Reagents

Agarose gel (1%)
Chloroform <!>
DNA, linearized or PCR-amplified and flanked with T3 and T7 promoters
DNase I
DTT <!>
Ethanol (100% and 80% in H_2O), cold <!>
H_2O, nuclease-free
Phenol:chloroform (50:50) <!>
Planarians of any species
Polymerases T3 and T7
Ribonucleotides
RNA transcription buffer <R>
RNasin®
Solution A <R>

Equipment

Dissecting microscope
Glass capillaries (3.5-inch micropipettes, Drummond Scientific)
Horizontal gel unit
Ice
Microcentrifuge tubes, nuclease-free
Microcentrifuges, at room temperature and 4°C
Microinjector (Nanoject, Drummond Scientific)
Mineral oil
Needle pipette puller
Petri dishes
Planarian water (see Protocol 1)
Tissue paper, cold and wet
Tweezers
Vortex
Water baths, set at 37°C and 68°C

METHOD

RNA Synthesis

1. Assemble two transcription reactions, one with T3 polymerase and another with T7 polymerase, in separate microcentrifuge tubes. To each tube, add the following:

DNA	1 µg
DTT	10 mM
Ribonucleotides	1%
RNA transcription buffer	20%
RNasin®	60 units
T3 or T7 polymerase	17 units
H$_2$O, nuclease-free	to 20 µl

    Alternatively, if the multiple cloning site of the vector that has the clone inserted is flanked with the same promoter (e.g., double T7 vector), reactions can be performed with T7 RNA polymerase only, but in separate tubes after the clone is linearized by digestion with appropriate DNA restriction enzymes.

2. Incubate the reactions in a water bath for 2 hours at 37°C.
3. Treat each reaction with DNase I by adding 1 unit to each tube.
4. Incubate the reactions in a water bath for 15 minutes at 37°C.
5. Transfer 1 µl from each reaction to fresh microcentrifuge tubes. Keep the tubes at –20°C.
6. Combine the remaining 19 µl of each reaction into one tube.
7. Add 360 µl of solution A to the tube. Leave it for 10 minutes at room temperature.
8. Add 200 µl of phenol:chloroform mix and vortex vigorously.
9. Microcentrifuge the tube at room temperature at 14,000 rpm for 2 minutes.
10. Transfer the aqueous phase to a fresh microcentrifuge tube.
11. Add 200 µl of chloroform and vortex vigorously.
12. Microcentrifuge at 14,000 rpm for 2 minutes at room temperature.
13. Transfer the aqueous phase to a fresh microcentrifuge tube.
14. Incubate in a water bath for 10 minutes at 68°C to denature the RNA.
15. Incubate in a water bath for 30 minutes at 37°C to reanneal the RNA.
16. Add 1 ml of cold 100% ethanol.
17. Microcentrifuge at 14,000 rpm for 15 minutes at 4°C.
18. Discard the supernatant.
19. Add 1 ml of cold 80% ethanol.
20. Microcentrifuge at 14,000 rpm for 10 minutes at 4°C.
21. Discard the supernatant.
22. Resuspend the pellet in 10 µl of nuclease-free H$_2$O. Keep the sample on ice.
23. Confirm single-stranded RNA (ssRNA) transcription and dsRNA formation by separating 1 µl of each ssRNA sample from Step 5 and 0.5 µl of dsRNA from Step 22 on a 1% agarose gel under nondenaturing conditions (see Fig. 1 for an example).

    See Troubleshooting.

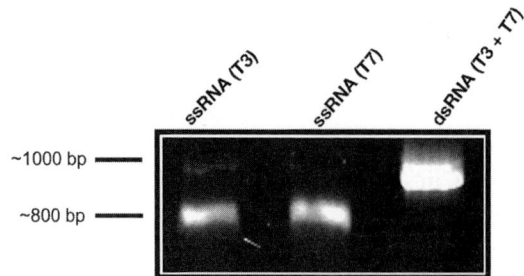

FIGURE 1. Confirmation of dsRNA formation. Electrophoretic migration patterns of single-stranded RNA (ssRNA) and double-stranded RNA (dsRNA) in an agarose gel under nondenaturing conditions. Similar amounts of ssRNA (1 µl) were loaded for each transcription reaction (T3 or T7), whereas for dsRNA, only 0.5 µl were loaded. Notice that dsRNA migration is shifted upward compared to ssRNA, confirming hybridization of the T3 and T7 ssRNA molecules. Notice that the approximate size of the bands (left) depends on the size of the DNA template used.

Microinjection of dsRNA

24. Prepare microinjection needles. Use a micropipette puller to form an elongate end. Then, while watching through a dissecting scope, break the tip of the needle to a diameter that will allow exit of liquid.

    Usually, eliminating 10–25% of the tip of the needle is enough. Important: Do not make the opening of the tip too wide because it will be difficult to prevent the liquid from coming out of the animal during the injection.

25. Fill the needle with mineral oil.

    Important: Do not allow air bubbles inside.

26. Attach the microinjector to the base of the dissecting scope and adjust it to a position that allows you to visualize the tip of the needle during the microinjection procedure. In addition, at this point, it is convenient to adjust the controlling box of the microinjector to dispense 32 nl per pulse.

27. Load the needle onto the microinjector and proceed to aspirate 1–2 µl of dsRNA.

28. Watching through the scope, place the worm on top of cold wet tissues.

    You may place ice underneath the tissues to keep the worm moist and cold.

29. Carefully introduce the needle into the worm. Use enough pressure to make sure the tip of the needle will be inside the worm.

    Different anatomical areas can be targeted, but usually, microinjections around the prepharyngeal area produce better results. Remember, planarians are flat and if the animal moves, it is easy to stick the needle out the other end.

30. Press the injection key, which will automatically dispense 32 nl. This should be repeated several times (three to five).

    Usually this amount (i.e., 32 × 3 nl) of liquid fills up the gastrovascular system of the worm, which confirms that dsRNA is getting into the worm. See Troubleshooting.

31. Transfer the injected worm to a Petri dish with fresh planarian water at room temperature.

32. To increase the strength of the phenotype, injection procedures (Steps 24–31) can be repeated several times (e.g., consecutive days or weeks).

    See Troubleshooting.

TROUBLESHOOTING

Problem (Step 23): ssRNA and/or dsRNA are not visualized on the gel.

Solution: Confirm that all reagents are good quality (i.e., not expired and have undergone few freeze-thaw cycles). Check that the DNA template included the T3 and T7 polymerase promoters. Make sure to work with clean equipment and surfaces (e.g., previously wiped with RNase and DNase decontaminants).

Problem (Step 30): No liquid comes out of the needle during injection.

Solution: Make sure that no air bubbles are in the needle. Also, confirm that the needle is properly attached to the microinjector.

Problem (Step 30): You are not sure if the injected liquid is getting inside the animal.

Solution: This requires practice. Sometimes, adding a few microliters of food coloring to the injection solution helps to ascertain the effectiveness of the microinjection.

Problem (Step 32): Animals fail to display any phenotype.

Solution: Keep in mind that not all genes produce observable phenotypes when tested by RNAi. Check whether the procedure is effectively reducing the expression of the target gene by performing in situ hybridization using probes to the target gene. For more sensitive evaluations, perform RT-PCR or quantitative RT-PCR (qRT-PCR). Sometimes it is possible to elicit a phenotype by adjusting the injection schedule (shorter or longer periods of time) or by simultaneously targeting two genes that may compensate for each other's activity.

Protocol 3

Live Imaging of Planarian Membrane Potential Using DiBAC$_4$(3)

This protocol describes how to use the anionic membrane voltage-reporting dye DiBAC$_4$(3) to generate images of cell membrane potential in live planarians. These images qualitatively reveal variations in time-averaged membrane potential across different regions of the organism. Changes in these images due to experimental treatments reveal how each particular treatment affects this physiological parameter. This method is a great improvement over standard electrophysiological techniques that cannot be used to gain an understanding of the electrical properties of an entire worm or a regenerating fragment due to small cell size and large cell number. When the proper controls are performed, this technique is a very powerful and simple way to gather physiologic data. In addition, it has been shown to be useful in the study of vertebrate regeneration (Adams et al. 2007) because of the known role of endogenous bioelectric signals in directing cell behavior and morphogenesis (McCaig et al. 2005; Levin 2007).

MATERIALS

The recipe for the item marked with <R> are on page 215.

CAUTION: See the Cautions Appendix for appropriate handling of materials marked with <!>.

Reagents

Cationic-membrane-reporting dye (see Step 14.iv)
Depolarizing agent (see Step 14.ii)
DiBAC$_4$(3) (1 mg/ml), prepared in 70% ethanol <R> <!>
Petroleum jelly (e.g., Vaseline) or other temporary sealant
Planarian water (see Protocol 1)
Worms with a strong *Smed-PC2(RNAi)* phenotype, ~10 days since first injection (see Protocol 2)

Equipment

Coverslips
Digital camera, attached to the microscope
Fluorescence filter set, suitable for λ_{ex} = 493, λ_{em} = 516 (e.g., fluorescein isothiocyanate [FITC] <!>)
Image analysis software with background correction, segmentation, and pixel quantification functions
Microscope with epifluorescence or confocal optics and 4x (or 5x) and 10x lenses
Paper towels
Petri dishes (35 mm) or 24-well plates
Press-to-Seal™ silicone spacers (each forms one well 20 mm in diameter and 0.5 mm deep)
Slides (25 x 75 x 1 mm)

METHOD

Staining

1. Dilute the $DiBAC_4(3)$ stock solution 1:10 in distilled H_2O and then dilute that $1:10^3$ in planarian water to a final concentration of approximately 0.1 ng/µl.
2. Fill a Petri dish or one well of a 24-well plate with the $DiBAC_4(3)$ solution, so that it is at least 2–3 mm deep.
3. Place the immobilized planarians (*Smed-PC2[RNAi]* worms) into the $DiBAC_4(3)$ solution.
4. Incubate the worms in the dark for at least 30 minutes at room temperature. Keep them in the solution until you are finished imaging.
   The dye has no lasting effect on worm behavior or regeneration.

Mounting

5. Remove the plastic from both sides of the silicone spacer. Place a thin layer of petroleum jelly on one side of the spacer and lay it, jelly-side up, on a paper towel or other clean surface.
6. Position the slide above the spacer and press down. Watch carefully to make sure that the seal is complete.
7. Flip the slide right-side up. Place a thin layer of jelly on the remaining face of the spacer.
8. Fill the pool created by the spacer with the diluted $DiBAC_4(3)$ solution (from Step 1).
9. Remove a *Smed-PC2(RNAi)* worm from the staining solution (from Step 4), position it on the slide, and cover it with a coverslip (Fig. 2).

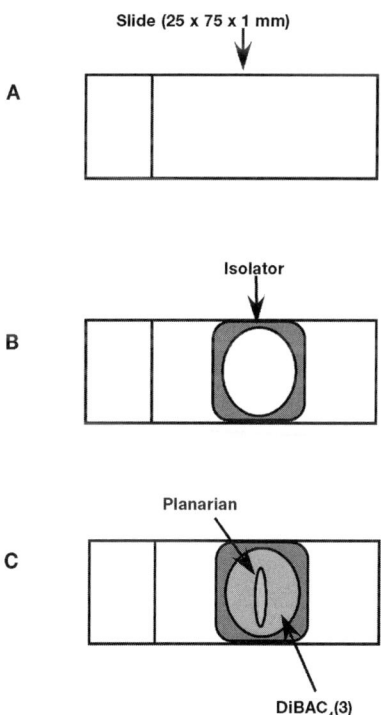

FIGURE 2. Sequential steps for mounting planarians for imaging with $DiBAC_4(3)$. (*A*) A slide recommended for the procedure. (*B*) Positioning of the isolator after petroleum jelly has been added and attached to the slide. (*C*) Final step involving positioning of the worm in the $DiBAC_4(3)$ solution.

10. Add more petroleum jelly along the junctions of the glass and silicone.

    This provides extra protection against fluid loss. Moreover, it minimizes changes in the $DiBAC_4(3)$ signal due to pressure on or contact with the worm. The concentration of $DiBAC_4(3)$ must stay constant. See Troubleshooting.

Imaging

11. Mount the slide on the microscope stage. Using bright-field light and the 4x or 5x lens, locate and focus on the animal.

    If the specimen is very small, the 10x lens can be used.

12. Confirm the focus on the $DiBAC_4(3)$ emission using the appropriate filter set.

    $DiBAC_4(3)$ will bleach; thus, it is important to minimize the amount of exposure. See Troubleshooting.

13. Take the picture.

    Because the $DiBAC_4(3)$ is constantly moving in and out of the cells, it can be useful to wait 20 to 30 seconds between exposures to give the unbleached dye time to replace the dye that was bleached in the previous exposure. See Troubleshooting.

14. Perform the following controls:

    i. Image unstained animals.

       This confirms that autofluorescence is not contributing to the signal.

    ii. Take one image, add a depolarizing agent, such as medium containing a high concentration of potassium gluconate (150 mM) and a potassium ionophore (e.g., 20 mM of salinomycin), and then take another image.

       This confirms that the depolarization of cells does cause an increase in emission intensity. The second image should be brighter than the first. The ionophore concentration must be titrated for each species. See Troubleshooting.

    iii. If possible, include a high-magnification image that illustrates the distribution of dye within individual cells.

       This shows whether it is the voltage across the cell membrane, or across the membranes of organelles, that is reported.

    iv. Repeat the imaging with a cationic-membrane-reporting dye such as one of the carbocyanine dyes (e.g., $DiSC_3[5]$).

       The intensity distribution should be inverted. Care should be taken with interpretation of cationic dye data because, unlike anionic dyes, cationic dyes will enter mitochondria. In addition, different dyes will have different ranges of potential to which they are most sensitive. Thus, the signal may not be entirely due to cell membrane potential. See Troubleshooting.

Image Processing and Analysis

15. Use the analysis software to open the image.

16. Correct the background using an appropriate function.

    In practice, this step, although required, usually has little effect on the data. The signal-to-noise ratio in $DiBAC_4(3)$ images tends to be very strong already; moreover, the result cannot be quantified on an absolute scale, and background noise will not affect the determination regarding which parts of the animal are emitting more intensely than others.

17. Examine the intensity of the pixels. To make areas of different intensity very obvious, pseudo-color the image with a red-green-blue look-up table (LUT).

    The pixel intensity indicates the degree of relative depolarization; brighter (more intense) pixels mean that that area of the specimen is depolarized relative to areas with less intense emission.

18. Segment the data using the analysis software.
    The intensity of the pixels can be quantified on an arbitrary scale, but it is important to make sure that only pixels with real data are counted.

    i. Exclude underexposed or overexposed pixels from the analysis.
        Underexposed or overexposed pixels will have values of 0 and 4095, respectively, for 12-bit images (or values of 0 and 255, respectively, for 8-bit images). Because the background correction may have artificially lowered the values of overexposed pixels to under 4095, it is a good idea to further limit the range of pixels values to be included (e.g., use values of 10 to 4000).

    ii. Use the segmentation function of the software to find pixels within the chosen range.
        Drawing regions of interest by hand is not recommended.

    iii. To yield additional information about the range of intensities and the sizes of areas showing different intensities, segment into multiple ranges.
        For example, use values of 10 to 1009, 1010 to 2009, 2010 to 3009, and 3010 to 4009. The number of pixels within each range provides information about the variability and the area of depolarization within the animal.

19. Choose a measure of intensity that will describe your data. A histogram of pixel-intensity frequencies will work for any data and should be examined first. If the histogram shows that pixel intensities are normally distributed, the mean and standard deviation (or the mean and confidence intervals) are the best measures to report and the simplest to use for comparisons. If intensities are uneven across the region of interest (i.e., blotchy), you may need somewhat more sophisticated statistical tests if you wish to make an inference about the general population; the histogram itself is fine for presentation of data.

    See Troubleshooting.

20. If the independent variable is nominal or ordinal (a category, such as treatment, with an arbitrary or unequal scale), compare the means using nonparametric statistics. To compare two means, use a Mann-Whitney test; for three or more means, use a Kruskal-Wallace test. If the independent variable is based on interval or ratio data, use a t-test or an ANOVA test.

TROUBLESHOOTING

Problem (Step 10): Fluid leaks out of well.
Solution: Return the animal to the original staining dish and give it at least 30 minutes to exchange $DiBAC_4(3)$ with the medium before trying again.

Problem (Step 12): Emission intensity is too high or too low.
Solution: Vary the concentration of $DiBAC_4(3)$ in the medium or use neutral density filters.

Problem (Step 13): $DiBAC_4(3)$ has bleached.
Solution: Return the animal to the original staining dish and give it at least 30 minutes to exchange $DiBAC_4(3)$ with the medium before trying again. The optimal solution is to image using a perfusion system that can completely replace all of the dye solution in the pool before each new image. Chambers for this purpose are available as Press-to-Seal™ silicone spacers.

Problem (Step 14.ii): K^+-gluconate plus ionophore has no effect on intensity.
Solution: Strong buffering, mucous, and other secretions can inhibit movement of ionophores or mute changes in membrane potential. Try a different ionophore or ionophore cocktail. Confirm that the K^+ concentration in the medium is higher than the intracellular concentration. If all else fails, kill the animal by adding a fast-acting toxin (e.g., sodium azide). As the cells die, the membrane potential will go to zero.

Problem (Step 14.iv): The cationic dye pattern is not the inverse of the anionic DiBAC$_4$(3) pattern.
Solution: This may indicate that the two dyes have entered different subcellular compartments or that they act by different mechanisms. There is no general solution to this problem, which makes the other controls even more important.

Problem (Step 19): A more quantitative measure is required.
Solution: Molecular Probes (now Invitrogen) has developed VSPs (voltage sensor probes), which are fluorescence resonance energy transfer (FRET)-based ratiometric methods for reporting membrane voltage. Because these are pairs of reporters, these kits tend to be much more expensive than DiBAC$_4$(3) alone. Nonetheless, because they are ratiometric, they are less affected by artifacts and can be used to generate more quantitative data.

Recipes

CAUTION: See the Cautions Appendix for appropriate handling of materials marked with <!>.

DiBAC$_4$(3) (1 MG/ML)

Tap out the smallest mass of DiBAC$_4$(3) (bis-[1,3-dibarbituric acid]-trimethine oxanol) solid that you can measure (usually 5×10^2 to 8×10^2 μg) and place it into a microcentrifuge tube. Add the volume of 70% ethanol <!> required to yield 1 mg/ml (i.e., 1 μl/μg). Centrifuge the tube to mix the solution. Store this stock solution at room temperature in the dark. It will last from 4 to 6 months.

1x MONTJUÏCH SALTS

Reagent	Quantity (for 1 liter)	Final concentration
CaCl$_2$ (2.5 M) <!>	0.4 ml	1.0 mM
KCl (1 M)	0.1 ml	0.1 mM
NaCl (5 M)	0.32 ml	1.6 mM
NaHCO$_3$ (1 M)	1.2 ml	1.2 mM
MgCl$_2$ (1 M)	0.1 ml	0.1 mM
MgSO$_4$ (1 M) <!>	1.0 ml	1.0 mM

Mix the reagents with H$_2$O that has 18 MΩ resistance (Milli-Q H$_2$O). Using HCl <!> and/or NaOH <!>, adjust the pH to 7.0. This recipe is from Cebrià and Newmark (2005).

RNA TRANSCRIPTION BUFFER

Reagent	Quantity (for 10 ml)	Final concentration
MgCl$_2$ (1 M)	300 μl	30 mM
NaCl (5 M)	100 μl	50 mM
Spermidine (1 M) <!>	100 μl	10 mM
Tris-HCl (1 M, pH 7.9)	2 ml	200 mM
H$_2$O, nuclease-free	7.6 ml	

SOLUTION A

Reagent	Quantity (for 10 ml)	Final concentration
Ammonium acetate (10 M)	1 ml	1 M
EDTA (0.5 M)	200 μl	10 mM
SDS (20%) <!>	100 μl	0.2%
H$_2$O, nuclease-free	8.7 ml	

ACKNOWLEDGMENTS

The authors thank Dr. W. Beane for comments and suggestions on the manuscript. N.J.O. is a National Institutes of Health (NIH) fellow supported under the Ruth L. Kirschstein National Research Service Award (F32 GM078774). D.S.A. is supported by the National Institute of Dental and Craniofacial Research (NIDCR) grant 1K22-DE016633. M.L is supported by the National Science Foundation (NSF) grant IBN 0347295, National Highway Traffic Safety Administration (NHTSA) grant DTNH22-06-G-00001, and NIH grant R21 HD055850. This review was prepared in a Forsyth Institute facility renovated with support from Research Facilities Improvement Grant Number CO6RR11244 from the National Center for Research Resources, NIH. We apologize to our colleagues whose work we could not cite due to space limitations.

REFERENCES

Adams, D.S., Masi, A., and Levin, M. 2007. H$^+$ pump-dependent changes in membrane voltage are an early mechanism necessary and sufficient to induce *Xenopus* tail regeneration. *Development* **134:** 1323–1335.

Agata, K. 2003. Regeneration and gene regulation in planarians. *Curr. Opin. Genet. Dev.* **13:** 492–496.

Baguñà, J. 1999. From morphology and karyology to molecules. New methods for taxonomical identification of asexual populations of freshwater planarians. A tribute to Professor Mario Benazzi. *Ital. J. Zool.* **66:** 207–214.

Baguñà, J. and Romero, R. 1981. Quantitative analysis of cell types during growth, degrowth and regeneration in the planarians *Dugesia mediterranea* and *Dugesia tigrina*. *Hydrobiologia* **84:** 181–194.

Benazzi, M., Baguñà, J., and Ballester, R. 1970. First report on an asexual form of the planarian *Dugesia lugubris* s.l. *Rend. Acc. Naz. Lincei* **48:** 282–284.

Benazzi, M., Ballester, R., Baguñà, J., and Puccinelli, I. 1972. The fissiparous race of the planarian *Dugesia lugubris* S.L. found in Barcelona (Spain) belongs to the biotype g: Comparative analysis of the karyotypes. *Caryologia* **25:** 59–68.

Benazzi, M., Baguñà, J., Ballester, R., Puccinelli, I., and Del Papa, R. 1975. Further contribution to the taxonomy of the "*Dugesia lugubris-polychroa* group" with description of *Dugesia mediterranea* n.sp. (Tricladida, Paludicola). *Boll. Zool.* **42:** 81–89.

Cebrià, F. 2007. Regenerating the central nervous system: How easy for planarians! *Dev. Genes Evol.* **217:** 733–748.

Cebrià, F. and Newmark, P.A. 2005. Planarian homologs of netrin and netrin receptor are required for proper regeneration of the central nervous system and the maintenance of nervous system architecture. *Development* **132:** 3691–3703.

De Vries, E. 1985. The biogeography of the genus *Dugesia* (Turbellaria, Tricladida, Paludicola) in the Mediterranean region. *J. Biogeogr.* **12:** 509–518.

Handberg-Thorsager, M. and Saló, E. 2007. The planarian *nanos*-like gene *Smednos* is expressed in germline and eye precursor cells during development and regeneration. *Dev. Genes Evol.* **217:** 403–411.

Hayashi, T., Asami, M., Higuchi, S., Shibata, N., and Agata, K. 2006. Isolation of planarian X-ray-sensitive stem cells by fluorescence-activated cell sorting. *Dev. Growth Differ.* **48:** 371–380.

Hicks, C., Sorocco, D., and Levin, M. 2006. Automated analysis of behavior: A computer-controlled system for drug screening and the investigation of learning. *J. Neurobiol.* **66:** 977–990.

Higuchi, S., Hayashi, T., Hori, I., Shibata, N., Sakamoto, H., and Agata, K. 2007. Characterization and categorization of fluorescence activated cell sorted planarian stem cells by ultrastructural analysis. *Dev. Growth Differ.* **49:** 571–581.

Levin, M. 2007. Large-scale biophysics: Ion flows and regeneration. *Trends Cell Biol.* **17:** 262–271.

McCaig, C.D., Rajnicek, A.M., Song, B., and Zhao, M. 2005. Controlling cell behavior electrically: Current views and future potential. *Physiol. Rev.* **85:** 943–978.

McConnell, J.V., ed. 1965. *A manual of psychological experimentation on planarians*. Worm Runner's Digest, Ann Arbor, Michigan.

McKay, R.D. 2004. Stem cell biology and neurodegenerative disease. *Philos. Trans. R. Soc. Lond. B Biol. Sci.* **359:** 851–856.

Morgan, T.H. 1898. Experimental studies of the regeneration of *Planaria maculata*. *Arch. Entw. Mech. Org.* **7:** 364–397.

Newmark, P.A. and Sánchez Alvarado, A. 2000. Bromodeoxyuridine specifically labels the regenerative stem cells of planarians. *Dev. Biol.* **220:** 142–153.

Newmark, P.A. and Sánchez Alvarado, A. 2002. Not your father's planarian: A classic model enters the era of functional genomics. *Nat. Rev. Genet.* **3:** 210–219.

Nicolas, C., Abramson, C., and Levin, M. 2008. Analysis of behavior in the planarian model. In *Planaria: A model for drug action and abuse* (ed. R.B. Raffa). RG Landes, Austin, Texas. (In press.)

Nogi, T. and Levin, M. 2005. Characterization of innexin gene expression and functional roles of gap-junctional communication in planarian regeneration. *Dev. Biol.* **287:** 314–335.

Novikov, V.V., Sheiman, I.M., Lisitsyn, A.S., Klyubin, A., and Fesenko, E.E. 2002. Dependence of the effect of weak combined low-frequency variable and constant magnetic fields on the intensity of asexual reproduction of planarians *Dugesia tigrina* on the magnitude of the variable field. *Biofizika* **47:** 564–567.

Ogawa, K., Wakayama, A., Kunisada, T., Orii, H., Watanabe, K., and Agata, K. 1998. Identification of a receptor tyrosine kinase involved in germ cell differentiation in planarians. *Biochem. Biophys. Res. Commun.* **248:** 204–209.

Orii, H., Sakurai, T., and Watanabe, K. 2005. Distribution of the stem cells (neoblasts) in the planarian *Dugesia japonica*. *Dev. Genes Evol.* **215:** 143–157.

Oviedo, N.J. and Levin, M. 2007. *smedinx-11* is a planarian stem cell gap junction gene required for regeneration and homeostasis. *Development* **134:** 3121–3131.

Oviedo, N.J. and Levin, M. 2008. The planarian regeneration model as a context for the study of drug effects and mechanisms. In *Planaria: A model for drug action and abuse* (ed. R.B. Raffa). RG Landes, Austin, Texas. (In press.)

Oviedo, N.J., Newmark, P.A., and Sánchez Alvarado, A. 2003. Allometric scaling and proportion regulation in the freshwater planarian *Schmidtea mediterranea*. *Dev. Dyn.* **226:** 326–333.

Oviedo, N.J., Pearson, B., Levin, M., and Sánchez Alvarado, A. 2008.

Planarian PTEN homologs regulate stem cells and regeneration through TOR signaling. *Dis. Mod. Mech.* (in press).

Pellettieri, J. and Sánchez Alvarado, A. 2007. Cell turnover and adult tissue homeostasis: From humans to planarians. *Annu. Rev. Genet.* **41:** 83–105.

Randolph, H. 1897. Observations and experiments on regeneration in planarians. *Arch. Entw. Mech. Org.* **5:** 352–372.

Reddien, P.W. and Sánchez Alvarado, A. 2004. Fundamentals of planarian regeneration. *Annu. Rev. Cell Dev. Biol.* **20:** 725–757.

Reddien, P.W., Oviedo, N.J., Jennings, J.R., Jenkin, J.C., and Sánchez Alvarado, A. 2005. SMEDWI-2 is a PIWI-like protein that regulates planarian stem cells. *Science* **310:** 1327–1330.

Reddien, P.W., Newmark, P.A., and Sánchez Alvarado, A. 2008. Gene nomenclature guidelines for the planarian *Schmidtea mediterranea*. *Dev. Dyn.* (in press).

Robb, S.M., Ross, E., and Sánchez Alvarado, A. 2007. SmedGD: The *Schmidtea mediterranea* genome database. *Nucleic Acids Res.* (database issue) **36:** D599–D606.

Rossi, L., Salvetti, A., Lena, A., Batistoni, R., Deri, P., Pugliesi, C., Loreti, E., and Gremigni, V. 2006. DjPiwi-1, a member of the PAZ-Piwi gene family, defines a subpopulation of planarian stem cells. *Dev. Genes Evol.* **216:** 335–346.

Rossi, L., Salvetti, A., Batistoni, R., Deri, P., and Gremigni, V. 2007a. Planarians, a tale of stem cells. *Cell. Mol. Life Sci.* **65:** 16–23.

Rossi, L., Salvetti, A., Marincola, F.M., Lena, A., Deri, P., Mannini, L., Batistoni, R., Wang, E., and Gremigni, V. 2007b. Deciphering the molecular machinery of stem cells: A look at the neoblast gene expression profile. *Genome Biol.* **8:** R62.

Sánchez Alvarado, A. 2003. The freshwater planarian *Schmidtea mediterranea*: Embryogenesis, stem cells and regeneration. *Curr. Opin. Genet. Dev.* **13:** 438–444.

Sánchez Alvarado, A. 2006. Planarian regeneration: Its end is its beginning. *Cell* **124:** 241–245.

Sánchez Alvarado, A. 2007. Stem cells and the planarian *Schmidtea mediterranea*. *C. R. Biol.* **330:** 498–503.

Sánchez Alvarado, A. and Kang, H. 2005. Multicellularity, stem cells, and the neoblasts of the planarian *Schmidtea mediterranea*. *Exp. Cell Res.* **306:** 299–308.

Sánchez Alvarado, A. and Newmark, P.A. 1999. Double-stranded RNA specifically disrupts gene expression during planarian regeneration. *Proc. Natl. Acad. Sci.* **96:** 5049–5054.

Sánchez Alvarado, A., Newmark, P.A., Robb, S.M., and Juste, R. 2002. The *Schmidtea mediterranea* database as a molecular resource for studying platyhelminthes, stem cells and regeneration. *Development* **129:** 5659–5665.

Sato, K., Shibata, N., Orii, H., Amikura, R., Sakurai, T., Agata, K., Kobayashi, S., and Watanabe, K. 2006. Identification and origin of the germline stem cells as revealed by the expression of *nanos*-related gene in planarians. *Dev. Growth Differ.* **48:** 615–628.

Stokely, P., Brown, T., Kuchan, F., and Slaga, T. 1965. The distribution of fresh-water triclad planarians in Jefferson County, Ohio. *Ohio J. Sci.* **65:** 305–318.

Umesono, Y., Watanabe, K., and Agata, K. 1997. A planarian orthopedic homolog is specifically expressed in the branch region of both the mature and regenerating brain. *Dev. Growth Differ.* **39:** 723–727.

Wang, Y., Zayas, R.M., Guo, T., and Newmark, P.A. 2007. *nanos* function is essential for development and regeneration of planarian germ cells. *Proc. Natl. Acad. Sci.* **104:** 5901–5906.

Zayas, R.M., Hernandez, A., Habermann, B., Wang, Y., Stary, J.M., and Newmark, P.A. 2005. The planarian *Schmidtea mediterranea* as a model for epigenetic germ-cell specification: Analysis of ESTs from the hermaphroditic strain. *Proc. Natl. Acad. Sci.* **102:** 18491–18496.

WWW RESOURCES

http://genome.wustl.edu/ The Genome Sequence Center at Washington University performed the sequence of the *S. mediterranea* genome that is currently being assembled. Additional details on the sequence process can be accessed at this website and the final assembly will be posted there for download.

http://www.ncbi.nlm.nih.gov/ The National Center for Biotechnology Information displays trace files generated during the sequencing of the *S. mediterranea* genome and will include EST data generated during the final assembly of the planarian genome.

http://smedgd.neuro.utah.edu/ The *S. mediterranea* Genome Data-base (SmedGD) integrates all available data associated with the *S. mediterranea* genome, including predicted and annotated genes, ESTs, protein homologies, gene expression patterns, and RNAi phenotypes.

http://www.ascb.org/ibioseminars/sanchez/sanchez1.cfm The American Society for Cell Biology library of seminars, iBioSeminars, features a lecture on basics of planarian regeneration.

http://planarian.bio.keio.ac.jp/go/EST/P/GO0007582/GO0006915.html This is an accessible portal from the Department of Biosciences and Informatics at Keio University (Japan) that allows for the search and retrieval of a large collection of *Dugesia japonica* ESTs.

http://turbellaria.umaine.edu The Turbellarian Taxonomic Database covers all turbellarian flatworms.

9 The Snail *Ilyanassa*
A Reemerging Model for Studies in Development

Maey Gharbiah,[1] James Cooley,[1] Esther M. Leise,[2] Ayaki Nakamoto,[1] Jeremy S. Rabinowitz,[3] J. David Lambert,[3,4] and Lisa M. Nagy[1,4]

[1]*Department of Molecular and Cellular Biology, University of Arizona, Tucson, Arizona 85721;*
[2]*Department of Biology, University of North Carolina, Greensboro, North Carolina 27402-6170;*
[3]*Department of Biology, University of Rochester, Rochester, New York 14627*

ABSTRACT

Ilyanassa obsoleta is a marine gastropod that is a long-standing and very capable model for studies of embryonic development. It is especially important as a model for the spiralian development program, a distinctive mode of early development shared by a large group of animal phyla, but poorly understood. *Ilyanassa* adults are readily obtainable and easy to keep in the laboratory, and they lay large numbers of embryos throughout most of the year. The embryos are amenable to classic embryological manipulation techniques as well as a growing number of molecular approaches. In this chapter, we present an overview of aspects of its biology and use as a model organism and describe protocols for rearing and handling *Ilyanassa* adults and embryos for examination of their development.

PROTOCOLS
1 Obtaining *Ilyanassa* Embryos, 223
2 Induction of Larval Metamorphosis, 227
3 Pressure Injection of Embyros, 230
4 Fixation of *Ilyanassa* Embryos and Larvae, 233
5 Isolation of Genomic DNA from *Ilyanassa* Larvae, 235
6 Isolating Protein from *Ilyanassa* Embryos, 237

BACKGROUND INFORMATION

Ilyanassa obsoleta (Caenogastropoda, Sorbeoconcha, Hypsogastropoda, Neogastropoda, Buccinoidea, Nassariidae) is also known as *Nassarius obsoletus, Nassa obsoleta,* and *Alectrion obsoleta*. It is a marine gastropod, with a shell of up to about 2.5 cm (Fig. 1A). It is abundant in near-shore environments along the east coast of North America. It has been introduced to various localities on the west coast, where it can be very common, and it has close relatives that are found all over the world. *Ilyanassa* live on intertidal mud or sand flats. They feed on detritus, bacteria, and algae found on the substrate as well as carrion, which they need for reproduction (Hurd 1985).

The *Ilyanassa* embryo was one of the first systems used by American embryologists to probe the mechanisms of early development. In 1896, H.E. Crampton demonstrated that the cleavage pattern could be significantly altered by removing the polar lobe—an anucleate protrusion that forms during the first cleavage. This embryo was also a favorite of T.H. Morgan (1933, 1935), who performed

[4]Corresponding authors.

This chapter, with full-color images, can be found online at cshprotocols.org/emo.

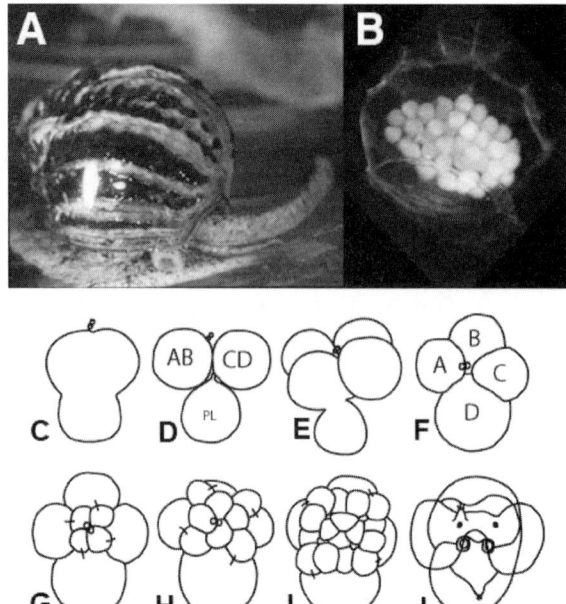

FIGURE 1. *Ilyanassa obsoleta.* (A) Adult snails are 0.5–2.5 cm. (B) Females deposit clutches of synchronously cleaving zygotes in transparent egg capsules (1–2 mm). The embryos in this capsule are about 48 hours old. (C–I) Early cleavage in *Ilyanassa* embryos. As the first mitotic division approaches (C), the vegetal pole of the zygote extrudes a lobe of cytoplasm called the polar lobe (PL), which is connected by a thin bridge of membrane to the CD cell at the ensuing cytokinesis (D). (The two tiny circles are polar bodies.) The lobe forms again at the next cleavage (E), and at the four-cell stage (F), one cell has inherited the polar lobe material and is thus specified to be the D cell. The four cells present at this stage are called macromeres. In subsequent cleavage cycles (G–I), the macromeres divide synchronously toward the animal pole (out of the page) to produce smaller cells called micromeres. A set of synchronously born micromeres is known as a quartet. (G) In the eight-cell embryo, the first quartet cells have just been born. (The hatch marks indicate sister-cell relationships from the previous division.) The second quartet cells are present at the 12-cell embryo stage (H), and the third quartet cells are present at the 24-cell stage (I). (J) Veliger larvae hatch about 7 days after zygote deposition.

various manipulations of *Ilyanassa* embryos. The *Ilyanassa* embryo became less popular as a model for general developmental biology after the rise of developmental genetic models such as *Drospohila melanogaster* and *Caenorhabditis elegans*, but is currently regaining some prominence because of its usefulness for understanding spiralian development and for studies of asymmetric cell division.

SOURCES AND HUSBANDRY

Adult *Ilyanassa* are abundant and easy to collect at low tide on sand or mud flats at many locations along the east and west coasts of North America. They can also be ordered from suppliers such as the Marine Resources Center at the Marine Biological Laboratory (Woods Hole, Massachusetts). The snails are hardy and can be maintained in simple benchtop saltwater aquaria and, unlike most marine invertebrates, they can be relied on to produce high-quality embryos nearly year-round, even though their normal spawning season is in the early summer (see Protocol 1).

RELATED SPECIES

The closely related species *Nassarius reticulatus*, found in Europe, is very similar to *Ilyanassa* in adult morphology, egg masses, and embryonic development (Pelseneer 1911). Other members of the genus *Nassarius* are also probably similar in these regards and are found throughout the world.

USES OF *ILYANASSA* AS A MODEL SYSTEM

Ilyanassa is one of the best models available for functional studies of spiralian development—the dominant mode of early development among a large multiphylum group of animals called the Lophotrochozoa. This clade includes roughly one third of the animal phyla (see Dunn et al. 2008). Of the three major clades of bilateral metazoans—deuterosomes, ecdysozoans, and lophotrochozoans—molecular aspects of lophotrochozoan embryos are the least understood, probably because there is no genetic model system in the group. Although they are morphologically diverse as adults, most lophotrochozoans share a conserved developmental program known as spiralian development. This is characterized by (1) a stereotyped pattern of early asymmetric cell division called spiral cleavage (see Fig. 1) (Wilson 1899; Schleip 1929), (2) similarities in the pattern of cell fates in the blastula (see, e.g., Boyer et al. 1996; Henry and Martindale 1998), and (3) similarities in larval morphology (Rouse 1999). Recent phylogenetic studies indicate that the common ancestor of the Lophotrochozoa had spiralian development (Dunn et al. 2008), highlighting the importance of this character for understanding the evolutionary history of animal development. There are also particular advantages for studying the evolution of development within the spiralians. The strong conservation of spiralian development means that homology can be precisely identified at the level of individual cells in the embryos of distantly related phyla, and many aspects of the common ancestor of spiralians can be reliably predicted. Against this backdrop of conservation, there is a wealth of evolutionary novelty, both subtle and dramatic (see, e.g. Freeman and Lundelius 1992; Henry and Martindale 1999).

Among the spiralians, *Ilyanassa* is a particularly powerful model for developmental biology for several reasons. The embryos undergo the classic pattern of spiral cleavage (Fig. 1C–J), and a large body of descriptive and experimental literature is available to guide molecular analyses (see Collier 2002). For instance, the larval fate maps of the micromeres of the 24-cell embryo have been determined (Render 1991, 1997), and the phenotypic effects of deleting each micromere at this stage have been reported (Clement 1967, 1986a,b; Sweet 1998). Perhaps most importantly, the functions of specific gene products can be probed in this embryo using a variety of approaches (Rabinowitz et al. 2008; A. Nakamoto et al., unpubl.). These experimental advantages are augmented by considerable practical advantages of this system. The adults are extremely abundant at many coastal locations and can be ordered from suppliers. The animals are hardy and can easily be maintained in inland laboratories using basic aquarium supplies and artificial seawater. Throughout the day, they lay large numbers of egg capsules, each containing 50–100 synchronously fertilized zygotes, so that embryos of a given stage can be obtained when desired. Thus, the *Ilyanassa* embryo combines important practical advantages, a large body of literature of descriptive and experimental studies, and a wealth of experimental methods, including specific gene knockdown studies; these factors make this system one of the very best for studies of the spiralian developmental program.

Recently, *Ilyanassa* has also emerged as a useful system for studying the cell biology of asymmetric cell division (Lambert and Nagy 2002; Kingsley et al. 2007; Rabinowitz et al. 2008). Almost all of the early cell divisions in the embryo are asymmetric, including the dramatic polar-lobe-associated first cleavage (Fig. 1; Clement 1952). These embryos are excellent material for imaging because of the large size of the cells during these divisions and the optical clarity of the embryos in the animal hemisphere where most cytoskeletal events take place.

In addition to the developmental studies that are the primary focus of this chapter, *Ilyanassa* is used for studies in a variety of fields of biology. It is an important model for studies of metamorphosis (Couper and Leise 1996; Froggett and Leise 1999; Leise et al. 2004) the ecology of parasitism (Curtis 1987, 1996, 1997, 2002; Curtis and Tanner 1999; Hendrickson and Curtis 2002), and imposex, a striking morphological disorder caused by the disruption of sexual endocrine systems by environmental contaminants (Oberdorster and McClellan-Green 2000, 2002; Gooding and LeBlanc 2001; Gooding et al. 2003; McClellan-Green et al. 2006; Sternberg and LeBlanc 2006; Finnegan et al. 2008; Sternberg et al. 2008). *Ilyanassa* is also useful for studies of comparative neurobiology (Dickinson and Croll 2003).

GENETICS AND GENOMICS

By convention, gene names in *Ilyanassa* are prefaced with "*Io*" or "*Iob*." On the basis of bulk fluorometric assays (Hinegardner 1974) and biochemical analysis (Collier and McCann-Collier 1962; Davidson et al. 1971), *Ilyanassa*'s genome size is estimated to be about 3.2 pg/haploid genome, which is large for a nonmammalian genome. Despite the considerable advantages of *Ilyanassa* as a model system, it has not yet been selected for whole-genome sequencing. At present, the most efficient use of sequencing in *Ilyanassa* is for transcriptome analysis. An initial set of about 9500 high-quality expressed sequence tags (ESTs) has been deposited into GenBank, and this effort continues. As the price of sequencing declines and the experimental strengths of *Ilyanassa* continue to improve, we expect that whole-genome sequencing will occur in the not too distant future.

TECHNICAL APPROACHES

A wide range of classical embryological manipulations have been demonstrated in the *Ilyanassa* embryo. These include cell deletion, which can be performed by pricking cells with a glass needle (Clement 1956, 1962, 1967, 1986a,b; Sweet 1998; Lambert and Nagy 2001; Goulding 2003); cell transplantation (Sweet 1998); and treatment with pharmacological agents that perturb signaling (Lambert and Nagy 2001, M. Gharbiah et al. unpubl.).

One of the most conceptually important embryological manipulations that can be performed in the *Ilyanassa* embryo, polar lobe deletion, is also one of the most straightforward. The polar lobes are anucleate extrusions that form during early meiotic and mitotic divisions in the embryo of *Ilyanassa* and some other spiralians. There are various ways to remove the polar lobe (see Clement 1952; Newrock and Raff 1975; Collier 1981). Deletion of the polar lobe that forms at first cleavage causes severe defects in development (Clement 1952). This experiment was one of the first demonstrations showing that the cytoplasm could regulate development (Crampton 1896).

In the protocols that follow, we present general methods for obtaining *Ilyanassa* embryos (Protocol 1), inducing larval metamorphosis (Protocol 2), fixing embryos and larvae (Protocol 4), and isolating DNA (Protocol 5) and protein (Protocol 6). Recently, a reliable and robust method for intracellular injection of aqueous solutions has been developed for *Ilyanassa* embryos (Protocol 3). This has enabled the development of a number of tools for functional analysis of individual genes. Injection of blocking antibodies, morpholinos, double-stranded RNA (dsRNA), small interfering RNA (siRNA), and messenger RNA (mRNA) fusions have been validated in this system (Rabinowitz et al. 2008; A. Nakamoto and L.M. Nagy; J.S. Rabinowitz and J.D. Lambert, both unpubl.). These approaches complement *Ilyanassa*'s existing strengths—the range of classical approaches in place, the rich literature, and the practical advantages—to make this system an excellent platform for examining spiralian development and the cell biology of asymmetric cell division.

Protocol 1

Obtaining *Ilyanassa* Embryos

Although the normal spawning season for *Ilyanassa* is during early summer, they can produce high-quality embryos nearly year-round in the laboratory. Snails collected in the late fall, winter, or spring can be induced to deposit zygotes before the natural spawning season by warming them to room temperature, and snails collected before the natural spawning season can be made to postpone zygote deposition until needed (up to at least 6 months) by maintaining them in tanks in a cold room at 4–8°C. This protocol describes how to induce embryo production in *Ilyanassa* snails, collect the embryos, and rear them to the stage required for study.

MATERIALS

The recipe for the item marked with <R> is on page 239.

Reagents

Artificial seawater, unfiltered (ASW) and filtered (FASW) <R>
FASW is required for rearing and for the manipulation of embryos and larvae. It is useful to keep a plastic squirt bottle containing FASW at each workstation.

Clams of various species, obtained from fish markets and frozen until needed

Ilyanassa snails
Collect snails from U.S. coastal regions or obtain them from a supplier such as the Marine Resources Center at the Marine Biological Laboratory (Woods Hole, Massachusetts).

Equipment

Aquarium
For the aquarium, use a 20-gallon tank with low sides containing calcareous gravel, an undergravel filter, and a hang-on power filter, as well as ASW to fill. Set up the intake of the hang-on power filter so that it will pull water through the undergravel filter. Use power filters with chemical and biological filtration that are rated for 50–60-gallon tanks.

Braking pipette
Construct a braking pipette by pulling a large (~100-µl) glass capillary into a fine tip over a flame and then fixing it into the end of a glass Pasteur pipette with hot glue (Fig. 2A,E). Attach this assembly to the tubing of a mouth pipette with a 1-ml plastic pipette tip. Alternatively, splice the mouth pipette tubing around a fine gel-loading pipette tip so that the airflow is constricted through the hole in the tip (Fig. 2F).

Bucket (5 gallon)

Dissecting forceps with fine tips (e.g., Dumont no. 5 Biologie; Fig. 2C)

Iridectomy scissors (Fine Science Tools, 15003-08; Fig. 2B)

Orbital shaker

Pasteur pipette, broken to an ~4-mm aperture and attached to a rubber bulb

Permanent marker

224 / Chapter 9

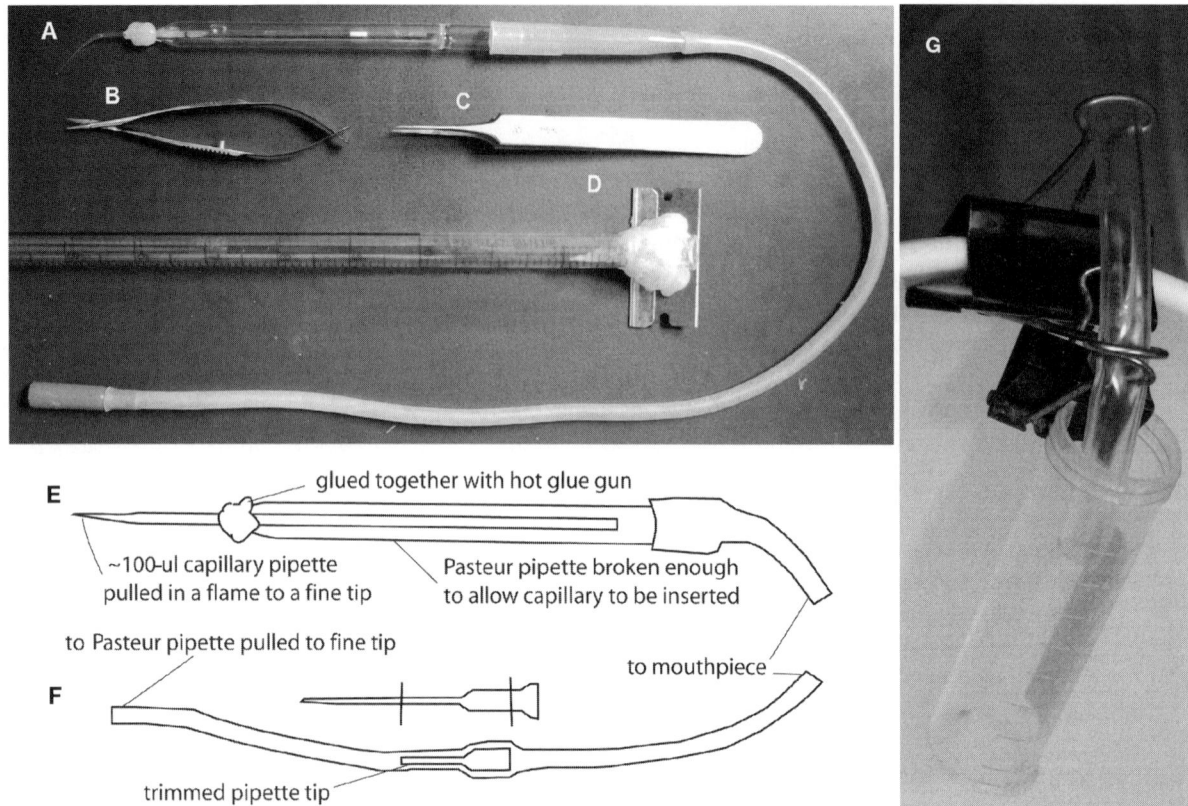

FIGURE 2. Tools used in handling *Ilyanassa* embryos. (*A*) Apparatus used to gently blow seawater into cut capsules to eject embryos. A fine-tipped, hand-pulled glass needle is wedged into the large opening of a 1-ml plastic pipette tip. The pipette tip is, in turn, tightly fit into 24 inches of rubber tubing, which is topped with a 200-μl plastic pipette tip. (*B*) Iridectomy scissors used to cut capsules. (*C*) Dissecting forceps used to handle capsules. (*D*) A razor, hot-glued to the end of a 10-ml plastic pipette, used to scrape capsules from tank walls. (*E*,*F*) Diagrams showing how to construct a braking pipette (see Protocol 1). (*G*) A capsule collection basket on the edge of a plastic bucket, with the siphon hose attached. The basket has a cut in the plastic that allows it to be clamped with the metal clip.

Petri dishes (35 mm)

For routine collection, sorting, and preparation of embryos, dishes can be reused after cleaning by repeated rinsing with H_2O or by washing in covered metal baskets in a laboratory dishwasher without detergent.

Razor blade capsule scraper

Attach a razor blade to a 10-ml plastic pipette with hot glue (Fig. 2D) or remove both ends of the pipette and combine with the siphoning apparatus (see below).

Siphoning apparatus

Construct by attaching one end of a 2-m length of Tygon tubing (ID = 7–8 mm) to the end of a 10-ml plastic pipette, with the cotton plug removed, or to the razor blade capsule scraper (see above). Attach the other end of the siphon tubing to a capsule collection basket made from a 50-ml polypropylene conical tube. To make the basket, remove the conical end from the tube and attach nylon mesh (about 0.5-mm aperture size) with a hot glue gun or make a large hole in the top of the cap and screw it on over a piece of the nylon mesh. Hold the siphon tube in the top of the basket with a 1-inch black metal paper clamp and attach the basket to the rim of a bucket with a 2-inch black metal paper clamp (Fig. 2G).

Stereomicroscope (optional; see Step 4)

METHOD

Embryo Production

To induce embryo production in cold-stored snails, begin at Step 1. To induce embryo production immediately in snails collected before the early summer spawning season, begin at Step 2.

1. Transfer 50–75 snails from storage at 4–8°C into a 5-gallon bucket filled with 2 gallons of ASW cooled to 4°C. Leave the bucket overnight at room temperature.
2. Move the snails to an aquarium at room temperature.
3. On day 2 and every second day thereafter, feed the snails one clam. Remove the clam after 6 hours to avoid microbial growth.

    Egg-laying will commence in 1–5 days and persist for 3–10 weeks, peaking at about 2–3 weeks.

4. When early stages are needed, regularly check the aquarium for females depositing capsules onto the tank wall (see Fig. 1B). Mark new capsules with a permanent marker on the glass and label with the time of capsule deposition.

    Capsule deposition is easily observed because the capsule gland on the foot has a characteristic ring appearance on the glass. For later stages, all of the capsules can be harvested, and the required stages can be selected under a stereomicroscope.

5. After the embryos have reached the required developmental stage, collect the capsules using one of the following methods:
    - To collect individual capsules, use a glass Pasteur pipette with an approximately 4-mm aperture to gently scrape off and retrieve individual capsules.
    - To collect many capsules simultaneously, use the razor blade tool to scrape the capsules off the tank wall (Fig. 2D). Use the siphoning apparatus to collect the detached capsules as they sink.

Removal of Embryos from Capsules

6. Transfer the capsules to a Petri dish containing FASW. Select a capsule containing embryos of the desired stage. Use the dissecting forceps to hold the capsule down onto the bottom of the dish so that the portion of the capsule that is free of embryos is facing upward.
7. Use the iridectomy scissors to cut a hole as large as possible in the capsule wall.

    Avoid contact with the embryos and strong deformation of the capsule wall because the embryos are easily damaged in this process. After cutting, the capsule can be moved to another dish using fine forceps if desired.

8. Allow the capsule to sit for at least 5 minutes or until needed, ideally with gentle agitation on an orbital shaker to help the transparent capsule jelly to dissolve.
9. Holding the capsule down with the fine forceps, use the braking pipette to direct a gentle stream of FASW into the opening of the capsule to *slowly* flush the embryos out.
10. Remove empty capsule with forceps. Allow the embryos to develop at room temperature until the desired developmental stage is reached.

    Ilyanassa embryos develop into veliger larvae in 5 days and normally hatch from the capsule 7–8 days after fertilization. The phenotypic effects of various experimental treatments are commonly scored at 7 days, when most of the major tissues are differentiated and morphologically distinct (Fig. 3). See Troubleshooting.

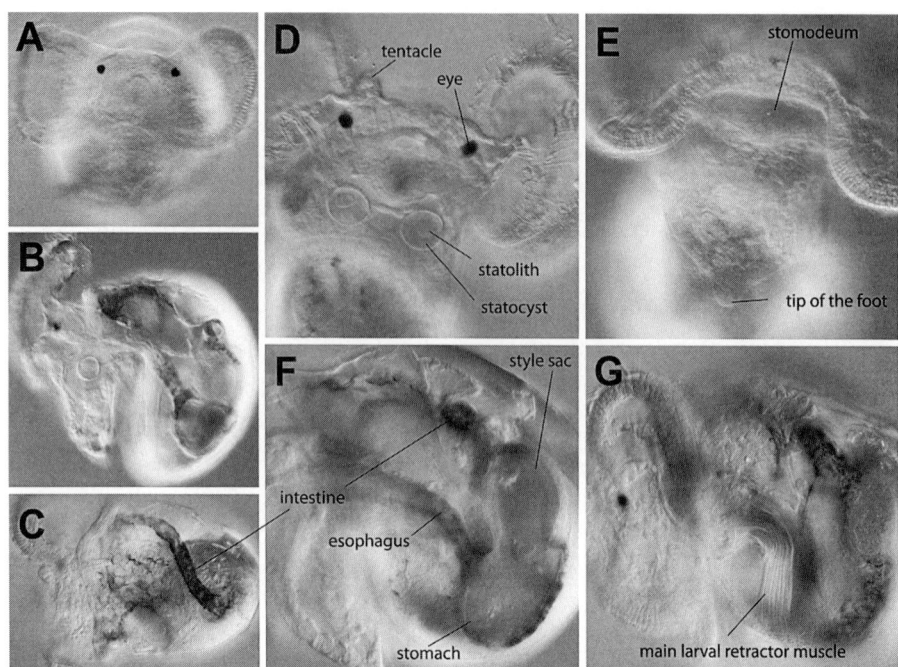

FIGURE 3. Larval structures of 7-day-old *Ilyanassa* larvae. (*A,D,E*) Frontal views. (*B,F,G*) Lateral views. (*C*) Dorsal view.

TROUBLESHOOTING

Problem (Step 10): Embryos die during development because of microbial growth.
Solution: Ensure that FASW, rather than ASW, is used and, optionally, add penicillin (10 units/ml) and/or streptomycin (10 μg/ml). Because embryo death can result in an increase in microbial activity, always remove damaged embryos from the culture dishes.

Problem (Step 10): Embryos die during development because of mechanical damage.
Solution: Avoid cutting capsules or flushing out embryos during the most fragile stages, such as meiosis and mitosis. If embryos die during culture, use Petri dishes coated with 1% agar prepared in H_2O to minimize damage from the plastic.

Protocol 2

Induction of Larval Metamorphosis

After hatching from capsules, larval *Ilyanassa* can be maintained in culture, feeding on single-celled algae. The larvae will become competent to undergo metamorphosis after about 3 weeks in culture. Metamorphosis can be induced artificially by treating with either the neurotransmitter serotonin (5-HT) or the nitric oxide synthase inhibitor 7-nitroindazole (7-NI) as described in this protocol. Both of these reagents have been shown to induce metamorphosis in more than 75% of larvae within 48 hours (Couper and Leise 1996; Froggett and Leise 1999; Leise et al. 2004).

MATERIALS

The recipes for items marked with <R> begin on page 239.

CAUTION: See the Cautions Appendix for appropriate handling of materials marked with <!>.

Reagents

Algal cultures of *Dunaliella tertiolecta* (~500,000 cells/ml) and *Isochrysis galbana* (~5,000,000 cells/ml)

> Obtain *D. tertiolecta* from the Provasoli-Guillard National Center for Culture of Marine Phytoplankton (no. 1320) and *I. galbana* from the Carolina Biological Supply Company. Culture the species in either Provasoli's E-S medium or Guillard's Medium F/2 (Hinegardner et al. 1981; Switzer-Dunlap and Hadfield 1981; Strathmann and Fernald 1987).

Artificial seawater, filtered (FASW) <R>
Ilyanassa larvae, obtained and reared as described in Protocol 1
Larval seawater <R>
7-Nitroindazole (7-NI; 0.2 mM) <!>

> To make 100 ml, mix 0.0032 g of 7-NI in 0.5 ml of DMSO <!>, vortex, and then mix with 99.5 ml of FASW.

Serotonin (0.1 mM, prepared in FASW) <!>

Equipment

Air pump and tubing
Alcohol lamp containing 95% ethanol <!> (optional; see Step 20)
Beakers (1 liter, glass)
Beakers (50 ml and 400 ml; plastic, tripour)
Connectors (three way; for plastic tubing)
Cork borer
Dissecting microscope (capable of 40x magnification) and light source
Glue gun and glue sticks (hot-melt type)
Nitex nylon mesh (130-μm and 250-μm mesh sizes)
Ocular micrometer
Pasteur pipettes and bulbs
Plastic tubing (1/8-inch ID, 1/16-inch wall; flexible and clear)
Sandwich wrap (plastic)

Spouts (glass, hand-fabricated from tubing)
Squirt bottle (500 ml; filled with FASW)
Stender dishes
Tissue culture plates (24 well, flat-bottomed, with lids) or glass fingerbowls (see Step 19)

METHOD

Construction of Droplet-stirred Culture Containers

See Figure 4. For further details, see Miller and Hadfield (1986) and Strathmann and Fernald (1987).

1. Remove the bottoms of the 400-ml plastic tripour beakers and replace them with 250- or 130-μm mesh Nitex using a hot-melt glue gun. (At the same time, replace the bottoms of the 50-ml tripour beakers with a 250-μm mesh for use in Step 15.)
2. Use a cork borer to create a 1-cm hole in the side of each beaker, just below the rim.
3. Insert the modified tripour beakers into 1-liter glass beakers.
4. Fill the 1-liter glass beaker with 625 ml of larval seawater.
5. Connect a glass spout to a 4-cm length of plastic tubing. Insert the spout into the hole below the rim of the modified tripour beaker.
6. Attach a three-way connector to the plastic tubing, at the opposite end of the glass spout.
7. To the opposite end of the three-way connector, attach the air pump tubing (which is, in turn, attached to the air pump).
8. Position the three-way connector to the side or bottom of the plastic beaker such that drops of the larval seawater are lifted through the spout to continuously mix the culture.

Larval Culture

9. Add 550–600 newly hatched larvae to a droplet-stirred culture container containing a tripour beaker with a 130-μm mesh bottom and larval seawater.
10. Add 20 ml each of the *D. tertiolecta* and *I. galbana* suspensions.
11. Cover the droplet-stirred culture container with plastic wrap and aerate at room temperature.
12. Feed the culture every day by adding 10 ml each of *D. tertiolecta* and *I. galbana*.
13. Change the larval seawater, glass beakers, and 130-μm plastic inserts weekly.
14. When the larvae are large enough, swap the tripour beaker with a 130-μm mesh bottom for a tripour beaker with a 250-μm mesh bottom.

FIGURE 4. Diagram showing how to construct droplet-stirred culture containers for the induction of metamorphosis in *Ilyanassa* larvae (see Protocol 2).

Induction of Metamorphosis

15. When the larvae have attained a shell length of between 575 and 600 μm, gently wash them into a 50-ml tripour beaker with a 250-μm mesh bottom.

    Determine larval shell length by measuring the longest shell length of 12–15 randomly selected larvae from a culture using an ocular micrometer. The larvae should reach this length by approximately 17 days posthatching.

16. Set up a series of five stender dishes, four half-full with FASW.

17. Gently immerse the larvae in the tripour beaker (from Step 15) into each of the first four dishes to remove excess algae from the larvae.

18. Back-wash the larvae into the fifth (empty) stender dish using a stream of FASW from a squirt bottle.

19. Under a dissecting microscope, use a Pasteur pipette to select healthy, swimming larvae. Transfer them to wells of a 24-well culture plate (five larvae per well).

    Alternatively, large batches of larvae can be induced to metamorphose in glass fingerbowls with approximately 250 larvae/100 ml inductive solution.

20. Use a Pasteur pipette to remove all FASW from each well of the culture plate.

    The tip of the pipette can be narrowed over the flame of an alcohol lamp to facilitate complete removal of the FASW.

21. Immediately add 2 ml of the chosen inductive solution, either 0.1 mM serotonin or 0.2 mM 7-NI, to each well.

    The larvae can remain moist in a "dry" well for a few minutes without damage.

22. Incubate the 24-well culture plate for 24 or 48 hours at room temperature.

    By 24 hours, about 50% of the larvae will have metamorphosed and by 48 hours, greater than 75% of the larvae should be juveniles (Gifondorwa and Leise 2006). Induction with 7-NI is slightly slower than induction with serotonin.

Protocol 3

Pressure Injection of Embryos

Intracellular microinjection is an important tool, especially for lineage tracing and perturbations of specific genes by knockdown approaches and synthetic mRNA injections. Two methods for the introduction of lineage tracers into particular cells are routine in *Ilyanassa*. Iontophoresis of charged molecules such as fluorophore-dextran conjugates (Render 1991, 1997) can be accomplished using a simply-built current generator (Hodor and Ettensohn 1998; X.Y. Chan and J.D. Lambert, unpubl.). Injection of an oil-based solution containing the fluorescent probe DiI (1,1-dioctadecyl-3,3,3′,3′-tetramethyl indocarbocyanine perchlorate) is also straightforward (Rabinowitz et al. 2008). However, injection of oil-based solutions and iontophoresis have not been useful for delivering water-soluble reagents to perturb gene function, and pressure injection of aqueous solutions has been more challenging. This protocol describes a recently optimized procedure for the pressure injection of aqueous solutions into *Ilyanassa* embryos and zygotes with high rates of survival and normal development. The key parameters seem to be the injection needles, injection media, and the stage of injected embryos. The use of this method to inject morpholino oligos, antisense RNA, siRNA, and synthetic mRNA to modulate gene expression has been demonstrated in *Ilyanassa* embryos, providing valuable new tools for functional studies in the *Ilyanassa* embryo (Rabinowitz et al. 2008; J.S. Rabinowitz and J. D. Lambert; M. Gharbiah et al., both unpubl.).

MATERIALS

The recipes for items marked with <R> begin on page 239.

CAUTION: See the Cautions Appendix for appropriate handling of materials marked with <!>.

Reagents

Agarose (1%) (optional; see Step 1)
Artificial seawater, filtered (FASW) <R>
Ficoll 400 (5%, prepared in FASW)
Ilyanassa zygotes (obtained as described in Protocol 1; see also Step 6)
10x Injection buffer <R>

Prepare 1x injection buffer containing 1% fluorescein isothiocyanate <!> (FITC)-dextran (10,000 MW), with or without the experimental molecule (e.g., morpholino oligo [0.05–3 mM], siRNA [1–10 µg/µl], or mRNA [100–1000 ng/µl]). A few microliters of this solution are sufficient to inject hundreds of embryos. Sulforhodamine 101 at a final concentration of 100 µM can be used as an alternative to 1% FITC dextran (10,000 MW).

Equipment

Microinjection capillaries (pulled from 1-mm OD thin-wall borosilicate glass with filament to a tip size of about 0.5-µm ID and 1-µm OD)

These can be purchased (e.g. Eppendorf Femtotip II) or pulled on a horizontal electrode puller (e.g. Sutter P-97).

Microinjector (pneumatic, with foot pedal control; e.g., Harvard Apparatus PL-100)

The microinjector must be capable of delivering at least 2000 hPa to the needle.

Microloader tips (Eppendorf)
Micromanipulator (fine scale, joystick controlled, mounted on a coarse manipulator; e.g. Narishige MO-202U)
Petri dishes (35 mm and 60 mm; polystyrene)
Slides (glass)
Spin column (with 0.22-μm cellulose acetate membrane; e.g., Costar® Spin-X®)
Stereomicroscope (e.g., Zeiss Stemi SV11)
  The stereomicroscope must have epifluorescent illumination and a minimum of 80x magnification.
Wire (approximately 175 μm) (optional; see Step 1)

METHOD

Preparation for Microinjection

Many of these steps can be performed while embryos develop to the correct stage as described in Step 6.

1. Using a glass slide, scratch a groove that is roughly as deep as the diameter of the embryo into the bottom of the 60-mm polystyrene Petri dish.

   Embryos will eventually be pipetted into this groove for the injections. Alternatively, line the dish with 1% agarose and the embryos can be housed in holes made in the agarose using a hot wire of approximately 175 μm.

2. Fill the bottom of the Petri dish with 5% Ficoll 400.

   Ficoll 400 suppresses cytoplasmic loss from the site of injection, perhaps due to increased viscosity.

3. Filter the injection solution through a spin column.

   When using small volumes, centrifuge the column with about 6 μl of 1x injection buffer to wet the filter and remove the flowthrough before applying the injection solution. This prevents the loss of the injection solution into the filter pad.

4. Back-fill the microinjection capillaries with the filtered injection solution using a Microloader tip.

   Filling only the tip of the needle will be sufficient to inject hundreds of embryos.

5. Attach the needle to microinjector, angle it at 30–45° from the horizontal, and perform a test injection into the 5% Ficoll 400 solution to verify that the correct puff size is being delivered.

   The puff of injection solution should be about one sixth of the diameter of the zygote at 2000 hPa.

Embryo Staging and Injection

6. Follow Protocol 1 to obtain and stage zygotes. Place the zygotes into 35-mm polystyrene Petri dishes containing FASW. Arrange the dishes according to the age of the embryos—oldest to youngest—and periodically check to verify their exact stage of development.

   Ilyanassa zygotes produce three polar lobes (PL), which are vegetal extrusions that accompany the meiotic divisions of the fertilized egg (PL1 and PL2) and the first mitosis (PL3). Zygotes are extremely fragile during the extended phase of each polar lobe. In our hands, consistently good survival rates for injections into zygotes are obtained only if they are performed less than 15 minutes after the retraction of PL2. Therefore, remove the embryos from their capsules at least 20 minutes before PL2 and inject immediately after PL2 retraction. For injection at cleavage stages, remove the embryos well before they have reached the desired developmental stage. Survival rates are highest if injection is performed when the cells have compacted following a cell division.

7. Immediately after PL2 has fully retracted in the prezygotic embryos or cells have compacted following a cell division in later-stage embryos, pipette the embryos into the groove scratched into the 60-mm Petri dish.

2. Add 300 µl of PEM fixation buffer and mix by flicking or gently tilting the tube.

    Embryos can also fixed by adding one-tenth volume of 37% formaldehyde to the FASW. The PEM fixation buffer is especially good for preserving the microtubule cytoskeleton and gives cleaner in situ hybridization staining.

3. Incubate for 30–120 minutes at room temperature.
    See Troubleshooting.

Fixation of Larvae

4. Add the selected larvae to a Petri dish containing trichlorobutanol solution and incubate for 8–12 minutes.

    Trichlorobutanol relaxes the larvae, preventing them from retracting into their shells and making larval structures difficult to see (Clement and Cather 1957). Incubating at 4°C may enhance the relaxation of the larvae. If trichlorobutanol is not available, larvae can also be relaxed in ice-cold 7% $MgCl_2$.

5. Use a disposable micropipette tip to transfer the larvae in 450 µl of solution to a microcentrifuge tube. Add 50 µl of 37% formaldehyde and immediately mix by flicking or gently inverting the tube.

6. Incubate for 30 minutes at room temperature or overnight at 4°C.

Washing and Storing Embryos and Larvae

7. Visually confirm that the embryos or larvae are at the bottom of the microcentrifuge tube (if not, flick the tube to dislodge them from the side and allow them to sink). Remove the liquid fixative with a pulled Pasteur pipette.

8. Wash the embryos or larvae twice with PBT and store at 4°C or wash twice with methanol and store at –20°C.

    Fixed embryos or larvae can last at least 1 month at 4°C in PBT and at least 2 years at –20°C in methanol. Extended storage in PBT will dissolve the mineral portion of the larval shell. This is often advantageous for microscopic observation of the larvae because the shell is highly birefringent and makes differential interference contrast (DIC) observation of internal structures difficult. The shell is retained longer in Tris-buffered saline (pH 7.0) with 0.1% Tween and 20 mM $CaCl_2$. The shell can be rapidly dissolved by adding EGTA (pH 8.0) to a final concentration of 5 mM.

TROUBLESHOOTING

Problem (Step 3): The embryos lyse or shrink during fixation.
Solution: After ensuring that the FASW has the correct salinity, vary the amount of sucrose up or down in the PEM fixation buffer until a concentration is found where the embryos look lifelike after fixation.

Protocol 5

Isolation of Genomic DNA from *Ilyanassa* Larvae

Ilyanassa is host to a several species of parasitic trematode worms, so care must be taken to avoid contamination of *Ilyanassa* genomic DNA with that of the parasites. The easiest way to avoid this contamination is to isolate DNA from veliger larvae, which are not parasitized. This also avoids other problems, such as the presence of large amounts of polysaccharides, encountered when isolating DNA from adult mollusc tissues. This protocol describes the isolation of genomic DNA from *Ilyanassa* larvae. RNA can also be isolated from the tissue with Trizol reagent (Gibco), but phenol:chloroform extraction is necessary to further purify Trizol-isolated RNA for use in downstream enzymatic reactions.

MATERIALS

The recipes for items marked with <R> begin on page 239.

CAUTION: See the Cautions Appendix for appropriate handling of materials marked with <!>.

Reagents

Artificial seawater, filtered (FASW) <R>
Chloroform <!>
DNA extraction buffer <R>
Ethanol (70% and 95%) <!>
Ilyanassa larvae, cultured to the veliger stage (see Protocol 1)
   Later stages are preferable because they have more cells.
Phenol:chloroform <R> <!>
Potassium acetate (8 M)
2-Propanol <!>
Proteinase K (1 mg/ml) <!>
RNase mix and buffer (e.g., Fermentas EN0551)
Sodium acetate (3 M, adjusted to pH 5.0 with glacial acetic acid <!>)
TE buffer <R>

Equipment

Ice
Microcentrifuge tubes (1.7 ml)
Microcentrifuge tube pestle (e.g., Kontes Pellet Pestle)
Microcentrifuges, with maximum speeds of about 15,000g
   Preferably, two microcentrifuges should be available: one at room temperature and one at 4°C. If this is not possible, perform the indicated 4°C centrifugation steps at room temperature.
Petri dish (35 mm)
Water baths, set at 37°C and 65°C

METHOD

Harvesting *Ilyanassa* Veliger Larvae

1. Select and open about 20 capsules into a 35-mm Petri dish containing FASW (as described in Protocol 1).
2. Fill a 1.7-ml microcentrifuge tube with the larval suspension using a 1-ml micropipette.
3. Chill the suspension on ice for 2 minutes and then centrifuge at maximum speed in a microcentrifuge for 2 minutes at 4°C.
4. Immediately remove the supernatant before the veligers begin to swim.
5. Repeat Steps 2–4 until most of the larvae are in the tube, contained in less than 20 µl of FASW.

Genomic DNA Extraction

6. Add 200 µl of DNA extraction buffer and 20 µl of proteinase K. Homogenize the larvae by grinding with a microcentrifuge tube pestle.
7. Incubate the homogenate for 15 minutes in a 65°C water bath.
8. Add 28 µl of 8 M potassium acetate and mix immediately and thoroughly by flicking and inverting the tube. Incubate the sample for 15 minutes on ice.
9. Centrifuge at maximum speed in a microcentrifuge for 10 minutes at 4°C.
10. Pipette the supernatant, which contains DNA, to a new tube. Add 100 µl of 2-propanol and mix immediately and thoroughly by flicking and inverting the tube.
11. Centrifuge at maximum speed in a microcentrifuge for 10 minutes at room temperature.
12. Discard the supernatant. Wash the pellet in 400 µl of 70% ethanol. Do not centrifuge again.
13. Discard the 70% ethanol, air-dry the pellet for 2–3 minutes, and resuspend the DNA in 100 µl of TE.

Genomic DNA Cleanup

14. To the 100-µl genomic DNA solution from Step 13, add the RNase mixture and the accompanying buffer according to manufacturer instructions. Incubate the sample in a 37°C water bath for 30 minutes.
15. Add 112 µl of phenol:chloroform, mix well, and centrifuge at maximum speed in a microcentrifuge for 10 minutes at room temperature.
16. Transfer the aqueous (upper) phase to a new tube and add an equal volume of chloroform. Mix well and centrifuge at maximum speed for 5 minutes at room temperature.
17. Add one-tenth volume of 3 M sodium acetate and mix and then add 2 volumes of 95% ethanol and mix. Incubate on ice for 30 minutes.
18. Centrifuge at maximum speed in a microcentrifuge for 30 minutes at 4°C.
19. Discard the supernatant. Wash the pellet in 400 µl of 70% ethanol. Do not centrifuge again.
20. Discard the 70% ethanol, air-dry the pellet for 2–3 minutes, and resuspend the DNA in 50 µl of TE buffer. Store the DNA at −20°C.

Protocol 6

Isolating Protein from *Ilyanassa* Embryos

This protocol describes the procedure for extracting protein from *Ilyanassa* embryos for use in techniques such as western blotting or two-dimensional (2D) gel electrophoresis. The protocol for 2D electrophoresis sample preparation was modified from a method previously published by Newrock and Raff (1975).

MATERIALS

The recipes for items marked with <R> begin on page 239.

CAUTION: See the Cautions Appendix for appropriate handling of materials marked with <!>.

Reagents

Acetone <!> (for 2D gel electrophoresis only; see Steps 6–25)
Ammonium acetate (0.1 M, prepared in methanol <!>) (for 2D gel electrophoresis only; see Steps 6–25)
Artificial seawater, filtered (FASW) <R>
Ilyanassa embryos at the desired developmental stage (see Protocol 1)
Liquid nitrogen <!>
2x Modified Newrock & Raff buffer A <R> (for 2D gel electrophoresis only; see Steps 6–25)
Modified rehydration buffer <R> (for 2D gel electrophoresis only; see Steps 6–25)

This modified rehydration buffer (Newrock and Raff 1975) is suitable for use with a polyacrylamide gel strip with an immobilized pH gradient of 3–10 (e.g., Immobiline DryStrip, GE Healthcare 17-6001-14). For gels with different properties, use the rehydration buffer suggested by the gel manufacturer.

Phenol (saturated) (for 2D gel electrophoresis only; see Steps 6–25)
100x Protease Inhibitor Cocktail for General Use (Sigma-Aldrich, P2714)

Make a 100x stock of protease inhibitor by dissolving the contents of one vial of Protease Inhibitor Cocktail for General Use powder in 1 ml of Phosphatase Inhibitor Cocktail 2 (Sigma-Aldrich, P5726), which is supplied at a concentration of 100x.

Equipment

Absorbent paper (for 2D gel electrophoresis only; see Steps 6–25)
Hydrophobic microtubes (BioExpress) or 1.7-ml microcentrifuge tubes (see Step 1)
Microcentrifuge (for 2D gel electrophoresis only; see Steps 6–25)

The microcentrifuge should be able to be cooled to 4°C and should have a maximum speed of about 15,000g.

Vortex (for 2D gel electrophoresis only; see Steps 6–25)

METHOD

Sample Collection and Crude Protein Extraction

1. Collect embryos:
   - For western blot samples, collect 25–75 embryos per western blot lane in a 1.7-ml centrifuge tube in sufficient FASW for transfer.

- For each 2D gel electrophoresis sample, collect 1000 embryos in a hydrophobic microtube in sufficient FASW for transfer.

    One thousand embryos is sufficient for a 13-cm polyacrylamide gel strip.

2. After the embryos have sunk to the bottom of the tube, remove as much FASW as possible while leaving the embryos covered (~10 µl of FASW should be sufficient).
3. Add the protease inhibitor to a final concentration of 2×.
4. Immediately plunge the tube into liquid nitrogen to snap-freeze.

    At this point, the embryos can be stored at –80°C.

5. Remove the tube from the liquid nitrogen (or from the –80°C freezer if stored). For western blotting, proceed with a standard western blotting protocol, loading 25–75 embryo equivalents on each gel lane. For 2D gels, proceed to Step 6.

Protein Purification for 2D Gel Electrophoresis

6. Add 50 µl of 2× Modified Newrock & Raff buffer A to each tube and incubate for 15 minutes on ice with occasional vortexing.
7. Centrifuge at maximum speed in a microcentrifuge for 5 minutes at 4°C.
8. Transfer the supernatant to a new hydrophobic microtube.
9. Add an equal volume of saturated phenol.
10. Centrifuge at maximum speed in a microcentrifuge for 5 minutes at room temperature.
11. Discard the clear upper phase (contains DNA and RNA), but keep the colored protein phase.
12. Add equal volumes of 2× Modified Newrock & Raff buffer A to each tube. Vortex vigorously.
13. Centrifuge at maximum speed in a microcentrifuge for 5 minutes at room temperature. Discard clear upper phase (contains DNA and RNA), but keep the colored protein phase.
14. Add 5 volumes of 0.1 M ammonium acetate and mix well by vortexing.
15. Store overnight at –20°C. The sample can be stored at –20°C for 2–3 days.
16. Centrifuge at maximum speed in a microcentrifuge for 15 minutes at 4°C.
17. Decant and discard the supernatant and then rinse the pellet with 1.5 ml of 0.1 M ammonium acetate.
18. Centrifuge at maximum speed in a microcentrifuge for 5 minutes at 4°C.
19. Decant and discard the supernatant and then rinse the pellet with 1.5 ml of acetone.
20. Centrifuge at maximum speed in a microcentrifuge for 5 minutes at 4°C.
21. Decant and discard the supernatant and then tap the tube face down on clean absorbent paper to remove any liquid.
22. Allow the pellet to air-dry for a maximum of 5 minutes.
23. Resuspend the pellet in 250 µl of modified rehydration buffer and vortex. Incubate the sample for at least 30 minutes at room temperature.
24. Vortex the sample and then add an additional 30 µl of modified rehydration buffer. Centrifuge the tube at maximum speed in a microcentrifuge for 1 minute at room temperature.
25. Load the supernatant onto the gel strip and proceed with a standard 2D gel electrophoresis protocol.

    Do not store the supernatant during this step. If the protocol must be stopped, it is best to store the sample at Step 15.

Recipes

The recipes for items marked with <R> in the text are listed here.
CAUTION: See the Cautions Appendix for appropriate handling of materials marked with <!>.

ARTIFICIAL SEAWATER, UNFILTERED (ASW) AND FILTERED (FASW)

Reagent	Quantity (for 5 gallons)	Final concentration
Instant Ocean	675 g	3.57% (w/v)
Tap water	to 5 gallons	

Other brands of artificial seawater, such as Reef Crystals, can also be used. Large volumes of ASW for aquaria can be prepared in 5-gallon buckets; mix using a power drill with a metal paint-mixing paddle. After mixing, the ASW should have a salinity of about 32 ppt. For FASW, filter the ASW through a 0.2-µm filter.

DNA EXTRACTION BUFFER

Reagent	Quantity (50 ml)	Final concentration
EDTA (0.5 M, pH 8.0)	10 ml	100 mM
SDS (10%) <!>	5 ml	1%
Tris-HCl (1 M, pH 9.0)	5 ml	100 mM
H_2O	30 ml	

Store indefinitely at room temperature.

10x INJECTION BUFFER

Reagent	Quantity (for 1 ml)	Final concentration
HEPES (0.5 M, pH 7.0)	0.2 ml	100 mM
KCl (3 M)	0.5 ml	1.5 M
H_2O	0.3 ml	

Store indefinitely at 4°C.

LARVAL SEAWATER

Reagent	Quantity (for ~500 ml)	Final concentration
FASW <R>	250 ml	0.5x
Natural seawater (filtered, 0.2 µm)	250 ml	0.5x
100x Penicillin/streptomycin <!> (Sigma-Aldrich)	5 ml	50 units/ml penicillin 50 µg/ml streptomycin

Store for 1 month at room temperature. See Miller and Hadfield (1986).

2x MODIFIED NEWROCK & RAFF BUFFER A

Reagent	Quantity (for 10 ml)	Final concentration
Magnesium acetate (1 M)	20 µl	0.002 M
100x Protease inhibitor	200 µl	2x
Sodium phosphate buffer (1 M, pH 7.2)	200 µl	0.02 M
Triton X-100 (10%) <!>	1 ml	1%
H_2O	8.58 ml	

Make the 100x stock of protease inhibitor by dissolving the contents of one vial of Protease Inhibitor Cocktail for General Use (Sigma-Aldrich, P2714) powder in 1 ml of Phosphatase Inhibitor Cocktail 2 (Sigma-Aldrich, P5726), which is supplied at a concentration of 100x. Prepare the Modified Newrock & Raff buffer A fresh before use. This recipe was modified from Newrock and Raff (1975).

MODIFIED REHYDRATION BUFFER

Reagent	Quantity (for 25 ml)	Final concentration
CHAPS <!>	0.5 g	2%
Thiourea <!>	3.8 g	2 M
Urea <!>	10.5 g	7 M
Just before use, add		
Bromophenol blue <!>	a dash	trace
DTT <!>	75 µg	3 µg/ml
IPG (immobilized pH gradient)-ampholyte containing buffer (e.g., GE Healthcare)	500 µl	20 µl/ml or 0.5%
H_2O	to 25 ml	

Buffer can be stored indefinitely at 4°C. This recipe was modified from Newrock and Raff (1975).

10x PBS

Reagent	Quantity (for 1 liter)	Final concentration (10x)
NaCl	10.2 g	1.75 M
Na_2HPO_4	11.94 g	84 mM
NaH_2PO_4	2.56 g	18.5 mM

Dissolve the reagents listed above in 800 ml of H_2O. Adjust the pH to 7.4 with HCl <!> and then add H_2O to 1 liter. Dispense the solution into aliquots and sterilize them by autoclaving for 20 minutes at 15 psi (1.05 kg/cm^2) on liquid cycle or by filter-sterilization. Store at room temperature.

PBT

Reagent	Quantity (for 500 ml)	Final concentration
10x PBS <R>	50 ml	1x
Tween 20 (10%)	25 ml	0.5%
H_2O	to 500 ml	

Store indefinitely at room temperature.

PEM FIXATION BUFFER

Reagent	Quantity (for 40 ml)	Final concentration (4x)
EGTA (500 mM)	0.8 ml	10 mM
MgSO$_4$ (1 M) <!>	40 µl	1 mM
PIPES (500 mM, pH 6.9)	8 ml	100 mM
Sucrose (2 M)	1.5 ml	75 mM
Triton X-100 (10%) <!>	400 µl	0.1%
H$_2$O	to 30 ml	
Just before use, add		
Paraformaldehyde <!> (16%, in 10-ml ampule; Polysciences)	10 ml	4%

Store 4x PEM without paraformaldehyde indefinitely at 4°C. PEM fixation buffer (with paraformaldehyde) can be stored at 4°C for at least 1 week.

PHENOL:CHLOROFORM

Reagent	Quantity (for 50 ml)	Final concentration
Phenol (pH 8.0) <!>	25 ml	0.5x
Chloroform <!>	25 ml	0.5x

Store at 4°C in a dark container for up to 1 month.

TE BUFFER

Reagent	Quantity (for 50 ml)	Final concentration
Tris-HCl (1 M, pH 8.0)	5 ml	10 mM
EDTA (0.5 M, pH 8.0)	10 ml	2 mM
H$_2$O	35 ml	

Store indefinitely at room temperature.

ACKNOWLEDGMENTS

The preparation of this manuscript and the methods reported herein were supported by National Science Foundation grants to J.D.L. (IOB0544220), E.M.L. (IBN-9604516 and IBN-0130677), and L.M.N. (IOB0820564).

REFERENCES

Boyer, B.C., Henry, J.Q., and Martindale, M.Q. 1996. Dual origins of mesoderm in a basal spiralian: Cell lineage analyses in the polyclad turbellarian *Hoploplana inquilina*. *Dev. Biol.* **179:** 329–338.

Clement, A.C. 1952. Experimental studies on germinal localization in *Ilyanassa*. I. The role of the polar lobe in determination of the cleavage pattern and its influence in later development. *J. Exp. Zool.* **121:** 593–626.

Clement, A.C. 1956. Experimental studies on germinal localization in *Ilyanassa*. II. The development of isolated blastomeres. *J. Exp. Zool.* **132:** 427–445.

Clement, A.C. 1962. Development of *Ilyanassa* following removal of the D-macromere at successive cleavage stages. *J. Exp. Zool.* **149:** 193–215.

Clement, A.C. 1967. The embryonic value of the micromeres in *Ilyanassa obsoleta*, as determined by deletion experiments. I. The first quartet cells. *J. Exp. Zool.* **166:** 77–88.

Clement, A.C. 1976. Cell determination and organogenesis in molluscan development: A reappraisal based on deletion experiments in *Ilyanassa. Am. Zool.* **16**: 447–453.

Clement, A.C. 1986a. The embryonic value of the micromeres in *Ilyanassa obsoleta*, as determined by deletion experiments. II. The second quartet cells. *Int. J. Invertebr. Reprod. Dev.* **9**: 139–153.

Clement, A.C. 1986b. The embryonic value of the micromeres in *Ilyanassa obsoleta*, as determined by deletion experiments. III. The third quartet cells and the mesentoblast cell, 4d. *Int. J. Invertebr. Reprod. Dev.* **9**: 155–168.

Clement, A.C. and Cather, J.N. 1957. A technic for preparing whole mounts of veliger larvae. *Biol. Bull.* **113**: 340.

Collier, J.R. 1981. Methods of obtaining and handling eggs and embryos of the marine mud snail *Ilyanassa obsoleta*. In *Laboratory animal management: Marine invertebrates*, pp. 217–232. National Academy Press, Washington, D.C.

Collier, J.R. 2002. A bibliography of the marine mud snail *Ilyanassa obsoleta*. *Invertebr. Reprod. Dev.* **42**: 95–110.

Collier, J.R. and McCann-Collier, M. 1962. The deoxyribonucleic acid content of the egg and sperm of *Ilyanassa obsoleta*. *Exp. Cell Res.* **27**: 553–559.

Couper, J.M. and Leise, E.M. 1996. Serotonin injections induce metamorphosis in larvae of the gastropod mollusc *Ilyanassa obsoleta*. *Biol. Bull.* **191**: 178–186.

Crampton, Jr., H.E. 1896. Experimental studies on gastropod development. *Arch. Entw.-mech.* **3**: 1–19.

Curtis, L.A. 1987. Vertical distribution of an estuarine snail altered by a parasite. *Science* **235**: 1509–1511.

Curtis, L.A. 1996. The probability of a marine gastropod being infected by a trematode. *J. Parasitol.* **82**: 830–833.

Curtis, L.A. 1997. *Ilyanassa obsoleta* (Gastropoda) as a host for trematodes in Delaware estuaries. *J. Parasitol.* **83**: 793–803.

Curtis, L.A. 2002. Ecology of larval trematodes in three marine gastropods. *Parasitology* (suppl.) **124**: S43–S56.

Curtis, L.A. and Tanner, N.L. 1999. Trematode accumulation by the estuarine gastropod *Ilyanassa obsoleta*. *J. Parasitol.* **85**: 419–425.

Davidson, E.H., Hough, B.R., Chamberlin, M.E., and Britten, R.J. 1971. Sequence repetition in the DNA of *Nassaria (Ilyanassa) obsoleta*. *Dev. Biol.* **25**: 445–463.

Dickinson, A.J. and Croll, R.P. 2003. Development of the larval nervous system of the gastropod *Ilyanassa obsoleta*. *J. Comp. Neurol.* **466**: 197–218.

Dunn, C.W., Hejnol, A., Matus, D.Q., Pang, K., Browne, W.E., Smith, S.A., Seaver, E., Rouse G.W., Obst M., Edgecombe, G.D., et al. 2008. Broad phylogenomic sampling improves resolution of the animal tree of life. *Nature* **452**: 745–749.

Finnegan, M.C., Pittman, S., and Delorenzo, M.E. 2008. Lethal and sublethal toxicity of the antifoulant compound Irgarol 1051 to the mud snail *Ilyanassa obsoleta*. *Arch. Environ. Contam. Toxicol.* (in press). doi: 10.1007/s00244-008-9166-x.

Freeman, G. and Lundelius, J. 1992. Evolutionary implications of the mode of D quadrant specification in coelomates with spiral cleavage. *J. Evol. Biol.* **5**: 205–247.

Froggett, S.J. and Leise, E.M. 1999. Metamorphosis in the marine snail *Ilyanassa obsoleta*, yes or NO? *Biol. Bull.* **196**: 57–62.

Gifondorwa, D.J. and Leise, E.M. 2006. Programmed cell death in the apical ganglion during larval metamorphosis of the marine mollusc *Ilyanassa obsoleta*. *Biol. Bull.* **210**: 109–120.

Gooding, M.P. and LeBlanc, G.A. 2001. Biotransformation and disposition of testosterone in the eastern mud snail *Ilyanassa obsoleta*. *Gen. Comp. Endocrinol.* **122**: 172–180.

Gooding, M.P., Wilson, V.S., Folmar, L.C., Marcovich, D.T., and LeBlanc, G.A. 2003. The biocide tributyltin reduces the accumulation of testosterone as fatty acid esters in the mud snail (*Ilyanassa obsoleta*). *Environ. Health Perspect.* **111**: 426–430.

Goulding, M. 2003. Cell contact-dependent positioning of the D cleavage plane restricts eye development in the *Ilyanassa* embryo. *Development* **130**: 1181–1191.

Hendrickson, M.A. and Curtis, L.A. 2002. Infrapopulation sizes of co-occurring trematodes in the snail *Ilyanassa obsoleta*. *J. Parasitol.* **88**: 884–889.

Henry, J.J. and Martindale, M.Q. 1998. Conservation of the spiralian developmental program: Cell lineage of the nemertean, *Cerebratulus lacteus*. *Dev. Biol.* **201**: 253–269.

Henry, J.J. and Martindale, M.Q. 1999. Conservation and innovation in spiralian development. *Hydrobiologia* **402**: 255–265.

Hinegardner, R. 1974. Cellular DNA content of the Mollusca. *Comp. Biochem. Physiol. A* **47**: 447–460.

Hinegardiner, R.T., Atz, J.W., Fay, R.C., Fingerman, M., Josephson, R.K., Meinkoth, N.A., Miller, J.W., and Rice, M.E. 1981. Foods and feeding. In *Marine invertebrates: Laboratory animal management*. National Academy Press, Washington, D.C.

Hodor, P.G. and Ettensohn, C.A. 1998. The dynamics and regulation of mesenchymal cell fusion in the sea urchin embryo. *Dev. Biol.* **199**: 111–124.

Hurd, L.E. 1985. On the importance of carrion to reproduction in an omnivorous estuarine neogastropod, *Ilyanassa obsoleta* (Say). *Oecologica* **65**: 513–515.

Kingsley, E.P., Chan, X.Y., Duan, Y., and Lambert, J.D. 2007. Widespread RNA segregation in a spiralian embryo. *Evol. Dev.* **9**: 527–539.

Lambert, J.D. and Nagy, L.M. 2001. MAPK signaling by the D quadrant embryonic organizer of the mollusc *Ilyanassa obsoleta*. *Development* **128**: 45–56.

Lambert, J.D. and Nagy, L.M. 2002. Asymmetric inheritance of centrosomally localized mRNAs during embryonic cleavages. *Nature* **420**: 682–686.

Leise, E.M., Kempf, S.C., Durham, N.R., and Gifondorwa, D.J. 2004. Induction of metamorphosis in the marine gastropod *Ilyanassa obsoleta*: 5HT, NO and programmed cell death. *Acta Biol. Hung.* **55**: 293–300.

McClellan-Green, P., Romano, J., and Rittschof, D. 2006. Imposex induction in the mud snail, *Ilyanassa obsoleta* by three tin compounds. *Bull. Environ. Contam. Toxicol.* **76**: 581–588.

Miller, S.E. and Hadfield, M.G. 1986. Ontogeny of phototaxis and metamorphic competence in larvae of the nudibranch *Phestilla sibogae* Bergh (Gastropoda: Opisthobranchia). *J. Exp. Mar. Biol. Ecol.* **97**: 95–112.

Morgan, T.H. 1933. The formation of the antipolar lobe in *Ilyanassa*. *J. Exp. Zool.* **64**: 433–467.

Morgan, T.H. 1935. The rhythmic changes in form of the isolated antipolar lobe of *Ilyanassa*. *Biol. Bull.* **68**: 296–299.

Newrock, K.M. and Raff, R.A. 1975. Polar lobe specific regulation of translation in embryos of *Ilyanassa obsoleta*. *Dev. Biol.* **42**: 242–261.

Oberdorster, E. and McClellan-Green, P. 2000. The neuropeptide APGWamide induces imposex in the mud snail, *Ilyanassa obsoleta*. *Peptides* **21**: 1323–1330.

Oberdorster, E. and McClellan-Green, P. 2002. Mechanisms of imposex induction in the mud snail, *Ilyanassa obsoleta*: TBT as a neurotoxin and aromatase inhibitor. *Mar. Environ. Res.* **54**: 715–718.

Pelseneer, P. 1911. Recherches sur l'embryology des Gastropodes. *Mem. Acad. Belg.* (*Class de sciences*) **3**: 1–167.

Rabinowitz, J.S., Chan, X.Y., Kingsley, E.P., Duan, Y., and Lambert, J.D. 2008. Nanos is required in somatic blast cell lineages in the posterior of a mollusk embryo. *Curr. Biol.* **18**: 331–336.

Render, J. 1991. Fate maps of the first quartet micromeres in the gastropod *Ilyanassa obsoleta*. *Development* **113**: 495–501.

Render, J. 1997. Cell fate maps in the *Ilyanassa obsoleta* embryo beyond the third division. *Dev. Biol.* **189**: 301–310.

Rouse, G.W. 1999. Trochophore concepts: Ciliary bands and the evolution of larvae in spiralian Metazoa. *Biol. J. Linn. Soc.* **66**: 411–464.

Schleip, W. 1929. Die determination der primitiventwicklung. Leipzig: Akad. Verlags.

Sternberg, R.M. and Leblanc, G.A. 2006. Kinetic characterization of the inhibition of acyl coenzyme A: Steroid acyltransferases by tributyltin in the eastern mud snail (*Ilyanassa obsoleta*). *Aquat. Toxicol.* **78:** 233–242.

Sternberg, R.M., Hotchkiss, A.K., and Leblanc, G.A. 2008. Synchronized expression of retinoid X receptor mRNA with reproductive tract recrudescence in an imposex-susceptible mollusc. *Environ. Sci. Technol.* **42:** 1345–1351.

Strathmann, M.F. and Fernald, R.L. 1987. *Reproduction and development of marine invertebrates of the Northern Pacific coast. Data and methods for the study of eggs, embryos, and larvae.* University of Washington Press, Seattle.

Sweet, H.C. 1998. Specification of first quartet micromeres in *Ilyanassa* involves inherited factors and position with respect to the inducing D macromere. *Development* **125:** 4033–4044.

Switzer-Dunlap, M. and Hadfield, M.G. 1981. Laboratory culture of *Aplysia*. In *Marine invertebrates: Laboratory animal management.* National Academy Press, Washington, D.C.

Wilson, E.B. 1899. Cell-lineage and ancestral reminiscence. In *Biological lectures 1898; The Marine Biological Laboratory, Woods Hole, Massachusetts.* Athenaeum Press, Boston.

10 Helobdella (Leech)
A Model for Developmental Studies

David A. Weisblat and Dian-Han Kuo
Department of Molecular and Cell Biology, University of California, Berkeley, California 94720-3200

ABSTRACT

Helobdella is a genus of freshwater leeches, several species of which have been used for developmental studies since the 1970s. *Helobdella* embryos have been used for cell-lineage tracing and for dye-mediated photoablation, and they have also been very useful for studies in cellular neurobiology. In this chapter, we discuss the reasons that *Helobdella* is used for studying development and some of the questions that are addressed through the use of this organism.

PROTOCOLS
1. Handling of Embryos, 249
2. Microinjection, 251
3. Devitellinization of Living Embryos, 256
4. Arnold's Silver Staining, 258
5. Immunostaining, 260
6. In Situ Hybridization, 263
7. Whole-mount Preparation for Microscopy, 268

BACKGROUND INFORMATION

Helobdella designates a genus of small freshwater glossiphoniid leeches (*Lophotrochozoa, Annelida, Clitellata, Hirudinida, Glossiphoniidae*), several species of which are useful for developmental studies. Investigations to date have centered on questions related to cell lineage and cell-fate decisions, segmentation, stem cell biology, and neurogenesis (Stent et al. 1992; Bissen 1997; Shankland and Savage 1997; Weisblat et al. 1999). Beginning in the 1870s, C.O. Whitman, the first director of Woods Hole Marine Biological Laboratory, used glossiphoniid leech embryos in pioneering studies of cell lineage to test Haeckel's prediction that the early embryo of any modern species should consist of interchangeable cells, recapitulating the presumed early stages of metazoan evolution via colonial organisms (i.e., "ontogeny recaptitulates phylogeny"). Contrary to the prediction, Whitman described stereotyped and unequal cell divisions giving rise to individually identified blastomeres beginning with first cleavage (Whitman 1878, 1887).

Developmental analyses of glossiphoniid leeches resumed in the 1970s when G.S. Stent and colleagues sought a leech species that would be useful for neurodevelopment. By about 1975, Roy T. Sawyer brought a population of *H. triserialis* from Golden Gate Park, San Francisco into laboratory culture. Embryos from this colony were used for the first modern studies of glossiphoniid leech development, including the first demonstrations of microinjectable enzymes and fluorescent compounds as tools for cell-lineage tracing in complex embryos (Weisblat et al. 1978, 1980; Gimlich and Braun 1985) and the first application of dye-mediated photoablation as a microsurgical tool for living embryos (Shankland 1984).

This chapter, with full-color images, can be found online at www.cshprotocols.org/emo.

The biogeography of *Helobdella* has been studied extensively by Marc Siddall and colleagues, who conclude that this genus has its origins in temperate South America. Their work and that of others indicates that the genus comprises three species complexes, designated *H. triserialis*, *H. robusta*, and *H. stagnalis* (Siddall and Borda 2003; Bely and Weisblat 2006).

Some of the main rationales for studying *Helobdella* development are as follows: (1) As an experimentally tractable member of the superphylum Lophotrochozoa, *Helobdella* can be studied and compared to the animal models belonging to the other two superphyla (Ecdysozoa and Deuterostomia). (2) The embryos of annelids, mollusks, and some other lophotrochozoan taxa exhibit a highly conserved series of early cell divisions called spiral cleavage. Studying *Helobdella* development can contribute to understanding cell biological mechanisms by which this ancient cleavage pattern is achieved and how diverse body plans have evolved from the ancestor of this group. (3) Each of the three superphyla contains a mixture of segmented and unsegmented taxa. Thus, whether segmentation was already present in the ancestral bilaterian or has evolved more than once (e.g., by independent exaptation of terminal addition mechanisms present in an unsegmented ancestor) is a matter of both interest and controversy (Davis and Patel 1999). Thorough comparisons of segmentation processes in annelids, arthropods, and vertebrates are required if this issue is to be resolved. (4) Comparisons of leeches that cannot generate additional segments or regenerate segments and oligochaetes, which are capable of these processes, should lead to better understanding of regeneration. (5) Medicinal leeches (genus *Hirudo*) are very useful for studying neurogenesis, but earlier stages are less accessible in *Hirudo*. Many individual neuron types can be identified as homologous between species, allowing *Helobdella* to be used to study the cell-lineage and molecular mechanisms by which these neurons arise during early embryogenesis.

SOURCES AND HUSBANDRY

Helobdella species of the *robusta* and *triserialis* complexes have been maintained in laboratory culture for numerous generations, living and breeding in artificial pond water at temperatures of 16–24°C on a diet of freshwater snails. Obtaining adequate supplies of snails can be difficult or expensive; casual attempts to devise an artificial substitute have demonstrated that the leeches will readily ingest such materials as gelatin enriched with yeast extract, but they are incapable of reproducing or growing on any concoction tried to date. In contrast, species of the *H. stagnalis* complex feed on the oligochaetes ("black worms") that are widely sold as pet food, but no species from this complex has yet been maintained in large-scale culture.

Lab colonies of most *Helobdella* isolates are prone to die off, for reasons that are unclear. Difficulties in maintaining the original laboratory population of *H. triserialis* and in recollecting them from the wild (the *H. triserialis* species complex is now recognized as a Latin American clade that was probably introduced to the United States via birds or water plants) led to the identification and culture of other *Helobdella* species or strains and eventually to the appreciation (based first on subtle developmental differences and then on cytochrome-oxidase-1 sequence analyses) of the fact that the three species complexes may be more diverse than previously thought. Identification of *Helobdella* species by DNA bar coding is described in Bely and Weisblat (2006).

The species *H. robusta* was collected from Sacramento, California as a replacement for *H. triserialis*; it was eventually recognized and described as a new species, but it suffers from the same propensity for dying off as did *H. triserialis*. Happily, a closely related species of the *robusta* complex collected from Austin, Texas by M. Shankland is apparently stable in lab culture and is now the focus of most developmental studies. This yet-to-be-named species is designated provisionally as *H*. "austin" (Hau).

Species of the *H. robusta* and *H. triserialis* species complexes can be maintained in aquaria of glass containers at 23°C in chorine-free spring water (natural or 1:100 dilution of reconstituted Instant Ocean in deionized water). Up to 200 animals can be maintained in a single bowl (1.75

liters). The water is changed daily (the glass surface should be rubbed everyday to remove accumulated mucus and small life forms) or more often after feeding or if the animals start to accumulate near the air-water interface. Some investigators find that colony health is improved by including duckweed in each bowl. Animals are fed freshwater snails (physid and/or planorbid), whose shells may be cracked open to facilitate the attack of the leeches. To maintain a peak reproductive output, feeding should occur every other day. At a cooler temperature (e.g., 16°C), however, a healthy colony can survive without feeding or water change for a few weeks, although reproductive activity will diminish. Full reproductive capacity can be restored by returning the colony to 23°C and a regular maintenance schedule. An individual *Helobdella* leech can produce up to five or more clutches of eggs during its life span. The length of each reproduction cycle is 30–40 days, starting at the age of 2 months. Detailed reproductive habits vary from species to species. For example, under laboratory conditions, *H. triserialis* is strictly limited to producing five clutches of embryos, the third of which is clearly the largest, whereas *H. robusta* is capable of generating up to seven clutches that exhibit a more uniform size distribution (Wedeen et al. 1990; D.A. Weisblat, unpubl.). In a properly maintained colony, one can routinely obtain several batches of embryos daily.

RELATED SPECIES

Leeches are a monophyletic annelid group arising within Clitellata, a larger monophyletic group comprising leeches and oligochaetes (Siddall and Burreson 1998; Erséus and Källersjö 2004). Clitellata arises within the polychaetes, which are now recognized as a paraphyletic group. *Helobdella* is among the more experimentally tractable representatives of the superphylum Lophotrochozoa. This taxon includes roughly half of the currently recognized animal phyla (annelids, molluscs, nonacoel flatworms, brachiopods, bryozoans, phoronids, and others), but the intensively studied animal models (i.e., fruit fly, nematode, and various vertebrates) belong to the two other superphyla, namely, Ecdysozoa (arthropods, nematodes, and their allies) and Deuterostomia (chordates, echinoderms, and hemichordates). Thus, examining the development of lophotrochozoan species in comparison with the other taxa is essential for a more complete understanding of how changes in developmental processes have permitted the evolution of the diverse body plans present in extent species.

USES OF THE *HELOBDELLA* MODEL SYSTEM

Stem Cell Biology and Regeneration

In *Helobdella* and other leeches, segmental mesoderm and ectoderm arise by stereotyped lineages from a posterior growth zone consisting of five bilateral pairs of stem cells (teloblasts), each of which is restricted in the types of cells to which it gives rise (Shankland and Savage 1997). Moreover, leeches produce exactly 32 segments during embryogenesis and no more, indicative of a further restriction on the potency of their segmentation stem cells. They cannot generate additional segments nor are they capable of regenerating segments in response to amputation. In contrast, many of the other clitellate annelid species (oligochaetes) produce segments continuously throughout life and are capable of regenerating segments in response to amputation (Bely 2006), indicating the presence of stem cell populations of much greater potency. Remarkably, the early embryos of oligochaetes and leeches are sufficiently similar that individual blastomeres and teloblast lineages can be recognized as unmistakably homologous throughout the Clitellata (Fernández and Olea 1982; Shimizu 1982; Dohle 1999). Thus, comparisons of leeches and oligochaetes should be fruitful for understanding the cellular and molecular mechanisms of regeneration processes and the mechanisms by which regenerative capability may be lost in evolution.

Neurogenesis

Medicinal leeches (genus *Hirudo*) provide one of the best preparations for cellular neurobiology: The constellation of properties that gives different classes of neurons distinct functional identities can be studied in individually identified neurons both in vivo and in vitro (Muller et al. 1981). Since the 1960s, the leech nervous system has been used to address a wide set of topics, including the elucidation of neuronal networks underlying a variety of behaviors (swimming, crawling, heartbeat, local bending, and even behavioral choice), synapse formation, coding and integration of sensory information, neuromodulation, transmitter release, and regeneration (Briggman et al. 2005; Briggman and Kristan 2006). The morphological and functional development of these well-characterized neural circuits can also be studied in *Hirudo*, beginning at the stage at which postmitotic neurons are extending their processes. However, the earlier stages of neurogenesis in *Hirudo* are less accessible because early embryos of the hirudinid species are much smaller and less easily cultured than those of glossiphoniid leeches. Yet many individual neuron types can be identified as homologous between species. Hence, *Helobdella* can be used to study the cell-lineage and molecular mechanisms by which these neurons arise from teloblasts during early embryogenesis.

GENETICS AND GENOMICS RESOURCES

All clitellate annelids are hermaphroditic, and at least some species of *triserialis* and *robusta Helobdella* species complexes are capable of both cross- and self-fertilization. An extensively inbred strain of *H. robusta* was the source of DNA for a National Science Foundation (NSF)-funded bacterial artificial chromosome (BAC) library generated by Kazutoyo Osoegawa and Chung-Li Shu in Pieter de Jong's laboratory at the Children's Hospital Research Institute (http://bacpac.chori.org/library.php?id=211) and for most of the recently completed *Helobdella* genome project (http://genome.jgi-psf.org/Helro1/Helro1.home.html), funded by the Department of Energy (DOE) and performed at the Joint Genome Institute with leadership from Dan Rokhsar and Jeff Boore.

As a self-fertile hermaphrodite with an 8–10-week egg-to-egg life cycle, *Helobdella* offers some features amenable to forward genetics. However, the small clutch sizes (typically 20–100 per individual laying) and difficulty in maintaining individual strains mitigate this possibility.

TECHNICAL APPROACHES

We describe methods for handling and microinjecting *Helobdella* embryos, as well as devitellinizing living embryos. Also included are protocols for silver staining, immunostaining, in situ hybridization, and whole-mount preparation for microscopy.

Protocol 1

Handling of Embryos

Gravid adults are identified by white egg masses that are visible through the ventral body wall. For convenience, gravid animals are removed to smaller bowls so that animals with newly laid embryos can be more readily identified. Zygotes (fertilized internally) turn pink when they are deposited into transparent cocoons on the ventral surface of the parent. A typical clutch consists of 20–100 embryos. Embryos can be collected after zygote deposition and before hatching (stage 10). Staging of embryos is summarized in Weisblat and Huang (2001). The following protocol entails the methods of embryo collection and culture.

MATERIALS

The recipe for the item marked with <R> is on page 271.

Reagents

Artificial spring H_2O
   Dilute 1 part of reconstituted Instant Ocean (32 g/liter in H_2O) in 100 parts of H_2O.
HL saline (sterilized by autoclaving and/or filtering) <R>
Leech (*Helobdella*) carrying cocoons
Snails (small, freshly killed) (optional; see Step 6)
   Kill the snails by crushing their anterior ends with forceps. Partially remove the crushed shells.

Equipment

Abrasive paper or whetstone
Forceps (two blunted Dumont No. 5)
Insect pins (size 0 or 00)
Microscope (dissecting)
Pasteur pipette (disposable glass)
Petri dish (60 × 10 mm, glass or plastic)
Tissue culture dish (35 × 10 mm; Falcon 35-3001)
Wire cutter (optional; see Step 1)

METHOD

1. To avoid injuries to the animal, use blunt forceps for handling the leech; blunted Dumont No. 5 forceps are useful for this purpose. Worn forceps with damaged tips are particularly suitable starting materials. Use a wire cutter if necessary to cut off tips from these forceps and remove the rough edges with abrasive paper or whetstone. The resulting forceps should have a tip diameter of 0.5–1 mm without any sharp angles on the surface.

2. Under a dissecting microscope, place the leech carrying cocoons in a 60 × 10-mm Petri dish filled with spring H_2O. Allow the leech to settle down with suckers firmly attached to the dish bottom. Use forceps to stimulate the anterior end until the anterior sucker is released from the

substrate while the posterior sucker remains attached to the substrate. Once the anterior end is free, gently grab it with one forceps and lift it up so that the cocoons on the ventral surface are exposed. Peel away the cocoons from the parent with the other pair of forceps.

3. Tear apart the isolated cocoons with insect pins or forceps to release the embryos.

4. Collect the embryos and transfer them to a tissue culture dish filled with HL saline (sterilized by autoclaving and/or filtering).

    We use disposable glass Pasteur pipettes to transfer the embryos because the embryos do not stick to the glass material. Glass or Falcon 35-3001 dishes are recommended to avoid stickiness. Pipettes and Petri dishes used for routine embryo cultures are rinsed and reused multiple times.

5. Maintain embryos at 23°C. Replace HL saline daily. Transfer hatching embryos (late stage 9) into artificial spring H_2O gradually during the course of 1–2 days.

6. (Optional) If desired, rear individual embryos to adulthood by assisting them in their first few feedings.

    i. Place juvenile animals (as judged by the exhaustion of yolk from the gut and adult-like locomotory behavior) on or near small, freshly killed snails. Pierce the soft bodies of the snails with an insect pin to facilitate the leech feeding.

    ii. When the leech has fed, it will become quiescent and its gut should be outlined with the pigmented material from the snail. At this point, remove the carcass of the snail promptly and change the medium to prevent contamination.

        After two–five assisted feedings, the young leech should be able to feed on its own, as do adult leeches.

Protocol 2

Microinjection

One major advantage of using *Helobdella* embryo as an experimental system is its amenability for microinjection. Individual blastomeres ranging in size from the zygote (400 µm diameter) down to individual micromeres and primary blast cells (~20 µm diameter) can be injected by pressure under a dissecting microscope. Smaller cells can be injected by iontophoresis under a compound microscope.

Microinjection is a useful tool for studying embryonic development. For example, developmental fates of a cell can be followed by injecting a lineage tracer. A specific cell can be killed by injecting a toxic substance (e.g., DNase or ricin A chain). Furthermore, cells can be killed at a given developmental stage by directing intense fluorescence illumination or a blue laser beam on fluorescein-labeled cells. Finally, it is feasible to manipulate gene expression in leech embryos by injecting zygotes or selected blastomeres with synthetic mRNA, morpholino antisense oligo, or a plasmid construct.

Due to the presence of gap junctions, molecules with MW (molecular weight) of less than roughly 1500 D can diffuse freely among early blastomeres. For purposes in which intercellular diffusion of an injected substance is undesirable, small molecules should be conjugated to larger molecules such as dextran (10 kD). Because there are many commercially available microinjection devices, there are just as many ways to perform microinjection. These different setups can be adopted for *Helobdella* embryos. Here, we describe a versatile homemade pressure injection system used in our lab. In brief, under a dissecting microscope, embryos are immobilized by suction in a custom-fabricated chamber with the target cell facing upward. Cells are visualized using transillumination via a long-working-distance dark-field condenser. The tip of a micropipette is brought into the target cell with a micromanipulator, and the injectant is delivered into the cell by pressure.

MATERIALS

The recipes for the items marked with <R> begin on page 270.

CAUTION: See the Cautions Appendix for appropriate handling of materials marked with <!>.

Reagents

Antibiotics (tetracycline 0.05 mg/ml <!>, penicillin 50 units/ml, streptomycin 0.05 mg/ml <!>) (optional; see Step 10)
Fast Green (4% in 200 mM KCl or H_2O; Sigma-Aldrich) <!>
Fluorescent dextran conjugate (10 kD; 100 mg/ml in 200 mM KCl <!>; Molecular Probes)
Helobdella embryos
Phenol Red solution (0.5%; Sigma-Aldrich) <!>
Zero-divalent HL saline <R>, HL saline <R>, or H_2O

Equipment

Air supply (pressurized: wall outlet or compressed air tank)
Alcohol lamp

Condenser (long working distance, dark field)
Culture dish (plastic, 35 x 10 mm)
Electrode holder
EPON mix <R>
Equipment for microinjection setup, in addition to items listed (see Fig. 1)
Lamp (tungsten halogen, for microinjection of smaller cells; see Step 11)
Micromanipulator
Micropipette (Drummond microcaps, for microinjection of smaller cells; see Step 11)
Micropipette puller
Microscope (dissecting)
Microscope (inverted or upright compound, with water-immersion objective lenses [10x and 20x], for microinjection of smaller cells; see Step 11)
Regulator
Slide (custom-constructed large silicon well, glass, for microinjection of smaller cells; see Step 12)
Tubing (capillary, thin walled; Frederick Haer & Co, 30-30-0)
Tubing (vacuum)
Vacuum chamber (optional; see Step 5.ii)

METHOD

Equipment Setup

See Figure 1.

1. **Pressure module.** Attach a regulator (b in Fig. 1A) to an air outlet (a in Fig. 1A) or a compressed air tank to provide stable air pressure during injection. If using an in-house air line, place a filter in the line to prolong the life of the regulator. Adjust the regulator to maintain a pressure of 5–15 psi. Attach a three-way connector to the regulator at one end (c in Fig. 1A). Connect one remaining end to the electrode holder via thin tubing and leave the other open (the open end serves as the control valve for injection pressure). Apply a finger to this opening to increase the pressure to the micropipette.

    All connections in the pressure module should be hard rubber or plastic vacuum tubing that can withstand a substantial amount of pressure.

2. **Suction module.** Figure 1D details the construction of a suction chamber (h in Fig. 1A, B) from a plastic 35 x 10-mm dish, polyethylene tubing, and epoxide casting resin (EPON). Attach the open end of the tubing to a syringe (d in Fig. 1A) to provide suction for immobilizing the embryo. For more precise control of the suction used to immobilize the embryo, modify this syringe so that the plastic shaft of the piston is replaced by a threaded metal shaft passing through a threaded lucite or metal block for one-handed operation (Fig. 1C).

3. **Illumination and other details.** During microinjection, visualize the embryo in the suction crater by dark-field transillumination. Attach the suction chamber to a stage with a long-working-distance dark-field condenser underneath it (i in Fig. 1A, B). A lamp and a reflective mirror (j in Fig. 1A, B) provide transillumination. Mount the electrode holder (e in Fig. 1A, B) connected to the air pressure on a micromanipulator and place it above the suction chamber. Place a dissecting microscope (g in Fig. 1A, B) above this setup so that it is possible to monitor the suction chamber and micropipette under the microscope.

    An additional lamp (k in Fig. 1A,B) can be placed above the suction chamber to provide epiillumination.

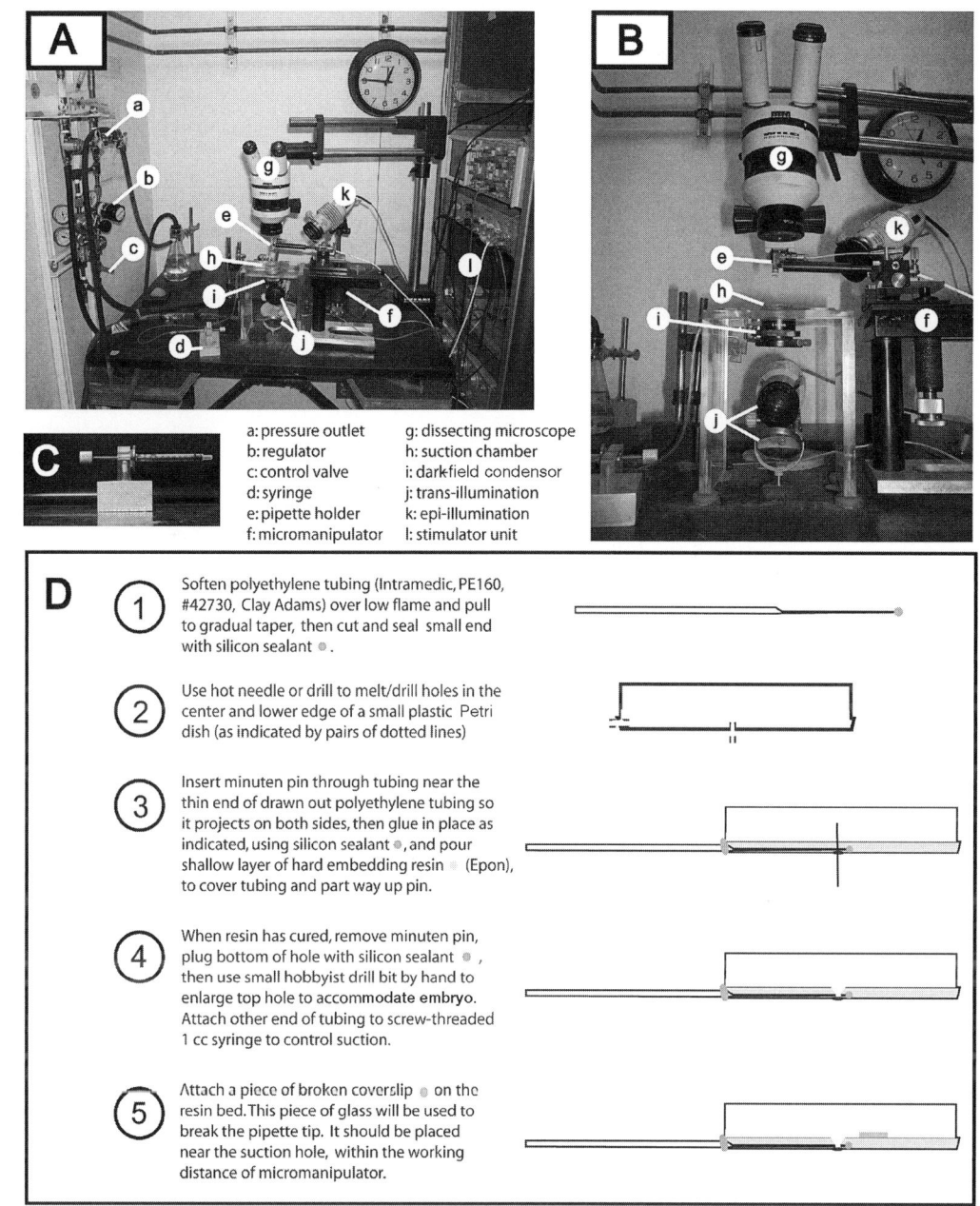

FIGURE 1. Equipment setup for pressure microinjection of the leech embryos. (*A*) Overview of the entire setup. (*B*) Close-up view of the dark-field transillumination unit. (*C*) A custom-built precision-control syringe that is connected to the suction chamber and provides suction force to immobilize the embryo during injection. (*D*) Step-by-step instructions for the construction of a suction chamber.

Micropipette Preparation

4. Draw micropipettes on a standard puller from thin-walled capillary tubing. Determine the setting of the puller empirically to obtain an optimal pipette shape for injecting a specific cell type.

   In Helobdella *embryos, it is possible to come up with a single setting that can be used for fabricating all-purpose micropipettes suitable for injecting large, yolky, early blastomeres and teloblasts as well as smaller yolk-free micromeres.*

5. Use Fast Green or Phenol Red dyes to monitor the tip of the micropipette and the course of the injection as follows.

    i. Mix the monitor dye with the substance intended for injection. For example, we routinely use a 3:1 mixture of 100 mg/ml fluorescent dextran and 4% Fast Green solution as a lineage tracer.

    ii. Back-load the mixture into the micropipette. If necessary, briefly place the loaded micropipette in a vacuum chamber to remove any air bubbles in the micropipette. Store the loaded micropipette in a humid chamber until ready to use.

    Fast Green is more visible and thus a better dye for many purposes. However, it can produce a low-to-middle level of fluorescence in living tissue. Thus, for live imaging, we often use Phenol Red as a monitor dye.

Injection

For microinjection of smaller cells, proceed to Step 11.

6. Before injection, fill the suction chamber with the medium of choice. Depending on the property of the injectant, incubate embryos in normal HL saline, zero-divalent HL saline, or H_2O during the course of injection.

    Helobdella embryos can survive for at least 1 hour in zero-divalent HL saline or H_2O.

7. Break the pipette tip by slowly and gently lowering it against the small glass plate fixed onto the chamber bed. Once the pipette tip is broken, one can see monitor dye flowing from the pipette tip immediately upon applying pressure on the control valve. The flow of dye should terminate as soon as the pressure is relieved.

    Often, depending on the size of the opening and the viscosity of the injectant, breaking the tip will result in fluid from the bath being drawn into the tip by capillary action, as judged by the retreat of the monitor dye from the tip. Clogged tips can sometimes be remedied by rebreaking the tip. If the broken tip is too large in diameter (as judged by continuous dye flow or the inability to successfully impale the cells of interest), discard the pipette and start over.

8. Briefly flame the tip of a micropipette in an alcohol lamp to make a smooth glass probe. (Discarded or broken injection pipettes can be recycled for this purpose.) Use both the probe and the threaded suction controller in combination to push the embryo to be injected into the crater and immobilize it with the target cell facing upward toward the micropipette. Adjust the dark-field transillumination so that the outline of the target cell is clearly visible.

9. Move the pipette tip over the target cell and lower the micropipette to impale the target cell. If bathing media has been drawn into the tip of the pipettor, displace it immediately, just before impaling the cell, by applying pressure until the monitor dye has returned to the tip. Once the cell is penetrated, as judged by the "jumping" movement of the pipette tip relative to the cell membrane, briefly apply pressure to the pressure valve. Observe the color of the dye to monitor the injection. Using Fast Green to monitor the injection of fluorescent dextrans, the presence of barely visible monitor dye around the tip of the pipette within the cell is sufficient.

    With a successful injection, the Fast Green stain should diffuse quickly from the pipette tip when pressure is applied and it should be barely or not at all visible several minutes after injection. If the monitor dye forms a nondiffusing clump at the tip of the pipette, the injection was probably unsuccessful. Any visible swelling of the cell during the injection can be taken as an indication that the cell is overinjected and will probably die or develop abnormally.

10. After injection, release the suction, transfer the embryo back into HL saline, and continue the experimental program. If necessary, add antibiotics to the culture medium (tetracycline 0.05 mg/ml, penicillin 50 units/ml, streptomycin 0.05 mg/ml).

Microinjection of Smaller Cells by Iontophoresis under a Compound Microscope

11. Smaller cells (e.g., primary blast cells, their daughters, and granddaughters; 5–20 µm in diameter) can be microinjected by iontophoresis using an inverted compound microscope or an upright compound microscope equipped with a water-immersion objective lens (20x and 40x).

    i. Immobilize the embryo by suction in a fire-polished Drummond microcups micropipette connected to a syringe, similar to those used for microinjecting mammalian embryos. Use a micromanipulator attached to the suction pipette to position the embryo so that the target cell is pointing away from the suction pipette but is visible under the microscope.

    ii. At the opposite side of the suction pipette, mount a back-loaded micropipette pulled from thin-wall capillary tubing (tip resistance = 30~100 MΩ) on an electrode holder connected to a stimulator (e.g., l in Fig. 1A). Immerse the ground electrode and tips of the suction and injection pipettes in zero-divalent HL saline that fills a custom-constructed large silicone-well glass slide.

    iii. Gradually bring the micropipette tip close to the target cell by using a micromanipulator attached to the injection pipette. Once the tip of the injection pipette is within about 1 µm of the surface of the target cell, penetrate the cell by thrusting the pipette tip against the cell. Pulses of negative DC current (~5 nA) deliver the injectant (which is usually negatively charged) into the target cell.

    iv. Monitor the progress of these injections by epifluorescence of fluorescent dextran mixed into the injectant. To minimize photo damage, use a tungsten halogen lamp in place of a high-intensity mercury or xenon arc lamp as the source of epi-illumination.

Protocol 3

Devitellinization of Living Embryos

Embryos of glossiphoniid leeches are enclosed in a thin vitelline envelope until "hatching" (stage 10). To perform microsurgical manipulation on early embryos (see Symes and Weisblat 1992; Isaksen et al. 1999; Huang et al. 2001), it is necessary to remove the vitelline envelope. The following protocol describes the procedure of removing the vitelline envelope from *H. robusta* and *H.* sp. *(Austin)* embryos. This protocol is applicable to embryos of stages 1–9 but is probably more useful for early stages. The devitellinized embryos can develop normally with careful culturing.

MATERIALS

The recipes for the items marked with <R> are on page 271.

CAUTION: See the Cautions Appendix for appropriate handling of materials marked with <!>.

Reagents

Agar
Dithiothreitol (DTT) solution (1 M in H_2O) <!>
Helobdella embryos
High-calcium HL saline <R>
HL saline <R>
HL saline supplemented with antibiotics <R>
Trypsin (type II; Sigma-Aldrich) <!>

Equipment

Alcohol lamp or bunsen burner
Culture dishes
Micropipette (flamed; see Protocol 2, Step 8)
Pasteur pipettes (disposable glass)

METHOD

1. Prepare a 0.1% trypsin solution in high-calcium HL saline. Aliquot and store at –20°C.

2. Prepare at least four fire-polished glass pipettes by briefly placing the tips of the pipettes in a flame.

3. Prepare melted 1% agar in high-calcium HL saline. Prepare at least four agar dishes by pouring the agar into culture dishes.

4. Thaw the trypsin solution immediately before use and add 1/100 volume of 1 M DTT solution to the trypsin solution.

5. Fill one agar dish with the trypsin/DTT solution and the other three with high-calcium HL saline (these will be the washing agar dishes).

6. Place embryos into the trypsin/DTT solution; watch them closely. The vitelline envelope should start to wrinkle, break open, and lift off from the embryo within a few minutes. The time required for each embryo varies significantly and ranges from 1 to 30 minutes. Use a fire-polished glass pipette to transfer an individual embryo to the washing agar dish containing high-calcium HL saline as soon as its vitelline envelope breaks loose.

7. Often, the vitelline envelope does not completely detach from the embryo. If this is the case, use a flamed micropipette to remove the loosened vitelline envelope once the embryo is in the first washing agar dish.

8. Once the vitelline envelope is removed from every embryo, transfer the embryos to a new washing agar dish with a new flamed glass pipette. Incubate for 5 minutes.

9. Repeat Step 8 twice.

10. Transfer the embryos back to the HL saline supplemented with antibiotics (tetracycline 0.05 mg/ml, penicillin 50 units/ml, streptomycin 0.05 mg/ml). They are now ready for further experimentation.

    Devitellinized embryos are fragile and sticky and should always be kept on an agar bed. The devitellinzed embryos should be moved to a new agar dish containing fresh HL saline and antibiotics daily if long-term culture is desired.

Protocol 4

Arnold's Silver Staining

Arnold's silver staining is used to visualize the boundaries of epidermal cells at the surface of embryos. This method, adapted from Arnolds (1979), is particularly useful for examining the transparent micromere-derived epithelium that covers the animal half of the embryo during gastrulation (see, e.g., Smith and Weisblat 1994; Smith et al. 1996)

MATERIALS

The recipes for the items marked with <R> begin on page 270.

CAUTION: See the Cautions Appendix for appropriate handling of materials marked with <!>.

Reagents

Carnoy's fixative <R>
Ethanol <!>
Formaldehyde <!> (4% in 0.25x PBS <R> or other buffered solution)
Formaldehyde <!> (0.8% in 0.1 M sodium cacodylate buffer [pH 7.4] <!>)
Helobdella embryos
Silver methenamine solution <R> <!>
Sodium borate solution (30 mM, pH 7.5; pH adjusted with boric acid)

Equipment

Culture dish (10 x 35 mm)
Fiber optic illuminators (2x)
Microscope (dissecting)

METHODS

Silver staining can be performed on either living or fixed embryos. To preserve the cell junction in fixed embryos, a special fixation procedure is used.

I. Living Embryos

1. In a culture dish, incubate embryos in H_2O for at least 5 minutes to remove excess ions, especially Cl^-.

2. Replace H_2O with sodium borate solution and incubate for 1 minute.

3. Replace sodium borate solution with freshly prepared silver methenamine solution and briefly incubate.

4. Place the dish under a dissecting microscope and direct fiber optic illumination onto the embryos at the highest intensity. Two illuminators can be used in this step. Watch the reaction closely. Cell outlines should turn dark brown within 10–15 minutes. Terminate the reaction

by turning off the illumination and replacing the silver methenamine solution with sodium borate solution.

5. Fix the embryos in 4% formaldehyde in 0.25x PBS or other buffered solution.

II. Fixed Embryos

1. Fix embryos in 0.8% formaldehyde in 0.1 M sodium cacodylate buffer (pH 7.4) for 30 minutes or less at room temperature.
   Overfixation results in a failure to visualize the epithelial junctions.

2. Briefly wash the embryos several times with H_2O.

3. Incubate the embryos in H_2O for 5 minutes, then follow Steps 2–4 of Method I above.

4. Fix the embryos overnight at 4°C in Carnoy's fixative and store in 100% ethanol.

Protocol 5

Immunostaining

The purpose of this staining protocol is to detect the localization of specific antigens in the embryo. The immunostaining protocols must be optimized for each antibody and embryonic stage. Here, as a starting point, we present a general-purpose immunostaining protocol as described by Goldstein et al. (2001). Immunostaining of the leech embryo can be performed on the intact embryo or on dissected parts. For postgastrulation embryos (stages 9 and 10), if the area of interest is in the germinal plate, immunostaining can be performed on the dissected germinal plate. It can also be performed on the ventral nerve cords dissected from late-stage embryos or juveniles.

MATERIALS

The recipes for the items marked with <R> begin on page 270.

CAUTION: See the Cautions Appendix for appropriate handling of materials marked with <!>.

Reagents

Bovine serum albumin (BSA) (optional; see Step 6)
Detergent (Tween 20 or Triton X-100 <!>)
Diaminobenzidine tetrahydrochloride (DAB; 2 mg/ml in H_2O; Sigma-Aldrich) <!> (for horseradish peroxidase (HRP)-conjugated secondary antibody; see Step 11.i)
Formaldehyde (4% in 0.25x PBS) <!>

> For convenience, we use 10-ml ampules of 16% formaldehyde solution packed in nitrogen (EM Science or Polysciences). Use the formaldehyde solution within 1–2 weeks of opening.

Helobdella embryos
Hoechst 33258 (optional; see Step 1.iii)
Hydrogen peroxide (H_2O_2; 30%) <!> (for HRP-conjugated secondary antibody; see Step 11.ii)
Methanol (100% and series of 70, 50, and 30%) (optional; see Step 4) <!>
Nickel salt solution (1% $NiCl_2$ or $NiSO_4$) <!> (for HRP-conjugated secondary antibody; optional; see Step 11.i)
Phosphate-buffered saline (PBS) <R>
Primary antibody
Relaxant (ice cold) <R>
Secondary antibody (fluorescence or HRP-conjugated)
Serum (normal; from the same species in which the secondary antibody is generated)

Equipment

Alcohol lamp or bunsen burner
Dish (Sylgard 184 silicone resin)
Microcentrifuge tubes (1.5 or 0.6 ml)
Minuten pins (or short lengths of sharpened, fine tungsten wire)
Pasteur pipette (disposable glass)
Rocking table

METHOD

1. Fix the embryos as follows.

   Early-stage Embryos

   i. Fix early-stage embryos (stages 1–9; before muscle formation) in 4% formaldehyde in 0.25× PBS.

   Late-stage Embryos

   To prevent late-stage embryos with developed nervous system and musculature (stages 9–11) from hypercontraction during fixation, paralyze and relax them before fixation.

   i. Place late-stage embryos in ice-cold relaxant for several minutes. The muscle contraction should completely diminish and the body should straighten up.

   ii. After the embryos are completely relaxed, fix them in 4% formaldehyde in 0.25× PBS for 1 hour at room temperature or overnight at 4°C.

   For embryos labeled with fluorescent dextran lineage tracer, overnight fixation at 4°C is recommended.

   If Hoechst 33258 DNA counterstaining is desired, the dye can be included in the fixative at a concentration of 2.5 µg/ml.

2. After fixation, wash the embryos several times in PBS.

   Embryos can be stored in PBS for up to several days at 4°C.

3. Remove the vitelline envelope to allow antibody penetration.

   Manual Method

   i. Use minuten pins or sharpened tungsten wire to remove the vitelline envelope.

   Broken Pipette Method

   i. Pull a glass Pasteur pipette to a smaller diameter over an alcohol lamp or a gas flame in a bunsen burner.

   ii. Break the tip of the pipette so that the size of the borehole is only slightly larger than the diameter of the embryos.

   iii. Remove the vitelline envelope by repeatedly pipetting the embryos in and out through the borehole.

   Both methods are best performed in a dish with a Sylgard silicone resin bed. The first method is more time consuming. Although the second method is fast and suitable for large numbers of embryos, it can sometimes cause significant damage to the embryos. Examine the devitellinized embryos individually and discard those that are damaged.

4. (Optional) For some epitopes, methanol treatment can enhance the staining.

   i. Wash fixed, devitellinized embryos once in a 1:1 mixture of PBS and methanol.

   ii. Wash the embryos twice in methanol. Incubate the embryos overnight or longer at –20°C.

   iii. Transfer the embryos back to PBS through a methanol series (70, 50, and 30%) and briefly wash several times in PBS.

   In general, embryos can be stored for up to 1 year in methanol at –20°C.

5. Wash the embryos in PBS supplemented with detergent(s) (detergent PBS) several times.

   In terms of detergent concentration, a good starting point is 0.1% Tween 20 (a milder condition). For nuclear proteins (such as transcription factors or histone), start with 1% Triton X-100. The optimal detergent concentrations can range anywhere from 0.1% Tween 20 to 2% Tween 20 + 2% Triton X-100 for various antibodies and embryonic stages.

6. Prepare blocking solution with PBS and detergent(s) at the same concentration as in Step 5 and normal serum from the same species in which the secondary antibody is generated. Place the embryos into the blocking solution and incubate on a rocking table for 1–3 hours at room temperature or up to overnight at 4°C if necessary.

   The optimal serum concentration must be determined for each specific antibody and embryonic stage. We have used concentrations between 1% and 10%. In general, for early-stage embryos, a higher concentration of serum is required, and a lower concentration can be used for dissected germinal plate or nerve cord.

   In some staining protocols, the blocking solution is suppplemented by the addition of BSA to a final concentration of 1–3%. For early-cleavage-stage embryos, adding BSA is often required. As a starting point, for embryos of stages 1–6, we recommend beginning with a blocking solution with 10% normal serum, 3% BSA, and 1% Tween 20. For embryos of stages 7–11, we recommend beginning with 3% normal serum and 0.1% Tween 20.

7. Add primary antibody to the blocking solution at an appropriate dilution factor. Incubate the embryos overnight or longer with gentle rocking at 4°C. Remove the primary antibody solution. Rinse the embryos several times in detergent PBS and then wash in detergent PBS for 6 hours with frequent buffer changes (30–60-minute intervals). In some cases, blocking solution can be used in place of detergent PBS.

   The optimal length of primary antibody incubation is also determined empirically for each individual antibody and embryonic stage. The time required ranges from 1 to 10 days in our experience, with the longest incubations required for zygote and early cleavage stages.

8. Transfer the embryos into blocking solution and add secondary antibody at an appropriate dilution factor. Incubate the embryos overnight or longer with gentle rocking at 4°C.

   Usually, the length of time for the secondary antibody incubation is the same as that used for the primary antibody incubation. However, in many cases, it can be shorter.

9. Remove the secondary antibody solution and rinse the embryos several times in detergent PBS. Wash the embryos in detergent PBS for 6–12 hours with frequent buffer changes.

10. If a fluorescent secondary antibody was used, rinse the embryos in PBS; otherwise, proceed to Step 11 (for an HRP-conjugated secondary antibody).

    The embryos are now ready for clearing and observation.

11. For an HRP-conjugated secondary antibody, detect the secondary antibody as follows.

    i. Prepare DAB working solution (1 part of 2 mg/ml DAB stock solution and 4 parts of PBS or detergent PBS). The color of DAB staining is reddish brown. If black staining is desired, add a small amount of 1% nickel salt solution (1/50 of the total volume of DAB working solution to a final concentration of 0.02%) to the DAB solution. Mix the working solution well.

    ii. Add H_2O_2 at a final concentration of 0.01% to give catalyzed DAB solution.

    iii. Incubate the embryos in catalyzed DAB solution. The color should develop within minutes.

    iv. Once the desired intensity is reached, terminate the reaction with several washes in PBS or detergent PBS.

    v. Rinse the embryos in PBS.

       The embryos are now ready for clearing and observation.

Protocol 6

In Situ Hybridization

The purpose of this protocol is to reveal the localization of transcripts in the embryo. In situ hybridization protocols for *Helobdella* embryos are derived from those used for zebrafish and *Xenopus*. The protocols are different for early- and late-stage embryos (stages 1–8 and stages 9–11, respectively) due to a difference in tissue permeability.

MATERIALS

The recipes for the items marked with <R> begin on page 270.

CAUTION: See the Cautions Appendix for appropriate handling of materials marked with <!>.

Reagents

Acetic anhydride <!>
Agarose gel
Antidigoxigenin antibody conjugate (alkaline phosphatase or peroxidase)
AP Buffer <R>
Blocking solution (2% normal sheep serum and 2% BSA in PBT)
Formaldehyde (4% in 0.25x PBS) <!>

> *For convenience, we use 10-ml ampules of 16% formaldehyde solution packed in nitrogen (EM Science or Polysciences). Use the formaldehyde solution within 1–2 weeks of opening.*

Formaldehyde (4% in PBT) <!>
Glycine (2 mg/ml in PBS)
H_2O (RNase-free)
Helobdella embryos
Lithium chloride (LiCl) <!>
Methanol (100%, 60% in PBS, 50% in PBS, 30% in PBS) <!>
4-Nitro blue tetrazolium chloride/5-bromo-4-chloro-3-indolyl-phosphate (NBT/BCIP) stock solution (Roche) <!>
PBT (0.1% Tween 20 in PBS)
Phosphate-buffered saline (PBS) <R>
Prehybridization buffer <R>
Pronase E (20 mg/ml stock solution; diluted to 0.5 mg/ml in PBS before use)
Relaxant (ice cold) <R>
RNA synthesis kit (e.g., MEGAscript by Ambion)
SSC (2, 0.2, 0.1x) <R>
Template DNA for antisense riboprobe
Triethanolamine buffer (0.1 M, pH 8) <!>

Equipment

Burner
Dish (Sylgard 184 silicone resin)

Equipment for agarose gel electrophoresis
Microcentrifuge tubes (1.5 or 0.6 ml)
Minuten pins
Oven (hybridization) or water bath
Pasteur pipette (disposable glass)
Rocking table
Vortex mixer

METHODS

Probe Synthesis

1. Synthesize antisense riboprobe using a commercial in vitro transcription kit (e.g., Ambion MEGAscript) with linearized plasmid or PCR product as the template.

    i. For the Ambion MEGAscript kit, use the following recipe (applicable to both T7 and SP6 kits):

      1.5 µl of RNA polymerase enzyme
      1.5 µl of 10x reaction buffer
      1.5 µl each of ATP, CTP, and GTP
      1 µl of UTP
      2.5 µl of digoxigenin-UTP (10 mM solution; Roche)
      1–2 µg of template DNA
      H_2O to a final volume of 15 µl

      Based on our experience, the probe should be between 600 and 2000 bp in length and is best at ~1 kb. We do not find it necessary to hydrolyze probe of this length.

    ii. Precipitate the probe with LiCl and then resuspend and quantify in RNase-free H_2O.

    iii. Run a small aliquot of probe on an agarose gel to ensure the probe quality.

    iv. Dilute the probe in prehybridization buffer to a final concentration of 100 ng/µl.

      The probe stock in prehybridization buffer can be stored for at least 1 year at –20°C or –80°C. The above reaction condition routinely yields 10–70 µg of probe per reaction.

Fixation and Devitellinization

2. Fix the embryos as follows.

Early-stage Embryos

   i. Fix the embryos in freshly prepared 4% formaldehyde in 0.25x PBS for 1 hour at room temperature or overnight at 4°C.

Late-stage Embryos

   i. Paralyze the embryos by bathing them in ice-cold relaxant solution for several minutes.

   ii. Fix the embryos immediately after paralyzing them, the same as for early-stage embryos (Step 2.i).

3. After fixation, wash the embryos several times in PBS to remove excess formaldehyde.

4. For embryos that are enclosed in the vitelline envelopes of stages 1–9, remove the vitelline envelopes from the embryos using a broken pipette or minuten pins as described in Protocol 5.

    For in situ hybridization, perform devitellinization with a broken pipette in PBS. However, when minuten pins are used, a methanol/PBS mixture can be used in the rehydration step following methanol treatment. Methanol treatment weakens the vitelline envelope and makes devitellinization with pins easier.

5. Briefly wash the embryos with 50% methanol in PBS. Then, wash the embryos at least twice in 100% methanol, and incubate in 100% methanol for at least 1 hour.

    For late-stage embryos, a long methanol treatment is desirable. Methanol treatment is essential for tissue penetration and can reduce endogenous alkaline phosphatase background, but there are some cases in which the absence of methanol treatment does not influence the outcome of in situ hybridization staining in early-cleavage-stage embryos. Embryos can be stored in methanol at –20°C for at least a year.

Tissue Penetration

6. Rehydrate the embryos as follows.

    i. Return the embryos to room temperature.

    ii. Incubate the embryos in 60% methanol in PBS, 30% methanol in PBS, and PBS for 5 minutes each.

    iii. Prepare the embryos for hybridization.

       a. For embryos of stages 1–8, wash the embryos five times in PBT for 5 minutes each and proceed to Step 12.

       b. For embryos of stages 9–11 (after the coelom has formed), proceed directly to Step 7 to perform tissue penetration by enzyme digestion.

7. For embryos of stages 9–11, rinse the embryos three times in PBS for 5 minutes each. Treat the embryos with 0.5 mg/ml Pronase E in PBS for 15–30 minutes at 37°C.

    Tissue penetration is the trickiest step in the in situ hybridization for late-stage embryos; it is very important not to overdigest or underdigest the embryos. The length of incubation that yields optimal tissue penetration depends on the age of embryos and enzyme activity. The optimal time may vary among batches of Pronase E and should be thus determined for each new batch of Pronase E. A good starting point is 15 minutes for stage 9 or 25 minutes for stages 10 and 11.

    Although enzyme digestion is not necessary for in situ hybridization on embryos younger than stage 9, it may sometimes help to yield better quality staining. For younger embryos, Pronase E digestion should be limited to less than 5 minutes.

8. Stop the Pronase E digestion reaction by rinsing the embryos twice with 2 mg/ml glycine in PBS for 1 minute each.

9. Wash the embryos once with PBS.

10. Rinse the embryos twice in 0.1 M triethanolamine buffer (pH 8) for 5 minutes each. Add 3 µl of acetic anhydride per each 1 ml of triethanolamine buffer and incubate with agitation for 5 minutes. Add another 3 µl of acetic anhydride and incubate with agitation for an additional 5 minutes. Wash the acetylated embryos three times in PBS for 5 minutes each.

    This acetylation step reduces nonspecific probe binding.

11. Postfix the embryos with 4% formaldehyde in PBT for 20 minutes at room temperature. Wash the embryos three times in PBT.

12. Prepare a 1:1 mixture of PBT and prehybridization buffer. Rinse the embryos with this mixture for 5 minutes. Wash the embryos twice with prehybridization buffer for 5 minutes each.

Hybridization

13. Incubate the embryos with fresh prehybridization buffer overnight at 60–67°C, depending on the length and guanine-cytosine (GC) content of the probe as expected.

    The optimal hybridization temperature can vary among different probes and should be determined empirically. Hybridization can be performed in a hybridization oven or a water bath. We recommend using a hybridization oven because it minimizes water condensation and allows for better handling during the washing steps. Gentle agitation during prehybridization, hybridization, and probe removal also improves the quality of staining.

14. Dilute the 100 ng/μl probe stock solution (from Step 1) 1:50 in the prehybridization buffer to make a hybridization solution with approximately 2 ng/μl final probe concentration. Denature the probe by incubating the probe solution for 5 minutes at 90°C and then allow it to cool to the hybridization temperature. Incubate the embryos in the hybridization solution containing the probe at the hybridizing temperature.

    Usually, overnight hybridization is sufficient, but it can be extended for up to several days when necessary. The probes can be reused several times.

    The optimal probe concentration may be anywhere between 0.1 and 2 ng/μl, and must be determined empirically.

Probe Removal

It is important to maintain the embryos at the hybridization temperature throughout the washing steps to avoid nonspecific staining. Perform all of the buffer changes in a hybridization oven or water bath.

15. Wash the embryos once with prewarmed prehybridization buffer for 5 minutes.

16. Prepare prewarmed 2:1, 1:1, and 1:2 mixtures of prehybridization buffer and 2× SSC buffer. Rinse the embryos sequentially in these mixtures for 5 minutes each.

17. Wash the embryos once in prewarmed 2× SSC for 10 minutes, twice in prewarmed 0.2× SSC for 15–30 minutes each, and then twice in prewarmed 0.1× SSC for 15–30 minutes. After the final 0.1× SSC wash, remove the buffer from the embryos and return the embryos to room temperature.

18. Prepare 2:1, 1:1, and 1:2 mixtures of 0.1× SSC and PBT. Wash the embryos sequentially with these mixtures for 5 minutes each at room temperature. Then, wash the embryos once in PBT for 5 minutes.

Antidigoxigenin Antibody Labeling

19. Place the embryos into a new tube, briefly rinse twice in antibody blocking solution, and then incubate in the blocking solution for 1–3 hours at room temperature.

20. Add to the blocking solution the alkaline phosphatase– (for NBT/BCIP coloration) or peroxidase– (for fluorescence labeling) conjugated antidigoxigenin antibody at a 1/5000 dilution factor. Incubate the embryos in the antibody solution overnight at 4°C.

21. Remove the unbound antibody with PBT washes on a rocking table at room temperature. First, briefly wash the embryos three times with PBT. Follow this with three 15-minute washes and another three 1-hour washes.

    The embryos are now ready for the color reaction. For detection of alkaline phosphatase activity, proceed to Step 22.

    For fluorescence labeling, we have had some success with TSA fluorescence labeling (Perkin Elmer). It is performed following manufacturer instructions and is not described here. Note that TSA fluorescence labeling is not quite as sensitive as NBT/BCIP for this application, and thus, its use is limited to more abundant transcripts.

NBT/BCIP Coloration

We routinely use standard NBT/BCIP substrate for alkaline phosphatase. NBT/BCIP coloration generally has the best sensitivity.

22. Briefly rinse the embryos twice in AP buffer and then wash twice in AP buffer for 5 minutes each.

23. Immediately before starting the color reaction, dilute the NBT/BCIP stock solution with 50× volumes of AP buffer to make the NBT/BCIP working solution. Mix the working solution thoroughly by vortexing.

24. Start the color reaction by adding an appropriate amount of working solution to the embryos. Perform the reaction in the dark at room temperature or at 4°C. If performing at room temperature, examine the reaction every 15–30 minutes under the dissecting microscope.

25. Once the desired staining intensity is reached, terminate the coloration reaction by rinsing the embryos several times in PBT.

    Depending on the transcripts and probe, the reaction time can range from as short as 10 minutes to as long as several days. If a long reaction time is necessary, replace the NBT/BCIP solution daily.

    For most probes, an alternative method is to perform the color reaction overnight at 4°C. Generally, the shorter the color reaction time, the less background staining is introduced. Determine the optimal color reaction conditions for each different probe.

26. After terminating the color reaction, store the embryos in PBS.

    The embryos are now ready to be cleared and observed under a microscope.

Protocol 7

Whole-mount Preparation for Microscopy

Due to the high yolk content and relatively large size of the leech embryo, it is preferable to examine the embryos in a cleared whole-mount preparation after the staining procedure. Three whole-mount procedures suitable for the leech embryos are described below. The first, glycerol whole mount, is quick and convenient because it does not require dehydration. However, prolonged incubation in buffered glycerol could cause the yolk to turn dark purple. Thus, it is not suitable for yolk-containing specimens that are intended for light microscopy. For fluorescence microscopy, antifading reagent can be added to the buffered glycerol. A recipe that works quite well for leech embryos is given here as Method I.

Method II relies on the use of benzyl benzoate:benzyl alcohol (BBBA), a solution that provides the best clearing of yolk-containing tissue in the leech embryo. It is suitable for both fluorescence and light microscopy. Embryos in EPON (Method III) can be observed as whole mount or, alternatively, transferred to catalyzed EPON and embedded for sectioning. Compared to BBBA, EPON is not quite as effective in clearing the yolk, but the viscosity of EPON facilitates orienting the specimens for imaging. As with BBBA, EPON is suitable for both light and fluorescence microscopy.

MATERIALS

The recipes for the items marked with <R> are on page 270.

CAUTION: See the Cautions Appendix for appropriate handling of materials marked with <!>.

Reagents

Benzyl benzoate:benzyl alcohol (BBBA; for BBBA whole mount) <R> <!>
Buffered glycerol with 4% *n*-propyl gallate (for glycerol whole mount) <R>
EPON 812 resin (PolyBed 812; Polysciences; for EPON whole mount) <!>
Ethanol (50, 70, 80, 90, 95, and 100%) (for BBBA whole mount and EPON whole mount) <!>
Helobdella embryos
Propylene oxide (PPO; for EPON whole mount) <!>

Equipment

Chemical hood or vacuum dessicator (for EPON whole mount)
Container (open, for EPON whole mount)

METHODS

I. Glycerol Whole Mount for Fluorescence Microscopy

1. Transfer fixed embryos directly from buffered saline into the buffered glycerol with 4% *n*-propyl gallate.

2. Observe the embryos directly as whole-mount preparations.

II. BBBA Whole Mount

1. Dehydrate the embryos by transferring them from buffered saline to ethanol through the ethanol series (50, 70, 80, 90, 95, and 100% twice).
2. Briefly store the embryos in 100% ethanol to ensure complete dehydration.
3. Transfer the embryos to BBBA immediately before observation.
    BBBA can "wash out" NBT/BCIP staining deposit over time. Hence, NBT/BCIP-stained specimens cleared by BBBA should be documented as soon as possible and BBBA is not suitable for long-term storage of NBT/BCIP-stained embryos.

    Embryos can be returned from BBBA to ethanol. This can be followed by rehydration for storing in buffered saline or by EPON embedding.

III. EPON Whole Mount

1. Dehydrate the embryos through the ethanol series (50, 70, 80, 90, 95, and 100% twice).
2. Briefly wash the embryos in PPO.
    PPO makes it possible to transfer embryos from ethanol to EPON, which are not miscible with each other.
3. Prepare a 1:1 mixture of EPON and PPO and transfer the embryos from PPO into this mixture (or simply replace the PPO with EPON:PPO). Allow the embryos to sit in the EPON:PPO mixture overnight in an open container in a chemical hood or vacuum dessicator, which allows the PPO to evaporate.
4. Replace the evaporated EPON/PPO mixture with fresh EPON the next day.
    The embryos in EPON are now ready for observation as a whole-mount preparation. Alternatively, the embryos can then be transferred into catalyzed EPON and embedded into plastic blocks for sectioning.

Recipes

CAUTION: See the Cautions Appendix for appropriate handling of materials marked with <!>.

AP BUFFER

Reagent	Quantity	Final concentration
Tris (1 M, PH 9.5) <!>	5 ml	0.1 M
$MgCl_2$ (1 M) <!>	2.5 ml	0.05 M
NaCl (5 M)	1 ml	0.1 M
Tween 20 (10%)	0.5 ml	0.1%
Sterile H_2O	to a total volume of 50 ml	

BENZYL BENZOATE:BENZYL ALCOHOL (BBBA)

Reagent	Quantity (10 ml)	Final concentration
Benzyl benzoate <!>	6 ml	60%
Benzyl alcohol <!>	4 ml	40%

BUFFERED GLYCEROL WITH 4% N-PROPYL GALLATE

Reagent	Quantity (10 ml)	Final concentration
Glycerol	80 ml	80%
n-Propyl gallate (Sigma-Aldrich) <!>	4 g	4%
Tris (0.1 M, pH 9) <!>	20 ml	0.02 M

Dissolve the *n*-propyl gallate powder (an antifading reagent) in the glycerol by slowly stirring in the dark, overnight at room temperature. *n*-Propyl gallate is not very soluble in aqueous solvent. Mix the 0.1 M Tris (pH 9) into the solution the next day. The buffered glycerol can be kept for at least 1 year when stored at 4°C.

CARNOY'S FIXATIVE

Reagent	Quantity (10 ml)	Final concentration
Ethanol <!>	6 ml	60%
Chloroform <!>	3 ml	30%
Glacial acetic acid <!>	1 ml	10%

EPON MIX (reagents available at Polysciences)

Reagent	Quantity	Final concentration
EPON (PolyBed 812) resin <!>	25 ml	52%
Methyl nadic anhydride (MNA) <!>	22.25 ml	46%
Dimethyl pimelimidate (DMP) <!>	0.69 ml	1.4%

Mix well by slowly stirring and store in a 50-ml syringe at –20°C.

HIGH-CALCIUM HL SALINE

Reagent	Quantity (1 liter)	Final concentration
NaCl (1 M)	4.8 ml	4.8 mM
KCl (1 M) <!>	1.2 ml	1.2 mM
MgCl$_2$ (1 M) <!>	2 ml	2 mM
CaCl$_2$ (1 M) <!>	18 ml	18 mM
Maleic acid (1 M) <!>	1 ml	1 mM

Adjust pH to 8.2.

HL SALINE

Reagent	Quantity (1 liter)	Final concentration
NaCl (1 M)	4.8 ml	4.8 mM
KCl (1 M) <!>	1.2 ml	1.2 mM
MgCl$_2$ (1 M) <!>	2 ml	2 mM
CaCl$_2$ (1 M) <!>	8 ml	8 mM
Maleic acid (1 M) <!>	1 ml	1 mM

Adjust pH to 6.6. HL saline may be supplemented with antibiotics as follows: tetracycline 0.05 mg/ml <!>, penicillin 50 units/ml, streptomycin 0.05 mg/ml <!>.

PHOSPHATE-BUFFERED SALINE (PBS)

Reagent	Quantity (1 liter)	Final concentration
NaCl <!>	8 g	137 mM
KCl <!>	0.2 g	2.7 mM
Na$_2$HPO$_4$	1.44 g	10 mM
KH$_2$PO$_4$	0.24 g	1.8 mM

Dissolve the ingredients in 800 ml of H$_2$O. Adjust the pH to 7.4 (or 7.2, if required) with HCl <!> and then add H$_2$O to 1 liter. Dispense into aliquots and sterilize by autoclaving for 20 minutes at 15 psi on liquid cycle or by filter-sterilization. Store at room temperature.

PREHYBRIDIZATION BUFFER

Reagent	Quantity	Final concentration
Deionized formamide <!>	25 ml	50%
Heparin (10 mg/ml) <!>	250 µl	0.05 mg/ml
Torula RNA	25 mg	0.5 mg/ml
Tween 20 (10%)	0.5 ml	0.1%
50x Denhardt's solution (optional)	1 ml	1x
CHAPS (10%) (optional)	0.5 ml	0.1%
20x SSC buffer <R>	12.5 ml	5x
Citric acid (1 M) <!>	460 µl	9.2 mM
RNase-free H$_2$O	to total volume of 50 ml	

RELAXANT

Reagent	Quantity (1 liter)	Final concentration
NaCl (1 M)	4.8 ml	4.8 mM
KCl (1 M) <!>	1.2 ml	1.2 mM
MgCl$_2$ (1 M) <!>	10 ml	10 mM
Ethanol (100%) <!>	80 ml	8%

SILVER METHENAMINE SOLUTION

Reagent	Quantity (30 ml)	Final concentration
AgNO$_3$ (0.3%) <!>	10 ml	0.1%
Hexamethylenetetramine (3%) <!>	10 ml	1%
Sodium borate (90 mM, pH 7.5)	10 ml	30 mM

Prepare immediately before use. Remove particles in the silver methenamine by filtration if necessary.

20x SSC

Reagent	Quantity (1 liter)	Final concentration
NaCl	175.3 g	3 M
Sodium citrate <!>	88.2 g	0.3 M

Dissolve the components in 800 ml of H$_2$O. Adjust the pH to 7.0 with a few drops of 14 N HCl <!>. Adjust the volume to 1 liter with H$_2$O. Dispense into aliquots. Sterilize by autoclaving. To make more dilute solutions of SSC, dilute the 20x stock with appropriate volumes of H$_2$O.

ZERO-DIVALENT HL SALINE

Reagent	Quantity (1000 ml)	Final concentration
NaCl (1 M)	1.98 ml	19.8 mM
KCl (1 M) <!>	0.12 ml	1.2 mM

REFERENCES

Arnolds, W.J.A. 1979. Silver staining methods for the demarcation of superficial cell boundaries in whole mounts of embryos. *Mikroskopie* **35:** 202–206.

Bely, A.E. and Weisblat, D.A. 2006. Lessons from leeches: A call for DNA barcoding in the lab. *Evol. Dev.* **8:** 491–501.

Bissen, S.T. 1997. Developmental control of cell division in leech embryos. *BioEssays* **19:** 201–207.

Briggman, K.L. and Kristan, Jr., W.B. 2006. Imaging dedicated and multifunctional neural circuits generating distinct behaviors. *J. Neurosci.* **26:** 10925–10933.

Briggman, K.L., Abarbanel, H.D., and Kristan, Jr., W.B. 2005. Optical imaging of neuronal populations during decision-making. *Science* **307:** 896–901.

Davis, G.K. and Patel, N.H. 1999. The origin and evolution of segmentation. *Trends Genet.* **15:** M68–M72.

Dohle, W. 1999. The ancestral cleavage pattern of the clitellates and its phylogenetic deviations. *Hydrobiologia* **402:** 267–283.

Erséus, C. and Källersjö, M. 2004. 18S rDNA phylogeny of Clitellata (Annelida). *Zool. Scr.* **33:** 187–196.

Fernández, J. and Olea, N. 1982. Embryonic development of glossiphoniid leeches. In *Developmental biology of freshwater invertebrates* (ed. F.W. Harrison and R.R. Cowden), pp. 317–361. Liss, New York.

Gimlich, R.L. and Braun, J. 1985. Improved fluorescent compounds for tracing cell lineage. *Dev. Biol.* **109:** 509–514.

Goldstein, B., Leviten, M.W., and Weisblat, D.A. 2001. Dorsal and snail homologs in leech development. *Dev. Genes Evol.* **211:** 329–337.

Huang, F.Z., Bely, A.E., and Weisblat, D.A. 2001. Stochastic WNT signaling between nonequivalent cells regulates adhesion but not fate in the two-cell leech embryo. *Curr. Biol.* **11:** 1–7.

Isaksen, D.E., Liu, N.-J.L., and Weisblat, D.A. 1999. Inductive interactions regulate cell fusion in leech. *Development* **126:** 3381–3390.

Muller, K.J., Nicholls, J.G., and Stent, G.S., eds. 1981. *Neurobiology of the leech.* Cold Spring Harbor Laboratory Press, Cold Spring Harbor, New York.

Shankland, M. 1984. Positional determination of supernumerary blast cell death in the leech embryo. *Nature* **307:** 541–543.

Shankland, M. and Savage, R.M. 1997. Annelids, the segmented worms. In *Embryology: Constructing the organism* (ed. S.F. Gilbert and A.M. Raunio), pp. 219–235. Sinauer, Sunderland, Massachusetts.

Shimizu, T. 1982. Development in the freshwater oligochaete *Tubifex*. In *Developmental biology of freshwater invertebrates* (ed. F.W. Harrison and R.R. Cowden), pp. 283–316. Liss, New York.

Siddall, M.E. and Borda, E. 2003. Phylogeny and revision of the leech genus *Helobdella* (*Glossiphoniidae*) based on mitochondrial gene sequences and morphological data and a special consideration of the *triserialis* complex. *Zool. Scr.* **32:** 23–33.

Siddall, M.E. and Burreson, E.M. 1998. Phylogeny of leeches (Hirudinea) based on mitochondrial cytochrome *c* oxidase subunit I. *Mol. Phylogenet. Evol.* **9:** 156–162.

Smith, C.M. and Weisblat, D.A. 1994. Micromere fate maps in leech embryos: Lineage-specific differences in rates of cell proliferation. *Development* **120:** 3427–3438.

Smith, C.M., Lans, D., and Weisblat, D.A. 1996. Cellular mechanisms of epiboly in leech embryos. *Development* **122:** 1885–1894.

Stent, G.S., Kristan, Jr., W.B., Torrence, S.A., French, K.A., and Weisblat, D.A. 1992. Development of the leech nervous system. *Int. Rev. Neurobiol.* **33:** 109–193.

Symes, K. and Weisblat, D.A. 1992. An investigation of specification of unequal cleavages in leech embryos. *Dev. Biol.* **150:** 203–218.

Wedeen, C.J., Price, D.J., and Weisblat, D.A. 1990. Analysis of the life cycle, genome and homeo box genes of the leech, *Helobdella triserialis*. In *The cellular and molecular biology of pattern formation* (ed. D.L. Stocum and T.L. Karr), pp. 145–167. Oxford University Press, New York.

Weisblat, D.A. and Huang, F.Z. 2001. An overview of glossiphoniid leech development. *Can. J. Zool.* **79:** 218–232.

Weisblat, D.A., Sawyer, R.T., and Stent, G.S. 1978. Cell lineage analysis by intracellular injection of a tracer enzyme. *Science* **202:** 1295–1298.

Weisblat, D.A., Zackson, S.L., Blair, S.S., and Young, J.D. 1980. Cell lineage analysis by intracellular injection of fluorescent tracers. *Science* **209:** 1538–1541.

Weisblat, D.A., Huang, F.Z., Isaksen, D.E., Liu, N.J., and Chang, P. 1999. The other side of the embryo: An appreciation of the non-D quadrants in leech embryos. *Curr. Top. Dev. Biol.* **46:** 105–132.

Whitman, C.O. 1878. The embryology of *Clepsine*. *Q.J. Microsc. Sci.* **18:** 213–315.

Whitman, C.O. 1887. A contribution to the history of the germ-layer in *Clepsine*. *J. Morphol.* **1:** 105–182.

WWW RESOURCES

http://calabreselx.biology.emory.edu/ Describes the research in the Calabrese laboratory at Emory University, Georgia, which focuses on the neural circuit that controls the heart of the medicinal leech.

http://mcb.berkeley.edu/labs/weisblat/ Describes the research on the leech in the Weisblat laboratory at the University of California, Berkeley. Includes its publications and protocols.

http://research.amnh.org/~siddall/ From the Siddall laboratory at the American Museum of Natural History, New York, describes research and expeditions concerning the leech, related primarily to taxonomy.

11 | *Pristionchus pacificus*
A Genetic Model System for the Study of Evolutionary Developmental Biology and the Evolution of Complex Life-history Traits

Robbie Rae, Benjamin Schlager, and Ralf J. Sommer

Max-Planck Institute for Developmental Biology, Department of Evolutionary Biology, Spemannstrasse 37, D-72076 Tübingen, Germany

ABSTRACT

Pristionchus pacificus is a nematode that has been established as a model system for evolutionary developmental biology. Initially, *P. pacificus* was used as a convenient nematode with which to compare the processes of vulva and gonad development as well as sex determination to *Caenorhabditis elegans*, one of

PROTOCOLS
1 Isolation of *Pristionchus* Nematodes from Beetles, 284
2 Assessment of the Olfactory Response to Chemicals or Bacteria in *Pristionchus* Nematodes, 286

the best-studied animal models. *P. pacificus* shares many features with *C. elegans*, including a short generation time, simple laboratory culture, and self-fertilization as a mode of reproduction. These features allowed forward and reverse genetic tools to be developed for this species. The application of these tools for genetic and molecular analysis of vulva formation revealed substantial differences between *P. pacificus* and *C. elegans*. The genome of *P. pacificus* has recently been sequenced and showed an expansion of protein-coding genes when compared to *C. elegans*. Interestingly, the *P. pacificus* genome encodes some genes, such as cellulases, that are known to be present in only plant-parasitic nematodes. Many of the putative functions of the predicted genes in the genome are related to the ecology of *P. pacificus* and other *Pristionchus* species. *Pristionchus* nematodes can be isolated from beetles and soil, indicating that the ecology of *P. pacificus* is strikingly different from that of *C. elegans*. Generally, *Pristionchus* species show an unexpected level of species specificity in their beetle associations, providing a unique opportunity to study the genetic and molecular mechanisms underlying the interactions of organisms in the environment. Thus, *P. pacificus* is not only an established model system for evolutionary developmental biology but an emerging model system for the evolution of complex life-history traits.

BACKGROUND INFORMATION

The phylum Nematoda has more than 1 million nematode species and consists of free-living as well as plant, vertebrate, and insect parasites (Lambshead 1993). One of the best-studied model organisms, *Caenorhabditis elegans*, is a nematode that can be cultured in the laboratory in a fast

This chapter, with full-color images, can be found online at www.cshprotocols.org/emo.

and inexpensive manner. Given the species richness and its ease of manipulation in the laboratory, nematodes are attractive candidates for basic and applied research, and currently there are more than ten ongoing nematode sequencing projects that focus on animal and plant parasites.

The search for a suitable nematode species to compare to *C. elegans* was initiated in the early 1990s (Sommer et al. 1994). *P. pacificus* (Fig. 1A) was selected as a comparative nematode model system for two reasons. First, there are important differences in postembryonic development, particularly vulva development, between *P. pacificus* and *C. elegans* (Sommer and Sternberg 1996). Second, techniques that had been developed for *C. elegans* could be easily transferred to *P. pacificus*. The life cycle of *P. pacificus* is largely the same as that of *C. elegans*: The egg develops through four larval stages (designated J1–J4 or L1–L4) and, finally, reaches the adult stage. If the food supply is depleted, the *P. pacificus* J2 larva develops into a dauer instead of a J3 larva. Dauer larvae retain their cuticle from stage J2, are resistant to a number of abiotic factors, persist in soil, and infect insect hosts.

The first strain of *P. pacificus* was isolated from soil under a tree in downtown Pasadena, California, in 1988, but was not described as a new species until 1996 (Sommer et al. 1996). Intensive studies during the last 5 years indicate that *P. pacificus* is a cosmopolitan species, having been isolated from Japan, the eastern and western United States, Bolivia, South Africa, mainland Asia, and a small number of locations in Europe (Herrmann et al. 2007; Zauner et al. 2007; M. Herrmann et al. unpubl.). Before 2006, it was not known whether *P. pacificus* was associated with beetles, as is the case with other *Pristionchus* species. Studies in Japan revealed that *P. pacificus* is

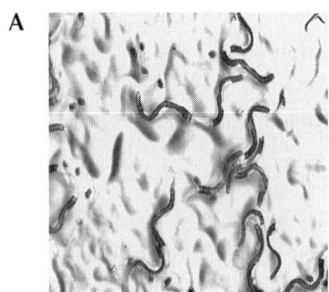

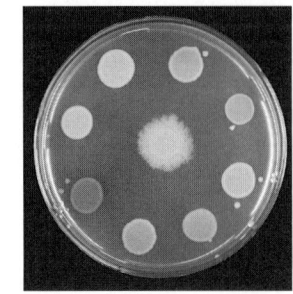

FIGURE 1. *Pristionchus* and its interactions. (*A*) *Pristionchus pacificus* nematodes feeding on *E. coli* OP50. (*B*) The Oriental beetle (*A. orientalis*) is a host for *P. pacificus*. (*C*) Range of bacteria isolated from the cuticle and intestine of *P. maupasi*.

closely associated with the Oriental beetle (*Anomala orientalis*) (Fig. 1B) (Herrmann et al. 2007). The molecular phylogeny of *Pristionchus* shows that *P. pacificus* is part of an ancient cluster of Asian species (Mayer et al. 2007). Therefore, *Pristionchus* nematodes are present both in the soil and on beetles, but the proportions of nematodes in each of these environments are unknown.

Although originally established as a model system in evolutionary developmental biology, *P. pacificus* is currently becoming an important model for the evolution of life-history traits. In contrast to *C. elegans*, the ecological niche of *P. pacificus* is well defined, given its close association with beetles. *P. pacificus* has a necromenic relationship with beetles; dauer larvae penetrate into the beetles and remain there until the death of their hosts, after which development is resumed and the nematodes feed on microbes on the beetle carcasses. The costs and benefits of these nematode-beetle interactions are currently being investigated.

SOURCES AND HUSBANDRY

P. pacificus can be obtained from the *Caenorhabditis* Genetics Center (CGC) at the University of Minnesota (http://www.cbs.umn.edu/CGC/). In addition, *P. pacificus* strains can be requested from the Sommer lab (http://www.pristionchus.org or http://www.eb.tuebingen.mpg.de). The CGC distributes the wild-type strain PS312 originally isolated in Pasadena, California and the polymorphic reference strain PS1843 from Port Angeles, Washington (Sommer et al. 1996; Srinivasan et al. 2002). The Sommer lab can distribute more than 80 additional isolates of *P. pacificus* from around the world. All of these strains can be cultured in the laboratory on plates that contain nematode growth medium (NGM) and the food source *Escherichia coli* OP50. *Pristionchus* strains can also be isolated from soil samples and beetles from the wild, whereby topsoil samples or beetles are placed on NGM plates seeded with *E. coli* OP50 (see Protocol 1). For more information on culturing *Pristionchus* and for other *Pristionchus* protocols, see Pires da Silva (2006).

RELATED SPECIES

Pristionchus is one of more than 20 known genera of the family Diplogastridae, many of which are associated with insects (Fürst von Lieven and Sudhaus 2000; Sudhaus and Fürst von Lieven 2003). Within the genus *Pristionchus*, we have currently isolated 22 species from beetles and soil (Mayer et al. 2007). All of these species can be maintained in the laboratory with *E. coli* OP50 and are available from the Sommer lab upon request (ralf.sommer@tuebingen.mpg.de).

USES OF THE *P. PACIFICUS* MODEL SYSTEM

Initially, *P. pacificus* was established as a model nematode species to compare developmental processes to *C. elegans*. Besides vulva formation, which is described in detail below, developmental genetic studies in *P. pacificus* have concentrated on sex determination (Pires da Silva and Sommer 2004), gonad development (Rudel et al. 2008), and dauer formation (A. Ogawa and R.J. Sommer, unpubl.). In addition, the associations of *Pristionchus* species with beetles, their interactions with bacteria, and their behavior in response to olfactory cues are well-defined complex life-history traits that provide a unique platform to investigate the genetic basis of ecologically relevant organismic interactions.

Vulva Development

The complete cell lineage of *C. elegans* has been known for 30 years (Sulston and Horvitz 1977), and the ability to perform cell-lineage analysis and cell-ablation experiments has prompted several comparative studies initially focusing on vulva development (Sommer 2006; Kiontke et al. 2007). The vulva is the egg-laying organ of nematode females and hermaphrodites, and the comparison of more than 50 different nematode species has indicated that the vulva is a homologous organ that is formed from homologous cells. However, these cells show species-specific cell-to-cell interactions during the formation of this conserved structure (Sommer 2008).

In all species studied to date, the vulva is formed from descendants of P cells. P cells are lateral hypodermal cells in embryos that migrate and adopt ventral positions in late embryogenesis. At the onset of postembryogenesis, they form a linear array of 12 ventral cells. In both *C. elegans* and *P. pacificus*, these cells are called P1 to P12 (from anterior to posterior, respectively). Each of these P cells goes on to divide along the anterior-posterior axis to generate a Pn.a and a Pn.p cell (where n is a number from 1 to 12). The Pn.a cells go on to divide and form neurons of the ventral nerve cord, whereas some Pn.p cells become precursors of the vulva (Fig. 2A).

LIN-39/HOX and the Vulva Equivalence Group

In both *C. elegans* and *P. pacificus*, a subset of Pn.p cells is used to form the hermaphrodite vulva. The Pn.p cells in the midbody region (P[3–8].p) form the vulva equivalence group (VEG), which is the group of cells competent to form vulva tissue (Fig. 2B). The HOX gene *lin-39* confers competence to these midbody Pn.p cells. The Pn.p cells not selected by *Cel-lin-39*, P(1,2,9–11).p, fuse with the hypodermal syncytium hyp7. In *Cel-lin-39* mutant hermaphrodites, all midbody Pn.ps undergo ectopic cell fusion (Fig. 2C). Thus, *Cel-lin-39* rescues midbody Pn.p cells from cell fusion, thereby establishing their competence (Shemer and Podbilewicz 2002).

In *P. pacificus*, nonvulval Pn.p cells undergo programmed cell death (Fig. 2B). Specifically, P(1–4).p in the anterior region and P(9–11).p in the posterior region die of apoptosis (Sommer and Sternberg 1996). In the late 1990s, genetic and molecular studies indicated that in *Ppa-lin-39* mutants, the midbody Pn.ps undergo ectopic cell death (Fig. 2C) (Eizinger and Sommer 1997). These early studies can be summarized by citing three important findings. First, Pn.p cells have different "default" cell fates in *C. elegans* and *P. pacificus* (cell fusion vs. apoptosis, respectively). Second, the HOX gene *lin-39* has a conserved role in establishing the VEG by rescuing the midbody Pn.ps from their default cell fate. Third, the genes regulated by the LIN-39 protein to rescue Pn.p cells must be different because the default cell fates are different. Directly or indirectly, LIN-39 suppresses cell-fusion effector genes in *C. elegans* and cell-death executors in *P. pacificus*. Indeed, genetic follow-up studies have placed LIN-39 in distinct genetic networks in *P. pacificus* and *C. elegans*. For example, the *Ppa-pax-3* gene interacts with *Ppa*-LIN-39 and is a HOX target during the formation of the VEG (Yi and Sommer 2007).

Reduction in Size of the VEG

The *C. elegans* VEG comprises six cells, P(3–8).p, all of which are competent to form part of the vulva. Two mechanisms reduce the size of the VEG in *P. pacificus* (Fig. 2B). First, the posteriormost cell P8.p is not fully competent to adopt a vulval fate. Second, the two anterior cells P(3,4).p die of apoptosis. These two mechanisms effectively reduce the VEG of *P. pacificus* to exactly those three cells that form the adult vulva in wild-type animals.

The molecular mechanisms involved in the reduction of the VEG were identified by genetic analysis of vulva-defective mutants in *P. pacificus*. In retrospect, an unbiased genetic analysis was crucial because the molecular mechanisms that were identified in the course of these studies in *P. pacificus* were fundamentally different from cell-fate specification in *C. elegans* and involved genes that are not even present in the *C. elegans* genome. Therefore, a candidate gene approach, as often used in evolutionary developmental biology, would have been completely misleading.

In *ped-5* and *ped-6* mutant animals, P3.p and P4.p do not undergo apoptosis but instead survive and form competent vulva precursor cells (VPCs) (Fig. 2D) (Sommer and Sternberg 1996).

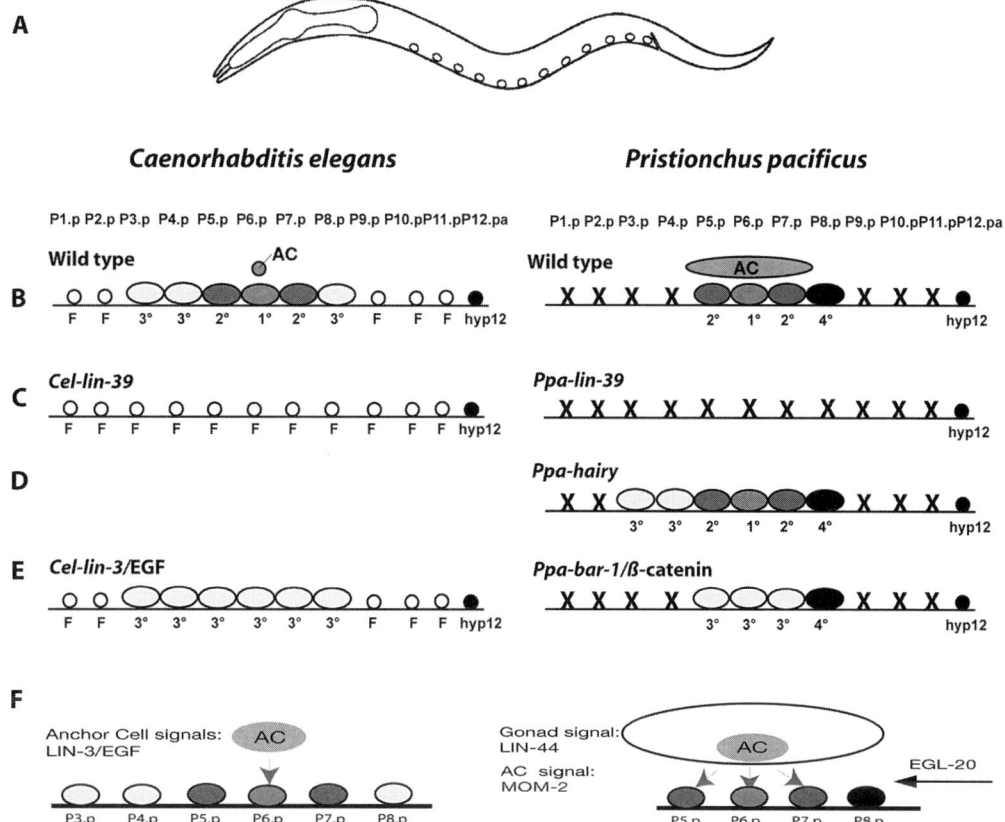

FIGURE 2. Schematic summary of ventral epidermal cell-fate specification in C. elegans and P. pacificus. (A) The ventral epidermis of hermaphrodites derives from 12 ectoblasts, named P(1–12).p according to their anterior-posterior position. The 12 cells are equally spaced between the pharynx and the rectum. (B) Wild-type animals. In C. elegans, the vulva is formed from the progeny of P(5–7).p. P6.p has 1° cell fate and generates eight progeny, and P(5,7).p cells have 2° cell fates and form seven progeny each. P(3,4,8).p cells have 3° cell fates; they are competent to form vulval tissue, but remain epidermal under wild-type conditions. P(1,2,9–11).p cells fuse with the hypodermis and are not competent to form part of the vulva. P12.pa is a special cell called hyp12 and forms part of the rectum. In P. pacificus, P(1–4,9–11).p cells die of programmed cell death (indicated by "X") and reduce the size of the vulva equivalence group (VEG). P(5–7).p cells have the 2°-1°-2° pattern seen in C. elegans, and P8.p, a special epidermal cell, has 4° cell fate. Whereas the single anchor cell (AC) of the somatic gonad induces vulva formation in C. elegans, several cells do so in P. pacificus. (C) In Cel-lin-39 mutants, positional information for the formation of the VEG is missing, and P(3–8).p fuse with the hypodermis like their lineage counterparts in the anterior and posterior body regions. In Ppa-lin-39 mutants, the VEG is not formed, and P(5–8).p die of programmed cell death. (D) Ppa-hairy is involved in regulating the size of the VEG. In Ppa-hairy mutants, P(3,4).p survive and form a VEG that is reminiscent of the pattern in C. elegans wild-type animals. There is no 1:1 ortholog of Ppa-hairy in the C. elegans genome. (E) In C. elegans, mutations in the EGF family member lin-3 result in a vulva-less phenotype. In P. pacificus, mutations in the β-catenin-like gene Ppa-bar-1 result in phenocopies of gonad-ablated animals. (F) Model of vulva induction for C. elegans and P. pacificus. See text for details.

The *ped-6* gene was shown to be homologous to the transcriptional corepressor *Drosophila groucho*, whereas the *ped-5* gene encodes a homolog of the basic helix-loop-helix (bHLH) transcription factor *hairy* (Schlager et al. 2006). Interestingly, *C. elegans* does not possess a true *hairy* ortholog and most *C. elegans* bHLH proteins do not have a *groucho* interaction domain (Schlager et al. 2006). Indeed, molecular studies in *P. pacificus* revealed that *Ppa*-HAIRY and *Ppa*-GROUCHO can physically interact and that *Ppa*-HAIRY binds to the promoter of the HOX gene *Ppa-lin-39*, indicating that the HAIRY/GROUCHO dimer regulates the size of the VEG by inhibiting the activity of *Ppa-lin-39* (Schlager et al. 2006).

Vulva Induction

One of the most important aspects of *C. elegans* vulva formation is the induction of the P(5–7).p cells to form vulval tissue by the anchor cell (AC) (Sternberg 2005). The AC is located dorsally to P6.p and is part of the somatic gonad. After cell ablation of the AC, the VPCs do not take on a vulva fate and remain epidermal (Fig. 2E,F). The induction of the *C. elegans* vulva is molecularly very complex and at least three major pathways are involved in this process (Sternberg, 2005). The P(5–7).p cells are induced by the AC through an epidermal growth factor (EGF)-like ligand encoded by the gene *lin-3*. In addition to EGF/RAS signaling, LIN-12/Notch and Wnt signaling were also shown to have a role in *C. elegans* vulva induction.

Comparative studies in *P. pacificus* showed that vulva induction requires multiple cells of the somatic gonad. Early work in *P. pacificus* has shown that the Frizzled-type Wnt receptor LIN-17 represses vulva formation and that *Ppa-lin-17* mutants result in a multivulva phenotype and vulva polarity defects (Zheng et al. 2005). These results implied that Wnt signaling is involved in vulva formation. Large-scale reverse genetic deletion screens were performed to obtain mutants for additional Wnt-signaling components.

Ppa-bar-1/β-catenin mutants show a fully penetrant vulva-less phenotype (Tian et al. 2008). This strongly implicated Wnt signaling as the inductive signaling pathway for vulva formation and revealed an antagonism between at least two Wnt pathways: an inhibitory pathway involving *Ppa-lin-17* and an inductive pathway involving *Ppa-bar-1*. In contrast to *Ppa-bar-1*, Wnt ligands and receptors do not show strong vulval phenotypes in single-mutant animals. However, some double- and triple-mutant animals are vulva-less, indicating that several Wnt pathways act in parallel during vulva induction (Tian et al. 2008). Together, these studies reveal that a complex network of Wnt-signalling pathways is involved in vulva induction and repression (Fig. 2E,F). At the same time, there is no evidence thus far for a role of EGF/RAS signaling in *P. pacificus* vulva formation.

The example of the nematode vulva with its unbiased genetic analysis has important conceptual consequences for evolutionary developmental biology. The distinct functions of homologous genes in *C. elegans* and *P. pacificus* strongly indicate that sequence conservation does not necessarily imply functional conservation. Any attempt to deduce function from structure can be misleading. This separation of structure and function presents a strong caveat to expression studies, an experimental approach very common in evolutionary developmental biology (Sommer 2008).

Pristionchus-beetle Interactions and Biogeography

The unexpected genetic and molecular differences in the developmental mechanisms of *P. pacificus* and *C. elegans* initiated an interest in microevolutionary and ecological studies of *P. pacificus* and its close relatives. Specifically, a search to identify the exact ecological niche of this organism was initiated in 2004. Original field studies concentrated on isolating *Pristionchus* by trapping and collecting beetles from 25 sampling sites across western Europe (Herrmann et al. 2006a). In total, 4242 beetles were collected, and six *Pristionchus* species, including *P. entomophagus*, *P. lheritieri*, *P. maupasi*, *P. uniformis*, *P.* sp. 4, and *P.* sp. 6, were isolated and characterized using morphological, molecular, and mating experiments. Near-species-specific nematode-beetle relationships were discovered: *P. maupasi* was isolated from cockchafers (*Melolontha melolontha* and *M. hippocastani*), *P. entomophagus* from dung beetles (*Geotrupes stercorosus*), and *P. uniformis* from Colorado potato beetles (CPBs) (*Leptinotarsa decemlineata*).

In a second survey concentrating on the eastern United States, 1241 beetles were collected and 285 were recorded as having *P. aerivorous*, *P. pseudoaerivorous* n. sp., *P. pauli* n. sp., *P. marianneae* n. sp., or *P. americanus* n. sp. present (Herrmann et al. 2006b). Only *P. aerivorous* had previously been described in North America (Sudhaus and Fürst von Lieven 2003); the rest were new species of *Pristionchus*. Strikingly, all of these species were gonochoristic (males and females) but the Western European species were hermaphroditic.

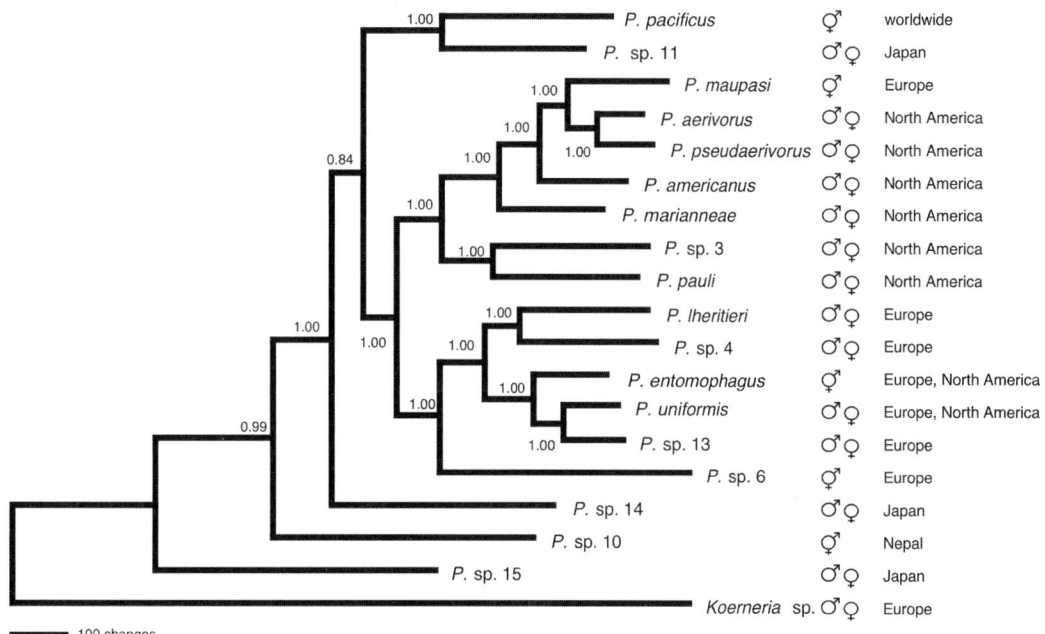

FIGURE 3. Phylogenetic relationships of 18 *Pristionchus* species. The Bayesian tree results from the analysis of 10,971 base pairs of concatenated ribosomal protein cDNA sequences. (Redrawn, with permission, from Mayer et al. 2007.)

In 2006, 1650 Oriental beetles (*A. orientalis*) were collected from more than 60 sampling sites in Japan (Herrmann et al. 2007). Of these individuals, 37 were infested with *P. pacificus*, thus showing that *A. orientalis* is an important host for *P. pacificus* in Japan. *A. orientalis* has also been shown to harbor *P. pacificus* on Long Island (U.S.).

Phylogenetic analysis shows that all of the American species as well as *P. maupasi* (from Europe) are grouped into one clade, whereas European species (*P. lheritieri*, *P. entomophagus*, *P. uniformis*, *P.* sp. 4, and *P.* sp. 13) are in another clade (Fig. 3) (Mayer et al. 2007). *P. pacificus* and *P.* sp. 11 fall into a separate clade as sister species. The phylogenetic position of *Pristionchus* in the Diplogastridae family is currently under investigation (Sudhaus and Fürst von Lieven 2003; Kiontke et al. 2007; W.E. Mayer et al., unpubl.).

Although it is thought that *Pristionchus* has a necromenic relationship with beetles, it has been reported that *P. uniformis* can infect and kill the CPB under laboratory and field conditions (Fedorko and Stanuszek 1971). These investigators speculated that *P. uniformis* may act as a vector of pathogenic bacteria and were able to isolate *Streptococcus durans* and one *Bacillus* species resembling *B. pumilis* or *B. subtilis* from dauer larvae of this species (Sandner et al. 1972). Clearly, the relationship among *Pristionchus*, beetles, and bacteria is not fully understood, and a number of factors are currently being investigated (R. Rae et al., unpubl.).

Pristionchus-bacterial Interactions

Pristionchus is grown on *E. coli* OP50 in the laboratory, but it is not well understood which bacteria these nematodes associate with in nature. Rae et al. (2008) analyzed bacteria from the cuticle and intestine of *P. pacificus*, *P. entomophagus*, and *P. maupasi* individuals that had emerged from their associated beetles (Fig. 1C). Using 16S sequence analysis, it was shown that one *Pristionchus* individual harbors more than 40 different bacterial species from many different eubacterial taxa. More than 20 bacteria isolated from cockchafers, dung beetles, and oriental beetles were established as laboratory cultures and were exposed to *P. pacificus*. A continuum of interactions from dissemination of bacteria to reduction in brood size and nematode mortality was observed (Rae et al. 2008). In olfactory experiments, it was found that *P. pacificus* avoids insect pathogens such as *Serratia marcescens* and *B. thuringiensis* and is resistant to human pathogenic bacteria (*Pseudomonas aeruginosa* and *Staphylococcus aureus*), which was surprising because *C. elegans* is highly susceptible to these pathogens (Ewbank 2002). Whether the reasons for this are due to differences in morphology or gene machinery remains to be investigated. It is interesting to note in this context that the *P. pacificus* genome shows a strong expansion of genes encoding detoxification enzymes such as cytochrome P450 enzymes, sulfotransferases, and ABC transporters (Dieterich et al. 2008).

Pristionchus can not only feed on bacteria, but are also able to use fungi and even other nematodes as food sources. Thus, *P. pacificus* is an omnivorous nematode. The different feeding strategies in *P. pacificus* and *C. elegans* are reflected in morphological differences between these species. Although *C. elegans* has a so-called grinder (a structure in the terminal part of the pharynx that disrupts all bacteria under laboratory conditions), no such structure is known in *Pristionchus* (Rae et al. 2008). Instead, *Pristionchus* nematodes have denticles in the mouth that are used to disrupt the hyphae of fungi and/or the cuticle of other nematodes. Interestingly, *Pristionchus* exhibits a mouth dimorphism: Animals can have a broad, shallow eurystomatous mouth form or a long, narrow stenostomatous mouth form (Fürst von Lieven and Sudhaus 2000; Hong and Sommer 2006a).

Behavior and Chemoattraction

Pristionchus must rely on a number of long- and short-range chemical cues to locate potential beetle hosts and to ensure host specificity. In olfaction experiments using insect sex pheromones and insect and plant semiochemicals, *Pristionchus* species display unique chemoattraction profiles (Hong and Sommer 2006b; Herrmann et al. 2007; Hong et al. 2008a). For example, *P. pacificus* is strongly attracted to the *A. orientalis* sex pheromone (Z)-7-tetradecen-2-1 (ZTDO) and *P. maupasi* is strongly attracted to phenol (a sex attractant of cockchafers) (Herrmann et al. 2007; Hong et al. 2008a). To decipher the molecular and genetic factors involved in *Pristionchus* olfaction, 19 strains of *P. pacificus* were assessed for their response to the moth sex pheromone (E)-11-tetradecenyl acetate (ETDA) (Hong et al. 2008b). The greatest difference in olfactory response to ETDA was in the California (insensitive) and Washington (attractive) strains. Using introgression lines, exogenous cGMP treatment, and null alleles, it was discovered that the cGMP-dependent protein kinase EGL-4 was responsible for this difference. EGL-4 is part of the cGMP-signaling pathway. Homologs of the *egl-4* gene are present in *C. elegans* and *Drosophila* and are known to affect olfaction and foraging behavior, respectively, in these species.

GENETICS AND GENOMICS RESOURCES

P. pacificus has been established as a genetic system based on its self-fertilizing hermaphroditic propagation. Hermaphrodites are modified females that produce sperm during a short period of larval development before becoming mature adult females. As long as males are not present, hermaphrodites will use the self-sperm to fertilize their oocytes. Males can be easily obtained and maintained under laboratory conditions and are used in genetic experimentation and classical genetic studies.

Genetic markers and mutants associated with specific morphological traits are available for *P. pacificus*. These are primarily mutants with a short body form (Dumpy) or uncoordinated movement (Unc) (Kenning et al. 2004). In addition, more than 50 mutants with various developmental and behavioral phenotypes have been isolated (Sommer 2006).

The gene nomenclature rules for *P. pacificus* basically follow those established for *C. elegans*. *P.*

pacificus genes are distinguished by a species prefix "*Ppa*." Genes that have been identified based on sequence similarity to *C. elegans* genes are named by simply adding this prefix to the known *C. elegans* name (e.g., *Ppa-lin-39*). Genes identified based on genetic screens are named by using a novel three-letter code; for example, *Ppa-pdl* describes the *P. pacificus* "*p*ristionchus *d*umpy-*l*ike genes." This system ensures that molecularly unidentified mutants are unique when compared to *C. elegans* (Kenning et al. 2004).

Two bacterial artificial chromosome (BAC) libraries have been produced for *P. pacificus* PS312, with a total of more than 30,000 clones (HindIII and EcoRI) (Srinivasan et al. 2002). A large set of molecular genetic markers has been generated, primarily by identifying single-strand conformation polymorphism (SSCP) and single-nucleotide polymorphism (SNP) markers in end-sequenced BAC clones (Srinivasan et al. 2002). More than 600 such markers are currently available. A genetic linkage map was built by mapping these markers on a meiotic mapping panel of 42 animals that were produced by crossing the PS312 wild-type strain with the polymorphic PS1843 reference strain (Srinivasan et al. 2002). This genetic linkage map is integrated with the physical map (Srinivasan et al. 2003) and the *P. pacificus* genome (Dieterich et al. 2008). In addition, several expressed sequence tag (EST) sequencing projects have been performed for *P. pacificus* PS312. All of this information is accessible online at http://www.pristionchus.org.

The genome of *P. pacificus* PS312 has been sequenced by whole-genome shotgun (WGS) with approximately ninefold coverage (Dieterich et al. 2008). The size of the genome is approximately 169 Mb and is thus substantially larger than the *C. elegans* genome. The reference strain *P. pacificus* PS1843 has also been sequenced by WGS with onefold coverage. All sequence data can be accessed at http://www.pristionchus.org. The *P. pacificus* PS312 genome is fully assembled and annotated and is currently being incorporated into WormBase (http://www.wormbase.org). The genomes of the PS312 and PS1843 strains differ by more than 4% because of numerous short insertions and deletions (indels) and SNPs that can be used as molecular markers. Additional SNPs in these strains can be obtained using BLAST searches.

TECHNICAL APPROACHES

Genetic and genomic tools for *P. pacificus* have been described in great detail, with associated protocols, in WormBook. For example, protocols for mutagenesis, deletion library screening for gene knockouts, and gene knockdown using RNA interference (RNAi) and morpholinos are available at http://www.wormbook.org/chapters/www_ppageneticprotocols/ppageneticprotocols.html. Here, we add two specific protocols: One describes the isolation of *Pristionchus* nematodes from beetles and soil (Protocol 1) and the other outlines ways to study behavior in response to insect pheromones and bacteria by so-called "chemotaxis assays" (Protocol 2).

Protocol 1

Isolation of *Pristionchus* Nematodes from Beetles

In this procedure, nematodes disembark from a beetle carcass and feed on *E. coli* OP50. The nematodes are then monitored for a few days and identified using simple morphological characteristics. This method, based on Herrmann et al. (2006a,b), is rapid, easy, and biased for *Pristionchus* species.

MATERIALS

The recipe for the item marked with <R> is on page 288.

Reagents

Beetles, collected from anywhere in the wild
E. coli OP50, grown overnight in Luria-Bertani (LB) broth ($OD_{600} \approx 0.5$–1.0)

Alternatively, soil samples (about 10 g of topsoil) from the natural environment can be used (see Step 2).

Equipment

Dissecting microscope
Flame (e.g., from Bunsen burner)
Petri dishes (6 cm), half filled with nematode growth medium (NGM) <R>
Pick, metal
Scissors
Tweezers

METHOD

1. Hold a live beetle with a tweezers and cut it transversely using a scissors.

   The purpose of this process is to kill the beetle so that the nematodes will disembark.

2. Place the two pieces of beetle on a Petri dish containing NGM and seeded with 300 µl of *E. coli* OP50.

   Instead of a beetle, 10 g of soil can be placed on the NGM plate with E. coli OP50. Nematodes can be isolated from the soil according to the same processes described below for beetles.

3. During a 2-week period, check daily using a dissecting microscope for any emerging and reproducing nematodes in the *E. coli* OP50. Identify the nematodes to the family level using morphological characteristics (Sudhaus and Fürst von Lieven 2003).

   For molecular confirmation of nematode identification, isolate DNA using the single-worm lysis procedure described by Floyd et al. (2002), perform polymerase chain reaction (PCR) using the primers created by Blaxter et al. (1998), and then sequence the PCR product.

4. Using a metal pick, place any nematodes that resemble *Pristionchus* (i.e., lack of grinder, flat mouth, striations on cuticle) on fresh *E. coli* OP50–seeded NGM plates to create isogenic lines.

While moving and isolating worms, flame-sterilize the metal pick between transfers to prevent cross-contamination.

5. Identify hermaphroditic and gonochoristic species.

   Any nematodes that produce offspring without a mate are hermaphroditic or parthenogenetic species; all other worms are gonochoristic species. For hermaphroditic species, eggs and offspring should be visible after 48 hours when stored at 25°C.

6. Establish cultures of the worms of interest by allowing them to grow extensively on NGM plates with *E. coli* OP50 at 20–25°C. These worms can then be used for experiments.

   For additional guidance on culturing Pristionchus, *see Pires da Silva (2006).*

Protocol 2

Assessment of the Olfactory Response to Chemicals or Bacteria in *Pristionchus* Nematodes

In the soil environment, nematodes must rely on a number of host-specific chemical cues in order to find potential beetle hosts. They must also discriminate among different food choices (i.e., bacteria), which is important because if the nematodes concentrate on an unsuitable food source, they may die. To detect the bacteria and host-specific chemicals, nematodes use sensory structures called amphids and phasmids that are located on the head and tail, respectively. The olfactory response of nematodes can be studied in the laboratory using the very simple agar-based assay described here. This assay determines the attractiveness of *Pristionchus* to a range of beetle-associated compounds. It has allowed detailed molecular mechanisms of olfaction to be studied in *Pristionchus* and has demonstrated that these nematodes respond to beetle cues in an almost species-specific manner (Hong and Sommer 2006b; Hong et al. 2008a). It can also be used to examine the food choice of these nematodes when they are faced with a number of bacteria.

MATERIALS

The recipes for the items marked with <R> are on page 288.

CAUTION: See the Cautions Appendix for appropriate handling of materials marked with <!>.

Reagents

Bacteria or insect pheromone of choice
> For example, the olfaction response of Pristionchus can be assessed by exposing the worms to beetle sex pheromones such as phenol <!>, (E)-11-tetradecenyl acetate (ETDA) (Sigma T2143) <!>, or (Z)-7-tetradecen-2-one (ZTDO) (Bedoukian Research P6140-95) at concentrations ranging from 1% to 10% (v/v). Alternatively, bacteria such as the human pathogen Pseudomonas aeruginosa or the insect pathogens Serratia marcescens or Photorhabdus luminescens can be used in olfaction assays if grown overnight in LB broth to an OD_{600} of 0.5–1.0.

E. coli OP50, grown overnight in LB broth ($OD_{600} \approx 0.5$–1.0) (for assays with bacteria only; see Step 4)

Ethanol (100%) <!> (for assays with insect pheromones only; see Step 4)

M9 buffer <R>

Pristionchus cultures, growing healthily on *E. coli* OP50–seeded 6-cm NGM plates (see Protocol 1)

Sodium azide (1 M) <!>

Equipment

Dissecting microscope
Marker pen
Petri dishes (9 cm), half filled with NGM <R>
Ruler

METHOD

1. Using a marker and a ruler, score the underside of a half-filled NGM plate 0.5 cm from each side of the dish.

   One mark will serve as the test spot (the location of the chemical or bacteria of choice), and the other mark will serve as the control spot.

2. Flip over the dish and add 10 µl of sodium azide to each mark.

   Sodium azide is used to anesthetize worms and to allow them to be counted.

3. To the test spot, add 20–50 µl of the insect pheromone or bacteria of choice.

4. To the control spot, add the same volume of *E. coli* OP50 (for assays with bacteria) or ethanol (for assays with insect pheromones).

5. Rinse 50–200 *Pristionchus* nematodes three times in M9 buffer at room temperature.

6. Add the worms from Step 5 to the middle of the plate.

7. After an appropriate length of time (e.g., 2.5, 12, or 24 hours), score the plates by counting the amount of nematodes present in each spot. View the worms using a dissecting microscope.

8. Calculate the chemotaxis index:

   i. Subtract the total number of nematodes in the control spot from the total number of nematodes in the test spot.

      Usually, the majority of worms are in the test spots, but some are found on other areas of the plate. These other worms are not used in the calculation.

   ii. Divide the result from Step 8.i by the total number of nematodes in the control and test spots.

      This gives a value ranging from –1.0 to +1.0. The closer to +1.0, the more attractive the substance; the closer to –1.0, the more unattractive the substance.

Recipes

CAUTION: See the Cautions Appendix for appropriate handling of materials marked with <!>.

M9 BUFFER

Reagent	Quantity (for 1 liter)	Final concentration
KH_2PO_4	3 g	22 mM
Na_2HPO_4	6 g	42 mM
NaCl	5 g	86 mM

NEMATODE GROWTH MEDIUM (NGM)

Reagent	Quantity (for 1 liter)	Final concentration
Agar	17.0 g	1.7%
NaCl	2.9 g	50 mM
Peptone	2.5 g	0.25%
$CaCl_2$ (1 M) <!>	1 ml	1 mM
Cholesterol (5 mg/ml)	1 ml	5 µg/ml
KH_2PO_4 (1 M)	25 ml	25 mM
$MgSO_4$ (1 M) <!>	1 ml	1 mM

Mix the first three reagents in H_2O and autoclave. After the mixture is cool, add the last four reagents.

REFERENCES

Blaxter, M.L., de Ley, P., Garey, J.R., Liu, L.X., Scheldeman, P., Vierstraete, A., Vanfleteren, J.R., Mackey, L.Y., Dorris, M., Frisse, L.M., et al. 1998. A molecular evolutionary framework for the phylum Nematoda. *Nature* **392:** 71–75.

Dieterich, C., Clifton, S.W., Schuster, L., Chinwalla, A., Delehaunty, K., Dinkelacker, I., Fulton, R., Godfrey, J., Minx, P., Mitreva, M., et al. 2008. The genome sequence of the nematode *Pristionchus pacificus* and the evolution of nematode parasitism. *Nat. Genet.* (in press).

Eizinger, A. and Sommer, R.J. 1997. The homeotic gene *lin-39* and the evolution of nematode epidermal cell fates. *Science* **278:** 452–455.

Ewbank, J.J. 2002. Tackling both sides of the host-pathogen equation with *Caenorhabditis elegans*. *Microbes Infect.* **4:** 247–256.

Fedorko, A. and Stanuszek, S. 1971. *Pristionchus uniformis* sp. P. (Nematoda, Rhabditida, Diplogasteridae) a facultative parasite of *Leptinotarsa decemlineata* Say and *Melolontha melolontha* L. in Poland. Morphology and biology. *Acta Parasitol. Pol.* **19:** 95–112.

Floyd, R., Abebe, E., Papert, A., and Blaxter, M. 2002. Molecular barcodes for soil nematode identification. *Mol. Ecol.* **11:** 839–850.

Fürst von Lieven, A.F. and Sudhaus, W. 2000. Comparative and functional morphology of the buccal cavity of Diplogastrina (Nematoda) and a first outline of the phylogeny of this taxon. *J. Zool. Syst. Evol. Res.* **38:** 37–63.

Herrmann, M., Mayer, W.M., and Sommer, R.J. 2006a. Nematodes of the genus *Pristionchus* associated with scarab beetles and the Colorado potato beetle in Western Europe. *Zoology* **109:** 96–108.

Herrmann, M., Mayer, W.M., and Sommer, R.J. 2006b. Sex, bugs and Haldane's rule: The nematode genus *Pristionchus* in the United States. *Front. Zool.* **3:** 1–15.

Herrmann, M., Mayer, W.M., Hong, R.L., Kienle, S., Minasaki, R., and Sommer, R.J. 2007. The nematode *Pristionchus pacificus* (Nematoda: Diplogastridae) is associated with the Oriental beetle *Exomala orientalis* (Coleoptera: Scarabaeidae) in Japan. *Zool. Sci.* **24:** 883–889.

Hong, R.L. and Sommer, R.J. 2006a. *Pristionchus pacificus*: A well-rounded nematode. *BioEssays* **28:** 651–659.

Hong, R.L. and Sommer, R.J. 2006b. Chemoattraction in *Pristionchus* nematodes and implications for insect recognition. *Curr. Biol.* **16:** 2359–2365.

Hong, R.L., Svatos, A., Herrmann, M., and Sommer, R.J. 2008a. Species-specific recognition of beetle cues by the nematode *Pristionchus maupasi*. *Evol. Dev.* **10:** 273–279.

Hong, R.L., Witte, H., and Sommer, R.J. 2008b. Natural variation in *P. pacificus* insect pheromone attraction involves the protein kinase EGL-4. *Proc. Natl. Acad. Sci.* **105:** 7779–7784.

Kenning, C., Kipping, I., and Sommer, R.J. 2004. Isolation of mutations with dumpy-like phenotypes and of collagen genes in the nematode *Pristionchus pacificus*. *Genesis* **40:** 176–183.

Kiontke, K., Barriere, A., Kolotuev, I., Podbilewicz, B., Sommer, R., Fitch, D.H., and Felix, M.A. 2007. Trends, stasis and drift in the evolution of nematode vulva development. *Curr. Biol.* **20:** 1925–1937.

Lambshead, P.J.D. 1993. Recent developments in marine benthic biodiversity research. *Oceanis* **19:** 5–24.

Mayer, W.E., Herrmann, M., and Sommer, R.J. 2007. Phylogeny of the nematode genus *Pristionchus* and implications for biodiversity, biogeography and the evolution of hermaphroditism. *BMC Evol. Biol.* **7:** 104–117.

Pires da Silva, A. 2006. *Pristionchus pacificus* genetic protocols (July 17, 2006), *WormBook* (ed. The *C. elegans* Research Community), WormBook, doi/10.1895/wormbook.1.114.1, http://www.wormbook.org/chapters/www_ppageneticprotocols/ppageneticprotocols.html

Pires da Silva, A. and Sommer, R.J. 2004. Conservation of the global sex determination gene *tra-1* in distantly related nematodes. *Genes Dev.* **18:** 1198–1208.

Rae, R., Riebesell, M., Dinkelacker, I., Wang, Q., Herrmann, M., Weller, A.M., Dieterich, C., and Sommer, R.J. 2008. Isolation of naturally associated bacteria of necromenic *Pristionchus* nematodes and fitness consequences. *J. Exp. Biol.* **211:** 1927–1936.

Rudel, D., Tian, H., and Sommer, R.J. 2008. Wnt signaling in *Pristionchus pacificus* gonadal arm extension and the evolution of organ shape. *Proc. Natl. Acad. Sci.* **105:** 10826–10831.

Sandner, H., Seryczynska, H., and Kamionek, M. 1972. Preliminary microbiological and ultrastructural investigations of bacteria isolated from *Pristionchus uniformis*. *Bull. Pol. Acad. Sci. Earth Sci.* **20:** 567–569.

Schlager, B., Röseler, W., Zheng, M., Gutierrez, A., and Sommer, R.J. 2006. HAIRY-like transcription factors and the evolution of the nematode vulva equivalence group. *Curr. Biol.* **16:** 1388–1394.

Shemer, G. and Podbilewicz, B. 2002. LIN-39/Hox triggers cell division and represses EFF-1/fusogen-dependent vulval cell fusion. *Genes Dev.* **16:** 3136–3141.

Sommer, R.J. 2006. *Pristionchus pacificus* (August 14, 2006), *WormBook* (ed. The *C. elegans* Research Community), WormBook, doi/10.1895/wormbook.1.102.1, http://www.wormbook.org/chapters/www_genomes Pristionchus/genomesPristionchus.html

Sommer, R.J. 2008. Homology and hierarchy of biological systems. *BioEssays* **30:** 653–658.

Sommer, R.J. and Sternberg, P.W. 1996. Apoptosis and change of competence limit the size of the vulva equivalence group in *Pristionchus pacificus*: A genetic analysis. *Curr. Biol.* **6:** 52–59.

Sommer, R.J., Carta, L.K., and Sternberg, P.W. 1994. The evolution of cell lineage in nematodes. *Dev. Suppl.* **1994:** 85–95.

Sommer, R.J., Carta, L.K., Kim, S.Y., and Sternberg, P.W. 1996. Morphological, genetic and molecular description of *Pristionchus pacificus* sp. N. (Nematoda: Neodiplogastridae). *Fundam. Appl. Nematol.* **19:** 511–521.

Srinivasan, J., Sinz, W., Lanz, C., Brand, A., Nandakumar, R., Raddatz, G., Witte, H., Keller, H., Kipping, I., Pires da Silva, A., et al. 2002. A bacterial artificial chromosome-based genetic linkage map of the nematode *Pristionchus pacificus*. *Genetics* **162:** 129–134.

Srinivasan, J., Sinz, W., Jesse, T., Wiggers-Perebolte, L., Jansen, K., Buntjer, J., van der Meulen, M., and Sommer, R.J. 2003. An integrated physical and genetic map of the nematode *Pristionchus pacificus*. *Mol. Genet. Genomics* **269:** 715–722.

Sternberg, P.W. 2005. Vulval development (June 25, 2005), *WormBook* (ed. The *C. elegans* Research Community), WormBook, doi/10.1895/wormbook.1.6.1, http://www.wormbook.org/chapters/www_vulvaldev/vulvaldev.html

Sudhaus, W. and Fürst von Lieven, A. 2003. A phylogenetic classification and catalogue of the Diplogastrina (Secernentea, Nematoda). *J. Nem. Morph. Syst.* **6:** 43–90.

Sulston, J.E. and Horvitz, H.R. 1977. Post-embryonic cell lineages of the nematode *Caenorhabditis elegans*. *Dev. Biol.* **56:** 110–156.

Tian, H., Schlager, B., and Sommer, R.J. 2008. Wnt signaling by differentially expressed Wnt ligands induces vulva development in *Pristionchus pacificus*. *Curr. Biol.* **18:** 142–146.

Yi, B. and Sommer, R.J. 2007. The *pax-3* gene is involved in vulva formation in *Pristionchus pacificus* and is a target of the Hox gene *lin-39*. *Development* **134:** 3111–3119.

Zauner, H., Mayer, W.E., Herrmann, M., Weller, A., Erwig, M., and Sommer, R.J. 2007. Distinct patterns of genetic diversity in *Pristionchus pacificus* and *Caenorhabditis elegans*, two partially selfing nematodes with cosmopolitan distribution. *Mol. Ecol.* **16:** 1267–1280.

Zheng, M., Messerschmidt, D., Jungblut, B., and Sommer, R.J. 2005. Conservation and diversification of Wnt signaling function during the evolution of nematode vulva development. *Nat. Genet.* **37:** 300–304.

WWW RESOURCES

http://www.cbs.umn.edu/CGC The *Caenorhabditis* Genetics Center is responsible for maintaining and distributing mutant strains of *C. elegans* and *P. pacificus* to the worm research community.

http://www.pristionchus.org This website allows anyone to browse the *Pristionchus* genome, perform BLAST searches, view physical and genetic maps, and retrieve sequences.

http://www.wormbase.org WormBase has a huge number of resources dedicated to *C. elegans*, including information on *Caenorhabditis* genomes, physical and genetic maps, and BLAST, as well as literature searches and community postings.

12 The African Butterfly *Bicyclus anynana*
A Model for Evolutionary Genetics and Evolutionary Developmental Biology

Paul M. Brakefield, Patrícia Beldade, and Bas J. Zwaan
Institute of Biology, Leiden University, 2300 RA Leiden, The Netherlands

ABSTRACT

The butterfly model based on laboratory stocks of the African species *Bicyclus anynana* provides a special system for the following reasons:

- A range of phenotypes has proven to be amenable to study in this system. These include wing color patterns (including eyespots), seasonal forms, male androconia (secondary sexual traits), and a range of life-history traits (relevant to ageing research).

- These phenotypes have a clear ecological relevance that is associated with dramatic differences in ecological environments represented by the dry and wet seasons in East Africa. In addition, the *Bicyclus* genus, as well as closely related genera from independent radiations in Asia and Madagascar, are highly speciose, providing opportunities to explore diversity among species for wing patterning, life histories, and male secondary sexual traits. There are also rich opportunities to examine the interactions among all of these phenotypes and both natural and sexual selection.

PROTOCOLS

1. Culture and Propagation of *Bicyclus* Laboratory Populations, 299
2. Surgical Manipulations on Pupal Wings: Damage and Cauteries, 301
3. Surgical Manipulations on Pupal Wings: Grafts, 303
4. Fixation and Dissection of Embryos, 305
5. Dissection of Larval and Pupal Wings, 307
6. In Situ Hybridization of Embryos and Larval and Pupal Wings, 309
7. Immunohistochemistry Staining of Embryos, 312
8. Immunohistochemistry Staining of Wing Discs, 314
9. Extraction and Gas Chromatography Analysis of Adult Pheromones, 316
10. Fresh Weight, Dry Weight, and Fat Content of Adult Butterflies, 318
11. Constant Volume Respirometry, 320
12. Hemolymph Extraction from Larvae, Pupae, and Adults, 322
13. Injection of Chemicals into Pupae, 324

- Organismal size also provides important practical advantages. *B. anynana* individuals are small enough to be readily reared in large numbers, but large enough to enable marking and tracking, and to facilitate such manipulations as microsurgical procedures on developing wing discs and the noninvasive sampling of hemolymph.

Here, we explore these properties that enable integrative research that links variation among genotypes, via development and physiology to variation in the phenotype, to variation in fitness in natural environments.

BACKGROUND INFORMATION

Natural History of *Bicyclus* butterflies

The tropical butterfly *B. anynana* (Butler 1879) was first established as a laboratory stock in Leiden in 1988. It was founded with more than 80 gravid females collected by P.M. Brakefield in the understory of a rubber plantation adjacent to primary forest near Nkhata Bay on the western shore of Lake Malawi in Malawi, Africa. *B. anynana* is a small, brown butterfly of the nymphalid tribe Satyrinae (Lepidoptera; Nymphalidae). Condamin (1973) published an important monograph in on the genus *Bicyclus* (Kirby 1871). This species-rich genus includes about 80 species distributed throughout sub-Saharan Africa (excluding Madagascar). Most species are forest or woodland dwellers, normally flying at or close to the ground. The larvae feed on different species of grasses. Adults feed on fallen fruit and these can be collected readily by hand-netting or with the use of fruit-baited traps (Windig et al. 1994; Molleman et al. 2006).

The male secondary sexual traits represented by wing androconia are a crucial taxonomic character together with features of the wing venation and the genitalia. The androconia distribute sex pheromones over the female antennae during courtship (Robertson and Monteiro 2005; Nieberding et al. 2008), and they vary in number, position, and morphology among species. The wing patterns are also diverse (Roskam and Brakefield 1996) and include marginal eyespots that can function in deflecting the attacks of birds away from the vulnerable body (Lyytinen et al. 2004). Species range from those with a wide distribution to narrowly distributed endemics found in a single forest. Many species occur in wet equatorial forests, whereas others inhabit highly seasonal environments (Roskam and Brakefield 1996). A molecular phylogeny for about two thirds of the species, based on both mitochondrial and nuclear gene sequences, has been produced by Monteiro and Pierce (2001).

The Story of *B. anynana*

As usual, the story of why the community initiated research using *Bicyclus* is a mixture of scientific logic and serendipity. Paul Brakefield had studied the ecological genetics of the marginal wing spots of the meadow brown butterfly in Europe. Presenting this work at the first conference on Butterfly Biology in London in 1981, Torben Larsen suggested that the tropical satyrids that express seasonal polyphenism would provide far better opportunities to analyze how natural selection works on butterfly eyespots (Brakefield and Larsen 1984). A Kenyan field trip and some pilot studies in Cardiff with *B. safitza* (which proved less suitable for laboratory culture) led to collecting the original stock of *B. anynana* in Malawi. Then, following on Fred Nijhout's classic microsurgical studies on eyespot formation in *Junonia (Precis) coenia*, Vernon French at Edinburgh helped to set up the *B. anynana* research program in evolutionary developmental biology that became based in Leiden. Sean Carroll's team at the University of Wisconsin at Madison analyzed the expression of key developmental pathways from *Drosophila* wings in the eyespot organizer in *J. coenia*; this analysis stimulated the development of molecular approaches in *B. anynana*. More recently, genomic and transgenesis tools have brought the system up to be a front-player in modern research.

B. anynana as the Butterfly "Lab Rat"

The *B. anynana* system was first established for studies in ecological and evolutionary genetics. An overall aim has been to link genetic variation via developmental and physiological mechanisms to the variation in phenotype that is screened by natural selection (Beldade and Brakefield 2002). Research on the processes that generate phenotypic variation in this species can be done in the context of the functional significance of the variation in natural environments. This effort is beginning to explore how those processes, together with natural selection, influence the directions taken in evolution and the patterns of diversity observed within lineages (Beldade et al. 2002b; Zijlstra et al 2004; Allen et al. 2008). This work on adaptive evolution is now being extended to examining the processes of speciation in the *Bicyclus* lineage, especially those involving diversification in the wing androconia and the associated male sex pheromone system (Nieberding et al. 2008).

B. anynana Stock Center

Stocks kept in laboratories reflect different "flavors" of variation, including phenotypic plasticity, mutations of large effect, and segregating quantitative variant traits. More than 30 spontaneous mutations identified through their discrete effects on morphology, usually in wing pattern, have been isolated in Leiden and used to establish genetic stocks (Fig. 1). X-ray mutagenesis has been applied to adult males to screen for phenotypic variants in their offspring, but it proved impossible to establish any variants as breeding stocks (Monteiro et al. 2003). In addition, artificial selection has

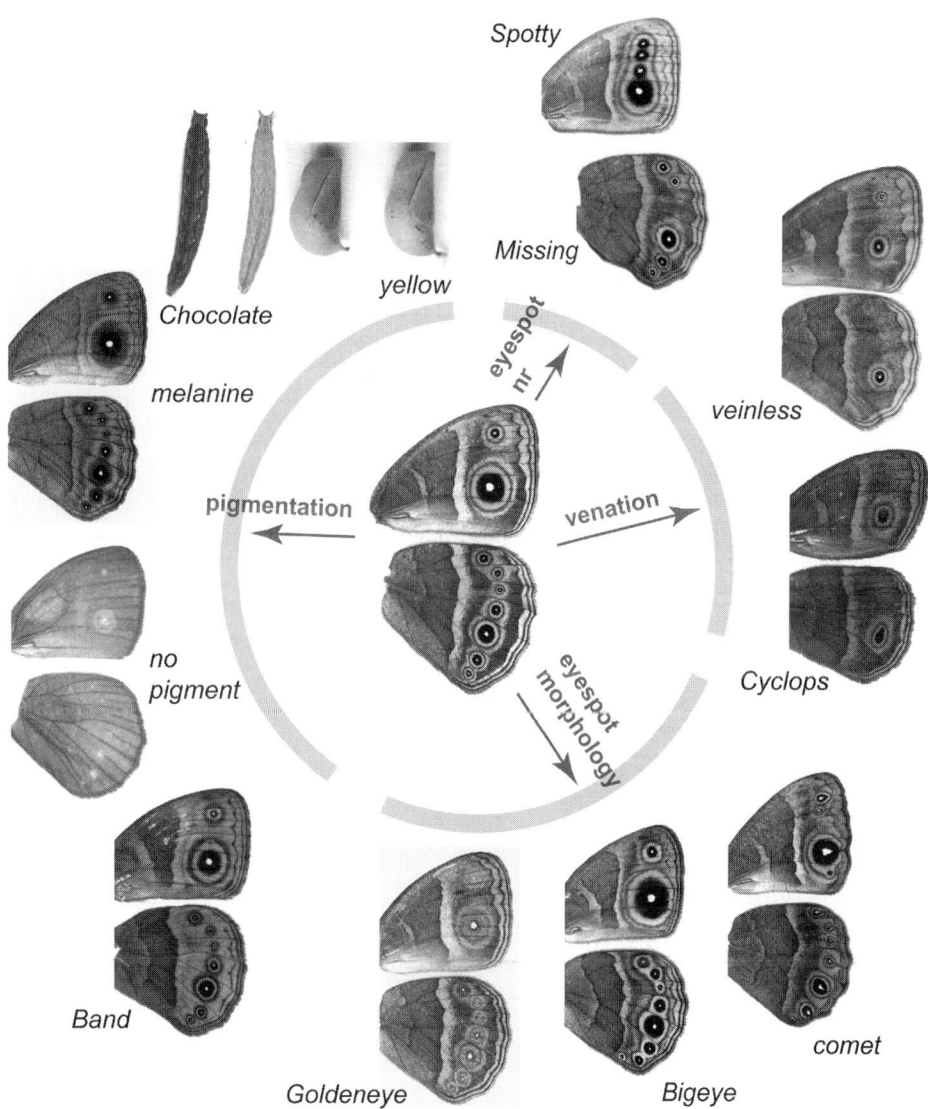

FIGURE 1. Mutant stocks illustrating variation in *B. anynana* color patterns. At the center, the "wild-type" phenotype of the ventral surface of the adult forewing (*top*) and hindwing (*bottom*) shows the series of marginal eyespots characteristic of *B. anynana* adults. The surrounding images are examples of spontaneous mutants with large effect on different aspects of adult wing pattern (variation in eyespot number and morphology, in wing venation coupled with eyespot pattern, and in overall pigmentation) and preadult pigmentation (the *Chocolate* mutant with dark larval integument is shown next to a typical wild-type final instar larva). Full-color images can be found online at www.cshprotocols.org/emo (http://www.cshprotocols.org/emo).

provided a major tool used to explore patterns of existing genetic variation for morphological and life-history traits (discussed below).

SOURCES AND HUSBANDRY

The laboratory stock of *Bicyclus* has been maintained in Leiden with several hundred adults in each (nonoverlapping) generation. The stock is now also kept in several other laboratories in Europe and North America. Effective population sizes are about two thirds of the census number under our standard culturing conditions in the laboratory (Brakefield et al. 2001).

Larvae are typically fed and maintained on pot-grown maize plants (*Zea mays*). Different-sized groups of larvae are raised in net sleeves or in larger cages. Prepupae and pupae can be readily removed from the foliage and will eclose (emerge from the pupal stage) successfully when layered in Petri dishes. For further details on the care and propagation of *Bicyclus*, see Protocol 1.

RELATED SPECIES

B. anynana is a member of the subtribe Mycalesini that are distributed from northeast Australia through Asia and into sub-Saharan Africa. Mycalesina butterflies have shown three of the most spectacular radiations of butterflies in the Old World tropics: in mainland Africa, on Madagascar, and in Asia, respectively. These expansions appear to have occurred in concert with those of their larval food-plants, the Poaceae, some 30–25 million years ago in the Oligocene (Peña and Wahlberg 2008).

Several species of *Bicyclus* have been brought to the laboratory in Leiden, but none have been maintained successfully for more than a few generations. The species from tropical wet forest habitats have comparatively long larval developmental times and are very difficult to culture. *B. safitza* and some other species from highly wet-dry seasonal environments may eventually prove possible to culture continuously in the laboratory. In addition, we have recently successfully established a sister species from the radiation on Madagascar, *Heteropsis iboina*, for comparative studies with *Bicyclus*.

STUDIES IN EVOLUTION AND DEVELOPMENT

Adaptive Phenotypic Plasticity

Research on *B. anynana* concentrated initially on the evolution of developmental plasticity, which is expressed as a seasonal polyphenism in the adult stage. An analysis of the daily trap captures of five species of *Bicyclus*, including *B. anynana*, in a forest-edge environment in Malawi provided base-line information on this phenomenon (Windig et al. 1994). A wet season form with eyespots and a medial band flies during the hot rainy season. This form alternates with a generation of the uniformly brown colored, cryptic, dry season form that survives throughout a long, cool dry season (Fig. 2). The ways in which this polyphenism is influenced by natural selection, and differences in resting backgrounds and predation, have been investigated by use of capture-recapture experiments (Brakefield and Frankino 2008) and laboratory studies with potential predators (Lyytinen et al. 2004). The phenotypic plasticity is induced by ambient temperature in the late larval period and is mediated by the ecdysteroid hormones (Brakefield et al. 1998; Kooi and Brakefield 1999; see further below). Declining temperatures in the early dry season act as a cue for larvae that will develop into the dry-season-form generation (Windig et al. 1994). The seasonal polyphenism provides an adaptive response to the alternating seasonal environments that differ in mortality factors

FIGURE 2. Phenotypic plasticity in *Bicyclus*. (*Top row*) Photographs of the habitat of the butterflies at the end of the wet season (shown online in color as a green background in *a*) and in the early dry season (shown online in color as brown, dry leaves in *b*). Photographs were taken at exactly the same spot near Zomba, Malawi. (*Bottom row*) (*c*) A mating pair of the wet season form (*B. safitza*); (*d*) an individual of the dry season form of *B. cottrelli* that has just alighted on leaf litter, thus hiding the ventral eyespot; (*e*) both forms of *B. safitza*. Photographs *c* and *d* were taken in the wild at Zomba, whereas *e* is of two sisters raised in the laboratory. In *e*, the wet-season-form individual was reared throughout development at 27°C, whereas the late larva of the dry-season-form butterfly was switched to a lower temperature (from Brakefield and Frankino 2008).

and reproductive opportunities. Lines produced by artificial selection now yield only one or the other of the alternative seasonal forms across the whole range of rearing temperatures (Brakefield et al. 1996). A set of crosses between these lines has suggested that from five to ten genes are involved in producing the highly divergent phenotypes (Wijngaarden and Brakefield 2000). Work on the wing patterns of the seasonal forms has also been extended to cover differences in life history-traits (see below).

Eyespot Evolutionary Developmental Biology

Following the earlier work by Fred Nijhout (for review, see Nijhout 1991), the evolution and development of wing eyespot formation became an important research focus (for reviews, see Beldade and Brakefield 2002; McMillan et al. 2002). Each eyespot is made up of concentric rings of color formed by specialized epithelial cells (scale cells) that have synthesized a particular color pigment shortly before adult eclosion. The decisions that specify pigment identity are made earlier in development, shortly before and immediately after pupation. Each eyespot is formed around an inductive organizer known as an eyespot focus. Some of the genetic pathways and developmental mechanisms involved in pattern determination have been identified by using a combination of microsurgical manipulations and gene expression studies (see, e.g., Brakefield et al. 1996; Brunetti et al. 2001; Reed and Serfas 2004; Saenko et al. 2008). An eyespot focus can be transplanted to a different site in very early pupae, an experiment that results in an ectopic eyespot being formed around the grafted tissue. A series of genes from well-known signaling pathways has been shown to be expressed at different stages in the eyespot focus and, later, in the surrounding epithelial cells. An early study has mapped a component of phenotypic variation in eyespot size to the *Distal-less* locus (Beldade et al. 2002a). A modeling approach has also begun to be applied to eyespot pattern determination in *Bicyclus* (Dilão and Sainhas 2004; Marcus and Evans 2008). The *Bicyclus* community as a whole is now developing the tools for gene mapping, genomics, transgenics (Marcus et al. 2004; Beldade et al. 2006, 2008; Ramos et al. 2006), and those required for studies of embryonic development, which has been shown to be associated to eyespot development (Saenko et al. 2008).

STUDIES IN BEHAVIOR AND LIFE-HISTORY EVOLUTION

Behavioral Studies

The two wing androconia of male *B. anynana* have been found to produce three sex pheromone components that enhance male mating success via receptors on female antenna (Nieberding et al. 2008). Six consecutive steps have been detected by video analysis in a typical successful courtship: location, orientation, flickering, thrust, attempting, and copulation. The flickering phase, during which the male initiates rapid flapping wing movements in front of the female, causes the androconia to fan out and presumably spread the pheromone on to the female. We have developed a useful tool for assaying male mating success and studying sexual selection (Joron and Brakefield 2003). Males painted with a fluorescent dust on their genitalia transfer the dust during copulation. Thus, males with different phenotypes can be marked with a group-specific color of dust and released in a flight cage or greenhouse. Receptive females released into the same greenhouse will mate during a period of 24–48 hours and can then be recaptured and inspected under UV light to identify their mating partner by group. This technique was first used to demonstrate a mating advantage for outbred over inbred males under free-flying conditions. The genetic load of *B. anynana* is higher than that of *Drosophila melanogaster* and is inherited mainly through the paternal line (Saccheri et al. 2005).

Life-history Evolution

Research studies have also been focused on conservation genetics and inbreeding depression (Saccheri et al. 2005), on fluctuating asymmetry (Breuker and Brakefield 2003), and, especially, on life-history evolution, including the basis of variation in rates of aging (www.lifespannetwork.nl). The latter area of interest in evolutionary medicine has followed on from the recognition that although adults of the wet season form are short-lived and reproduce rapidly in a favorable environment, those of the dry season form are long-lived and allocate more resources to survival in a stressful environment. The dry-season-form individuals must survive as active adults for several months of reproductive dormancy in a cool climate before they can reproduce with the rains and increased temperatures of the early wet season. These differences in life histories have been the focus for work on the evolutionary genetics of key life-history traits, including egg size, developmental time, size at maturity, life span, and starvation resistance (e.g., Zijlstra et al. 2004; Fischer et al. 2007; Pijpe et al. 2008).

Physiology, Plasticity, and Life Histories

The roles of ecdysteroid hormones in regulating the phenotypic plasticity of the wing pattern and in preadult developmental time have been investigated (Brakefield et al. 1998; Zijlstra et al. 2004). Microinjections or infusions of 20-OH ecdysone into early pupae fated to develop the dry season form yield adults with wing patterns more characteristic of the wet season form. By studying the ecdysone receptor using immunological staining against the receptor protein, P.B. Koch et al. (unpubl.) showed that (1) the location of ecdysone receptors coincides with the location of the developing eyespot, (2) upon injection of ecdysone, the number of ecdysone receptors increases, and (3) butterflies artificially selected for smaller eyespots on the ventral wing surface have fewer ecdysone receptors at the position of the developing eyespot.

Correlated responses are regularly observed in artificial experiments on phenotypic plasticity or life-history traits. These can be explained in part by the involvement of ecdysone in the regulation of the different traits (Zijlstra et al. 2004). For example, lines that have been selected for short or long preadult developmental time show corresponding changes toward larger or smaller ventral eyespots, respectively. Fast-developing butterflies have higher levels of ecdysone shortly after pupation than do slow-developing butterflies. In addition, the slow-selected butterflies show a much smaller response to ecdysone injection in the pupal stage (increase in pupal developmental time; increase in ventral eyespot size) than the fast-selected butterflies. In general, measuring ecdysone levels in larval, pupae, or adult butterflies will provide information on ecdysone production, whereas the responses to ecdysone injection will provide information on the sensitivity of the target tissues to ecdysone (likely to be at least partly caused by differences in the number of ecdysone receptors). We have also begun to explore the involvement of juvenile hormone (JH). We have used several JH forms and analogs (JH-III; methoprene; periproxyfen) to study the influence of this hormone on several life-history traits such as egg size and life span. Generally, the effects of these manipulations are very small (B.J. Zwaan et al., unpubl.), probably because we lack the basic knowledge of JH titer dynamics in the life history of *Bicyclus*. Only very recently were we able to measure the titers of JH, because of the very low levels of JH normally present in insects.

GENETIC RESOURCES: POPULATIONS

Artificial selection is routinely applied for a series of morphological and life-history traits in *B. anynana*. The following are among the traits that we have targeted (usually in upward and downward directions) and for which we have retained selected lines in the laboratory.

- **Seasonal polyphenism:** A dry-season-form line and a wet-season-form line (Brakefield et al. 1996); the elevation and shape of the norm of reaction for eyespot size against rearing temperature (Brakefield and Frankino 2008).
- **Eyespot patterns:** The size, color composition, shape, and position of individual eyespots (McMillan et al. 2002) and the pattern of relative size and color composition among particular eyespots (Beldade et al. 2002b; Allen et al. 2008).
- **Allometric growth patterns:** Wing size relative to body size, and of forewing size relative to hindwing size (Frankino et al. 2005, 2007).
- **Life-history traits:** Egg size, developmental time (both as a single trait and in combination with eyespot plasticity), pupal weight, protandry, adult starvation resistance, and life span (e.g., Zijlstra et al. 2004; Fischer et al. 2007; Pijpe et al. 2008).

Mutant stocks are available that carry alleles arising from spontaneous mutations which yield a discrete change in morphology. These stocks include the following as grouped by the principal phenotypic change:

- **Eyespot patterns (see Fig. 1):** Eyespot size, color, or shape and the presence or absence of specific subsets of eyespots. One of these mutants also has vestigial development of the wing androconia, although such pleiotropic effects on other wing or adult traits are the exception. Five of more than 20 of the stocks are, however, homozygous lethal in embryonic development (Saenko et al. 2008).
- **Medial band:** The pale band across the ventral wings has a dent in it on both wings rather than being nearly straight.
- **Venation:** Pattern including *Cyclops* (Brakefield et al. 1996). Individual wing veins or trachea are either vestigial or additional to the normal pattern, or the whole venation system is weakly formed (Saenko et al. 2008).
- **Eye color:** The *pearl* mutant with cream-colored eyes (the ultrastructure of the eye and the opsins have been studied by D.G. Stavenga et al., unpubl.).

- **Pigmentation:** Several wing color mutants make the brown background color either darker or paler, or yield a distal wing portion that is a pale cream color (Brakefield et al. 2001). In addition, spontaneous mutations have also been isolated that affect other tissues and/or developmental stages, including a larval mutant with a much darker brown color in the final instar (*Chocolate*) and a *yellow* pupal mutant whose color is due to loss of synthesis of a blue, pterobilin pigment (yellow pigments are grass-derived carotenoids).
- **Size:** Pygmy-sized adults with an extended larval developmental time.

GENETIC RESOURCES: GENOMICS AND TRANSGENESIS

Recent and ongoing efforts to develop genomic and transgenic tools for *B. anynana* are opening up a new generation of questions concerning linking genotypes to phenotypes and to fitness (Beldade et al. 2008). The complex repetitive genome of *B. anynana* is now subject to analysis by newly developed resources and tools including (1) sequence information, (2) linkage map, (3) microarrays, and (4) germ-line transformation. A recent project that involved sequencing approximately 10,000 expressed sequences tags (ESTs) from *B. anynana* developing wings has identified more than 4000 new genes expressed in these tissues (Beldade et al. 2006). We continue to sequence ESTs and expect to add to the increasing number of genes expressed in developing wings and other tissues. We have created cDNA libraries from diverse tissues in order to sequence those ESTs, and we have also selected bacterial artificial chromosome (BAC) clones containing candidate genes of interest, enabling sequence analysis of potential regulatory noncoding regions of those genes. Thus far, we have sequenced 11 BAC clones and more than 100,000 ESTs (in collaboration with the Department of Energy Joint Genome Institute; see http://jgi.doe.gov/sequencing/why/ 3112.html), which were assembled into more than 17,000 unigenes. Various types of DNA sequence polymorphisms have been identified, including microsatellites, amplified-fragment-length polymorphisms (AFLPs), and single-nucleotide polymorphisms, in expressed genes that can be used in gene mapping. A linkage map covering all 28 chromosome pairs of *B. anynana* is now available, and the microsatellite and AFLP markers (Van't Hof et al. 2005) are currently being enriched with markers in genes expressed during wing development (P. Beldade et al., in prep.). Moreover, all approximately 17,000 *B. anynana* UniGenes identified to date from ESTs have been used to design the first generation of *B. anynana* microarrays (using NimbleGen-Roche technology), which are currently being tested (P. Beldade and A.D. Long, unpubl.). The use of these microarrays will enable high-throughput expression profiling analysis, which complements the single-gene approach possible with quantitative real-time polymerase chain reaction. Finally, *B. anynana* is the first (and, to date, the only published) butterfly for which germ-line transformation has been developed (Marcus et al. 2004). Such techniques, coupled with an elegant laser-mediated method that enables precise transcriptional activation of transgenes in developing pupal wings (Ramos et al. 2006), will enable fine-scale functional analysis of candidate genes. These studies hold great promise for furthering our understanding of the mechanisms of morphological evolution, beautifully represented in butterfly wing patterns.

TECHNICAL APPROACHES

The following protocols describe the preparation and manipulation of *Bicyclus* embryos and developing wing discs for studies in gene expression and the development of wing patterns. Additional protocols describe methods used in insect physiology and to isolate male sex pheromones.

Protocol 1

Culture and Propagation of *Bicyclus* Laboratory Populations

This protocol describes the methods required for the culture of *Bicyclus*. The larvae are typically fed and maintained on pot-grown maize plants (*Zea mays*) that are about 50 cm high. Males and females can be separated as pupae or adults. Adults are fed on moist banana and will readily mate in the lab. Females lay eggs on available grass plants. We use a standard temperature of 27°C (because an approximate rule developmental time is twice as long at 20°C), a high relative humidity of about 60–70% (not critical), and a 12:12 photoperiod similar to the climate experienced in the wet season near the equator. Both larval and pupal moults are gated by photoperiod and can be readily timed (e.g., by use of time exposure filming). To set up cohorts of standard developmental stages, an appropriate timing for this photoperiod is chosen: Pupation usually occurs shortly after lights out and larval molts also occur during the night. At around 27°C, egg development takes 4 days and the total generation time is 5–6 weeks, yielding about eight generations a year in selection experiments.

MATERIALS

Reagents

Bicyclus stocks, established using eggs, larvae, or pupae
   All life stages can be delivered by mail.
Maize plants (*Zea mays*) for propagating *Bicyclus*
   Grow in about 9-cm pots, at a density of approximately 12–15 plants per pot to a height of about 50 cm.
   A natural larval host plant, Oplismenus africanus, is also available. It yields comparable developmental times and pupal weights, but the plant grows much more slowly than maize. Artificial diets have been used for the larvae, but larvae show extended developmental time and reduced pupal weight.

Equipment

Cages (hanging)
   These should be cylindrical or cubic, made of black or green net, and in sizes large enough to enclose up to several hundred butterflies or groups of 12 or 15 pots.
Net sleeves, large enough to fully enclose two pots of maize
Plastic cups to isolate mating pairs

METHOD

Propagation and Collection of Larvae

1. Feed and maintain the larvae on *Zea mays* plants, grown in pots.
   Young larvae that have eaten out their food supply will crawl on new plants made available.
2. To collect the older larvae, shake the foliage above a tray. When the majority of larvae drop off the foliage, transfer them to fresh plants.

Older larvae are generally quite resistant to handling and can literally be piled up on the surface of a pot of plant food. They will crawl up and find foliage by themselves.

3. To assess the developmental staging, observe the color differences in the larvae. The five different larval molts can be distinguished by the size of their head capsule, and larvae that are also about to molt can be readily picked out.

   Young larvae, crawlers that have ceased feeding, hanging prepupae, and pupae are all green (except for those pupating on a dark surface), whereas the fifth (final) instar larvae are brown. This color change is useful in tracking developmental stages.

4. Transfer the pupae into hanging net cages for eclosion, mating, and oviposition on young maize plants or *Oplismenus* cuttings kept in tubes with water.

   Prepupae and their silk holding pads can be removed carefully from foliage or stems and placed on filter paper to pupate. Pupae can be even more easily removed from stems or foliage and placed together in Petri dishes for eclosion into adults.

5. Feed the emerging adults on slices of banana, which should be renewed twice a week. Place the banana either on the netting at the top of the cage. Keep moist by covering with wet cotton wool or place on wet cotton wool in a Petri dish at the base of the cage.

Mating of *Bicyclus*

6. Separate the male from female pupae or collect virgins of each sex within 24 hours of eclosion to be used for mating.

   Pupae can be sexed by inspection of genital pores under a binocular microscope.

   Individual butterflies can be marked using an overhead pen to write on the ventral surface (the one exposed) of their wings.

   Butterflies do not mate on the day of eclosion, enabling a straightforward collection of virgins of each sex; females usually mate when 2 or 3 days old.

7. Mix the males and females in a mating cage, usually a hanging cylindrical net, or set up to mate and lay eggs in single pairs in small plastic pots. Copulation lasts about 30 minutes.

8. Remove pairs *in copula* by picking the female out by her closed forewings or carefully enclosing the pair in a plastic cup.

   Pairings or individual females after separation can then be set up separately in small plastic pots to obtain families. Well-watered cuttings of grass provide egg-laying substrates.

   Eggs are laid out singly, in large numbers from about 2–3 days after mating for 2–3 weeks. Eggs can also be collected en masse in the hanging cages by adding a pot of young maize plants. Synchronized cohorts of eggs can be obtained by giving females access to egg-laying plants for only one period of light (or an even shorter period of time).

   Individual females can lay more than 300 eggs, although a brood size of 100 is more typical. In addition, under the conditions described for stock rearing, about two thirds of the adults in a population contribute offspring to the next generation (Brakefield et al. 2001).

9. Allow eggs to hatch on leaves to establish a new generation or, if individuals eggs need to be analyzed or manipulated, remove the eggs without damage from leaves by lightly rubbing with the fingers.

   The eggs can be assessed or further analyzed, for example, measured for size (they are spherical and can be assayed by image analysis; Fischer et al. 2007), set up for embryo dissection (Protocol 4), or transported dry en masse in Eppendorf tubes.

Protocol 2

Surgical Manipulations on Pupal Wings: Damage and Cauteries

This protocol describes cautery experiments with young pupal wings that can be used to examine the potential role of groups of cells in the developing wing in wing pattern determination. Experiments can be performed on one of the (left or right) wings, leaving the other wing as a control. The pupal wing case overlies a stack of four sheets of epithelial cells that will form in turn the dorsal and ventral surfaces of the forewing and then the respective surfaces of the hindwing. The dorsal surface of the forewing is immediately underneath the cuticle. The trachea or wing veins of the forewing are visible through the pupal cuticle under a binocular microscope providing landmarks. In addition, raised "bumps" or irregularities on the pupal cuticle indicate where the eyespot organizers (or foci) are present on the underlying forewing. The wild-type pattern of the dorsal forewing in *B. anynana* consists of a small anterior eyespot and a large forewing eyespot; each eyespot has a central white pupil, a black inner ring, and an outer gold ring of color. Experiments that involve a short insertion of sharpened needles through the pupal cuticle can explore the effects of damage to groups of cells at different times from shortly after pupation until pattern determination has occurred (at ~24 hours after pupation). This type of experiment originally helped to characterize the eyespot organizers, because early focal damage to the location of these cells can result in shrinkage or complete elimination of the corresponding eyespot in the adult wing (relative to the control wing). Curiously, damage to other sites on the distal wing shortly before pattern determination can produce nonectopic eyespots (with no white pupil) where no eyespot would occur normally (Brakefield and French 1995). The size and timing of the effect on pattern formation are dependent on the amount of damage applied; e.g., when heat is applied to the needle tip (i.e., with cautery), the effect is magnified and tends to be advanced in timing. A deep(er) insertion of the needle can also produce damage to lower cell layers that can yield effects on adult wing surfaces, other than that of the dorsal forewing.

MATERIALS

Reagents

Bicyclus pupae of desired age

Equipment

Dissecting microscope
Heating element with variable power source
Tungsten Dissecting Probe (~0.25 mm diameter; World Precision Instruments)

METHOD

Cauteries

1. Collect pupa of known age between 1 hour and 24 hours after pupation.
2. Select target sites by examining characteristic vein pattern and cuticle marks, visible on the pupal cuticle.

3. Pierce the cuticle and epidermis by inserting the tungsten dissecting probe approximately 0.1 mm past the hardened cuticle. Immediately withdraw the tungsten dissecting probe after insertion.

    Severe cauteries can be applied by using a heated tungsten dissecting probe. To do this, connect the tungsten dissecting probe to the heating element and adjust the current to heat the tip to about 100°C. Damage (without cautery) to the developing wing epithelial cells is also obtained by simply inserting an unheated needle tip.

    Clean the tungsten dissecting probe after each cautery with 70% ethanol to prevent infections.

    See Troubleshooting.

4. Return the pupa, with the operated side up, to a temperature of choice until eclosion occurs.

    Pupae can be kept in small plastic cups; however, if the surface is too smooth, insert a piece of paper for the butterfly to crawl up on.

5. After wing expansion, sacrifice the adult by freezing at –20°C for approximately 20 minutes and remove both forewings (and hindwings, with deep damage).

    See Troubleshooting.

6. Inspect and compare experimental and control wings under a dissection microscope to assess effects on pattern determination.

    The properties of the ectopic eyespot formed around a wound site by standardized damage can give information about the properties of the damaged tissue.

TROUBLESHOOTING

Problem (Step 3): Bleeding.
Solution: Remove hemolymph with a piece of filter paper.

Problem (Step 5): Hole in wing or crippled wing after eclosion.
Solution: Try to damage wings less severely during the cautery.

Protocol 3

Surgical Manipulations on Pupal Wings: Grafts

This protocol describes the procedure of transplanting tissue of the eyespot organizer in early pupae of *B. anynana*. In very young pupae, before apolysis, the pupal wing case is attached to the sheet of cells that will form the dorsal surface of the adult forewing (see also Protocol 2). Therefore, if a piece of pupal cuticle is removed and transplanted to a novel site on the developing wing (where a corresponding patch has been removed), the underlying cells of the dorsal forewing are transferred together with it. The pattern determination effects of the transplanted wing tissue on the surrounding cells can be examined by comparing the experimental and control wings of the adult butterfly. In crucial experiments to explore eyespot formation on butterfly wings, an eyespot focus transplanted to a novel site in the distal part of the forewing was shown to yield an ectopic eyespot pattern in the tissue surrounding the graft (French and Brakefield 1995). Square grafts can be turned 180° (or 90°) before implantation to distinguish donor and host tissue because the orientation of the scale cells is determined before grafting.

MATERIALS

Reagents

B. anynana pupae (3–4 hours old)

Equipment

Dissecting microscope
Filter paper
Fine forceps (no. 55; World Precision Instruments)
Microsurgical knife 15° (5 mm depth; MSP)
   Alternatively, use a self-made knife with broken chips of a razor blade.
Tungsten dissecting probe (~0.25 mm diameter; World Precision Instruments)

METHOD

1. Collect 3–4-hour old pupa.
   Start the graft 3.5–4 hours after pupation, when the cuticle is hardened sufficiently, but before the underlying epidermis has separated during apolysis (at 5–6 hours).

2. Use a tungsten needle, forceps, and knives to prepare a square hole on the cuticle above the forewing of the host pupa to which the graft will be moved.
   The hole should be the same size as the graft will be.

3. Use the same instruments to cut out the graft square (cuticle plus epidermis attached) from the donor pupa.
   The graft is usually a small square of cuticle plus epidermis, the size of one third to one fourth of wing cell width.

4. Move the graft to the hole in the host, keeping the same orientation or rotating it 180° to reverse proximal-distal polarity.

   If there is a good fit between the graft and host site, the graft will quickly secure by dried hemolymph, with no need for adhesives or wax. No sterile precautions or antibiotics are required. Dry any bleeding with a small square of filter paper.

5. Use the integument from the host site that was previously removed to seal the donor site and prevent desiccation.

6. Leave the grafted pupae at room temperature until the graft is well-secured (15 minutes to 1 hour).

7. Return pupa to temperature of choice until eclosion.

8. After wing expansion, sacrifice the adult by freezing at −20°C for approximately 20 minutes and remove both forewings.

   See Troubleshooting.

9. Score the grafts under the dissecting microscope.

   The morphology of the ectopic eyespot formed around the graft will depend on the properties of the donor eyespot focus and of the epidermis of the host tissue.

   Because wounding of pupal wings results in pigment production (see Protocol 2), take care when interpreting graft results involving poor healing of tissue damage around the graft.

TROUBLESHOOTING

Problem (Step 8): Holes in wings after eclosion.

Solution: Try to perform grafts quickly so that the grafted epidermis does not dry out and die under the cuticle. In addition, make sure that the graft and host hole are the same size.

Protocol 4

Fixation and Dissection of Embryos

This protocol describes the fixation and dissection of *B. anynana* embryos that can subsequently be used in in situ hybridization (ISH, Protocol 6) and immunohistochemistry (IHC, Protocol 7) analysis (Saenko et al. 2008).

MATERIALS

The recipes for items marked with <R> begin on page 325.

CAUTION: See the Cautions Appendix for appropriate handling of materials marked with <!>.

Reagents

B. anynana eggs
Bleach, 50% <!> (final concentration of sodium hypochlorite is <2.5%) in Milli-Q H_2O
Block buffer embryos (BBE) <R>
Fix buffer embryos (FBE) <R>
H_2O_2, 4–5%
Methanol, 100% (Merck) <!>
Methanol, 25, 50, and 75% in PBS <!>
Modified phosphate-buffered saline (PBS) <R>, precooled on ice
Modified phosphate-buffered saline with detergent (PBST) <R>, precooled on ice

Equipment

24-well plate (Techno Plastic Products)
   If embryos will be used for ISH, pretreat plate with 4–5% H_2O_2 to destroy RNases.
Baskets for embryos
   Remove the top 1 cm of a 5-ml pipette tip and glue a piece of mesh to one side. These baskets fit into a well of a 24-well plate.
Dissection microscope
Fine forceps (no. 55; World Precision Instruments)
Glass dish, small
Glass slide (Menzel-Glaser)

METHOD

Perform all steps at room temperature unless otherwise indicated.

Dechorionation and Fixation of Embryos

Precool PBS and PBST on ice.

1. Collect eggs of desired age into a glass dish.

   At 27°C, it takes about 100 hours to complete embryonic development (Saenko et al. 2008).

2. Rinse the eggs in water to remove any dirt.
3. In a small glass dish, dechorionate the embryos with 50% bleach for 1–2 minutes.
4. Rinse well with water to remove any bleach.
5. Transfer eggs to an Eppendorf tube.
6. Fix the eggs in FBE for 30–60 minutes.
7. Wash three times, 5 minutes each in PBST.
8. Dehydrate the embryos:

    5 minutes in PBS
    5 minutes in 25% methanol
    5 minutes in 50% methanol
    5 minutes in 75% methanol.

    See Troubleshooting.

9. Fix embryos in 100% methanol for a few days at –20°C.

    Omitting this step makes dissection of the embryos much harder. Fixed embryos can be stored long term in methanol at –20°C.

    See Troubleshooting.

Dissection

If embryos will be used for ISH, work under RNase-free conditions. Optional: If they are to be used for IHC, they can first be rehydrated and subsequently dissected in BBE.

10. Place the embryos in a droplet of 100% methanol on a glass slide.

    See Troubleshooting.

11. Use very fine forceps and/or dissecting needles to make a hole in the vitelline membrane surrounding the embryo.

    See Troubleshooting.

12. Carefully remove the vitelline membrane piece by piece without damage to the embryo itself.

    The embryo can be easily distinguished from yolk and membrane. See Troubleshooting.

13. Remove as much yolk as possible.
14. Transfer the embryo into a basket placed in a well of a 24-well plate containing 100% methanol.

    A maximum of 100 embryos will fit in one basket.

15. Gradually rehydrate the dissected embryos:

    5 minutes in 75% methanol
    5 minutes in 50% methanol
    5 minutes in 25% methanol
    Then soak in PBS

16. Rinse the embryos twice briefly with PBST and wash three times for 5 minutes each in PBST.

    The embryos are now ready for use in ISH (Protocol 6) or IHC (Protocol 7).

TROUBLESHOOTING

Problem (Steps 8–12): Dissection is difficult.
Solution: Leave embryos for a longer time in methanol. This works as an extra fixation step and should make dissection easier. Practice also helps to improve dissection skills.

Protocol 5

Dissection of Larval and Pupal Wings

This protocol describes the dissection of *B. anynana* larval and pupal wings. The wings can subsequently be used in in situ hybridization (ISH, Protocol 6) and immunohistochemistry (IHC, Protocol 8). Dissected pupal wings can also be analyzed with stereomicroscopy to study scale maturation and pigment deposition.

MATERIALS

Reagents

The recipes for items marked with <R> begin on page 325.

B. anynana larval and pupae
Modified phosphate-buffered saline (PBS) <R>, precooled on ice

Equipment

24-well plate (Techno Plastic Products)
If wings will be used for ISH, pretreat plate with 4–5% H_2O_2 to destroy RNases.
Baskets (prepared as described in Protocol 4)
Curved forceps (no. 5; World Precision Instruments)
Dissection dish
Prepare Sylgard 186 (Dow Corning) as described on the product, but add a pinch of activated charcoal powder to the Sylgard while it is still fluid and mix well. Fill a Petri dish with the Sylgard mixture and allow it to harden.
Dissection microscope
Dissection scalpel and blade
Fine forceps (no. 55; World Precision Instruments)
Fine scissors (no. 5; F.S.T. Fine Science Tools)
Insect pins

METHODS

I. Dissection of Larval and Prepupal Wings

1. Collect larva or prepupa of desired age.
   Wings in prepupae are generally much larger, surrounded by a thick peripodial membrane that is hard to remove and more fragile.

2. Place larva in ice-cold PBS to mildly anesthetize it.

3. Pin the caterpillar to a dissection dish filled with ice-cold PBS so that it is on its side.
   Forewing and hindwing discs are located under the second and third pair of larval legs, respectively, midway between the dorsal midline (visible as a white line running laterally along the larval body) and the larval legs.

4. With fine scissors, make a small superficial cut of approximately 1 mm just dorsally from the legs.

5. The disc will generally pop out, due to the pressure inside the larval body. If not, pull it gently with fine forceps.

   The wing disc is covered with a transparent peripodial membrane that can be removed.

6. Once the disc is fully exposed, cut off the trachea that connects the disc to the larval body.

   All four wings can be removed, and removing one single hindwing will not kill the larvae or prevent full development (typically resulting in a three-winged adult).

7. Transfer the wings to a basket in a well of a 24-well plate, containing ice-cold PBS.

   A maximum of 20 larval wings will fit in one basket.

8. Continue immediately with ISH (Protocol 6) or IHC (Protocol 8).

II. Dissection of Pupal Forewings and Hindwings

1. Collect pupa that are at least 5 hours old.

   Before 5 hours, the cuticle is too soft and the dissection is technically very difficult. Before 10 hours, it is hard to separate the cuticle from the forewing still attached to it.

2. In a dissection dish, submerge the pupa in ice-cold PBS to mildly anesthetize it.

3. Pin the pupa through the abdomen and thorax laterally so that the wing is accessible.

4. Use a scalpel or fine scissors to cut the cuticle around the wing margin.

   Forewing shape is visible on each side of the pupa.

5. Lift the cuticle and cut through the trachea that connects the forewing to the thorax.

   Until about 25 hours postpupation, the forewing is more or less loosely attached to the cuticle and will be lifted with it. Separate forewing tissue from the cuticle by gently scraping with fine forceps. See Troubleshooting.

6. Transfer the forewings to a basket in a well of a 24-well plate containing ice-cold PBS and leave it there while proceeding with hindwing dissection, or continue.

   A maximum of ten pupal wings will fit in one basket.

7. Remove the peripodial membrane that covers the hindwing with fine forceps.

   The hindwing (including characteristic veins) is easy to distinguish through the transparent peripodial membrane made visible after forewing removal.

8. Cut the trachea that connects the hindwing to the thorax.

9. Transfer the hindwing to a basket in a well of a 24-well plate containing ice-cold PBS.

   All four wings from a pupa (left and right forewings and hindwings) can be dissected. However many wings are removed, pupae generally do not survive wing extraction.

10. Continue immediately with ISH (Protocol 6) or IHC (Protocol 8).

TROUBLESHOOTING

Problem (Step 5): Damaged wing.
Solution: Be very careful with cutting and manipulation, especially with prepupae and young pupae.

Protocol 6

In Situ Hybridization of Embryos and Larval and Pupal Wings

The modified in situ hybridization (ISH) protocol can be used to localize RNA transcripts in developing tissues of *B. anynana*, including larval and pupal wings and embryos. A slightly different version of this protocol has been published elsewhere and is available as a online-accessible video (Diane Ramos and Antónia Monteiro; http://www.jove.com/index/details/stp?ID=208).

MATERIALS

The recipes for items marked with <R> begin on page 325.

CAUTION: See the Cautions Appendix for appropriate handling of materials marked with <!>.

Reagents

Anti-DIG antibody (Roche), 1:2000 in PBST

B. anynana larval and pupal wings, dissected as described in Protocol 5 *or* embryos, fixed as described in Protocol 4

0.5x Blocking reagent (Roche) in PBST

10x Blocking reagent (Roche)

EDTA (50 mM in PBST)

Formaldehyde, 4.5% and 9% in PBST (prepared fresh) <!>

Glycerol

Glycine, 2 mg/ml in PBST

H_2O_2, 4–5%

Hybridization buffer (HB) <R>

$MgCl_2$, 50 mM in staining buffer (SB + $MgCl_2$) <R>

Modified phosphate-buffered saline (PBS) <R>

Modified phosphate-buffered saline with detergent (PBST) <R>

PBST:preHB mix

Combine equal volumes of PBST and preHB.

Prehybridization buffer (preHB) <R>

Proteinase K, 12.5 µg/ml (QIAGEN), in ice-cold PBST

Sense and antisense probe for ISH, 50–100 ng/µl

The probes are synthesized using the DIG RNA–labeling kit (Roche) with the desired DNA sequence inserted in the pCRII-TOPO vector as template (Invitrogen).

20x SSC <R>

Staining buffer (SB) <R>

Staining solution (SS) <R>

Equipment

24-well plate (Techno Plastic Products)

Pretreat plate with 4–5% H_2O_2 to destroy RNases.

Baskets (see Protocol 3)

Coverslips (Menzel-Glaser)

Glass dishes

Glass slides (Menzel-Glaser)
Heating block, set at 80°C
Pasteur pipette
Oven, set to 55°C

METHOD

If using dissected larval and pupal wings (from Protocol 5), fix as described beginning with Step 1 before starting with section Prehybridization. For fixed embryos, prepared as described in Protocol 4, begin the method immediately with section Prehybridization.

Fixation of Larval and Pupal Wings

1. To dissected wings (from Protocol 5), add approximately 1 ml of 9% formaldehyde and fix for 45–60 minutes.
2. Stop the fixation by rinsing twice and washing once for 5 minutes in PBST.

Prehybridization

Perform all steps under RNase-free conditions and on ice unless otherwise indicated. Washes are performed for 5 minutes unless otherwise indicated; volumes per well are approximately 1 ml unless otherwise indicated.

3. Transfer the tissue to a basket.

   The tissue will remain in the basket throughout the protocol. Move the basket from well to well between the steps to avoid direct handling of tissue and to minimize damage. A maximum of 100 embryos, 20 larval wings, or 10 pupal wings will fit in each basket.

4. Incubate the tissue for 4 minutes in ice-cold proteinase K with gentle rocking.

   When using embryos, allow the embryos to settle for 1 minute.

5. Immediately rinse twice with 2 mg/ml glycine.
6. Wash twice in PBST.
7. Postfix in 4.5% formaldehyde for 15–20 minutes.
8. Rinse in PBST.
9. Wash in PBST.
10. Wash for 10 minutes in PBST.
11. Wash in PBST:preHB.
12. Transfer the basket to a well containing preHB and incubate with rocking for at least 1 hour at 55°C.
13. Prepare the probe solution:

    i. To a 1.5-ml Eppendorf tube, add 49 µl of HB + 1 µl of 50–100 ng/µl RNA probe.

    ii. Heat-denature for 5 minutes at 80°C.

    iii. Add the probe mixture to a well containing 0.95 ml of prewarmed HB.

    iv. Homogenize by swirling.

    Prepare both a sense (control) probe and an antisense probe for each target transcript.

14. Transfer the baskets to the HB-probe wells and immediately place back at 55°C.

15. Incubate for 48 hours at 55°C with rocking.

    Depending on probe size and/or desired stringency, temperature can be increased to up to 65°C.

Posthybridization

16. Wash six times for 30 minutes each in prewarmed (55°C) preHB.
17. Transfer the 24-well plate to room temperature and allow it to cool.
18. Wash in PBST:preHB.
19. Wash twice in PBST.
20. Incubate in 0.5x blocking reagent for 1 hour with rocking.
21. Incubate in 1:2000 anti-DIG antibody with rocking overnight at 4°C.

Staining

22. Wash ten times for 15 minutes each in PBST.
23. Rinse and wash twice in SB.
24. Wash in SB + $MgCl_2$.
25. Transfer the samples to clean glass dishes with SB.
26. Replace the SB with SS using a Pasteur pipette. Observe the staining reaction occasionally; staining may continue for less than 1 hour to overnight.

    Protect the staining reaction from light by keeping it in the dark.
27. When sufficient coloring can be detected, stop the staining reaction by removing the SS.

    No coloring should be apparent with the sense probe. Stop both the sense reaction and the antisense reaction at the same time.
28. Wash three times in H_2O (pH 7) or in 50 mM EDTA.
29. Mount samples in glycerol on a glass slide and cover with a coverslip.
30. Observe the samples under a light microscope.

Protocol 7

Immunohistochemistry Staining of Embryos

This protocol describes the detection of proteins in developing embryos of *B. anynana* (Saenko et al. 2008).

MATERIALS

Reagents

The recipes for items marked with <R> begin on page 325.

B. anynana embryos, fixed and dissected as described in Protocol 3
Glycerol, 70% in PBS
Block buffer embryos (BBE) <R>
Fix buffer embryos (FBE) <R>
Fluorescently labeled secondary antibody against primary antibody (commercially available)
Modified phosphate-buffered saline (PBS) <R>
Primary antibody against protein of interest (commercially available *or* provided by another research group)
Wash buffer embryos (WBE) <R>

Equipment

24-well plate (Techno Plastic Products)
Baskets, prepared as described in Protocol 3
Coverslips (Menzel-Glaser)
Fine forceps (F.S.T. Fine Science Tools)
Fluorescence microscope
Glass slide (Menzel-Glaser)

METHOD

Perform all steps at room temperature unless otherwise indicated. Incubations and washes are preferably done with gentle rocking.

1. Transfer the embryos to a basket in which they will remain throughout the protocol.

   Move the basket from well to well between the steps to avoid direct handling and damaging of embryos. A maximum of 100 embryos can be processed in one basket.

2. Incubate embryos in BBE for 1–2 hours.

3. Incubate the embryos in 400 µl of BBE with primary antibody overnight at 4°C.

   The final concentration of the antibody ranges between 1:10 and 1:1000, depending on the antibody. Usually, the lab or company that provides the antibody suggests a working concentration.

4. Wash embryos ten times for 5 minutes each in WBE.

5. Incubate embryos in 400 µl of WBE with secondary antibody for 1–2 hours at 4°C.

   Consult the manufacturer's information sheet for the appropriate concentration of secondary antibody. Keep all material hereafter in the dark.

6. Wash ten times for 15 minutes each with WBE.
7. Wash in PBS for 10 minutes.
8. Incubate the embryos in 500 µl of 70% glycerol for 60 minutes at room temperature or overnight at 4°C.

   The embryos can be stored in 70% glycerol in sealed tubes in the dark at either 4°C or −20°C for a maximum of 1 month.

9. Mount embryos on slides in 100% glycerol and immediately observe under the fluorescence microscope.

Protocol 8

Immunohistochemistry Staining of Wing Discs

This protocol describes the method for detection of proteins in developing wings.

MATERIALS

The recipes for items marked with <R> begin on page 325.

Reagents

B. anynana larval or pupal wings, dissected as described in Protocol 4
Glycerol, 80% in PBS
Block buffer wing discs (BBW) <R>
Fix buffer wing discs (FBW) <R>
Modified phosphate-buffered saline (PBS) <R>
Permount SP15 Mounting medium (Fisher)
Primary antibody against protein of interest (commercially available *or* provided by another research group)
Fluorescently labeled secondary antibody against primary antibody (commercially available)
Wash buffer wing discs (WBW) <R>

Equipment

24-well plate (Techno Plastic Products)
Baskets (prepared as described in Protocol 4)
Clear nail polish
Coverslips (Menzel-Glaser)
Fluorescence microscope
Glass slides (Menzel-Glaser)

METHOD

Perform all steps at 4°C or on ice unless otherwise indicated. Incubations and washes are preferably done with gentle rocking.

Fixation

1. Transfer the dissected wings to a basket in which they will remain throughout the protocol.
   Move the basket from well to well between the steps to avoid direct handling and damaging of embryos. A maximum of 100 embryos can be processed in one basket.

2. Fix the wing discs in FBW for 30 minutes.

3. Stop fixation by washing four times for 5 minutes each with PBS.

Antibody Staining

4. Incubate the wings in BBW for 1–2 hours.

5. Incubate further in 400 µl of WBW with primary antibody for 18–24 hours at 4°C.

    The final concentration of the antibody ranges between 1:10 and 1:1000, depending on the antibody. Usually, the lab or company that provides the antibody suggests a working concentration.

6. Wash four times for 15 minutes each in WBW.

7. Incubate in 400 µl of WBW with secondary antibody for 1–2 hours (or overnight).

    Consult the manufacturer's information sheet for the appropriate concentration. After fluorescent secondary antibody is added, keep in the dark to prevent fading of fluorescence.

8. Wash four times for 15 minutes each in WBW.

9. Incubate in 80% glycerol for 1–5 hours.

10. Mount discs on glass slides in 100% glycerol. Seal coverslips with nail polish.

    The slides can be kept at –20°C for several months.

11. Observe the preparations under the fluorescence microscope.

Protocol 9

Extraction and Gas Chromatography Analysis of Adult Pheromones

Pheromones are used for communication among individuals of the same species, including attracting mates. A blend of chemical components is recognized as a pheromone if it elicits a behavioral or physiological response. Gas chromatography helps to extract, separate, and, when the extract is associated to an internal standard, quantify the amounts of the chemical components present on a target tissue. This protocol describes the preparation of pheromone extracts of *B. anynana* wings and the subsequent use of these extracts in gas chromatography (GC) analysis (Nieberding et al. 2008).

MATERIALS

CAUTION: See the Cautions Appendix for appropriate handling of material marked with <!>.

Reagents

B. anynana adult butterflies of target age
  Use fresh butterflies or dead individuals stored at –80°C to minimize pheromone evaporation.
Hexane, G.C. grade (Merck) <!>
Palmityl acetate (Sigma-Aldrich)

Equipment

Gas chromatography (GC) system 6890 Series (Agilent)
GC vial + cap, 12 x 32 mm wide mouth (Alltech)
Limited volume insert, no. 98024 (Alltech)
Scissors (F.S.T. Fine Science Tools)

METHOD

Perform all steps at room temperature unless otherwise specified.

Preparation of Pheromone Extracts

1. Depending on the question under study, cut off whole wings, part of the wings, or androconical pockets and place in the GC vial.

2. Soak the tissue in hexane for 5 minutes, using 500 µl for two hindwings or 250 µl for small parts of wings or androconical pockets. Ensure that the whole tissue is submerged in hexane.
   Optional: Add 2 ng/µl palmityl acetate as an internal standard. The internal standard serves as a quality control and/or concentration measure. Extracts can be stored at –20°C until running the G.C. analysis.

3. Fill a limited volume insert with approximately 200 µl of the extract and place the insert back in the G.C. vial.

GC Analysis

4. Run a G.C. analysis with the following settings:

Column:	HP-1, 30 m; 0.32 mm (Hewlett-Packard)
Injector:	240°C
Detector:	Flame ionization detector (FID) 295°C
Program:	Start at 50°C. Increase to 295°C with steps of 15°C per minute. Hold at 295°C for 6 minutes.
Flow carrier gas:	3 ml/minute
Pressure:	85 kPa
Carrier gas:	N_2
Injection:	2 µl in splitless mode

5. Collect and analyze the data produced by the G.C.

Protocol 10

Fresh Weight, Dry Weight, and Fat Content of Adult Butterflies

For many biological processes, such as metabolic rate, starvation resistance, and fecundity, knowledge of body composition is important. This protocol describes how to measure fresh weight (FW), dry weight (DW), and fat-free dry weight (FFDW) of *B. anynana* adults. From this, the water and fat weight can be calculated, as well as fat and water fractions.

MATERIALS

CAUTION: See the Cautions Appendix for appropriate handling of materials marked with <!>.

Reagents

B. anynana adults of desired age
Dichloromethane (Merck) <!>
Dichloromethane:methanol solution (2:1, v/v) <!>
Methanol (Merck) <!>

Equipment

Amber vial with silicone lining in cap (Grace Davison Discovery Science)
Balance, accuracy 0.01 mg
Forceps (F.S.T. Fine Science Tools)
Orbital shaker
Scissors (World Precision Instruments)
Stove, set at 40°C
Urine cup
Weighing paper

METHOD

Perform all steps at room temperature unless otherwise specified. Depending on the data required, all or only parts of the protocol below can be followed. For example, if fresh-weight information is not required, Step 1 can be skipped.

Determining Fresh Weight

1. To obtain FW, measure the butterfly alive or immediately after being sacrificed, with an accuracy of 0.01 mg.

    If measuring fresh weight of a butterfly while it is alive, place the butterfly in a urine cup that has been used to tar the balance beforehand.

    If measuring fresh weight of a recently sacrificed butterfly, place the butterfly directly on the balance using a tarred weighing paper.

    It is generally not a good idea to determine fresh weight of butterflies that have been stored in the freezer for a long period of time, because they tend to lose weight through water loss.

Determining Dry Weight

2. Freeze the butterflies at −20°C.

   Adult butterflies can be stored at −20°C for later analysis.

3. Remove the butterfly from −20°C and cut off the wings, legs, and antennae.

   Wings, legs, and antennae do not contain any fat that is used in metabolic processes. Therefore, these can be removed when determining dry weight and fat content. If, however, water content is of interest, do not cut off the wings, legs, and antennae because of weight differences compared to FW. Leaving wings, legs, and antennae intact is generally not a good idea because drying the animal will make it very brittle and these body parts are easily lost between the different procedures.

4. Transfer the body to an amber vial without cap and dry for 48 hours at 40°C.

5. To obtain DW, weigh the body with an accuracy of 0.01 mg.

Measuring Fat Content

6. Transfer the body back to the vial and add 2 ml of dichloromethane:methanol solution (2:1).

7. Close the vial with a cap and place on a shaker. Shake at 100 rpm for 48 hours at room temperature.

8. Refresh the solvent with another 2 ml of dichloromethane:methanol solution (2:1).

9. Repeat Step 7.

10. Remove the solvent and dry the body for 24 hours at 40°C.

11. Obtain FFDW by weighing the body (accuracy of 0.01 mg).

Calculations

Water content (H_2O in mg): H_2O = FW − DW

Fat content (FAT in mg): FAT = DW − FFDW

Various ratios can be calculated with these numbers depending on the question being addressed.

Protocol 11

Constant Volume Respirometry

This protocol describes the use of constant volume respirometry to measure CO_2 production. The setup used is the TR-2 system from Sable Systems International. In this setup, dry CO_2-free air is sequentially pumped through 16 sealed chambers. The CO_2 produced by the butterflies is subsequently measured with a Li-Cor 6251 infrared CO_2 detector. CO_2 levels (μl/ hour) respired from individual butterflies is an index of the resting metabolic rate (RMR). The RMR of each individual is corrected for weight, preferably after removal of water and fat mass, to be able to compare the RMRs among individuals.

MATERIALS

Reagents

B. anynana adults of desired age

Equipment

Climate chamber with temperature control
Custom-made chambers and caps (volume = 35 cm^3) that fit *B. anynana* (Leiden University)
ExpeData™ software (Sable Systems International, www.sablesys.com)
LiCor6251 infrared CO_2 detector (Now: LiCor7000) (Sable Systems International, www.sablesys.com)
TR-2 system (Now: SI-1 [small insects]) (Sable Systems International, www.sablesys.com)

METHOD

Measuring CO_2 Respiration (RMR)

1. Transfer individual adults to a chamber and close the chamber.
   Leave one chamber empty as a control.
2. Place the chambers in the incubator set at the preferred measuring temperature. Close the incubator and make sure that the light is off inside the incubator when measuring RMR.
   Performing the measurements in the dark will ensure a resting state of the butterfly.
3. Measure CO_2 output using the ExpeData software connected to the TR-2 setup. Use the following settings:

Push-through flow:	100 ml/minute
Run time:	24 minutes
Measuring interval:	0.5 seconds, 180 samples per chamber (= 90 seconds)
Number of runs:	Three. The initial run is a flush run to allow the animals to settle and is discarded in the analysis. The second and third runs are averaged in the analysis.

   See Troubleshooting.

Measuring Fresh, Dry, or Fat-free Dry Weight

4. After measuring CO_2 respiration, remove butterflies from the chambers and measure any of these weights as described in Protocol 9.

Calculating RMR/mg of Tissue

5. Use the individual's weight to convert its RMR (µl/ hour) to RMR per body weight (µl/hour/mg).

    This ratio is the most appropriate measure for representation of the data. For statistical analysis, it is preferable to use the weight measure as a covariate.

TROUBLESHOOTING

Problem (Step 3): Irregular or blocked flow.

Solution: Usually this is caused by small leaks in the tubing. This problem can be solved by checking the tubing and connections and using parafilm around the chamber connections. Replace tubing and/or connectors if the leakage problem remains. In addition, the airflow may also be compromised if air dryers have become too wet. The cotton wool placed at the end caps will become wet and obstruct the airflow. A similar problem occurs when this cotton wool is compacted too much.

Protocol 12

Hemolymph Extraction of Larvae, Pupae, and Adults

This protocol describes the extraction of hemolymph from *B. anynana* larvae, pupae, and adults. The hemolymph can subsequently be used for various purposes and must be treated accordingly. Different purposes and treatments of the hemolymph are described in the last part of the protocol.

MATERIALS

Reagents

B. anynana larvae, pupae or adults

Equipment

Glass capillary, 20 µl (for larvae and pupae) or 1 µl (for adults)
Microsurgical knife, 15° (5-mm depth, MSP)

METHOD

Hemolymph Extraction of Larvae, Pupae, and Adults

1. Collect larvae, pupae, or adult at desired age.
2. Puncture the sample as follows, according to the stage of specimen:
   - For fifth instar larvae, carefully puncture with a microsurgical knife at the dorsal vessel immediately above the third pair of legs. Early fifth instar larvae are difficult.
   - For pupae, carefully puncture with a microsurgical knife at the posterior ridge of the wing.
   - For adults, carefully puncture the aorta and adjacent pulsatile organ that lie just under the dorsal juncture of the scutum and scutelum of the mesothorax.
3. Touch the wound with a capillary and press lightly to fill it.
   - For larvae and pupae, use one 20-µl capillary and allow it to fill to approximately 20 µl. If it is difficult to get 20 µl from pupae, especially from older pupae, make another puncture at a different point or use more individuals.
   - For adults, use single 1-µl capillaries until the desired volume is reached; the maximum is usually about 5 µl.

   A protocol for extraction of larger volumes of adult hemolymph is currently being developed.
4. Transfer the hemolymph to the Eppendorf tube by carefully blowing out the capillary.

Processing Extracted Hemolymph

5. Treat the hemolymph as described, according to the question under study:

   - Blow the hemolymph from the capillary into a glass tube prefilled with 150 µl of methanol and 150 µl of isooctane to dissolve/extract nonpeptide hormones that can later be measured with liquid chromatography mass spectrometry (LCMS) (Westerlund and Hoffmann 2004).

   - Blow the hemolymph from the capillary in a prechilled tube with phenyl-thiocarbamide and keep on ice immediately after extraction to prevent melanization. Hemolymph treated this way can be used to quantify antibacterial peptides (Armitage and Siva-Jothy 2005). If the hemolymph turns black, it cannot be used any longer.

   - Mix the hemolymph with sodium cacodylate buffer for counting cells, using a hemocytometer (Armitage and Siva-Jothy 2005).

   - Immediately freeze the hemolymph to destroy the hemocyte membranes. The hemolymph can later be used to measure phenoloxidase activity (Armitage and Siva-Jothy 2005).

Protocol 13

Injection of Chemicals into Pupae

This protocol describes the injection of chemicals in *B. anynana* pupae. The effect of the injection of the chemical can be monitored for a number of target traits. For example, in our lab, we looked at the effect of injecting the hormone 20-hydroxyecdysone (20E) on development time (Koch et al. 1996).

MATERIALS

Reagents

B. anynana pupae

Chemicals of interest, e.g., hormone (Sigma-Aldrich)
  The desired concentration of the chemical is typically obtained from the literature or experimental trials.

Grace's medium (Life Technology)

Equipment

Injection needle (~0.3 mm diameter; Hamilton)
Petri dishes
Syringe, 10 µl (Hamilton)

METHOD

1. Collect pupa of desired age.

2. Dissolve the chemical at the desired concentration in Grace medium and use 3 µl per injection.
   Injecting less than 3 µl can be difficult, whereas injecting more than 3 µl results in higher mortality.

3. Inject the pupa at the desired age at the lateral posterior region of the fifth abdominal segment.

4. Return injected pupa to temperature of choice until eclosion occurs.

5. Measure the effect of the hormone injection on a selected trait(s).

Recipes

CAUTION: See the Cautions Appendix for appropriate handling of materials marked with <!>.

BLOCK BUFFER EMBRYOS (BBE)

Reagent	Quantity	Final concentration
Wash buffer embryos (WBE) <R>	100 ml	1x
BSA	2 g	2%

Store at 4°C

BLOCK BUFFER WING DISCS (BBW)

Reagent	Quantity (for 100 ml)	Final concentration
Tris-Cl (1 M, pH 6.8)	5 ml	50 mM
NaCl (1 M)	150 µl	150 mM
IGEPAL	500 µl	0.5%
BSA	500 mg	5 mg/ml

FIX BUFFER EMBRYOS (FBE)

Reagent	Quantity (for 100 ml)	Final concentration
10x modified PBS <R>	10 ml	1x
1 M EGTA	5 ml	50 mM
Formaldehyde <!>	10 ml	10%

Prepare fresh for each use. The solution can be filtered-sterilized and stored for approximately 1 year.

FIX BUFFER WING DISCS (BBD)

Reagent	Quantity (for 100 ml)	Final concentration
PIPES (1 M, pH 6.9)	10 ml	0.1 M
EGTA (100 mM, pH 6.9)	1 ml	1 mM
Triton X-100 <!>	1 ml	1.0%
$MgSO_4$ (1 M) <!>	200 µl	2 mM

Just before use, add 1 part 37% formaldehyde <!> to 3 parts of the solution.

HYBRIDIZATION BUFFER (HB)

Reagent	Quantity (for 50 ml)	Final concentration
Prehybridization buffer <R>	50 ml	1x
Yeast tRNA (20 mg/ml; Roche), heat-denatured for 5 minutes at 80°C	250 µl	100 µg/ml
Heparin (25 mg/ml) <!> or	100 µl or	50 µg/ml or
Glycine (100 mg/ml)	0.5–1 ml	1–2 mg/ml (glycine)

1x MODIFIED PHOSPHATE-BUFFERED SALINE (1x PBS)

Reagent	Quantity	Final concentration
NaCl	8.175 g	140 mM
K_2HPO_4	1.07 g	6.1 mM
KH_2PO_4	0.47 g	3.9 mM

Adjust the final volume to 1 liter with DEPC-treated H_2O; adjust pH to 7, distribute into aliquots, and autoclave.

1x MODIFIED PHOSPHATE-BUFFERED SALINE WITH TWEEN 20 (PBST)

Prepare 1x modified phosphate-buffered saline, adjust to pH 7 and autoclave. After autoclaving, add 1 ml of Tween 20 to a final concentration of 0.1%.

10x MODIFIED PHOSPHATE-BUFFERED SALINE (10x PBS)

Reagent	Quantity	Final concentration
NaCl	81.75 g	1.40 M
K_2HPO_4	10.7 g	61 mM
KH_2PO_4	4.7 g	39 mM

Adjust the final volume to 1 liter with DEPC-treated H_2O; adjust pH to 7, distribute into aliquots, and autoclave.

PREHYBRIDIZATION BUFFER (preHB)

Reagent	Quantity (for 100 ml)	Final concentration
Formamide (deionized) <!>	50 ml	50%
20x SSC (pH 5.0) <R>	25 ml	5x
Tween 20	20 µl	0.02%

Adjust the final volume to 100 ml with DEPC-treated H_2O; adjust pH to 5.5 with HCl.

20x SSC STOCK

Reagent	Quantity (for 1 liter)	Final concentration
NaCl	175.3 g	3 M
Sodium citrate	88.2 g	0.3 M

Dissolve the reagents in 800 ml of distilled or deionized H_2O and adjust pH to 7 with HCl <!>. Add water to a final volume of 1 liter and dispense into aliquots.

STAINING BUFFER (SB)

Reagent	Quantity (for 100 ml)	Final concentration
Tris-Cl (1 M, pH 9.5)	10 ml	100 mM
NaCl (100 mM)	100 µl	0.1 mM
Tween 20	100 µl	0.1%

Prepare fresh for each use from stock solutions.

STAINING BUFFER WITH MgCl$_2$ (SB + MgCl$_2$)

Reagent	Quantity (for 100 ml)	Final concentration
Tris-Cl (1 M, pH 9.5)	10 ml	100 mM
NaCl (100 mM)	100 µl	0.1 mM
Tween 20	100 µl	0.1%
1 M MgCl$_2$ <!>	5 ml	50 mM

Prepare fresh for each use from stock solutions.

STAINING SOLUTION (SS)

Reagent	Quantity	Final concentration
Staining buffer <R>	1 ml	1x
NBT/BCIP Stock Solution (Roche)	10 µl	

Prepare fresh and protect from the light.

WASH BUFFER EMBRYOS (WBE)

Reagent	Quantity (for 100 ml)	Final concentration
1x modified PBS <R>	100 ml	1x
Triton X-100 <!>	100 µl	0.1%

Store at 4°C.

WASH BUFFER WING DISCS (BBW)

Reagent	Quantity (for 100 ml)	Final concentration
Tris-Cl (1 M, pH 6.8)	5 ml	50 mM
NaCl (1 M)	150 µl	150 mM
IGEPAL	500 µl	0.5%
BSA	100 mg	1 mg/ml

ACKNOWLEDGMENTS

We are indebted to all who have worked with *B. anynana* from research staff and students to laboratory technicians and maize growers. Torben Larsen was the initial catalyst to begin work with the genus, and Vernon French supplied the expertise in developmental biology that was crucial to shaping the evo-devo work (following a chance teatime conversation with Linda Partridge). Cornell Dudley and John Wilson were most helpful in Malawi. Sean Carroll, Antónia Monteiro, and Tony Long have been especially influential in the modern genomics era, and Bernd Koch and Klaus Fischer in the areas of physiology and life histories. We thank all of the funding bodies for their contributions, especially the Human Frontiers Science Program, NWO, Leiden University, NSF, and JGI. We thank Suzanne Saenko and Martin Brittijn for the photos and arwork and Nicolien Pul for the organization of all protocols. We are also grateful to the people of Malawi for the early support of their research authorities and for providing access to their exciting biodiversity.

REFERENCES

Allen, C.E., Beldade, P., Zwaan B., and Brakefield, P.M. 2008. Development explains differences in evolvability of serially repeated color pattern elements on butterfly wings. *BMC Evol. Biol.* **8:** 94

Armitage, S.A.O. and Siva-Jothy, M.T. 2005. Immune function responds to selection for cuticular colour in *Tenebrio molitor*. *Heredity* **95:** 650–656.

Beldade, P. and Brakefield, P.M. 2002. The genetics and evo-devo of butterfly wing patterns. *Nat. Rev. Genet.* **3:** 442–452.

Beldade, P. and Brakefield, P.M. 2003. Concerted evolution and developmental integration in modular butterfly wing patterns. *Evol. Dev.* **5:** 169–179.

Beldade, P., Brakefield, P.M., and Long, A.D. 2002a. Contribution of *Distal-less* to quantitative variation in butterfly eyespots. *Nature* **415:** 315–318.

Beldade, P., Koops, K., and Brakefield, P.M. 2002b. Developmental constraints versus flexibility in morphological evolution. *Nature* **416:** 844–847.

Beldade, P., Rudd, S., Gruber, J.D., and Long, A.D. 2006. A wing expressed sequence tag resource for *Bicyclus anynana* butterflies, an evo-devo model. *BMC Genomics* **7:** 130.

Beldade, P., McMillan, W.O., and Papanicolaou, A. 2008. Butterfly genomics eclosing. *Heredity* **100:** 150–157.

Brakefield, P.M. and Frankino, W.A. 2008. Polyphenisms in Lepidoptera: Multidisciplinary approaches to studies of evolution. In *Phenotypic plasticity of insects: Mechanisms and consequences* (ed. D.W. Whitman and T.N. Ananthakrishnan), pp. 121–152. Science, Plymouth, United Kingdom.

Brakefield, P.M. and French, V. 1995. Eyespot development on butterfly wings: The epidermal response to damage. *Dev. Biol.* **168:** 98–111.

Brakefield, P.M. and Larsen, T.B. 1984. The evolutionary significance of dry and wet season forms in some tropical butterflies. *Biol. J. Linn. Soc.* **22:** 1–12.

Brakefield, P.M., Gates, J., Keys, D., Kesbeke, F., Wijngaarden, P.J., Monteiro, A., French V., and Carroll, S.B. 1996. Development, plasticity and evolution of butterfly eyespot patterns. *Nature* **384:** 236–242.

Brakefield, P.M., Kesbeke, F., and Koch, P.B. 1998. The regulation of phenotypic plasticity of eyespots in the butterfly *Bicyclus anynana*. *Am. Nat.* **152:** 853–860.

Brakefield, P.M., El Jilali, E., van der Laan, R., Breuker, C., Saccheri, I., and Zwaan, B.J. 2001. Effective population size, reproductive success and sperm precedence in the butterfly, *Bicyclus anynana*, in captivity. *J. Evol. Biol.* **14:** 148–156.

Breuker, C.J. and Brakefield, P.M. 2003. Lack of response to selection for lower fluctuating asymmetry of mutant eyespots in the butterfly *Bicyclus anynana*. *Heredity* **91:** 17–27.

Brunetti, C.R., Selegue, J.E., Monteiro, A., French, V., Brakefield, P.M., and Carroll, S.B. 2001. The generation and diversification of butterfly eyespot color patterns. *Curr. Biol.* **11:** 1578–1585.

Butler, A.G. 1879. *Mycalesis anynana*. *Ann. Mag. Nat. Hist.* **5:** 187.

Condamin, M. 1973. Monographie du genre *Bicylus* (Lepidoptera: Satyridae). *Mem. Inst. Fond. Afr. Noire* **88:** 1–324.

Dilão, R. and Sainhas, J. 2004. Modelling butterfly wing eyespot patterns. *Proc. Biol. Sci.* **271:** 1565–1569.

Fischer, K., Zwaan, B.J., and Brakefield, P.M. 2007. Realized correlated responses to artificial selection on pre-adult life-history traits in a butterfly. *Heredity* **98:** 157–164.

Frankino, W.A., Zwaan, B.J., Stern, D.L., and Brakefield, P.M. 2005. Natural selection and developmental constraints in the evolution of allometries. *Science* **307:** 718–720.

Frankino, W.A., Zwaan, B.J., Stern, D.L., and Brakefield, P.M. 2007. Internal and external constraints in the evolution of morphological allometries in a butterfly. *Evolution* **61:** 2958–2970.

French, V. and Brakefield, P.M. 1995. Eyespot development on butterfly wings: The focal signal. *Dev. Biol.* **168:** 112–123.

Joron, M. and Brakefield, P.M. 2003. Captivity masks inbreeding effects on male mating success in butterflies. *Nature* **424:** 191–194.

Kirby, W.F. 1871. *A synonymic catalogue of the diurnal Lepidoptera.* Van Vorst, London.

Koch, P.B., Brakefield, P.M., and F. Kesbeke. 1996. Ecdysteroids control eyespot size and wing color pattern in the polyphenic butterfly *Bicyclus anynana* (Lepidoptera: Satyridae). *J. Insect. Physiol.* **42:** 223–230.

Kooi, R.E. and Brakefield, P.M. 1999. The critical period for wing pattern induction in the polyphenic tropical butterfly *Bicyclus anynana* (Satyrinae). *J. Insect Physiol.* **45:** 201–212.

Lyytinen, A., Brakefield, P.M., Lindstrom, L., and Mappes, J. 2004. Does predation maintain eyespot plasticity in *Bicyclus anynana*? *Proc. R. Soc. Lond. B Biol. Sci.* **271:** 279–283.

Marcus, J.M. and Evans, T.M. 2008. A simulation study of mutations in the genetic regulatory hierarchy for butterfly eyespot focus determination. *Biosystems* **93:** 250–255.

Marcus, J.M., Ramos, D.M., and Monteiro, A. 2004. Germline transformation of the butterfly *Bicyclus anynana*. *Proc. R. Soc. Lond. B Biol. Sci.* (suppl. 5) **271:** S263–S265.

McMillan, W.O., Monteiro, A., and Kapan, D.D. 2002. Development and evolution on the wing. *Trends Ecol. Evol.* **17:** 125–133.

Molleman, F., Kop, A., Brakefield, P.M., De Vries, P.J., and Zwaan, B.J. 2006. Vertical and temporal patterns of biodiversity of fruit-feeding butterflies in a tropical forest in Uganda. *Biodivers. Conserv.* **15:** 107–121.

Monteiro, A. and Pierce, N.E. 2001. Phylogeny of *Bicyclus* (Lepidoptera: Nymphalidae) inferred from COI, COII, and EF-1α gene sequences. *Mol. Phylogenet. Evol.* **18:** 264–281.

Monteiro, A., Prijs, J., Bax, M., Hakkaart, T., and Brakefield, P.M. 2003. Mutants highlight the modular control of butterfly eyespot patterns. *Evol. Dev.* **5:** 180–187.

Monteiro, A., Chen, B., Scott, L.C., Vedder, L., Prijs, H.J., Belicha-Villanueva, A., and Brakefield, P.M. 2007. The combined effect of two mutations that alter serially homologous color pattern elements on the fore and hindwings of a butterfly. *BMC Genet.* **8:** 22.

Nieberding, C.M., de Vos, H., Schneider, M.V., Lassance, J.-M., Estramil, N., Andersson, J., Bång, J., Hedenström, E., Löfstedt, C., and Brakefield, P.M. 2008. The male sex pheromone of the butterfly *Bicyclus anynana*: Towards an evolutionary analysis. *PLoS ONE* **3:** e2751.

Nijhout, H.F. 1991. *The development and evolution of butterfly wing patterns.* Smithsonian Institution Press, Washington, D.C.

Peña, C. and Wahlberg, N. 2008. Prehistorical climate change increased diversification of a group of butterflies. *Biol. Lett.* **4:** 274–278.

Pijpe, J., Brakefield, P.M., and Zwaan, B.J. 2008. Increased lifespan in a polyphenic butterfly artificially selected for starvation resistance. *Am. Nat.* **171:** 81–90.

Ramos, D.M., Kamal, F., Wimmer, E.A., Cartwright, A.N., and Monteiro, A. 2006. Temporal and spatial control of transgene expression using laser induction of the hsp70 promoter. *BMC Dev. Biol.* **6:** 55.

Reed, R.D. and Serfas, M.S. 2004. Butterfly wing pattern evolution is associated with changes in a Notch/Distal-less temporal pattern formation process. *Curr. Biol.* **14:** 1159–1166.

Robertson, K.A. and Monteiro, A. 2005. Female *Bicyclus anynana* butterflies choose males on the basis of their dorsal UV-reflective eyespot pupils. *Proc. R. Soc. B Biol. Sci.* **272:** 1541–1546.

Roskam, J.C. and Brakefield, P.M. 1996. A comparison of temperature-induced polyphenism in African *Bicyclus* butterflies from a seasonal savannah-rainforest ecotone. *Evolution* **50:** 2360–2372.

Saccheri, I.J., Lloyd, H.D., Helyar, S.J., and Brakefield, P.M. 2005. Inbreeding uncovers fundamental differences in the genetic load affecting male and female fertility in a butterfly. *Proc. R. Soc. Lond. B Biol. Sci.* **272:** 39–46.

Saenko, S.V., French, V., Brakefield, P.M., and Beldade, P. 2008. Conserved developmental processes and the formation of evolutionary novelties: Examples from butterfly wings. *Philos. Trans. R. Soc. Lond. B Biol. Sci.* **363:** 1549–1555.

Van't Hof, A.E., Zwaan, B.J., Saccheri, I.J., Daly, D., Bot, A.N.M., and Brakefield, P.M. 2005. Characterization of 28 microsatellite loci for the butterfly *Bicyclus anynana*. *Mol. Ecol. Notes* **5:** 169–172.

Westerlund, S.A., and Hoffmann, K.H. 2004. Rapid quantification of juvenile hormones and their metabolites in insect haemolymph by liquid chromatography–mass spectrometry (LC-MS). *Anal. Bioanal. Chem.* **379:** 540–543.

Wijngaarden, P.J. and Brakefield, P.M. 2000. The genetic basis of eyespot size in the butterfly *Bicyclus anynana*: An analysis of line crosses. *Heredity* **85:** 471–479.

Windig, J.J., Brakefield, P.M., Reitsma, N., and Wilson, J.G.M. 1994. Seasonal polyphenism in the wild: Survey of wing patterns in five species of *Bicyclus* butterflies in Malawi. *Ecol. Entomol.* **19:** 285–298.

Zijlstra, W.G., Steigenga, M.J., Koch, P.B., Zwaan, B.J., and Brakefield, P.M. 2004. Butterfly selected lines explore the hormonal basis of interactions between life histories and morphology. *Am. Nat.* **163:** E76–E87.

13 The Two-spotted Cricket *Gryllus bimaculatus*
An Emerging Model for Developmental and Regeneration Studies

Taro Mito and Sumihare Noji

Department of Life Systems, Institute of Technology and Science, The University of Tokushima, 2-1 Minami-Jyosanjima-cho, Tokushima 770-8506, Japan

ABSTRACT

The two-spotted cricket *Gryllus bimaculatus* De Geer (Orthoptera: Gryllidae) is one of the most abundant cricket species, and it inhabits the tropical and subtropical regions of Asia, Africa, and Europe. *G. bimaculatus* can be easily bred in the laboratory and has been widely used to study insect physiology and neurobiology. Recently, this species has become established as a model animal for studies on molecular mechanisms of development and regeneration because its mode of development is more typical of arthropods than that of *Drosophila melanogaster*, and the cricket is probably ancestral for this phylum. Moreover, the cricket is a hemimetabolous insect, in which nymphs possess functional legs with a remarkable capacity for regeneration after damage. Because RNA interference (RNAi) works effectively in this species, the elucidation of mechanisms of development and regeneration has been expedited through loss-of-function analyses of genes. Furthermore, because RNAi-based techniques for analyzing gene functions can be combined with assay systems in other research areas (such as behavioral analyses), *G. bimaculatus* is expected to become a model organism in various fields of biology. This offers the possibility of establishing the cricket as a simple model system for exploring more complex organisms such as humans.

BACKGROUND INFORMATION

The two-spotted cricket *G. bimaculatus* is a hemimetabolous insect that is abundant in tropical and subtropical regions of Asia, Africa, and Europe. Its adult body length is about 30 mm. The chromosome number is $2n = 28 + XX$ (female)/XO (male) (Yoshimura et al. 2006) and the genome size is estimated to be a few gigabases (T. Mito and S. Noji, unpubl.). The cricket is a popular food source for insectivorous animals such as spiders and reptiles, and thus, *G. bimaculatus* is readily available from pet stores. *Gryllus* can also be easily bred in the laboratory with general insect food, and, unlike some other cricket species, it does not require prolonged exposure to cold temperatures in order to complete its life cycle. With such advantages as an experimental animal, this species has been widely used to study insect physiology and neurobiology (see, e.g., Engel and Hoy 1999; Paydar et al. 1999; Wenzel and Hedwig 1999; Hedwig and Poulet 2004).

Recently, *G. bimaculatus* has become established as a model animal for studies on molecular mechanisms of development and regeneration. Since the early 1990s, Sumihare Noji's group at the University of Tokushima in Japan has used *G. bimaculatus* to investigate mechanisms of development and regeneration. There are several reasons why this species is an emerging model organism. First, although developmental mechanisms have been extensively studied in the fruit fly *Drosophila melanogaster*, this organism exhibits an evolutionarily derived mode of development and the underlying molecular mechanisms may be unrepresentative for arthropods, even for insects. On the other hand, the mode of development in *Gryllus* is more general and is probably ancestral for arthropods. Thus, *Gryllus* is a promising model system for comparing developmental mechanisms with those of *Drosophila*. Second, in hemimetabolous insects such as cockroaches and crickets, nymphs possess functional legs with a remarkable capacity for regeneration after damage. Although the phenomenon of insect leg regeneration in cockroach nymphs is well studied, the underlying molecular mechanisms remain unclear.

In the course of studies on development and regeneration, Noji and coworkers found that RNAi works effectively in this species, and they developed a number of RNAi-based techniques for analyzing gene functions (Miyawaki et al. 2004; Mito et al. 2005; Nakamura et al. 2008b). As a consequence, there have been remarkable advances in this field (Mito and Noji 2006; Nakamura et al. 2008b). Furthermore, because these RNAi techniques can be combined with assay systems in other research areas, it is expected that *G. bimaculatus* will become more popular in various fields of biology. Recent studies with this organism have considerably extended our understanding of circadian rhythms as well as of learning and memory (Moriyama et al. 2008; T. Takahashi et al., unpubl.).

SOURCES AND HUSBANDRY

G. bimaculatus can be obtained inexpensively at pet stores. A mutant strain with white eyes (autosomal recessive; *gwhite*) was isolated by Isao Nakatani and coworkers at Yamagata University in 1989. At present, this strain is continuously maintained in S. Noji's laboratory at the University of Tokushima. The mutant cricket embryos are often preferentially used for whole-mount in situ hybridization or immunohistochemistry because their bodies are more transparent than those of the wild type during late embryogenesis, which results in clearer staining results.

Nymphs and adults of *G. bimaculatus* can be easily reared in plastic cases at 26–30°C under a 10-hour light/14-hour dark photoperiod. They are usually fed on artificial diets for insects or artificial fish foods. Fertilized eggs can be collected with a wet kitchen towel and are incubated in plastic dishes at 28°C and 70% humidity. Under these incubation conditions, most of the eggs hatch on the 13th day. The total developmental span of *G. bimaculatus* embryos has been divided into 16 stages based on the morphological features of the developing embryos and their appendages, and these stages are summarized in Table 1. The nymphs molt eight times before becoming adults. Adults lay a maximum number of eggs 1 week after the final molt. The generation time of the cricket is about 50 days.

RELATED SPECIES

For a comparative view of developmental studies, the following short- and intermediate-germ insects other than crickets are also studied: the beetle *Tribolium castaneum* (Coleoptera), the milkweed bug *Oncopeltus fasciatus* (Hemiptera), and the grasshoppers *Schistocerca americana*, *S. gregaria*, and *Locusta migratoria* (Orthoptera). In *T. castaneum*, *O. fasciatus*, and *L. migratoria*, parental and/or embryonic RNAi techniques have proved to be successful (see, e.g., Bucher et al. 2002; Liu and Kaufman 2004; He et al. 2006). *T. castaneum* is an established model system whose genome has been sequenced. It is a holometabolous insect and therefore more evolutionarily derived (or "advanced") than hemimetabolous insects such as *G. bimaculatus*, *O. fasciatus*, and the grasshoppers.

Nymphal or larval RNAi techniques have been shown to be applicable in *S. americana* (Dong and Friedrich 2005), *T. castaneum* (Tomoyasu and Denell 2004), and the cockroach *Blattella germanica* (Marie et al. 2000). However, regenerative RNAi has not been reported in insect species other than *G. bimaculatus*.

TABLE 1. Developmental stages for *Gryllus bimaculatus* embryos

Stage	Day	hAEL	Characteristics
1.0	~1		Freshly laid eggs are light yellow in color and 2.88 ± 0.12 mm long. The anterior pole of the egg is slightly pointed and the posterior is blunt. The concave and convex sides are dorsal and ventral, respectively. The yolk is finely granular.
2.0	1	24	A uniform array of nuclei is visible on the surface of the yolk.
2.3		27	The bipartite germ anlage is formed.
2.4		28	Merging of the bipartite germ anlage from the posterior side begins.
2.6		30	The unitary germ anlage is formed.
3.0		30–36	The germ anlage contracts in a left-right direction.
4.0	1.5	36	The germ band (from the head to the T2 segment) is formed.
4.3		39	The T3 segment is formed.
4.4		40	Elongation of the germ band to form the abdominal segments begins, as does *wg* expression in the A1 segmen; A/E = 21%
4.9		45	*wg* expression in the A2 segment begins; A/E = 28%
5.0		46	*wg* expression in the A3 segment begins; A/E = 30%
5.2	2	48	*wg* expression in the A4 segment begins; A/E = 40%
6.0	2.5	60	Formation of the limb bud begins; A/E = 42%
6.5		66	A/E = 44%
7.0	3	72	*wg* expression in the A8/9 segments begins; A/E = 45%. The limb bud is formed without any segmentation boundary.
8.0	4	93	The boundary between the trochanter and femur is formed.
9.0	5	120	The boundary between the femur and tibia is formed.
10.0	5.5	132	The embryo undergoes katatrepsis. The tibial segment of the limb bud elongates.
11.0	6	144	After completion of katatrepsis, dorsal closure begins. The boundary between the tibia and tarsus is formed.
12.0	8		The femoral, tibial, and tarsal segments of the metathoracic leg are clearly distinguishable. At the distal end of the tibia, two pairs of spur primordia are generated. The tarsus is divided into three segments and its distal end possesses the claw primordium.
13.0	9		Dorsal closure is completed. The embryo attains its full length and fills the entire egg. Cuticle secreted by the embryo covers the entire body. Pigmentation in the compound eye is visible. The legs begin to twitch. Rapid growth of the leg segments starts at this stage.
14.0	10		Each leg is fully grown. The tibial spurs and tarsal claws become sharp.
15.0	11		The embryo is light yellow in color. Regularly arranged black bristles appear on the leg and the posterior ridge of each tergum.
16.0	12–13		Most of the eggs are hatched.

(hAEL) Hours after egg-laying at 28°C. A/E = 100 × (length of the abdominal region)/(full length of the embryo).

In orthopteran insects other than *G. bimaculatus*, large-scale genomic resources have been developed in only two species. An expressed sequence tag (EST) analysis from the whole body and dissected organs of *L. migratoria* has produced 12,161 unique sequences (Kang et al. 2004). More recently, an EST resource containing 8607 unique sequences has been established using a nerve cord cDNA library of the Hawaiian trigonidiine cricket *Laupala kohalensis* (Danley et al. 2007).

USES OF THE *G. BIMACULATUS* MODEL SYSTEM

Development

Description of Embryogenesis

Several studies have provided a morphological description of *G. bimaculatus* embryogenesis (Niwa et al. 1997; Sarashina et al. 2003, 2005; Miyawaki et al. 2004; Zhang et al. 2005). The *Gryllus* egg is approximately 0.5–0.6 mm in diameter and 2.88 ± 0.12 mm long (Fig. 1A). After fertiliza-

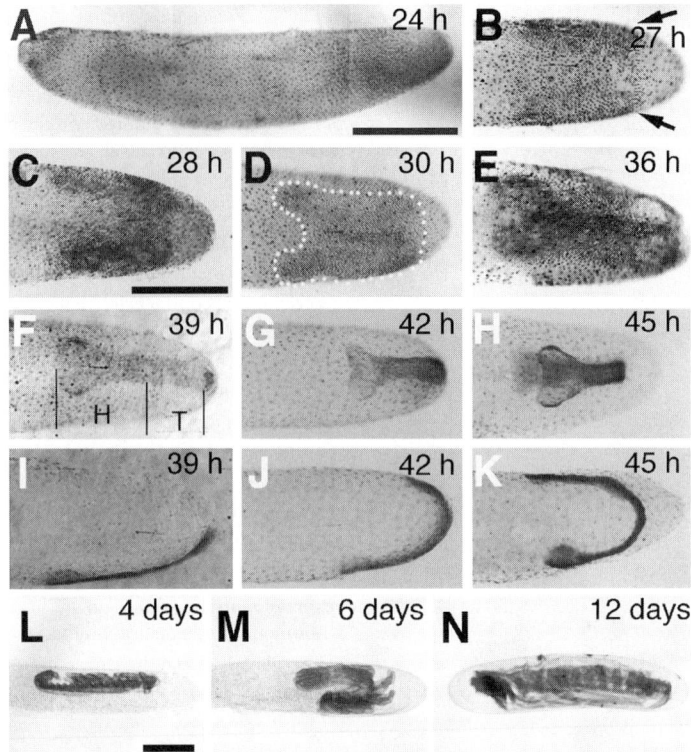

FIGURE 1. Formation of the *G. bimaculatus* embryo in eggs as revealed by distributions of nuclei stained with fuchsin at 24–45 hours after egg-laying (hAEL) (*A–K*). (*A* and *I–K*) Lateral views; (*B–H*) ventral views of the posterior regions of eggs. Arrows indicate prospective germ bands. (*L–N*) Lateral views of the embryo at 4, 6, and 12 days AEL, respectively. In all photographs, the anterior of the egg is to the left. Bars, (*A, L–N*) 0.7 mm; (*B–K*) 0.5 mm.

tion, male and female pronuclei fuse near the center of the egg (Sato and Tanaka-Sato 2002). Mitoses take place within the yolk about 3 hours after egg-laying (hAEL); then, a uniform array of nuclei appear on the surface of the yolk. At 27 hAEL, the number of nuclei per unit area increases in the posterolateral region (bipartite germ anlage as indicated by the two arrows in Fig. 1B), and a large patch of closely adjacent nuclei is observed on each egg flank. At 28 hAEL, the bipartite germ anlage starts to merge from the posterior side (Fig. 1C), and at 30 hAEL, the unitary germ anlage is observed in the posteroventral region of the egg (Fig. 1D). The germ anlage then starts to contract in a left-right direction (Fig. 1E). The posterior region of the delineated germ anlage, the germ band, elongates slowly to form the prospective abdomen (Fig. 1F–K). Until 96 hAEL, the abdomen of the germ band is fully segmented (Fig. 1L). At 5.5–6 days AEL, the embryo undergoes katatrepsis, i.e., the embryo reorientates by moving head first around the posterior pole from the dorsal to ventral side of the egg (Fig. 1M), and then dorsal closure begins. At 10–11 days AEL, embryogenesis is complete (Fig. 1N).

Developmental stages are defined on the basis of morphological features as shown in Table 1 and Figure 2. Anterior segmentation occurs almost simultaneously by stage 4.0, at least to the level of the segment polarity genes, because *Gryllus wingless* (*Gb'wg*) is expressed in five vertical stripes corresponding to the mandibular through second thoracic segments (Miyawaki et al. 2004). The remaining posterior segments are then sequentially produced through germ-band elongation from the posterior growth zone. The specification of the posterior segments can be tracked by the appearance of *Gb'wg* stripes, which appear one by one in the third thoracic segment at stage 4.3 and then in abdominal segment 1 at stage 4.4. At stage 7.5, the posteriormost stripe appears in abdominal segment 10.

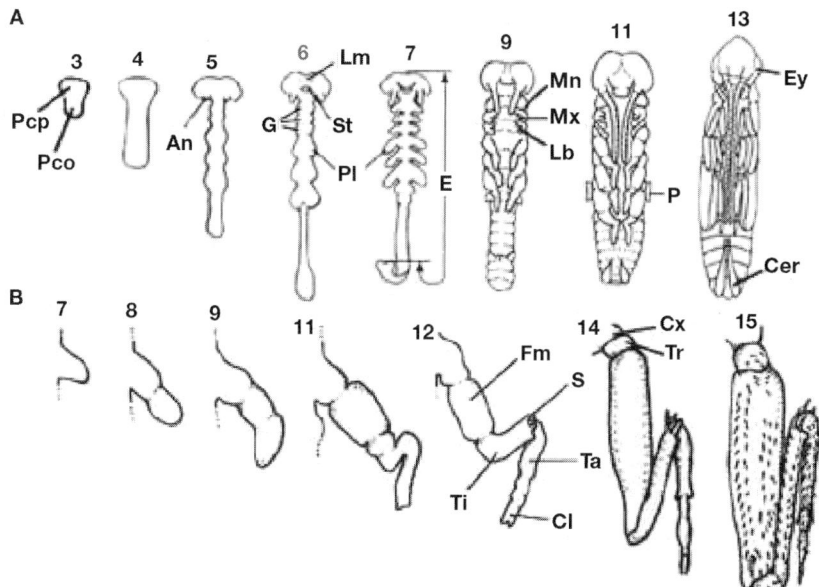

FIGURE 2. (A) Series of illustrations showing G. bimaculatus embryogenesis (ventral view). The numbers indicate developmental stages. In a hemimetabolous insect such as the cricket, the adult leg develops from the limb bud formed in the embryo. Gryllus is a short-germ insect: The protocephalon (Pcp) and protocorm (Pco) of presumptive anterior segments comprise most of the embryonic rudiment, which lengthens as posterior segments are added during development. (B) Illustrations depicting metathoracic leg formation. (An) Antenna; (Cer) cercus; (Cl) claw; (Cx) coxa; (E) full length of embryo; (Ey) eye; (Fm) femur; (G) gnathal segments; (Lb) labrum; (Lm) labium; (Mn) mandible; (Mx) maxilla; (P) pleuropodium; (Pl) prothoracic leg; (S) tibial spur; (St) stomodium; (Ta) tarsus; (Ti) tibia; (Tr) trochanter.

Segment Patterning

G. bimaculatus is a model system that is amenable to functional analyses of genes via RNAi, by introducing double-stranded RNA (dsRNA) into either an egg (embryonic RNAi) or an individual. Application of parental RNAi, a phenomenon in which female adult crickets injected with dsRNA produce progeny showing an RNAi knockdown phenotype, has drastically improved the efficiency of functional analysis of *Gryllus* genes involved in developmental processes, in particular those involved in the segmentation process.

Although segmentation mechanisms have been studied extensively in *D. melanogaster*, this organism exhibits an evolutionarily derived mode of development, and the molecular mechanisms underlying *Drosophila* segmentation may not be representative for arthropods, even for insects. In *Drosophila*, which is a long-germ insect, simultaneous segment patterning takes place through the genetic cascade of maternal, gap, pair-rule, and segment polarity genes. Early steps of segment patterning progress in a syncytial mode, so that products of the maternal and gap genes can diffuse in the egg. Bicoid (Bcd) has an especially important role in early steps of the anterior-posterior (AP) pattern formation via its morphogenetic gradient.

Because of the ancestral relationship between *G. bimaculatus* and other arthropods, Noji and coworkers are using the developmental system of the cricket as a model for studying the general segmentation mechanisms for insects other than *Drosophila*. The *bcd* gene is thought to be absent in *G. bimaculatus*, and in *Gryllus* eggs, the germ band is formed in the posterior region. The germ band elongates as development proceeds. Anterior segmentation occurs almost simultaneously, whereas posterior segmentation takes place sequentially during elongation of the posterior portion of the germ band. In contrast to *Drosophila*, most of segments are specified in a cellular mode.

How is the AP axis determined in *Gryllus*, which is thought not to possess a *bcd* gene? To address this question, Shinmyo et al. (2005) focused on the *caudal* (*cad*) gene, which is involved in the formation of abdominal structures in *Drosophila*, by analyzing the function of its ortholog in *Gryllus* using RNAi. It was determined that reduction of *Gryllus cad* (*Gb'cad*) expression by RNAi results in deletion of the gnathum, thorax, and abdomen in embryos, leaving only the anterior head, suggesting that, in *Gryllus*, *cad* is required for formation of a surprisingly larger part of the body than in *Drosophila* (Shinmyo et al. 2005, 2006). Such *cad* function may be conserved in short- and intermediate-germ insects, because a similar *cad* phenotype was reported in the short-germ beetle *T. castaneum* (Copf et al. 2004). For the formation of the head region, *Gryllus ortho-denticle* (*otd*) also appears to be essential (T. Mito et al., unpubl.).

In studies of the roles of the gap gene orthologs *Gryllus hb* (*Gb'hb*) and *Gryllus Kr* (*Gb'Kr*), it was noted that RNAi knockdown of *Gb'hb* results in a gap-like phenotype that lacks gnathal and thoracic segments. Detailed analyses of the *Gb'hb* phenotype revealed that it results from transformation of both gnathal and thoracic regions into abdominal identity and reduction of the number of abdominal segments (Mito et al. 2005). This differs from the typical gap phenotypes caused by *hb* depletion in *Drosophlia* and *Tribolium*, which exhibit deletion of contiguous segments in the gnathal and thoracic regions (Lehmann and Nüsslein-Volhard 1987; Schröder 2003), but resemble the phenotype of the intermediate-germ insect *O. fasciatus* (Liu and Kaufman 2004). On the other hand, *Gb'Kr* apparently acts as a gap gene in a manner similar to *Drosophila*, because *Gb'Kr* RNAi results in several contiguous segments in the thorax and abdomen (Mito et al. 2006). These results suggest that the gap patterning system, a system of patterning several contiguous segments, operates in *Gryllus*. However, it is unclear how the *Gryllus* gap genes act in the cellular environment, where transcription factors cannot diffuse to produce morphogenetic gradients as they do in *Drosophila*. *Gryllus extradenticle* may function upstream of the gap genes *Gb'hb* and *Gb'Kr* (Mito et al. 2008).

Conservation in *Gryllus* has been examined at the level of pair-rule patterning, which is downstream from gap patterning in the *Drosophila* segmentation hierarchy. Using RNAi for the pair-rule gene ortholog *Gryllus even-skipped* (*Gb'eve*) resulted in segment fusion in alternative segmental units in the anterior region of the embryo (Mito et al. 2007), which suggests that *Gb'eve* acts as a pair-rule gene at least during anterior segmentation.

In later stages of *Gryllus* segmentation, abdominal segments are sequentially determined as the posterior region of the embryo elongates. Using RNAi, Miyawaki et al. (2004) demonstrated that the Wingless (Wg)/Armadillo (Arm)-signaling pathway (a canonical Wnt-signaling pathway) is required for posterior segmentation. Because a Wnt-signaling pathway is involved also in vertebrate segmentation, a common mechanism for sequential segmentation involving Wnt/Wg could exist in *Gryllus* and in vertebrates. In addition, *Gb'eve* may also be involved in posterior elongation because its RNAi depletion results in severe shortening of the abdomen (Mito et al. 2007).

On the basis of the results of our RNAi analyses, we have proposed a model for a regulatory network among segmentation genes in *Gryllus*, which is shown in Figure 3. Our present knowledge of the *Gryllus* segmentation process can be summarized as follows. First, the AP axis is determined by *Gb'cad* and probably *Gb'otd* without the Bcd system. Next, in the process of anterior segmentation, gap and pair-rule patterning occur despite a cellularized environment. The remaining posterior segments are then formed sequentially under the control of the Wg/Arm-signaling pathway and *Gb'eve*.

Exploring Development in Hemimetabolous Insects

Insects can be grouped into two main categories, holometabolous and hemimetabolous, according to the extent of their morphological change during metamorphosis. In holometabolous insects, the larvae are usually quite unlike the adult and a pupal stage is present between the last larval stage and the adult. In contrast, in hemimetabolous insects, the larva hatches in a form that generally resembles the adult, except for its small size and absence of wings and genitalia. In holometabolous insects, new adult organs develop from the imaginal discs, whereas in hemimetabolous insects, the legs, eyes, and most of the other adult structures are formed by bud-

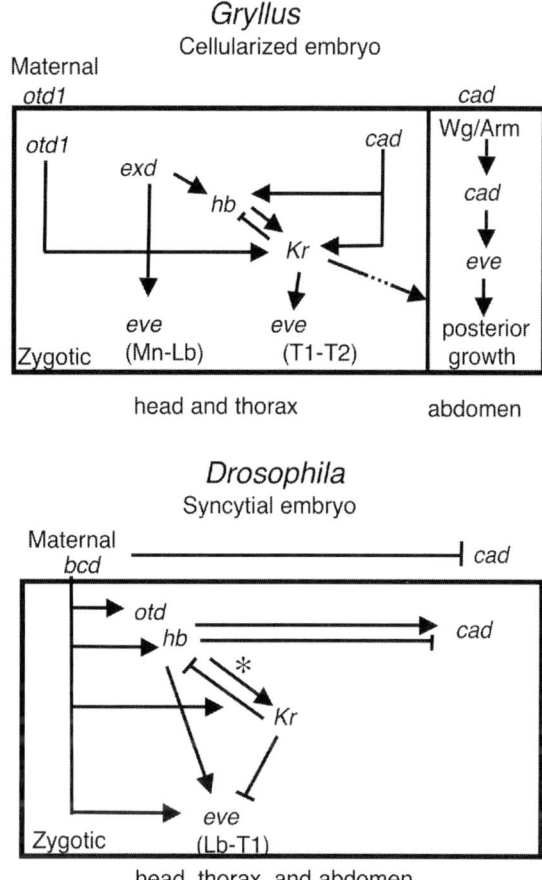

FIGURE 3. Models for regulatory networks of segmentation genes in *Gryllus* and *Drosophila*. Relationships between genes do not necessarily show direct regulation. (*Top*) In *Gryllus*, *caudal* (*cad*) and *orthodenticle* (*otd*) organize patterning in gnathal and thoracic regions by activating the gap genes such as *hunchback* (*hb*) and/or *Krüppel* (*Kr*). *Gb'Kr* is activated by *Gb'cad* and *Gb'hb*, although it remains unclear whether *Gb'cad* activates *Gb'Kr* independently of *Gb'hb* or only through *Gb'hb* function. *Gb'Kr* represses *Gb'hb* expression in the thoracic region. *Gb'Kr* is required for *Gb'even-skipped* (*eve*) thoracic stripes that correspond to T1 and T2 segments. *Gb'Kr* is also involved in segmentation in the posterior region, possibly via expression of other gap genes. *Gb'extradenticle* (*exd*) may directly or indirectly regulate the expression of *Gb'hb* and *Gb'eve* in the gnathal region and *Gb'Kr* expression in the thoracic region (*dotted arrow*), although further analyses are needed to clarify precise regulatory relationships among these genes. (*Bottom*) In *Drosophila*, gap genes such as *hb* and *Kr* are activated by *bicoid* (*bcd*) in the anterior region of the embryo. *bcd* and *hb* activate *eve* (the pair-rule stripe 2, Lb-T1), whereas *Kr* represses *eve*. *Kr* is activated by *bcd* and *hb*. *hb* activates *Kr* at low levels and represses it at high levels (*asterisk*). *hb* also acts as a primary regulator of the abdominal expression of *cad*.

ding and are already present at hatching. How the molecular mechanisms governing these modes of development differ from one another is thus an intriguing question.

Leg development. In contrast to the *Drosophila* leg, which develops from a specific imaginal disc, the cricket leg (like a vertebrate limb) develops from a limb bud (Fig. 2B) in the absence of any leg imaginal disc. Nymphal legs then develop, looking like small adult legs. As discussed later, although the leg developmental processes differ considerably in cricket and *Drosophila*, the signaling molecules involved in development and regeneration are essentially similar. This suggests that the genetic information gathered about leg development in *Drosophila* can be used as a starting point to understand the molecular mechanisms controlling leg development in hemimetabolous insects.

On the basis of experimental data from *Drosophila*, Meinhardt (1983) proposed the boundary model to explain the formation of the proximodistal (P/D) axis of the leg. In this model, the leg is divided into three compartments: posterior, dorsal/anterior, and ventral/anterior. The site where the boundaries between these compartments intersect defines the presumptive distal tip of the leg and is the source of a diffusible morphogen that induces outgrowth and specifies cell fate along the P/D axis. Ten years later, three reports demonstrated that *Drosophila* limbs are indeed subdivided into three domains that derive from adjacent cell populations founded early in development (Campbell et al. 1993; Diaz-Benjumea et al. 1994; Basler and Struhl 1994). Posterior cells organize growth and cell patterning in the three domains by secreting the Hedgehog (Hh) protein, which acts indirectly by inducing neighboring anterior cells to secrete Decapentaplegic (Dpp) or Wingless (Wg) proteins at the dorsal AP and the ventral AP boundaries, respectively (Campbell et al. 1993: Basler and Struhl 1994; Diaz-Benjumea et al. 1994). A P/D axis is initiated at the site where *dpp*-expressing cells come into close association with those expressing *wg* (Campbell et al. 1993). The distal region is patterned by a distal-to-proximal gradient of epidermal growth factor receptor (EGFR) tyrosine kinase activity (Campbell 2002), and this graded activity is established according to the concentration of EGF-related ligands. The source of these ligands is located at the presumptive tip (the most distal region that requires the highest level of EGFR activity), whereas more proximal regions require progressively less EGF (Campbell 2002). This indicates that EGF corresponds to the predicted morphogen produced by the distal organizer and that the tarsus may be patterned by a distal-to-proximal gradient of EGFR activity (Galindo et al. 2002). These findings thus provided the molecular basis of the boundary model, now termed the molecular boundary model.

We examined whether the molecular boundary model could be applied to developing hemimetabolous insect legs. For this purpose, we observed the expression patterns of the *hh*, *wg*, and *dpp* homologs in *G. bimaculatus* (named *Gb'hh*, *Gb'wg*, and *Gb'dpp*, respectively) and found no striking differences between *Drosophila* and *Gryllus* expression patterns (Niwa et al. 2000). In fact, those genes were all expressed in the *Gryllus* limb bud, even though the *Gb'dpp* transcripts were detected as a spot in the dorsal tip of the limb bud, whereas *dpp* transcripts form a stripe along the dorsal side of the AP boundary in the *Drosophila* leg imaginal disc. The morphogenetic significance of this difference is currently unknown. However, these results allowed us to conclude that the molecular boundary model is a fundamental and universal mode for forming the P/D axis in the insect leg.

Determination of segment identity in the developing hemimetabolous insect leg is the next step in the leg patterning process. Nymphs of hemimetabolous insects have six leg segments, arranged along the P/D axis in the following order: coxa, trochanter, femur, tibia, tarsus, and pretarsus. These leg segments are formed not simultaneously, but sequentially from the presumptive trochanter/femur, femur/tibia, tibia/tarsus boundaries; the initial structure consists of the presumptive coxa/pretarsus (Miyawaki et al. 2002). During formation of the leg segments, a proximal leg segment appears to be intercalated from the proximal side between the most proximal and the most distal structures. Segment identity may be determined by the *homothorax* (*hth*), *dachshund* (*dac*), and *Distal-less* (*Dll*) genes in the cricket limb bud (Inoue et al. 2002), at least as revealed in the fly imaginal disc (Goto and Hayashi 1999; Kojima 2004). Their expression patterns are likely to be regulated by Hh, Wg, and Dpp, which are expressed in the limb bud (Kojima 2004). Because recent findings show that in *Drosophila* legs, the sharp discontinuity of Dpp signaling is required for forming the leg joint, expression patterns of *Gb'dpp* in the cricket leg bud may be also involved in the formation of the leg joint (Manjon et al. 2007). Thereafter, intrasegmental pattern formation, which is characterized by spines and hairs, follows the determination of segment identity. Thus far, we do not know the mechanisms that drive intrasegmental pattern formation, but information about the regulation of the pattern within a segment has been obtained from intercalary regeneration experiments in hemimetabolous insects (Meinhardt 1983).

Eye development. In *Drosophila*, the entire adult head is secondarily formed from imaginal discs that develop during larval and pupal stages. The compound eyes develop from the eye-antennal discs. Patterning of the eye field and adjacent head cuticle occurs with the formation of the retina. On the other hand, in *G. bimaculatus*, all eye components are formed from the eye lobes. Differentiation of the retina is initiated during early embryogenesis. Partitioning of the embryonic head anlagen precedes retina formation. A hatched cricket nymph already has small compound eyes, which continue to grow to form additional rows of ommatidia throughout nymphal stages. It was reported that the expression patterns of Decapentaplegic in retinal differentiation are significantly different between *Drosophila* and the hemimetabolous grasshopper *S. americana* (Friedrich and Benzer 2000). These facts reveal that there are fundamental differences in the timing and spatial context of eye development between fruit fly and cricket.

We have thus been studying expression patterns of *G. bimaculatus* homologs of the *Drosophila* genes required for eye patterning. Dachshund (Dac) is a transcriptional regulator required for eye, brain, and leg development in holometabolous *Drosophila* (Mardon et al. 1994; Martini et al. 2000). During brain development, *Gb'dac* is first expressed in the medial head region, which corresponds to part of the developing protocephalic region, and then in the primordial and adult Kenyon cells. During eye development, *Gb'dac* is first expressed in the lateral head region, which becomes the eye primordium and part of the deutocerebrum. *Gb'dac* is then expressed in the posterior region of the eye primordium before the formation of compound eyes. The expression domain shifts to the anterior domain concomitantly with the movement of morphogenetic furrows. *Gb'dac* is also expressed in the developing optic lobes during differentiation of the retina. These expression patterns were compared with those of *Drosophila dac*. We found that although the developmental processes of the *Gryllus* eye and brain differ from those of *Drosophila*, the expression patterns of *Gb'dac* are essentially similar to those of *Drosophila dac* (Inoue et al. 2004).

Analyses of other eye-patterning gene homologs, such as *eyeless/twin of eyeless* (Pax6 homologs), *sine oculis*, and *eyes absent*, are currently under way. These genes are expressed in the eye-forming region during *Gryllus* embryogenesis, and they therefore appear to be involved in eye development as they are in *Drosophila* (S. Noji et al., unpubl.).

Body Size Regulation

One of the fundamental problems in developmental biology is how organisms grow to be the correct body size with an individual lifespan. Body size is known to be controlled by both genetic and nongenetic (e.g., caloric restriction) cues through the insulin/insulin-like growth factor (I/IGF)-signaling pathway, in concert with short-range-signaling systems such as the EGF system in each organ. However, it is still unclear which signaling components are involved in the nongenetic control of body size.

It has been reported that mutation of an insulin receptor substrate (IRS) termed Chico causes small body size and a long lifespan in both *Drosophila* (Bohni et al. 1999; Clancy et al. 2001) and mouse (White 2002; Taguchi et al. 2007). On the other hand, mutations in a phosphatase suppressor of the I/IGF-signaling pathway PTEN have been reported to increase adult size in both *Drosophila* (Goberdhan et al. 1999; Oldham et al. 2002) and mouse (Nguyen et al. 2006). Noji and coworkers have examined *Gryllus chico*, *Pten*, and *Egf receptor* homologs using the nymphal RNAi technique. When expression of *chico* or *Egf receptor* was knocked down by RNAi at the early nymphal stages of *G. bimaculatus*, the affected crickets displayed small body size. However, in the case of RNAi for *Pten* (a lipid phosphatase gene), no significant effect on body size was observed. These results indicate that *chico* and *Egfr*, but not *Pten*, are involved in control of cricket body size (T. Bando et al., unpubl.).

Leg Regeneration

Nymphs of hemimetabolous insects such as cockroaches and crickets have three pairs of legs each consisting of six segments, which are arranged along the P/D axis in the following order: coxa, trochanter, femur, tibia, tarsus, and pretarsus (Fig. 2B). Damaged legs show a remarkable capacity for regenerating complex structures from a blastema. Because nymphal legs are relatively large, they are easy to manipulate surgically. Numerous grafting experiments have advanced our understanding of the mechanisms of pattern formation and tissue and organ regeneration, including vertebrate systems. Such experiments have shown that each leg segment has a similar set of P/D and circumferential positional values; furthermore, when positional values are missing in amputated or grafted legs, the legs can intercalate the missing values (French 1981; Meinhardt 1983; Campbell and Tomlinson 1995). In this intercalary regeneration, a rule, designated as the shortest intercalation rule, was discovered, which states that the confrontation of cells with different positional values would result in intercalation of intermediate values via the shortest route. Until 1986, the nymphal leg system was used extensively by scientists interested in the mechanisms of appendage regeneration, but very few groups are currently working on this system. This loss of focus was mainly due to the paucity of molecular data and a lack of tools for functional analyses. However, the situation has gradually changed as techniques for analyzing gene expressions and functions have become available.

One interesting phenomenon in leg regeneration is the formation of supernumerary legs, which occurs after grafting the distal part of a leg onto the contralateral leg stump, a procedure that brings into contact at the amputation plane two structures with inverted AP axes (French 1981; Meinhardt 1983; Campbell and Tomlinson 1995). The formation of supernumerary legs in this context means that additional P/D axes were induced upon grafting. In 1976, the polar coordinate model was proposed to interpret the regeneration phenomena (French et al. 1976). The model has two rules. First, the rule of shortest intercalation states that the confrontation of cells with different positional values would result in the intercalation of intermediate values via the shortest route. Second, the full circle rule states that leg outgrowth and ultimately a P/D axis would be generated at the site where there was a full complement of circumferential positional values (Bryant et al. 1981). This model can account for the formation of supernumerary legs following grafting experiments, because a full circumference of positional values will be intercalated at the two positions of maximal circumferential misalignment (French et al. 1976). The model, based on local interactions, provides a simple and unified interpretation for a wide range of experimentally produced and naturally occurring regenerative phenomena, such as limb regeneration in insects, crustaceans, and amphibians; duplication of structures; and formation of complete, tapering, or branching supernumeraries. Meinhardt (1983) interpreted the formation of supernumerary legs with the boundary model. In 1995, the molecular boundary model was proposed to explain the formation of supernumerary legs and may thus provide a molecular basis for the polar coordinate model (Campbell and Tomlinson 1995).

To verify that the molecular boundary model would apply during regeneration, we examined the expression patterns of *Gryllus hedgehog* (*Gb'hh*), *wingless* (*Gb'wg*), and *decapentaplegic* (*Gb'dpp*) in regenerating legs (Mito et al. 2002) and silenced those genes through RNAi. We have shown that these genes are expressed in the regeneration blastema of a leg amputated at the distal tibia. These genes are expressed in a pattern comparable to that of the leg bud of the *Gryllus* embryo and the *Drosophila* leg imaginal disc (Niwa et al. 2000; Mito et al. 2002). This finding suggests that the initiation mechanism for the regenerating leg P/D axis is similar to that for normal leg development. In addition, the formation of a regenerating leg P/D axis is triggered at a site where ventral *wg*-expressing cells abut dorsal *dpp*-expressing cells in the AP boundary, similar to *Drosophila* leg development (Campbell and Tomlinson 1995; Mito et al. 2002).

When dsRNA for a given gene is injected into a cricket nymph, in some cases a phenotype appears after molting and lasts to the adult stage (nymphal RNAi). To examine the effect of RNAi silencing during leg regeneration, the metathoracic leg of third instar nymphs is amputated at the distal tibia, immediately after injection of the dsRNA. In nymphs silenced for *Gb'hh*, regeneration was abnormal, with formation of a supernumerary leg, whereas no phenotype was observed in the absence of amputation (Nakamura et al. 2008a). In nymphs silenced for *Gb'dpp* and *Gb'wg*, no phenotype was found either during development or during regeneration (Nakamura et al. 2008a). However, when expression of *Gb'arm* (the β-catenin ortholog) was knocked down through RNAi, no regeneration was observed, indicating that the canonical Wnt pathway may be involved in initiation of regeneration (Nakamura et al. 2007). During leg regeneration, *Gb'Egfr* is expressed in the blastema, as observed in the limb bud (Nakamura et al. 2008b). In the nymphs exposed to *Gb'Egfr* dsRNAs, the leg was formed normally without amputation, but the distal structures of the tarsus and claw were not normal after tibial amputation (Nakamura et al. 2008b). Furthermore, during formation of the supernumerary legs upon grafting, the genes encoding the three main signaling factors involved in leg development, Hh, Dpp, and Wg, were expressed as predicted from the molecular boundary model (Mito et al. 2002). However, the *Egfr* RNAi nymphs were unable to form the distal structures of the supernumerary legs, indicating that EGFR is indeed involved in the regeneration of distal structures (Nakamura et al. 2008a). This is consistent with the role of EGF signaling during *Drosophila* leg development (Campbell 2002). Thus, these data indicate that the molecular boundary model is the most plausible hypothesis for explaining the various phenomena observed during the formation of the P/D axis of both the developing and regenerating insect leg. Furthermore, the model may be applicable to the regenerating vertebrate limb, because supernumerary legs can be formed by grafting in a manner similar to that described above (Bryant and Iten 1976).

To identify the molecules that may be involved in intercalary regeneration, we investigated *Gb'hh* and *Gb'wg* expression patterns during regeneration (Nakamura et al. 2007). We found *Gb'hh* and *Gb'wg* induced in the host, proximally to the amputation plane, but not in the structure grafted distally to the amputation plane (Nakamura et al. 2007). This directional induction occurs even in the reversed intercalation. Because these results are consistent with a distal-to-proximal respecification of the regenerate, *Gb'wg* may be involved in the reestablishment of the positional values in the regenerate. Furthermore, we found that no regeneration occurs when *Gb'arm* was knocked down by RNAi (Nakamura et al. 2007). These results indicate that the canonical Wnt/Wingless-signaling pathway is involved in the process of leg regeneration and determination of positional information in the leg segment.

Two directions in future work on the cricket leg should provide original contributions to the general field of regeneration. First, this robust model system will help to identify new molecules whose function in regeneration can only be discovered through powerful genome-wide RNAi screens. Second, the experimental versatility of the cricket leg makes possible a precise dissection of the molecular mechanisms underlying intercalary regeneration. Ultimately, this system will tell us which signaling pathways and what genetic circuitry define positional information in regenerating structures.

Calling Song as a Behavioral Output in Gene Function Analysis

Human hereditary disorders include single-gene disorders, which are caused by the mutation of one gene. *Drosophila* has been used as a model organism to study the pathogenic mechanisms of such diseases because it has 61% of the genes responsible for human single-gene disorders. Recent progress in techniques of RNAi-based functional analyses of *G. bimaculatus* genes has offered the possibility of establishing the cricket as a novel model system for exploring human pathogenic mechanisms through analyses of higher-order functions of the orthologs of the genes responsible for human diseases.

Noji and coworkers have focused on a *G. bimaculatus* ortholog of the gene responsible for the human disorder fragile-X mental retardation (*Gb'fmr*). Fragile-X syndrome is the most common form of inherited mental retardation and is caused by mutations in the *fragile X mental retardation 1* (*fmr1*) gene. In *Drosophila*, functional analyses of the *fmr1* gene have suggested that FMR1 has a role in regulating the development and plasticity of neuronal synaptic connections (for review, see Zhang and Broadie 2005). Furthermore, it has been shown that *Drosophila fmr1* mutants display behavioral defects, such as arrhythmic circadian activity, and have erratic patterns of locomotor activity, probably caused by defects in neurons (Dockendorff et al. 2002).

To assay the behavioral outputs of knocking down the *fmr1* ortholog in the cricket, we attempted to combine a loss-of-function technique using RNAi on the nymph and an analysis of the cricket song. Cricket songs involve species-specific signals and are important for species recognition. In many species, only male crickets produce songs and their repertoire includes a calling song, a courtship song, and an aggressive song. The calling song serves to both attract conspecific mature females and advertise a male's presence to other males. The calling song of *G. bimaculatus* consists of several phrases, each of which consists of four or five pulses. Songs have several parameters such as pulse duration, phrase duration, pulse period, dominant frequency, and bandwidth. We focused on the two temporal parameters, pulse duration and pulse period, of the cricket song. We recorded the calling songs of individuals and compared the parameters between the songs of *Gb′fmr*-RNAi and control crickets (dsRNA injected at the seventh instar). The pulse duration in the calling song of *Gb′fmr*-RNAi adult crickets was significantly reduced compared with that of the control, whereas the pulse period was not affected. This suggests that *Gb′fmr* may be involved in proper formation of the nerve/muscle system controlling the stridulation of the forewings (A. Hamada et al., unpubl.).

Circadian Rhythms

Circadian rhythms are endogenous oscillations with a period of about 24 hours driven by a timing machinery called a circadian clock. In insects, the rhythm can be observed in a variety of physiological functions, such as eclosion (Pittendrigh et al. 1958), locomotion (Tomioka and Chiba 1984; Tomioka et al. 1998), and stridulation (Loher 1972). Neurophysiological studies have revealed that the circadian clock driving overt rhythms is located in specific loci in the central nervous system. In crickets and cockroaches, the optic lobe contains the circadian clock that regulates overt behavioral and electroretinographic rhythms (Nishiitsutsuji-Uwo and Pittendrigh 1968; Tomioka and Chiba 1982, 1984, 1992; Wills et al. 1985).

Molecular studies revealed that the oscillation is regulated by cyclical expressions of so-called clock genes and their product proteins through feedback mechanisms (Dunlap 1999; Stanewsky 2002). In *Drosophila*, the clock genes *period* (*per*) and *timeless* (*tim*) encode key components of the clock. CLOCK (CLK) and CYCLE (CYC) proteins, encoded by *Clock* (*Clk*) and *cycle* (*cyc*) genes, respectively, form heterodimers to promote transcription of *per* and *tim* during the late subjective day. The product proteins PERIOD (PER) and TIMELESS (TIM) increase during the subjective night, form heterodimers, enter the nucleus during the late subjective night, and bind to the CLK–CYC complex to suppress their transcriptional activity during the late night. This negative action reduces PER and TIM levels to initiate the clock's next cycle. In the negative-feedback loop, posttranslational regulations of PER and TIM are provided by DOUBLE-TIME, CASEIN KINASE 2, and SHAGGY, which control the stability and the nuclear transport of PER and TIM by phosphorylation (Price et al. 1998; Martinek et al. 2001; Akten et al. 2003).

The proposed molecular mechanism, however, is not fully supported by the results obtained in other insect species. For example, in the silkmoth *Antheraea pernyi*, expression of PER is restricted to the cytoplasm of presumptive clock neurons and PER does not enter the nucleus (Sauman and Reppert 1996). In the crickets *Teleogryllus commodus* and *T. oceanicus*, no rhythmic expression of PER can be observed (Lupien et al. 2003). There might be diverse clock mechanisms among insects and therefore it is necessary to compare the clock mechanisms across the diverse classes to understand the insect clock mechanism.

Kenji Tomioka's group has addressed this issue using *G. bimaculatus*, which is an appropriate model for this purpose because the circadian clock has been localized in the optic lobe and the role of the optic lobe in regulation of the locomotor rhythm has been well demonstrated (Tomioka and Abdelsalam 2004). Recently, Tomioka and coworkers examined the daily expression pattern and function of a *Gryllus* ortholog of *period* (*Gb′per*). They found that *Gb′per* mRNA was expressed cyclically with peaks at the early night, and its knockdown by RNAi disrupted circadian rhythms, suggesting that *Gb′per* has an important role in the circadian system in the cricket (Moriyama et al. 2008).

Learning and Memory

The nervous systems of vertebrates and invertebrates store information for short-term memory (STM) and long-term memory (LTM) by changing the strength of their synaptic connections. Olfactory learning in insects is a useful model for studying neural mechanisms that underlie learning and memory. Crickets are capable of quickly learning olfactory signals and memorizing them for practically a lifetime (Matsumoto and Mizunami 2000, 2002, 2004), and they can be easily used for detailed pharmacological (Matsumoto et al. 2003; Unoki et al. 2005) and electrophysiological (Paydar et al. 1999) studies. Makoto Mizunami's group has been studying the biochemical pathway involved in the formation of LTM in associative olfactory learning by using *G. bimaculatus*. They performed behavioral and pharmacological studies to clarify the biochemical pathway involved in the formation of LTM associative olfactory learning and demonstrated that the cAMP pathway is a downstream target of the nitric oxide (NO)-cGMP pathway for the formation of LTM and that the cyclic nucleotide-gated channel and calcium–calmodulin intervene between the NO-cGMP pathway and the cAMP pathway. More recently, their assay system of learning and memory has been combined with the nymphal RNAi technique to reveal gene functions in LTM formation. Consequently, it has been shown that knocking down the *NOS* (nitric oxide synthase) gene led to complete impairment of LTM formation in crickets, suggesting that NO signaling in the mushroom body participates in LTM formation (T. Takahashi et al., unpubl.).

GENETICS

The genetic background of *G. bimaculatus* has not been well established. Only the above-mentioned white-eye mutant (autosomal recessive; *gwhite*) is continuously maintained in a laboratory.

GENOMICS/ASSOCIATED RESOURCES

S. Noji and his colleagues have established the following genomic resources.

1. EST projects are in progress. So far, 1056 cDNA sequences from regenerating legs of third instar nymphs and 9600 cDNA sequences from early embryos have been deposited with the DNA databank of Japan (DDBJ).

2. A high-throughput cDNA (HTC) analysis using pyrosequencing technology has been performed and has produced 32,010 cDNA sequences from the whole bodies of first to eighth instar nymphs (which were deposited with DDBJ).

3. A BAC (bacterial artificial chromosome) genomic library of *G. bimaculatus* has been constructed at the Noji laboratory. Using this library, analyses of gene structures and *cis*-regulatory elements are ongoing.

TECHNICAL APPROACHES

Whole-mount In Situ Hybridization and Immunohistochemistry

Standard protocols for whole-mount in situ hybridization using a digoxigenin (DIG)-labeled antisense RNA probe (Wilkinson 1992) or double probes (one DIG-labeled and one fluoroscein isothiocyanate [FITC]-labeled) (Dietrich et al. 1997) for chick embryos and immunohistochemistry for grasshopper embryos (Patel et al. 1989) are applicable to *G. bimaculatus* embryos and regenerating blastemata (epithelial tissue isolated from the cuticle [Mito et al. 2002]).

RNAi

To analyze gene functions in *G. bimaculatus*, Noji and coworkers have developed four types of RNAi techniques: embryonic, parental, nymphal, and regenerative RNAi (Miyawaki et al. 2004; Mito et al. 2005; Nakamura et al. 2007; Ronco et al. 2008). Embryonic RNAi involves injection of dsRNA into eggs. In parental RNAi, dsRNA is injected into the body cavity of adult female crickets. With this method, one can observe RNAi effects on many eggs laid after injection (a few thousand per injected individual). To analyze gene functions during postembryonic development, dsRNA is injected into the nymphal body cavity, and this is designated nymphal RNAi. The RNAi effect continues in general through subsequent molts. Regenerative RNAi is an application of nymphal RNAi. To study regeneration of the leg, a leg is amputated after injection of dsRNA. During regeneration of the lost part of the leg, RNAi effects can be observed. Classical operations that induce intercalary regeneration can be combined with this method.

Transgenesis

Generation of transgenic crickets is critical for analyzing the functions of genes and *cis*-regulatory elements. However, this attempt is still under way. Although we succeeded in introducing genes in the cricket genome of a part of somatic cells using *piggyBac* and *Minos* transposon elements (Zhang et al. 2002; Shinmyo et al. 2004), germ-line transmission of exogenous sequences has not yet been achieved.

ACKNOWLEDGMENTS

The authors thank Hideyo Ohuchi, Tetsuya Bando, and other members of the Noji laboratory for their contributions to many parts of the cricket studies described in this chapter. T.M. and S.N. were supported by a grant from the Ministry of Education, Culture, Sports, Science, and Technology of Japan.

REFERENCES

Akten, B., Jauch, E., Genova, G.K., Kim, E.Y., Edery, I., Raabe, T., and Jackson, F.R. 2003. A role for CK2 in the *Drosophila* circadian oscillator. *Nat. Neurosci.* **6:** 251–257.

Basler, K. and Struhl, G. 1994. Compartment boundaries and the control of *Drosophila* limb pattern by *hedgehog* protein. *Nature* **368:** 208–214.

Bohni, R., Riesgo-Escovar, J., Oldham, S., Brogiolo, W., Stocker, H., Andruss, B.F., Beckingham, K., and Hafen, E. 1999. Autonomous control of cell and organ size by CHICO, a *Drosophila* homolog of vertebrate IRS1-4. *Cell* **97:** 865–875.

Bryant, S.V. and Iten, L.E. 1976. Supernumerary limbs in amphibians: Experimental production in *Notophthalmus viridescens* and a new interpretation of their formation. *Dev. Biol.* **50:** 212–234.

Bryant, S.V., French, V., and Bryant, P.J. 1981. Distal regeneration and symmetry. *Science* **212:** 993–1002.

Bucher, G., Scholten, J., and Klingler, M. 2002. Parental RNAi in *Tribolium* (Coleoptera). *Curr. Biol.* **12:** R85–R86.

Campbell, G. 2002. Distalization of the *Drosophila* leg by graded EGF-receptor activity. *Nature* **418:** 781–785.

Campbell, G. and Tomlinson, A. 1995. Initiation of the proximodistal axis in insect legs. *Development* **121:** 619–628.

Campbell, G., Weaver, T., and Tomlinson, A. 1993. Axis specification in the developing *Drosophila* appendage: The role of *wingless*, *decapentaplegic*, and the homeobox gene *aristaless*. *Cell* **74:** 1113–1123.

Clancy, D.J., Gems, D., Harshman, L.G., Oldham, S., Stocker, H., Hafen, E., Leevers, S.J., and Partridge, L. 2001. Extension of lifespan by loss of CHICO, a *Drosophila* insulin receptor substrate protein. *Science* **292:** 104–106.

Copf, T., Schröder, R., and Averof, M. 2004. Ancestral role of caudal genes in axis elongation and segmentation. *Proc. Natl. Acad. Sci.* **101:** 17711–17715.

Danley, P.D., Mullen, S.P., Liu, F., Nene V., Quackenbush, J., and Shaw, K.L. 2007. A cricket Gene Index: A genomic resource for studying neurobiology, speciation, and molecular evolution. *BMC Genomics* **8:** 109.

Diaz-Benjumea, F.J., Cohen, B., and Cohen, S.M. 1994. Cell interaction between compartments establishes the proximal-distal axis of *Drosophila* legs. *Nature* **372:** 175–179.

Dietrich, S., Schubert, F.R., and Lumsden, A. 1997. Control of dorsoventral pattern in the chick paraxial mesoderm. *Development* **124:** 3895–3908.

Dockendorff, T.C., Su, H.S., McBride, S.M., Yang, Z., Choi, C.H., Siwicki, K.K., Sehgal, A., and Jongens, T.A. 2002. *Drosophila* lacking *dfmr1* activity show defects in circadian output and fail to maintain courtship interest. *Neuron* **34:** 973–984.

Dong, Y. and Friedrich, M. 2005. Nymphal RNAi: Systemic RNAi mediated gene knockdown in juvenile grasshopper. *BMC Biotechnol.* **5:** 25.

Dunlap, J.C. 1999. Molecular bases for circadian biological clocks. *Cell* **96:** 271–290.

Engel, J.E. and Hoy, R.R. 1999. Experience-dependent modification of ultrasound auditory processing in a cricket escape response. *J. Exp. Biol.* **202:** 2797–2806.

Friedrich, M. and Benzer, S. 2000. Divergent *decapentaplegic* expression patterns in compound eye development and the evolution of insect metamorphosis. *J. Exp. Zool.* **288:** 39–55.

French, V. 1981. Pattern regulation and regeneration. *Philos. Trans. R. Soc. Lond. B Biol. Sci.* **295:** 601–617.

French, V., Bryant, P.J., and Bryant, S.V. 1976. Pattern regulation in epimorphic fields. *Science* **193:** 969–981.

Galindo, M.I., Bishop, S.A., Greig, S., and Couso, J.P. 2002. Leg patterning driven by proximal-distal interactions and EGFR signaling. *Science* **297:** 256–259.

Goberdhan, D.C., Paricio, N., Goodman, E.C., Mlodzik, M., and Wilson, C. 1999. *Drosophila* tumor suppressor *PTEN* controls cell size and number by antagonizing the Chico/PI3-kinase signaling pathway. *Genes Dev.* **13:** 3244–3258.

Goto, S. and Hayashi, S. 1999. Proximal to distal cell communication in the *Drosophila* leg provides a basis for an intercalary mechanism of limb patterning. *Development* **126:** 3407–3413.

He, Z.B., Cao, Y.Q., Yin, Y.P., Wang, Z.K., Chen, B., Peng, G.X., and Xia, Y.X. 2006. Role of *hunchback* in segment patterning of *Locusta migratoria manilensis* revealed by parental RNAi. *Dev. Growth Differ.* **48:** 439–445.

Hedwig, B. and Poulet, J.F. 2004. Complex auditory behaviour emerges from simple reactive steering. *Nature* **430:** 781–785.

Inoue, Y., Mito, T., Miyawaki, K., Matsushima, K., Shinmyo, Y., Heanue, T.A., Mardon, G., Ohuchi, H., and Noji, S. 2002. Correlation of expression patterns of *homothorax*, *dachshund*, and *Distal-less* with the proximodistal segmentation of the cricket leg bud. *Mech. Dev.* **113:** 141–148.

Inoue, Y., Miyawaki, K., Terasawa, T., Matsushima, K., Shinmyo, Y., Niwa, N., Mito, T., Ohuchi, H., and Noji, S. 2004. Expression patterns of dachshund during head development of *Gryllus bimaculatus* (cricket). *Gene Expr. Patterns* **4:** 725–731.

Kang, L., Chen, X., Zhou, Y., Liu, B., Zheng, W., Li, R., Wang, J., and Yu, J. 2004. The analysis of large-scale gene expression correlated to the phase changes of the migratory locust. *Proc. Natl. Acad. Sci.* **101:** 17611–17615.

Kojima, T. 2004. The mechanism of *Drosophila* leg development along the proximodistal axis. *Dev. Growth Differ.* **46:** 115–129.

Lehmann, R. and Nüsslein-Volhard, C. 1987. hunchback, a gene required for segmentation of an anterior and posterior region of the *Drosophila* embryo. *Dev. Biol.* **119:** 402–417.

Liu, P.Z. and Kaufman, T.C. 2004. hunchback is required for suppression of abdominal identity, and for proper germband growth and segmentation in the intermediate germband insect *Oncopeltus fasciatus*. *Development* **131:** 1515–1527.

Loher, W. 1972. Circadian control of stridulation in the cricket *Teleogryllus commodus* Walker. *J. Comp. Physiol.* **79:** 173–190.

Lupien, M., Marshall, S., Leser, W., Pollack, G.S., and Honegger, H.-W. 2003. Antibodies against the PER protein of *Drosophila* label neurons in the optic lobe, central brain, and thoracic ganglia of the crickets *Teleogryllus commodus* and *Teleogryllus oceanicus*. *Cell Tiss. Res.* **312:** 377–391.

Manjon, C., Sanchez-Herrero, E., and Suzanne, M. 2007. Sharp boundaries of Dpp signalling trigger local cell death required for *Drosophila* leg morphogenesis. *Nat. Cell Biol.* **9:** 57–63.

Mardon, G., Solomon, N.M., and Rubin, G.M. 1994. *dachshund* encodes a nuclear protein required for normal eye and leg development in *Drosophila*. *Development* **120:** 3473–3486.

Marie, B., Bacon, J.P., and Blagburn, J.M. 2000. Double-stranded RNA interference shows that Engrailed controls the synaptic specificity of identified sensory neurons. *Curr. Biol.* **10:** 289–292.

Martinek, S., Inonog, S., Manoukian, A.S., and Young, M.W. 2001. A role for the segment polarity gene *shaggy*/GSK-3 in the *Drosophila* circadian clock. *Cell* **105:** 769–779.

Martini, S.R., Roman, G., Meuser, S., Mardon, G., and Davis, R.L. 2000. The retinal determination gene, *dachshund*, is required for mushroom body cell differentiation. *Development* **127:** 2663–2672.

Matsumoto, Y. and Mizunami, M. 2000. Olfactory learning in the cricket *Gryllus bimaculatus*. *J. Exp. Biol.* **203:** 2581–2588.

Matsumoto, Y. and Mizunami, M. 2002. Lifetime olfactory memory in the cricket *Gryllus bimaculatus*. *J. Comp. Physiol. A* **188:** 295–299.

Matsumoto, Y. and Mizunami, M. 2004. Context-dependent olfactory learning in an insect. *Learn. Mem.* **11:** 288–293.

Matsumoto, Y., Noji, S., and Mizunami, M. 2003. Time course of protein synthesis-dependent phase of olfactory memory in the cricket *Gryllus bimaculatus*. *Zool. Sci.* **20:** 409–416.

Meinhardt, H. 1983. Cell determination boundaries as organizing regions for secondary embryonic fields. *Dev. Biol.* **96:** 375–385.

Mito, T. and Noji, S. 2006. Evolution of developmental systems underlying segmented body plans of bilaterian animals: Insights from studies of segmentation in a cricket. *Paleont. Res.* **10:** 337–344.

Mito, T., Inoue, Y., Kimura, S., Miyawaki, K., Niwa, N., Shinmyo, Y., Ohuchi, H., and Noji, S. 2002. Involvement of *hedgehog*, *wingless*, and *dpp* in the initiation of proximodistal axis formation during the regeneration of insect legs, a verification of the modified boundary model. *Mech. Dev.* **114:** 27–35.

Mito, T., Sarashina, I., Zhang, H., Iwahashi, A., Okamoto, H., Miyawaki, K., Shinmyo, Y., Ohuchi, H., and Noji, S. 2005. Non-canonical functions of *hunchback* in segment patterning of the intermediate germ cricket *Gryllus bimaculatus*. *Development* **132:** 2069–2079.

Mito, T., Okamoto, H., Shinahara, W., Shinmyo, Y., Miyawaki, K., Ohuchi, H. and Noji, S. 2006. *Krüppel* acts as a gap gene regulating expression of *hunchback* and *even-skipped* in the intermediate germ cricket *Gryllus bimaculatus*. *Dev. Biol.* **294:** 471–481.

Mito, T., Kobayashi, C., Sarashina, I., Zhang, H., Shinahara, W., Miyawaki, K., Shinmyo, Y., Ohuchi, H., and Noji, S. 2007. *even-skipped* has gap-like, pair-rule-like, and segmental functions in the cricket *Gryllus bimaculatus*, a basal, intermediate germ insect (Orthoptera). *Dev. Biol.* **303:** 202–213.

Mito, T., Ronco, M., Uda, T., Nakamura, T., Ohuchi, H., and Noji, S. 2008. Divergent and conserved roles of *extradenticle* in body segmentation and appendage formation, respectively, in the cricket *Gryllus bimaculatus*. *Dev. Biol.* **313:** 67–79.

Miyawaki, K., Inoue, Y., Mito, T., Fujimoto, T., Matsushima, K., Shinmyo, Y., Ohuchi, H., and Noji, S. 2002. Expression patterns of *aristaless* in developing appendages of *Gryllus bimaculatus* (cricket). *Mech. Dev.* **113:** 181–184.

Miyawaki, K., Mito, T., Sarashina, I., Zhang, H., Shinmyo, Y., Ohuchi, H., and Noji, S., 2004. Involvement of Wingless/Armadillo signaling in the posterior sequential segmentation in the cricket, *Gryllus bimaculatus* (Orthoptera), as revealed by RNAi analysis. *Mech. Dev.* **121:** 119–130.

Moriyama, Y., Sakamoto, T., Karpova, S.G., Matsumoto, A., Noji, S., and Tomioka, K. 2008. RNA interference of the clock gene *period* disrupts circadian rhythms in the cricket *Gryllus bimaculatus*. *J. Biol. Rhythms* **23:** 308–318.

Nakamura, T., Mito, T., Tanaka, Y., Bando, T., Ohuchi, H., and Noji, S. 2007. Involvement of the canonical Wnt/Wingless signaling in determination of the proximodistal positional values within the leg segment of the cricket *Gryllus bimaculatus*. *Dev. Growth Differ.* **49:** 79–88.

Nakamura, T., Mito, T., Bando, T., Ohuchi, T., and Noji, S. 2008a. Dissecting insects leg regeneration through RNA interference. *Cell. Mol. Life Sci.* **65:** 64–72.

Nakamura, T., Mito, T., Miyawaki, K., Ohuchi, H., and Noji, S. 2008b. EGFR signaling is required for re-establishing the proximodistal axis during distal leg regeneration in the cricket *Gryllus bimaculatus* nymph. *Dev. Biol.* **319:** 46–55.

Nguyen, K.T., Tajmir, P., Lin, C.H., Liadis, N., Zhu, X.D., Eweida, M., Tolasa-Karaman, G., Cai, F., Wang, R., Kitamura, T., et al. 2006. Essential role of *Pten* in body size determination and pancreatic β-cell homeostasis in vivo. *Mol. Cell Biol.* **26:** 4511–4518.

Nishiitsutsuji-Uwo, J. and Pittendrigh, C.S. 1968. Central nervous system control of circadian rhythmicity in cockroach. III. The optic lobes, locus of the driving oscillation? *Z. Vgl. Physiol.* **58:** 14–46.

Niwa, N., Saitoh, M., Ohuchi, H., Yoshioka, H., and Noji, S. 1997. Correlation between *Distal-less* expression patterns and structures of appendages in development of the two-spotted cricket, *Gryllus bimaculatus*. *Zool. Sci.* **14:** 115–125.

Niwa, N., Inoue, Y., Nozawa, A., Saito, M., Misumi, Y., Ohuchi, H., Yoshioka, H., and Noji, S. 2000. Correlation of diversity of leg morphology in *Gryllus bimaculatus* (cricket) with divergence in *dpp* expression pattern during leg development. *Development* **127:** 4373–4381.

Oldham, S., Stocker, H., Laffargue, M., Wittwer, F., Wymann, M., and Hafen, E. 2002. The *Drosophila* insulin/IGF receptor controls growth and size by modulating PtdIns *P*3 levels. *Development* **129:** 4103–4109.

Patel, N.H., Martin-Blanco, E., Coleman, K.G., Poole, S.J., Ellis, M.C., Kornberg, T.B., and Goodman, C.S. 1989. Expression of *engrailed* proteins in arthropods, annelids, and chordates. *Cell* **58:** 955–968.

Paydar, S., Doan, C.A., and Jacobs, G.A. 1999. Neutral mapping of direction and frequency in the cricket cercal sensory system. *J. Neurosci.* **19:** 1771–1781.

Pittendrigh, C.S., Bruce, V.G., and Kaus, P. 1958. On the significance of transients in daily rhythms. *Proc. Natl. Acad. Sci.* **44:** 965–973.

Price, J.L., Blau, J., Rothenfluh, A., Abodeely, M., Kloss, B., and Young, M.W. 1998. *double-time* is a novel *Drosophila* clock gene that regulates PERIOD protein accumulation. *Cell* **94:** 83–95.

Ronco, M., Uda, T., Mito, T., Minelli, A., Noji, S., and Klingler, M. 2008. Antenna and all gnathal appendages are similarly transformed by *homothorax* knock-down in the cricket *Gryllus bimaculatus*. *Dev. Biol.* **313:** 80–92.

Sarashina, I., Shinmyo, Y., Hirose, A., Miyawaki, K., Mito, T., Ohuchi, H., Horio, T., and Noji, S. 2003. Hypotonic buffer induces meiosis and formation of anucleate cytoplasmic islands in the egg of the two-spotted cricket *Gryllus bimaculatus*. *Dev. Growth Differ.* **45:** 103–112.

Sarashina, I., Mito, T., Saito, M., Uneme, H., Miyawaki, K., Shinmyo, Y., Ohuchi, H., and Noji, S. 2005. Location of micropyles and early embryonic development of the two-spotted cricket *Gryllus bimaculatus* (Insecta, Orthoptera). *Dev. Growth Differ.* **47:** 99–108.

Sato, M. and Tanaka-Sato, H. 2002. Fertilization, syngamy, and early embryonic development in the cricket *Gryllus bimaculatus* (De Geer). *J. Morph.* **254:** 266–271.

Sauman, I. and Reppert, S.M. 1996. Circadian clock neurons in the silkmoth *Antheraea pernyi*: Novel mechanisms of period protein regulation. *Neuron* **17:** 889–900.

Schröder, R. 2003. The genes *orthodenticle* and *hunchback* substitute for *bicoid* in the beetle *Tribolium*. *Nature* **422:** 621–625.

Shinmyo, Y., Mito, T., Matsushita, T., Sarashina, I., Miyawaki, K., Ohuchi, H., and Noji, S. 2004. *piggyBac*-mediated somatic transformation of the two-spotted cricket, *Gryllus bimaculatus*. *Dev. Growth Differ.* **46:** 343–349.

Shinmyo, Y., Mito, T., Matsushita, T., Sarashina, I., Miyawaki, K., Ohuchi, H., and Noji, S. 2005. *caudal* is required for gnathal and thoracic patterning and for posterior elongation in the intermediate-germband cricket *Gryllus bimaculatus*. *Mech. Dev.* **122:** 231–239.

Shinmyo, Y., Mito, T., Uda, T., Nakamura, T., Miyawaki, K., Ohuchi, H., and Noji, S. 2006. *brachyenteron* is necessary for morphogenesis of posterior gut but not for AP axial elongation from the posterior growth zone in the intermediate-germband cricket, *Gryllus bimaculatus*. *Development* **133:** 4539–4547.

Stanewsky, R. 2002. Clock mechanisms in *Drosophila*. *Cell Tiss. Res.* **309:** 11–26.

Taguchi, A., Wartschow, L.M., and White, M.F. 2007. Brain IRS2 signaling coordinates life span and nutrient homeostasis. *Science* **317:** 369–372.

Tomioka, K. and Abdelsalam, S.A. 2004. Circadian organization in hemimetabolous insects. *Zool. Sci.* **21:** 1153–1162.

Tomioka, K. and Chiba Y. 1982. Persistence of circadian ERG rhythms in the cricket with optic tract severed. *Naturwissenschaften* **69:** 355–356.

Tomioka, K. and Chiba, Y. 1984. Effects of nymphal stage optic nerve severance or optic lobe removal on the circadian locomotor rhythm of the cricket, *Gryllus bimaculatus*. *Zool. Sci.* **1:** 385–394.

Tomioka, K. and Chiba, Y. 1992. Characterization of optic lobe circadian pacemaker by in situ and in vitro recording of neuronal activity in the cricket *Gryllus bimaculatus*. *J. Comp. Physiol. A* **171:** 1–7.

Tomioka, K., Sakamoto, M., Harui, Y., Matsumoto, N., and Matsumoto, A. 1998. Light and temperature cooperate to regulate the circadian locomotor rhythm of wild type and *period* mutants of *Drosophila melanogaster*. *J. Insect Physiol.* **44:** 587–596.

Tomoyasu, Y. and Denell, R.E. 2004. Larval RNAi in *Tribolium* (Coleoptera) for analyzing adult development. *Dev. Genes Evol.* **214:** 575–578.

Unoki, S., Matsumoto, Y., and Mizunami, M. 2005. Participation of octopaminergic reward system in insect olfactory learning revealed by pharmacological study. *Eur. J. Neurosci.* **22:** 1409–1416.

Wenzel, B. and Hedwig, B. 1999. Neurochemical control of cricket stridulation revealed by pharmacological microinjections into the brain. *J. Exp. Biol.* **202:** 2203–2216.

White, M.F. 2002. IRS proteins and the common path to diabetes. *Am. J. Physiol. Endocrinol. Metab.* **283:** E413–E422.

Wilkinson, D.G. 1992. A practical approach. In *In situ hybridization* (ed. D.G. Wilkinson), pp. 75–83. IRL Press, Oxford.

Wills, S.A, Page, T.L., and Colwell, C.S. 1985. Circadian rhythms in the electroretinogram of the cockroach. *J. Biol. Rhythms* **1:** 25–37.

Yoshimura, A., Nakata, A., Mito, T., and Noji, S. 2006. The characteristics of karyotype and telomeric satellite DNA sequences in the cricket, *Gryllus bimaculatus* (Orthoptera, Gryllidae). *Cytogenet. Genome Res.* **112:** 329–336.

Zhang, H., Shinmyo, Y., Hirose, A., Mito, T., Inoue, Y., Ohuchi, H., Loukeris, T.G., Eggleston, P., and Noji, S. 2002. Extrachromosomal transposition of the transposable element *Minos* in embryos of the cricket *Gryllus bimaculatus*. *Dev. Growth Differ.* **44:** 409–417.

Zhang, H., Shinmyo, Y., Mito, T., Miyawaki, K., Sarashina, I., Ohuchi, H., and Noji, S. 2005. Expression patterns of the homeotic genes *Scr, Antp, Ubx*, and *abd-A* during embryogenesis of the cricket *Gryllus bimaculatus*. *Gene Expr. Patterns* **5:** 491–502.

Zhang, Y.Q. and Broadie, K. 2005. Fathoming fragile X in fruit flies. *Trends Genet.* **21:** 37–45.

14 The American Wandering Spider *Cupiennius salei*
A Model for Behavioral, Evolutionary, and Developmental Studies

Nikola-Michael Prpic,* Michael Schoppmeier,† and Wim G.M. Damen
Institute for Genetics, Evolutionary Genetics, University of Cologne, 50674 Köln, Germany

ABSTRACT

The spider *Cupiennius salei* is used as a laboratory model for embryological and physiological studies. Their highly developed sensory organs make this spider an excellent model for behavioral studies. Furthermore, *Cupiennius* has contributed greatly to answering evolutionary developmental questions. This chelicerate arthropod is a very useful model for such studies also due to its phylogenetic position and the availability of tools to study and manipulate its embryonic development.

PROTOCOLS
1. Collection and Fixation of Spider Embryos, 351
2. Whole-mount In Situ Hybridization of Spider Embryos, 353
3. Detection of Cell Death in Spider Embryos using TUNEL, 356
4. Gene Silencing via Embryonic RNAi in Spider Embryos, 359
5. Detection of Cell Proliferation using BrdU Labeling, 362
6. Dissecting Embryos for Light Microscopy, 366

BACKGROUND INFORMATION

C. salei is a large species of spider. Its common name is American Wandering Spider, and it was described first by Keyserling in 1877. The body size of adult animals is up to 4.5 cm and the span of the legs can be up to 15 cm (Fig. 1A). Spiders belong to the chelicerates, which are one of four extant arthropod classes, the remaining three being myriapods, crustaceans, and insects (Fig. 2). Chelicerates are a basally branching arthropod taxon with a fossil record that dates back to the Cambrian. The chelicerates also include, apart from spiders, scorpions, mites, ticks, and horseshoe crabs. *Cupiennius* is systematically placed in the family *Ctenidae* that belongs to the Araneomorphae, the largest group of spiders, comprising approximately 90% of all extant spider species.

C. salei is found in Mexico, Guatemala, and Honduras (it has also been recorded in Haiti). The natural habitat of *Cupiennius* is the leaf bases of bananas, bromeliads, and similar plants. They hide during the daytime and hunt insects at night. *C. salei* was introduced as a research organism

Present addresses: *Georg-August-Universität, Johann-Friedrich-Blumenbach-Institut für Zoologie und Anthropologie, Abteilung Entwicklungsbiologie, GZMB, Justus-von-Liebig-Weg 11, D-37077 Göttingen, Germany. †Friedrich-Alexander University Erlangen, Institute for Biology, Department of Developmental Biology, Staudtstrasse 5, D-91058 Erlangen, Germany

This chapter, with full-color images, can be found online at www.cshprotocols.org/emo.

348 / Chapter 14

FIGURE 1. The American Wandering Spider *Cupiennius salei*. (A) Adult female specimen of *C. salei*. (B) The spiders are kept separate in 2- to 3-liter jars, with a substrate of flower ground.

in the early 1960s, when some specimens were found as stowaways in a banana transport to Munich (Germany), and brought to Mechthild Melchers at the Zoological Institute of the University of Munich (Barth 2001). Starting from these initial spiders, Barth established a spider culture in Munich and introduced *Cupiennius* as a research animal.

Cupiennius has primarily served as a physiological model for work on sensing and behavior (Barth 2001). However, in the late 1990s, *Cupiennius* developed into a model for studies in evolutionary developmental biology (McGregor et al. 2008). It was considered an appropriate organism for these studies because of the phylogenetic position of the chelicerates (Fig. 2). The chelicerates form a monophyletic group that had already split from the other arthropods in the Cambrian era.

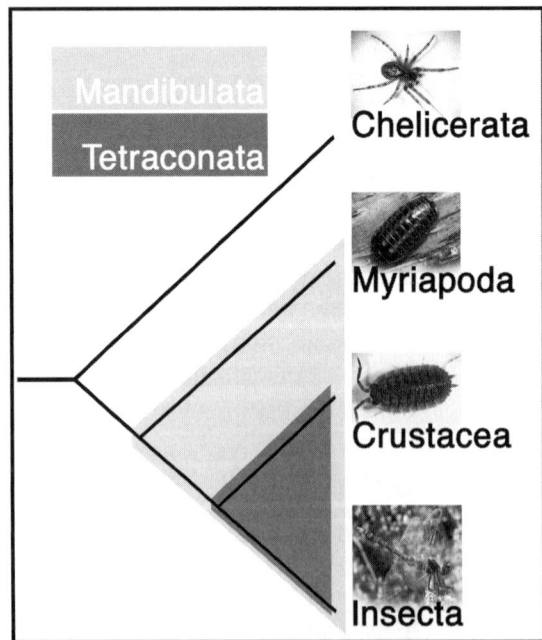

FIGURE 2. Arthropod phylogeny. Depicted are relationships of major arthropod groups. Insects and crustaceans are sister taxa (Tetraconata); the Tetraconata together with the myriapods form the Mandibulata. Chelicerates, to which the spiders belong, are a basally branching arthropod group.

Comparisons between spiders and insects show us the degree of conservation and divergence of developmental mechanisms during arthropod evolution. Any embryological feature conserved between spiders and insects is likely to represent an ancestral feature for arthropods. Comparative molecular embryological work in insects and spiders should eventually allow us to define a molecular archetype for the phylum Arthropoda. This in itself will be a necessary cornerstone for comparing the different metazoan phyla, including chordates.

SOURCES AND HUSBANDRY

Cupiennius can easily be kept in the lab. They require a day/night cycle and moderate temperature (up to 25°C). Adult spiders are kept separately in 2–3-liter jars, with a substrate of flower-ground that is kept moist to ensure high humidity (Fig. 1B). They feed on insects such as crickets (*Acheta domesticus*; commercially available, e.g., in pet shops or via mail order). After the final molt, females are mated with a male, and after 10–14 days produce a cocoon filled with up to 1500 synchronously developing embryos. Females can produce up to four cocoons successively at about 1-month intervals. The female spider carries the cocoon with her for about 4 weeks, until the spiderlings hatch from the cocoon. The young spiders are initially kept together and feed on fruit flies, but after two to three more molts, they are separated to avoid cannibalism. The generation time of *C. salei* is 9 months. There are several lab cultures of *C. salei*, for example, in Cologne, Germany (W. Damen), Frankfurt am Main, Germany (E.A. Seyfarth), and Vienna, Austria (F. Barth), from which specimens may be obtained. Usually, no special precautions are required when handling *Cupiennius*. However, a bite may be painful (like the sting of a bee) and cause allergic reactions in sensitive individuals. Therefore, we recommend avoiding direct contact with the spiders and wearing gloves if direct contact cannot be avoided.

RELATED SPECIES

Two other spider species have been used in developmental studies: *Tegenaria atrica* (a member of the *Agelenidae* [Superfamily Amaurobioidea]) and *Achaearanea tepidariorum* (member of the *Theridiidae* [Superfamily Araneoidea]). The protocols provided here can also be used for these spider species.

USES OF THE *C. SALEI* MODEL SYSTEM: STUDIES IN DEVELOPMENT AND BEHAVIOR

Work in *C. salei* has contributed greatly to answering evolutionary developmental questions. The tools available make this spider an excellent laboratory model for studying development and for addressing questions in the evolution of developmental mechanisms in arthropods. *Cupiennius* is used to study axis formation, segmentation, appendage development, neurogenesis, and silk production. These studies contribute to our understanding of the evolution of these processes, but they also help us to understand the origin and diversification of evolutionary novelties (e.g., see McGregor et al. 2008).

C. salei has also been used to study sensory organs and behavior. Spiders have developed their sensory organs to a fascinating technical perfection and complexity. The highly developed sensory systems of spiders and their varied and complex range of behavior are excellently adapted to environmental conditions. *Cupiennius* has significantly contributed to our understanding of the interplay among sensory organs, environment, and behavior (Barth 2001).

GENETICS

C. salei is not directly amenable for classical genetic studies. The generation time is 9 months and spiders must be kept in separate jars. Forward genetics is therefore difficult and not yet possible. However, gene silencing experiments using embryonic RNA interference (RNAi) have been performed successfully, allowing functional genetic studies of the organism. The haploid genome size of *C. salei* is approximately 2500 Mb. There are currently no initiatives to sequence the genome, but expressed sequence tag (EST) projects are under way (McGregor et al. 2008).

TECHNICAL APPROACHES

The following protocols describe the preparation and manipulation of *Cupiennius* embryos for studies in gene expression, apoptosis, gene silencing, and cell proliferation. The embryos may also be dissected for microscopic observation and imaging.

Protocol 1

Collection and Fixation of Spider Embryos

This protocol describes the collection and fixation of embryos of the spider *Cupiennius salei*. The fixed embryos can be stored at –20°C for prolonged periods and used for in situ hybridization (Protocol 2), in studies of apoptosis using TUNEL (Protocol 3), and for immunohistochemistry (Protocol 5).

MATERIALS

CAUTION: See the Cautions Appendix for appropriate handling of materials marked with <!>.

Reagents

Bleach (2.8% Natriumhypochlorite) <!>
CO_2 gas <!>
Cupiennius female with cocoon, maintained in 2–3-liter glass jar (as described in the Husbandry section)
Formaldehyde (37%; Sigma-Aldrich) <!>
Heptane (100%; Merck) <!>
Methanol (100%; Merck) <!>

Equipment

Dissection microscope
Eppendorf tubes (2 ml), round bottom
Forceps (long)
Rocking table or rotating wheel
Small scissors
Tubes (12 or 15 ml), with lid and round bottom
Watchmaker forceps (Dumont 5)

METHOD

Perform all steps at room temperature.

Collection of Embryos

1. Paralyze a female spider that carries a cocoon with eggs by introducing some CO_2 gas into the jar. Wait for 10–20 seconds, until the spider is not moving (if spider still moves, add more CO_2), and remove the cocoon from the female with forceps. Blow fresh air into the jar with the spider for about 5 seconds to remove the CO_2 and wake up the spider.
   Females take care of their cocoons and carry them attached with silk threads to their opisthosoma.

2. Open the cocoon by making a small (5–10 mm) incision in the silk cover of the cocoon with small scissors. Collect the eggs/embryos, which will roll out of the cocoon, in a 15-ml tube with cap.

   Big cocoons contain more than 1000 synchronously developing eggs. The embryonic stage of the embryos can be estimated by putting a few eggs in methanol and observing them under the dissection microscope.

3. Collect about 100 eggs in one 12- or 15-ml round-bottom tube with cap.

   Embryos can be kept in these tubes for several days until they have reached the desired development stage. In this way, a developmental series of embryos from a single cocoon can be fixed.

Fixation of Embryos

4. Add 10 ml of bleach to the tube to remove the chorion from the eggs. Gently rock the eggs in the tube for 2–3 minutes. Once the chorion is dissolved, the clear vitelline membrane is visible.

5. Remove the bleach (withdrawn with a pipette) and rinse three times with water. Meanwhile, prepare a mixture consisting of 200 µl of formaldehyde and 10 ml of heptane. Mix well on a shaker for at least 5 minutes.

6. Remove all of the liquid from the eggs and add the formaldehyde/heptane mixture to the eggs. Fix on a rocking table or wheel for several hours or overnight in such a way that there is only gentle movement of eggs and fixative. Do not shake. The duration of fixation depends on the intended application. For in situ hybridizations, fix overnight (16 hours); for immunohistochemistry, fix for 1 hour to a few hours.

7. Remove all of the fixative from the eggs and add 10 ml of methanol. Mix gently. Due to an osmotic shock, the vitelline membranes of most of the eggs will break.

8. Replace the methanol with 10 ml of fresh methanol. The embryos now can be stored at –20°C.

9. Remove the vitelline membranes from the embryos using two sharp Dumont-5 forceps under the dissection microscope. Transfer the devitellinized embryos into a 2-ml Eppendorf tube with a round bottom (up to 30 embryos per tube) and store at –20°C until needed. Embryos can be stored –20°C for more than 1 year.

Protocol 2

Whole-mount In Situ Hybridization of Spider Embryos

This protocol describes the detection of transcripts in embryos of the spider *Cupiennius salei*.

MATERIALS

The recipes for the items marked with <R> begin on page 370.

CAUTION: See the Cautions Appendix for appropriate handling of materials marked with <!>.

Reagents

20x SSC <R>
Alkaline-phosphatase-conjugated antidigoxigenin antibody (Roche)
Blocking reagent (10% in water; Roche)
Cupiennius embryos (fixed as described in Protocol 1)

The starting material is fixed devitellinized embryos, preserved in methanol in 2-ml round-bottom Eppendorf tubes stored at –20°C (Protocol 1).

Formaldehyde (37%; Sigma-Aldrich) <!>
HYB-A <R>
HYB-B <R>
Labeled antisense RNA probe complementary to the gene of interest

Prepare an antisense RNA probe labeled with either digoxigenin or fluorescein, using a standard in vitro transcription protocol (e.g., Roche). The length of the probe should be at least 300 nucleotides. For each hybridization reaction, dissolve 50–100 ng of probe in 50 µl of HYB-A.

Methanol (100%; Merck) <!>
Methanol:PBST

Prepare solutions of 50:50 and 30:70 methanol:PBST.

4-Nitro blue tetrazolium chloride/5-bromo-4-chloro-3-indolyl-phosphate (NBT/BCIP substrate mixture; Roche) <!>
Phosphate-buffered saline with Tween 20 (PBST) <R>
Proteinase K solution (Roche)
Staining buffer (prepare fresh) <!>

Equipment

Dissection microscope
Eppendorf tubes (2 ml), round bottom
Incubator or hybridization oven, set at 65°C
Rocking table or rotating wheel

METHOD

Perform all steps at room temperature unless otherwise specified.

Postfixation and Proteinase K Digestion

1. Bring the embryos to room temperature. Remove as much methanol as possible (with a pipette), leaving just enough liquid to cover the embryos. Add 2 ml of 50:50 methanol:PBST

mixture. After 5 minutes, replace the 50:50 methanol:PBST with a 30:70 methanol:PBST mixture and let stand for another 5 minutes.

2. Replace 30:70 methanol:PBST mixture with 2 ml of PBST. Wash embryos three times in PBST for 5 minutes each. After each wash, remove the PBST and replace with 2 ml of fresh PBST.

3. Fix the embryos in 5% formaldehyde in PBST (1 ml of PBST + 170 μl of 37% formaldehyde) for 20 minutes. Place the tubes on a rocking table or on a rotating wheel in such a way that there is a slight movement of the liquid in the tubes.

4. Wash the embryos three times in PBST for 5 minutes each. After each wash, remove the PBST and replace with 2 ml of fresh PBST.

5. Remove the PBST and add 1 ml of PBST supplemented with 5 μg/ml of proteinase K. Incubate for 4 minutes.

    The exact concentration of proteinase K depends on the batch of proteinase K and must be estimated empirically. Too long an incubation may lead to overdigestion.
    See Troubleshooting.

6. Remove the PBST/proteinase K solution and immediately replace with 2 ml of PBST. Wash another three times in PBST for 5 minutes each. After each wash, remove the PBST and replace with 2 ml of fresh PBST.

7. Fix the embryos in 5% formaldehyde in PBST (1 ml of PBST + 170 μl of 37% formaldehyde) for 20 minutes. Place the tubes on a rocking table or on a rotating wheel in such a way that there is a slight movement of the liquid in the tubes.

8. Remove the PBST/formaldehyde solution and replace with 2 ml of PBST. Wash another three times in PBST for 5 minutes each. After each wash, remove the PBST and replace with 2 ml of fresh PBST.

Prehybridization and Hybridization

9. Replace PBST with 1 ml of PBST:HYB-B (1:1) and incubate for 5 minutes.

10. Remove as much liquid as possible and add 500 μl of HYB-B. Incubate for 5 minutes at 65°C.

11. Replace HYB-B with 500 μl of HYB-A and incubate the embryos for prehybridization for at least 1 hour at 65°C.

    This prehybridization step can be extended for to up to 3 hours.
    See Troubleshooting.

12. Remove the 500 μl of HYB-A and add the labeled antisense RNA probe (50–100 ng) in 50 μl of HYB-A. Incubate overnight (16 hours) at 65°C.

    See Troubleshooting.

Probe Removal

13. Remove as much hybridization solution from the tube as possible and add 500 μl of HYB-B. Incubate for 10 minutes at 65°C.

14. Replace liquid with 500 μl of HYB-B and incubate for another 30 minutes at 65°C.

15. Replace HYB-B with 1 ml of PBST:HYB-B (1:1) and incubate for 10 minutes at room temperature.

16. Wash the embryos twice in PBST for 5 minutes each. After each wash, remove the PBST and replace with 2 ml of fresh PBST.

17. Wash embryos twice in PBST for 15 minutes each. After each wash, remove the PBST and replace with 2 ml of fresh PBST. Place the tubes on a rocking table or on a rotating wheel in such a way that there is a slight movement of the liquid in the tubes. Additional wash steps are possible.

18. Incubate embryos in PBST supplemented with 1% blocking solution for 1 hour. Place the tubes on a rocking table or on a rotating wheel in such a way that there is a slight movement of the liquid in the tubes.

Detection of the Probe

19. Remove the PBST-blocking solution mix and replace with 1 ml of PBST with 0.5 µl of alkaline-phosphatase-coupled antibody directed against the label of the probe. If the probe is labeled with digoxigenin, this will be an anti-DIG antibody conjugated to alkaline phosphatase. Incubate for 4 hours at room temperature. Place the tubes on a rocking table or on a rotating wheel in such a way that there is a slight movement of the liquid in the tubes.

20. Remove antibody solution and wash embryos twice in PBST for 5 minutes each. After each wash, remove the PBST and replace with 2 ml of fresh PBST.

21. Wash embryos twice in PBST for 30 minutes each. After each wash, remove the PBST and replace with 2 ml of fresh PBST. Place the tubes on a rocking table or on a rotating wheel in such a way that there is a slight movement of the liquid in the tubes. Additional wash steps are possible.

22. Store embryos overnight at 4°C. No rocking is required.
    This step is an additional wash step and reduces unspecific background staining.

23. Remove PBST and wash embryos three times in staining buffer for 5 minutes each. After each wash, remove the staining buffer and replace with 2 ml of fresh staining buffer.

24. Remove staining buffer and add 1 ml of fresh staining buffer supplemented with the substrate (e.g., 20 µl of NBT/BCIP mixture if the antibody is coupled to alkaline phosphatase). Keep embryos in the dark.

25. Observe the staining reaction, which may take hours or even overnight; stop the reaction at an appropriate moment by removing the staining solution. Wash several times with PBST to remove all substrates. Store embryos in PBST at 4°C.
    See Troubleshooting.

26. The embryos now can be observed either as whole mounts or as flat mounts (Protocol 6).

TROUBLESHOOTING

Problem (Step 5): Optimal proteinase K concentration.
Solution: The exact duration and concentration of the proteinase K must be estimated empirically because there are differences in the activity in different batches of proteinase K. Proteinase K is required to make the transcripts in the embryos accessible for probes and antibodies. Too much proteinase K activity causes the embryos to fall apart, whereas too low a concentration may cause low specificity and sensibility. To find optimal proteinase K incubation, run a series of test reactions, varying both the concentration and duration of incubation time.

Problem (Steps 11 and 12): Evaporation at 65°C.
Solution: The 65°C incubations are best done in a 65°C incubator or hybridization oven instead of a water bath. Due to the low amounts of liquid especially during the hybridization in Step 12, evaporation of water that then condenses at the lid of the tube may be a problem if the 65°C incubation is done in a water bath. Incubations of the tubes in a 65°C incubator solve this problem because there is then no condensation of the water on the lid. Too much water evaporation results in significant changes in concentrations.

Problem (Step 25): High background.
Solution: Keep embryos in the dark during the staining reaction. Observe the progress of the staining reaction quickly. Too much light during the staining reaction in the presence of the NBT/BCIP substrate may cause an unspecific pinkish color in the embryo and yolk.

Protocol 3

Detection of Cell Death in Spider Embryos using TUNEL

One of the features of apoptosis or cell death is the cleavage or fragmentation of DNA that occurs in dead or dying cells. This protocol describes the detection of fragmented DNA in whole-mount *Cupiennius* embryos (Prpic and Damen 2005). The 3′-OH ends of these DNA fragments can be labeled with the terminal deoxynucleotidyl-transferase-mediated dUTP–digoxygenin nick-end labeling (TUNEL) technique. The present protocol uses a terminal deoxynucleotidyl transferase to add labeled dUTP to the fragmented DNA, and this label is then detected by immunocytochemistry. The TUNEL technique is a relatively easy way to obtain a reliable picture of the cell death pattern during normal and abnormal development.

MATERIALS

The recipes for the items marked with <R> begin on page 370.

CAUTION: See the Cautions Appendix for appropriate handling of materials marked with <!>.

Reagents

Alkaline phosphatase buffer <R>
Antidigoxigenin antibody coupled with alkaline phosphatase (anti-dig-AP; Roche) <!>
Blocking solution <R>
Cupiennius embryos, fixed and dehydrated as whole mounts (see Protocol 1; up to 30 embryos per tube)
Digoxygenin-11-2′-deoxy-uridine-5′-triphosphate (DIG-UTP; Roche) <!>
DNAse I (RNAse-free; Roche) <!>
DNAse I buffer <!>
Formaldehyde (37%; Sigma-Aldrich) <!>
Ice
Methanol <!>
4-Nitro blue tetrazolium chloride/5-bromo-4-chloro-3-indolyl-phosphate (NBT/BCIP stock solution; Roche) <!>
Phosphate-buffered saline with detergent (PBST-T) <R>
Proteinase K solution (Roche)
Sodium borohydride <!>
Terminal deoxynucleotidyl transferase (TdT; Sigma-Aldrich)
Terminal deoxynucleotidyl transferase buffer (TdT buffer) <R>
Tris-buffered saline with Tween 20 for TUNEL (PBST-T) <R>

Equipment

Eppendorf tubes (2 ml), round bottom
Fume hood
Glass staining dishes (standard)
Incubation oven, set at 37°C
Rotating wheel

METHOD

Keep the embryos in 2-ml Eppendorf tubes with a round bottom (up to 30 embryos per tube; in tubes with a pointed bottom, the embryos pile up and may be squashed). Perform all washing steps at room temperature unless otherwise specified, on a rotating wheel set to a shallow angle, and at low rotation speed. Important: Include proper negative and positive controls (Steps 18 and 19–23) along with the TUNEL experiment.

Rehydration and Refixation

1. Rehydrate the embryos stepwise in a series of successive incubations in 90, 75, 50, and 25% methanol in PBST-T (1.5-ml total volume) for 5 minutes each.

   Important: Treat enough embryos also for the negative and positive controls described in Steps 18 and 19–23.

2. Wash twice with PBST-T for 5 minutes.

3. Remove PBST-T and add 1 ml of PBST-T supplemented with 5 µg/ml of proteinase K. Incubate for 4 minutes at room temperature. Wash with PBST-T three times to stop the digest.

   Important: For the positive control, go to Step 19 and return here after completing Step 23.

4. Refix in 1 ml of PBST-T with 125 µl of formaldehyde for 20 minutes and wash three times with PBST-T for 5 minutes each.

DNA Labeling

Perform the next steps under the fume hood and with sufficient safety precautions, because solutions will develop explosive or foul-smelling gases.

5. Incubate embryos in 0.1% sodium borohydride ($NaBH_4$) in water (1.5-ml volume) for 20 minutes at room temperature.

   Important: Sodium borohydride solution will develop hydrogen gas. Do not close the tubes (open tubes prevent explosion). Do not expose open tubes to naked light or fire. During the 20-minute incubation, rock the tubes gently from time to time.

6. Quickly rinse embryos with TdT buffer.

   Important: Try to remove as much sodium borohydride solution as possible from the previous step before adding TdT buffer. This is difficult, because many hydrogen bubbles have formed, but this is nevertheless recommended, because mixing the TdT buffer with too much of the sodium borohydride solution leads to a chemical reaction of an unclear nature, the product of which is a foul-smelling gas that should not be inhaled.

   See Troubleshooting.

7. Wash with fresh TdT buffer for 5 minutes.

   The tubes containing the embryos can be removed from the fume hood after this washing step.

8. Incubate in TdT buffer containing 20 µM of DIG-UTP and 0.3 U/µl of terminal deoxynucleotidyl transferase (TdT) for 2 hours at 37°C. TdT and DIG-UTP are very expensive; keep the volume in which the embryos are incubated in this step to a minimum by adjusting it to the number of embryos in the tube. All embryos must be covered by the solution.

   Important: For the negative control (Step 18), omit TdT and replace it with an equal volume of water.

Blocking and Detection

9. Wash three times with TBST (5 minutes each) and incubate in TBST for 20 minutes at 70°C.

10. Wash three times (5 minutes each) in PBST-T and then block in 1.5-ml blocking solution for 1 hour.
11. Incubate in fresh 1.5-ml blocking solution containing 1:2000 anti-DIG-AP antibody for 3 hours.
12. Wash several times with PBST-T (all together, at least 2 hours).
13. Wash overnight with PBST-T at 4°C.
14. Return to room temperature. Wash once with PBST-T for 30 minutes.
15. Wash three times with AP buffer for 5 minutes each.
16. Transfer all embryos to glass staining dishes (about 30 embryos per dish). Stain in AP buffer with NBT/BCIP stock solution (20 μl per 1 ml of AP buffer).
    Important: Color reaction develops rapidly. Be prepared to stop the reaction within minutes after addition of NBT/BCIP solution.
17. Stop staining reaction with several quick washes in PBST-T and refixation in formaldehyde (125 μl in 1 ml of PBST-T) for about 30 minutes.
    Embryos can be stored in the refixation solution for several months at 4°C.

Negative Control

18. Follow the protocol as specified above, but omit the TdT from the enzymatic reaction in Step 8 and replace it with the same volume of water.

Positive Control

19. Execute Steps 1–3 from the main protocol above.
20. Wash three times with DNAse I buffer for 5 minutes each.
    See Troubleshooting.
21. Incubate in DNAse I buffer containing 0.06 units of DNAse I per 1 μl of buffer for 30 minutes at 37°C.
22. Wash three times with PBST-T for 5 minutes each.
23. Proceed with Step 4 of the main protocol above.

TROUBLESHOOTING

Problem (Steps 6 and 20): Embryos stick together and to the wall of the tubes.
Solution: Tween 20 may be added to the DNAse I buffer and TdT buffer. Try to start with 0.02% Tween 20 in the buffers. If this does not solve the problem, increase the amount of Tween 20 to a maximum of 0.1%.

Protocol 4

Gene Silencing via Embryonic RNAi in Spider Embryos

The discovery of RNA interference (RNAi), in which double-stranded RNA (dsRNA) suppresses the translation of homologous mRNA, has had a huge impact on evolutionary developmental biology by enabling the analysis of loss-of-function phenotypes in organisms in which classical genetic analysis is laborious or not possible. This protocol describes the application of RNAi in embryos of the spider *C. salei* (Schoppmeier and Damen 2001).

MATERIALS

CAUTION: See the Cautions Appendix for appropriate handling of materials marked with <!>.

Reagents

Cupiennius embryos, collected as described in Protocol 1
Deionized and RNase-free water
dsRNA targeted against the gene of interest (for synthesis, use, e.g., Ambion T7 kit)
Glue

> *Incubate Tesa tape (Beiersdorf) in heptane for several days; pour off. Supernatant is used as glue and can be stored at room temperature for several months.*

Heptane (100%; Merck) <!>
Phenol red solution (0.5% in DPBS; Sigma-Aldrich) <!>
Reagents for spider embryo dechorionization (see Protocol 1)
Reagents for spider embryo fixation (see Protocol 1)
Voltalef 10S Oil (Arkema)

Equipment

3D micromanipulator (e.g., Eppendorf Micromanipulator 5174)
Agar plate (2%, Agar-Agar, Merck)
Capillaries (e.g., Hilgenberg Borosilicate glass capillary, 1 mm diameter with filament)
Capillary-filling filament tip (e.g., Eppendorf Microloader)
Capillary puller (e.g., Sutter P9 Micropipette Puller)
Coverslips (24 x 24 mm)
Dissecting microscope
Equipment for spider embryo fixation (see Protocol 1)
Equipment for spider embryo dechorionization (see Protocol 1)
Fine brush for lining up embryos
Forceps
Incubator, set at 28°C
Microinjector (e.g., Eppendorf Femto Jet)
Paper tissues
Petri dishes
Plastic Pasteur pipette
Razor blade
Small disposable weighing dishes

METHOD

Perform all steps at room temperature unless otherwise specified.

Embryo Collection and Prearrangements

1. Collect spider embryos within the early embryonic stages (blastula stage or later; 2–3 days after egg laying at 23°C).

   The yolk granules should have sunk down, leaving a space between the embryonic tissue and vitelline membrane (perivitelline space). Earlier stages are difficult to inject, and injections lead to artifacts.
   See Troubleshooting.

2. Dechorionate embryos according to Protocol 1, Step 4, and wash well but not too rigidly with deionized water.

3. Transfer embryos on a 2% agar plate using a plastic Pasteur pipette. Transfer as little water as possible. Remove residual water with a paper towel.

4. Use a fine brush to line up approximately 100 embryos on the agar plate, in arrays of about ten embryos per row and per column. Make sure that embryos do not touch one another.

5. Place a drop of glue on a coverslip and allow the heptane to evaporate (takes about 1–2 minutes).

6. Place the coverslip with the sticky side down onto the embryos and apply slight pressure on the coverslip. As you pick up the coverslip; the embryos will lift onto and stick to it. Let embryos dry for 25 to 35 minutes at 28°C.

   This step is crucial and will enhance injection efficiency.

7. Transfer the coverslip into a small weighing dish that is only slightly larger than the coverslip itself. Add Voltalef 10S oil until the embryos are almost but not completely covered.

   The uppermost part of the egg should be free from oil. The smaller the weighing dish, the less oil that is required.

Injection

8. Dilute dsRNA to a concentration of choice by adding RNase-free water (we recommend a final concentration of between 1 and 4 µg/µl). Add 1/20 volume of phenol red solution to monitor injections. Keep dsRNA on ice.

9. Draw the dsRNA solution into the capillary using a capillary-filling filament tip. Break the capillary tip by squeezing it with a razor blade under a dissecting microscope.

10. Place the weighing dish with embryos onto the dissecting microscope stage and bring the embryos and capillary tip into focus.

11. Position the capillary to inject dsRNA into the space between the embryonic tissue and vitelline membrane (perivitelline space); this works best if the needle angle is at about 65°. Because the vitelline membrane is rather flexible, it will be necessary to provide some pressure to penetrate it. After penetrating the vitelline membrane, pull the capillary back a bit and inject the dsRNA. You can monitor the injections as a red drop appears in the perivitelline space, and the egg may appear to be "growing" by turgor. Pull out the capillary, but beware if too much fluid leaks because this may cause the embryo to die. In addition, avoid penetrating the embryonic tissue or the yolk because this may result in injection artifacts.

    See Troubleshooting.

12. Modulate injection time and pressure according to the volume to be injected.

    dsRNA delivery depends on the capillary (i.e., the settings of the capillary puller and the opening after breaking the capillary) as well as on the injection device. Therefore, it will be necessary to

adjust injection time and pressure for each capillary such that the volume of injected dsRNA remains constant.
See Troubleshooting.

Rearing of the Injected Embryos

13. Cover embryos completely with Voltalef 10S oil after finishing injections.

14. Place the weighing dish with embryos into a Petri dish, place a piece of wet paper towel around the weighting dish (to create a humid atmosphere), close the dish, and incubate at 28°C.
    Because the embryos stick to the coverslip, their ability to perform morphogenetic movements is limited. Therefore, it is necessary to fix the embryos before dorsal closure is completed or to detach them from the coverslip mechanically. Check the injected embryos daily to monitor their development. Ensure that the embryos stay covered with oil.

Fixation

15. When the embryos have reached the desired stage, remove the coverslip from the oil using forceps and carefully decant the Voltalef 10S oil.
    The recovered oil may be reused.

16. Transfer the coverslip into a Petri dish and add heptane until the embryos are completely covered.

17. Rinse carefully in heptane until the oil is washed out and all embryos have detached from the cover slide (this may take up to 5 minutes).

18. Transfer embryos with a plastic Pasteur pipette into a fixation vial, wash twice with heptane, and proceed with the fixation (Protocol 1, Step 6).

TROUBLESHOOTING

Problem (Steps 1 and 11): High lethality rate and/or unspecific phenotypes.
Solution: It is crucial that the embryos are not collected before blastula stage. Early *Cupiennius* embryos are quite yolky and therefore very fragile. Make sure that you do not apply too much pressure during the manipulations. Avoid injections into embryonic tissue or yolk, because this will cause development of unspecific phenotypes and/or high lethality. If too much fluid is leaking from the vitelline membrane, reduce injection pressure to avoid fluid leakage or choose a capillary with a smaller opening.

Make sure that your dsRNA does not cause off-target effects as a result of high sequence similarity to other genes. As a control, set up independent injections with two different dsRNAs that are targeted against two independent areas of the gene of interest. As both dsRNAs are directed against different areas of the same gene, their injection should result in identical phenotypes.

As general controls, use uninjected embryos from the same cocoon that are treated in the same way as the injected ones and also include embryos from the same cocoon injected with dsRNA prepared from unrelated genes such as *GFP* or *lacZ*. Always use embryos from the same cocoon for the experimental and control groups.

Problem (Step 12): Penetrance of RNAi phenotypes.
Solution: There is a direct correlation between dsRNA concentration and penetrance of RNAi phenotypes. Thus, try to increase the dsRNA concentration and/or the volume of injection (time).

Protocol 5

Detection of Cell Proliferation using BrdU Labeling

This protocol describes the detection of proliferating cells in whole-mount *Cupiennius* embryos. When labeled nucleotides are introduced into mitotically dividing cells, these cells incorporate the labels into the newly synthesized DNA. Thus, only cells that have synthesized DNA after the addition of the label will be detected. The present protocol uses 5-bromo-2′-deoxy-uridine (BrdU) as a label that is subsequently detected by immunocytochemistry. BrdU labeling is a relatively easy way to detect cells that have recently synthesized DNA. The main advantage of this technique is that the label accumulates over time and, by varying the incubation time before fixation, an increasingly cumulative picture of cell proliferation activity can be obtained.

MATERIALS

The recipes for the items marked with <R> begin on page 370.

CAUTION: See the Cautions Appendix for appropriate handling of materials marked with <!>.

Reagents

Alkaline phosphatase buffer (AP buffer) <R>
Antidigoxigenin antibody coupled with alkaline phosphatase (anti-DIG-AP; Roche) <!>
5-Bromo-2′-deoxy-uridine Labeling and Detection Kit II (Roche)
  This kit includes BrdU labeling reagent <!>, antimouse antibody (alkaline phosphatase [AP] coupled) <!>, anti-BrdU solution <!>, incubation buffer, and washing buffer.
Cupiennius embryos, live (see Protocol 1)
Formaldehyde (37%; Sigma-Aldrich) <!>
Heptane <!>
Hydrochloric acid (HCl; 2 N) <!>
Ice
Methanol <!>
4-Nitro blue tetrazolium chloride/5-bromo-4-chloro-3-indolyl-phosphate (NBT/BCIP stock solution; Roche) <!>
Phosphate-buffered saline with Triton-X (PBS-Triton) <R>
Reagents for spider embryo dechorionization (see Protocol 1)
Reagents for spider embryo injection (see Protocol 4)

Equipment

Centrifuge tubes (12 or 15 ml), round bottom
Equipment for spider embryo devitellinization (see Protocol 1)
Equipment for embryo injection (see Protocol 4)
Glass staining dishes
Incubation oven, set at 28°C
Microcentrifuge tubes (2 ml), round bottom

Paper pH indicator sticks
Shaking platform or vortex shaker

METHOD

Perform all steps at room temperature unless otherwise specified.

Preparation of the Embryos

1. Obtain live spider embryos from female spiders according to Steps 1–3 of Protocol 1. The developmental stage depends on the developmental process that you wish to study; all embryonic stages can be used for this protocol.

2. Add 10 ml of bleach to the 12- or 15-ml round-bottom tube to remove the chorion from the eggs. Gently rock the eggs in the tube for 2 to 3 minutes. Once the chorion is dissolved, the clear vitelline membrane is visible.

3. Remove the bleach and rinse very carefully three times with water. Check to make sure that no bleach remains (you should not smell bleach anymore); if necessary, perform additional wash steps with water.

4. Prepare the embryos for injection as described in Steps 3–7 of Protocol 4.

Injection of the Embryos

5. Dilute BrdU labeling reagent 1:50 with distilled water (working concentration is 200 µmol/l) (e.g., mix 0.5 µl of BrdU labeling reagent and 25 µl of water). Keep on ice at all times.

6. Add 1/20 volume of phenol red solution to the diluted BrdU labeling reagent. The dye makes it easier to monitor the amount of injected liquid.

7. Fill the capillary with the diluted BrdU labeling reagent and open the capillary by following Step 9 of Protocol 4.

8. Inject the embryos with BrdU labeling reagent according to Steps 10–12 of Protocol 4.

9. Cover embryos completely with Voltalef 10S oil after you have finished injections.

10. Place the weighing dish with embryos into a Petri dish, place a folded piece of wet paper towel into the Petri dish next to the weighing dish (for humidity), close the lid of the Petri dish, and incubate for at least 15 minutes at 28°C. Incubation time can be varied from 15 minutes to several hours, depending on the level of label accumulation desired.

Fixation of the Embryos

11. After incubation, lift the coverslip that carries the embryos with fine forceps and allow as much oil as possible to run off the coverslip.
    The recovered oil may be reused.
    See Troubleshooting.

12. Transfer the coverslip to a Petri dish and carefully add heptane to cover the embryos. Incubate in heptane until the embryos detach (detachment may take several minutes).

13. Carefully transfer the embryos to a round-bottom 12- or 15-ml centrifuge tube and wash three times with fresh heptane to completely remove the oil.

14. Prepare a solution of 10 ml of heptane supplemented with 200 µl of formaldehyde and vig-

orously shake this mixture on a shaking platform (or a vortex shaker) for 10 minutes.

15. As completely as possible, remove the heptane from the embryos with a pipette and add the heptane/formaldehyde mixture (prepared in Step 14). Incubate for 1 hour.
16. Replace the heptane/formaldehyde solution with 10 ml of methanol. Incubate for 5 minutes.
17. Wash twice with 10 ml of fresh methanol for 5 minutes each.
18. Incubate the embryos in methanol for at least 30 minutes at –20°C. This step can be carried out over night.
19. Remove the vitelline membranes from the embryos using two sharp Dumont-5 forceps under the dissection microscope. Transfer embryos to 2-ml round-bottom microcentrifuge tubes (up to 30 embryos per tube).

Immunocytochemistry

20. Wash the embryos three times with PBS-Triton for 5 minutes each.
21. Incubate in 1.5 ml of freshly prepared 2 N HCl for 1 hour.
22. Wash several times with PBS-Triton until the pH is neutral or basic. After each wash, check the waste wash liquid with a paper pH indicator stick.
23. Wash once with Roche incubation buffer for 15 minutes.
24. Dilute anti-BrdU solution 1:10 with incubation buffer and incubate the embryos in this solution for 1 hour. The volume of anti-BrdU solution plus incubation buffer added to the embryos depends on the amount of embryos used; all embryos must be covered with the solution.
25. Wash three times with PBS-Triton for 5 minutes each.
26. Wash three times with PBS-Triton for 15 minutes each.
27. Wash over night with PBS-Triton. This step should be performed at 4°C.
28. Wash once with PBS-Triton for 1 hour.
29. Dilute anti-mouse-AP antibody 1:10 with washing buffer from Roche and incubate embryos in this solution for 1 hour. The volume of anti-mouse-AP antibody plus washing buffer added to the embryos depends on the amount of embryos used; all embryos must be covered with the solution.
30. Wash three times with PBS-Triton for 5 minutes each.
31. Wash three times with PBS-Triton for 15 minutes each.
32. Wash once with PBS-Triton for 5 hours.
33. Wash once with PBS-Triton for 30 minutes.
34. Wash three times with AP buffer for 5 minutes each.
35. Transfer the embryos to glass staining dishes (approximately 30 embryos per dish). Stain in AP buffer with NBT/BCIP stock solution (20 µl per 1 ml of AP buffer) for only a few minutes while you observe color development.

    Important: Color reaction develops rapidly. Be prepared to stop the reaction within minutes after adding the NBT/BCIP solution.

    See Troubleshooting.

36. Stop the staining reaction with several quick washes in PBS-Triton and refix the embryos in formaldehyde (125 µl of formaldehyde in 1 ml of PBS-Triton) for about 30 minutes.

    Embryos can be stored in the refixation solution for several months at 4°C.

TROUBLESHOOTING

Problem (Step 11): Embryos do not detach from the coverslip.

Solution: Check that all embryos are completely covered with heptane and incubate for up to 10 minutes. Collect all detached embryos before forcing the remaining ones to detach. You can use a plastic Pasteur pipette for this: Aspirate the heptane and forcefully blow it out again next to the embryos that have not detached.

Problem (Step 35): Embryos stain completely and appear black or covered with black dots.

Solution: This problem can have two causes. First, the color reaction in Step 35 may have been carried out for too long a time. In most cases, the color precipitate forms very quickly and the reaction must be stopped within minutes after the addition of substrate. If the reaction is nevertheless allowed to continue, a large amount of precipitate will form that also is distributed within the tissue, and this produces a blackish background staining. Try to optimize the staining time by closely monitoring the development of the precipitate. The other possible cause may be that the incubation time in Step 10 is too long. Depending on the developmental stage at which the embryos have been injected with the label, there can be a high rate of cell proliferation; in these cases, an incubation time of several hours could be enough for virtually all cells to undergo mitosis (and incorporate the label). Under these conditions, the embryos appear black because every nucleus is labeled. To determine the incubation time that is best suited for the developmental stage under study, try to work with a small series of different incubation times in parallel.

Protocol 6

Dissecting Embryos for Light Microscopy

Spider embryos can be studied as whole mounts under the dissection microscope. Alternatively, the embryos can be dissected and observed under the compound microscope. Preparing and dissecting spider embryos for compound microscopy is difficult due to the high amount of yolk, which makes the embryos very fragile. The following protocols describe step by step how the necessary tools can be made and then used to obtain good preparations. Not all embryonic stages can be dissected and prepared; very young stages can only be examined as whole mounts (see Step 14).

MATERIALS

CAUTION: See the Cautions Appendix for appropriate handling of materials marked with <!>.

Reagents

Glycerine

Equipment

Bunsen burner
Compound microscope with camera
Coverslips, glass
Eyelashes (or beard-hair cuts)
Insect minutia needles (0.2-mm gauge or thinner)
These are normally used to pin microscopic insect species; obtain from insect collectors or entomology suppliers.
Lighter
Microscope slides, glass
Pasteur pipette, glass
Pipette tip, disposable (for 200-μl pipette)
Tungsten wire (0.125-mm gauge), cut to approximately 3-cm lengths (Agar Scientific Ltd., Stansted, Essex, U.K.)
Watchmaker forceps (Dumont 5)

METHOD

Perform all steps at room temperature.

Producing the Eyelash Tool (see also Fig. 3A, c and d)

1. Briefly hold a disposable pipette tip (preferably a tip for 200-μl pipettes) in the flame of a lighter or Bunsen burner. The tip should begin to melt but should not catch fire.

2. Quickly insert the eyelash with its root into the blob of melted plastic.

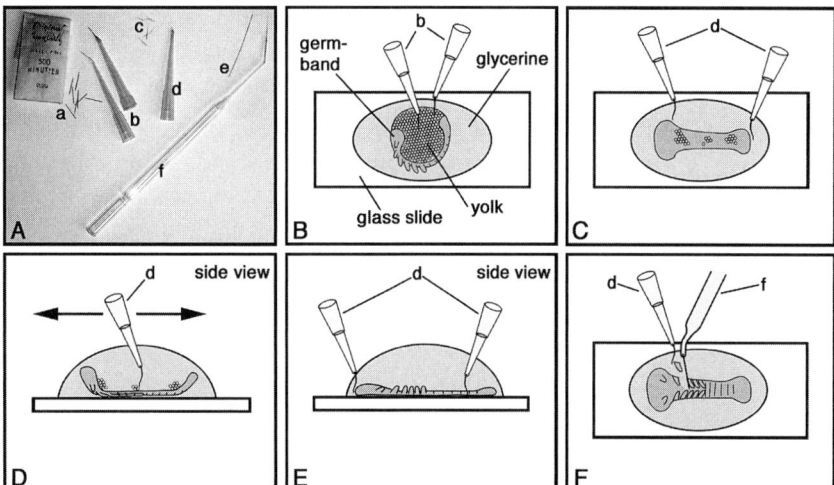

FIGURE 3. Dissection of *C. salei* embryos. (A) Tools needed for dissection: (a) Fine insect minutia needles; (b) microcutter tools with minutia of different strengths inserted at the tip; (c) eyelashes; (d) eyelash tool with an eyelash inserted at the tip; (e) piece of fine tungsten wire; (f) tungsten wire tool made from a glass Pasteur pipette and a piece of fine tungsten wire. (B–F) Illustration showing important steps during embryo preparation; tools are labeled as in A. (B) Embryo is placed on its side and microcutter tools (b) are used to puncture the yolk and remove most of it. (C) Embryo is then placed on its ventral side using the eyelash tools (d). (D) Yolk remaining on the back of the embryo is brushed away with the eyelash tool by moving it carefully back and forth (arrows). (E) Using the eyelash tools, embryo is then turned around and placed with its ventral side up (F). Using the tungsten wire tool (f), an organ of choice (in the example, it is one of the legs) is cut away. The dissected organ should be moved carefully using the eyelash tool.

> *Single lashes can be obtained by holding your lashes firmly between the finger and thumb and pulling abruptly. This loosens older lashes completely with their roots attached. Alternatively, and if more rigid hairs are required, beard hairs (approximately 1 cm long) may be used.*

3. Let the plastic bubble cool. During cooling, it is possible to continue to manipulate the eyelash to find the optimal position—the lash should be inserted straight. The resulting tool should be similar to a paintbrush with just a single hair.

Producing the Microcutter Tool (see also Fig. 3A, a and b)

4. Briefly hold a disposable pipette tip (preferably a tip for 200-µl pipettes) in the flame of a lighter or a Bunsen burner. The tip should begin to melt but should not catch fire.

5. Quickly insert the insect minutia with the stronger end into the blob of melted plastic.
   > *Even in the same batch, insect minutia differ in very much strength and sharpness. Produce several microcutter tools, each with different minutia, in order to find which work the best for your dissections.*

6. Let the plastic bubble cool. During cooling, it is possible to further manipulate the minutia to find the optimal position—the minutia should be inserted at an angle of about 130°. After the plastic hardens, the minutia can be bent to the ideal angle.

Producing the Tungsten Needle Tool (see also Fig. 3A, e and f)

7. Insert an approximately 3-cm-long piece of tungsten wire into the tip opening of a glass Pasteur pipette. The wire should be about halfway inside the pipette tip.

8. Using a Bunsen burner, melt the tip of the Pasteur pipette. As the tip melts, it will bend, but it should nevertheless remain as straight as possible. After melting, check that the wire is firmly inserted and does not come loose from the glass pipette tip.

9. Set the Bunsen burner to welding flame. Insert into the flame almost the entire length of the tungsten wire that sticks out of the glass tip, taking care to keep the glass tip itself away from the flame. Hold the wire in the hottest area of the flame (upper part of the flame) for several minutes. The wire should be white hot at all times.

10. After several minutes, part of the wire tip will break off, producing an extremely fine tip at the wire break point.

    This tool is very delicate and requires some practice for proper use. The fine tip of the melted tungsten wire will bend on the slightest contact with any hard material such as glass or plastic. However, it is flexible enough to withstand contact with embryonic tissue and even vitelline membranes. Tools that have bent tips can be either resharpened in the flame or modified to somewhat more rigid needles by cutting off the bent tip with fine scissors.

Embryo Dissection (see also Fig. 3B–F)

11. Depending on the protocol, immerse the embryos to be dissected in PBS, methanol, or another solution. Transfer the embryos from this solution into glycerine through a series of increasing concentrations of glycerine (10, 50, 80, and 100%). Incubate the embryos for 5 minutes at each step.

12. Transfer single embryos in a drop of glycerine to separate microscope slides.

13. Check that the embryos are covered completely with glycerin, but they should not float. Remove surplus glycerine with a paper towel.

14. Use the microcutter tool to puncture the yolk mass and remove larger chunks of yolk (Fig. 3B). The microcutter tool is very sharp, which is ideal for cutting and removing large chunks of yolk, but special care must be taken not to cut into the embryonic tissue.

    Very young embryos in the stages before formation of the germ band (the basic embryonic body tissue) cannot be dissected. They are composed of a large yolk mass that will burst if touched with any tool. These stages can only be viewed as whole mounts. Old stages after inversion (when the two halves of the germ band have fused dorsally) are difficult to prepare, because their back must be sliced open with the microcutter tool. In these cases, it is easier to cut the embryos into two halves along the anterior-posterior axis and prepare the two halves separately.

15. After the largest part of the yolk has been cut away with the microcutter tool, use the eyelash tool to carefully turn the embryo over (Fig. 3C). The ventral side of the embryo must face the microscope slide. Use the eyelash tool to gently brush the embryo from posterior to anterior and/or vice versa (Fig. 3D); this will remove all remaining yolk granules.

    The brushing may take a while until all yolk granules have been removed. It also requires some practice, because with the brushing movement the tissue of the germ band is stretched. Unpracticed users will often tear off the head or the posterior end, but the ability to use the tool properly soon improves with practice.

16. Using the eyelash tool, turn the embryo over extremely carefully so that its ventral surface faces you (Fig. 3E). You can also use the eyelash tool to carefully flatten out and spread the germ band onto the slide.

17. If you wish to produce organ preparations, proceed to the next step of this protocol. Otherwise, proceed to Step 20.

18. Using the tungsten needle tool, carefully cut away those organs from the germ band that you want to study (e.g., legs, pedipalps) (Fig. 3F). Try to avoid touching the glass slide with the needle during dissection, because this will bend the needle. In older embryos, the tissue can

be very thick; in these cases. it may be helpful to use both the microcutter and the tungsten needle tool to cut away the organ of choice.

19. Use the tungsten needle tool to chip away excess tissue that you do not want in your preparation. After some practice, the needle tool can be used to very precisely dissect any organ of choice and spread it on the microscope slide.

20. Use a blunt object (e.g., the back end of a pen) to break several glass coverslips into smaller shards. Pick up a shard with watchmaker forceps and place the shard over your dissected preparation to cover it.

    The capillary forces will press the shard onto your preparation and flatten it without squashing it. Alternatively, you can use a complete coverslip equipped with wax or plastiline feet, but there is a risk of squeezing the preparation too much when lowering the wax/plastiline feet as you press on the slip. Wear protective clothing (gloves, safety goggles) when producing the glass shards.

21. Store the slide in a horizontal position and at 4°C. Document the dissected preparation as soon as possible by microscopic photography.

Recipes

CAUTION: See the Cautions Appendix for appropriate handling of materials marked with <!>.

ALKALINE PHOSPHATASE BUFFER (AP BUFFER)

Reagent	Quantity (for 100 ml)	Final concentration
Tris-Cl (1 M, pH 9.5)	2.5 ml	100 mM
$MgCl_2$ (1 M)	5 ml	50 mM
NaCl (5 M)	2 ml	100 mM
Tween 20	100 µl	0.1%

Adjust the final volume to 100 ml with DEPC-treated H_2O.

BLOCKING SOLUTION

Prepare PBST-T <R> containing 10 mg/ml bovine albumin and 2% sheep serum. Prepare fresh, do not store, and keep on ice.

DNASE I BUFFER

Reagent	Quantity (for 100 ml)	Final concentration
Tris-Cl (1 M, pH 7.5)	4 ml	40 mM
$MgCl_2$ (100 mM)	6 ml	6 mM
Dithiothreitol (DDT; 100 mM) <!>	0.1 ml	0.1 mM

HYB-A

Reagent	Quantity (ml)	Final concentration
Formamide <!> (Roche)	50	50%
20x SSC (pH 5.0)	25	5x
Tween 20	0.2	0.2%
Salmon testis DNA (20 mg/ml) (Sigma-Aldrich)	1	0.2 mg/ml
tRNA (20 mg/ml) (Sigma-Aldrich)	0.5	0.1 mg/ml
Heparin (50 mg/ml) <!>	0.1	50 µg/ml

Adjust the final volume to 100 ml with DEPC-treated H_2O; adjust pH to 6.5.

HYB-B

Reagent	Quantity (ml)	Final concentration
Formamide <!> (Roche)	50	50%
20x SSC (pH 5.0)	25	5x
Tween 20	0.2	0.02%

Adjust the final volume to 100 ml with DEPC-treated H_2O; adjust pH to 6.5.

PHOSPHATE-BUFFERED SALINE WITH TWEEN-20 (PBST)

Reagent	Quantity	Final concentration
NaCl	7.9 g	135 mM
KCl	0.2 g	2.7 mM
KH_2PO_4	0.180 g	1.5 mM
Na_2HPO_4	1.15 g	8.1 mM
Tween 20	0.2 ml	0.02%

Adjust the final volume to 1 liter with DEPC-treated H_2O; adjust to pH 7.4.

PHOSPHATE-BUFFERED SALINE WITH TWEEN-20 FOR TUNEL (PBST-T)

Reagent	Quantity	Final concentration
NaCl	81.75 g	1.4 M
KCl	2.22 g	30 mM
KH_2PO_4	1.8 g	15 mM
Na_2HPO_4	11.5 g	80 mM
Tween 20	1 ml	0.1%

Adjust the final volume to 1 liter with DEPC-treated H_2O; adjust to pH 7.4.

PHOSPHATE-BUFFERED SALINE WITH TRITON-X (PBS-Triton)

Reagent	Quantity	Final concentration
NaCl	81.75 g	1.4 M
KCl	2.22 g	30 mM
KH_2PO_4	1.8 g	15 mM
Na_2HPO_4	11.5 g	80 mM
Triton-X	1 ml	0.1%

Adjust the final volume to 1 liter with DEPC-treated H_2O; adjust to pH 7.4.

20x SSC

Reagent	Quantity (for 1 liter)	Final concentration
NaCl	175.3 g	3 M
Sodium citrate	88.2 g	0.3 M

Dissolve the reagents in 800 ml of distilled or deionized H_2O and adjust the pH to 7 with a few drops of 14 N HCl <!>. Add water to a final volume of 1 liter; dispense into aliquots.

STAINING BUFFER

Reagent	Quantity (for 100 ml)	Final concentration
Tris-Cl (1 M, pH 9.5)	10 ml	100 mM
$MgCl_2$ (1 M)	5 ml	50 mM
NaCl (5 M)	2 ml	100 mM
Tween 20	200 μl	0.2%

Prepare fresh for each use.

TERMINAL DEOXYNUCLEOTIDYL TRANSFERASE BUFFER (TDT BUFFER)

Reagent	Quantity (for 10 ml)	Final concentration
Tris-HCl (1 M, pH 7.2)	300 µl	30 mM
Sodium cacodylate (1 M) <!>	1.4 ml	140 mM
Cobalt chloride (100 mM) <!>	100 µl	1 mM

TRIS-BUFFERED SALINE WITH DETERGENT (TBST)

Reagent	Quantity (for 100 ml)	Final concentration
Tris-Cl (1 M, pH 7.5)	2.5 ml	25 mM
NaCl (5 M)	2.8 ml	140 mM
KCl (1 M) <!>	270 µl	2.7 mM
Tween 20	100 µl	0.1%

Adjust the volume to 100 ml with H_2O. Dispense into aliquots and sterilize the solution by autoclaving for 20 minutes on liquid cycle. Store at room temperature.

REFERENCES

Barth, F.G. 2001. *Sinne und Verhalten: Aus dem Leben einer Spinne*, pp. 1–424. Springer, Berlin.

McGregor, A.P., Hilbrant, M., Pechmann, M., Schwager, E.E., Prpic, N.-M., and Damen, W.G.M. 2008. My favorite animal: *Cupiennius salei* and *Achaearanea tepidariorum*: Spider models for investigating evolution and development. *BioEssays* **30:** 487–498.

Prpic, N.M. and Damen, W.G. 2005. Cell death during germ band inversion, dorsal closure, and nervous system development in the spider *Cupiennius salei*. *Dev. Dyn.* **234:** 222–228.

Schoppmeier, M. and Damen, W.G. 2001. Double-stranded RNA interference in the spider *Cupiennius salei*: The role of *Distal-less* is evolutionarily conserved in arthropod appendage formation. *Dev. Genes Evol.* **211:** 76–82.

15 The Crustacean *Parhyale hawaiensis*
A New Model for Arthropod Development

E. Jay Rehm, Roberta L. Hannibal, R. Crystal Chaw,
Mario A. Vargas-Vila, and Nipam H. Patel

*Department of Molecular and Cell Biology, Department of Integrative Biology, and
Howard Hughes Medical Institute, University of California, Berkeley, California 94720-3140*

ABSTRACT

The great diversity of arthropod body plans, together with our detailed understanding of fruit fly development, makes arthropods a premier taxon for examining the evolutionary diversification of developmental patterns and hence the diversity of extant life. Crustaceans, in particular, show a remarkable range of morphologies and provide a useful outgroup to the insects. The amphipod crustacean *Parhyale hawaiensis* is becoming established as a "new" model organism for developmental studies within the arthropods. In addition to its phylogenetically strategic position, *P. hawaiensis* has proven to be highly amenable to experimental manipulation, is straightforward to rear in the laboratory, and has large numbers of embryos that are available year-round. A detailed staging system has been developed to characterize the entirety of *P. hawaiensis* embryogenesis. Robust protocols exist for the collection and fixation of all embryonic stages, in situ hybridization to study mRNA localization, and immunohistochemistry to study protein localization. Microinjection of blastomeres enables detailed cell-lineage analyses, transient and transgenic introduction of recombinant genetic material, and targeted knockdowns of gene function using either RNA interference (RNAi) or morpholino methods. Directed genome sequencing will generate important data for comparative studies aimed at understanding *cis*-regulatory evolution. Bacterial artificial chromosome (BAC) clones containing genes of interest to the developmental and evolutionary biology communities are being targeted for sequencing. An expressed sequence tag (EST) database will facilitate discovery of additional developmental genes and should broaden our understanding of the genetic controls of body patterning. A reference genome from the related amphipod crustacean *Jassa slatteryi* will shortly be available.

PROTOCOLS

1. Fixation and Dissection of *P. hawaiensis* Embryos, 384
2. Injection of *P. hawaiensis* Blastomeres with Fluorescently Labeled Tracers, 389
3. Antibody Staining of *P. hawaiensis* Embryos, 393
4. In Situ Hybridization of Labeled RNA Probes to Fixed *P. hawaiensis* Embryos, 396

BACKGROUND INFORMATION

The amphipod *P. hawaiensis* (Dana 1853) is a crustacean species that is particularly well suited for developmental, genetic, and evolutionary analyses, and it has the potential to fill an important taxonomic gap in current comparative studies. Commonly referred to as beachhoppers or scuds, amphipods are malacostracan crustaceans and are therefore closely affiliated with more familiar crustacea such as krill, lobsters, and crabs. Within the taxonomic group Crustacea, amphipods rank as one of the most ecologically successful and speciose extant orders, and they occur in nearly all known marine, fresh, and brackish water environments as well as in high-humidity terrestrial ecosystems such as tidal zones, coastal flood plains, and forest leaf litter (Barnard and Karaman 1991; Lindeman 1991; Vinogradov et al. 1996; Kamaltynov 1999; Sherbakov et al. 1999; Vainola and Kamaltynov 1999; Poltermann et al. 2000; Sheader et al. 2000; Gasca and Haddock 2004). This ecological diversity is matched by a high level of morphological diversity. Several thousand amphipod species have been described, with several new species descriptions occurring every year.

Several aspects of the life history of *P. hawaiensis* make this particular species amenable to many types of classical and modern laboratory analyses and techniques. *P. hawaiensis* is a detritovore that has a circumtropical, worldwide, intertidal, and shallow-water marine distribution (Shoemaker 1956; Barnard 1965), and it may occur as a species complex (Myers 1985). It has been reported to aggregate in large populations ($>3000/m^2$) on decaying mangrove leaf material in environments subjected to rapid changes in salinity (Poovachiranon et al. 1986). The ability to tolerate rapid temperature and osmotic changes characteristic of its preferred shallow-water habitat allows this species to thrive under typical laboratory conditions.

Several characteristics of *P. hawaiensis* have made it particularly amenable to embryological and molecular genetic manipulation. Females produce embryos every 2 weeks once they reach sexual maturity. Embryogenesis is relatively short, lasting approximately 10 days at 26°C. Close examination of the embryonic development of *P. hawaiensis* has produced the most detailed staging system for any crustacean (Browne et al. 2005). Complete embryogenesis has been divided into 30 discrete stages, which are readily identifiable in living animals or by means of common molecular markers in fixed specimens. As direct developers, hatchlings possess a complete complement of segments and appendages morphologically similar to those of adult animals. Females normally brood the embryos in a ventral brood pouch. Embryos can be rapidly and easily removed from the brood pouch and maintained in seawater. Eggs can be collected and hatched individually, and the mature animals can subsequently be used in pairwise sister–brother or mother–son matings to generate inbred lines. Fertilized eggs can be removed from females before their first cleavage and are sufficiently large to perform microinjections (Gerberding et al. 2002) and blastomere isolations (Extavour 2005) with relative ease. Developing *P. hawaiensis* embryos are optically clear, allowing for both detailed microscopic analyses in situ and the use of fluorescently tagged tracer molecules in live embryos. The yolk, although opaque, is sequestered early in development at the center of the developing egg and then later to the developing midgut of the embryo. In addition, early cleavage is holoblastic (total), allowing the fates of individual early cells to be explored through experimental manipulation (Gerberding et al. 2002; Extavour 2005).

In addition to being robust laboratory performers amenable to a wide variety of experimental manipulations, the phylogenetic position of *P. hawaiensis* has important consequences for our understanding of the evolutionary processes responsible for the remarkable radiation of insects and crustaceans. Several recent studies have reexamined evolutionary relationships among the major groups of arthropods. The data suggest two possible relationships between insects and crustaceans. One possibility is that the two groups are sister taxa (Boore et al. 1995, 1998; Friedrich and Tautz 1995; Eernisse 1997; Giribet et al. 2001). The other possibility is a "Pancrustacea" clade, in which the insects branch from within branchiopod crustacea (Regier and Shultz 1997; Hwang et al. 2001). In this scenario, insects represent a terrestrialized branch of crustaceans, and branchiopod crustaceans, which include *Artemia* and *Daphnia*, are more closely related to insects than are malacostracan crustaceans such as crabs, isopods, and amphipods. In either case, malacostracan crustaceans such as amphipods provide a key group for understanding arthropod evolution.

Many of the preceding general considerations of *P. hawaiensis* as well as the following sketches of morphology and embryonic development were extracted and condensed from a more extensive staging paper, which can be consulted for further details (Browne et al. 2005).

SOURCES AND HUSBANDRY

A population of *P. hawaiensis* was originally obtained in 1997 from the marine filtration system of the John G. Shedd Aquarium in Chicago. From this founding population, a laboratory culture of *P. hawaiensis* was successfully established and has subsequently been dispersed to other laboratories in the United States and Europe. Laboratory breeding colonies are maintained in shallow plastic trays (47 × 33 × 12 cm) at 26°C. The containers should remain covered, and the water should be at least 4 inches deep to prevent rapid changes in salinity due to evaporation. A thin (1/4 inch) layer of calcium carbonate gravel spread over the bottoms of tanks serves as a pH buffer as well as a substrate for the animals. Adequate circulation is essential to allow proper biological filtration and gas exchange. This is best achieved by using an airstone and/or a submersible water pump. Illumination provided by fluorescent strip lighting is placed on a day/night cycle of 14/10 hours. Artificial saltwater (Tropic Marin) is prepared to a specific gravity of 1.018–1.022, and after the salt has fully dissolved, it is aerated overnight with an airstone before use. Filtered natural seawater can also be used. Trays should receive a 50% water change each week. Optimal growth is obtained by heating the aquarium room to 26°C, although animals can be grown at 20–26°C. *P. hawaiensis* is maintained on a diet of *Ascophyllum nodosum* kelp granules (Starwest Botanicals) soaked in seawater and a 1:1:1 liquid mixture of fatty acids (Selcon), plankton (MarineSnow), and vitamins (MultiVit, Hawaiian Marine).

Once a tank is established, it is typically only necessary to monitor salinity and temperature. When a tank is first set up, however, it is important to check a number of other parameters. To maintain an ideal pH range of 8.0–8.4, artificial seawater should be made from deionized water and changed weekly, and a calcium carbonate substrate should be used. Failure to clean the tank sufficiently and/or poor gas exchange will lower the pH. Ammonia and nitrites are the natural products of waste protein decomposition and even small amounts are toxic to crustaceans. Test kits are available from aquarium supply stores and detectable levels of either substance indicate that the biological load of the tank exceeds its capacity to process waste. Decreasing the population density and/or changing the seawater should solve the problem. Although nitrates are far less toxic than other nitrogenous compounds, they fuel algae growth, which can be a nuisance. In a well-maintained tank, nitrates should be close to zero and always maintained at less than 10 ppm.

Smaller tanks can be maintained on individual laboratory benches. Lidded plastic food storage containers aerated with small airstones and external aquarium pumps are adequate for this task. Typically, benchtop working populations of *P. hawaiensis* are sustained solely on a diet of baby carrots. The lower population densities in these benchtop colonies together with carrots as a food source results in very large animals. Sexually mature pairs collected from these benchtop tanks routinely result in gravid females with broods of 20–30 embryos. Large broods facilitate the collection of synchronous populations of embryos. For experimental manipulation of embryos, sexually mature pairs in amplexus are isolated in small plastic food storage containers with seawater.

Within 12–24 hours, most of the animals will have separated. The rate of separation may be decreased by keeping pairs at 18°C. Separation occurs as females shed their old cuticle. Once the molt is complete, mature oocytes are released through the oviducts, fertilized, and deposited in the female's ventral brood pouch. Gravid females are anesthetized with clove oil and the embryos are removed from the brood pouch (see Protocol 1). Following embryo collection, females are allowed to awaken and are then returned to a benchtop tank. Low-density working populations can be supplemented, as needed, from the larger tanks.

Embryos should be serially rinsed in three to four washes of seawater in medicine cups, after which they may be placed in filter-sterilized seawater (.22-μm Millipore filter [Bedford, Massachusetts]) in 35-mm tissue culture dishes (Falcon) and maintained in a high-humidity chamber at 26°C until they reach specific time points for manipulation, dissection, and/or fixation. Under these conditions, embryonic development will proceed as described by Browne et al. (2005). Adult females should be allowed to awaken for at least 30 minutes in 500 ml of seawater before being returned to an aerated benchtop tank. Amphipods and their embryos must not be left too long in clove oil.

RELATED SPECIES

Jassa slatteryi and *P. hawaiensis* are relatively closely related amphipods. From cDNA sequence comparisons, we estimate that about the same molecular distance separates these two amphipods as separates *Drosophila melanogaster* and *D. virilis*. The embryonic development patterns of *J. slatteryi* and *P. hawaiensis* appear to be largely identical; if not for the smaller size of early *J. slatteryi* embryos, they would be difficult to distinguish morphologically from those of *P. hawaiensis*. What does distinguish *J. slatteryi* is the relatively small size of its genome. At 690 Mb, the *J. slatteryi* genome is only 10 Mb larger than the smallest recorded genome within the malacostracan crustaceans. Unlike *P. hawaiensis*, however, *J. slatteryi* has proved to be difficult to culture in the laboratory. Obtaining wild-caught *J. slatteryi* adults is reasonably straightforward, because they live along the Northern California coastline and in the San Francisco Bay. Efforts are under way to develop an inbred laboratory population of *J. slatteryi*.

J. slatteryi is an approved sequencing target of the National Human Genome Research Institute's (NHGRI) Large-Scale Genome Sequencing Program (http://www.genome.gov/10002154 and http://www.genome.gov/Pages/Research/Sequencing/SeqProposals/EcdysozoaProposalFinalPDF.pdf). The *J. slatteryi* genome will be sequenced with fivefold coverage by the Broad Institute at the Massachusetts Institute of Technology. In addition, *J. slatteryi* BAC and cDNA libraries are being constructed. The Broad Institute will generate paired-end reads from 100,000 embryonic cDNAs and sixfold BAC end sequencing to complement the *J. slatteryi* genome sequence.

USES OF THE *P. HAWAIENSIS* MODEL SYSTEM

Although *P. hawaiensis* has been used sporadically for toxicological and ecological studies throughout the years, a concerted effort has recently been made to establish the amphipod *P. hawaiensis* as a new crustacean model organism for studying the relationship between development and evolution (Browne et al. 2005).

Body Plan of *P. hawaiensis*

The body plan of *P. hawaiensis* (Fig. 1) is organized around a series of repeating segmental units along the anterior-posterior (AP) axis. Several synapomorphic characteristics clearly unite the Amphipoda (Schram 1986; Schmitz 1992). The orientation (pointed anterior) of T4 and T5 walking appendages relative to those of T6–T8 (pointed posterior) is responsible for the name of the group: amphipod. Other recognizable characteristics are lateral compression of the body, sessile compound eyes, and large coxal plates attached dorsally to the base of thoracic appendages. As is the case in most amphipods, the *P. hawaiensis* cephalon (head) is composed of the six anteriormost segments. The anteriormost preantennal segment bears no paired appendage. The remaining five segments do possess paired appendages. From anterior to posterior, the paired appendages of the head are the uniramous first antenna (An1), uniramous second antenna (An2), gnathoba-

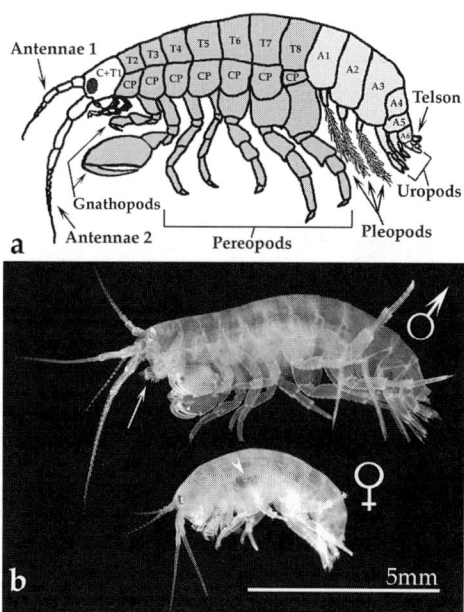

FIGURE 1. The *P. hawaiensis* body plan. (a) Schematic diagram of the adult body plan. By convention, the head (*white*) consists of the anteriormost six segments (termed the cephalon [C]) plus the first segment of the thorax (T1). All head segments posterior to the second cephalic segment bear a pair of appendages. From anterior to posterior, these appendages are as follows: antennae 1, antennae 2, mandibles, first maxillae, second maxillae, and the maxillipeds of the first thoracic segment (T1). (*Dark gray*) The pereon, composed of T2–T8. Each thoracic segment of the pereon possesses a pair of appendages, and the most proximal element of each appendage, the coxa, has a dorsal branch that is compressed and expanded into a structure called the coxal plate (CP). Appendages T2 and T3 are termed gnathopods and are distinctly subchelate (clawed). The T3 gnathopod is sexually dimorphic: In mature males, it is greatly enlarged (compare male to female in *b*). Segments T4–T8 possess appendage pairs termed pereopods that function primarily for locomotion. (*Light gray*) Abdominal segments A1–A6. A1–A3 form the pleon, and each bears a pair of appendages termed pleopods. The final three segments of the abdomen (A4–A6) form the urosome, and each of these segments bears a pair of appendages termed uropods. At the very posterior is the telson, which is a cleft flap of cuticle just posterior and dorsal to the anus. (*b*) Sexually mature animals possess a number of dimorphic characters. Males are larger than females, and T3 appendages are greatly enlarged in males. Females possess a ventral brood pouch in which they incubate their eggs until hatching (*arrowhead*). All amphipods retain a highly compact arrangement of mouthparts termed the buccal mass (*top, arrow*).

sic mandibles (Mn), biramous first maxillae (Mx1), and the biramous second maxillae (Mx2). In addition, the first thoracic segment (T1) is fused to the cephalon. The T1 appendage pair, the maxillipeds, are triramous, fused at their base and extensively modified to assist in feeding. There is a close arrangement of the gnathal appendages, including the maxillipeds, in a basket shape around the mouth to form a highly compact buccal mass. The next seven segments of the thoracic region, T2–T8, articulate independently and bear a pair of appendages. Appendages T2–T5 of sexually mature females possess an additional ventral appendage branch, termed an oosteogite, which interlocks to form the protected ventral brood pouch into which fertilized eggs are shed and incubated until hatching. The T2 and T3 appendages, termed gnathopods, are subchelate and function in grasping and mating. The T3 appendage is sexually dimorphic (larger on mature males). The T4–T8 appendages, functioning in locomotion, are termed pereopods.

The abdominal region is composed of the next six segments and is divided into two regions: segments A1–A3 constitute the pleon and A4–A6 the urosome. Each abdominal segment bears a pair of biramous appendages. A1–A3 appendages are termed pleopods and are used both for swimming and for drawing water across the ventrally located thoracic gills. Each pair of pleopods is fused at its base along the ventral midline of the animal and is highly setose. The appendages of

A4–A6 are termed uropods. These appendages are reduced and thickened and bear a number of stout spikes along their proximodistal axis. The terminal structure, the telson, is cleft and reduced in size relative to comparable structures of other types of crustaceans.

Developmental Staging of *P. hawaiensis*

Parhyale embryogenesis is briefly described here with particular attention to morphological signposts that are helpful for staging embryos. Key time points of *P. hawaiensis* development are further illustrated by means of live bright-field and matching postfixation DAPI (4′-6-diamidino-2-phenylindole, a fluorescent stain for nuclei) images, as well as bright-field anti-Engrailed stains. For more detailed descriptions and additional imagery, consult Browne et al. (2005). All developmental times refer to embryos maintained at 26°C.

Mating and Fertilization

Sexually mature male and female *P. hawaiensis* form mating pairs in which males grasp and hold smaller females with their second thoracic appendages (T2, gnathopods) until mating occurs. The pairs remain in premating amplexus until the female molts, at which point the male deposits sperm into the female's paired oviducts and releases her. Before the female's new cuticle hardens, she sheds her eggs into a ventral brood pouch through two bilaterally symmetric oviducts, fertilizing them in the process.

Stages 1–4: Oocyte to Eight-cell Embryo (Fig. 2)

Within 1 hour of being deposited into the brood pouch, eggs assume an elliptical shape as their outer chorion shells harden. The oocyte/yolk mass usually appears purplish and is consistent for all embryos within a given brood. After 4 hours (26°C), a total first cleavage occurs that is meridional and perpendicular to the egg's long axis, resulting in a two-cell embryo. A second total cleavage that is meridional and parallel to the long axis of the egg results in a four-cell embryo. At 8 hours, a highly unequal and equatorial third cleavage occurs, perpendicular to the first two planes of cleavage, resulting in an eight-cell embryo with four distinct macromeres (Mav, Er, El, and Ep) and four micromeres (g, mr, ml, and en). An interesting and unusual feature of *P. hawaiensis* embryogenesis is that each cell of an eight-cell embryo already bears a distinct fate (Gerberding et al. 2002). At the eight-cell stage, individual identities of each macromere and micromere can be determined by size, shape, refringency, and relative position of the blastomeres.

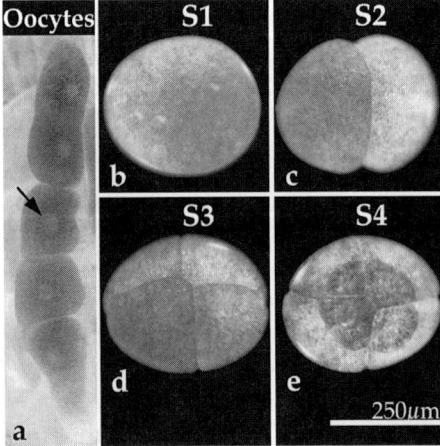

FIGURE 2. *P. hawaiensis* oocytes and stages 1–4 (S1–S4). (a) Right ovary with six mature oocytes (anterior up, posterior down). The oocyte nucleus (arrow) is visible as a white oval in the yolk-laden oocyte. (b) Stage 1 (S1); single uncleaved cell. (c) Stage 2 (S2); the first cleavage plane is meridional and slightly asymmetric. (d) Stage 3 (S3); the second cleavage plane is meridional and also slightly asymmetric. (e) Stage 4 (S4); vegetal view, the third cleavage plane is equatorial and highly asymmetric, generating an eight-cell embryo composed of four macromeres and four micromeres.

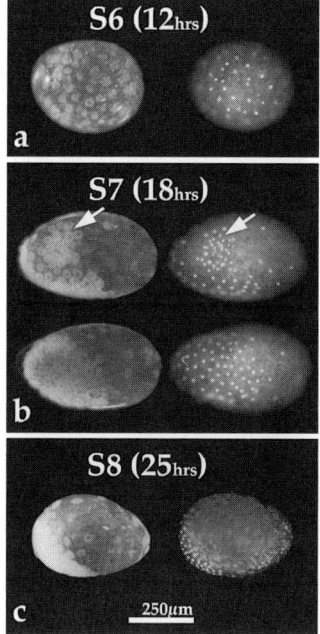

FIGURE 3. *P. hawaiensis* stages 6–8. Incident light bright-field photos of living embryo (*left*) and matching DAPI image after fixation (*right*). (a) Stage 6 (S6), "soccerball" stage; most cells are approximately the same size and evenly distributed around the egg periphery. (b) Stage 7 (S7), "rosette" stage; same embryo shown in two orientations. Top image pair (lateral view) shows position and distribution of cells and their nuclei in the rosette. *White arrows* indicate the position of rosette cells and nuclei. Lower image pair (animal view) shows the large cluster of cells and nuclei migrating into the position of the future germ disc (there is a slight rotation between bright-field and DAPI images). (c) Stage 8 (S8); formation of the germ disc at the anterior end of the egg. Gastrulation and the establishment of multiple germ layers occur as the germ disc is condensing.

Stages 6–8: Blastoderm, Gastrulation, and Germ-disc Condensation (Fig. 3)

By 12 hours of development, embryos reach the "soccerball" stage (stage 6). The yolk has been displaced toward the center of the egg, which is largely devoid of nuclei. Similarly sized cells with distinctive whitish cytoplasm are distributed evenly along the egg periphery. The first significant cell migrations are associated with early gastrulation, and they occur between 12 and 18 hours of development. By stage 7 (18 hours), a majority of the cells have aggregated in the presumptive ventral end of the embryo, where they assume a hexagonal shape and begin forming a tight sheet. A second distinct cluster of cells, the "rosette," derives from Mav and g descendants. The rosette marks the position of the first group of cells that will move beneath the peripheral surface of the embryo. The rosette further indicates the future anterior end of the embryonic AP axis and, once it has migrated under the ectoderm, provides the first evidence of multiple germ layers. A third group of cells remain relatively stationary, scattered across the remaining embryo. As these cells continue to proliferate, their descendants either migrate into the condensing germ disc or remain stationary outside of the embryo proper. At stage 8 (25 hours), embryos are defined by germ-disc condensation, the gradual aggregation of cells along the anterior ventral region of the egg. Rosette cells are now positioned under the ectoderm anlagen and begin to move to their final positions. A depression in the germ disc becomes visible, directly above the rosette cells. This is shortly followed by the migration of ml and mr micromere descendants, under the ectoderm anlagen regions, that flank the rosette. The further development of the germ disc is characterized by continued cell proliferation and a general reduction in cell size.

Stages 11–18: Germ-band Formation and Elongation (Fig. 4)

The germ disc continues to grow and organize via proliferation as well as the recruitment of cells from both posterior and lateral positions. By stage 11 (60 hours), cells within the germ disc begin organizing into head lobes and germ band, fixing the relative positions of both anterior/posterior and dorsal/ventral axes relative to the egg axis. Patterning of the embryonic ectoderm posterior to the head results from the condensation of a grid-like array of cells, the parasegment precursor rows (PSPR), which subsequently undergo two rounds of wave-like, mediolateral, mitotic division

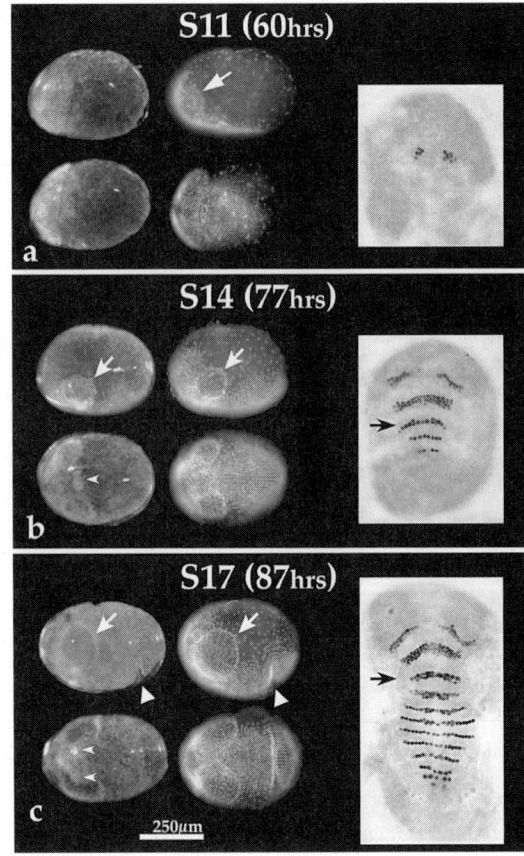

FIGURE 4. *P. hawaiensis* stages 11, 14, and 17 with Engrailed expression. Live bright-field and matching postfixation DAPI images showing lateral (upper image pairs) and ventral (lower image pairs) views. Engrailed (En) expression for the corresponding stage is shown at the right, where all embryos are oriented anterior-up. Anti-En staining (with MAb 4D9) is black in these bright-field images. The black arrows mark the position of the mandibular (Mn) segment En stripe. (a) Stage 11 (S11); formation of ectodermal cell rows. The top image pair (lateral view) shows the aggregation of cells forming the midgut anlagen (*white arrow* indicates midgut anlagen nuclei in the DAPI field). The bottom image pair (dorsal view) shows the organization of the ectoderm into transverse cell rows. First expression of En initiates in An1. (b) Stage 14 (S14); bilobed germ-cell cluster. Midgut anlagen becomes ovoid (*white arrow*). Germ-cell clusters at the midline begin to separate from one another (*white arrowhead*). Organization of the ectoderm into transverse cell rows is clearly visible in the DAPI image of the ventral view. En expression is throughout the cephalon, with the fifth stripe corresponding to the second maxillae segment (Mx2). (c) Stage 17 (S17); ventral flexure visible. *White triangles* are positioned where posterior ventral flexure initiates (T4/T5). *White arrows* indicate the position of the expanding midgut anlagen. *White arrowheads* flanking the midline in the ventral view (lower) indicate position of the migrating germ-cell clusters. Ventral flexure is clearly visible in the DAPI images. En expression is in 12 stripes; the 12th stripe corresponds to T7.

to generate four transverse rows of cells that constitute a parasegment. Further divisions, although reproducible, occur in much more complex patterns and begin to define primordia of ectodermal derivatives such as appendages and the central nervous system. Unlike the ectoderm, the underlying segmental mesoderm is generated via mesoteloblastic growth, like in other crustaceans. Mesoteloblasts pass through highly ordered series of asymmetric divisions to generate progeny called mesoblasts. As they do so, the mesoteloblasts shift posteriorly beneath the developing ectodermal grid. The resulting mesoblasts are organized into rows beneath the posterior compartment of each of the forming parasegments, ultimately producing the somatic mesodermal derivatives of the associated segment. By stage 14 (77 hours), the bilateral midgut anlagen have become two well-defined circular structures flanking segments T2 and T3. These will dramatically increase in size until they meet each other along their most ventral margins and begin to fuse. During development of the embryonic germ band, the germ cells become clearly visible underneath the developing maxillae-1 (Mx1) and maxillae-2 (Mx2) segments. Initially, they are present as a single medial cluster of large whitish cells underneath the ectoderm. By stage 17 (87 hours), the germ cells have separated into two bilaterally opposed whitish clusters, which then begin to move laterally and posteriorly away from the embryonic midline. Anterior appendage primordium have begun to elongate, whereas thoracic appendages are visible as buds. As the germ band elongates, it begins to fold inward toward the egg interior at segment T5. This posterior (ventral) flexure then expands to include more anterior and posterior segments, appearing as a clear furrow in the germ band when viewed from the side.

Stages 19–30: Appendage Formation, Organogenesis, Neurogenesis, and Hatching (Fig. 5)

The posterior (ventral) flexure becomes increasingly pronounced, broadening into a distinctive paddle-shaped structure by 96 hours. By 112 hours, the posterior paddle has significantly narrowed, limb buds are becoming apparent along the abdomen, and the developing telson is clearly visible and oriented toward the anterior. By stage 21 (120 hours), all abdominal segments possess visible

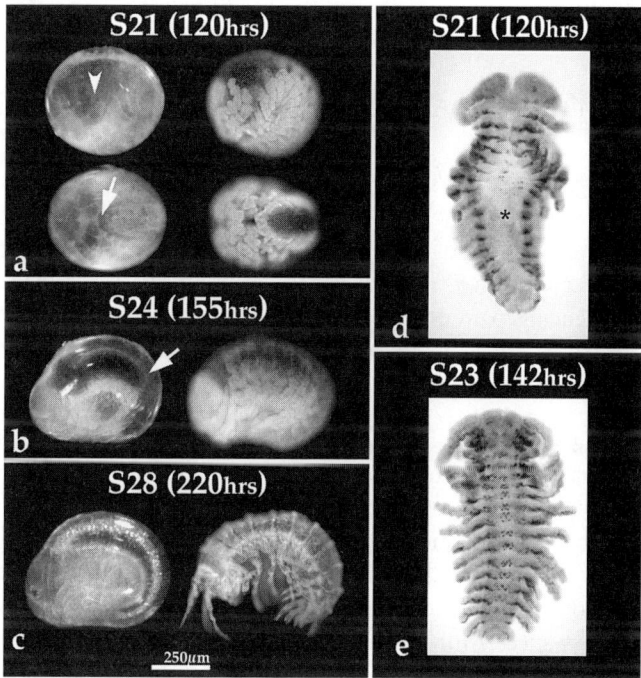

FIGURE 5. *P. hawaiensis* stages 21, 24, and 28 with Engrailed (en) expression. Live bright-field and matching postfixation DAPI images showing lateral (upper image pairs) and ventral (lower image pairs) views in a and only lateral views in b–c. In d–e, the bright-field images of En expression are oriented such that anterior is up. (a) Stage 21 (S21); hindgut proctodeum becomes visible. *White arrowhead* on lateral view (*upper*) shows position of the germ-cell clusters. In ventral view (*lower*), the hindgut proctodeum (*white arrow*) is visible as a wedge-shaped ridge of cells at the posterior end of the embryo. (b) Stage 24 (S24); midpoint of digestive cecae extension. The *white arrow* shows the posterior extent of extending cecum (~T7). Yolk granules have cleared from the cephalon. Body tergites and coxal plates are visible. (c) Stage 28 (S28); yellow cuticle. As the cuticle continues to accumulate, the embryo acquires a yellow-gold appearance. Eye rhabdomere fields appear intensely red, limbs have acquired their final morphology, and the digestive cecae are significantly depleted of yolk stores. (d) En expression, stage 21 (S21); all segments are En positive. *White asterisk* indicates the prominent and transient ventral midline split. (e) En expression, stage 23 (S23); the ventral midline cleft has closed and all segments possess a subset of En-positive neurons.

limb buds, and the hindgut proctodeum is just becoming visible at the posterior terminus. The developing midgut can be seen as an organized epithelial sheet spreading over the yolk. The germ-cell clusters are visible in the lower hemisphere of the midgut anlagen and are beginning to migrate dorsally in the embryo. By 132 hours, the midgut completely envelops the remaining yolk, and thoracic limb buds have begun elongation. At 144 hours, dorsal closure of the ectoderm over the gut appears to be nearly complete, and the digestive cecae first become apparent as bud-like posterior projections from the now wedge-shaped midgut anlagen. Digestive cecae are major secretory organs consisting of a pair of blind tubes that open directly into the anterior end of the medial midgut. By stage 24 (155 hours), the digestive cecae have extended approximately midway and are nearing the abdomen. By the end of digestive cecae extension (168 hours), most yolk has relocated from the developing medial midgut to the digestive cecae. Eye fields become distinctly visible as two bilaterally symmetric, whitish hued clusters of rhabdomeres by 180 hours. At this time, the tubular heart, located along the dorsal thoracic region, begins to visibly beat. At 192 hours, pigment cells that surround the whitish rhabdomeres begin to develop a deep red color. By stage 28 (216 hours), the embryo cuticle acquires a yellow-gold appearance. Beyond this point, detection of gene expression by in situ hybridization becomes problematic because the cuticle has thickened. The digestive cecae are significantly depleted of yolk stores, and the heart is beating strongly. Hatching occurs at approximately 250 hours of development. Immediately before hatching, the embryo is in a constant state of coordinated motion. In addition, the embryo increases substantially in size immediately before and during hatching, most likely via active uptake of water. Soon after breaking free of the eggshell, hatchlings undergo a molt and revert to a smaller size.

GENETICS, GENOMICS, AND ASSOCIATED RESOURCES

The genome of *P. hawaiensis* is estimated to be 3.6 Gb in size (A. Aboobaker and N. Patel, unpubl.). The Joint Genome Institute (JGI) has undertaken both EST and directed BAC sequencing of *P. hawaiensis*. Both of these sequencing projects are nearing conclusion. In addition, the related amphipod crustacean *J. slatteryi*, whose genome size is 690 Mb, has been selected for fivefold genome sequencing, paired-end reads from 100,000 cDNA clones, and sixfold BAC end sequencing.

EST Sequencing

Using a collection of *P. hawaiensis* embryos spanning the first two thirds of embryogenesis (0–156 hours), a normalized total embryonic cDNA library with an average insert size of 2.2 kb was constructed for EST sequencing; 29,566 cDNA clones were end-sequenced using the traditional methods of Sanger sequencing. Of the sequences passing quality control to cluster, 90% yielded a total of 47,732 unique ESTs with a mean read length of 670 bp. These ESTs were parsed into 13,366 clusters of highly related sequence. The resulting cluster distribution suggests that the source cDNA library successfully normalized gene coverage of the embryonic transcriptome: 68% of the total clusters contained sequence from only a single clone and only 66% of passing clones formed clusters with greater than 1 clone/cluster. A searchable web-based database of *P. hawaiensis* EST sequences is being developed at JGI, which will provide the scientific community ready access to these data.

To supplement the EST sequences, JGI will perform an additional single 454-FLX run on sheared cDNA to generate about 80 Mb of data in approximately 200-bp reads (400,000–600,000 clones). JGI will cluster the 454 ESTs with the ESTs already run on the Sanger platform.

Directed BAC Sequencing

A *P. hawaiensis* BAC library (F. Poulin and N. Patel, in collaboration with the C. Amemiya laboratory, unpubl.) was constructed from nuclear DNA isolated from a isofemale line "iso2" that was established approximately 6 years ago. With an average insert size of 120–150 kb, the library was designed to cover the entire genome approximately five times. Because the size of *Parhyale* genome is estimated to be 3.6 Gb, an array of 129,024 clones was plated onto seven high-density filters. Screens with single-copy genes confirmed fivefold to sevenfold coverage.

Probes made from cDNA sequence from about 60 developmentally important genes were used to screen the BAC library filters, using standard techniques for radioactive hybridization. Positive BAC clones for each probe were sized, screened by polymerase chain reaction (PCR), and restriction-mapped to ensure that the regions of interest were centered before being sent for sequencing. The Stanford Human Genome Center is in the process of sequencing 70 BAC clones.

TECHNICAL APPROACHES

The following four protocols describe essential laboratory techniques for the study of embryogenesis in *P. hawaiensis*.

Protocol 1

Fixation and Dissection of *P. hawaiensis* Embryos

This protocol describes the dissection and fixation of *P. hawaiensis* embryos. Embryonic tissue fixed in the following manner is suitable for in situ hybridization experiments to study mRNA expression or for immunocytochemistry to study protein localization.

MATERIALS

The recipes for the items marked with <R> begin on page 400.

CAUTION: See the Cautions Appendix for appropriate handling of materials marked with <!>.

Reagents

Clove oil (Oshadhi)
Formaldehyde (37%; Fisher) <!>
Methanol (100, 90, 70, and 50%; all dilutions in PT <R>) <!>
NaOH (1 M) <!>
P. hawaiensis (gravid females)
PT <R>
Seawater (filter-sterlized) <R>

Equipment

Alligator clamp leads (two) that can be plugged into the transformer
Autotransformer (variable type 3PN1010, 0–120 V, 10A; Staco Energy Products Co.)
Beaker (50 ml)
Bunsen burner
Centrifuge tubes (50 ml; BD Biosciences)
Forceps
Humidity chamber
> *This chamber is a small, lidded plastic box either lined with moistened paper towels or containing an inverted pipette tip holder and filled with 2 mm of water.*

Incubator, set at 26°C
Medicine cups
Microcentrifuge tubes
Needle (26G1/2; BD Biosciences)
Pasteur pipettes (5.75 inch; VWR)
Petri dish (100 x 15 mm) coated with Sylgard 184
> *Dissection dishes are prepared by coating a plastic Petri dish with 2–3 mm of Sylgard 184 following instructions for the Silicone Elastomer Kit (World Precision Instruments).*

Pyrex spot plate (3 or 9 well)
Syringe (1-ml tuberculin slip tip; BD Biosciences)
Timer (optional; see Step 18)
Tissue culture dish (35 x 10 mm; Falcon)
Transfer pipettes (Samco Scientific Corp.)
> *Clip the tip of a plastic transfer pipette for use in Step 13.*

Tungsten wire (0.005-inch diameter; Ted Pella)

METHOD

Making Dissection Tools

Pasteur Pipettes

Drawn and blunted Pasteur pipettes are used to extract embryos from brood pouches of gravid female P. hawaiensis. Wear eye protection to protect against small pieces of flying glass when breaking drawn pipettes.

1. Heat a glass pipette, about 1 inch from the end, over a Bunsen burner flame until the glass softens. Remove from the heat and pull gently on the end with forceps, drawing the glass to a finer point.
2. Break the drawn end carefully by flexing it with forceps. Discard the small piece.
3. Heat the newly formed end so that it rounds up.
   The objective is to obtain a small and rounded end, which fits easily into the brood pouch and does no harm to the female.

Tungsten Needles

Fine tungsten needles are used for dissecting P. hawaiensis embryos away from their surrounding membranes. They are honed to a fine point electrochemically. Wear protective eyewear and gloves while sharpening the needles. Care should be taken to avoid electric shock.

4. Thread a short piece of 0.005-inch diameter tungsten wire through a 26G1/2 needle. Crook the back end of the wire so that it stays attached to the needle. Attach the needle to a 1-ml syringe.
5. Set up a 50-ml beaker with 1 M NaOH.
6. Attach an alligator clamp from one lead to the beaker lip, ensuring that the lead makes contact with the NaOH solution. Attach the clamp from the other lead to the base of the needle.
   Take care to avoid touching the two leads together because this may cause a shock and damage the transformer.
7. Make sure that the variable autotransformer is turned off. Plug both leads into the transformer. Adjust the output knob to no more than 5% of maximum voltage output (~6 V).
8. Turn on the transformer while holding the needle with its attached lead by the plastic barrel of the syringe.
9. Dip the tip of the needle into the NaOH solution, which completes an electrical circuit.
   Small bubbles at the needle's point of contact with the solution will indicate that the tungsten is dissolving. Typically, less than 1 minute of steady up and down motion produces a sufficiently sharp needle tip.
10. If unsure of the proper voltage setting, attach the leads and dip the needle into the NaOH solution with the transformer turned off. While holding the plastic syringe, turn on the transformer with the output voltage knob positioned at zero. Slowly increase the transformer's voltage until bubbles begin to form.

Extracting *P. hawaiensis* Embryos from Gravid Females

Gravid females are anesthetized and the embryos are scooped out of the brood pouch.

11. Add 10 µl of clove oil to 50 ml of filtered seawater in a centrifuge tube. Shake vigorously for 5 seconds.

12. Collect gravid *P. hawaiensis* in a medicine cup. Remove as much seawater as possible before replacing it with the clove oil/seawater mixture.

    The animals should be fully asleep within 10–15 minutes.

    Typically, up to 20 gravid females can be anesthetized in 10 ml of the clove oil/seawater mixture. The animals should not be left in the clove oil/seawater mixture for longer than 1 hour.

13. Transfer the anesthetized amphipods in a drop of clove oil/seawater mixture to a plastic Petri dish that has been coated with Sylgard.

    Females may be moved using forceps or a plastic transfer pipette whose tip has been clipped to allow the animals to pass freely.

14. Under a dissecting microscope, expose the brood pouch by using forceps to gently hold the animal on its back. Gently plunge forcep tips into the Sylgard surface of the dish, straddling the gravid female, to ensure that it rests on its back and the brood pouch remains exposed. Extract the embryos by sweeping the blunted end of a drawn Pasteur pipette through the brood pouch in a posterior-to-anterior direction.

    One-cell embryos recently deposited into the brood pouch are particularly soft and difficult to remove without damage. The one-cell stage lasts for about 4 hours at 26°C. It is best to wait 1 hour following deposition before extracting one-cell embryos from the brood pouch.

15. Wash the embryos with three or four changes of sterile seawater in medicine cups. Store the embryos in 35-mm tissue culture dishes in sterile seawater until ready for use. Keep in small humidity chambers to avoid evaporation. Periodically change the seawater to prevent fungal infections.

    At 26°C, embryos will develop in accord with the established staging system (Browne et al. 2005). Embryos incubated at 18°C will develop significantly more slowly, but otherwise normally.

16. Revive the females by transferring them to a medicine cup of seawater. Ensure that they are fully awake before returning them to tanks with other amphipods.

Dissection and Fixation of *P. hawaiensis* Embryos

Embryos are hand-dissected away from membranes in a seawater/formaldehyde fix.

17. Place a few embryos (start with two or three and increase with experience) in one well of a 3- or 9-well Pyrex spot plate containing fixative (9 parts seawater to 1 part 37% formaldehyde).

    Glass pipettes can be used to move embryos from one well to another, but designate different pipettes for the different solutions so that fixative is not inadvertently introduced into a dish of living embryos. See Troubleshooting.

18. Poke a hole in the egg shell with a dissecting needle, choosing a region of the egg with little embryonic tissue. Fixative will begin to enter the egg, so start a timer if fix duration is critical (see Step 22). If necessary, stabilize the embryo with either a pair of forceps or a second dissecting needle while poking the hole.
    - Very young embryos (0–18 hours) are quite yolky and contain only a single membrane. It is difficult to maintain the embryo's shape when dissecting away this membrane. Begin by poking a shallow hole. It may help to keep the embryo in fix for 5 minutes before making the hole. Start the timer, as usual, only AFTER making the hole.
    - Older embryos (1–2 days) are relatively easy to dissect because they are surrounded by only a single membrane and the embryo has condensed to one side of the egg. Target the side of the egg with little embryonic tissue for the initial hole and subsequent dissection. This allows the membrane to be easily removed without damaging many embryonic cells.

19. Allow the fixative to enter the embryo for 1 minute. This will assist in the subsequent dissection.

    Do not fix too long before completing dissections, however, because the membranes will begin to irreversibly attach to the embryo.

20. Remove the outer membrane by carefully reentering the initial hole with two dissecting needles and tearing a larger opening.

    This often results in loose flaps of membrane that can then be peeled away using forceps or dissecting needles. Alternatively, the outer membrane pops off the embryo as a more or less intact capsule.

21. If the inner membrane has not been removed with the outer, repeat Step 20 to remove the inner membrane.

    At various stages, the inner membrane routinely remains snugly wrapped around the embryo/yolk mass after removal of the outer eggshell. The inner membrane is visible as a slight sheen over the opaque, whitish embryo. This inner membrane is often quite challenging to remove. If left more than a few minutes in fix, the inner membrane tends to fuse to the embryo and yolk, making it difficult or impossible to remove.

    - After 2.5–3 days, the embryo has developed an inner membrane that frequently does not come off with the outer membrane. Poke the initial hole along the dorsal midline of the embryo, which is readily identified by the small number of large cells that will give rise to the dorsal organ. Most of the embryo lies on the other, ventral side of the egg. Do not attempt to dissect too many embryos at once because the inner membrane can quickly become permanently fixed to the embryo. Quick and efficient removal of the outer membrane enables the inner membrane to be peeled away before becoming overfixed.

    - Once the embryo has developed appendages (after 4 days), it can be difficult to remove the membranes without dismembering the animal. Poke a hole along the dorsal midline, just behind the embryo's head. It may help to wait 2 minutes before continuing to dissect the membranes. Often, both membranes will come off together.

    - By day 7, cuticle deposition begins causing significant levels of background, especially where limbs emerge from the developing body. Beyond day 9, as the embryonic cuticle attains a yellowish coloration, detection of gene expression by in situ hybridization becomes impossible.

22. Remove the dissected embryos from the fixative after sufficient fixation time has elapsed. Collect the fixed and dissected embryos in a medicine cup filled with 20 ml of PT. Store them in this solution until all dissections are completed.

    If the dissected embryos are to be used for in situ hybridization, they can be fixed for 15 minutes or up to a few hours. However, embryos that are to be used for antibody staining must not be overfixed. Embryos dissected for antibody staining are typically exposed to fixative for no longer than 15–20 minutes, although these conditions may require optimization depending on the antibodies to be used. Underfixation typically causes high background, whether the embryos are used for in situ hybridization or antibody staining.

23. Rinse the embryos by gathering them from the medicine cup and transferring them into a microcentrifuge tube. Fill the tube with about 1 ml of PT. Swirl the contents by spinning the tube as it rests in a tube rack. Allow the embryos to settle and reduce the PT down to about 25 µl. Repeat two to three times.

Dehydration of P. hawaiensis Embryos

P. hawaiensis embryos can be dehydrated and subsequently stored in 100% methanol at –20°C. Dehydrated embryos work fine for in situ hybridizations and many antibody-staining reactions. However, some antibodies do not stain tissue that has been dehydrated. In these cases, it is important to begin antibody reactions directly after dissection and fixation.

24. Transfer the embryos to a microcentrifuge tube containing 1 ml of 50% methanol.

    i. Swirl the tubes to mix and allow them to sit for 5 minutes.

    ii. Carefully remove the methanol solution and replace it with 1 ml of 70% methanol.

    iii. Swirl and wait 5 minutes.

    iv. Remove and replace the solution with 1 ml of 90% methanol. Swirl and wait 5 minutes.

v. Complete the dehydration with three quick washes of 100% methanol.

25. Store the dehydrated embryos in 100% methanol at –20°C.

TROUBLESHOOTING

Problem (Step 17): Dissected embryos readily and permanently stick to the inner surfaces of glass pipettes.

Solution: This can be prevented by passing yolk from the first group of dissected embryos into pipettes that will be used to move the embryos, coating them with a nonstick layer of yolk proteins. Alternatively, rinse pipettes with PT before using them to move embryos. It also helps to avoid drawing embryos past the neck of the pipette—embryos tend to stick more frequently to the wider part of the pipette.

Protocol 2

Injection of *P. hawaiensis* Blastomeres with Fluorescently Labeled Tracers

This protocol describes the injection of *P. hawaiensis* blastomeres with fluorescently labeled tracers for the purpose of cell-lineage analysis. The total (holoblastic) cleavages that characterize early embryogenesis in *P. hawaiensis* generate an eight-cell embryo with a stereotypical arrangement of blastomeres, each of which already possesses an invariant cell fate. Fluorochrome-conjugated dextran solutions, mRNAs encoding fluorescent proteins, and biotin-dextran have all proven to be useful lineage markers. The relative merits of various tracers will be considered. Although not explicitly discussed here, it should be noted that microinjection has also been used to generate genetically transformed lines of *P. hawaiensis* by coinjecting one-cell embryos with a Minos transposable element together with mRNA encoding the Minos transposase (see Pavlopoulos and Averof 2005).

MATERIALS

The recipe for the item marked with <R> is on page 403.

CAUTION: See the Cautions Appendix for appropriate handling of materials marked with <!>.

Reagents

Fluorescent dyes (50 mg/ml stocks):
  Fluorescein isothiocyanate <!> (FITC)-dextran (MW = 250,000)
  Tetramethylrhodamine isothiocyanate <!> (TRITC)-dextran (MW =155,000)
  Biotin-dextran (MW =70,000)
  See "Tracer Choice" in the Method section.
Halocarbon oil 700
mMessage mMachine SP6 Kit (Ambion)
mRNA encoding fluorescent proteins (see Step 2)
Nitrogen gas supply <!>
P. hawaiensis pairs
  Materials for collecting embryos from gravid females are also required (see Protocol 1).
Phenol red solution (0.5%; Sigma-Aldrich, P0290) <!>
Seawater (filter-sterilized) <R>
Sylgard 184 Silicone Elastomer Kit (World Precision Instruments)

Equipment

Capillary glass tubing with filament (1-mm OD; World Precision Instruments, TW100F4)
Dissecting microscope with light source (20x–50x total magnification)
  An inverted compound microscope with a stage-mounted micromanipulator may also be necessary (see "Injection Rig Set-up" in the Methods section).
Forceps (blunt)
Glass microscope slide
Injection holder (for 1-mm OD tube; Narishige HI-7)
Iron plate (Narishige IP)
Joystick micromanipulator (Narishige MN-151)
Magnetic base (Narishige GJ-1)

Microinjector (Narishige IM300)
Micropipette puller (Sutter P-97)
Modeling clay
Pasteur pipettes (5.75 inch; VWR)
Petri dish (large, plastic, 150 x 15 mm)
Petri dish (medium, plastic, 60 x 15 mm)
Petri dishes (small, plastic, 35 x 10 mm)
Pipette bulb
Plastic containers with lids (e.g., 14 x 10 x 10-cm plastic food storage containers)
Razor blade
Silicone tubing (1/32-inch ID; 3/32-inch OD; Cole-Parmer, 06411-60)

METHOD

Tracer Choice

Both fluorochrome-conjugated dextran solutions and mRNAs encoding fluorescent proteins have proven to be useful for continuous in vivo observation of clones following microinjection of early blastomeres. Biotin–dextran injections require subsequent fixation and enzymatic development and thus provide data for only a single time point. However, biotin–dextran tracers result in a superior signal-to-noise ratio, are permanent, and allow for simultaneous DAPI staining. For details on the fixation, dissection, and histochemical reaction of embryos injected with biotin–dextran, see Gerberding et al. (2002). Fluorescent tracers provide excellent spatial resolution through gastrulation, and diminishing resolution thereafter. FITC–dextran injections have higher background because of embryonic tissue autofluorescence. Injection of high concentrations of FITC–dextran (25–50 mg/ml) results in photoablation when the tissue is subsequently irradiated with blue light. TRITC–dextran suffers from fewer autofluorescence issues and does not cause cell death. Injection of mRNA encoding DsRed.T1 results in detectable fluorescence after 1.5 hours. DsRed.T1 mRNA injections result in very strong fluorescence and can also be detected with anti-DsRed antibodies for permanent preparations. Enhanced green fluorescent protein (EGFP) mRNA injections typically result in weaker fluorescence, which is detectable approximately 3 hours after injection.

Preparing Reagents for Injection

Fluorochrome-coupled dextrans must persist stably as high-molecular-weight molecules in a cellular environment. Some dextrans break down rather quickly following injection and subsequently pass readily across cell membranes. The dextrans described here have all proved to be effective tracers.

1. Prepare the dextrans at the following working concentrations:

FITC–dextran	2.0 mg/ml
TRITC–dextran	2.0 mg/ml
Biotin–dextran	1.0 mg/ml

2. Prepare mRNAs for the fluorescent proteins EGFP-1 (Clontech) and DsRed.T1 (Bevis and Glick 2001) in the pSP64T expression vector that contains 5′- and 3′-flanking sequences from a *Xenopus laevis* β-globin gene and includes a poly(dA) region (Krieg and Melton 1984; Gerberding et al. 2002). Generate capped transcripts by run-off in vitro transcription using the mMessage mMachine SP6 Kit from Ambion. Inject them at a concentration of 1.0 mg/ml.

Injection Rig Set-up

Injections of P. hawaiensis *embryos from the one- to eight-cell stages can be accomplished using dissecting microscopes with sufficient magnification (20x–50x). When attempting to inject micromeres at the eight-cell stage, some researchers prefer to use an inverted compound microscope with a stage-*

mounted micromanipulator. This section describes a tabletop micromanipulator used in conjunction with a dissecting microscope. Fitting the dissecting microscope with an ocular reticle that has been scaled with a stage micrometer allows the calculation of injection volumes and enables the microinjector to be adjusted such that various needles deliver consistent injection volumes. The microinjector must be capable of precise and reproducible delivery of very small volumes (~50 pl).

3. Position the iron plate adjacent to a dissecting microscope and secure the magnetic stand to the plate by activating the magnet. Connect the microinjector to a source of nitrogen gas. Mount the joystick micromanipulator onto the secured magnetic stand. Use silicone tubing to connect the injection holder to the output of the microinjector, ensuring that the tubing is free of kinks.

4. Prepare a Sylgard-coated injection dish by coating a 60-mm plastic Petri dish with 2–3 mm of Sylgard 184 following instructions for the Silicone Elastomer Kit. After the Sylgard hardens, use a razor blade to make a shallow V-shaped trough in the Sylgard that is approximately the width of a *P. hawaiensis* embryo (0.25–0.75 mm). Rinse the dish twice in distilled H_2O to prevent the embryos from sticking.

5. Pull the needles by placing glass tubing in the micropipette puller and selecting an appropriate program.

    A good needle has a tip sturdy enough to penetrate the eggshell and maintains a large enough bore to prevent frequent clogging, but is small enough in diameter to avoid excessive damage to the embryo. Needles can be stored by pressing them into a strip of clay in a large Petri dish.

    Typical program parameters for pulling needles suitable for injecting P. hawaiensis blastomeres using a Sutter P-97 micropipette puller are heat 615, pull 100, velocity 20, time 250. Trial and error will be required to optimize these settings. The Sutter "Pipette Cookbook" is a good source for program design help.

Collecting *P. hawaiensis* Pairs

6. Collect *P. hawaiensis* pairs in amplexus the night before injecting.

    Pairs can be placed in lidded containers (such as 14 x 10 x 10-cm plastic food storage containers) and left overnight at room temperature without aeration.

Collecting and Staging Embryos

For details concerning embryo collection, see Protocol 1.

7. The following morning, pairs will have separated and gravid females can be removed for embryo collection. Collect and stage embryos and plan injection schedule accordingly.

    Note that freshly shed eggs lack a hard outer chorion and are extremely fragile: Do not remove the eggs from the brood pouch until they have hardened for at least 1 hour. The following are elapsed times of early embryonic stages at 26°C:

Stage 1	one cell	0–4 hours
Stage 2	two cell	4–6 hours
Stage 3	four cell	6–7.5 hours
Stage 4	eight cell	7.5–9 hours

    The rate of embryogenesis can be modulated by adjusting the incubation temperature between 18°C and 32°C. If one-cell embryos are required, for instance, they can be kept at 18°C until injections begin. If eight-cell embryos are needed, incubating at higher temperatures will speed development.

Injecting Embryos

Tracers are diluted and phenol red is added to assist needle filling. Embryos are injected with 30–70 pl of tracer solution.

8. Prepare a small (~5 µl) volume of injection solution using the desired tracer. Use 4.5 µl of tracer at the appropriate working concentration (see above) and 0.5 µl of 0.5% phenol red solution.

9. Back-fill an injection needle. Pipette 0.5 μl of injection solution into the nonpulled end of a needle.
   Capillary action will draw the solution into the needle.

10. Insert the back-filled needle into the injection holder and lower it onto a glass microscope slide with a drop of halocarbon oil on it. Adjust the micromanipulator and stand so that the needle enters the oil at an angle of approximately 45° and the needle tip is centered under the dissecting microscope. Do not overtighten the collar on the injection holder; this will constrict the enclosed silicon gasket and prevent injection.
    The back end of the needle sits in the gasket, requiring little pressure to make a tight seal. This gasket should be inspected and replaced regularly.

11. Turn on the microinjector and open the nitrogen gas supply valve. Use a regulator to adjust input line pressure to 80–90 psi. Balance the needle and try to inject using an injection pressure of 25–40 psi with a duration of 20–50 msec. If no solution comes out of the needle, use a razor blade to carefully remove the tip of the needle while it remains in the injection holder.

12. Repeat Step 11 until the injections produce a consistent drop size of 3–5 μm (~33 pl) in halocarbon oil. Leave the needle tip in oil while assembling the embryos and between rounds of injections.
    Microinjector settings for injection pressure and duration can be adjusted to fine-tune the drop size. Injection durations of less than 20 msec are not reliably reproduced by the Narishige IM300. Injection pressures greater than 40 psi can cause damage to the embryo and thus are also not recommended.

13. Place embryos into the trough of a Sylgard-coated injection dish filled with sterile seawater. Position the injection dish under the dissecting microscope.

14. Use blunt forceps to maneuver an embryo into position so that the needle is aimed at the desired cell.

15. Gently push the needle into the embryo, using forceps if necessary to prevent the embryo from rolling.
    See Troubleshooting.

16. Inject.
    In general, one-cell embryos are easier to inject, so practice with them first. One-cell embryos can handle larger injection volumes, which are best achieved by multiply injecting a small volume. When injecting micromeres, try to inject just under the cell membrane to avoid penetrating the macromere underneath. See Troubleshooting.

17. Following injections, transfer the embryos to a 35-mm Petri dish filled with sterile seawater and store at the appropriate temperature.

TROUBLESHOOTING

Problem (Step 15): The needle does not penetrate the eggshell.
Solution: The needle is too thin. To make it thicker, cut it with a razor blade. If the problem persists, adjust the needle puller program to make stouter needles.

Problem (Step 16): The needle clogs.
Solution: Use the Clear function on the microinjector to drive the debris from the needle. Gently flicking the needle tip with forceps while clearing may help. These steps can be done in seawater next to the embryos. If the needle remains clogged, position the needle tip in a drop of halocarbon oil on a glass slide and cut the end with a razor blade.

Problem (Step 16): The needle leaves a large hole in the embryo that leaks yolk when the needle is withdrawn.
Solution: Try another needle with a smaller-diameter tip. Leaking embryos, especially at the one- and two-cell stages, will often live, so keep damaged embryos as well.

Protocol 3

Antibody Staining of *P. hawaiensis* Embryos

Below is a simplified protocol for antibody staining of *P. hawaiensis* embryos. It also works well for other arthropods and phyla. Fixed embryos are rehydrated, washed, blocked with normal goat serum, and incubated overnight with primary antibody. Embryos are then washed and incubated with a peroxidase-conjugated secondary antibody that binds to the primary antibody. A subsequent histochemical reaction produces a black stain in those cells where antibodies have localized.

MATERIALS

The recipes for the items marked with <R> begin on page 400.

CAUTION: See the Cautions Appendix for appropriate handling of materials marked with <!>.

Reagents

DAB + Ni <!> <R>
  Prepare immediately before use.
DAPI (4′,6-diamidino-2-phenylindole) solution <!> <R>
Glycerol solutions (50% and 70%) <R>
Hydrogen peroxide (3%) <!>
Methanol (70, 50, and 30%; all dilutions in PT <R>) <!>
P. hawaiensis embryos
  This protocol assumes that the P. hawaiensis *embryos have been fixed for 15–20 minutes and dehydrated for storage in 100% methanol as described in Protocol 1. D. melanogaster embryos are often included in the same tubes as internal controls and to help reduce overall background. Large volumes of dechorionated, fixed, and dehydrated* D. melanogaster *embryos are prepared and stored at –20°C (Patel 1994) for this purpose.*
10x Phosphate buffered saline (PBS) <R>, diluted to 1x before use
Primary antibody
PT <R>
PT + NGS <R>
Secondary antibody, conjugated with horseradish peroxidase (HRP)

Equipment

Dissecting microscope
Light source (see Step 1)
Microcentrifuge tubes

METHOD

Perform all incubations and washes in microcentrifuge tubes at room temperature.

Rehydration of Embryos

1. Rehydrate *P. hawaiensis* embryos stored in 100% methanol by incubating them for 5 minutes each in 70, 50, and 30% methanol. During rehydration and all subsequent wash steps, gently mix the embryos by spinning the microcentrifuge tubes as they rest in a tube rack. When changing solutions, pipette away as much as possible without allowing the embryos to dry. Avoid shaking or flicking the tubes because the embryos will splash up the tube sides, dry out, and not stain properly.

   P. hawaiensis embryos often sink slowly (or not at all) to the bottom of the tube. Be careful to avoid pipetting away these embryos when changing solutions. It is often necessary to illuminate the tube with a strong localized light source (such as that provided by a typical fiber optic lamp designed for a dissecting microscope) to ensure that the translucent embryos are not inadvertently withdrawn. Note the Troubleshooting note in Protocol 1 (Step 17) concerning the affinity of fixed embryos for glass pipettes.

   Fly embryos in methanol require only a series of 100% PT washes (three times for 5 minutes) for rehydration.

Primary Antibody Reaction

2. Wash the *P. hawaiensis* and fly embryos with PT three times for 5 minutes. Following the washes, add about 10 µl of fly embryos to each tube of *P. hawaiensis* embryos.

3. Incubate in 300 µl of PT + NGS for 30 minutes at room temperature to block nonspecific antibody-binding sites.

4. Add the appropriate amount of primary antibody to achieve the desired final concentration.

5. Gently swirl the embryos in the antibody solution and incubate overnight at 4°C.

Secondary Antibody Reaction

6. Wash the embryos with PT three times for 1 minute.

   Before the washes, diluted primary antibody can be recovered for reuse. Store diluted antibody at 4°C. It can typically be reused several times.

7. Wash the embryos with PT three times for 30 minutes.

8. Incubate in 300 µl of PT + NGS for 10–30 minutes at room temperature.

9. Add an appropriate peroxidase-conjugated secondary antibody to the proper final concentration.

10. Gently swirl the embryos in the secondary antibody solution and incubate for 2 hours at room temperature.

11. Wash with PT three times for 1 minute.

12. Wash with PT three times for 30 minutes.

Histochemical Development Reactions

The following section describes a histochemical reaction for use with secondary antibodies that have been conjugated with HRP. The colorless substrate DAB becomes brown when oxidized by the free oxygen that is formed by the action of peroxidase on the hydrogen peroxide. The presence of nickel ions results in a darker purplish-black. Staining the embryos with DAPI following the HRP reaction enables the nuclei of all cells to be visualized in the ultraviolet spectrum. Moving reacted embryos into 70% glycerol increases the optical clarity of the tissue and makes it easier to mount.

13. Prepare DAB + Ni solution immediately before use.

    1 ml of DAB is sufficient for three reactions.

14. Pipette away the final wash from Step 12 to within 1 mm of the embryos.

15. Add 300 µl of DAB + Ni. Mix very gently, ensuring that no bubbles are present.

16. Prepare a 0.3% solution of hydrogen peroxide by mixing 10 µl of 3% hydrogen peroxide with 90 µl of 1× PBS.

    This solution will last only about 30 minutes.

17. Add 15 µl of 0.3% hydrogen peroxide to each tube of embryos in DAB + Ni. Mix. Swirl the tubes gently, mixing the reagents quickly and thoroughly. Avoid introducing bubbles. Remove any bubbles that appear.

    The addition of hydrogen peroxide initiates the reaction. At this point, the reaction should be monitored by opening the lid of the microcentrifuge tube and observing the embryos under a dissecting microscope. Allow the reaction to proceed until satisfied with the signal-to-noise ratio. Reactions typically continue for 5–10 minutes.

18. Stop the reaction by removing the DAB and washing several times with PT.

    See Troubleshooting.

19. If DAPI staining is desired, place the reacted embryos into 200 µl of DAPI solution for 30 minutes or longer. If no DAPI staining is required, simply put the embryos into 50% glycerol for 30 minutes.

20. Move the embryos to 70% glycerol.

    Embryos can be stored in 70% glycerol for several weeks at room temperature, several years at 4°C, and several decades at –20°C.

TROUBLESHOOTING

Problem (Step 18): A weak signal (understaining).
Solution: There are several possible causes of a weak signal. The most common in *P. hawaiensis* are stopping the staining reaction too soon and overfixation. If the signal remains weak after 10 minutes of reaction, add another 15 µl of 0.3% hydrogen peroxide to the tube. Monitor the reaction closely because it may react quickly and result in overstaining. Make sure that specimens are fixed for no longer than 15–20 minutes.

Problem (Step 18): High background (overstaining).
Solution: The most common reasons for high background in *P. hawaiensis* are underfixation of the tissue or the fact that the reactions were allowed to proceed for too long. Be sure to prepare the DAB + Ni solution just before use. It is possible to reduce excess background by washing the embryos overnight in PT before placing them in glycerol.

Protocol 4

In Situ Hybridization of Labeled RNA Probes to Fixed *P. hawaiensis* Embryos

This protocol describes in situ hybridization of fluorescein- or DIG-labeled RNA probes to fixed *P. hawaiensis* embryos. Standard techniques of molecular biology should be used to produce an appropriate template for generation of antisense RNA probes. RNA-labeling mixes designed to produce fluorescein- or DIG-labeled RNA probes using T3, T7, or SP6 RNA polymerases are commercially available. Probes should be purified using QIAGEN RNeasy columns or similar means. Considerations for double-labeling experiments using both fluorescein- and DIG-labeled RNA probes are included.

MATERIALS

The recipes for the items marked with <R> begin on page 400.

CAUTION: See the Cautions Appendix for appropriate handling of materials marked with <!>.

Reagents

Alkaline phosphatase (AP) reaction buffer <R>
  Prepare immediately before use.
Antibodies (anti-DIG and/or antifluorescein)
BCIP/NBT <!> solution <R> and/or Fast Red <!> reaction solution <R>
DAPI solution <!> <R> (for DAPI staining only; see Step 33)
Formaldehyde (37%) <!>
Glycerol solutions (50% and 70%) <R>
Glycine buffer <R>
Methanol (30, 50, and 70%; all dilutions in PT <R>) <!>
P. hawaiensis embryos
  This protocol assumes that P. hawaiensis embryos have been fixed for 15 minutes or longer and dehydrated for storage in 100% methanol as described in Protocol 1. P. hawaiensis embryos should always be dehydrated before using them for in situ hybridizations, even if they have been freshly dissected. Dehydration and rehydration mitigates against their tendency to float in solution, greatly speeding up the protocol. Fixed D. melanogaster embryos can be added to tubes of P. hawaiensis embryos to reduce general background and help prevent floating. See Protocol 3 for further details.
PT <R>
PT + BSA <R>
RNA probe(s)
SDS hybridization (Hyb) buffer <R>
20x SSC <R>, diluted to 2x before use

Equipment

Centrifuge tubes (50 ml; BD Biosciences)
Dissecting microscope
Light source
Microcentrifuge tubes
Water bath at 65°C

METHOD

Perform all incubations and washes in microcentrifuge tubes at room temperature using about 1 ml of solution, unless otherwise indicated. Wash volumes should be added gently down the sides of the tubes, rather than directly onto the embryos. No agitation is necessary during washes or incubations.

Day 1

Rehydration of Embryos

1. Rehydrate *P. hawaiensis* embryos stored in 100% methanol by incubating them for 5 minutes each in 70, 50, and 30% methanol. During rehydration and all subsequent wash steps, gently mix the embryos by spinning the microcentrifuge tubes as they rest in a tube rack. When changing solutions, pipette away as much because possible without allowing the embryos to dry. Avoid shaking or flicking the tubes because the embryos will splash up the tube sides, dry out, and not stain properly.

   P. hawaiensis *embryos often sink slowly (or not at all) to the bottom of the tube. Be careful to avoid pipetting away these embryos when changing solutions. It is often necessary to illuminate the tube with a strong localized light source (such as that provided by a typical fiber optic lamp designed for a dissecting microscope) to ensure that the translucent embryos are not inadvertently withdrawn. Note the Troubleshooting note in Protocol 1 (Step 17) concerning the affinity of fixed embryos for glass pipettes.*

Postfixation

2. Wash the *P. hawaiensis* embryos with PT three times for 5 minutes.
3. Fix the embryos in 9 parts PT to 1 part 37% formaldehyde for 30 minutes.
4. Wash with PT twice for 5 minutes.
5. (Optional) Add about 10 µl of rehydrated fly embryos to each tube of *P. hawaiensis* embryos.

   The fly embryos help to reduce the background by increasing the tissue mass in the tube.

6. Wash with 500 µl of Hyb buffer once for 5 minutes.

   Amphipods tend to float in Hyb buffer. Be particularly careful when placing a pipette into a tube with Parhyale *embryos in Hyb buffer.*

Hybridization

7. Add 500 µl of fresh Hyb buffer and prehybridize the embryos by incubating for 10–60 minutes (or longer if preferred) at 65°C.
8. Dilute the probe to a final concentration of 1 ng/µl in Hyb buffer.

   If using two differently labeled probes, they should be mixed together at this point.

9. (Optional) Heat the Hyb buffer containing the probe for 5 minutes at 85°C. Cool to 65°C.

   This will relax RNA probes that contain high degrees of secondary structure, allowing them to hybridize more effectively to targets in situ. The presence of 50% formamide in Hyb buffer typically makes this step unnecessary.

10. Add 300–500 µl of Hyb buffer containing probe to each tube of embryos.
11. Hybridize without shaking in a water bath for 20–24 hours at 65°C.

Day 2

Washes

Initial washes are performed at 65°C, with subsequent washes at room temperature. Hyb buffer and 2x SSC can be stored in 50-ml screw-cap tubes in a water bath set at 65°C. All wash volumes are approximately 1 ml, unless otherwise indicated.

12. Recover the probe and store at –20°C.

    Probes can be reused many times.

13. Wash with Hyb buffer once for 30 minutes at 65°C.
14. Wash with 2x SSC four times for 30 minutes at 65°C.
15. Wash with 2x SSC twice for 10 minutes at room temperature. Remove only about 500 µl of the 2x SSC from the final wash.
16. Wash once for 20 minutes by adding 500 µl of PT at room temperature to tubes containing 500 µl of 2x SSC.
17. Wash with PT three times for 20 minutes at room temperature.
18. Wash with PT + BSA once for 30 minutes at room temperature.

Probe Visualization I

Antibody reactions are performed to detect localized RNA probes. React the strongest probe first when performing double-labeling reactions. Antibodies should only be added for one probe at a time. The antifluorescein antibody typically results in weaker staining than the anti-DIG antibody for a given probe. BCIP/NBT reactions produce stronger staining than Fast Red.

19. Dilute the anti-DIG (1:3000) or antifluorescein (1:4000) antibodies in PT + BSA.
20. Add 300–500 µl of diluted antibody per tube and incubate overnight at 4°C.

Day 3

Probe Visualization II

Room temperature washes are followed by visualization reactions.

21. Remove the antibody and wash with PT four times for 30 minutes at room temperature.
22. Wash with AP reaction buffer three times for 5 minutes at room temperature.
23. React by adding 500 µl of the appropriate reaction solution (BCIP/NBT or Fast Red) to each tube. Store the tubes in the dark at room temperature.

    Reactions can be monitored by briefly examining the embryos under a dissecting microscope. Color frequently develops within 1 hour, although the reactions may need to be left overnight. Reaction solutions can be replaced after 3–4 hours, if necessary.

24. Stop the reactions by washing with PT three times for 5 minutes at room temperature.
25. For single-labeling experiments, wash the embryos in PT overnight at room temperature and continue with Steps 32–34.

Day 4

Double-labeling Experiment: Second Probe Visualization

To continue with a double-labeling experiment, the washes are followed by the addition of an antibody specific to the second probe.

26. Wash with PT three times for 5 minutes at room temperature.
27. Rinse the embryos briefly with glycine buffer at room temperature.
28. Wash with glycine buffer once for 10 minutes at room temperature.
29. Wash with PT three times for 5 minutes at room temperature.
30. Wash with PT + BSA once for 1 hour at room temperature.
31. React by adding 500 µl of the appropriate reaction solution (BCIP/NBT or Fast Red) to each tube. Store the tubes in the dark at room temperature.

    Reactions can be monitored by examining the embryos under a dissecting scope and may need to be left overnight. Reaction solutions can be replaced after 3–4 hours, if necessary.

32. Stop the reactions by washing with PT three times for 5 minutes at room temperature.
33. If DAPI staining is desired, place the reacted embryos into 200 µl of 50% glycerol with DAPI for 30 minutes or longer. If no DAPI staining is required, simply place the embryos into 50% glycerol for 30 minutes.
34. Transfer the embryos to 70% glycerol.

    Embryos can be stored in 70% glycerol for several weeks at room temperature, several years at 4°C, and several decades at –20°C.

Recipes

The recipes for the items marked with <R> are also listed here.

CAUTION: See the Cautions Appendix for appropriate handling of materials marked with <!>.

ALKALINE PHOSPHATASE (AP) REACTION BUFFER

Reagent	Final concentration
$MgCl_2$ <!>	5 mM
NaCl	100 mM
Tris (pH 9.5) <!>	100 mM
Triton X-100 <!>	0.1%

Prepare just before use: Buffer that has been sitting for a few hours at room temperature will not work as well. AP reaction buffer with a pH of 9.5 is used for BCIP/NBT reactions. For Fast Red and Vector Red reactions, a pH of 8.2 is optimal. This can be made by mixing 7 volumes of 1 M Tris (pH 9.5) with 13 volumes of 1 M Tris (pH 7.5) for a final concentration of 100 mM Tris (pH 8.2).

BCIP/NBT SOLUTION

Reagent	Quantity (for 1 ml)	Final concentration
BCIP (50 mg/ml in 70% DMF) <!>	6.8 µl	0.68%
NBT (50 mg/ml in 70% DMF) <!>	8.1 µl	0.81%
AP reaction buffer <R>	1 ml	1x

Make 50 mg/ml stock solutions of BCIP (5-bromo-4-chloro-3-indolyl-phosphate) and NBT (nitro-blue tetrazolium chloride) in 70% dimethylformamide (DMF) <!>. These stocks last indefinitely when stored at 4°C or –20°C. Mix the reagents just before use and store in the dark.

DAB SOLUTION

Reagent	Quantity (for 33 ml)	Final concentration
1x PBS	33 ml	1x
Tween 20	16.5 µl	0.05%
3,3'-Diaminobenzidine (DAB) (10-mg tablets) <!>	10 mg	0.3 mg/ml

Add one 10-mg DAB tablet to a 50-ml tube containing 33 ml of 1x PBS and 16.5 µl of Tween 20. Rock gently in the dark for about 30 minutes. Filter through a 0.22-µm filter to remove particulate matter. Store aliquots at –70°C or in a nondefrosting –20°C freezer. Aliquots should be used immediately after thawing.

DAB + Ni

Reagent	Quantity (for 1 ml)	Final concentration
DAB solution (0.3 mg/ml) <!>	1 ml	0.3 mg/ml
NiCl$_2$ • 6H$_2$O (8%) <!>	8 µl	0.064%

Make an 8% stock solution of nickel chloride in deionized H$_2$O and store indefinitely at room temperature. Prepare DAB + Ni by thawing a 1-ml aliquot of 0.3 mg/ml DAB solution and adding 8 µl of 8% nickel chloride stock solution just before use. Mix by inverting the tube several times. Do not store DAB + Ni because the nickel eventually precipitates as nickel phosphate.

DAPI SOLUTION

Reagent	Quantity (for 200 ml)	Final concentration
DAPI (1 mg/ml) <!>	200 µl	1 µg/ml
Glycerol (50% in 1x PBS <R>)	200 ml	1x

Prepare a 1 mg/ml stock solution of DAPI (4′,6-diamidino-2-phenylindole) in deionized H$_2$O. Make 200 ml of a 50% glycerol solution (ultrapure) (Invitrogen) in 1x PBS. Add 200 µl of the DAPI to the glycerol/PBS. Store in the dark at 4°C.

FAST RED REACTION SOLUTION

Dissolve one Fast Tris tablet (Sigma-Aldrich T9043) in 1 ml of deionized H$_2$O in a microcentrifuge tube. Once this has completely dissolved, dissolve one Fast Red Naphthol <!> tablet (Sigma-Aldrich F0775) in the same tube. With both tablets in solution, filter through a 0.22-µm syringe filter.

GLYCEROL SOLUTIONS (50% and 70%)

Glycerol (ultrapure)
1x PBS <R>

Prepare 50% and 70% glycerol solutions by mixing the appropriate volumes of ultrapure glycerol with 1x PBS. Use pH paper to ensure a pH of about 7.4. Acidic glycerol will cause rapid fading of DAB reaction products.

GLYCINE BUFFER

Reagent	Quantity (for 50 ml)	Final concentration
Glycine (pH 2.0)	0.375 g	125 mM
Triton X-100 <!>	0.05 ml	0.1%

Dissolve 0.375 g of glycine in 40 ml of deionized H$_2$O and adjust the pH to 2.0 with concentrated HCl <!>. Add Triton X-100 and adjust the volume to 50 ml with deionized H$_2$O.

NORMAL GOAT SERUM (NGS)

Heat inactivate the serum for 30 minutes at 56°C. Filter through a 0.22-μm filter while still warm. Aliquot into sterile tubes and store the aliquots at –20°C. Once thawed, aliquots are stable for several months at 4°C.

10x PHOSPHATE-BUFFERED SALINE (PBS)

Reagent	Quantity (for 1 liter)	Final concentration (10x)
NaH_2PO_4	2.56 g	18.6 mM
Na_2HPO_4	11.94 g	84.1 mM
NaCl	102.2 g	1.75 M

Adjust the pH to 7.4 using NaOH <!> or HCl <!> as necessary. Prepare 1x PBS by diluting with deionized H_2O. Both 1x and 10x PBS can be kept indefinitely at room temperature.

PT

Reagent	Quantity (for 1 liter)	Final concentration
10x PBS <R>	100 ml	1x
100% Triton X-100 <!>	1 ml	0.1%

Store at 4°C or room temperature.

PT + BSA

Add 0.5 g of BSA to 500 ml of PT <R>. Store at 4°C or room temperature.

PT + NGS

Mix 4.75 ml of PT <R> with 0.25 ml of NGS <R> and store at 4°C. PT + NGS will usually last for 2–3 weeks. Discard if the solution looks cloudy.

SDS HYBRIDIZATION (HYB) BUFFER

Reagent	Final concentration
Formamide <!>	50%
20x SSC (pH 4.5) <R>	5x
Triton X-100 <!>	0.25%
SDS <!>	1%
Salmon sperm DNA	100 μg/ml

Prepare a 10 mg/ml stock solution of salmon sperm DNA in deionized H_2O. Autoclave for 20 minutes to shear and denature the DNA. Store at 4°C or in aliquots at –20°C. Prepare 40 ml of SDS hybridization buffer in distilled H_2O and test the pH. If the pH is not between 5 and 6, check the starting solutions and remake. A slightly acidic pH is essential to prevent the embryos from disintegrating. Important: Just before preparing Hyb buffer, remove a working volume of 20x SSC and use concentrated HCl <!> to adjust its pH to 4.5. Take precautions to ensure that the Hyb buffer is RNase-free. Hyb buffer can be stored at –20°C but should be warmed before use in order to resuspend the SDS. It can also be made fresh and kept at 65°C for the duration of the procedure.

SEAWATER

Dissolve sufficient sea salt mixture (Tropic Marin) in deionized H₂O to result in a specific gravity of 1.018–1.022, as measured by a hydrometer. Large volumes of seawater can be prepared in plastic garbage cans. Smaller volumes can be removed and filter-sterilized (22-μm) for use with embryos.

20× SSC

Reagent	Final concentration
NaCl	3 M
Sodium citrate, dehydrate <!>	0.3 M

ACKNOWLEDGMENTS

The authors thank P. Liu, R. Parchem, and M. Protas for helpful comments and C. Reiss for incisive edits.

REFERENCES

Barnard, J.L. 1965. Marine Amphipoda of atolls in Micronesia. *Proc. U.S. Natl. Mus.* **117:** 459–551.

Barnard, J.L. and Karaman, G.S. 1991. The families and genera of marine gammaridean Amphipoda (except marine gammaroids). *Rec. Aust. Mus. Suppl.* **13:** 1–866.

Bevis, B.J. and Glick, B.S. 2001. Rapidly maturing variants of the *Discosoma* red fluorescent protein (DsRed). *Nat. Biotechnol.* **20:** 83–87.

Boore, J.L., Collins, T.M., Stanton, D., Daehler, L.L., and Brown, W.M. 1995. Deducing the pattern of arthropod phylogeny from mitochondrial DNA rearrangements. *Nature* **376:** 163–165.

Boore, J.L., Lavrov, D.V., and Brown, W.M. 1998. Gene translocation links insects and crustaceans. *Nature* **392:** 667–668.

Browne, W.E., Price, A.L., Gerberding, M., and Patel, N.H. 2005. Stages of embryonic development in the amphipod crustacean, *Parhyale hawaiensis*. *Genesis* **42:** 124–149.

Dana, J.D. 1853. Crustacea. Part II. *United States Exploring Expedition* **14:** 689–1618.

Eernisse, D.J. 1997. Arthropod and annelid relationships re-examined. In *Arthropod relationships* (ed. R.A. Fortey and R.H. Thomas), pp. 43–56. Chapman and Hall, London.

Extavour, C.G. 2005. The fate of isolated blastomeres with respect to germ cell formation in the amphipod crustacean *Parhyale hawaiensis*. *Dev. Biol.* **277:** 387–402.

Friedrich, M. and Tautz, D. 1995. Ribosomal DNA phylogeny of the major extant arthropod classes and the evolution of myriapods. *Nature* **376:** 165–167.

Gasca, R. and Haddock, S.H.D. 2004. Associations between gelatinous zooplankton and hyperiid amphipods (Crustacea: Peracarida) in the Gulf of California. *Hydrobiologia* **530/531:** 529–535.

Gerberding, M., Browne, W.E., and Patel, N.H. 2002. Cell lineage analysis of the amphipod crustacean *Parhyale hawaiensis* reveals an early restriction of cell fates. *Development* **129:** 5789–5801.

Giribet, G., Edgecombe, G.D., and Wheeler, W.C. 2001. Arthropod phylogeny based on eight molecular loci and morphology. *Nature* **413:** 157–161.

Hwang, U.W., Friedrich, M., Tautz, D., Park, C.J., and Kim, W. 2001. Mitochondrial protein phylogeny joins myriapods with chelicerates. *Nature* **413:** 154–157.

Kamaltynov, R.M. 1999. On the evolution of Lake Baikal amphipods. *Crustaceana* **72:** 921–931.

Krieg, P.A. and Melton, D.A. 1984. Functional messenger RNAs are produced by SP6 *in vitro* transcription of cloned cDNAs. *Nucleic Acid Res.* **12:** 7057–7070.

Lindeman, D. 1991. Natural history of the terrestrial amphipod *Cerrorchestia hyloraina* Lindeman (Crustacea: Amphipoda; Talitridae) in a Costa Rican cloud forest. *J. Nat. Hist.* **25:** 623–638.

Myers, A.A. 1985. Shallow-water, coral reef and mangrove Amphipoda (Gammaridea) of Fiji. *Rec. Aust. Mus. Suppl.* **5:** 1–143.

Patel, N.H. 1994. Imaging neuronal subsets and other cell types in whole mount *Drosophila* embryos and larvae using antibody probes. *Methods Cell Biol.* **44:** 445–487.

Pavlopoulos, A. and Averof, M. 2005. Establishing genetic transformation for comparative developmental studies in the crustacean *Parhyale hawaiensis*. *Proc. Natl. Acad. Sci.* **102:** 7888–7893.

Poltermann, M., Hop, H., and Falk-Peterson, S. 2000. Life under Arctic sea ice—Reproduction strategies of two sympagic (ice-associated) amphipod species, *Gammarus wilkitzkii* and *Apherusa glacialis*. *Mar. Biol.* **136:** 913–920.

Poovachiranon, S., Boto, K., and Duke, N. 1986. Food preference studies and ingestion rate measurements of the mangrove amphipod *Parhyale hawaiensis* (Dana). *J. Exp. Mar. Biol. Ecol.* **98:** 129–140.

Regier, J.C. and Shultz, J.W. 1997. Molecular phylogeny of the major arthropod groups indicates polyphyly of crustaceans and a new hypothesis for the origin of hexapods. *Mol. Biol. Evol.* **14:** 902–913.

Schmitz, E.H. 1992. Amphipoda. In *Microscopic anatomy of invertebrates* (ed. F.W. Harrison), pp. 443–528. Wiley, New York.

Schram, F.R. 1986. *Crustacea*. Oxford University Press, Oxford.

Sheader, M., Van Dover, C.L., and Shank, T.M. 2000. Structure and function of *Halice hesmonectes* (Amphipoda: Pardaliscidae) swarms from hydrothermal vents in the eastern Pacific. *Mar. Biol.* **136:** 901–911.

Sherbakov, D.Y., Kamaltynov, R.M., Ogarkov, O.B., Vainola, R.,

Vainio, J.K., and Verheyen, E. 1999. On the phylogeny of Lake Baikal amphipods in the light of mitochondrial and nuclear DNA sequence data. *Crustaceana* **72:** 911–919.

Shoemaker, C.R. 1956. Observations on the amphipod genus *Parhyale. Proc. U.S. Natl. Mus.* **106:** 345–358.

Vainola, R. and Kamaltynov, R.M. 1999. Species diversity and speciation in the endemic amphipods of Lake Baikal: Molecular evidence. *Crustaceana* **72:** 945–956.

Vinogradov, M.E., Volkov, A.F., and Semenova, T.N. 1996. *Hyperiid amphipods (Amphipoda, Hyperiidea) of the world oceans.* Smithsonian Institution Libraries, Washington, D.C.

WWW RESOURCES

http://www.genome.gov/10002154 The NHGRI's list of approved sequencing targets. *J. slatteryi* is an approved sequencing target of the NHGRI's Large-scale Genome Sequencing Program.

http://www.genome.gov/Pages/Research/Sequencing/SeqProposals/ EcdysozoaProposalFinalPDF.pdf The Ecdysozoan Sequencing Proposal. The five species chosen for analysis include *J. slatteryi*.

http://sutter.com/contact/faqs/pipette_cookbook.pdf The Sutter "Pipette Cookbook." Program parameters for pulling needles suitable for injecting *P. hawaiensis* blastomeres can be designed using this cookbook; see Protocol 2 of this chapter.

16 The Sea Lamprey *Petromyzon marinus*
A Model for Evolutionary and Developmental Biology

Natalya Nikitina, Marianne Bronner-Fraser, and Tatjana Sauka-Spengler
Division of Biology, California Institute of Technology, Pasadena, California 91125

ABSTRACT

Sea lampreys (*Petromyzon marinus*) are cyclostomes, the most basal extant group of vertebrates, and are thought to have existed largely unchanged for more than 500 million years. They are aquatic, eel-shaped animals that spend a major part of their life as filter-feeding larvae called ammocoetes, inhabiting many freshwater bodies in the northern hemisphere. After metamorphosis, sea lampreys migrate to the ocean (or to the Great Lakes),

PROTOCOLS
1. Culturing Lamprey Embryos, 410
2. Microinjection of RNA and Morpholino Oligos into Lamprey Embryos, 413
3. DiI Cell Labeling in Lamprey Embryos, 415
4. Whole-mount In Situ Hybridization on Lamprey Embyros, 417
5. Immunostaining of Whole-mount and Sectioned Lamprey Embyros, 421

where they feed on the blood and bodily fluids of salmonid fish and ultimately return to freshwater streams and rivers to spawn and die. The unique evolutionary position of lampreys and the relative ease of obtaining mature adults and embryos make this animal an ideal model for investigations into early vertebrate evolution. Studies of features shared between lampreys and jawed vertebrates, but distinct from those in nonvertebrate chordates, have informed on the origin and evolution of hallmark vertebrate characteristics such as the neural crest, ectodermal placodes, and jaw. In addition, studies of features that are unique to lampreys (e.g., the variable lymphocyte receptor [VLR]-mediated immune system) provide insights into mechanisms of parallel evolution (e.g., the adaptive immune system). With the establishment of techniques for the extended maintenance and spawning of lampreys in the laboratory, the sequencing of the lamprey genome, and the adaptation and optimization of many established molecular biology and histochemistry techniques for use in this species, *P. marinus* is poised to become an evolutionary developmental model of choice.

BACKGROUND INFORMATION

Petromyzon marinus (commonly known as sea lamprey but sometimes called lake lamprey or lamprey eel) is an aquatic animal with an elongated snake-like body reaching, on average, 45–60 cm in length and weighing up to 900 g. Adults are mottled brown in color with a lighter ventral side that becomes yellow during the breeding season.

This chapter, with full-color images, can be found online at www.cshprotocols.org/emo.

Lamprey eggs are laid in freshwater streams during the summer, and after spawning, the adults usually die. The embryos hatch after about 2 weeks, and the larvae (called ammocoetes) spend the first 7–9 years of their life burrowed in mud in freshwater streams, filter-feeding on small mud-inhabiting organisms. After this period of time, ammocoetes undergo an extensive metamorphosis, during which they develop suctorial discs, eyes, and dorsal fins, and they migrate into the ocean, where they spend the next 18–24 months feeding on fish blood. When lampreys reach their full adult size, they become sexually mature, stop feeding, and, like other anadromous fish, migrate back to the streams and rivers where they hatched to repeat the cycle of reproduction and death (Hardisty and Potter 1971b; Hardisty 1979).

Sea lampreys are generally marine, inhabiting the Atlantic Ocean and migrating into freshwater rivers and streams of Europe and the United States Atlantic coast to reproduce. The U.S. also has two landlocked populations of sea lampreys, one in the Great Lakes and the other in Lake Champlain. These populations originated in 1835 when lampreys were inadvertently introduced into Lake Ontario. It is quite possible that they migrated by attaching themselves to the hulls of ships going through the canal systems of Lake Erie or the St. Lawrence River. By 1946, sea lampreys had spread throughout the Great Lakes and the freshwater streams of the Great Lakes region. Because these landlocked lampreys have not been able to enter the ocean to feed on marine fish, they have become parasitic, preying on freshwater fish and causing considerable damage to the ecosystem, which has necessitated government-controlled measures. Freshwater lampreys inhabiting the Great Lakes are distinct from Atlantic lampreys in size and color, with the former being smaller and more brown and yellow in color and the latter much larger and silvery in color.

SOURCES AND HUSBANDRY

Each year, from late April through mid-July, adult sea lampreys are collected from tributaries of Lake Huron and Lake Michigan by the U.S. Fish and Wildlife Service Marquette Biological Station (Marquette, Michigan). The maturing adults, migrating upstream, are trapped, and spawning adults are removed directly from their nests. After collection, the animals are transported to laboratories at the U.S. Geological Survey Hammond Bay Biological Station (Millersburg, Michigan). Adult prespermiating males and preovulating females are then shipped to our laboratory where they are kept in a custom-made lamprey aquatic housing system (Aquaneering).

This housing system features four durable, insulated, polypropylene tanks, each with an individual tank valve for precisely metering the water flow rates. The system also features a mechanism for aeration and uses a four-stage filtration system that includes a fluidized bed biological filter and single-canister charcoal filter. Wastewater first flows through a Dacron pad mechanical filter, trapping particles greater than 10 μm, and then through the fluidized bed biological filter. The fluidized bed biological filters are composed of a fine medium that provides a large surface area on which hundreds of different types of beneficial bacteria can grow. This ensures a biologically stable ecology in the tank system that has undetectable levels of ammonia and nitrites. Each tank has an individual temperature- and light-control system that can be set to manually or automatically manipulate the temperature and the light/dark cycle in the lamprey rack. Each also has an automatic water-exchange system that can be programmed to perform the daily change-out of water conditioned to the proper pH and salinity levels. In our laboratory, sea lampreys are held under conditions that mimic their natural cycle until they reach the spermiated/ovulating phase, when they are used for artificial fertilization.

RELATED SPECIES

Thirty-eight lamprey species have been described; four species inhabit the southern hemisphere and have been placed into the families Geotriidae and Mordaciidae, whereas the 34 remaining species are all from the northern hemisphere and have been assigned to six genera that are grouped together into one family, Petromyzonidae (Hubbs and Potter 1971). The phylogenetic relationships among the various lamprey genera have not yet been adequately resolved, but molecular phylogenetic analyses indicate that lampreys are paraphyletic in origin.

Three lamprey species, *P. marinus*, *Lampetra (Lethenteron) japonica*, and *L. fluviatilis* (European river lamprey), have historically been the "stars" of lamprey research; they remain, to this day, the preferred lamprey research models. These lampreys are relatively large, produce huge quantities of eggs, are widely distributed, and are easy to capture when they migrate into freshwater streams at the start of the breeding season. *L. japonica* is used by Japanese researchers; *L. fluviatilis* is studied in Europe; and *P. marinus*, the focus of this chapter, is a preferred model in European and North American labs. All of the techniques for laboratory maintenance, embryo culture, and embryo manipulation described here are applicable to all three species, although a different embryo culture medium (10% Steinberg's solution) has been used successfully with *L. japonica* embryos (Horigome et al. 1999).

USES OF THE *P. MARINUS* MODEL SYSTEM

The sea lamprey has been studied since the mid-19th century. At that time, lampreys were thought to be either primitive, degenerate vertebrates or direct ancestors of modern vertebrates. As a consequence, many of the first studies focused on comparing the morphology of the adult lampreys to that of higher vertebrates in an attempt to clarify the taxonomic position of this animal group (Huxley 1876). Adult lampreys used in these studies were wild-caught; the first attempt to maintain lamprey larvae in captivity was reported in 1900 (Reese 1900). Techniques for spawning, in vitro fertilization, and short-term maintenance of lamprey were first established in related species from the genus *Lampetra*, which allowed for a detailed analysis of early embryonic development to be performed in this species (Damas 1944). The embryonic development of *P. marinus* was first described by George Piavis (Piavis 1961, 1971); his developmental staging system is still widely used. More recently, Tahara (1988) described developmental stages for *Lampetra reissneri* and correlated them with Piavis stages, creating the developmental table that is probably the most commonly used today.

With the advent of modern molecular techniques, lamprey became one of the most important research models for understanding vertebrate origins (for review, see Kuratani et al. 2002; Osorio and Retaux 2008). Lampreys, together with hagfish, are the most basal animals in which many of the true vertebrate characteristics (e.g., neural crest, placodes, segmented brain, skull, paired sensory organs, pharyngeal skeleton) are present. By studying the molecular and developmental mechanisms responsible for the formation of these structures in lamprey and higher vertebrates, we can gain insight into how these vertebrate characteristics evolved.

Here, we concentrate on the aspects of lamprey biology that are most relevant to developmental and evolutionary biology. Readers interested in a more comprehensive description of lamprey physiology, morphology, behavior, and ecology may refer to *The Biology of Lampreys* (Hardisty and Potter 1971a,b; 1971–1982).

Embryonic Development

An excellent description of sea lamprey development was provided by Piavis (1971), who divided the embryonic developmental pattern into 18 stages—from fertilized egg to ammocoete—based on easily discernable external characteristic. Cleavage is holoblastic; animal blastomeres are visibly smaller than vegetal blastomeres. Gastrulation (Piavis stage 9 or 64–100 hours postfertilization [hpf]) occurs by a mechanism that is very similar to that of *Xenopus* gastrulation, and the first signs of neural plate formation are seen at 4.5 days of development, with the thickened layer of the neural ectoderm on the dorsal aspect of the embryo and the flattened neural plate border clearly distinguishable from the surrounding ectoderm. The raised edges of the neural plate fuse to form a solid neural rod that later cavitates to form a hollow neural tube. The embryos begin to move inside their chorions by days 9–10 (Piavis stage 13). On day 11 (Piavis stages 13–14), they hatch, their hearts start to beat, and gill clefts form and start to function by day 17 (Piavis stage 15). Eye development in the lamprey proceeds in a unique fashion: The formation of the eye primordium, with an undifferentiated retina and lens, is accomplished during embryonic life (days 17–20; Piavis stages 15–16), but the growth of the eye and the maturation of the retinal receptors arrest during the long larval life and resume only after metamorphosis (Rubinson 1990).

Lamprey is one of the slowest-developing vertebrates, and its embryos develop abnormally and often arrest in development at temperatures even slightly below the optimal temperature of 18°C. Thus, the pace of their development cannot be adjusted by modulating the rearing temperature.

Immunology

It has been known since the mid-20th century that lampreys possess an efficient adaptive immune system (for review, see Cooper and Alder 2006); they demonstrate accelerated rejection of secondary skin allografts and increased agglutination after repeated immunization with particulate antigens (e.g., anthrax bacteria, sheep blood cells). Interestingly, lampreys do not appear to respond to soluble antigens. Despite the presence of lymphocytes and the obvious ability to mount an immune response, lampreys do not possess either immunoglobulin-type or T-cell receptors, which are the hallmark of the adaptive immune systems of higher vertebrates. Instead, lampreys rely on a completely different type of variable lymphocyte receptor (VLR) that is composed of a highly variable leucine-rich repeat (LRR) and a stalk segment of conserved sequence that connects the VLR to the surface of the lymphocyte (Pancer et al. 2004). Like the vertebrate immunoglobulin-type receptors, diverse VLRs are generated in lymphocytes by genome recombination. A combination of variable gene segments, including many LRRs, is assembled into a VLR, and the intervening genomic sequence is permanently excised from the genome of that lymphocyte lineage (Nagawa et al. 2007). Vast numbers of lymphocyte receptors, rivaling the diversity of the mammalian immune repertoire, are produced in this way (Alder et al. 2005).

Lampreys do not possess an anatomical equivalent of the mammalian thymus. In ammocoetes, blood cell formation and lymphocyte maturation occur in the typhlosole (an invagination of the intestinal epithelium) and the nephric fold (Amemiya et al. 2007). In adult lampreys, lymphocyte formation occurs in the gills and the ventral kidney.

Phylogeny

Phylogenetic relationships among the lampreys, hagfishes, and gnathostomes have been the subject of many molecular analyses, and the current consensus, based on molecular analyses of ribosomal RNA and nuclear genes, is that the lampreys and hagfishes form a monophyletic sister group to the craniates (Stock and Whitt 1992; Takezaki et al. 2003). The 38 currently recognized species of extant lampreys are considered to be paraphyletic in origin and are grouped into three families on the basis of morphological characteristics, including dentition and the shape of the suctorial disc (Gill et al. 2003). From an ecological perspective, lampreys can be divided into three groups on the basis of their feeding strategies as adults: anadromous lampreys, freshwater parasitic lampreys, and brook lampreys. Anadromous lampreys, which includes the two species most commonly used for research (*P. marinus* and *L. japonica*), spend their adult lives in saltwater habitats, feeding on the blood of marine fish. Parasitic freshwater species spend their entire lives in the streams and rivers in which they hatch and, due to the more limited food supply, they tend to be smaller than their anadromous relatives and have a shorter adult stage. Twenty of the extant lamprey species are dwarf, nonparasitic brook lampreys, which never feed after metamorphosis. The brook lampreys most likely evolved from parasitic freshwater ancestors by delaying metamorphosis and shortening the adult stage (Hardisty and Potter 1971a).

GENETICS, GENOMICS, AND ASSOCIATED RESOURCES

The sequencing of the 2.1–2.4-Gb genome of *Petromyzon marinus*, financed by the National Human Genome Research Institute (NHGRI), is well under way, and the project aims to produce a high-quality assembly. So far, the genome has been sequenced to 5.9x whole-genome coverage, and the current assembly primarily consists of whole-genome, shotgun, plasmid-end sequences, with a small sampling (0.005x) of fosmid-end sequences. This assembly can be viewed and downloaded from the website maintained by Washington University in the St. Louis Genome Sequencing Center (http://genome.wustl.edu/genome.cgi?GENOME=Petromyzon%20marinus&SECTION=assemblies), and raw sequence data in NCBI's Trace Archive can be searched using BLAST (http://www.ncbi.nlm.nih.gov/BLAST/Blast.cgi?PAGE=Nucleotides&PROGRAM=ST_SPEC=TraceArchive&BLAST_PROGRAMS=megaBlast&PAGE_TYPE=BlastSearch).

A bacterial artificial chromosome (BAC) library using the genomic material of the specimen being sequenced was constructed, arrayed, and spotted onto nylon macroarray filters, which are available from BACPAC Resources, Children's Hospital Oakland Research Institute (Oakland, CA; http://bacpac.chori.org/library.php?id=199). Because initial studies indicated that the genome is A+T-rich and repetitive and that the heterozygosity rate for a single animal is very high, additional sequencing of BAC and fosmid ends is planned. Conventional expressed sequence tag (EST) sequencing of cDNA libraries is also under way; this should help with gene annotation and will provide clones of interest for the community. Because the sequencing project is ongoing, the quality of the assembly should improve and the number of partially or fully sequenced cDNA clones from different stages and tissue-specific libraries will increase.

TECHNICAL APPROACHES

The following protocols describe some basic techniques for culturing and studying lamprey embryos in the laboratory.

Protocol 1

Culturing Lamprey Embryos

This protocol describes how to produce lamprey embryos by collecting sperm and eggs from mature lamprey, performing fertilization, and culturing the embryos through to the desired developmental stage. The embryos produced in this protocol can be used for all of the investigative procedures described in Protocols 2–5.

MATERIALS

The recipe for the item marked with <R> is on page 427.

Reagents

H_2O, distilled and spring (Sparkletts) (both equilibrated to 18°C)
Lampreys (mature spermiating males and ovulating females)
> Obtain lampreys and maintain actively spermiating males (Fig. 1A) and ovulating females (Fig. 1B) in an aquarium as described in the "Sources and Husbandry" section. Because lampreys tend to die after spawning, we usually maintain the actively spermiating males and ovulating females at 12°C in order to extend their lives and slow down their natural tendency to die.

0.1x MMR, equilibrated to 18°C <R>

Equipment

Crystallization dishes
Incubator, preset to 18°C

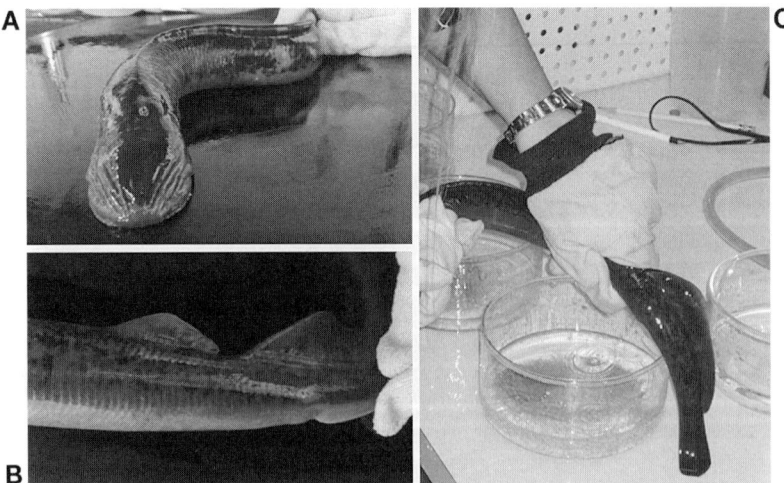

FIGURE 1. (A) Mature spermiating males are recognizable by their prominent dorsal ridge. (B) Ovulating females lack the dorsal ridge but have enlarged, soft abdomens. (C) In vitro fertilization is performed in spring water at 18°C. Eggs are obtained by gently squeezing the abdomen of a mature female and are then deposited into a crystallization dish. Subsequently, sperm from mature males are expressed in a similar manner and are added to the dish. The dish of eggs and sperm is gently swirled several times and then allowed to rest for about 15 minutes.

METHOD

1. Obtain eggs by gently squeezing the abdomens of gravid females (Fig. 1C). Deposit the eggs into a crystallization dish containing approximately 250–350 ml of spring H_2O. Obtain sperm from spermiating males using the same technique.

   Mature females (Fig. 1B) can be distinguished from immature specimens by their distended soft abdomen and the absence of a dorsal ridge, whereas mature males (Fig. 1A) have a very thick, prominent, raised dorsal ridge (Vladykov 1949). See Troubleshooting.

2. To fertilize the eggs, add the sperm to the crystallization dish, swirl the solution of sperm and eggs several times, and allow the dish to stand for at least 15 minutes.

   To allow cortical rotation to occur normally, it is important to avoid moving the dish with fertilized eggs too much immediately after adding sperm. The success of fertilization can be monitored by observing the fraction of eggs whose fertilization envelopes lift.

3. Wash out the excess sperm with four changes of distilled H_2O. Do so gently to avoid introducing air bubbles.

   Air bubbles adhere to the membranes of the eggs, causing the eggs to float and the embryos to be lost during the washes.

4. Replace the distilled H_2O with fresh spring H_2O and incubate the embryos at 18°C until the first cell division is under way, which normally takes about 6–6.5 hours.

   See Troubleshooting.

5. Before cleavage (within 10 hours of fertilization), replace the spring water with sterile 0.1× MMR.

   MMR contains $CaCl_2$. Cleaving embryos require Ca^{2+} in order to form intercellular junctions and will die if left in spring H_2O for more than 10 hours.

6. Incubate the embryos in the incubator until the desired developmental stage is reached. On a daily basis, inspect the culture dishes, remove any dead or arrested embryos, and replace the 0.1× MMR. To avoid fungal infections and massive embryonic death, carefully maintain the embryos in sterile, nonconfluent conditions. After the third day of incubation, ensure that the embryos are spread over the dish such that every embryo is at least 1 cm away from its neighbors. Leave the embryos undisturbed until day 4.5.

   See Troubleshooting.

TROUBLESHOOTING

Problem (Step 1): Females do not release eggs easily or blood is present during the extraction procedure.
Solution: The females have not yet matured. Allow them to develop for another 2–4 days at 18–19°C before attempting to extract eggs again.

Problem (Step 4 or 6): Abnormal cleavage or arrested embryonic development.
Solution: Consider the following:

- Abnormal cleavage and arrested development are evidence of polyspermy, so avoid using too much sperm, especially in June or early July. The quality of sperm progressively decreases throughout the mating season. In August, use sperm from several males in order to ensure complete fertilization.

- Ensure the highest egg quality by obtaining immature female lampreys and allowing them to develop for 2–3 weeks before extracting their eggs. Do not use older females that are already ovulating.
- Wash the eggs extensively after fertilization (Step 3).
- Ensure that the incubator temperature does not fluctuate from 18°C.

Problem (Step 6): Embryos do not survive past days 4–5 of development.
Solution: Disturbing or moving embryos between days 3–5 of incubation can result in massive embryonic death. On day 3, be sure that each embryo is at least 1 cm away from its neighbors, and leave it undisturbed until gastrulation is completed (day 4.5). During this vulnerable stage, embryos can be raised individually in 48-well plates.

Problem (Step 6): Embryos cleave and gastrulate normally, but show slowed development at later stages (i.e., the head does not begin to form by day 5).
Solution: Ensure that spermiating males and ovulating females are not exposed to temperatures below 10°C. Lower temperatures seem to affect the viability of the eggs.

Protocol 2

Microinjection of RNA and Morpholino Oligos into Lamprey Embryos

Lamprey embryos are particularly amenable to injection techniques. They have the same advantages as both zebrafish and *Xenopus* embryos in that, due to their double chorion, they are not prone to surface-tension-induced explosion when removed from liquid and can therefore be injected in a dry dish. This eliminates the need to support the embryo while performing injections, making the procedure very rapid. Also, a single ovulating female may contain up to 100,000 eggs, so the numbers of injectable embryos per fertilization is not a limiting factor, which is a great advantage over *Xenopus* and zebrafish. Finally, the second division lasts for several hours, providing a very large injection window. This protocol describes how to microinject RNA and morpholinos into lamprey embryos using previously described techniques (McCauley and Bronner-Fraser 2006; Sauka-Spengler et al. 2007).

MATERIALS

The recipe for the item marked with <R> is on page 427.

CAUTION: See the Cautions Appendix for appropriate handling of materials marked with <!>.

Reagents

Lamprey embryos (at the two-cell stage; see Protocol 1)
0.1x MMR, equilibrated to 18°C <R>
Morpholino oligos (FITC <!> labeled; e.g., Gene Tools)
Sense RNA (100 µg/µl)

The full-length sense RNA for the gene of interest can be prepared by in vitro transcription (e.g., using one of Ambion's mMessage mMachine kits). Before injection, a small amount of RNase-free vegetable dye should be added to the RNA solution to assist with visualizing the injection (the vegetable dye is not required for the FITC-labeled morpholinos because their slightly yellow color allows direct visualization).

Equipment

Glass needles, fine-pulled
Incubator, set at 18°C
Injection dish

Prepare an injection dish by gluing a fine nylon mesh (Segar Filtration Inc.) to the bottom of a standard Petri dish. Use a few drops of acetone <!> to bind the mesh to the dish.

Micrometer
Microscope (dissecting)
Microscope (fluorescent)
Picospritzer
Transfer pipettes (wide bore, plastic)

METHOD

1. Use a wide-bore, plastic, transfer pipette to transfer 100–200 two-cell-stage embryos (Fig. 2A) into an injection dish. Choose viable embryos that are cleaving normally; avoid those that are not dividing or are dividing irregularly.

2. Remove the liquid from the dish and allow the embryos to settle into the holes in the mesh.

3. Back-fill the glass needle with 6–10 µl of morpholino (or sense RNA) solution. For each two-cell-stage embryo, inject about 5 nl into one blastomere. Use the dissecting microscope to aid in visualization.

    The exact volume required should be determined by a standard calibration method or by measuring the droplet diameter with the help of a micrometer. As a rough guide, the diameter of the drop that is injected should not exceed about one-fourth the diameter of a single blastomere. For most embryos, this volume corresponds to approximately 5 nl of aqueous injection solution. If required, adjust the picospritzer until the desired drop size is achieved. See Troubleshooting.

4. Incubate the injected embryos at 18°C in 0.1x MMR until the desired developmental stage is reached.

    See Protocol 1 for general information on culturing lamprey embryos.

5. When the embryos reach the desired stage of development, use the fluorescent microscope to select those that integrated the morpholino on one side only (Fig. 2B).

    See Troubleshooting.

TROUBLESHOOTING

Problem (Step 3): The injection needle is blocked.
Solution: Clean or replace the needle. The morpholino solution and egg cytoplasm have a tendency to block the needle, so it should be checked regularly and cleaned or replaced as necessary.

Problem (Step 5): The injected agent leaks into the noninjected side of the embryo.
Solution: Leakage occurs when two-cell-stage embryos are injected because, in lamprey, the first cell division is not completed until the middle of the second cell division. To avoid leakage, either (1) inject two blastomeres at the four-cell stage or (2) inject during the middle of the second cell division. If you choose the latter, perform the injection at the end of the dividing blastomere, opposite the furrowing site, to ensure an even distribution of the injected agent throughout the dividing cell. This approach will necessarily narrow the injection time window.

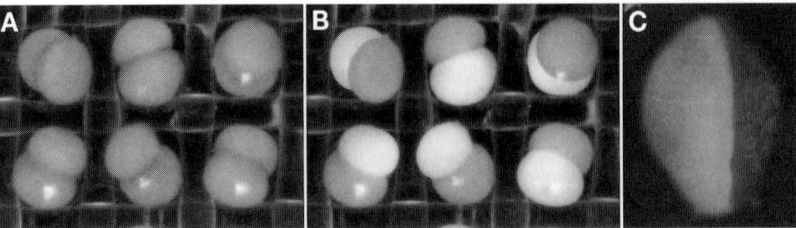

FIGURE 2. (*A*) Lamprey embryos at the two-cell stage. (*B*) Embryos injected with an FITC-labeled morpholino antisense oligo. Each injection was into a single blastomere at the two-cell stage. (*C*) Dorsal view of a 5-day-old neurula, which had been injected at the two-cell stage, showing the unilateral integration of the fluorescent morpholino.

Protocol 3

DiI Cell Labeling in Lamprey Embryos

This protocol describes how to label lamprey embryo cells by microinjecting the fluorescent dye DiI (1,1′-dioctadecyl-3,3,3′,3′-tetramethylindocarbocyanine perchlorate) to study cell fate during development. It was adapted from McCauley and Bonner-Fraser (2003).

MATERIALS

The recipes for the items marked with <R> are on page 427.

CAUTION: See the Cautions Appendix for appropriate handling of materials marked with <!>.

Reagents

DiI (0.5 µg/µl), prepared in 0.3 M sucrose
Lamprey embryos (5–6 days old; see Protocol 1)
1x MMR <R>
Paraformaldehyde (4%) <!>
10x Phosphate-buffered saline (PBS) <R>, diluted to 1x before use

Equipment

Dissecting microscope
Fluorescence microscope
Forceps (sharp; two pairs)
Glass electrodes (pulled)
Incubator, preset to 18°C
Petri dishes

Coat some dishes with 1% agarose and create 1-mm depressions that will hold the embryos in place (see Step 2).

Picospritzer
Vibratome (optional; see Step 8)

METHOD

1. Submerge the 5–6-day-old embryos in 1x MMR. Remove the chorions from the embryos using sharp forceps. Use a dissecting microscoipe to aid in visualization.

    The 1x MMR solution is used during chorion removal and for subsequent culturing to prevent the embryos from exploding.

2. Place the dechorionated embryos into 1-mm depressions in the agarose-coated Petri dishes.

3. Back-fill the glass electrodes with the DiI solution. Using the picospritzer, inject a single bolus under the ectoderm to label a discrete region of the embryo (e.g., the dorsal neural tube).

4. View the injected embryos under a fluorescence microscope to confirm that the labeling is confined to a single discrete location.

5. Transfer the embryos to Petri dishes containing 1× MMR. Incubate the dishes at 18°C until they reach the desired stage of development.

   See Protocol 1 for general information on culturing lamprey embryos.

6. Once the embryos have reached the desired age, replace the 1× MMR with 4% paraformaldehyde and incubate for 1–2 hours at room temperature to fix the tissue.

7. Rinse the embryos thoroughly with room-temperature 1× PBS to remove the fixative.

8. Examine the embryos under the fluorescence microscope to observe the final destinations of the labeled cells (e.g., the neural crest cells).

   To determine the location of the labeled cells with greater accuracy, the embryos can be embedded in 5% agarose and sectioned with a vibratome.

Protocol 4

Whole-mount In Situ Hybridization on Lamprey Embryos

This protocol describes an optimized procedure for RNA in situ hybridization in lamprey embryos. It is based on the protocol previously described by Sauka-Spengler et al. (2007).

MATERIALS

The recipes for the items marked with <R> begin on page 425.

CAUTION: See the Cautions Appendix for appropriate handling of materials marked with <!>.

Reagents

Anti-DIG FAB fragments (alkaline-phosphatase conjugated [1:2000 dilution], prepared in blocking solution)

> *If using FITC <!>-labeled probes, use anti-FITC FAB fragments instead of anti-DIG FAB fragments, conjugated to alkaline phosphatase.*

Bleaching solution (freshly prepared) <R>
Blocking solution <R>
Color development solution <R>
Glycerol (75%, prepared in PBST)

> *To prevent mold growth, include 0.2% sodium azide in the 75% glycerol solution.*

Glycine (2 mg/ml, freshly prepared in PBST-DEPC)
Hybridization mix <R>
Hybridization mix:MABT (1:1)
Lamprey embryos at desired developmental stage (see Protocol 1)
1× MABT <R>
MEMFA <R>
Methanol <!>
NTMT <R>
Paraformaldehyde (4%) <!>
Proteinase K (14–22 mg/ml) <!>, prepared in PBST-DEPC
PBST <R>

> *In addition, prepare three separate solutions of PBST containing 25, 50, and 75% methanol.*

PBST-DEPC <R>

> *In addition, prepare three separate solutions of PBST-DEPC containing 25, 50, and 75% methanol.*

Postfix solution <R>
RNA probe (DIG labeled; final concentration 2–4 µg/ml)

> *Both antisense and sense (control) probes should be synthesized for the genes of interest using a standard in vitro transcription kit containing DIG-labeled nucleotides. FITC-labeled probes can also be used.*

Equipment

Aluminum foil
Forceps (sharp, two pairs)
Hybridization oven set at 70°C

Light box
Microcentrifuge tubes (2 ml, transparent)
Microscope (fluorescence)
Nutator

METHOD

All steps should be performed at room temperature unless otherwise noted. Throughout the protocol, use approximately 2 ml of each solution for each group of 20–30 embryos. All steps should be performed using a nutator, except where specifically mentioned otherwise.

Fixation and Dehydration

1. Place embryos in a 2-ml microcentrifuge tube. Fix the embryos in MEMFA for 1 hour with shaking on the nutator.

2. Wash the embryos in PBST-DEPC for 15 minutes with shaking on the nutator. Discard the solution. Repeat this step three more times.

3. With shaking on the nutator, wash the embryos for 15 minutes in each of the following solutions:

    PBST-DEPC containing 25% methanol
    PBST-DEPC containing 50% methanol
    PBST-DEPC containing 75% methanol
    100% methanol

    This slow, gradual dehydration of embryos is important to prevent the chorions from sticking to the surface of the embryo.

4. Rinse the embryos several times in methanol.

    At this point, embryos can be stored for up to several years at –20°C.

5. With shaking on the nutator, wash the embryos for 30 minutes in each of the following solutions:

    Once with PBST-DEPC containing 75% methanol
    Once with PBST-DEPC containing 50% methanol
    Once with PBST-DEPC containing 25% methanol
    Three times with PBST-DEPC

    This slow rehydration procedure ensures that the chorions separate from the surface ectoderm to facilitate chorion removal postfixation. Although lamprey chorions are softer and easier to remove manually before the embryos are fixed, it is not practical to remove the chorions from the embryos at early stages of development (before days 5.5–6) because the embryos are very fragile. (Also, we have found that enzymatic dechorionation methods do not work in lamprey.)

Dechorionation and Pretreatment

6. Remove the chorions from the embryos using forceps.

    After chorion removal, embryos can be dehydrated and stored at –20°C, once again, as described in Steps 3 and 4. When being brought out of storage, wash the embryos three times for 5–10 minutes each in PBST-DEPC before proceeding with Step 7.

7. Replace the PBST-DEPC with freshly made bleaching solution. Place the tubes containing the embryos on the light box for 7–10 minutes.

    This bleaching step increases the signal in the detection step. If a stronger signal is desired, the incubation time can be increased to 15 minutes, but overbleaching will damage the ectoderm in younger embryos. Four-day embryos should not be subjected to bleaching for more than 7 minutes. Note that the embryos must be in transparent tubes to allow the light to penetrate.

8. Rinse the embryos three times for 5 minutes each with PBST-DEPC.
9. Treat the embryos with 14–22 µg/ml proteinase K in PBST-DEPC at room temperature. Incubation time depends on the age of the embryo:
   - For 4.5-day embryos and younger, incubate for 5 minutes.
   - For 4.5–6-day embryos, incubate for 7–8 minutes.
   - For 6-day and older embryos, incubate for 9–10 minutes.

   The proteinase K treatment makes the embryos very fragile, so they should not be agitated until they are refixed. Therefore, do not use the nutator from now until Step 11.

10. Replace the proteinase K solution with freshly prepared 2 mg/ml glycine. Incubate the embryos for 10 minutes.
11. Rinse the embryos twice for 5 minutes each with PBST-DEPC.
12. Incubate the embryos in postfix solution for 20 minutes with mixing on the nutator.
13. Rinse the embryos four times for 5 minutes each with PBST-DEPC.
14. Wash the embryos two or three times for 10 minutes each with hybridization mix.

    At this point, the embryos can be stored indefinitely at –20°C in hybridization mix.

Prehybridization and Hybridization

15. Add fresh hybridization mix to the embryos and prehybridize by incubating for at least 3 hours at 70°C in the hybridization oven.
16. Add fresh, prewarmed (70°C) hybridization mix containing 1–10 µl/ml labeled RNA probe. To hybridize the probe, incubate the embryos overnight (or for at least 16 hours) at 70°C in the hybridization oven with agitation on the nutator, to hybridize the probe.

Posthybridization Washes

17. Remove the hybridization mix containing the probe and store it at –20°C for reuse.

    Probes can be reused five to eight times.

18. Using prewarmed (70°C) hybridization mix, wash the embryos in the 70°C hybridization oven as follows:

    Twice for 15 minutes each (change the hybridization mix after each wash)
    Four times for 30–45 minutes each (change the hybridization mix after each wash)

19. Wash the embryos with hybridization mix:MABT (1:1) for 30 minutes in the 70°C hybridization oven.
20. Wash the embryos four times with MABT for 30 minutes each at room temperature.

Blocking and Color Development

21. Block the embryos in blocking solution for 1 hour at room temperature.
22. Add the diluted anti-DIG-alkaline phosphatase antibody solution to the embryos and shake on the nutator overnight at 4°C.
23. Wash the embryos with MABT as follows:

    Twice for 5 minutes each at room temperature
    Eight times for 30–60 minutes each at room temperature
    Overnight at 4°C

24. Wash the embryos four times with NTMT for 15 minutes each at room temperature.

25. Incubate the embryos in color development solution in the dark (e.g., by covering in aluminum foil) for 1–72 hours.

    The incubation should continue until the color has developed, which will depend on the abundance of the target RNA in the tissue. Slow-developing embryos (those being stained for RNA that is expressed at very low levels) can be incubated overnight.

Postcolor Washes and Postfix

26. Wash the embryos three times with PBST for 5 minutes each in the dark.
27. Incubate the embryos in 4% paraformaldehyde for 2 hours at room temperature.

    Alternatively, incubate the embryos in 4% paraformaldehyde overnight at 4°C.
28. Wash the embryos three times with PBST for 10 minutes each.
29. With shaking on the nutator, wash the embryos as follows:

    With PBST containing 25% methanol for 15 minutes
    With PBST containing 50% methanol for 15 minutes
    With PBST containing 75% methanol for 15 minutes
    With 100% methanol for 2–3 hours

    This step removes background staining. At this stage, embryos can be stored at –20°C.
30. With shaking on the nutator, wash the embryos for 10 minutes in each of the following solutions:

    Once with PBST containing 75% methanol
    Once with PBST containing 50% methanol
    Once with PBST containing 25% methanol
    Three times with PBST
31. Replace the PBST with the 75% glycerol solution and store the embryos at 4°C until ready for photographing.

    When excited with a 633-nm laser, the NBT/BCIP precipitate fluoresces intensely (wavelengths of emitted light are 650 nm) and will quench autofluorescence in lamprey tissues. These properties make it possible to obtain single-cell resolution in whole-mount in situ hybridization and permit multichannel combination of in situ staining with other visible fluorophores (Trinh et al. 2007).

Protocol 5

Immunostaining of Whole-mount and Sectioned Lamprey Embryos

This protocol describes how to immunostain whole-mount or sectioned lamprey embryos using an antibody raised against the protein of interest and detected with an HRP-conjugated secondary antibody. Enzyme-conjugated rather than fluorochrome-conjugated secondary antibodies are used for antigen detection in lamprey embryos because, after fixation, yolk platelet-rich lamprey embryos exhibit a strong autofluorescence in all three channels. Some lamprey antigens (e.g., activated caspase-3 and phospho-histone H3) are recognized by commercially available antibodies raised against antigens from other species.

MATERIALS

The recipes for the items marked with <R> begin on page 426.

CAUTION: See the Cautions Appendix for appropriate handling of materials marked with <!>.

Reagents

3,3-Diaminobenzidine tetrahydrochloride (DAB) <!>
   Prepare a stock solution of 2 mg/ml DAB in H_2O, filter it through a 0.22-μm membrane, and store it at –20°C, protected from light. Before the procedure, dilute the DAB solution to 1:20 in PBST for whole-mount embryos or 1:40 for sections. Staining can be enhanced to give a black signal, which is most useful for cell counting, by adding $CoCl_2$ <!> and $NiCl_2$ <!> to final concentrations of 0.1%.

Donkey serum, heat inactivated (5% or 10% [see Step 6], prepared in PBST)

Eukitt <!> (for embryo sections only)

Gelatin (20%), prewarmed for 1 hour at 37°C before use (for embryo sections only)

Glycerol (25, 50, and 75%; prepared in PBST) (for whole-mount embryos only)

Goat–antirabbit antibody, HRP conjugated (1:500, prepared in PBST)

H_2O, distilled (for embryo sections only)

Hydrogen peroxide (H_2O_2; 0.5%, prepared in PBST) <!>

Hydrogen peroxide (30%) <!>

Lamprey embryos, cultured to the desired stage (see Protocol 1)

Liquid nitrogen <!> (for embryo sections only)

MEMFA (for whole-mount embryos only) <R>

Methanol <!> (for whole-mount embryos only)

Paraformaldehyde (4%) <!>

10x PBS <R>, diluted to 1x before use (for embryo sections only)

PBST <R>

PBST-DEPC <R> (for whole-mount embryos only)
   In addition, prepare three separate solutions of PBST-DEPC containing 25, 50, and 75% methanol.

Rabbit or mouse antibody, raised against protein of interest (1:250, prepared in PBST)
   The required dilution will vary for different antibodies. The stated 1:250 dilution is suitable for rabbit anti-activated caspase-3 (Promega) and should be used as a guide only.

Sucrose (5% and 15%) (for embryo sections only)

Triton X-100 <!> (for embryo sections only)

Equipment

Centrifuge tubes (15 ml)
Coplin jars or plastic slide holders (for embryo sections only)
Cryostat (for embryo sections only)
Dissection needle (for embryo sections only)
Microcentrifuge tubes (2 ml; for whole-mount embryos only)
Molds, rubber (for embryo sections only)
> Use rubber molds rather than plastic molds to prevent them from cracking when frozen.

Nutator (for whole-mount embryos only)
Superfrost plus slides (for embryo sections only)
Water bath, preset to 42°C (for embryo sections only)

METHOD

Perform all steps at room temperature unless otherwise noted. Agitate whole-mount embryos on the nutator for all incubations. Incubate embryo sections in Coplin jars without agitation.

Gelatin Embedding and Sectioning

1. Cryoprotect the lamprey embryos by incubating them in 5% sucrose for 5–6 hours at room temperature and then 15% sucrose for 5–6 hours at room temperature or overnight at 4°C.

2. Incubate the embryos in prewarmed 20% gelatin for 4–7 hours at 37°C.

3. Embed lamprey embryos in the desired orientation in 20% gelatin as follows:

    i. Prepare the mold by filling it halfway with 20% gelatin and allow it to stand for 1–2 minutes to create a solid gelatin "cushion."

    ii. Place the embryo in liquid 20% gelatin on top of the "cushion" and adjust the position of the embryo quickly using a hot dissection needle. Allow the gelatin to set for 5–10 minutes.

    iii. Immerse the entire mold into liquid nitrogen.

    iv. Store the blocks at –80°C.

4. Make 10–14-µm sections of the blocks using a cryostat and mount the sections on Superfrost plus slides.

5. Store the slides at –20°C.
    > When the slides are taken out of –20°C storage, allow them to warm to room temperature before using.

Fixation and Blocking

6. Prepare whole-mount embryos or embryo sections for immunostaining as follows.

To Prepare Whole-mount Embryos

   i. Transfer embryos into 2-ml microcentrifuge tubes (up to 20 embryos per tube). Fix the embryos in MEMFA for 1 hour.

   ii. Wash the embryos in PBST-DEPC for 15 minutes. Discard the solution. Repeat this step three more times.

iii. Wash the embryos for 15 minutes in each of the following four solutions:

>   PBST-DEPC containing 25% methanol
>   PBST-DEPC containing 50% methanol
>   PBST-DEPC containing 75% methanol
>   100% methanol

iv. Rinse the embryos several times in methanol.
    At this point, embryos can be stored for up to several years at –20°C.

v. To block endogenous peroxidase activity, treat the embryos with 0.5% H_2O_2 in PBST for 1 hour at room temperature.

vi. Wash the embryos in PBST three times for 15 minutes each.

vii. To block nonspecific proteins, incubate the embryos in 10% donkey serum in PBST for 1 hour at room temperature.

To Prepare Embryo Sections

i. Incubate the slides containing sectioned embryos in preheated (42°C) 1x PBS for 10 minutes in the 42°C water bath to remove the gelatin.

ii. Wash once for 10 minutes in 1x PBS and then twice for 10 minutes each in PBST.

iii. To block endogenous peroxidase activity, treat the slides with 0.5% H_2O_2 in PBST for 30 minutes at room temperature.

iv. Permeablize the cells by incubating the slides in 0.2% Triton X-100 for 5 minutes then rinse three times in 1x PBS.

v. Incubate the embryos in 5% donkey serum in PBST for 1 hour at room temperature.

Antibody Binding

7. Incubate the embryos or sections in the antibody raised to the protein of interest overnight at 4°C.

8. Wash the antibody from the samples as follows:
   - For whole-mount embryos, wash five to eight times with PBST for 1 hour each. Then incubate overnight in PBST at 4°C.
   - For sections, wash twice with PBS for 10 minutes each. Then wash twice with PBST for 10 minutes each.

9. Incubate the samples in HRP-conjugated goat–antirabbit antibody for 4 hours (for whole-mount embryos) or 2 hours (for sections) at room temperature.

10. Remove unbound antibodies as follows:
    - For whole-mount embryos, rinse five times in PBST. Then wash overnight in PBST at 4°C.
    - For sections, wash twice in PBS for 10 minutes each. Then wash twice with PBST for 10 minutes each.

Color Development

11. Add 10 ml of DAB solution to the embryos in a 15-ml plastic centrifuge tube.
    If using embryo sections, skip Step 7 and proceed to Step 8 to allow the DAB to penetrate.

12. For whole-mount embryos, incubate for 30 minutes at room temperature.

13. Add 3.3 µl of 30% hydrogen peroxide and incubate for 30–45 minutes, checking regularly for color development.

    If the color has not yet developed, the DAB/hydrogen peroxide mix can be replaced after 30 minutes.

14. When sufficient color formation has occurred, stop the reaction by washing whole-mount embryos in several changes of PBST or by dipping embryo section slides several times in a 1-liter beaker containing distilled H_2O.

15. Postfix whole-mount embryos and sections in 4% paraformaldehyde for 20–30 minutes at room temperature.

16. Prepare the stained samples for photographing.

To Prepare Stained Whole-mount Embryos

    i. Rinse three–four times for 5 minutes each in PBST

    ii. Process through glycerol/PBST series: 25%, 50%, 75%.

    iii. Photograph in 75% glycerol.

        Whole-mount embryos can be processed to gelatin and sectioned as described on page 422.

To Prepare Stained Embryo Sections

    i. Dehydrate the slides.

    ii. Mount slides in Eukitt.

    iii. View and photograph.

Recipes

The recipes for items marked with <R> are also listed here.

CAUTION: See the Cautions Appendix for appropriate handling of materials marked with <!>.

BLEACHING SOLUTION

Reagent	Quantity (for 10 ml)	Final concentration
Formamide <!>	500 µl	5%
H_2O_2 <!>	2.8 ml	28%
20x SSC	250 µl	0.5x
H_2O, DEPC-treated	to 10 ml	

Prepare fresh. To avoid foaming, add the ingredients in the following sequence: formamide, DEPC-treated H_2O, 20x SSC, and H_2O_2. After use, dilute the bleaching solution with an equal amount of H_2O and store it in the fume hood before disposal.

BLOCKING SOLUTION

Reagent	Quantity (for 100 ml)	Final concentration
Blocking reagent, 10% stock solution in 1x MAB (Roche)	20 ml	2%
Sheep serum, heat inactivated	20 ml	20%
MABT <R>	to 100 ml	

This solution can be stored frozen for several months at –20°C.

COLOR DEVELOPMENT SOLUTION

Reagent	Quantity (for 10 ml)	Final concentration
4-Nitro blue tetrazolium chloride (NBT) <!> (100 mg/ml)	45 µl	0.45 mg/ml
5-Bromo-4-chloro-3-indolyl-phosphate (BCIP) <!> (50 mg/ml)	35 µl	0.175 mg/ml
NTMT <R>	to 10 ml	

Prepare just enough color development solution for the reaction. Excess can be stored for several days in foil-covered tubes at 4°C.

HYBRIDIZATION MIX

Reagent	Quantity (for 500 ml)	Final concentration
CHAPS <!> (10% in DEPC-treated H_2O)	25 ml	0.5%
EDTA (0.5 M, pH 8.0) <!>	5 ml	5 mM
Formamide, deioinized <!> (Ambion)	250 ml	50%
Heparin <!>	50 mg	100 µg/ml
20x SSC	32.5 ml	1.3x
tRNA from baker's yeast	100 mg	200 µg/ml
Tween 20	1 ml	0.2%
H_2O	to 500 ml	

Hybridization mix can be stored for several months at –20°C.

1x MABT

Reagent	Quantity (for 1000 ml)	Final concentration
Maleic acid <!>	11.6 g	0.4 M
NaCl	8.77 g	150 mM
Tris base <!>	20–30 g	0.2 M
Tween 20	1 ml	0.1%
H_2O	to 1000 ml	

Adjust pH to 7.5 with Tris base. Filtered MABT can be stored for a few weeks to several months.

MEMFA

Reagent	Quantity (for 10 ml)	Final concentration
Formaldehyde, ultrapure <!> (16%)	2.5 ml	4%
10x MEM salts <R>	1 ml	1x
H_2O, distilled	to 10 ml	

Store for 5–12 hours at 4°C, protected from light.

10x MEM SALTS

Reagent	Quantity (for 500 ml)	Final concentration
MOPS (pH 7.4) <!>	104.65 g	1 M
EGTA	3.804 g	20 mM
$MgSO_4$ <!>	0.602 g	10 mM
H_2O, distilled	to 500 ml	

This solution can be stored for several months at room temperature, protected from light.

10x MMR

Reagent	Quantity (for 1000 ml)	Final concentration
$CaCl_2 \cdot 2H_2O$	2.94 g	20 mM
EDTA (0.5 M)	2 ml	1 mM
HEPES (1 M)	50 ml	50 mM
KCl	1.491 g	20 mM
$MgSO_4$	1.204 g	10 mM
NaCl	58.44 g	1 M
H_2O	to 1 liter	

Mix all ingredients well before adding $CaCl_2$. Adjust pH to 7.8 with 1 M NaOH. MMR is prepared as a 10x stock, then diluted for 1x or 0.1x with distilled H_2O, and autoclaved before use. The 10x stock can be stored for several months to years at room temperature.

NTMT

Reagent	Quantity (for 200 ml)	Final concentration
$MgCl_2$ (2 M)	5 ml	50 mM
NaCl (5 M)	4 ml	100 mM
Tris-Cl (1 M, pH 9.5)	20 ml	100 mM
Tween 20	200 µl	0.1%
H_2O	to 200 ml	

There is no need to adjust pH if stock Tris-HCl solution is used. This solution should be prepared fresh before use. It can be stored for 12–24 hours at room temperature. Precipitates form when stored for longer periods.

10x PBS

Reagent	Quantity (for 1 liter)	Final concentration
NaH_2PO_4	2.75 g	23 mM
NaCl	90 g	1.54 M
$Na_2HPO_4 \cdot 7H_2O$	21.45 g	80 mM
H_2O	to 1 liter	

Adjust pH to 7.4 with NaOH. Store for several months at room temperature.

PBST

Reagent	Quantity (for 1 liter)	Final concentration
10x PBS <R>	100 ml	1x
Tween 20	1 ml	0.1%
H_2O	to 1 liter	

Store at room temperature.

PBST-DEPC

Reagent	Quantity (for 1 liter)	Final concentration
10x PBS <R>	100 ml	0.1x
Diethylpyrocarbonate (DEPC) <!>	1 ml	
Tween 20	1 ml	0.1%
Distilled H₂O	to 1 liter	0.1%

Stir at room temperature for 1 hour and then autoclave for 1 hour to remove the DEPC. Add Tween 20 after autoclaving. Store for months to years at room temperature.

POSTFIX SOLUTION

Reagent	Quantity (for 10 ml)	Final concentration
Glutaraldehyde <!> (25%)	80 μl	0.2%
Paraformaldehyde <!>	400 mg	4%
PBST-DEPC <R>	to 10 ml	

Prepare solution just before use. Do not store for more than 1–2 hours at 4°C.

REFERENCES

Alder, M.N., Rogozin, I.B., Iyer, L.M., Glazko, G.V., Cooper, M.D., and Pancer, Z. 2005. Diversity and function of adaptive immune receptors in a jawless vertebrate. *Science* **310:** 1970–1973.

Amemiya, C.T., Saha, N.R., and Zapata, A. 2007. Evolution and development of immunological structures in the lamprey. *Curr. Opin. Immunol.* **19:** 535–541.

Cooper, M.D. and Alder, M.N. 2006. The evolution of adaptive immune systems. *Cell* **124:** 815–822.

Damas, H. 1944. Recherches sur le developpement de Lampetra fluviatilis L. Contriution a l'etude de la cephalogenese des vertebres. *Arch. Biol.* **55:** 3–282.

Gill, H.S., Renaud, C.B., Chapleau, F., Mayden, R.L., and Potter, I.C. 2003. Phylogeny of living parasitic lampreys (Petromyzontiformes) based on morphological data. *Copeia* **2003:** 687–703.

Hardisty, M.W. 1979. *Biology of the cyclostomes*. Chapman and Hall, London.

Hardisty, M.W. and Potter, I.C., eds. 1971a. Paired species. In *The biology of lampreys* (ed. M.W. Hardisty and I.C. Potter), pp. 249–278. Academic, London.

Hardisty, M.W. and Potter, I.C., eds. 1971b. The general biology of adult lampreys. In *The biology of lampreys* (ed. M.W. Hardisty and I.C. Potter), pp. 127–206. Academic, London.

Hardisty, M.W. and Potter, I.C., eds. 1971–1982. *The biology of lampreys*. Academic, London.

Horigome, N., Myojin, M., Ueki, T., Hirano, S., Aizawa, S., and Kuratani, S. 1999. Development of cephalic neural crest cells in embryos of *Lampetra japonica*, with special reference to the evolution of the jaw. *Dev. Biol.* **207:** 287–308.

Hubbs, C.L. and Potter, I.C. 1971. Distribution, phylogeny and taxonomy. In *The biology of lampreys* (ed. M.W. Hardisty and I.C. Potter), vol. 1, pp. 1–65. Academic, London.

Huxley, T.H. 1876. The nature of the craniofacial apparatus of *Petromyzon*. *J. Anat. Physiol.* **10:** 412–429.

Kuratani, S., Kuraku, S., and Murakami, Y. 2002. Lamprey as an evo-devo model: Lessons from comparative embryology and molecular phylogenetics. *Genesis* **34:** 175–183.

McCauley, D.W. and Bronner-Fraser, M. 2003. Neural crest contributions to the lamprey head. *Development* **130:** 2317–2327.

McCauley, D.W. and Bronner-Fraser, M. 2006. Importance of SoxE in neural crest development and the evolution of the pharynx. *Nature* **441:** 750–752.

Nagawa, F., Kishishita, N., Shimizu, K., Hirose, S., Miyoshi, M., Nezu, J., Nishimura, T., Nishizumi, H., Takahashi, Y., Hashimoto, S., et al. 2007. Antigen-receptor genes of the agnathan lamprey are assembled by a process involving copy choice. *Nat. Immunol.* **8:** 206–213.

Osorio, J. and Retaux, S. 2008. The lamprey in evolutionary studies. *Dev. Genes Evol.* **218:** 221–235.

Pancer, Z., Amemiya, C.T., Ehrhardt, G.R., Ceitlin, J., Gartland, G.L., and Cooper, M.D. 2004. Somatic diversification of variable lympho-cyte receptors in the agnathan sea lamprey. *Nature* **430:** 174–180.

Piavis, G.W. 1961. Embryological stages in the sea lamprey and effect of temperature on development. *U.S. Fish Wildl. Serv. Fish. Bull.* **61:** 111–143.

Piavis, G.W. 1971. Embryology. In *The biology of lampreys* (ed. M.W. Hardisty and I.C. Potter), vol. 1, pp. 361–400. Academic, London.

Reese, A.M. 1900. Lampreys in captivity. *Science* **11:** 555.

Rubinson, K. 1990. The developing visual system and metamorphosis in the lamprey. *J. Neurobiol.* **21:** 1123–1135.

Sauka-Spengler, T., Meulemans, D., Jones, M., and Bronner-Fraser, M. 2007. Ancient evolutionary origin of the neural crest gene regulatory network. *Dev. Cell* **13:** 405–420.

Stock, D.W. and Whitt, G.S. 1992. Evidence from 18S ribosomal RNA sequences that lampreys and hagfishes form a natural group. *Science* **257:** 787–789.

Tahara, Y. 1988. Normal stages of development in the lamprey *Lampetra reissneri* (Dybowski). *Zool. Sci.* **5:** 109–118.

Takezaki, N., Figueroa, F., Zaleska-Rutczynska, Z., and Klein, J. 2003. Molecular phylogeny of early vertebrates: Monophyly of the agnathans as revealed by sequences of 35 genes. *Mol. Biol. Evol.* **20:** 287–292.

Trinh le, A., McCutchen, M.D., Bonner-Fraser, S.E., Bumm, L.A., and McCauley, D.W. 2007. Fluorescent in situ hybridization employing the conventional NBT/BCIP chromogenic stain. *Biotechniques* **42:** 756–759.

Vladykov, V.D. 1949. Quebec lampreys (Petromyzonidae). Department of Fisheries, Province of Quebec, Canada. Contribution 26, pp. 1–6.

WWW RESOURCES

http://www.invasivespeciesinfo.gov/aquatics/lamprey.shtml The National Invasive Species Information Center. This website has numerous links to and information about *P. marinus*, its biology, and efforts to control its impact on native species.

17 The Dogfish *Scyliorhinus canicula*
A Reference in Jawed Vertebrates

Marion Coolen,[1] Arnaud Menuet,[1] Danièle Chassoux,[2]
Claudia Compagnucci,[3] Sébastien Henry,[4] Laurent Lévèque,[4]
Corinne Da Silva,[5] Frédérick Gavory,[5] Sylvie Samain,[5]
Patrick Wincker,[5] Claude Thermes,[6] Yves D'Aubenton-Carafa,[6]
Isabel Rodriguez-Moldes,[7] Gavin Naylor,[8] Michael Depew,[3]
Pascal Sourdaine,[9] and Sylvie Mazan[1]

[1]*CNRS UMR 6218 Immunologie et Embryologie Moléculaires, Université Sciences et Techniques d'Orléans, 45071 Orléans, France;* [2]*INSERM U565, F-75005 Paris, France, CNRS UMR5153, F-75005 Paris, France, and Muséum National d'Histoire Naturelle, F-75005 Paris, France;* [3]*Department of Craniofacial Development, Guy's Hospital, King's College London, United Kingdom;* [4]*CNRS FR 2424, Station Biologique de Roscoff, 29682 Roscoff, France;* [5]*Genoscope (CEA), 91057 Evry, France, CNRS UMR 8030, F-91057 Evry, France, and Université d'Evry, 91057 Evry, France;* [6]*CNRS UPR 2167, Centre de Génétique Moléculaire, F-91198 Gif sur Yvette, France;* [7]*Department of Cell Biology and Ecology, Faculty of Biology, University of Santiago de Compostela, 15706 Santiago de Compostela, Spain;* [8]*150-a Dirac Science Library, School of Computational Science, Florida State University, Tallahassee, Florida 32306;* [9]*UMR 100 IFREMER, Université de Caen Basse-Normandie "Physiologie et Ecophysiologie des Mollusques Marins," 14032 Caen, France*

ABSTRACT

Due to their large size and overly long generation times, chondrichthyans have been largely ignored by geneticists during past decades. However, their key phylogenetic position makes them ideal subjects for the study of the molecular bases of the important morphological and physiological innovations that characterize jawed vertebrates. Such analyses are crucial to understanding the origin of the complex genetic mechanisms unraveled in osteichthyans. The small spotted dogfish *Scyliorhinus canicula*, a representative of the largest order of extant sharks, presents a number of unique advantages in this context. Due to its relatively small size among sharks, its abundance, and easy maintenance, the dogfish has been an important model in comparative anatomy and physiology for more than a century. During the past few years, revived interest in the dogfish has occurred with the development of large-scale transcriptomic and genomic resources, together with the establishment of facilities allowing massive egg and embryo production. These new tools open the way to molecular analyses of the elaborate physiological and sensory systems used by sharks. They also make it possible to take advantage of unique characteristics of these species, such as organ zonation, in analyses of cell proliferation and differentiation. Finally, they offer important perspectives to the field of evolutionary developmental biology that will provide a better understanding of the origin and diversifications of jawed vertebrates. The dogfish whole-genome sequence, which may shortly become accessible, should establish this species as an essential shark reference, complementary to other chondrichthyan models. These analyses are likely to reveal an organism of an underestimated complexity, far from the primitive prototypical gnathostome anticipated in gradistic views.

This chapter, with full-color images, can be found online at www.cshprotocols.org/emo.

BACKGROUND INFORMATION

The lesser spotted dogfish, *Scyliorhinus canicula*, is among the smallest catshark species. Also termed small spotted catshark or simply dogfish, it is found at depths from a few meters to more than a hundred meters, and it is widely distributed in the northeast Atlantic, from Norway to Senegal. *S. canicula* is, as are most cartilaginous fish, poikilothermic (i.e., its body temperature varies with the temperature of its surroundings), and it feeds opportunistically on a wide range of benthic fauna (Lyle 1983). Its reproduction is oviparous, with internal fertilization, and egg-laying occurs throughout the year with peak frequency in the spring (Rodriguez-Cabello et al. 1998). *S. canicula* has been reported to be the most abundant catshark species in European inshore waters (Ellis and Shackley 1997). Although it is commercially fished and often caught as by-catch in demersal fisheries, no decline in populations has been recorded thus far, contrary to the case with most large sharks, skates, and rays.

The phylogenetic position of the dogfish, together with its abundance, largely explains the interest that it has raised in the scientific community. *S. canicula* is member of the chondrichthyan clade, which, as the sister group to osteichthyan fish, provides an outgroup to the major osteichthyan taxa (Fig. 1A). This key position makes the dogfish an essential comparative reference, providing insight into the origin of the major characteristics of gnathostomes (jawed vertebrates). It is likewise ideally placed to distinguish between ancestral and derived characteristics in comparative analyses between sarcopterygians and actinopterygians, which contain the usual vertebrate model organisms (Fig. 2). Its phylogenetic position among chondrichthyans confirms *S. canicula* as an informative and interesting representative of the clade. The dogfish belongs to Scyliorhinidae, the largest family in the largest order of extant sharks, the Carcharhiniformes, which form a clade with the Lamniformes, Orectolobiformes, and Heterodontiformes. This grouping is often termed the Galeomorph superorder and is contrasted with the Squalimorph superorder, comprising the remaining four orders of extant sharks (Squaliformes, Hexanchiformes, Pristiophoriformes, and Squatiniformes) (Fig. 1B). The earliest fossil dates ascribed to the Carcharhiniformes are described as Tithonian (upper Jurassic, 144–151 million years ago [MYA]); however, molecular-based estimates deduced from sequence comparisons suggest a date of 226 MYA (this estimate remains prone to large margins of error; see Heinicke et al. 2008). Inside Carcharhiniformes, phylogenetic analysis based on molecular data suggests that the family Scyliorhinidae is paraphyletic, comprising two distinct groups (Iglésias et al. 2005). The first, most basal group contains the genera Scyliorhinus and Cephaloscylium, whereas the second

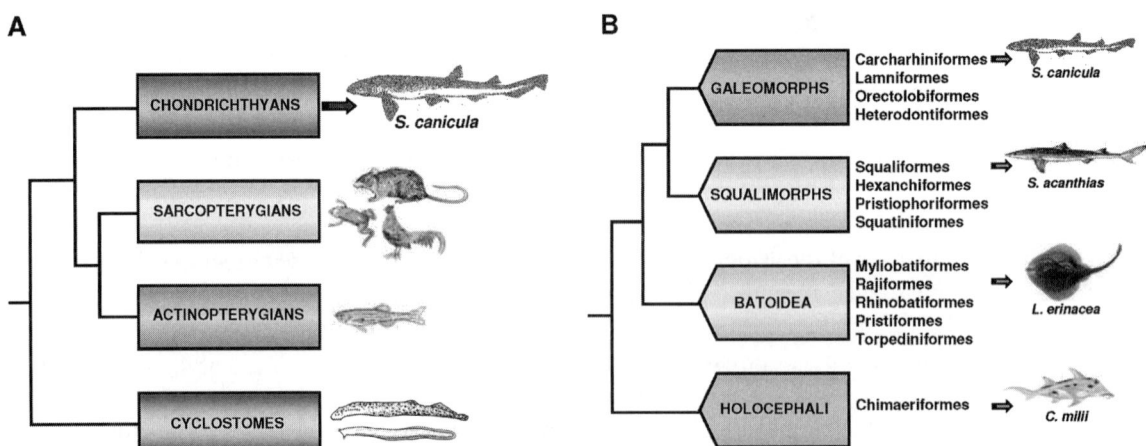

FIGURE 1. Phylogenetic position of the dogfish among jawed vertebrates (*A*) and inside chondrichthyans (*B*). (*A*) Chondrichthyans form a sister group to osteichthyans (sarcopterygians and actinopterygians), a group containing the usual vertebrate model organisms (mouse, frog, chick). (*B*) The most commonly studied chondrichthyan models provide a good phylogenetic sampling of the group. Shown are relative positions of the major chondrichthyan groups.

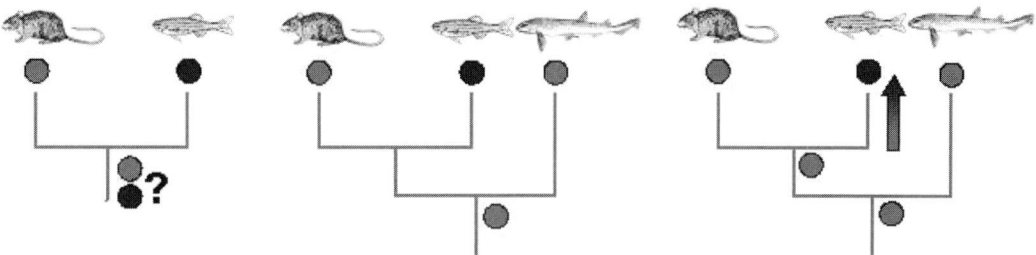

FIGURE 2. Use of an outgroup to infer the gnathostome ancestral state; figure explicitly shows the rationale used. When a given character displays different states in two taxa (tetrapods, *light gray*; actinopterygians, *dark gray*), one way to infer the ancestral state and polarize characters is to determine the character state in an outgroup (here, chondrichthyans). In this example, the character state observed in the outgroup allows us to conclude that the ancestral state is the one observed in tetrapods and that the situation observed in actinopterygians is derived. The underlying rationale in no way implies that the outgroup should systematically reflect the primitive condition, because the inference of the ancestral state relies simply on stepwise inferences.

contains the genera Apristurus, Asymbolus, Cephalurus, Galeus, and Parmaturus that collectively fall as the sister group to the remaining families in the Carcharhiniformes. *S. canicula* therefore holds the important distinction of being one of the most evolutionarily basal members of the Carcharhiniformes, a group that has become the most dominant group of sharks alive today. This placement makes it particularly well suited for studies in estimating the ancestral condition of the order Carcharhiniformes. More broadly, as a representative of Galeomorphes, *S. canicula* fills a gap in the sampling of cartilaginous fish recently considered for the development of large-scale genomic or transcriptomic resources. Its study should be important and complementary to that of *Squalus acanthias*, *Raja erinacea*, and *Callorhinchus milii*, representatives of the three other major chondrichthyan taxa (Squaliformes, Batoidea, and Holocephali, respectively), to infer the chondrichthyan ancestral state (Fig. 1B).

A distinctive feature of the dogfish among other candidate chondrichthyan models is the relatively easy accessibility of its eggs and embryos. Historically, this fact made it the prevailing chondrichthyan model at the height of the era of comparative embryology, during the end of the 19th century. The work of this time led to detailed histological studies of embryonic development and even pioneering attempts to establish fate maps at the blastula stages (for review, see Sauka-Spengler et al. 2004). Although these data should be considered cautiously and revisited using modern cell-labeling techniques, they clearly established the dogfish as a candidate chondrichthyan experimental model in embryology. Beginning in the 1950s, *S. canicula* also became an important model in comparative physiology. These studies, involving a number of different research groups, addressed such varied issues as endocrine control of reproduction, control of osmolarity, characterization of lymphomyeloid organs, and physiology of nutrition, respiration, and muscle contraction. Another intense field of investigation was aimed at the description of the dogfish central nervous system, brain projection patterns, and distribution of neurotransmitters. By the end of the 20th century, the expansion of molecular biology techniques engendered the emergence of a novel evolutionary-development field that focused on the dogfish; this may become the primary chondrichthyan model in this field. These studies have generally revealed remarkably elaborate physiological and developmental processes that contrast with classical views of chondrichthyans as primitive, prototypical gnathostomes.

SOURCES AND HUSBANDRY

The lesser-spotted dogfish is well known as being relatively easy to rear in captivity, and for this reason, it is very commonly maintained in public aquaria. However, probably due to its low commercial value, there has been no attempt to rear this species on a larger scale for aquaculture purposes. The emergence of this species as a new genomic model has created a fast-growing demand for the mass production of embryos, particularly for large-scale gene expression analyses and emerging functional approaches. Given the relatively low fertility of this species, availability of embryos throughout the year has become a real bottleneck in the development of these studies.

Currently in Europe, the main provider of *S. canicula* eggs is the Station Biologique de Roscoff in France, which since 2006, has offered a specific program to develop and optimize its facilities for the maintenance of dogfish to support genomic research. As a result, production has risen from a few hundred eggs in 2005 to 8000 eggs in 2007, which are shipped to various laboratories in Europe, and adult mortality has been dramatically reduced. Within a few years, ongoing development should reach a target egg production of about 20,000 per year, sufficient to meet the expected needs of the international scientific community working on related genomic topics. The Station Biologique de Roscoff is thus becoming the first structured resource center dedicated to this species.

A major challenge for maintenance facilities in rearing the dogfish is its relatively large size and low fertility. This species also exhibits a slow growth rate and a long life cycle; its maximum reported life span is 12 years. The age at which 50% of females reach maturity is estimated to be from 6 to 8 years (Ivory et al. 2004). It is therefore likely that it will be necessary to rely on natural populations to obtain reproductive adults. Optimized maintenance facilities are thus crucial to keep the need to collect wild specimens to a minimum. Finally, as pointed out by Sims et al. (2001), relatively little is known about the behavior of this species in the wild. Such basic knowledge, in particular, of reproductive and feeding behavior, will be important for improving maintenance conditions.

RELATED SPECIES

Three other chondrichthyan species that have aroused interest in the scientific community during the past few years are the spiny dogfish shark *Squalus acanthias* (squaliformes), the little skate *Raja erinacea* (rajiformes), and a chimera *Callorynchus milii* (holocephali). All of these species exhibit distinct advantages. *S. acanthias* is a traditional shark model in physiology, but embryos remain difficult to obtain, largely because of its relatively large adult size. *C. milii*, characterized by a particularly compact genome among chondrichthyans (about 0.9 Gb), was selected for a low-coverage (1.4x) genome sequencing project (Venkatesh et al. 2007). Although an important fraction of the genome remains to be determined, its analysis has already yielded important results, such as the identification of more than 150 gene-loss events in osteichthyans. An important limitation, however, is that this species is virtually inaccessible to experimentation. Finally, the little skate, another classic model in physiology, has been selected for whole-genome sequencing (see the Evolution of the Proteome project at http://www.genome.gov). This species is considered a good compromise model for the group because it has a moderate genome size among chondrichthyans and is likely amenable to various experimental approaches (C-value = 3.5 pg). Considering their respective relative phylogenetic distribution, it is noteworthy that these species together with the dogfish provide an excellent representation of chondrichthyans and therefore appear fully complementary in comparative approaches. Despite other constraints, the possibility to obtain eggs relatively easily remains a highly advantageous and distinctive feature of the dogfish.

USES OF THE *S. CANICULA* SYSTEM: STUDIES IN DEVELOPMENT, BEHAVIOR, ECOLOGY, MORPHOLOGY, AND PHYSIOLOGY

The dogfish has been an invaluable resource for identifying developmental processes and physiological systems conserved between chondrichthyans and osteichthyans and has thus contributed significantly to our understanding of the origin of major gnathostome characteristics. Molecular resources now provide new tools to address the bases of the unity, as well as the important underlying morphological or physiological diversity, of gnathostomes. Their application in the dogfish has opened active fields of investigation in evolutionary developmental biology and comparative physiology and may also contribute significantly to our understanding of genome and proteome evolution in vertebrates.

From Comparative Morphology to Evolutionary Developmental Biology: Origin and Diversification of the Gnathostome Body Plan

Despite the diversity of animal forms observed in the taxon, gnathostomes are characterized by a highly conserved body plan, including important morphological innovations such as the jaw, true teeth, gill arches lying internally to the gills and branchial blood vessels, paired appendages, a horizontal semicircular canal in the inner ear, and myelinated nerve fibers. During the past few years, the dogfish has become a reference model for assessing conservation (at the gnathostome level) of genetic programs thought to control the formation during ontogeny of these conserved morphological characteristics. A well-documented example is the emergence of paired appendages, which opened the way to important diversifications and adaptations. Comparative analysis between the dogfish and osteichthyans highlighted strong similarities in the genetic programs controlling limb positioning along the dorsoventral body axis, its anterior or posterior identity, and its proximodistal (PD) patterning (Tanaka et al. 2002). Unexpectedly, these studies suggested a conservation with tetrapods, which share with the dogfish, but not zebrafish, a late phase of *Hoxd* expression (typically required for proper limb development). This phase of expression, previously thought to be a tetrapod innovation associated with the formation of digits, thus appeared to correspond to an ancestral condition that was lost or diverged in actinopterygians (Freitas et al. 2007). As shown by this example, extending comparisons to a chondrichthyan outgroup can be important for correctly inferring the ancestral gnathostome condition and its possible modifications in the actinopterygian or sarcopterygian lineages.

Similarly, recent analyses in the dogfish provided new insights into common mechanisms specifying underlying early axis mechanisms across gnathostomes. In this case, both the phylogenetic position and the morphological characteristics of the dogfish embryo were crucial in the comparative analyses. In *S. canicula*, the egg is telolecithal (i.e., the yolk is concentrated at one end of the egg) and, at early development stages (blastula to gastrula), the embryo develops as a large, flat blastodisc, just like in the chick. This embryonic morphology lends itself remarkably well to comparisons with amniotes and, together with the protracted development of the dogfish, allows excellent spatial and temporal resolutions of expression patterns. By taking advantage of these characteristics, we produced an exhaustive molecular characterization of the early dogfish embryo, which led to the identification of its remarkable similarity to all vertebrate model organisms, including the zebrafish, *Xenopus*, and amniotes (Coolen et al. 2007). Unexpectedly, we identified territories in the dogfish embryo that share similar molecular properties and comparable relative organization to the extraembryonic territories of amniotes. Furthermore, the location of these territories at either side of the blastoderm was found to define an early blastoderm polarity, reflecting the future rostrocaudal axis of the embryo proper, consistent with the essential roles of extraembryonic tissues in axis specification reported in amniotes. Together, these data led for the first time to a unifying model of early axis specification in jawed vertebrates, suggesting in particular a relationship between the early dorsoventral axis of teleosts and amphibians and the embryonic-extraembryonic organization of mammals (Fig. 3). At a very general level of interpretation, these findings suggest that analyses conducted in novel model organisms, selected on criteria radically different from those traditionally taken into account in developmental genetics (anthropomorphism, small size, fast development), can contribute significantly to an integrated understanding of developmental processes.

Deciphering molecular mechanisms underlying the unity of morphological structures also provides a basis for addressing their diversification. In this respect, chondrichthyans offer a wealth of model systems that should benefit greatly from both detailed anatomical descriptions available

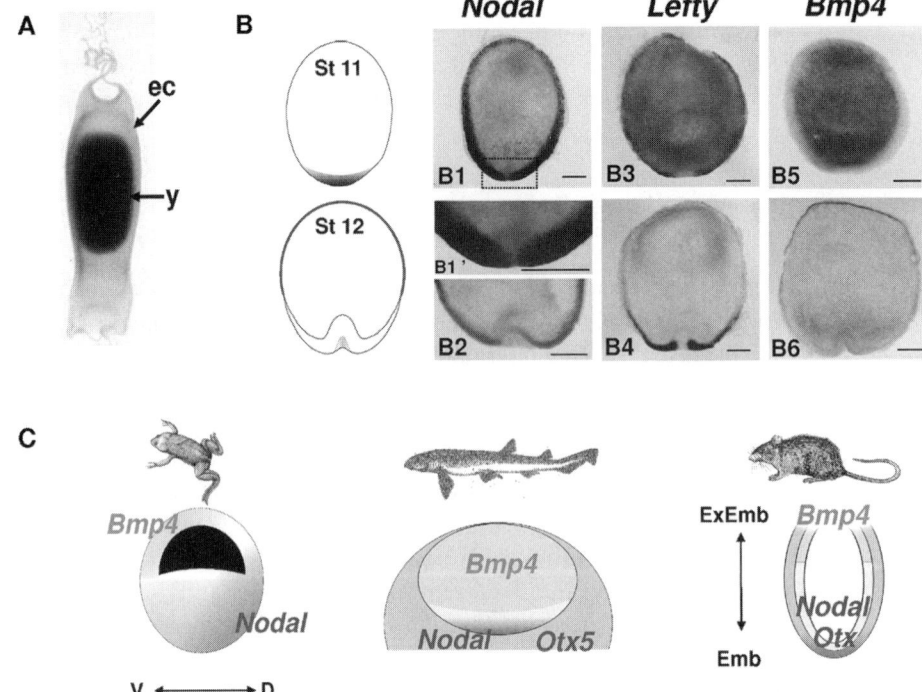

FIGURE 3. Broad characteristics of the dogfish early embryo and similarities with model organisms. (*A*) Lateral view of a dogfish egg. The embryo develops inside a collagen egg case (ec) at the surface of a large yolk mass (y). (*B*) Whole-mount in situ hybridizations of the dogfish embryo at pregastrulation (stage 11: B1, B3, B5) and early gastrulation (stage 12: B2, B4, B6). At these stages, the embryo develops as a relatively large (>1 mm), flat blastoderm that facilitates comparisons with amniotes, particularly the chick. The examples show the expression profiles or transcript distributions of signaling molecules, three members of the transforming growth factor β (TGF-β) superfamily, known to be involved in axis specification. The results, obtained with the use of *Nodal* (B1–2), *Lefty* (B3–4), and *Bmp4* (B5–6) probes, underscore the excellent spatial and temporal resolution of expression patterns achieved in the dogfish. All embryos are shown in dorsal views. B1′ is a higher magnification of B1 at the level of the *dotted square*, showing a *Nodal* negative territory in the midline of the posterior margin. (*C*) Similarities between the early polarities observed in the dogfish (*center*) and those described in *Xenopus* (*left*, dorsoventral [D-V]) and amniotes (*right*, embryonic-extraembryonic [Emb-ExEmb] in the mouse) at blastula stages. Comparisons among the dogfish, amphibians, and teleosts on one hand and amniotes on the other suggest that the genetic interactions taking place along the dorsoventral polarity of the former may be evolutionary related to those taking place between embryonic and extraembryonic poles in mammals. The photographs in *B* and the scheme in *C* are taken or redrawn from Coolen et al. 2007.

for several representatives of the taxon and the development of evolutionary developmental approaches in the dogfish. The analysis of jaw development provides an excellent paradigm. The acquisition of jaws was a landmark event in vertebrate evolution, one that led, in large part, to vertebrate diversification and success. This momentous event was achieved in coordination with a reorganization of the vertebrate head, which included the expansion of the forebrain, an elaboration of cephalic placodes, and the formation of a neural crest population supporting structures of the rostral nervous system. During the last 400+ million years—essentially, the entirety of the existence of jaws—gnathostome skulls have been characterized by two basic traits (Goodrich 1958; De Beer 1985; Depew and Simpson 2006). First, they exhibit a high degree of fidelity to an initial basic structural design. This seminal gnathostome basic body plan (or bauplan) includes a chondrocranium composed of a number of basic yet fundamental neurocranial and splanchnocranial units, including the palatoquadrate (PQ) and Meckel's cartilage (MC) cores of the upper and lower jaws, respectively (Fig. 4). Second, they manifest amazing end-point phenotypes. In particular, the jaws

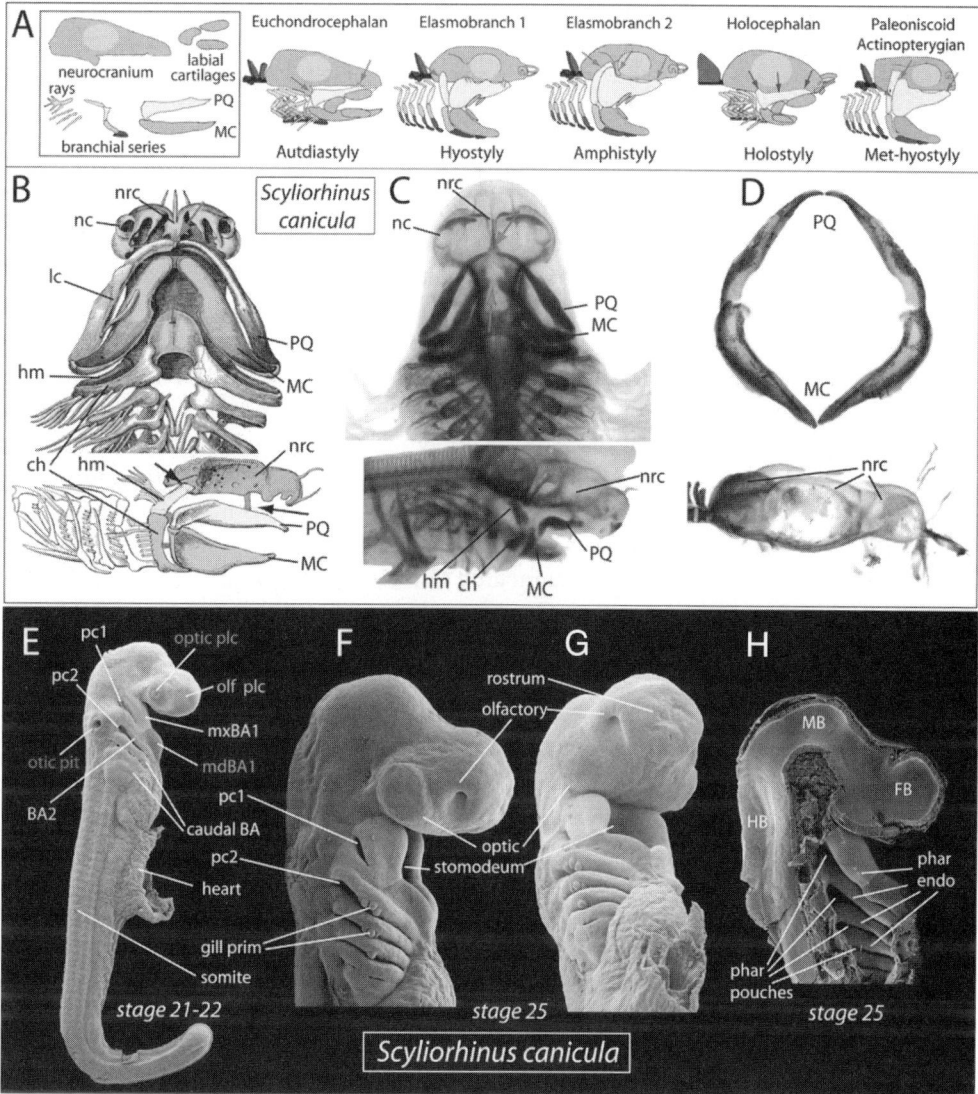

FIGURE 4. Craniofacial anatomy of *S. canicula*. (*A*) The chondrichthyan suspensorium, the structural mechanism by which the jaws are articulated with the neurocranium, is extremely varied and has long been the subject of phylogenetic theories. The principle components of the suspensorium (*box at left*) include (1) the placement, number, and cytodifferentiation state (e.g., synchondrotic, synovial, ligamentous) of the contact points between the palatoquadrate (PQ) and the neurocranium and (2) the relative involvement of the second, or hyoid, arch cartilages (caudal to the PQ and Meckel's cartilage [MC]). Various strategies for jaw suspension are depicted in the series of five images (*right*), showing four related groups of chondrichthyans and (*far right*) a fossil actinopterygian. *Arrows* indicate contact points between the PQ and the neurocranium (in the Holocephalan, *arrows* indicate synchondrotic fusion of the PQ and the neurocranium). The lines in the second image (Elasmobranch 1) indicate free-moving ligamentous attachments. (*B*) The chondrocranium of *Scyliorhinus canicula*. *Arrows* indicate the midline articulations of the bilateral PQ and MC elements. (*C*) Alcian blue–stained *S. canicula* fetus with principle chondrocranium components indicated. *Arrows* are as in *B*. (*D*) Alizarin red– and alcian blue–stained cranium of a late fetal *S. canicula*. (*E–G*) Scanning electron micrographs of stages 21–22 (*E*) and 25 (*F, G*) *S. canicula* embryos; the craniofacial primodia have been pseudoshaded to aid identification. (*H*) Hemisected embryo revealing the endodermal components of the branchial arches and the internal aspect of the developing central nervous system. (BA) Branchial arch; (BA2) second (hyoid) branchial arch; (ch) ceratohyal; (FB) forebrain; (HB) hindbrain; (hm) hyomandibula; (lc) labial cartilages; (MB) midbrain; (MC) Meckel's cartilage; (mdBA1) mandibular first branchial arch; (mxBA1) maxillary first branchial arch; (nc) nasal capsule; (nrc) neurocranium; (olf) olfactory; (pc1) first pharyngeal cleft; (pc2) second pharyngeal cleft; (phar endo) pharyngeal endoderm of the arches; (phar pouches) pharyngeal pouches; (plc) placode; (PQ) palatoquadrate.

of vertebrates have been greatly modified to meet new ontogenetic and functional demands. Jaws can be structurally defined as articulated, prehensile oral apparatuses principally derived from the skeletal and dental elements arising in the embryo from the anteriormost branchial arch (BA1), with a small yet significant contribution (in all but the chondrichthyans) from the frontonasal prominences (FNP). Polarity is an inherent character state of jaws: Minimally, there is the "hinge" and all "other" portions. Subsequently, within all other portions of the appositional units are those closest to the hinge and those furthest away. A consequence of such polarity is the potential for modularity within the appositional units. Notably, the articulations of gnathostome jaws are found in the same relative positions with regard to their ontogeny and their relationship as functional units with the rest of the skull. A number of strategies for suspending and bracing the jaws against the neurocranium have evolved, including autdiastyly, hyostyly, holostylic, amphistylic, met-hyostyly, and dermatostylic (Goodrich 1958; De Beer 1985; Grogan et al. 1999; Maisey 2005, 2008) (these strategies are illustrated in Fig. 4). Regardless of the type of suspensorium (suspended jaws), the PQ cartilage of the upper jaw is always closer to, and substantially more intimate with, the neurocranium. This, then, further generates PD polarity within the jaws. In particular, it is with the chondrichthyan that experimentation with the type of suspensorium has reached its greatest breadth. The bone, dentine, and cartilage of jaws are derived from the Hox-negative "trigeminal" cranial neural crest (CNC) that migrates to BA1 and the FNP. A conserved pattern of this jaw-yielding CNC is observed throughout the vertebrates, including chondrichthyans. The developmental history of the CNC informs its capacity to interpret spatial/patterning information in the jaw primordia and in elaborate reciprocal inductive and reactive epithelial-mesenchymal interactions. Current models of jaw development and evolution have been built on molecular and cellular evidence gathered mostly in amniotes such as mice and chicks. However, little is understood of the molecular and cellular foundations of the basal jaw bauplan represented by the chondrichthyans. Despite their possession of the oldest, least modified jaws, we are only now beginning to address similarities or differences in chondrichthyan spatiotemporal expression patterns of genes known to be critical for mouse or chick jaw development.

Chondrichthyans are also important in the analysis of the gnathostome ancestral condition of brain organization. The development of the telencephalon has been of special interest because, in cartilaginous fish, like in sarcopterygians (both lobe-finned fish and tetrapods), this anterior encephalic vesicle develops by a morphogenetic process known as evagination, which implies protrusion toward the outside and expansion. In contrast, in actinopterygians (ray-finned fish), the telencephalon develops by a different process called eversion, which makes comparison less straightforward. Among cartilaginous fish, other chondrichthyans such as *Squalus acanthias*, have also been largely used for neuroanatomical studies. However, the dogfish appears closer to teleosts and amniotes than any other shark species in its brain complexity, and it thus is a particularly relevant model for comparative studies of brain organization and development. Like teleosts and amniotes, galeomorph sharks in general present elaborate brains, in which, for example, the neuronal cell bodies migrate away from the periventricular matrix. This situation contrasts with that in the laminated brains of squalomorphs such as *S. acanthias*, in which most neurons are periventricularly located (see Buttler and Hodos 2005 for a more extensive characterization of these two types of brain organization). The neurochemical organization of the brain is also better known in the dogfish than in other elasmobranch, and immunohistochemical analyses have provided a detailed picture of the organization of neuronal systems expressing modulatory neuroactive substances such as monoamines and neuropeptides (for review, see Smeets 1998). Our current knowledge of the organization of cholinergic and GABAergic systems in the elasmobranch brain is thus based essentially on studies conducted in the dogfish (Anadón et al. 2000; Carrera et al. 2006, 2008; Rodríguez-Moldes et al. 2008). Similarly, most of our knowledge of fiber connection patterns in the chondrichthyan brain came from experimental studies conducted in the dogfish (for review, see Smeets et al. 1983). These analyses have highlighted a common pattern of brain organization shared by all jawed vertebrates as well as a number of elasmobranch-specific, and possibly derived, characteristics. More recently, analyses of the developing dogfish brain have suggested an ancient origin for major morphogenetic processes, long thought to be characteristics of the mammalian brain. A tangential migration of ventral telencephalic (subpallial) GABAergic cells to the dorsal telencephalon (pallium) thus probably also occurs in the dogfish (Carrera et al. 2008). Similarly, the existence of cerebellar granule-cell precursors expressing Pax6 (a transcription factor important in early developmental programs, highly conserved across many species) during their migration from the external germinal layer is observed in the dogfish as well as in tetrapods (Rodríguez-Moldes et al. 2008). These examples point to the dogfish as an important model for use in gaining insight into the origin of the complex organization of the gnathostome brain.

Finally, sharks possess highly specialized and elaborate sensory organs, some of them absent in the usual vertebrate model organisms. An excellent example is provided by electroreceptors, found only in so-called "primitive fish" (sharks, rays, lampreys, bichirs, lungfish, coelacanths, sturgeons) and among mammals such as monotremes. Sharks are sensitive to extremely weak electric fields; understanding the molecular and cellular mechanisms underlying the formation of the electroreceptive organs (Lorenzini ampullae) is important for understanding how such organs have been modified, or lost, in different taxa. Molecular evidence recently obtained from the dogfish has provided the first insights into the genetic mechanisms controlling the formation of extrasensory markers. Expression analyses of electrosensory cell markers suggest that they may be derived from the neural crest and that ephrin signaling may be involved early in electrosensory organ formation (Freitas et al. 2006).

Other dogfish sensory organs may provide excellent systems for evolution-development analyses. For example, analyses of the olfactory epithelium have recently revealed the presence of crypt receptor neurons expressing the olfactory marker protein (OMP), also present in bony fishes but not in tetrapods (Ferrando et al. 2007). Similarly, as shown in the dogfish, hydrostatic pressure sensing in chondrichthyans involves a modulation of the activity of vestibular hair cells, a system that differs radically from that used by bony fish (Fraser and Shelmerdine 2002). Highly detailed molecular analyses of the formation and differentiation of placodes and neural crest derivatives, using the dogfish as a chondrichthyan reference, will be crucial to our understanding of the developmental bases of these diversifications. The conservation across gnathostomes of the genetic programs controlling these processes is only now starting to emerge (O'Neill et al. 2007).

Origin and Evolution of Physiological Systems

Chondrichthyans are readily comparable to osteichthyans, not only in morphology, but in physiology as well. Studies of several members of this group, including *S. canicula, S. acanthias, R. erinacea*, and *Triakis scyllia*, have provided insights into the origin of most of the broad physiological systems used by osteichthyans. These studies have also generally highlighted a remarkable complexity in elasmobranch-specific organs or adaptations.

Analyses in the dogfish, focused on the hormonal control of reproduction, glucose metabolism, and heart and muscle physiology, similarly underscored the conservation of important physiological responses between chondrichthyans and osteichthyans (D'Antonio et al. 1995; Quérat et al. 2001; West et al. 2004; Thébault et al. 2005; Gemballa et al. 2006). Unlike lampreys, chondrichthyans possess an adaptative immune system (for review, see Bartl 1998; Marchalonis et al. 1998). *S. canicula* is particularly well suited for analyses of the origin of the complex immune system typical of all jawed vertebrates. It serves as an experimental model of infection and inflammation, and the morphology and development of the complex lymphomyeloid organs have been well characterized. In particular, the dogfish possesses lymphomyeloid epigonal tissue also found in other elasmobranchs and closely associated with male and female gonads, therefore representing a unique model for the study of interactions between the immune and reproductive systems (Fänge and Pulsford 1983). Inflammatory cytokines such as interleukin-1β (IL-1β) and nonspecific defenses involving inducible nitrous oxide synthetase (iNOS) were characterized in the dogfish, showing for the first time their presence in a cartilaginous fish (Bird et al. 2002; Reddick et al. 2006).

Analyses of other physiological processes, such as the control of osmoregulation or spermatogenesis, have been greatly facilitated by the size, specialization, or cellular organization of organs. Elasmobranch fish, which have the ability to regulate their body-fluid volume when exposed to different environmental salinities, have thus developed elaborate mechanisms to control osmoregulation through the regulation of ion transport, body-fluid volume, and external and internal osmotic pressure sensing. Some of these mechanisms appear very similar to those used by mammals. For instance, exposure to different pharmacological agents or variations in the physiological and environmental conditions in the dogfish highlighted an involvement of the arginine-vasotocin and renin-angiotensin systems well characterized in mammals (Wells et al. 2005, 2006). Like in osteichthyans, these functions are performed in different tissues, including the intestinal epithelium, kidneys, and the gills, which share a common origin and role in calcium homeostasis with the parathyroid gland, long thought to be a tetrapod-specific organ (Okabe and Graham 2004). In addition, a distinctive feature of chondrichthyans is their rectal gland, which has no equivalent in osteichthyans (Anderson et al. 2007). This highly specialized organ, composed of a single-cell-type transporting epithelium, is an important source of ion transport molecules and their regulators (MacKenzie et al. 2002). The rectal gland provides a unique experimental system for the study of regulation of the expression and activity of ion transport molecules in response to physiological modifications. As an example, dietary sodium loading in the dogfish was shown to increase the activity of Na/K-ATPase without changes in mRNA levels, pointing to an as-yet-unknown post-transcriptional activation event (MacKenzie et al. 2002).

Spermatogenesis is another process that may benefit greatly from analyses conducted in chondrichthyans and, in particular, the dogfish. This process, which leads to the differentiation of diploid germ-line stem cells into mature haploid functional spermatozoa, involves complex endocrine, paracrine, and autocrine controls of cell proliferation, specification, and differentiation. However, the genetic control of this process is poorly known in vertebrates other than mammals. The behavior of spermatogonial stem cells is tightly linked to their microenvironment, partly defined by neighboring somatic cells including Sertoli cells. In the dogfish, spermatogenesis occurs within spermatocysts, in which germ cells show synchronous development. Their genesis occurs in a germinative zone, running along the dorsal length of the testis, and spermatocysts move progressively, in maturational order, toward the opposite margin of the gonad where spermiation occurs (Fig. 5). This zonation allows the observation of the main stages of spermatogenesis on a single transverse section of the testis (Loir et al. 1995; Loppion et al. 2008). The cellular organization and size (~15 cm long, weight = ~13 g) of its testes thus make the dogfish a unique animal model for the study of genetic programs controlling the different steps of germ-cell differentiation. The development of transcriptomic and genomic resources currently in progress for this species should make it possible to exploit these characteristics for additional germ-cell differentiation studies.

GENETICS

Considering the generation time of the dogfish (discussed above), establishing collections of mutants and the use of classical genetic studies are at present unrealistic.

GENOMICS/ASSOCIATED RESOURCES

Comparative Genomics: The Dogfish Reference

The close relationship observed in morphology and physiology between chondrichthyans and osteichthyans also applies at the molecular level. The availability of a chondrichthyan reference is important to assess sequence conservation at the gnathostome scale, identify their innovations, and detect derived characteristics of actinopterygians or sarcopterygians. The availability of

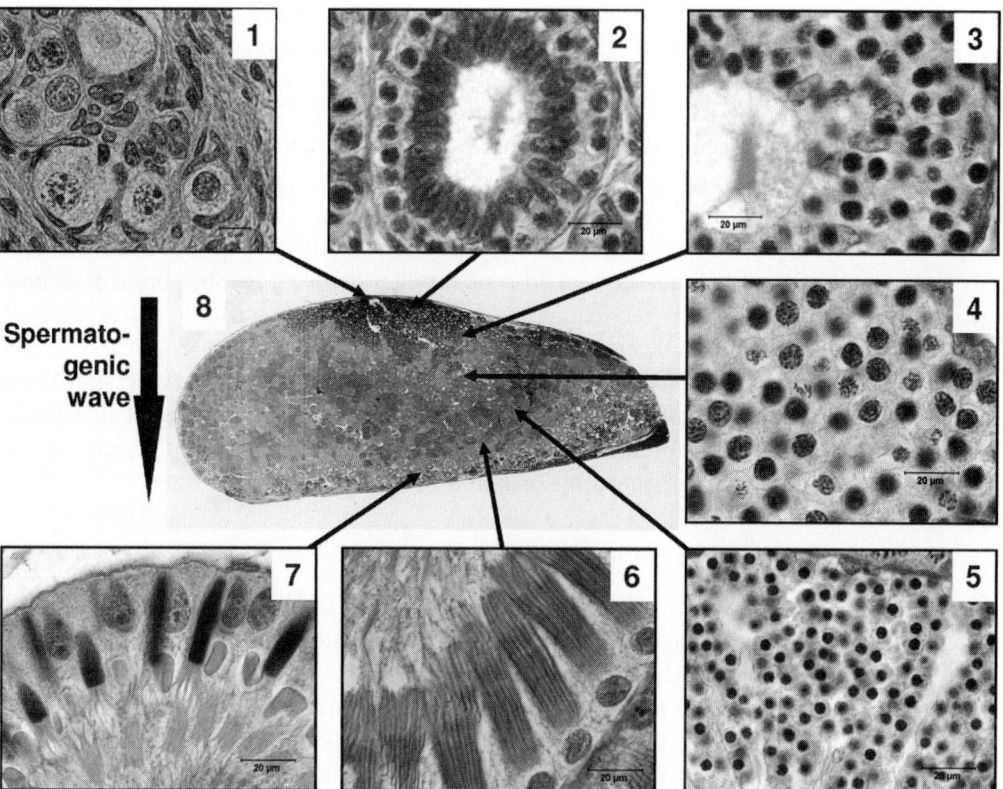

FIGURE 5. Zonation of the testis in *S. canicula*; zonation makes this species an interesting model for the study of spermatogenesis. In *S. canicula*, spermatogenesis occurs within spermatocysts composed of Sertoli cells associated with stage-synchronized germ cells. (*1*) Germinative area; (*2, 3*) spermatogonia; (*4*) spermatocyte; (*5–7*): spermatids). As shown from testicular cross sections (*8*), cysts radiate in a maturational order from the germinative area, where they are formed, to the opposite margin of the testis where spermiation occurs.

genomic or transcriptome resources (discussed below) in either chondrichthyan model is important for addressing these points. As a representative of the major group of extant sharks, the dogfish is an attractive model. In considering its divergence time and relative phylogenetic position with respect to the skate *R. erinacea* and chimera *C. milii* (Fig. 1B), we expect the dogfish to be fully complementary to these chondrichthyan models in comparative genomics. The ready access to eggs and embryos, which makes it possible to connect sequence evolution and expression characteristics, is a definitive advantage of the dogfish, particularly for studies in evolutionary developmental biology.

Evolution of the Proteome: Vertebrate Multigene Families and their Evolution

Although regulatory changes have been considered a driving force in evolution, a large body of evidence has also underscored the important role of protein modifications. Identifying the modifications of the proteome at all vertebrate nodes (e.g., gene losses and duplications, accelerated evolutionary rates) is important for correlating these genetic events with the evolution of developmental or physiological processes. Our understanding of the relationships between protein structure and function can also benefit from comparative analyses. These studies, aimed at reconstructing ancestral protein sequences, would reveal the stepwise process of modifications that occurred at each vertebrate node, including the agnathan-gnathostome transition, which is marked by important morphological and physiological innovations. Chondrichthyans are invaluable references in such approaches, especially with respect to the chronology of the two phases of polyploidization, which occurred early in the vertebrate lineage. It is now clear that these duplications predated the split between chondrichthyans and osteichthyans. As a result, orthologs of human genes are typically easily recognized in cartilaginous fish, whereas this is generally not the case in lampreys. This observation is important for correct annotation of paralogous genes, particularly in situations marked by gene losses or duplications in osteichthyans, and for gaining insight into the dynamic of evolution of vertebrate gene families. The availability of the genome will be an essential step toward fully exploiting the dogfish, and several studies focused on developmental genes underscore the potential of this model. Exhaustive characterizations in the dogfish of two homeodomain gene families involved in early brain regionalization or specification, *Emx* and *Otx*, have contributed significantly to establishing correct orthological relationships and our understanding of their mode of evolution (Germot et al. 2001; Derobert et al. 2002). These studies have highlighted several examples of unexpected expression shuffling processes and also the fixation of class-specific expression territories (Plouhinec et al. 2005; Coolen et al. 2007). Such analyses are prerequisite to understanding the mechanisms responsible for the maintenance and fixation of gnathostome orthology classes.

Evolution of Noncoding Regions

Heterologous transgenesis experiments and comparative sequence analyses have recently pointed to an unexpectedly high divergence rate of noncoding sequences in actinopterygians. Extended segments of sequence similarity possibly corresponding to *cis*-regulatory signals have thus been identified between tetrapods and other chondrichthyans, such as the skate or chimera, but they remain undetectable in teleosts (Kurokawa et al. 2006; Venkatesh et al. 2006). These similarities have also been observed in comparisons with the dogfish, as shown by analyses of the flanking regions of the *Lhx1* homeodomain gene (Fig. 6). This example additionally suggests that, in some cases, this paradox may be related to the partitioning of candidate regulatory sequences between the duplicates, as often encountered in actinopterygians (Fig. 6). Such situations considerably complicate comparisons and underscore the interest of chondrichthyan sequences in detecting long-range conservations across jawed vertebrates. Among chondrichthyans, the advantage of the dogfish will be to allow parallel expression characterizations and correlate the presence of a given expression territory in in situ analyses with the detection of the corresponding *cis*-acting control elements. However, because the dogfish and chondrichthyans in general are unlikely to be amenable to transgenesis analyses (which remains an obvious major limitation), a direct functional characterization of such candidate regulatory signals will have to rely on analyses in heterologous systems.

Transcriptome Resources

Large-scale expressed sequence tag (EST) resources have been developed during the past few years by the French National Sequencing Center, Genoscope (project coordinator, S. Mazan). A total of 250,000 arrayed ESTs consisting of single 5′ reads have been obtained from a total of three cDNA libraries: two embryonic libraries (stages 6–15 and 18–25) and one juvenile library. Approximately 50,000 contiguous sequences (contigs) were obtained from this collection of ESTs. Their annotation is under way and will be completed before public release. Current projects include extending these ESTs to adult organs of interest and spotting selected cDNAs on high-density microarrays.

Genomic Resources

The currently available genomic resources for *S. Canicula* include one arrayed bacterial artificial chromosome (BAC) library, consisting of 200,000 clones with an average insert size of 150 kb (construction A. Billault, Genoscope; requests to S. Mazan). A sequencing project of 40 selected BACs is under way, in collaboration with Genoscope. Whole-genome sequencing, an essential step

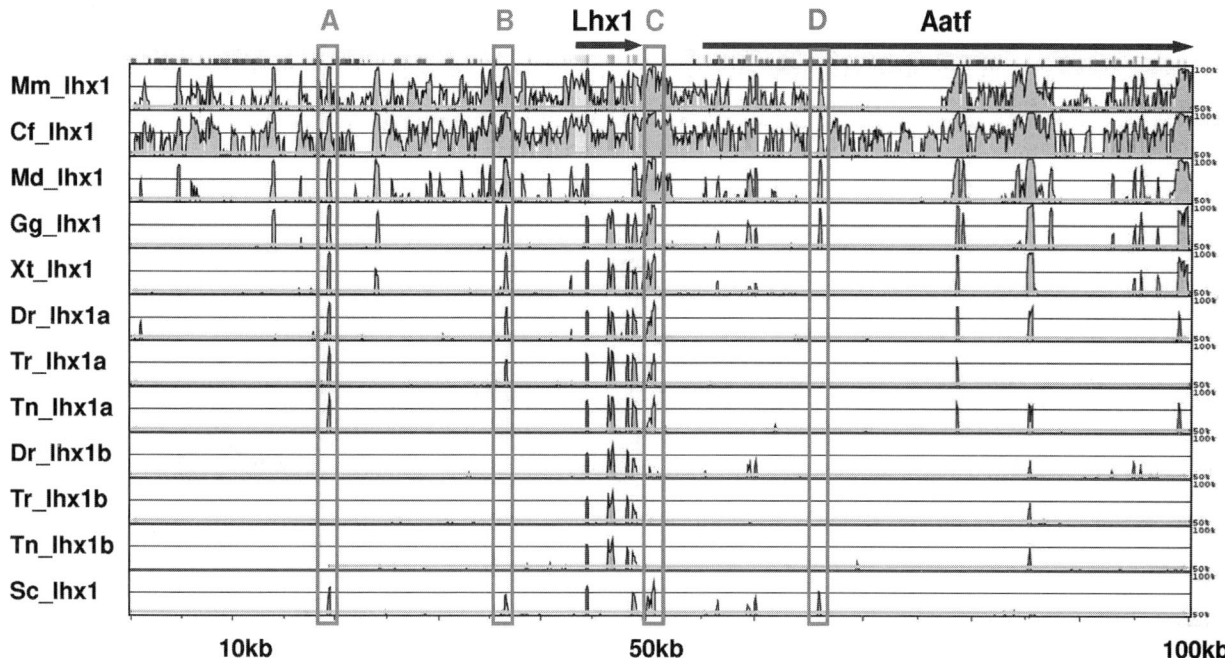

FIGURE 6. Sequence comparison of the *Lhx1* locus across vertebrates. The DNA sequence under study spans a region of 100 kb that includes the *Lhx1* gene, a member of the LIM homeobox gene family that encodes prototypical patterning factors, and the *Aatf* gene, encodes the apoptosis antagonizing transcription factor. Results were obtained using the LAGAN global alignment tool and the human locus (shown at the top of the figure) as reference. Species included in the comparison are shown at left. (Mm) *Mus musculus*; (Cf) *Canis familiaris*; (Md) *Monodelphis domestica*; (Gg) *Gallus gallus*; (Xt) *Xenopus tropicalis*; (Dr) *Danio rerio*; (Tr) *Takifugu rubripes*; (Tn) *Tetraodon nigroviridis*; (Sc) *Scyliorhinus canicula*. Two paralogous loci are present and were included in each teleost studied. Each peak in the graphic corresponds to a segment of at least 100 bp, with an identity rate higher than 50%. The peaks are shaded when the identity rate exceeds 70%, with noncoding regions in *light gray*, coding regions in *black*, and 5′ and 3′ UTR in *dark gray*. Four regions of similarity (boxed and termed A–D) are found conserved between the dogfish and human outside coding regions. D is not detected in teleosts, whereas A–C are only found in one paralogous locus of the teleosts studied.

required to establish the dogfish as a reference chondrichthyan model, has been considered out of reach until recently. This situation is changing with the rapid development of sequencing techniques. Furthermore, the genome size, although certainly as large as that of most chondrichthyans, may have been largely overestimated. Two different estimates of dogfish genome content are available online, with C-values of 5.65 and 7.50 pg (http://www.genomesize.com). Recent tests based on quantitative fluorescent imaging studies indicate that these estimates should be revised and that the dogfish genome size is substantially lower (C-value = 3.56 pg; D. Chassoux, unpubl.). A high-coverage whole-genome sequencing project has been recently submitted to Genoscope by an international consortium.

TECHNICAL APPROACHES

Technical approaches have been essentially restricted to descriptive analyses, such as in situ (whole-mount or sections) hybridizations or immunohistochemistry. Using standard protocols (e.g., see Fig. 7), both techniques yield excellent results in the dogfish.

An important goal is to establish functional approaches in the dogfish. This should become possible with the development of large-scale embryo productions. Preliminary tests of in ovo

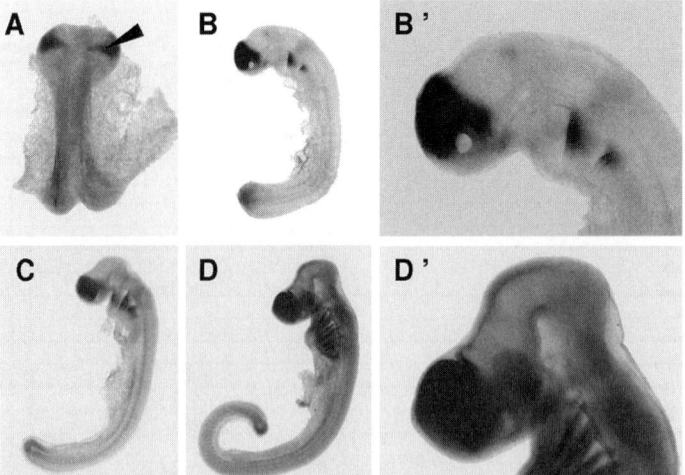

FIGURE 7. Examples of a series of *S. canicula* embryos (stages 15, 18, 20, 25) following whole-mount in situ hybridizations. Eggs of *S. canicula* can be easily maintained in laboratory conditions until hatching, making it possible to obtain an embryonic series. The probe in this case is *Emx3*, which codes for a homeodomain gene expressed in the dorsal pallium and branchial arches. (*A*) Dorsal view, stage 15; (*B–D*) lateral views; stages 18, 20 and 25, respectively. *B'* and *D'* are higher magnifications of *B* and *D* in the cephalic region.

pharmacological treatments currently give promising results (M. Coolen, A. Menuet, and S. Mazan, unpubl.). In the longer term, establishing conditions for embryo culture and electroporation will be essential to broaden the spectrum of issues addressed.

ACKNOWLEDGMENTS

We thank Bernard Kloareg and the Service Mer et Observation (Station Biologique de Roscoff) for their support in the development of the dogfish as a model organism. Large-scale EST databases and molecular resources were obtained in collaboration with Genoscope (Evry) and with the support of the Université d'Orléans, Université Paris 11, CNRS, Région Centre and the Groupement d'Intérêt Scientifique Génomique Marine (S.M., coordinator).

REFERENCES

Anadón, R., Molist, P., Rodríguez-Moldes, I., López, J.M., Quintela, I., Cerviño, M.C., Barja, P., and González, A. 2000. Distribution of choline acetyltransferase (ChAT) immunoreactivity in the brain of an elasmobranch, the lesser-spotted dogfish (*Scyliorhinus canicula*). *J. Comp. Neurol.* **420:** 139–170.

Anderson, W.G., Taylor, J.R., Good, J.P., Hazon, N., and Grosell, M. 2007. Body fluid volume regulation in elasmobranch fish. *Comp. Biochem. Physiol. A Mol. Integr. Physiol.* **148:** 3–13.

Bartl, S. 1998. What sharks can tell us about the evolution of MHC genes. *Immunol. Rev.* **166:** 317–331.

Bird, S., Wang, T., Zou, J., Cunningham, C., and Secombes, C.J. 2002. The first cytokine sequence within cartilaginous fish: IL-1β in the small spotted catshark (*Scyliorhinus canicula*). *J. Immunol.* **168:** 3329–3340.

Butler, A.B. and Hodos, W. 2005. In *Comparative vertebrate neuroanatomy: Evolution and adaptation*, 2nd ed. Wiley, New York.

Carrera, I., Sueiro, C., Molist, P., Holstein, G.R., Martinelli, G.P., Rodríguez-Moldes, I., and Anadón, R. 2006. GABAergic system of the pineal organ of an elasmobranch (*Scyliorhinus canicula*): A developmental immunocytochemical study. *Cell Tissue Res.* **323:** 273–278.

Carrera, I., Ferreiro-Galve, S., Sueiro, C., Anadón, R., and Rodríguez-Moldes, I. 2008. Tangentially migrating GABAergic cells of subpallial origin invade massively the pallium in developing sharks. *Brain Res. Bull.* **75:** 405–409.

Coolen, M., Sauka-Spengler, T., Nicolle, D., Le-Mentec, C., Lallemand, Y., Da Silva, C., Plouhinec, J.L., Robert, B., Wincker, P., Shi, D.L., and Mazan, S. 2007. Evolution of axis specification mechanisms in jawed vertebrates: Insights from a chondrichthyan. *PLoS ONE* **2:** e374.

D'Antonio, M., Vallarino, M., Lovejoy, D.A., Vandesande, F., King, J.A., Pierantoni, R., and Peter, R.E. 1995. Nature and distribution of gonadotropin-releasing hormone (GnRH) in the brain, and GnRH and GnRH binding activity in serum of the spotted dogfish *Scyliorhinus canicula*. *Gen. Comp. Endocrinol.* **98:** 35–49.

De Beer, G. 1985. *The development of the vertebrate skull*. University of Chicago Press, Chicago.

Depew, M. and Simpson, C.A. 2006. 21st century neontology and the comparative development of the vertebrate skull. *Dev. Dyn.* **235:** 1256–1291.

Derobert, Y., Plouhinec, J.L., Sauka-Spengler, T., Le Mentec, C., Baratte, B., Jaillard, D., and Mazan, S. 2002. Structure and expression of three *Emx* genes in the dogfish *Scyliorhinus canicula*: Functional and evolutionary implications. *Dev. Biol.* **247:** 390–404.

Ellis, J.R. and Schackley, S.E. 1997. The reproductive biology of *Scyliorhinus canicula* in the Bristol Channel, U.K. *J. Fish Biol.* **51:** 361–372.

Fänge, R. and Pulsford, A. 1983. Structural studies on lymphomyeloid tissues of the dogfish, *Scyliorhinus canicula* L. *Cell Tissue Res.* **230:** 337–351.

Ferrando, S., Bottaro, M., Gallus, L., Girosi, L., Vacchi, M., and Tagliafierro, G. 2007. First detection of olfactory marker protein (OMP) immunoreactivity in the olfactory epithelium of a cartilaginous fish. *Neurosci. Lett.* **413:** 173–176.

Fraser, P.J. and Shelmerdine, R.L. 2002. Dogfish hair cells sense hydrostatic pressure. *Nature* **415:** 495–496.

Freitas, R., Zhang, G., Albert, J.S., Evans, D.H., and Cohn, M.J. 2006. Developmental origin of shark electrosensory organs. *Evol. Dev.* **8:** 74–80.

Freitas, R., Zhang, G., and Cohn, M.J. 2007. Biphasic *Hoxd* gene expression in shark paired fins reveals an ancient origin of the distal limb domain. *PLoS ONE* **2:** e754.

Germot, A., Lecointre, G., Plouhinec, J.-L., Le Mentec, C., Girardot, F., and Mazan, S. 2001. Structural evolution of *Otx* genes in craniates. *Mol. Biol. Evol.* **18:** 1668–1678.

Gemballa, S., Konstantinidis, P., Donley, J.M., Sepulveda, C., and Shadwick, R.E. 2006. Evolution of high-performance swimming in sharks: Transformations of the musculotendinous system from subcarangiform to thunniform swimmers. *J. Morphol.* **267:** 477–493.

Goodrich, E.S. 1958. *Studies on the structure and development of vertebrates*. Dover, New York.

Grogan, E.D., Lund, R., and Didier, D. 1999. Description of the Chimaerid jaw and its phylogenetic origins. *J. Morphol.* **239:** 45–59.

Heinicke, M.P., Naylor, G.J.P., and Hedges, S.B. 2008. Chondrichthyes. In *The timetree of life* (ed. S.B. Hedges and S. Kumar). Oxford University Press, United Kingdom. (In press.)

Iglésias, S.P., Lecointre, G., and Sellos, D.Y. 2005. Extensive paraphylies within sharks of the order Carcharhiniformes inferred from nuclear and mitochondrial genes. *Mol. Phylogenet. Evol.* **34:** 569–583.

Ivory P., Jeal F., and Nolan C.P. 2005. Age determination, growth and reproduction in the lesser-spotted dogfish, *Scyliorhinus canicula* (L.). *J. Northwest Atl. Fish. Sci.* **35:** 89–106.

Kurokawa, D., Sakurai, Y., Inoue, A., Nakayama, R., Takasaki, N., Suda, Y., Miyake, T., Amemiya, C.T., and Aizawa, S. 2006. Evolutionary constraint on *Otx2* neuroectoderm enhancers-deep conservation from skate to mouse and unique divergence in teleost. *Proc. Natl. Acad. Sci.* **103:** 19350–19355.

Loir, M., Sourdaine, P., Mendis-Handagama, S.M., and Jegou, B. 1995. Cell-cell interactions in the testis of teleosts and elasmobranchs. *Microsc. Res. Tech.* **32:** 533–552.

Loppion, G., Crespel, A., Martinez, A.S., Auvray, P., and Sourdaine, P. 2008. Study of the potential spermatogonial stem cell compartment in dogfish testis, *Scyliorhinus canicula* L. *Cell Tissue Res.* **332:** 533–542.

Lyle, J.M. 1983. Food and feeding habits of the lesser spotted dogfish, *Scyliorhinus canicula* (L.), in Isle of Man waters. *J. Fish. Biol.* **23:** 725–737.

MacKenzie, S., Cutler, C.P., Hazon, N., and Cramb, G. 2002. The effects of dietary sodium loading on the activity and expression of Na,K-ATPase in the rectal gland of the European dogfish (*Scyliorhinus canicula*). *Comp. Biochem. Physiol. B. Biochem. Mol. Biol.* **131:** 185–200.

Maisey, J.G. 2005. Braincase of the Upper Devonian shark *Cladodoides wildungensis* (Chondrichthyes, Elasmobranchii), with observations on the braincase in early chondrichthyans. *Bull. Am. Mus. Nat. Hist.* **288:** 1–103.

Maisey, J.G. 2008. The postorbital palatoquadrate articulation in elasmobranchs. *J. Morphol.* **269:** 1022–1040.

Marchalonis, J.J., Schluter, S.F., Bernstein, R.M., and Hohman, V.S. 1998. Antibodies of sharks: Revolution and evolution. *Immunol. Rev.* **166:** 103–122.

Okabe, M. and Graham, A. 2004. The origin of the parathyroid gland. *Proc. Natl. Acad. Sci.* **101:** 17716–17719.

O'Neill, P., McCole, R.B., and Baker, C.V. 2007. A molecular analysis of neurogenic placode and cranial sensory ganglion development in the shark, *Scyliorhinus canicula*. *Dev. Biol.* **304:** 156–181.

Plouhinec, J.L., Leconte, L., Sauka-Spengler, T., Bovolenta, P., Mazan, S., and Saule, S. 2005. Comparative analysis of gnathostome *Otx* gene expression patterns in the developing eye: Implications for the functional evolution of the multigene family. *Dev. Biol.* **278:** 560–575.

Quérat, B., Tonnerre-Doncarli, C., Géniès, F., and Salmon, C. 2001. Duality of gonadotropins in gnathostomes. *Gen. Comp. Endocrinol.* **124:** 308–314.

Reddick, J.I., Goostrey, A., and Secombes, C.J. 2006. Cloning of iNOS in the small spotted catshark (*Scyliorhinus canicula*). *Dev. Comp. Immunol.* **30:** 1009–1022.

Rodriguez-Cabello, C, Velasco, F., and Olaso, I. 1998. Reproductive biology of lesser spotted dogfish *Scyliorhinus canicula* (L., 1758) in the Cantabrian Sea. *Sci. Mar.* **62:** 187–191.

Rodríguez-Moldes, I., Ferreiro-Galve, S., Carrera, I., Sueiro, C., Candal, E., Mazan, S., and Anadón, R. 2008. Development of the cerebellar body in sharks: Spatiotemporal relations of Pax6-expression, cell proliferation and differentiation. *Neurosci. Lett.* **43:** 105–110.

Sauka-Spengler, T., Plouhinec, J.L., and Mazan, S. 2004. Molecular and cellular aspects of gastrulation in a chondrichthyan, the dogfish *Scyliorhinus canicula*. In *Gastrulation: From cells to embryo* (ed. C.D. Stern), pp. 151–156. Cold Spring Harbor Laboratory Press, Cold Spring Harbor, New York.

Sims, D.W., Nash, J.P., and Morritt, D. 2001. Movements and activity of male and female dogfish in a tidal sea lough: Alternative behavioural strategies and apparent sexual segregation. *Mar. Biol.* **139:** 1165–1175.

Smeets, W.J. 1983. The secondary olfactory connections in two chondrichthyans, the shark *Scyliorhinus canicula* and the ray *Raja clavata*. *J. Comp. Neurol.* **218:** 334–344.

Smeets, W.J. 1998. Cartilaginous fishes. In *The central nervous system of vertebrates* (ed. R. Nieuwenhuys et al.), vol. 1, pp. 552–654. Springer-Verlag, Berlin/Heidelberg.

Tanaka, M., Münsterberg, A., Anderson, W.G., Prescott, A.R., Hazon N., and Tickle C. 2002. Fin development in a cartilaginous fish and the origin of vertebrate limbs. *Nature* **416:** 527–531.

Thébault, M.T., Izem, L., Leroy, J.P., Gobin, E., Charrier, G., and Raffin, J.P. 2005. AMP-deaminase in elasmobranch fish: A comparative histochemical and enzymatic study. *Comp. Biochem. Physiol. B. Biochem. Mol. Biol.* **141:** 472–479.

Venkatesh, B., Kirkness, E.F., Loh, Y., Halpern, A.L., Lee, A.P., Johnson, J., Dandona, N., Viswanathan, L.D., Tay, A., Venter, J.C., Strausberg, R.L., and Brenner, S. 2006. Ancient noncoding elements conserved in the human genome. *Science* **314:** 1892.

Venkatesh, B., Kirkness, E.F., Loh, Y., Halpern, A.L., Lee, A.P., Johnson, J., Dandona, N., Viswanathan, L.D., Tay, A., Venter, J.C., Strausberg, R.L., and Brenner, S. 2007. Survey sequencing and comparative analysis of the elephant shark (*Callorhinchus milii*) genome. *PLoS Biol.* **5**: e101.

Wells, A., Anderson, W.G., and Hazon, N. 2005. Glomerular effects of AVT on the in situ perfused trunk preparation of the dogfish. *Ann. N.Y. Acad. Sci.* **1040**: 515–517.

Wells, A., Anderson, W.G., Cains, J.E., Cooper, M.W., and Hazon, N. 2006. Effects of angiotensin II and C-type natriuretic peptide on the in situ perfused trunk preparation of the dogfish, *Scyliorhinus canicula*. *Gen. Comp. Endocrinol.* **145**: 109–115.

West, T.G., Curtin, N.A., Ferenczi, M.A., He, Z.H., Sun, Y.B., Irving, M., and Woledge, R.C. 2004. Actomyosin energy turnover declines while force remains constant during isometric muscle contraction. *J. Physiol.* **555**: 27–43.

18 The Genus *Polypterus* (Bichirs)
A Fish Group Diverged at the Stem of Ray-finned Fishes (Actinopterygii)

Masaki Takeuchi,[1] Masataka Okabe,[2] and Shinichi Aizawa[1]

[1]*Laboratory for Vertebrate Body Plan, Center for Developmental Biology, Riken Kobe, Chuo-ku, Kobe 650-0047, Japan;* [2]*Department of Anatomy, The Jikei University School of Medicine, Minatoku, Tokyo 105-8461, Japan*

ABSTRACT

Axis and germ-layer formations are central issues in vertebrate embryology that can be examined in the zebrafish, *Xenopus*, chick, and mouse. An intriguing question is how the mechanisms that existed in an ancestral vertebrate have been modified during vertebrate evolution. A major stream of vertebrates (Osteichthyes) evolved in two monophyletic lineages: the Sarcopterygii and Actinopterygii. The zebrafish is a well-known and well-studied teleost fish highly derived in the actinopterygian lineage, ray-finned fishes. The early cleavage pattern of teleosts is meroblastic in contrast to the holoblastic cleavage seen in amphibians, which diverged early in sarcopterygian lineage. There are many differences in the molecular and cellular mechanisms of embryogenesis between teleosts and amphibians, including the differences in bauplan (body plan) between teleosts and amphibians or actinoperigian and sarcopterigian. However, the lineage leading of the teleost fishes has undergone a whole-genome duplication before their radiation, which may have also caused changes in molecular usage independent of the teleost bauplan. *Polypterus* diverged from all other actinopterygians about 400 million years ago (Mya) during the Devonian period, soon after the divarication of an ancestral bony fish into Actinopterygii and Sarcopterygii. *Polypterus* is thus uniquely well suited for studies assessing the ancestral state or bauplan of Osteichthyes and Actinopterygii, as well as the divergence of embryogenetic processes in teleosts and amphibians.

PROTOCOLS
1 Microinjection and Animal Cap Assay, 458
2 Whole-mount In Situ Hybridization, 460

BACKGROUND INFORMATION

Polypterus (bichir) was once thought either to be the most primitive member of the Sarcopterygii or to constitute a group, Branchiopterygii, which is an outgroup of both actinopterygians and sarcopterygians. Now, however, *Polypterus* is regarded as a taxonomic order, whose phylogenetic position places it at the base of the Actinopterygii branch, ray-finned fishes (Fig. 1) (Nelson 2006), a view also supported by DNA sequencing evidence (Noack et al. 1996; Kikugawa et al. 2004). Adult bichirs have a slender body with a dorsal fin consisting of about ten spiny finlets (hence the name, poly-pterus) and with a pair of marked tubular nostrils; their heads are small and flat with small

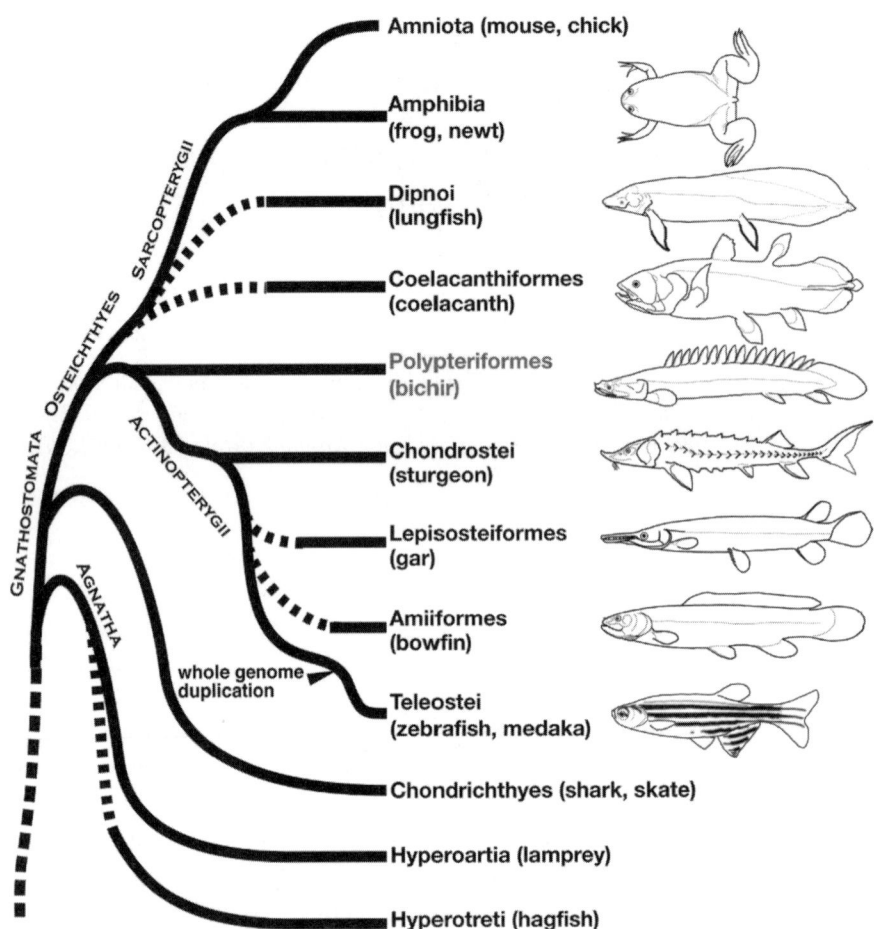

FIGURE 1. Phylogenic relationship of extant vertebrates with a focus on basal taxa of Actinopterygii and Sarcopterygii. Polypteriformes (bichir) is the most basal taxon in Actinopterygii (ray-finned fishes). Currently, model animals exist only in Amniota, Amphibia, and Teleostei.

eyes (Fig. 2a–c). *Polypterus* has many plesiomorphic characteristics; i.e., it shares several characteristics of cartilaginous fishes and basal bony fishes. Many bones are cartilaginous, the intestine has a spiral valve, and a spiracular opening appears at the rear of the eyes. The body is covered by rhombic ganoid scales composed of enamel, dentine, and bone, homologous to teeth. *Polypterus* has the dermal gular plate instead of the branchiostegal rays found in teleosts, the maxilla is firmly united to the skull, and the caudal fin is primitively heterocercal. But, strikingly, *Polypterus* also has characteristics of the primitive sarcopterygians—the coelacanth and lungfish—and moves along the riverbed using pectoral fins with fleshy bases similar to the lobe-fins of lungfish and coelacanth (Fig. 2d,e). A pair of highly vascularized swim bladders or lungs connect to the ventral side of the pharynx, like the lungs of terrestrial animals, and these are partially used for respiration. The larvae and juveniles have a pair of external gills, similar to those commonly present in amphibian larvae, as is widely known in neotenic axolotl (Fig. 2f,g). These external gills are composed of ectodermal tissue and have no supporting skeleton, which distinguishes them from the normal or internal gills that derive from pharyngeal pouches and have pharyngeal skeletons.

Ancestry and Phylogeny

A common ancestor between Actinopterygii and Sarcopterygii predates the Devonian period, before approximately 415 Mya. *Polypterus* is considered to have diverged from all other actinopterygians

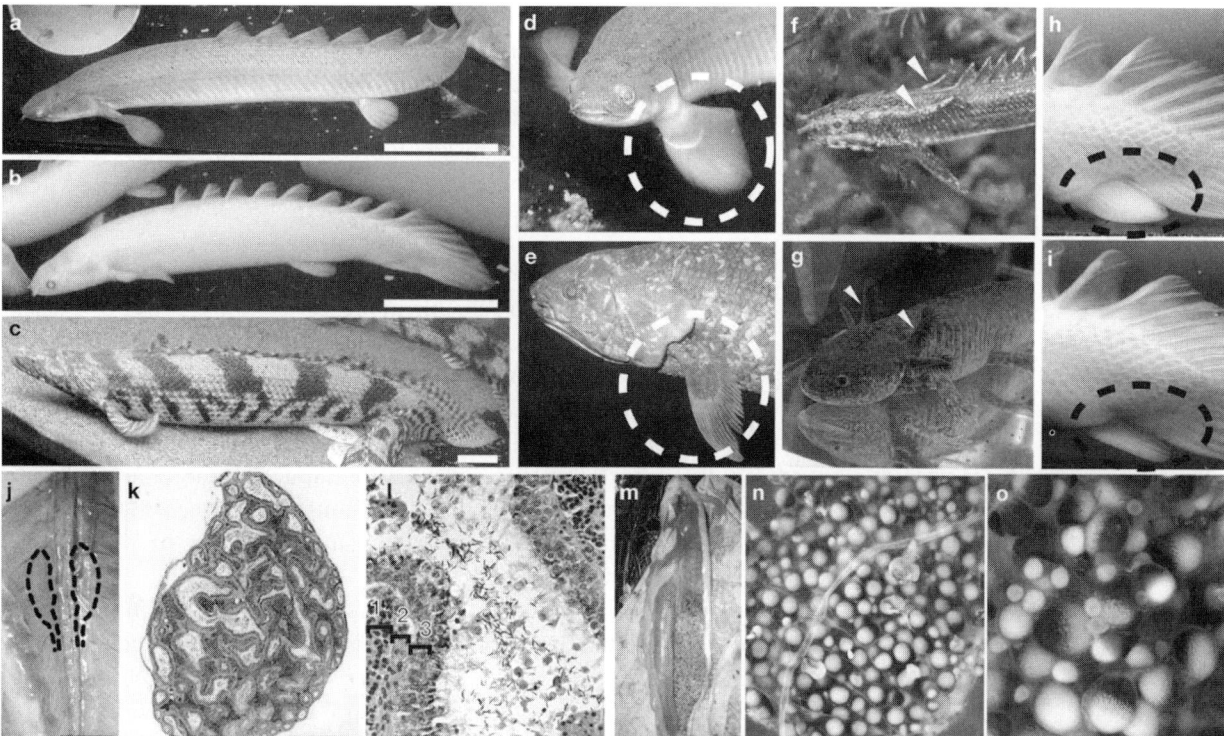

FIGURE 2. *Polypterus.* (a–c) Adult *P. senegalus* (a, wild type; b, albino) and *P. endlicheri* (c). Bars, (a–c) 5 cm. (d,e) Fleshy pectoral fins (*dotted circles*) of *P. senegalus* (d) and *Latimeria menadoensis* (e, coelacanth). (Photograph by Dr. Mark Erdmann.) (f,g) External gills (*arrowheads*) in juvenile *P. endlicheri* (f) and *Ambystoma mexicanum* (g, axolotl). (h,i) Anal fins (*dotted circles*) in *P. senegalus* male (h) and female (i). (j–l) Macroscopic (j) and histological views (k,l) of testis of mature male *P. senegalus*; (l) an enlarged view. Testis is located in the dorsal side of the abdominal cavity at midtrunk (*dotted lines* in j); left testis has been removed. (k,l) Stained with hematoxylin and eosin. Seminiferous tubules are well developed (k), and spermatozoa are abundant (l). (*Bracket 1*) Leydig cell layer; (*bracket 2*) myoid cell layer; (*bracket 3*) spermatocyte layer. (m–o) Macroscopic views of ovary of mature *P. senegalus* female with abundant eggs (m,n) of various sizes (n,o). These testis and ovary are widely found among mature male and female *P. senegalus*. However, either spawning is not yet controlled or artificial fertilization has been unsuccessful. Bars, (j) 10 mm; (k) 1 mm; (l) 0.1 mm; (m) 30 mm; (n) 2.5 mm; (o) 1 mm.

about 400 Mya, near the beginning of the Devonian period, soon after the divarication of an ancestral bony fish into Actinopterygii and Sarcopterygii. Subsequently, during the Carboniferous period, a new lineage of actinopterygian fishes radically transformed their mode of embryogenesis, and the Teleosts then appeared in the Permian. *Polypterus* fossils have been found in Africa and Bolivia, South America that date back to the middle Cretaceous period and all extant forms are carnivorous freshwater fishes living in tropical Africa. *Polypterus* belongs to Order Polypteriformes, Family Polypteridae. The family includes one more genus, *Erpetoichthys*; Genus *Polypterus* has about 10–15 species, whereas *Erpetoichthys* has only one (*E. calabaricus*). Bichir is a common name of the fishes belonging to *Polypterus*. The name "bichir" originated from one legendary species, *P. bichir bichir* (named by Lacépède in 1803), discovered by Geoffroy Saint-Hilaire when he accompanied the Egyptian campaign of Napoleon Bonaparte. *Erpetoichthys* is commonly called the reedfish.

Current model animals in the fish group (notably, zebrafish and medaka) belong to Teleostei. This successful taxon has undergone radiation into about 27,000 species, now comprising more than half of all extant vertebrates. Studies in genome evolution have shown that before radiation, whole-genome duplication (WGD) occurred in the Teleostei lineage (Amores et al. 1998; Jaillon et al. 2004). Most of the duplicated genes have subsequently undergone either degeneration of one of the duplicates or functional segregation (Ohno 1970; Force et al. 1999). This process likely brought about a variety of molecular events in embryogenesis peculiar to teleost fishes. There are also many differences between teleosts and amphibians in their cellular mechanisms of embryogenesis. In the actinopterygian lineage, *Acipenser*, *Lepisosteus*, and *Amia* diverged from the teleost lineage after the divergence of *Polypterus*, but before the WGD event; however, the phylogenetic relationship between *Lepisosteus* and *Amia* remains in dispute (Inoue et al. 2003; Kikugawa et al. 2004; Hurley et al. 2007). Figure 1 shows the phylogenetic relationships among the various groups discussed here.

Culture

Fishes in *Polypterus*, *Lepisosteus*, and *Amia* have been maintained successfully by amateur aquarists. *Amia* and/or *Lepidosteus* should serve as a valuable model for assessing the ancestral state of teleosts before the WGD event; however, there are no reports of successful breeding in captivity. Sturgeons (*Acipenser*) are industrially cultured; the pattern of their embryogenesis (Bolker 1993a,b) appears to be similar to that of *Polypterus* (see this chapter and Bartsch et al. 1997). Among sarcopterygian fishes, lungfish have been cultured by Australian scientists (Joss, J. et al.; see http://www.bio.mq.edu.au/dept/centres/lungfish/index.html) and should serve as a valuable model for use in assessing the ancestral state of tetrapods. In contrast, *Polypterus* should present a unique opportunity to assess the ancestral states or bauplan (body plan or "blueprint") for Osteichthyes as well as for that of Actinopterygii.

Difficulty in efficient and controlled spawning or fertilization in an aquarium tank has hampered ontogenetic studies with bichirs. Two species of bichirs, *P. senegalus senegalus* (named by Cuvier in 1829, called gray bichir) and *P. endlicheri endlicheri* (named by Heckel in 1847, called saddled bichir), are popular among amateur aquarists in Japan. Encouraged by several reports of successful breeding in these species in captivity, we have chosen them for our embryological studies. Mature *P. senegalus* and *P. endlicheri* are about 30–50 cm in total length, respectively (Fig. 2a–c). Thus, a larger tank space is required for their housing than is needed for zebrafish or medaka. Their eggs are approximately 1.4 mm (*P. senegalus*) (Figs. 2n,o and 3a) and 1.9 mm (*P. endlicheri*), respectively (compared to the eggs of *Xenopus*, 0.9 mm, and of zebrafish, 0.5 mm). Bichir eggs were reported to have a single micropyle (Bartsch and Britz 1997), but it remains to be determined whether this is indeed a true micropyle and thus whether the pattern of fertilization is the same as that in teleosts. In concordance with the presence of micropyles in eggs, teleost sperm do not have acrosomes and enter into eggs just beneath the micropyles. Bichir has spermatozoa of aberrant shape but with acrosomes (Björn 2006).

SOURCES AND HUSBANDRY

Almost all bichir species are cultured in southeastern Asia for commercial purposes and are available in tropical fish stores. Farmed *P. senegalus* is the most inexpensive and accessible. An albino variant, also available in *P. senegalus* (Fig. 2b), is useful for in situ hybridization and immunohistochemistry studies.

The sex of bichirs is easily determined by the size and shape of the anal fin; the male fin is larger and extended, whereas the female fin is spindly (Fig. 2h,i). The male is craggy and the female is plump. Bichirs are inactive nocturnal fishes; during the daytime, they mostly lie on the bottom and during the nighttime, they lurk on the bottom. *P. senegalus* and *P. endlicheri* can be easily maintained in 80-liter (bottom face: 60 × 45 cm) and 400-liter (bottom face: 120 × 60 cm) aquaria, respectively, heated to 28°C (live range: 23–33°C) and adjusted to neutral pH (6.5–7.5), with four to six fishes per tank. In a breeding tank, we provide PVC pipes (caliber: 5–10 cm) as hiding places and water plants such as willow moss.

To keep fishes in good condition, it is recommended that they be fed a variety of foods. Bichirs eat small live fishes (zebrafish, goldfish, medaka, etc.), fillets of any kinds of fishes and meats, pink mice, mealworms, and various dry or moist foods (dry shrimp, food pellets for carnivorous

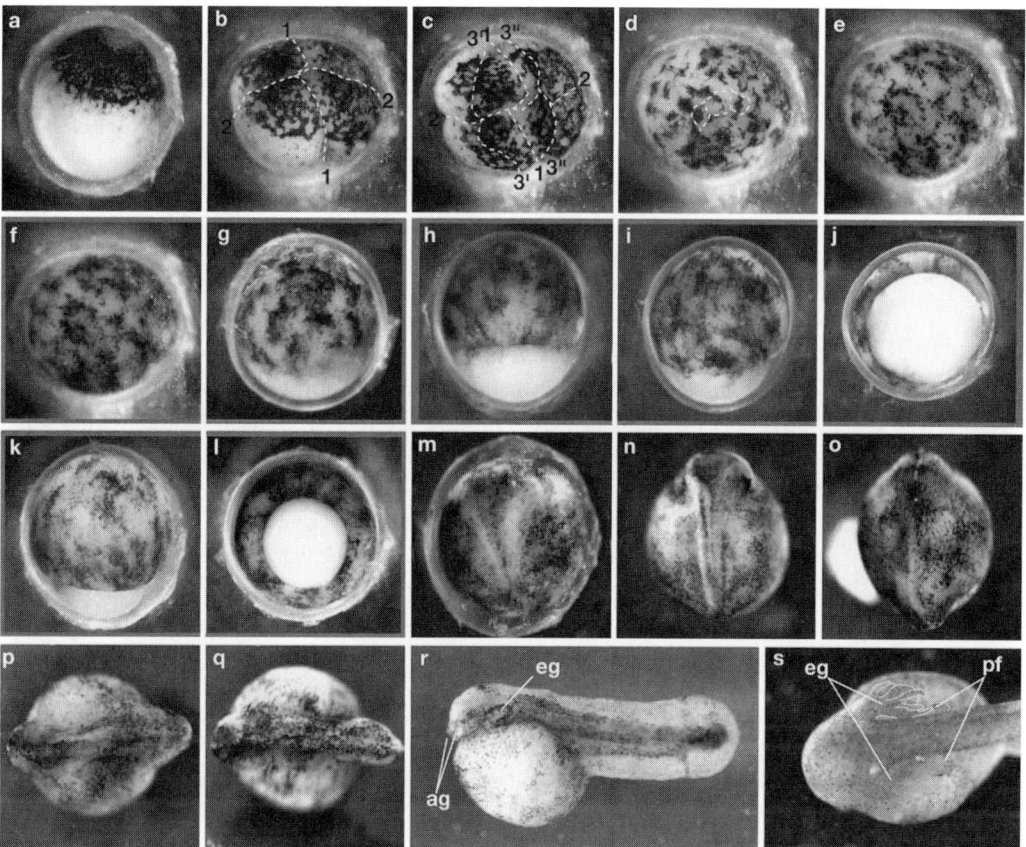

FIGURE 3. Embryogenesis in *P. senegalus*. (a) Fertilized egg (st. 1); (b) four-cell stage (st. 3); (c) eight-cell stage (st. 4); (d) morula (st. 6); (e) midblastula (st. 8); (f,g) late blastula (st. 9); (h–j) early gastrula (st. 10.5); (k,l) late gastrula (st. 12); (m) early neurula (st. 13); (n) midneurula (st. 15); (o) late neurula (st. 19); (p,q) tailbud stage (36 and 42 hpf); (r) prehatching tadpole (55 hpf), and (s) posthatching tadpole (4 dpf). At morula (d), each blastomere (*dotted circles*) is still distinguishable; at early neurula (m), the yolk plug is completely internalized; at midneurula (n), the neural groove is not distinct; at late neurula (o), neural folds meet one another at the midline; at early tailbud stage (p), attachment glands start to bud; at tailbud stage (q), the neural tube fuses and the dorsal midline disappears; in prehatching tadpole (r), the finfold, external gill bud, and urogenital opening are apparent; and in the posthatching tadpole (s), external gills are branched with active vascular flow. (a,g) Lateral views with animal pole at the top; (b–f) animal views; (h,m–q) dorsal views with the anterior at the top, (i,k) lateral views with the dorsal at the right; (j,l) vegetal views with the dorsal at the top; (r) a lateral view with the anterior at the left; (s) a dorsal view with the anterior at the left. (b,c) Numbers indicate the order of cleavage. (ag) Attachment gland bud; (eg) external gill bud; (pf) pectoral fin bud.

fishes). Our bichirs are routinely fed zebrafish, Trout Pellets (Oriental Yeast Co., Ltd.), and Sinking Carnivore Pellets (Kyorin Co., Ltd.).

Bichirs are not vulnerable to disease; however, the market fishes, regardless of whether they are wild or farmed, mostly live with parasitic worms, *Macrogyrodactylus polypteri*, classified as Trematoda. The worm, a parasite known only to bichirs, is a white, stringy worm parasitic on body surfaces, smaller than 2 mm and visible to the naked eye. The worm is not severely pathogenic and seldom kills the fish. It is a sign of infection when fishes rub their bodies against obstacles. To treat them, we include the following drugs in the aquarium tank water for 2 weeks: Refish and Green F Gold (Japan Pet Drugs Co., Ltd.): Trichlorfon (0.25 mg/l), Nitrofurazone (6.3 mg/l), and Sulfamerazine sodium (6.3 mg/l) each at the final concentration.

Spawning behavior is characteristic and usually starts 1 day before the spawning itself. First, a male bichir begins to track a female and peck at her abdomen with his rostrum. After continuous pecking by the male, they snuggle together and rub their anal fins together. These behaviors are repeated as the male and female swim through water plants, and finally, the female fish spawns eggs, releasing about ten eggs each time. During spawning, the male bends his anal fin into a cup shape under the genital opening of the female; spawned eggs are thus trapped and fertilized. By beating his caudal fin, the male distributes the fertilized eggs over the water plants and the eggs stick to the leaves. This behavior is repeated; spawning lasts for several days, and in successful cases, about 300 eggs are spawned. We are still not able to control spawning or artificially fertilize in vitro, although most mature males store numerous spermatozoa and many mature females have plenty of eggs (Fig. 2j–o). It is this difficulty that precludes *Polypterus* from becoming a laboratory animal.

Bichir larvae hatch on the sixth day after fertilization and start to ingest food on the eighth to tenth day when their yolk is consumed. We feed them newly hatched brine shrimps (Artemia) every day for about 1 month, and then for another several months, we feed them newly hatched brine shrimps and crushed sinking trout pellets (also sludgeworms when available). Juvenile bichirs prefer eating juvenile zebrafish and pellet foods. They are housed individually in small tanks to prevent cannibalization. The *P. senegalus* and *P. endlicheri* that we bred are now 2 years old but not yet mature; their body sizes are 20 cm and 30 cm, respectively, and their gonads, examined in several male and female fishes, are not yet mature. It is said to take 2–5 years, depending on the species, for bichirs to mature sexually.

Embryonic Culture

P. senegalus and *P. endlicheri* embryos are obtained by natural cross-breeding in our aquarium and are usually cultured at 28.5°C in embryonic medium (EM; see recipe on page 463) for zebrafish embryos. The developmental speed can be adjusted by lowering the water temperature to 23°C (in which case, development proceeds at about 1.5-fold lower speed). However, at below 20°C, development becomes abnormal.

Embryos are enveloped with chorion, which consists of sticky materials on the outside, translucent shell-like material in the middle, and vitelline membrane inside. The chorion is initially sticky and hard, but it becomes less and less so with development. The chorion of embryos later than neurula stage can be mechanically removed with tweezers, but it can hardly be detached in intact embryos at earlier stages. Neurula embryos free of the chorion develop normally in EM on a 2% agar dish.

Dissection of embryos and explant cultures are performed in Ringer's solution on a 2% agar dish, as with *Xenopus* embryos. The explants are cultured in EM on a 2% agar dish until the appropriate stages are reached. A variety of experimental procedures used in *Xenopus* embryos are applicable to studies with bichirs.

RELATED SPECIES

We have studied only two *Polypterus* species (*P. senegalus* and *P. endlicheri*), but other species as well as reedfish (*Erpetoichthys calabaricus*) are also available and could be handled similarly.

USES OF THE *POLYPTERUS* MODEL SYSTEM: STUDIES IN EARLY DEVELOPMENT

Early Cleavage Stages

In actinopterygians, teleosts undergo meroblastic, discoidal (or partial) cleavage; in contrast, bichir eggs undergo holoblastic (total or complete) cleavage. Holoblastic cleavage is also observed in sturgeon and dipnois as well as in amphibians, suggesting that ancestral Osteichthyes embryos must have undergone holoblastic cleavage. However, the details of holoblastic development in bichirs have not yet been reported.

We have performed comparisons of the developmental stages of *Polypterus* with teleosts and amphibians and observed many interesting features. *P. senegalus* and *P. endlicheri* eggs are pigmented in the animal hemisphere as are most amphibian eggs (Fig. 3a), whereas most teleost eggs are not pigmented. *Polypterus* eggs contain a much larger amount of yolk granules in the vegetal side than do *Xenopus* eggs. The first cleavage occurs at the meridian at about 1.5 hours after fertilization (hpf); this first cleavage occurs at about 1.5 hpf in *Xenopus* embryos, and at about 0.5 hpf in zebrafish embryos. The second cleavage occurs orthogonally to the first cleavage; the furrow starts at the animal pole and continues to the vegetal pole as the typical holoblastic cleavage (stage 3; Fig. 3b). Table 1 shows the stages of early embryogenesis. The third cleavage in bichir is also along the meridian, generating two furrows in parallel with the first cleavage plane (stage 4; Fig. 3c). In *Xenopus*, the third cleavage can be either equatorial or along the meridian, whereas in the newt, it is equatorial. Subsequent cleavages in bichir embryos are more unequal than those in *Xenopus* embryos and occur mostly in the animal hemisphere (stages 5–6; Fig. 3d), resulting in vegetal macromeres much larger than those in *Xenopus* embryos. Bichir embryos do not generate the yolk syncytial layer (YSL), which is produced by meroblastic cleavage in teleosts and gars (Long and Ballard 2001); the YSL has essential roles in axis and germ-layer formations. The blastocoel becomes apparent at about 4 hpf in bichir embryos (stages 6–7, Fig. 4a) and in *Xenopus* embryos. The relative size of the blastocoel in bichir embryos is much smaller than that in *Xenopus* embryos (Fig. 4a,d). The blastocoel roof is multicell-layered (Fig. 4g), and there are spherical cells loosely contacting one another in the blastocoel floor (Fig. 4d); these cells are not found in *Xenopus* but are present in newt embryos. Midblastula transition, when cell divisions slow down and zygotic gene transcription begins, occurs at 5 hpf in *Xenopus* and 3 hpf in zebrafish. Judging from the expression of molecular markers for each germ layer, such as orthologs of *sox2*, *brachyury/no-tail*, *sox17*, and others (data not shown), the midblastula transition occurs about 6 hpf in bichir embryos. No cortical rotation is apparent upon fertilization, and no morphological markers exist to assess the prospective dorsal side before gastrulation.

TABLE 1. Stages of early embryogenesis

Stage	State of the embryo	Timing	Notes
1	egg		
2	2-cell	(1.5 hpf)	
3	4-cell	(2 hpf)	
4	8-cell	(2.5 hpf)	
5	16–32 cell	(3 hpf)	
6	morula	(3.5 hpf)	
7	early blastula	(4 hpf)	
8	midblastula	(5–6 hpf)	
9	late blastula	(6–7 hpf)	
10	early gastrula	(8–9 hpf)	
11	midgastrula	(13–15 hpf)	
12	late gastrula	(15–20 hpf)	
13	early neurula	(20–22 hpf)	When yolk plug is completely internalized, and neural plate becomes apparent.
14	midneurula	(23–25 hpf)	When neural fold is formed.
15	late neurula	(24–26 hpf)	When neural plate starts to close at the posterior part.
16	late neurula	(27–28 hpf)	When neural tube is closed in the posterior half.
17	late neurula	(27–29 hpf)	When neural tube is closed, except for brain region that is wide-shaped.
18	late neurula	(28–30 hpf)	When neural tube is closed, except for brain region that is narrow-shaped.
19	late neurula	(29–31 hpf)	When entire neural tube is closed, but the midline where left and right neural folds meet is evident.
20	tailbud	(30–32 hpf)	When the neural folds are completely fused, losing the midline, and the tailbud is formed.

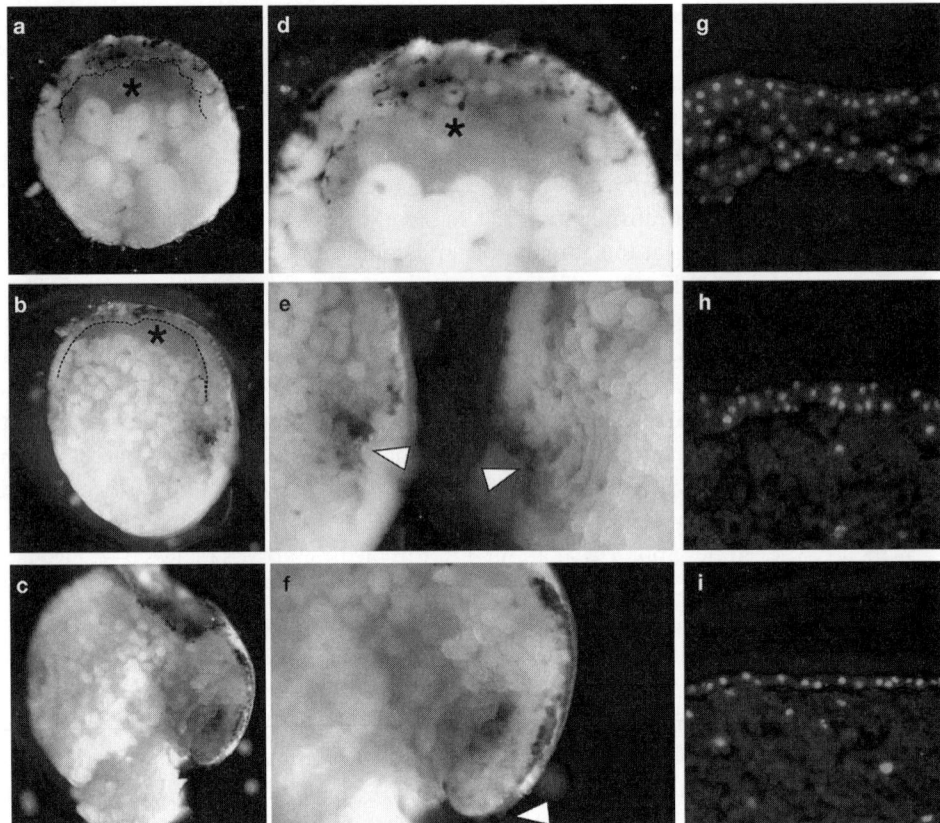

FIGURE 4. Gastrulation in *P. senegalus*. (a–f) Internal views of embryos dissected at the median plane; (g–i) DAPI-stained views of blastocoel roof. (d) Enlarged view of a at the animal pole region; (e,f) enlarged views of b and c at the blastopore regions, respectively. (a,d,g) Blastula embryos with distinct blastocoel (asterisks in a,b). The cells comprising the blastocoel floor are spherical and loosely attached to one another (d), and the blastocoel roof is multicell-layered (g). (b,e) Early gastrula embryo with distinct blastopore (*arrowheads*). (e) Both the left and right halves; the cells around the blastopore are elongated. The blastocoel roof now consists of a few cell layers (h). (c,f,i) At midgastrula, the pigmented cortical layer is involuted (c,f), suggesting that EVL or forerunner cells do not exist in *Polypterus* embryos as judged topographically by the gastrulation in teleost embryos. The blastocoel roof is now monolayered (i).

Gastrulation

Gastrulation in bichir embryos begins at about 9 hpf (stage 10.5; Fig. 3h–j) (*Xenopus*, 9 hpf; zebrafish, 6 hpf). By this stage, the blastocoel roof, multicell-layered at blastula stage, now consists of a few cell layers (Fig. 4h); it becomes monolayered at midgastrula (Fig. 4i). The blastopore in bichir embryos is closer to the equatorial plane than it is in *Xenopus* embryos. It remains to be demonstrated whether the bichir dorsal blastopore lip has Spemann organizer activity or expresses a series of organizer genes.

In bichir gastrulation, the involution movement occurs mostly in the dorsal half of the cells. This, together with the more equatorial location of the dorsal blastopore, results in the involution of part of the pigmented blastomeres (stages 11–13; Figs. 3k,l and 4c,f); whereas in *Xenopus*, nearly all pigmented cells remain as ectoderm. The involution of the most superficial pigmented cells indicates that bichir embryos do not have an enveloping layer (EVL). In teleosts and gars (Fig. 1), the most superficial EVL cells do not involute themselves, but rather overlay the cells fated to involute and become embryonic tissues. This involution also indicates that bichir embryos do not have cells homologous to forerunner cells that migrate vegetally at the leading edge of the shield, but do not involute themselves during teleost gastrulation. At the end of gastrulation, they migrate deep into the embryo and give rise to Kuppfer's vesicle that has an essential role in the development of left-right asymmetry (Essner et al. 2005). Bichir embryos have cells morphologically reminiscent of bottle cells that line the initial archenteron during *Xenopus* gastrulation (stage 10; Fig. 4e), but in teleosts, these cells are absent. Probably because of the more active cell movement in animal cells than in vegetal cells, externally the bichir early gastrula resembles *Xenopus* exogastrula embryos (stage 10) (Fig. 3i,k). Vegetal macromeres are finally internalized by epiboly movement and occupy the floor of the archenteron.

The archenteron becomes apparent at stages 12–13, and at the end of gastrulation when the yolk plug is completely internalized, a fan-shaped neural plate also becomes apparent at 20 hpf (stage 13) (Fig. 3m). In contrast, neurulation starts at 15 hpf in *Xenopus* and 10 hpf in zebrafish embryos. Lateral edges of the neural plate uplift to generate neural folds (Fig. 3n), and the entire neural plate closes to yield a neural tube by about 30 hours (stages 14–19) (Fig. 3o); this occurs at 20 hpf in *Xenopus* embryos. This pattern of neural tube formation is the same as that in *Xenopus* embryos, but it is different from the pattern in teleosts. In teleosts, the neural plate develops into the neural keel by infolding at the midline, and the keel in turn rounds into the cylindrical neural rod, whose formation is completed throughout the embryo by about 16 hpf (Kimmel et al. 1995). In bichir embryos, soon after closure of the neural tube, the tailbud is formed at the dorsal side of the blastopore slit at about 32 hpf (stage 20) (Fig. 3p); in *Xenopus*, the tailbud takes much longer to develop (at 32 hpf) after neural tube closure, and in zebrafish, it occurs much earlier (at 11 hpf), at early neurula.

In amphibians, the sperm entry point or the cortical rotation brought about by fertilization determines the future dorsal region, and the first cleavage demarcates the dorsoventral (DV) axis (Klein 1987; Heasman 1997). When horseradish peroxidase (HRP) is injected into the animal pole of one of two blastomeres at the two-cell stage and the fate of the HRP-positive cells is traced, most of the HRP-positive cells localize to either the left or the right side of the body. In teleosts, fertilization occurs just beneath the micropyle and causes no cortical rotation. Dorsal is also thought to be determined by the dorsal movement, along microtubules, of dorsalizing factors that originally exist in the vegetal pole (Mizuno et al. 1999). However, this movement occurs later, at about the four- to eight-cell stage. Moreover, blastomeres subsequently intermingle with one another intensively, and the first cleavage does not divide the embryo into left and right halves (Kimmel and Warga 1987). We also examined the cell fate of bichir blastomeres at the two-cell stage by injecting rhodamine dextran and tracing their fate to the tailbud stage (Fig. 5a). We found that almost all of the fluorescence was detected in either the left or right half of each embryo (Fig. 5b–e; 17/17, 100%). Thus, the first cleavage plane accords with the DV axis in bichirs, as it does in amphibians.

We also examined the double axis formation by β-catenin. Both blastomeres at the two-cell stage were injected with *Xenopus* β-catenin mRNA in the vicinity of the first cleavage furrow in the equatorial zone (Fig. 5f). In about half of these embryos (4/7), the β-catenin injection caused a secondary axis, duplicating the neural tube at the tailbud stage (Fig. 5g,h) or head structures together with attachment glands at the prehatching tadpole stage (Fig. 5i,j). Another half (3/7) of the embryos exhibited a normal phenotype. The embryos with secondary axis likely resulted from the β-catenin injection into the ventral side; normal embryos resulted from the injection into the dorsal side. The maternal canonical Wnt-signaling activity must thus function as a dorsal determinant in bichir embryos, in common with zebrafish and *Xenopus* embryos. Moreover, the β-catenin injection experiment indicates that the first cleavage plane defines the DV axis, in concordance with the fate-tracing experiment by dye injection. It must be determined whether the putative micropyle is indeed the real micropyle, whether fertilization takes place beneath the micropyle as in teleosts, or whether or not bichir acrosomes are functional. Cortical rotation has not been demonstrated in bichir, but it is possible that bichir eggs fertilize as *Xenopus* do, causing cortical rotation for DV axis determination.

In summary, these features of bichir early embryogenesis suggest that bichirs have mostly inherited frog-like development and established only minimal, if any, teleost-like development.

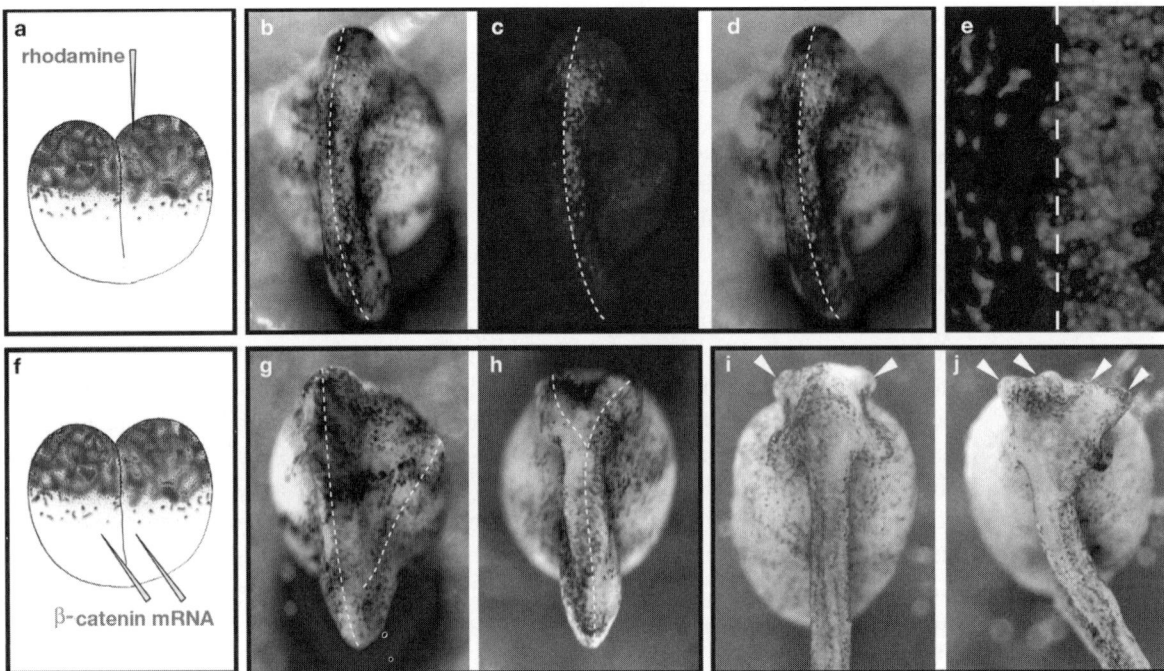

FIGURE 5. First cleavage and future axis in *P. senegalus* embryos. (a–e) When the two-cell stage embryo is injected with rhodamine into one of its blastomeres (a), the rhodamine fluorescence (c) is mostly detected in one half of the developing embryo at the tailbud stage. (d) Merged view of b and c. (e) Distribution of rhodamine-positive cells detected by confocal laser scanning microscopy. (f–j) Double axis formation. The two-cell stage embryos were injected with *Xenopus* β-catenin mRNA into each blastomere at the equatorial zone (in the vicinity of the first cleavage furrow) as illustrated in f. At the tailbud stage, embryos with secondary axis are either V (g)- or Y (h)-shaped with duplication of anterior structures. At the prehatching tadpole stage, they have four attachment glands (j; *arrowheads*). (i) Normal embryo with the two glands. Among 14 embryos analyzed at the neurula stage, two were dead, two abnormal, three normal, and six had the second axis. Among 14 embryos analyzed at the tailbud stage, seven were dead, three normal, and four had the second axis. (b–e,g,h) *Dotted lines* indicate the midline.

Holoblastic cleavage accompanying blastocoel and archenteron formation, neural tube formation via neural folds and their closure, correlation between body axis and first cleavage, and absence of EVL, YSL, and forerunner cells must therefore also be the pattern of early embryogenesis in ancestral Osteichthyes fish. Furthermore, it is likely that many characteristics found in teleosts arose during actinopterygian evolution after the divergence of Polypterioformes. These include meroblastic or discoidal cleavage, the presence of micropyle in eggs with the loss of acrosomes in sperms, dorsal determination at stages later than fertilization, absence of correlation between the first cleavage and the body axis, elimination of lumenal embryonic cavities such as blastocoele and archenteron, neural rod formation via neural keel, and the presence of EVL, YSL, and forerunner cells. The studies in amia and/or gar are of particular interest in determining how these various characteristics became established. At the same time, molecular analysis is essential to confirm that bichir early embryogenesis is indeed *Xenopus*-like. In this regard, yolky vegetal macromeres are especially a "hot spot." In the process of actinopterygian evolution, a cytoplasmic fusion in vegetal blastomeres may have brought about the teleostean yolk cell. The zebrafish YSL and the vegetal endoderm of *Xenopus* have striking similarities in both inductive function and morphogenesis. Both of these vegetal regions are involved in the establishment of the DV axis, as well as the induction of mesendoderm during the late blastula stages. Here, however, molecular and cellular mechanisms directing these processes diverge between the two animals. After the epiboly movement, the vegetal cells constitute the ventral part of archenteron. In *Xenopus*, the yolky cells express endodermal markers and become pancreas, liver, and gut, whereas in teleosts, the yolk cell does not express endoderm markers and serves as an extraembryonic nutritive. Do cells in bichirs express endodermal markers? Analysis of molecular machineries, morphogenetic domains, and cell identity specification within bichir vegetal cells would present unique insight into the evolution of gastrulation.

GENETICS AND GENOMICS

Bichirs are diploid organisms, and the chromosome number is $2n = 36$ in all reported *Polypterus* species including redfish (Cataudella et al. 1978), but not *P. weeksii* ($2n = 38$) (Vervoort 1980). Genome size appears to be similar among *Polypterus* species at a level slightly larger than the human genome; C values are 4–5 pg in bichirs and 3.5 pg in humans. It takes 2–5 years, depending on the species, for bichirs to mature sexually; the fishes are not amenable to genetic studies because there are no established mutant fish lines. However, several variants of *P. senegalus*, such as albino, short body, long-fin, and others, are available in the tropical fish market.

Gene overexpression studies by DNA injection work with bichir embryos. Therefore, knockdown studies of gene expression would also be practical with Morpholino oligonucleotide (MO), although these studies have not yet been conducted. The knockdown studies in bichirs should be productive, because the fish genes are less redundant than those in teleosts and *Xenopus*; bichirs did not undergo WGD nor are they pseudotetraploid.

Although the *P. senegalus* X gene would have been called *PsX* or *PosX*, *PsX* has already been used for genes in *Pelodiscus sinensis*, a soft-shelled turtle. We therefore propose to call the *P. senegalus* X gene *PosX* and the *P. endlicheri* X gene *PoeX*.

A good-quality bacterial artificial chromosome (BAC) library is available for *P. senegalus* from Chris T. Amemiya (Benaroya Research Institute at Virginia Mason, Seattle, Washington). We are now conducting an expressed sequence tag (EST) analysis. The *P. senegalus* cDNA library was constructed using the λ ZAP II vector (Uni-ZAP XR Vector Kits, Invitrogen) on poly(A)$^+$ RNAs from about 800 embryos at early stages (fertilized eggs~early neurula). The primary library consists of about 0.9×10^6 independent clones, and the average insert size is about 2.0 kb. Sequencing has been completed on about 10,000 clones at their 5'- and 3'-terminal ends. No plan for genome sequencing has been announced.

TECHNICAL APPROACHES

In this section, we describe the embryonic culture of *Polypterus*. We also include protocols for microinjection and in situ hybridization using *Polypterus* embryos.

Protocol 1

Microinjection and Animal Cap Assay

This protocol describes the collection of *Polypterus* embryos and the method for microinjection.

MATERIALS

The recipes for the items marked with <R> are on page 463.

Reagents

Embryonic medium (EM) <R>
Material to be microinjected
Polypterus
Ringer's solution <R>

Equipment

Culture dish containing 2% agar
Glass needle for microinjection
Incubator
Petri dish
Tweezers

METHOD

Embryo Collection

1. To collect one- to four-cell-stage embryos, observe the *Polypterus* carefully the day after spawning behavior is observed.
    Fertilized eggs and early embryos are sticky and adhere to water plants and tank walls; embryos at later stages detach and fall to the bottom.
2. Carefully collect early embryos manually.
    Embryos with chorion can be kept out of water for about 30 minutes.

Microinjection

3. Place the embryo to be injected in a Petri dish containing Ringer's solution.
4. Use tweezers to hold the embryo while performing microinjection.
5. Insert the glass needle containing the material to be injected through the chorion and inject.
    The glass needle easily clogs or breaks when inserted. The size of the needle tip is critical and must be determined empirically.

Embryo Culture

6. Culture the injected embryos in EM on a 2% agar dish until appropriate stages for analysis are reached.

   For an animal cap assay, inject animal cap explants from early blastula embryos with capped mRNAs. Rather small explants must be prepared; the bichir blastocoel is smaller than that of Xenopus.

Protocol 2

Whole-mount In Situ Hybridization

This protocol describes a method for in situ hybridization in *Polypterus* embryos. It is a minor modification of the whole-mount in situ hybridization using DIG-labeled antisense RNA probes described for *Xenopus* (Harland 1991).

MATERIALS

The recipes for the items marked with <R> begin on page 463.

CAUTION: See the Cautions Appendix for appropriate handling of materials marked with <!>.

Reagents

Antisense RNA probes, digoxigenin (DIG)-labeled (0.5–1.0 µg/ml) <!>
Blocking solution <R>
Coloring solution (BM purple; Roche)
Formaldehyde in MEM (MEMFA) <R>
Hybridization solution (HS) <R>
Hydrogen peroxide (H_2O_2; 10%, prepared in methanol) <!>
MAB containing Tween 20 (MABT) <R>
Maleic acid buffer (MAB) <R>
Methanol <!>
Paraformaldehyde <!>
Paraformaldehyde <!>, 4% in 1x PBS (4% PFA) <R>, prepared fresh before use
Phosphate-buffered saline (PBS) <R>
PBS containing Tween 20 (PBST) <R>
Proteinase K (1 mg/ml)
Polypterus embryos
Saline–sodium citrate buffer (20x SSC) <R>
Triethanolamine (TEA) in PBST <R>

Equipment

Cell culture dishes (~3 cm) to remove chorion
Equipment for in situ hybridization (see Harland 1991)
Polypropylene tubes of low-nucleotide and protein-binding quality for holding embryos to be dissected and fixed
Tweezers

METHOD

Unless otherwise indicated, one to five embryos can be placed in 1–1.5 ml of medium. Perform Step 1 in culture dishes and subsequent steps in polypropylene tubes.

Fixation

1. For embryos later than the tailbud stage, use tweezers to remove the chorion. For embryos earlier than the tailbud stage, remove the chorion after Step 7.
2. Fix the embryos in MEMFA overnight at 4°C.
3. Wash embryos in PBS for 5 minutes at room temperature.
4. Dehydrate the embryos through a graded series of methanol/PBS (25, 50, and 75% methanol in PBS) for 5 minutes each at room temperature.
5. Wash the embryos twice in 100% methanol for 30 minutes each at room temperature.
6. Store the fixed embryos in 100% methanol at −20°C until needed.

Pretreatment

7. Rehydrate the embryos through a graded series of methanol/PBS (75, 50, and 25% methanol in PBS) for 5 minutes at room temperature.
8. Wash the embryos in PBST for 5 minutes at room temperature.
9. Quench the embryos by dipping them in −20°C ethanol and maintain them in the ethanol for 7 minutes.
10. Wash the embryos three times in PBST for 5 minutes each at room temperature.

Proteinase K Treatment

11. Prepare a solution of 10 µg/ml proteinase K in PBST. Immerse the embryo in 1–1.5 ml of the proteinase K solution and perform digestion for 5–10 minutes at room temperature.
12. Rinse the embryos in PBST.
13. Treat the embryos twice with a solution of 0.1 M TEA in PBST for 5 minutes each at room temperature.
14. Add acetic anhydride to the TEA in PBST (final concentration 0.25%) and allow the embryos to incubate in this solution for 10 minutes at room temperature.
15. Wash the embryos twice in PBST for 5 minutes each time at room temperature.
16. Fix the embryos in 4% PFA for 20 minutes at room temperature.
17. Wash the embryos five times in PBST for 5 minutes each at room temperature.

Hybridization

18. Prepare a 1:1 mixture of PBST and HS and place the embryos in this mixture for 15 minutes at room temperature.
19. Replace the PBST/HS mixture with HS only and prehybridize the embryos for more than 6 hours at 60°C.
20. Hybridize the embryos with probe (0.5–1.0 µg) in HS for 1–2 overnight periods at 60°C.
21. Wash the embryos in HS for 15 minutes at 60°C.
22. Wash the embryos in a 3:1 mixture of HS and 2x SSC for 15 minutes at 60°C.
23. Wash the embryos in a 1:1 mixture of HS and 2x SSC for 15 minutes at 60°C.
24. Wash the embryos in a 1:3 mixture of HS and 2x SSC for 15 minutes at 60°C.
25. Wash the embryos three times in 2x SSC for 20 minutes each at 60°C.

26. Wash the embryos in 0.2x SSC for 30 minutes at 60°C.

Antibody Reaction

27. Wash the embryos three times in MABT for 15 minutes each at room temperature.
28. Replace the MABT with fresh blocking solution and block the reaction for 1–2 hours at room temperature.
29. Treat the embryos with anti-DIG antibody (1/1000) in fresh blocking solution overnight at 4°C.
30. Rinse the embryos in MABT at room temperature.
31. Wash the embryos six times in MABT for 1 hour each at room temperature.

Color Development

32. Wash the embryos in coloring solution (BM purple) for 5 minutes, using aluminum foil to shield the reaction from the light.
33. Allow the color to develop in BM purple, shielding from the light with aluminum foil. Check the color development first at 10–20 minutes, then at 30–40 minutes, 1–2 hours, 3–4 hours, etc.
34. Stop the reaction with 4% PFA when the embryos are appropriately stained.

Bleaching (optional)

These steps are performed to reduce pigments in pigmented blastomeres.

35. Transfer the embryos into 100% methanol.
36. Treat the embryos with 10% hydrogen peroxide in methanol under fluorescent light at room temperature.
37. Rinse the embryos with 100% methanol.

Recipes

CAUTION: See the Cautions Appendix for appropriate handling of materials marked with <!>.

BLOCKING REAGENT, 10% IN MAB

Reagent	Quantity (for 1 liter)	Final concentration
Blocking reagent (Roche)	5 g	10%

Dissolve blocking reagent in MAB with heating and adjust the final volume to 50 ml with MAB.

BLOCKING SOLUTION

Reagent	Quantity (for 1 liter)	Final concentration
Blocking reagent in MAB	1 ml	2% (blocking reagent)
Sheep serum (inactivated)	1 ml	20%
MAB	3 ml	

Adjust the final volume to 1 liter with H_2O and adjust the pH to 7.0–7.5.

EMBRYONIC MEDIUM (EM)

Reagent	Quantity (for 1 liter)	Final concentration
NaCl	0.87 g	15 mM
KCl <!>	0.04 g	0.5 mM
$CaCl_2 \cdot 2H_2O$ <!>	0.15 g	1 mM
$MgSO_4 \cdot 7H_2O$ <!>	0.25 g	1 mM
KH_2PO_4	0.02 g	0.15 mM
$Na_2HPO_4 \cdot 12H_2O$	0.02 g	0.05 mM
$NaHCO_3$	0.06 g	0.7 mM

Adjust the final volume to 1 liter with H_2O and adjust the pH to 7.0–7.5.

HYBRIDIZATION SOLUTION (HS)

Reagent	Quantity (for 10 ml)	Final concentration
Formamide	5 ml	50%
20x SSC	2.5 ml	5x
Denhardt's solution	200 µl	1x
EDTA (0.5 M, pH 8.0) <!>	200 µl	5 mM
10% Tween 20	100 µl	0.1%
10% CHAPS	100 µl	0.1%
Heparin (10 mg/ml)	100 µl	100 µl/ml
tRNA (10 mg/ml)	100 µl	100 µl/ml

Adjust the final volume to 10 ml with DEPC-H_2O.

MALEIC ACID BUFFER (MAB)

Reagent	Quantity (for 1 liter)	Final concentration
Maleic acid	11.6 g	100 mM
NaCl	8.77 g	150 mM

Adjust the pH to 7.5 with 10 N NaOH and adjust the final volume to 1 liter with H_2O.

MALEIC ACID BUFFER CONTAINING TWEEN 20 (MABT)

Reagent	Quantity (for 1 liter)	Final concentration
MAB	99 ml	99%
10% Tween 20	1 ml	0.1%

Adjust the final volume to 1 liter with DEPC-H_2O.

10x MEM

Reagent	Quantity (for 100 ml)	Final concentration
MOPS <!>	20.9 g	1000 mM
EGTA	0.76 g	20 mM
$MgSO_4$ <!>	0.25 g	10 mM

Adjust the final volume to 100 ml with H_2O, adjust the pH to 7.4, and sterilize by autoclaving. Store at 4°C, shielded from light.

MEMFA (FORMALDEHYDE IN MEM)

Reagent	Quantity (for 100 ml)
10x MEM	1 ml
10x PBS	1 ml
Formalin (37%) <!>	1 ml

Adjust the final volume to 10 ml with DEPC-H_2O; prepare fresh for each use.

PARAFORMALDEHYDE<!>, 4% IN 1x PBS (4% PFA)

Add 4 g of paraformaldehyde powder <!> to 80 ml of DEPC H_2O <!> and then add 2 drops 10 M NaOH <!>. Incubate the solution at 65°C, mixing occasionally until the powder is completely dissolved. Chill on ice until cold and then add 10 ml of 10x PBS and adjust the volume to 100 ml with DEPC-H_2O. Prepare fresh before use.

PBST (PBS CONTAINING TWEEN 20)

Reagent	Quantity (for 1 liter)	Final concentration
10x PBS	100 ml	1x
10% Tween 20	10 ml	0.1% Tween 20

Adjust the final volume to 1 liter with DEPC-H_2O.

10x PHOSPHATE-BUFFERED SALINE (PBS)

Reagent	Quantity (for 1 liter)	Final concentration
NaCl	80 g	1.37 M
KCl <!>	2 g	2.7 mM
Na_2HPO_4	11.5 g	81 mM
KH_2PO_4	2.0 g	14.7 mM

Adjust the final volume to 1 liter with H_2O and adjust the pH to 7.0 with 1 N HCl. Add 1 ml of DEPC, stir overnight, and sterilize by autoclaving.

RINGER'S SOLUTION

Reagent	Quantity (for 1 liter)	Final concentration
NaCl	6.8 g	116 mM
KCl <!>	0.2 g	2.9 mM
$CaCl_2$ <!>	0.2 g	1.8 mM
HEPES	1.2 g	5 mM

Adjust the final volume to 1 liter with H_2O and adjust the pH to 7.0–7.5.

20x SALINE-SODIUM CITRATE (SSC) BUFFER

Reagent	Quantity (for 1 liter)	Final concentration
NaCl	175.3 g	3 M
Sodium citrate	88.2 g	300 mM

Adjust the final volume to 1 liter with H_2O and adjust the pH to 7.0 with 1 N HCl. Add 1 ml of DEPC, stir overnight, and sterilize by autoclaving.

TRIETHANOLAMINE (TEA) IN PBST

Reagent	Quantity (for 100 ml)	Final concentration
TEA <!>	1.5 g	100 mM
10x PBS	10 ml	1x
10% Tween 20	1 ml	0.1%

Adjust the final volume to 100 ml with DEPC-H_2O and adjust the pH to 7–8.

ACKNOWLEDGMENT

We thank Dr. Mark Erdmann for the use of the coelacanth picture (Fig. 2e).

REFERENCES

Amores, A., Force, A., Yan, Y.L., Joly, L., Amemiya, C., Fritz, A., Ho, R.K., Langeland, J., Prince, V., Wang, Y.L., et al. 1998. Zebrafish hox clusters and vertebrate genome evolution. *Science* **282:** 1711–1714.

Ballard, W.W. 1986. Stages and rates of normal development in the holostean fish, *Amia calva*. *J. Exp. Zool.* **238:** 337–354.

Bartsch, P. and Britz, R. 1997. A single micropyle in the eggs of the most basal living actinopterygian fish, *Polypterus* (Actinopterygii, Polypteriformes). *J. Zool.* **241:** 589–592.

Bartsch, P., Gemballa S., and Piotrowski, T. 1997. The embryonic and larval development of *Polypterus senegalus* Cuvier, 1829: Its staging with reference to external and skeletal features, behavior and locomotory habits. *Acta Zool.* **78:** 309–328.

Björn, A.A. 2006. Preformed acrosome filaments. A chronicle. *Braz. J. Morphol. Sci.* **23:** 279–285.

Bolker, J.A. 1993a. Gastrulation and mesoderm morphogenesis in the white sturgeon. *J. Exp. Zool.* **266:** 116–131.

Bolker, J.A. 1993b. The mechanism of gastrulation in the white sturgeon. *J. Exp. Zool.* **266:** 132–145.

Cataudella, S., Sola, L., and Capanna, E. 1978. Remarks on the karyotype of the Polypteriformes. The chromosomes of *Polypterus delhezi*, *P. endlicheri congicus* and *P. palmas*. *Experientia* **34:** 999–1000.

Essner, J.J., Amack, J.D., Nyholm, M.K., Harris, E.B., and Yost, H.J. 2005. Kupffer's vesicle is a ciliated organ of asymmetry in the zebrafish embryo that initiates left-right development of the brain, heart and gut. *Development* **132:** 1247–1260.

Eycleshymer, A.C. and Wilson, J.M. 1906. The gastrulation and embryo formation in *Amia calva*. *Am. J. Anat.* **5:** 133–162

Force, A., Lynch, M., Pickett, F.B., Amores, A., Yan, Y.L., and Postlethwait, J. 1999. Preservation of duplicate genes by complementary, degenerative mutations. *Genetics* **151:** 1531–1545.

Harland, R.M. 1991. In situ hybridization: An improved whole-mount method for *Xenopus* embryos. *Methods Cell Biol.* **36:** 685–695.

Heasman, J. 1997. Patterning the *Xenopus* blastula. *Development* **124:** 4179–4191.

Hoegg, S., Brinkmann, H., Taylor, J.S., and Meyer, A. 2004. Phylogenetic timing of the fish-specific genome duplication correlates with the diversification of teleost fish. *J. Mol. Evol.* **59:** 190–203.

Hurley, I.A., Mueller, R.L., Dunn, K.A., Schmidt, E.J., Friedman, M., Ho, R.K., Prince, V.E., Yang, Z., Thomas, M.G., and Coates, M.I. 2007. A new time-scale for ray-finned fish evolution. *Proc. Biol. Sci.* **274:** 489–498.

Inoue, J.G., Miya, M., Tsukamoto, K., and Nishida, M. 2003. Basal actinopterygian relationships: A mitogenomic perspective on the phylogeny of the "ancient fish." *Mol. Phylogenet. Evol.* **26:** 110–120.

Jaillon, O., Aury, J.M., Brunet, F., Petit, J.L., Stange-Thomann, N., Mauceli, E., Bouneau, L., Fischer, C., Ozouf-Costaz, C., Bernot, A., et al. 2004. Genome duplication in the teleost fish *Tetraodon nigroviridis* reveals the early vertebrate proto-karyotype. *Nature* **431:** 946–957.

Kemp, A. 1982. The embryological development of the Queensland lungfish, *Neoceratodus forsteri* (Krefft). *Mem. Queensl. Mus.* **20:** 553–597.

Kerr, J.G. 1899. V. The external features in the development of *Lepidosiren paradoxa*, Fitz. 1899. *Philos. Trans. R. Soc. B Biol. Sci.* **192:** 299–330.

Khalil, L.F. 1964. On the biology of *Macrogyrodactylus polypteri* Malmberg, 1956, a monogenetic trematode on *Polypterus senegalus* in the Sudan. *J. Helminthol.* **38:** 219–222.

Kikugawa, K., Katoh, K., Kuraku, S., Sakurai, H., Ishida, O., Iwabe, N., and Miyata, T. 2004. Basal jawed vertebrate phylogeny inferred from multiple nuclear DNA-coded genes. *BMC Biol.* **2:** 3.

Kimmel, C.B. and Warga, R.M. 1987. Indeterminate cell lineage of the zebrafish embryo. *Dev. Biol.* **124:** 269–280.

Kimmel, C.B., Ballard, W.W., Kimmel, S.R., Ullmann B., and Schilling, T.F. 1995. Stages of embryonic development of the zebrafish. *Dev. Dyn.* **203:** 253–310.

Klein, 1987. The first cleavage furrow demarcates the dorsal-ventral axis in *Xenopus* embryos. *Dev. Biol.* **120:** 299–304.

Long, W.L. and Ballard, W.W. 2001. Normal embryonic stages of the longnose gar, *Lepisosteus osseus*. *BMC Dev. Biol.* **1:** 6.

Mizuno, T., Yamaha, E., Kuroiwa, A., and Takeda, H. 1999. Removal of vegetal yolk causes dorsal deficiencies and impairs dorsal-inducing ability of the yolk cell in zebrafish. *Mech. Dev.* **81:** 51–63.

Nelson, J.S. 2006. *Fishes of the world*, 4th ed. Wiley, New York.

Noack, K., Zardoya, R., and Meyer, A. 1996. The complete mitochondrial DNA sequence of the bichir (*Polypterus ornatipinnis*), a basal ray-finned fish: Ancient establishment of the consensus vertebrate gene order. *Genetics* **144:** 1165–1180.

Ohno, S. 1970. Evolution by gene duplication. Springer-Verlag, Heidelberg.

Olsson, L., Hossfeld, U., Bindl, R., and Joss, J.M.P. 2004. The development of the Australian lungfish (*Necoceratodus forsteri*): From Richard Semon's pioneering work to contemporary approaches. *Rudolstädter Naturhist. Schriften* **12:** 51–128.

Vervoort, A. 1980. Karyotypes and nuclear DNA contents of Polypteridae (Osteichthyes). *Experientia* **36:** 646–647.

FURTHER READING

This chapter focused on early embryogenesis; for larval development and anatomical and behavioral observations, see Bartsch et al. (1997), as well as the references. Among related animals (Fig. 1), development of white sturgeons is similar to that of bichirs (Bolker et al. 1993a,b). Development of gars and *Amia* is more similar to that of teleosts (Eycleshymer and Wilson 1906; Ballard 1986; Long and Ballard 2001). Development of Dipnoi (lungfish) is described by Kerr (1899), Kemp (1982), and Olsson et al. (2004).

WWW RESOURCES

http://www.answers.com/topic/polypteriformes-bichirs-biological-family This site provides information on the evolution, physical characteristics, distribution, habitat, behavior, diet, reproductive biology, conservation status, and significance to humans of Polypteriformes (bichirs).

http://www.aquaticcommunity.com/predatory/breedingbichir.php This site provides information on breeding bichirs, including a bibliography.

http://www.bio.mq.edu.au/dept/centres/lungfish/index.html This is the site for the Australian Lungfish Research Facility at Macquarie University. It includes information on lungfish biology, the research being performed at the lab, and a list of publications.

19 *Astyanax mexicanus*, The Blind Mexican Cave Fish
A Model for Studies in Development and Morphology

Richard Borowsky

Department of Biology, New York University, New York 10003

ABSTRACT

The perpetual darkness of caves has two important consequences for their permanent inhabitants. First, eyes and pigmentation lose their primary functions. Second, in the absence of photosynthesis, food is rare. For these reasons, cave-adapted species typically have reduced eyes and pigmentation and increased or more efficient metabolisms.

PROTOCOLS

1 Determining the Sex of Adult *Astyanax mexicanus*, 474
2 Breeding *Astyanax mexicanus* through Natural Spawning, 476
3 In Vitro Fertilization of *Astyanax mexicanus*, 477
4 Handling Eggs and Fry, 479

Additionally, other senses are usually augmented to compensate for the loss of vision. Identifying the genetic bases underlying these phenotypic changes will enhance our understanding of the specific pathways involved in control of these phenotypes and, in general, the evolutionary process. Unfortunately, the genetics of most cave animals cannot be studied because they are not easily bred. Blind Mexican tetras, *Astyanax mexicanus*, are the valuable exception to this rule because fish from the various cave populations are fully interfertile with one another and with eyed sister forms still living in nearby surface streams. Hybrids between surface and cave forms permit genetic analysis of their differences, and study of the pure forms as well as of hybrids allows study of their developmental differences. Quantitative trait loci (QTL) analysis has already identified some specific genes responsible for differences between cave and surface forms as well as other likely candidates; more will be added in the future. This system is a valuable addition to the array of existing models for the study of developmental and evolutionary genetics, because cave populations are repositories of numerous naturally occurring mutations affecting development of the eyes and other senses, pigmentation, bone structure, metamerism, and metabolism. These alleles have been prescreened by natural selection for high viability, which simplifies their study. In contrast, new alleles obtained through mutagenesis in other model species are typically burdened with lower viability.

BACKGROUND INFORMATION

The Mexican tetra, *Astyanax mexicanus* (de Filippi 1853; syn. *A. fasciatus*) is a member of the characidae, a family of fish of South American origin that also includes piranhas and pacu. Characins are intolerant of seawater ("primary freshwater origin") and were unable to reach

This chapter, with full-color images, can be found online at www.cshprotocols.org/emo.

Central America (CA) from South America (SA) until the end of the Tertiary period, with the establishment of a land bridge (Myers 1966). Most primary freshwater SA families that entered CA, therefore, have not progressed far, and only *A. mexicanus* has reached northern Mexico and southern Texas. Myers estimates that they reached the region with current cave populations about 1 million years ago.

Cave specimens differ from surface sisters most obviously in having only rudimentary eyes that are not even externally visible and in their greatly reduced pigmentation (Fig. 1). They also differ in a host of other less obvious features, such as metabolism and sensitivities of other sensory modalities.

The cave populations of northeast Mexico are apparently derived from two different stocks, one from the original invasion of the area and the other representing a reinvasion from southern *refugia* after the first invaders became locally extinct with climate change (Dowling et al. 2002; Strecker et al. 2003, 2004). Despite this diverse origin, all cave populations tested thus far have proven be to interfertile with one another and with the current surface species. Because of this history, the species offers the opportunity to study the genetic and developmental bases of recently evolved and extensive multitrait changes, driven by the repeated introduction of surface fish into a subterranean environment.

Sequence conservation between *Astyanax* and the zebrafish model *Danio* is apparently extensive, although their two lines diverged no later than the early Cretaceous period, approximately 100 million years ago (MYA) (Nelson 2006). Sequence comparison studies using the similarity search programs BLAT and BLAST of *Danio* sequence with queries from *Astyanax* microsatellite sequences yield numerous convincing matches (R. Borowski, unpubl.). Thus, *Danio* genomic tools should prove useful for many investigations in *Astyanax*. Developmental stages in both fish are similar and the transparent eggs of both make them equally easy to study (Jeffery 2001).

SOURCES AND HUSBANDRY

Sources

"Blind Mexican tetras" have been in the aquarium trade in the United States since the early 1940s. Commercial stocks were originally derived from the Cueva Chica population first discovered by Coronado and investigated by Breder (1942); the ancestry and purity of the existing commercial

FIGURE 1. Surface and cave (Pachon) individuals of *Astyanax mexicanus*.

stocks are uncertain because of the possibility of reestablishments from other localities. For this reason, they are an inappropriate choice for most scientific work. Nevertheless, they are hardy in the lab and very easily bred. Thus, they provide good material for pilot studies, including the honing of technical skills such as in vitro fertilization, embryonic staining, and transgenesis though egg injection. The author's laboratory maintains stocks of surface fish from 11 different caves spanning the full geographic range of cave forms. In addition, the Jeffery lab (Department of Biology, University of Maryland) maintains stocks that are potentially available to the community. Stocks and potential availability are listed in Table 1.

Collecting wildlife specimens in Mexico is forbidden without permits from the Mexican government. Importation into the U.S. is controlled by the U.S. Fish and Wildlife Service, which checks for Mexican permits and inspects stock at ports of entry. Currently, no blind Mexican tetra population is covered by the Convention on International Trade in Endangered Species of Wild Fauna and Flora (CITES); however, populations specifically named as endangered in document NOM-059-ECOL-2001 (Norma Oficial Mexicana of the Secretaría de Medio Ambiente y Recursos Naturales) are given special protection. It is possible to obtain permits for collecting NOM-listed populations, but only with difficulty, whereas permits for other populations are more easily obtained. Permits for any scientific work to be performed in Mexico may be obtained by application through the U.S. Department of State (by contacting the science liaison officer at the U.S. Embassy in Mexico City) or through collaborative arrangements with Mexican ichthyologists.

It should be stressed that the introduction of live fish from any source into any fish room carries the risk of cointroduction of parasites or pathogens. Newly introduced stock must be quarantined until it is determined that they exhibit no signs of disease. The safest practice is to introduce new stock as fertilized eggs, with chorion intact, because no known fish parasite can penetrate the chorion. Nets and other devices used in handling fish should never be shared among tanks without disinfection. Circulating water systems should incorporate water sterilization through UV radiation, ozonization, or otherwise. A good guide to general fish room practice is provided by Westerfield (2000).

Husbandry

In the wild, cave and surface forms of *A. mexicanus* live in running freshwater, above ground and below. Water temperatures in their caves average around 22°C, but the fish are comfortable in the

TABLE 1. Stocks of cave and surface *Astyanax mexicanus* currently maintained in U.S. laboratories

	Laboratory[1]	Successfully bred?
Cave localities		
Molino/Jineo	RB, WJ	No
Pachon	RB, WJ	Yes
Yerbaniz	RB	No
Japonese	RB	Yes
Los Sabinos	WJ	Yes
Arroyo	RB	No
Tinaja	RB, WJ	Yes
Piedras	RB	No
Curva	RB, WJ	Yes
Rio Subterraneo	RB, WJ	Yes
Surface localities		
Rio Sabinas Drainage	RB	Yes
Rio Valles Drainage	RB	Yes
Rio Choy Drainage	WJ	Yes
Balmorhea State Park, Texas	WJ	Yes

[1]Contact Richard Borowsky (RB; rborowsky@nyu.edu) or William Jeffery (WJ; jeffery@umd.edu.) for availability.

range of 19°C to 27°C. Generally, it is best to keep temperatures on the low side for maintenance because one stimulus for breeding is a rise in temperature. Also, lower temperatures help to control bacteria in the system and keep the fish in better health.

Proper feeding is essential for successful breeding. It is best to feed the fish several times a day with small amounts of food, rather than once a day with larger amounts. Food that is uneaten after 10 to 15 minutes will never be eaten and will decay if not removed from the tank. Fish from different cave populations differ in their food preferences, but useful foods include staple food flakes, freeze-dried crustacea (Mysis for high fat, krill, grass shrimp, etc.), frozen crustacea, or live foods. Crustacea should be ground or chopped to pieces 1–2 mm in size. Dried "blood worms" (dipteran insect larvae) are useful in the diet but are a potent allergen for some people and must be used with great caution. The stomach content of surface forms of *A. mexicanus* in the wild contains more than 50% plant material (Darnell 1962) and thus their diet in the lab should not be solely animal in origin. Commercial flake foods typically contain plant material. We supplement the diet with occasional feedings of small amounts of dried seaweed ("sushi nori"). The person in charge of fish feeding must monitor what the fish eat and what they ignore and then use this knowledge to tailor the diet to the stock.

Before being set up to mate, healthy females must be brought into breeding condition during the course of 10 to 14 days on a diet high in fats. Females can be bred as often as every two weeks, whereas males are always ready to breed. Under good conditions, a generation may be as short as 6 months. As many as 1000 fertilized eggs may be produced from a single mating.

Fry hatch as soon as 1 day postfertilization (dpf), but hatching may be delayed in some cases by as much as a day. Fry are free-swimming once they hatch, but they do not feed until 4 dpf. At this point, they can be fed any commercial fry food for egg layers. These typically consist of fine powders that must be suspended in water before adding to the tank or to prepared liquid suspensions. By 5 dpf, this food should be supplemented with brine shrimp nauplii, hatched in the fish room for that purpose. By the time fry are 9–10 dpf, they can feed exclusively on brine shrimp, and this diet can continue until they reach more than 12 mm in total length. Then they can be gradually weaned from nauplii and then fed more finely ground versions of an adult diet. As with most fish, social interactions amplify genetic differences in potential, and growth rates will vary widely among tank mates. Growth is stimulated by spreading the brood out into several tanks, and growth of smaller fish, in particular, is stimulated by moving out their larger siblings.

Fish rooms for *Astyanax* should house tanks larger than those used for zebra fish. Ideally, standard 5.5-gallon tanks (~17-liter water capacity) must be used for matings and maintenance of groups consisting of up to eight adults. Standard 15-gallon aquaria (~50-liter water capacity) can accommodate groups of up to 25 adults. Single animals can be maintained in isolation in 3.75-liter containers. The system should be recirculating.

Fish rooms and water systems can be designed and installed by many companies. In the U.S., these include Aquaneering Inc. (www.aquaneering.com), Marine Biotech (www.marinebiotech.com), and Thoren Aquatics (www.thorenaquatics.com). The author's fish rooms contain both Aquaneering and Thoren components, custom-designed racks, and water systems.

USES OF THE *A. MEXICANUS* SYSTEM: STUDIES IN DEVELOPMENT, BEHAVIOR, AND MORPHOLOGY

Extensive literature on cave forms of *A. mexicanus* has become available since their initial description more than 80 years ago (Hubbs and Innes 1936). A comprehensive bibliography of cave fish papers through 2004 (Proudlove 2006) is a good guide to the older literature; a thorough review covers the more modern literature, especially in development, through 2000 (Jeffery 2001). PubMed covers the most recent papers. The literature includes studies of feeding behavior, metabolism, morphology, neuroanatomy, pigmentation, eye development and structure, and tastebud and neuromast distributions. Nearly all of these studies are directed toward describing the differences between the cave and surface populations.

GENETICS AND GENOMICS

Genetic Markers and Maps

An extensive collection of microsatellite markers developed by Protas is available for mapping studies (Protas et al. 2006). Microsatellite sequences and primers for size typing are accessioned in GenBank under numbers BV67803–BV678968; additional primer sequences are published (Protas et al. 2008). In addition, approximately 100 coding sequences for various nuclear and mitochondrial genes are available in GenBank; most of these are genes involved in eye development or function, such as *PAX6*, crystallins, opsins, etc. Twelve genes are mapped in the Protas et al. 2008 publication. All public databases should be queried with all synonyms: *Astyanax mexicanus*, *Astyanax fasciatus*, and *Astyanax jordani*.

Preliminary linkage maps are published and have been used for mapping quantitative trait loci (QTL) controlling differences between surface fish and cave fish from two different populations (Protas et al. 2006, 2008). Traits studied include eye and lens size, melanophore numbers, albinism, chemoreception, body shape, and metabolism. The last published map (Protas et al. 2008) is approximately 10 cM resolution, but the latest revision (R. Borowsky, unpubl.) has 5 cM resolution.

Amenability to Classical Genetics

All populations of *A. mexicanus* cave fish and their surface sisters are fully interfertile. In some cases, interpopulation hybridization results in progeny with a skewed sex ratio (typically all female), but this problem is episodic rather than consistent. If one cross yields a single sex progeny, another cross with different parents from the same two populations could yield a mixed progeny. Thus, F_2 hybrids can be produced for mapping QTL responsible for phenotypic differences between different populations. One caveat is that not all populations breed readily, because the factors involved in bringing females into breeding condition appear to vary among populations and are not yet fully understood. In contrast, sperm can be obtained from males from all populations that have been tested. Females from surface populations, Pachon cave, and Tinaja cave are relatively easily bred. Thus, one can obtain hybrids between any cave population and surface fish by using males from the cave population. Because they are easy to breed, most studies have involved the use of Pachon cave fish.

Genomic Resources

The principal resource currently available is a bacterial artificial chromosome (BAC) library constructed from DNA from a surface *A. mexicanus* (Di Palma et al. 2007). The library vector is pCC1BAC, the average insert size is 104 kb, and the coverage is 3.6x. The library is available on 153 96-well plates and can be screened with the aid of a three-filter set, all available from the Hubbard Genome Center at the University of New Hampshire (http://hcgs.unh.edu/BAC/Astyanax.html). *A. mexicanus* genome sequencing applications have been submitted to the Broad Institute and Genoscope.

TECHNICAL APPROACHES

We describe here methods for natural breeding and in vitro fertilization of *A. mexicanus*, as well as the handling and manipulation of eggs and fry for their continued growth and maintenance.

Protocol 1

Determining the Sex of Adult *Astyanax mexicanus*

This protocol describes methods for determining the sex of an individual *A. mexicanus*. Adult males and females differ most obviously in body shape (deeper in females, as shown in Fig. 2) and in form of the anal fin (longer anterior anal fin rays in females, as in Fig. 3). If the sex of an adult cannot be determined by body shape or the form of the anal fin, it can be assessed by testing the anal fin for the presence (or absence) of denticle or hook-like bony elements on the anterior fin rays of the anal fin. These features are observed as an opacity toward the anterior half of the male's anal fin (as shown in Fig. 3), detected as described in the method presented here.

MATERIALS

CAUTION: See the Cautions Appendix for appropriate handling of materials marked with <!>.

Reagents

A. mexicanus adults
Tricaine methane sulfonate (MS222; 0.1% in buffered water, pH 7.5) <!>

Equipment

Cotton or nylon wool
Low-power dissecting microscope (10x to 40x)

FIGURE 2. Surface and Pachon cave F_1 hybrids. The female is above, the male below. The female has a fuller body shape and a more triangular shape to the anal fin. Note that the eyes of the hybrids are smaller than those of the surface fish in Figure 1.

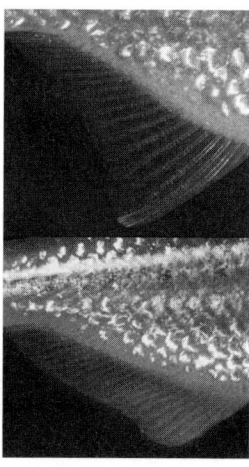

FIGURE 3. The anal fins of female (*above*) and male (*below*) are shown in greater detail. Note the differences in shape and the greater thickness and opacity of the male's anal fin in the anterior half.

METHOD

1. Anesthetize the individual by immersion in 200 ml of MS222 solution in a separate container for 30 seconds or until it becomes disoriented. It can then be held without damage in the folds of a wet paper towel.

2. Using a piece of fine cotton or nylon wool, swab the anterior portion of the anal fin distally. If the cotton catches, the fish is an adult male. If not, it is female or juvenile.

3. If the fish appears to be male, confirm this characteristic by microscopic examination under low power (20x–30x) with transillumination for the presence of hook-like bony elements on the anterior fin rays (see Fig. 3).

4. To assist recovery from MS222 anesthesia, place the fish in system water in an aquarium tank; observe it for signs of movement.

    Once the fish begins movement of the body or fanning of the opercula, it will revive fully on its own.

5. If the fish does not revive within 1 minute, aid its recovery by holding it gently at the midsection and directing a gentle stream of water at its mouth from a plastic disposable pipette.

    The mouth and opercula will open if the stream is directed from in front and above the animal. This movement of water past the gills will speed the loss of MS222 from the animal.

Protocol 2

Breeding *Astyanax mexicanus* through Natural Spawning

Male and female *A. mexicanus* can be bred successfully in tanks under appropriate conditions. As discussed in the "Husbandry" section of the introduction, females should be maintained on a diet high in fats for 10 to 14 days before breeding. The transfer of a male and female into clean water in a fresh tank and a change (increase) in water temperature are cues for breeding. Newly fertilized eggs may also be obtained through in vitro fertilization (described in Protocol 3).

Note that blind fish should never be paired with eyed fish in illuminated aquaria, because the eyed fish are aggressive and will kill even much larger blind fish. Such matings must be carried out in the dark or by using in vitro fertilization.

MATERIALS

Reagents

A. mexicanus adult males and females (sex determined as in Protocol 1)
Fish food for adult female: live food (preferable) or frozen crustacea (Mysis or other)
Fish food for fry (see "Husbandry" in the chapter introduction)

Equipment

Water tanks (one with the temperature set between 19°C and 21°C and the other [the mating tank] set at a temperature that is 4–5° higher)

METHOD

1. Bring the females into reproductive condition by feeding them well with appropriate fish food and keeping them uncrowded in a clean tank, with the temperature stably maintained and between 19°C and 21°C.

2. During the late morning, transfer the conditioned female and a male into a fresh tank with warmer (an increase of 4–5°C) clean water. Leave the fish together in this breeding tank overnight.

3. Remove the parents after the eggs are deposited.

4. The following morning, inspect the bottom of the tank for fertilized eggs. If no eggs are visible, keep the pair in the breeding tank for one more day.

   If no eggs are visible by the second day, return the fish to their original tank(s).

5. If fertilized eggs are needed in early stages of development, inspect the mating tanks regularly for signs of newly fertilized eggs.

   Depending on the stock, spawning may take place during the night or in early morning hours. The fry will hatch within 1 to 2 dpf. See the "Husbandry" section in the introduction for further information on the care and feeding of the fry.

6. Fertilized eggs prior to hatching can be handled without harm using wide-tip pipettors or even a fine net.

7. Posthatch fry are delicate but can be transferred from one tank to another by using a 250-ml beaker or other small container as a ladle. Do not use nets before the fry are approximately 2 weeks old.

Protocol 3

In Vitro Fertilization of *Astyanax mexicanus*

This protocol describes an alternative method for breeding *A. mexicanus*, using in vitro fertilization. Sperm collected from the male and eggs collected from the female are placed in a Petri dish, and sperm are activated by the addition of fresh system water. The eggs are observed under low magnification for signs of fertilization, usually marked by the onset of cleavage.

MATERIALS

The recipe for item marked with <R> is on page 480.

CAUTION: See the Cautions Appendix for appropriate handling of materials marked with <!>.

Reagents

A. mexicanus adult males and females (sex determined as in Protocol 1)
Hank's medium <R> (from Westerfield 2000)
Tricaine methane sulfonate (MS222; 0.1% in buffered water, pH 7.5) <!>

Equipment

Compound light microscope (400x to 1000x)
Pipettor (50-μl glass microcapillary, e.g., Fisher 21-164-2G, with appropriate fine controller [mouth tube or hand pipettor] or 0.5-ml fine plastic disposable pipettor, e.g., Denville P7231)

METHOD

1. Set up matings between male and female adult *A. mexicanus*, as described in Protocol 2 (Steps 1–3).

2. Select females from the mating tanks with newly fertilized eggs on the bottom.

    Depending on the stock, spawning may have occurred during the night or in the early morning.

3. Examine a sample of the newly fertilized eggs under low-power magnification to determine the stage of development.

    If the eggs appear to be at a stage earlier than the eight-cell stage, more (unfertilized) eggs can be obtained directly from the female for use in in vitro fertilization. But sperm must be collected first in anticipation of gathering viable eggs.

4. Anesthetize the male(s) in MS222 (see Protocol 1, Step 1) to facilitate handling.

5. Position the anesthetized individual ventral side up on the microscope stage under low-power magnification to facilitate positioning of the pipette.

6. Lavage the gonopore area of the selected male in Hank's medium, using the pipettor to ensure that the sperm are not activated by contact with residual tank water clinging to the fish.

7. Obtain the sperm by gently squeezing the male between the thumb and forefinger on the abdomen near the gonopore.

8. Using the pipettor from Step 2, collect any residual Hank's medium at the gonopore, now mixed with sperm, usually in a volume of 20 µl. The presence of sperm makes the liquid cloudy enough to be visible, even in the plastic pipettor.

9. If desired, verify the presence of sperm under high power in the compound microscope.

    i. Dilute a small portion of Hank's sample containing sperm >20-fold with water. Reserve the remainder in undiluted Hank's for Step 11.

    ii. Place a fraction of the sample to be tested onto a slide with a large drop of fresh water to activate the sperm.

    iii. Place a coverslip over the sample and observe active sperm (if present) under 400 or 1000 power magnification.

10. Squeeze the abdomen of the female just above and posterior to the insertion of the pelvic fins to recover unfertilized eggs, using a 50-mm plastic Petri dish or other container to collect and hold the egg mass.

    If eggs are not easily expelled, use another female. If the eggs are at the right stage, they will be expelled with gentle pressure. If firm pressure is required, it will not be possible to obtain usable eggs and the pressure will harm the female.

11. Place the Hank's medium with the bulk of the sperm sample as a small drop next to the egg mass in the Petri dish and then add 1 ml of clean system water to activate the sperm and the eggs.

    Note that the water must be added at the last moment. If added too soon, the eggs cannot be fertilized or the sperm will be exhausted. If too much water is added, the sperm will be too dilute and the fertilization efficiency will be low.

12. Five minutes after activating the eggs and sperm, increase the volume with system water to fill the dish.

13. During the next several hours, examine the eggs periodically under the microscope at low magnification.

    The chorion will rise in minutes and cytoplasm will subsequently stream from the yolky portion to the animal pole, whether the egg is fertilized or not. Unfertilized eggs pass through a pseudo-development stage that lasts for several hours and involves twisting movements of cytoplasm. The first reliable sign that the egg has been fertilized is the onset of cleavage.

Protocol 4

Handling Eggs and Fry

Fertilized eggs are transferred into larger dishes to reduce crowding and are observed until they begin to hatch into fry.

MATERIALS

Reagents

Eggs resulting from in vitro fertilization (Protocol 3)
Fish food for fry (see "Husbandry" section in the introduction)

Equipment

Compound light microscope (400x to 1000x)
Fresh water tank or aquaria
Petri dishes

METHOD

1. If the fertilization dish is crowded, transfer the eggs to larger, clean dishes. Eggs are adhesive but can be dislodged for transfer with gentle streams of water from a pipette or by swirling the water in the dish and then pouring it into the larger container.

2. Remove dead eggs from the dish as they become evident during the first day postfertilization (1 dpf).

    Developing eggs remain clear; dead eggs rapidly become opaque.

3. Continue to observe the eggs. Once the fry begin to hatch (1 dpf), transfer them to a larger dish or into an aquarium.

    Chorionated eggs are hardy and can be easily transferred from one dish to the next, but fry are fragile. One way to transfer fry successfully to a larger dish or an aquarium is to submerse the smaller container in the larger one and gently turn it over, washing the contents over.

4. At 4 dpf, begin to introduce food into the tank to feed the fry.

    See the "Husbandry" section in the introduction for further information on the care and feeding of the fry.

Recipe

CAUTION: See the Cautions Appendix for appropriate handling of materials marked with <!>.

HANK'S MEDIUM (adapted, with permission, from full protocol in Westerfield 2000)

Reagent	Quantity (for 1 liter)	Final concentration
NaCl (1 M)	137 ml	0.137 M
KCl (1 M) <!>	5.4 ml	5.4 mM
Na_2HPO_4 (1 M)	0.25 ml	0.25 mM
KH_2PO_4 (1 M)	0.44 ml	0.44 mM
$CaCl_2$ (1 M) <!>	1.3 ml	1.3 mM
$MgSO_4$ (1 M) <!>	1.0 ml	1.0 mM
$NaHCO_3$ (1 M)	4.2 ml	4.2 mM

REFERENCES

Breder, C.M. 1942. Descriptive ecology of la Cueva Chica, with special reference to the blind fish, *Anoptichthys*. *Zoologica* **27:** 7–15.

Darnell, R.M. 1962. Fishes of the Rio Tamesi and related coastal lagoons in east-central Mexico. *Publ. Inst. Mar. Sci. Univ. Texas* **8:** 299–365.

de Filippi, F. 1853. Nouvelles espèces de poissons. *Rev. Mag. Zool.* **5:** 164–171.

Di Palma, F., Kidd, C., Borowsky, R., and Kocher, T.D. 2007. Construction of bacterial artificial chromosome libraries for the Lake Malawi cichlid (*Metriaclima zebra*), and the blind cavefish (*Astyanax mexicanus*). *Zebrafish* **4:** 41–47.

Dowling, T.E., Martasian, D.P., and Jeffery, W.R. 2002. Evidence for multiple genetic forms with similar eyeless phenotypes in the blind cavefish, *Astyanax mexicanus*. *Mol. Biol. Evol.* **19:** 446–455.

Hubbs, C.L. and Innes W.T. 1936. The first known blind fish of the family Characidae: A new genus from Mexico. *Occas. Pap. Mus. Zool. Univ. Mich.* **342:** 1–7.

Jeffery, W.R. 2001. Cavefish as a model system in evolutionary developmental biology. *Dev. Biol.* **231:** 1–12.

Myers, G.S. 1966. Derivation of freshwater fish fauna of Central America. *Copeia* **1966:** 766–773.

Nelson, J.S. 2006. *Fishes of the world*, 4th ed. Wiley, New York.

Protas, M.E., Hersey, C., Kochanek, D., Zhou, Y., Wilkens, H., Jeffery, W.R., Zon, L.I., Borowsky, R., and Tabin C.J. 2006. Genetic analysis of cavefish reveals molecular convergence in the evolution of albinism. *Nat. Genet.* **38:** 107–111.

Protas, M., Tabansky, I., Conrad, M., Gross, J.B., Vidal, O., Tabin, C.J., and Borowsky R. 2008. Multi-trait evolution in a cave fish, *Astyanax mexicanus*. *Evol. Dev.* **10:** 196–209.

Proudlove, G.S. 2006. *Subterranean fishes of the world. An account of the subterranean (hypogean) fishes described to 2003 with a bibliography 1541–2004*. International Society for Subterranean Biology, Moulis, France.

Strecker, U., Bernatchez, L., and Wilkens H. 2003. Genetic divergence between cave and surface populations of *Astyanax* in Mexico (Characidae, Teleostei). *Mol. Ecol.* **12:** 699–710.

Strecker, U., Faundez, V.H., and Wilkens H. 2004. Phylogeography of surface and cave *Astyanax* (Teleostei) from Central and North America based on cytochrome *b* sequence data. *Mol. Phylogenet. Evol.* **33:** 469–481.

Westerfield, M. 2000. *The zebrafish book. A guide for the laboratory use of zebrafish* (Danio rerio), 4th ed. University of Oregon Press, Eugene (http://zfin.org/zf_info/zfbook/zfbk.html).

20 Darwin's Finches
Analysis of Beak Morphological Changes During Evolution

Arhat Abzhanov

Department of Organismic and Evolutionary Biology/FAS Biological Laboratories 4105, Harvard University, Cambridge, Massachusetts 02138

ABSTRACT

Finches of the Galápagos Islands were first described by Charles Darwin during his voyage on the HMS Beagle in 1835. Since then, through the subsequent work of many biologists, Darwin's finches have become a classic textbook example of many important processes in evolution. Today, this group of birds continues to be a significant source of information on such processes as speciation, niche partitioning, morphological adaptation, and species ecology. The approximately 14 species of Darwin's finches are closely related to one another and display a remarkable degree of diversity in bill shapes and sizes that are adapted for different food sources (e.g., seeds, insects, and even young leaves or blood from sea birds) in an otherwise scarce environment. For example, the deep and wide bills of the Ground Finches, one of the subgroups of Darwin's finches, are used to feed on seeds, whereas the Cactus Finches use their elongated and narrow bills to probe cactus fruit and flowers. These differences in bill shapes are not due to their differential usage or other external factors; rather, the differences are genetically and developmentally regulated and can be observed and studied during embryogenesis. Therefore, Darwin's finches are becoming a very useful nonmodel animal and avian system in which to investigate the molecular basis of morphological changes during evolution.

PROTOCOLS

1. Collection of Embryos from Darwin's Finches (Thraupidae, Passeriformes), 487
2. In Situ Hybridization Analysis of Embryonic Beak Tissue from Darwin's Finches, 490
3. Microarray Analysis of Embryonic Beak mRNA from Darwin's Finches, 494

BACKGROUND INFORMATION

Darwin's finches (also known as the Galápagos Finches) are a group of 14 (13–15 according to several different phylogenies) closely related species of passerine birds (Passeriformes). Most of Darwin's finches inhabit the Galápagos Islands, although one species is found on Cocos Island (Grant and Grant 2008). This group includes the following genera: *Certhidea*, *Geospiza*, *Camarhynchus*, *Platyspiza*, and *Pinaroloxias*. For a long time, taxonomists placed Darwin's finches in one family with the New World sparrows and Old World buntings (Emberizidae, Passeriformes) (Sulloway 1982). However, some of the more recent taxonomical references place Darwin's finches in the tangers group (Thraupidae, Passeriformes) (Sibley and Ahlquist 1990; Monroe and Sibley

This chapter, with full-color images, can be found online at www.cshprotocols.org/emo.

FIGURE 1. Morphology of adult Darwin's finches. Two species are shown: the Large Ground Finch *G. magnirostris* with deep and wide bills (*A*, male; *B*, female) and the Large Cactus Finch *G. conirostris* with probing elongated bills (*C*, male; *D*, female). The female *G. conirostris* (*D*) is feeding on a flower bud of *Opuntia* sp. cactus.

1993; Burns and Skutch 2003). Their closest relatives inhabit Central and South America and the Caribbean Islands (Burns et al. 2002).

Darwin's finches represent a well-established case of adaptive radiation where the entire existing diversity of modern species is derived from a single ancestral colonizing species (Grant and Grant 2002, 2008; Futuyma 2005). Most species of Darwin's finches are similar to one another in many respects—they have similar sizes, plumage, and songs—but are strikingly different in other details, such as the size and shape of their bills and feeding behaviors (Fig. 1). In fact, Charles Darwin noted several years after his famous exploration trip on the HMS Beagle that "[t]he most curious fact is the perfect gradation in the size of the beaks of the different species of *Geospiza*" and that "I have stated that in the thirteen species of ground-finches, a nearly perfect gradation may be traced, from a beak extraordinarily thick, to one so fine, that it may be compared to that of a warbler" (Darwin 1839). In an insightful effort to explain these observations, he wrote, "Seeing this gradation and diversity of structure in one small, intimately related group of birds, one might fancy that, from the original paucity of birds in this archipelago, one species has been taken and modified for different ends" (Darwin 1839). Thus, from the start, Darwin's finches were instrumental to the origin of the theory of adaptive evolution by natural selection (Grant and Grant 2008).

Important studies on Darwin's finches since Charles Darwin were conducted by David Lack, Richard Bowman (UCLA), and Peter and Rosemary Grant (Princeton) who extensively documented the morphology, ecology, behavior, phylogeny, and environment-induced evolutionary changes among the finches (Lack 1940, 1945; Bowman 1961; Grant 1999; Grant and Grant 2008). Of particular significance is the work by the Grants, who started their studies in 1973 and over time tracked many thousands of individual finches across several generations, showing numerous evolutionary phenomena. Their scientific efforts and accomplishments were described in *The*

Beak of the Finch by Jonathan Weiner (Weiner 1995). Numerous other research groups study Darwin's finches to explore various aspects of their biology, such as incipient speciation, introgression, character displacement, song behavior evolution, and adaptive beak biomechanics (Podos 2001; Herrell et al. 2005; Huber et al. 2007). Thus, besides historical interest, this group of animals provides unique advantages because a very large body of literature exists on its ecology, adaptive changes, selective pressures, and beak morphology and function, providing a superb context for interpreting results from molecular studies. This should ultimately allow for a better understanding of evolution where modern genetic and developmental information is integrated with population- and species-level ecological studies.

SOURCES AND HUSBANDRY

All of the species of Darwin's finches are restricted to the Galápagos archipelago and Cocos Island; no colonies of these species exist outside their native environment. The names and locations of species that have been collected for comparative analyses are as follows: the Warbler Finch (*Certhidea fusca*), Large Ground Finch (*G. magnirostris*), Large Cactus Finch (*G. conirostris*), and Sharp-beaked Finch (*G. difficilis*) are from Genovesa Island, which is located in the north of the archipelago, and the Small and Medium Ground Finches (*G. fuliginosa* and *G. fortis*, respectively) and the Cactus Finch (*G. scandens*) are from Santa Cruz Island.

Thus far, all material for developmental evolution work on Darwin's finches has been collected on site with full permission of the Galápagos National Park and the Charles Darwin Research Station. Special permits are required to conduct a scientific research project anywhere on Galápagos, move biological specimens from one island to another island within the archipelago, or export the research material from Galápagos/Ecuador. Even though many of the related species of songbirds are relatively easily kept in captivity and animal facilities, the Ecuadorian authorities do not allow the establishment of colonies of the Galápagos species away from the islands. None of the populations and species that were chosen for our projects are threatened or endangered; all are locally abundant birds.

USES OF DARWIN'S FINCHES AS A MODEL SYSTEM

Developmental Evolution of Beak Morphology

The diverse shapes and sizes of finch beaks are believed to be maximally effective for the exploitation of particular types of food including seeds, insects, and cactus flowers, especially under conditions of food scarcity (Grant 1999). The external differences in beak morphology among the species of Darwin's finches reflect differences in their respective craniofacial skeletons (Price and Grant 1985; Grant 1999). Bird beaks consist of a hard, albeit hollow and porous, bony structure formed from the fused premaxillary and maxillary bones, structures that have clear homologs in mammals (Hanken and Hall 1993). This bony structure gives the beak its shape and strength. The specialized shapes of beaks in Darwin's finches are apparent at hatching (Grant 1999; Grant and Grant 2003, 2008), indicating that the differences in their respective morphologies are genetically determined and established during embryogenesis. It was previously shown that species-specific differences in beak morphology in Darwin's finches could be observed by midstages of development (Abzhanov et al. 2004). These differences are also relatively evolutionarily recent (Grant 1986; Sato et al. 1999, 2001; Petren et al. 2000, 2005). It was thus hypothesized that differences in beak morphology in such a closely related group of species are primarily caused by changes in the regulation of key developmental genes, leading to differential patterns, timing, and/or levels of gene expression in the frontonasal and mandibular processes, the mesenchymal primordia of the upper and lower beaks, respectively. In turn, these regulatory changes bring about morphological alterations that have various adaptive consequences. Thus, the Darwin's finch model system is well suited to study the evolution of craniofacial development, particularly of the cranial skeletal structures (beak and skull).

The original studies on developmental craniofacial biology in Darwin's finches focused on beak development in six species from the genus *Geospiza* that represented three distinct morphologies in three different sizes. These included Small, Medium, and Large Ground Finches with deep and wide beaks adapted for crushing hard seeds, as well as (Medium) Cactus and Large Cactus Finches with long and pointed (low depth and width) beaks adapted for penetrating cactus flowers and fruit. Also included was the Sharp-beaked Finch, which has a small elongated symmetrical beak adapted for probing and seed-cracking; this represents the basal morphology for the genus (Petren et al. 2000; Abzhanov et al. 2004). All evolutionary developmental genetics work published to date for Darwin's finches has involved the *Geospiza* genus. However, current ongoing work involving Darwin's finches includes genera other than *Geospiza*.

Two different approaches, described below, have been used to identify genetic causes for morphological beak diversity in Darwin's finches. First, a set of candidate genes was studied to identify transcripts that correlated very closely with beak morphology and were functionally relevant for generating such morphology in embryos (Abzhanov et al. 2004; Grant et al. 2006). Subsequently, a less constrained cDNA microarray analysis of the transcripts expressed in the beak primordia was used to identify novel genes and pathways whose expression correlates with specific beak morphologies.

Candidate Gene Analysis

Thanks to recent advances in the understanding of craniofacial development, a large number of growth factors are known to be expressed during craniofacial development in chicken embryos (Francis-West et al. 1994, 1998, 2003; Schneider et al. 2001, Helms et al. 2005). The expression patterns of these growth factors were recently analyzed in different *Geospiza* species using in situ hybridizations on medial cranial sections of stages 26 (E5) and 29 (E6.5) embryos (Hamburger and Hamilton 1951; Abzhanov et al. 2004). Of particular interest were factor(s) whose expression in the mesenchyme of the beak primordium correlated with increasing depth and width of beaks seen as one compares the Sharp-beaked Finch *G. difficilis* to Small, Medium, and, finally, Large Ground Finches (*G. fuliginosa*, *G. fortis*, and *G. magnirostris*, respectively). To eliminate changes in expression that were merely related to the overall size of the bird and not to changes in beak morphology, expression patterns were also compared in Medium and Large Cactus Finches (*G. scandens* and *G. conirostris*, respectively), which are similar in size to the Medium and Large Ground Finches, respectively, but share the more pointed beak morphology of the Sharp-beaked Finch.

Most of the factors examined showed no differences among Darwin's finch species; however, *Bone morphogenetic factors 2* and *7* (*Bmp2* and *Bmp7*) correlated with beak size but not beak shape. In contrast, a striking correlation was observed between beak morphology and the expression of *Bmp4*. Once the cartilage condensation has occurred at stage 29, *Bmp4* continues to be expressed in mesenchymal cells surrounding the rostralmost part of the prenasal cartilage. When embryos of the three Ground Finch species were examined, a dramatic increase in the level of *Bmp4* expression was noted in the Large Ground Finch at stage 26, whereas all other species exhibited expression more or less equivalent to that of the Sharp-beaked Finch. By stage 29, however, all three Ground Finch species displayed elevated levels of *Bmp4* expression, with the Large Ground Finch being the strongest and the Small Ground Finch the weakest of these. Importantly, the Medium Cactus Finch, a relatively pointed-beaked species of similar size to the Medium Ground Finch, and the Large Cactus Finch, which is similar in size to the Large Ground Finch, did not show this increase in *Bmp4* expression. Thus, the species with deeper, wider beaks relative to their length express *Bmp4* in the mesenchyme of their beak primordia at higher levels and at earlier stages (heterochronic shift) than species with more narrow and shallow beak morphologies. Therefore, changes in expression of *Bmp4* represent examples of two important evolutionary changes: heterochrony (temporal alterations) and heterotopy (spatial alterations).

In a set of misexpression experiments designed to mimic the elevated levels of *Bmp4* seen in the Large Ground Finch, RCAS::*Bmp4* virus was injected into the mesenchyme of the frontonasal process of stages 23–24 chick embryos. Beaks resulting from infection of the mesenchyme were reminiscent of those of the Ground Finches: They were deep and wide. The more massive *Bmp4*-infected beaks had a corresponding increase in size of the skeletal core, analogous to the more massive beak skeleton of the Large Ground Finch. This skeletal phenotype was observed in the majority of infected embryos. In contrast, mesenchymal injection of RCAS::*Noggin* virus, which antagonizes BMP2/4/7 signaling, led to a dramatic decrease in the size of the upper beak and to much smaller skeletal elements inside the upper beak, further supporting the requirement for BMP signaling in the frontonasal mesenchyme for regulation of the depth and amount of chondrogenesis in the beak (Abzhanov et al. 2004). In short, the comparative candidate gene approach showed that *Bmp4* correlates with and regulates both beak depth and width.

Microarray Analysis

Another important morphological parameter is beak length. The candidate gene approach, which is limited to molecules with known functions in craniofacial and/or skeletal development, did not yield any pathways that could be involved in controlling this important dimension. Thus, a less constricted microarray approach was recently used to identify candidate molecule(s) associated with the elongated and pointed (low depth and width) beak of the Cactus Finch that is used for probing cactus flowers and fruit (Bowman 1961).

It was hypothesized that differences in beak morphology would be revealed by differential levels of gene expression in the frontonasal process, the ectomesenchymal progenitor that was previously shown to carry the morphogenic patterning information in developing chick beaks. To identify genes related to beak differences in finches, DNA chip technology was used to produce high-density cDNA microarrays corresponding to genes expressed during beak development and to analyze changes in gene expression.

The obtained data were clustered on signal ratios and intensities for different transcripts to find those that were up- or down-regulated in all individuals of a particular species. These species-specific clusters were further cross-compared to reveal transcripts that were consistently up-regulated in frontonasal processes of all individuals of the Cactus Finch beak morphology and that stayed unchanged or were down-regulated in beak primordia of Ground Finches. Comparisons with similarly sized birds allowed the separation of transcripts exhibiting size-specific regulation from those with morphology-specific regulation.

The resulting final cluster of genes expressed at higher levels in beaks of Cactus Finches contained multiple transcripts. Chosen for further validation were candidates that were expressed at moderate or high levels on the microarray and at least fivefold higher in Cactus Finches. Compared to the reference species (Sharp-beaked Finch, *G. difficilis*), *Calmodulin* (*CaM*) was up-regulated in Cactus Finch beaks and was unchanged or somewhat down-regulated in Ground Finch beaks.

To validate the microarray screen, comparative in situ hybridization was performed on stage 26 embryos of Darwin's finches. *CaM* was expressed at detectably higher levels in the distal-ventral mesenchyme of the frontonasal processes in Cactus and Large Cactus Finches compared to the similar-sized processes in Ground Finches. These observations indicated that differential levels of CaM-dependent signaling could be important in the development of distinct beak morphologies in different species of Darwin's finches. To functionally test this model, the level of CaM-dependent signaling was elevated in the developing chick beak primordium with a constitutively activated form of a downstream effector of CaM, the CaM-dependent protein kinase II (CaMKII). In chick embryos where activated CaMKII was misexpressed specifically in the distal-ventral mesenchyme, a significant increase in the length of the beaks was observed, whereas beak width and depth were not affected (Abzhanov et al. 2006).

In summary, a cDNA-microarray-based analysis of transcripts expressed in beak primordia of embryos of wild Darwin's finches revealed that *CaM* is a valuable new candidate gene associated with a previously developmentally uncharacterized beak-length dimension (Abzhanov et al. 2006).

GENETICS, GENOMICS, AND ASSOCIATED RESOURCES

Multiple microsatellite and amplified fragment length polymorphism (AFLP) genetic/genomic markers have been established for Darwin's finches for work on population genetics and evolutionary ecology (Petren et al. 2000, 2005). These are largely concordant with the corresponding genetic markers described for other birds, including other songbirds and chickens. No genetic maps currently exist for Darwin's finches but studies on other songbirds predict a high degree of synteny with the existing genetic maps for both chicken (*Gallus gallus*) and Zebra Finch (*Taennopogia guttata*). At this time, no species of Darwin's finches have been used for genome sequencing and no bacterial artificial chromosome (BAC) libraries exist for these species. Standard bird/vertebrate nomenclature is used for naming genes.

Some species of Darwin's finches display significant natural genetic variation, but they are not easily accessible and there are no colonies in captivity, making them less amenable for classical genetic studies. However, a considerable number of population genetic studies have been performed on various populations of Darwin's finches in their native habitat by several groups of researchers. For example, Peter and Rosemary Grant and collaborators used genetic markers (microsatellites and AFLPs) to study levels of heritability of beak morphology in Medium Ground Finches (*G. fortis*) on Daphne Major Island during changing environmental conditions, and they showed increasing introgression during hybridization of this population with that of Cactus Finch (*G. scandens*). Many of these genetic markers are used to confirm the identity of the embryos from our collections when there is any doubt as to the species classification of both parents.

To identify novel candidate genes associated with long and pointy beak morphology in Cactus Finches (*G. scandens* and *G. conirostris*), a cDNA library was generated that represents transcripts expressed in the craniofacial primordial (frontonasal mass and mandibular primordium) at stages that correspond to chicken developmental stages 26, 29, and 32 (Hamburger and Hamilton 1951; Abzhanov et al. 2006). The cDNA library was printed on a high-density microarray (about 21,000 spots) that was then used to screen individually labeled samples from various species of Darwin's finches. These microarrays can be used to probe any species of Darwin's finches.

TECHNICAL APPROACHES

The following protocols have been used to study developmental evolution in Darwin's finches: embryo collection and tissue preparation (Protocol 1), in situ hybridization analysis on embryonic tissue sections (Protocol 2), and microarray analysis of mRNA from individual beak tissue samples (Protocol 3). They were originally used for work on model organisms (mouse and chicken embryos) in the laboratory of Dr. Clifford Tabin (Harvard University Medical School), modified for work on songbird embryos, and specifically tested on Zebra Finches (*T. guttata*), Darwin's finches (*Geospiza*), and their allies. In principle, the protocols should be useful for embryos of any songbirds (Passeriformes) with little or no modification; all have been previously published (Abzhanov et al. 2004, 2006).

Protocol 1

Collection of Embryos from Darwin's Finches (Thraupidae, Passeriformes)

There are no breeding colonies of Darwin's finches anywhere in the world. Thus, all of the embryonic material is collected in the wild. This protocol describes how, in a field setting, fertilized eggs are collected and incubated at a precise temperature and how the resulting embryos are harvested and processed for in situ hybridization, antibody staining, and microarray analyses. In addition, the protocol includes steps for preparing the heads of older embryos for histological staining of bone and cartilage with alcian blue and alizarin red. It is likely that the same or similar methods can be used to obtain embryonic tissue from other species of songbirds. The main limitation of this protocol is that, when used in the field without external constant electricity sources, it requires power generators that need to run on a more or less constant basis, as well as stockpiles of supplies (fuel, oil, and fresh water).

MATERIALS

CAUTION: See the Cautions Appendix for appropriate handling of materials marked with <!>.

Reagents

Ethanol (25, 70, 75, and 90%) <!> (for histological staining only)
Paraformaldehyde (4%) <!> (for in situ hybridization, antibody staining, or histological staining only)
  Prepare immediately before use by diluting a 20% paraformaldehyde stock in 1x PBS.
1x PBS (for in situ hybridization, antibody staining, or histological staining only)
  Prepare immediately before use by diluting a 10x PBS stock with H_2O. Bottled water or a conventional travel water purification system can be used.
RNA*later*® (Ambion) (for microarray analysis, in situ hybridization, or antibody staining only)

Equipment

Bags (1–2 liters; plastic)
Box with a soft material lining
  A small plastic box for microcentrifuge tubes works well as long as it has holes for ventilation and heat exchange.
Centrifuge tubes (15 and 50 ml)
Cooler
  The Kool Auto Kaddy 12V cooler or the Compact 12V Kooler (both Koolatron) work well.
Electrical power generator (1000 W; portable)
Incubator with forced-air control (Hova-Bator 110 V)
Ladders (1 and 3 meter; folding)
Permit for egg collection from Galapagos National Park, Ecuador
Petri dishes (plastic; 6 and 10 cm diameter; sterile)
Tarp/tent (large; waterproof)
Tweezers (#5 and #55; Fine Science Tools)

METHOD

Egg Collection and Incubation

1. Set up the Hova-Bator incubator under a waterproof tarp or tent, using the portable generator as the power source. Set the incubator to 38°C and high (above 70%) humidity. Take care not to allow the incubator to cool down below 36°C at night and above 38°C during the day.

2. Set up the ladders, and using a box with a soft material lining, collect eggs from the nests of wild songbirds as specified by the permit.

   A strategy should be developed that minimizes the impact on the breeding pairs and the population as a whole. For example, we collect only the third egg (one egg per nest) because females re-lay until the normal clutch site is reached (Thomson 1964).

3. Incubate the egg(s) at 38°C until embryos reach the desired developmental stage:

   i. For stage 26 embryos, incubate the eggs for 6 days.

   ii. For stage 30 embryos, incubate the eggs for 7 days.

Embryo Harvest

4. Fill a Petri dish with a volume of 1x PBS that is 10 times greater than the estimated volume of the harvested embryo tissue. Use 6-cm Petri dishes for small embryos and 10-cm Petri dishes for large embryos.

5. Using tweezers, carefully open the egg and place the embryo into the Petri dish.

6. Remove all membranes surrounding the embryo and separate the head from trunk. Rinse the embryo well. If the embryos are to be used for in situ hybridization or antibody staining, allow the blood to drain before fixation. Dispose of all waste in a plastic bag.

   To prepare the embryos for microarray analysis, proceed to Step 7. To prepare the embryos for in situ hybridization, antibody staining, and/or histological staining, proceed to Step 9.

Sample Preparation for Microarray Analysis

7. Immediately place the embryo samples in 15- or 50-ml centrifuge tubes containing RNA*later*®. Use a 5:1 ratio of RNA*later*® volume:tissue volume.

8. Keep the tissue in RNA*later*® for 10–20 minutes at ambient temperature before placing the samples in a cooler for transport to the laboratory. According to the manufacturer, samples can be stored in RNA*later*® for one month at 4°C, one week at 25°C, or indefinitely at –20°C. Avoid rapid changes in temperature.

   These samples can be subsequently used for microarray analysis as described in Protocol 3. See Troubleshooting.

Sample Fixation for In Situ Hybridization and Antibody Staining

9. Place the samples into 15- or 50-ml centrifuge tubes containing 4% paraformaldehyde.

10. Incubate the samples for 3–7 hours (depending on the size of the tissue) at 10–20°C. For example, incubate a 0.5-cm-diameter head for 3–4 hours and a 1-cm-diameter head for 6–7 hours. Mix the tubes regularly (about once every 30 minutes) for better fixative diffusion.

11. Discard the fixative and rinse the samples twice in 1x PBS.

    To prepare the heads of older embryos (≥8 days of incubation) for histological staining, skip Step 12 and proceed to Step 13.

12. Discard the PBS and immerse the samples in RNA*later*® at ambient temperature for 10–20 minutes before placing them in a cooler for transport to the laboratory. According to the manufacturer, samples can be stored in RNA*later*® for one month at 4°C, one week at 25°C, or indefinitely at –20°C. Avoid rapid changes in temperature.

    These samples can subsequently be used for in situ hybridization as described in Protocol 2. See Troubleshooting.

Sample Dehydration for Histological Staining

13. After fixation (Steps 9–11), dehydrate the heads of older embryos (after 8 days of incubation) into a 90% ethanol solution via a series of washes with 25, 50, 75, and 90% ethanol (by volume). Wash the heads in each solution for 15–20 minutes. Use a ratio of 10:1 for the ethanol solution volume:tissue volume.
14. Store and transport the tissue in 90% ethanol at below 0°C, ideally at –20°C.

TROUBLESHOOTING

Problem (Steps 8 and 12): Tissues of early embryos desiccate and shrink when treated or stored in RNA*later*®.

Solution: Store these embryos in 90% methanol, ideally at a temperature below 0°C.

Protocol 2

In Situ Hybridization Analysis of Embryonic Beak Tissue from Darwin's Finches

The beaks of Darwin's finches develop their distinct shapes during embryogenesis. To visualize where and when mRNA transcripts of genes involved in beak development are present during embryogenesis, this protocol describes how to analyze embryonic tissue samples using in situ hybridization. The principle of this technique is to use probes specific to mRNAs coding for proteins of interest to reveal how the corresponding genes are expressed in the beaks of different species of Darwin's finches. This is a modified version of a protocol originally developed for work with muscle tissue in limbs and trunk in the laboratories of Dr. Andrew Lassar and Dr. Clifford Tabin at Harvard Medical School (Brent et al. 2003). It was modified for use on sections of cranial tissues of songbirds.

MATERIALS

The recipes for items marked with <R> begin on page 498.

CAUTION: See the Cautions Appendix for appropriate handling of materials marked with <!>.

Reagents

Acetic anhydride–TEA solution (freshly prepared) <!> <R>
Antidigoxygenin antibody conjugated to alkaline phosphatase (anti-DIG-AP; diluted 1:2500 in 5% HISS)
   The 5% HISS solution should be prepared with MABT.
BCIP-NBT solution <!> <R>
Finch heads (embryonic, preserved in RNA*later*®; see Protocol 1)
Formamide (50%, prepared in 1x SSC <R> and preheated to 37°C) <!>
Heat-inactivated (30 seconds at 56°C) sheep serum (HISS) (20%, prepared in MABT)
Hybridization solution for in situ hybridization analysis (preheated to 85°C) <R>
MABT <R>
NTMT (pH 9.5) <R>
Optimal cutting temperature (OCT) <!>
Paraformaldehyde (4%, prepared in 1x PBS) <!>
1x PBS
1x PBS (containing 0.1% Tween 20) (PBS-T)
Proteinase K <!>
RNA probe (DIG-labeled; 4–5 µg of probe per 1 µg of template) <!>
   Chicken probes have been used on passerine embryos with success (Abzhanov et al. 2004, 2006) and, except for certain rapidly diverging genes, most chicken probes are expected to work readily on Darwin's finches and other songbirds. In addition to target-specific antisense probes, sense probes should be used to provide a negative control.
2x and 0.2x SSC (preheated to 65°C) <R>
5x SSC <R>
Sucrose (10% and 30%)
TNE solution (preheated to 37°C) <R>

Equipment

Box (dark; see Step 34)
Coverslips (glass)
Cryostat
Dry ice <!>
Freezing chamber (soft plastic)
Humidified chamber (dark; see Steps 27 and 30)
Hybridization oven (preheated to 37°C and 65°C)
Microscope and imaging equipment
Mountant
Plastic chamber with lid and holder for 24 slides
Shaking incubator
Slide boxes
Slides (glass)
Tweezers (#5 and #55; Fine Science Tools)

METHOD

Tissue Sectioning

1. Remove the head samples from RNA*later*® and rehydrate them in 1× PBS for 1–2 hours at room temperature (depending on the size of the sample).

   Avoid repeated freezing and thawing of samples in RNAlater®; this will damage their morphology.

2. Wash the samples in 10% sucrose at room temperature for 4–6 hours with constant shaking and then equilibrate them in 30% sucrose overnight at 4°C.

   A fully equilibrated tissue sample in 30% sucrose should sink to the bottom of the tube.

3. Using tweezers, carefully place each head in a freezing chamber containing OCT. Orient each head to achieve the desired cutting angle; avoid air bubbles.

   The choice of section plane (sagittal, transverse, coronal) depends on the dimension of the bird beak (or other structure) that needs to be studied. See Troubleshooting.

4. Freeze the tissue samples in the OCT on dry ice for 15–20 minutes. Then, store the frozen blocks at –80°C until sectioning.

5. Place the frozen tissue sample blocks in a cryostat machine. Cut the tissue into 5- to 8-nm sections on glass slides.

6. Air-dry the tissue sections on slides for 2–3 hours at room temperature.

7. Store the slides with sectioned tissue in a sealed slide box at –80°C.

Tissue Fixation

Perform Steps 8–17 at room temperature.

8. Thaw and air-dry the slides for 15 minutes in a closed plastic chamber.

9. Postfix the slides in 4% paraformaldehyde for 10 minutes.

10. Wash the slides twice for 5 minutes each in PBS-T.

11. Treat the slides with 1 µg/ml of proteinase K for 5–15 minutes at room temperature.

    The duration of treatment should be optimized depending on the tissue type and age of embryos used for sections.

12. Wash the slides twice for 5 minutes each in PBS-T.
13. Postfix the slides in 4% paraformaldehyde for 10 minutes.
14. Wash the slides twice for 5 minutes each in PBS-T.
15. Treat the slides with acetic anhydride–TEA solution for 15 minutes.
16. Wash the slides twice for 5 minutes each in PBS-T.
17. Air-dry the slides for 10 minutes.

Probe Hybridization

18. Cover the slides with preheated (85°C) hybridization solution containing 1–2% (v/v) DIG-labeled in situ hybridization probe.
19. Cover the slides with coverslips and incubate overnight at 65°C in the hybridization oven.
    Depending on the signal and level of background, the hybridization temperature can vary from 60°C to 70°C.
20. During the next day, remove the coverslips and rinse the tissue sections with 5x SSC at room temperature.
21. Wash the slides in the 50% formamide solution for 30 minutes at 37°C.
22. Wash the slides twice in TNE for 10 minutes each at 37°C.
23. Wash the slides with 2x SSC for 20 minutes at 65°C.
24. Wash the slides with 0.2x SSC for 20 minutes at 65°C.

Signal Detection

25. Wash the slides with MABT for 10 minutes at room temperature.
26. Block the slides in 20% HISS for 2 hours.
27. Apply the diluted anti-DIG-AP to the slides. Incubate the slides overnight in a humidified chamber at 4°C.
    The humidified chamber could be a flat plastic box with wet paper towels on the bottom where the slides lie horizontally on plastic tubing. The slides should not touch the walls of the chamber nor the paper towels.
28. Wash the slides in MABT for 15 minutes at room temperature.
29. Wash the slides in NTMT (pH 9.5) for 10 minutes.
30. Incubate the slides in BCIP-NBT solution for 10 minutes. Develop the slides in a dark humidified chamber and change the solution as required to prevent the slides from drying out.
31. Rinse the slides briefly in NTMT (pH 9.5).
32. Wash the slides in 1x PBS for 15 minutes.
33. Fix the slides in 4% paraformaldehyde for 30 minutes to stop the reaction.
34. Mount and coverslip the slides; dry them in a dark box.
35. Photograph the slides.
    See Troubleshooting.

TROUBLESHOOTING

Problem (Step 3): It is difficult to position the head in OCT to achieve the proper sectioning angle.
Solution: Hold the embryo with a pair of tweezers until the OCT solidifies enough to prevent the head from rolling or shifting. Make sure that the head is oriented correctly relative to the bottom of the chamber.

Problem (Step 35): The tissue is damaged.
Solution: Reduce the duration of proteinase K treatment in Step 11. The optimal duration of the treatment is best determined empirically and depends on the thickness and composition of the embryonic tissue section, the incubation temperature, and the level of tissue fixation.

Protocol 3

Microarray Analysis of Embryonic Beak mRNA from Darwin's Finches

In this protocol, microarray technology is used as a very sensitive and rapid method to identify genes that are potentially involved in beak development and morphology in Darwin's finches. The method allows for the direct comparison between cDNA targets from two different species (each labeled with a different dye). The prevalence of one of the dyes for any of the genes on the resulting scan indicates a higher level of accumulation of transcripts from that gene in a particular beak morphology/species. The obtained expression profiles can be clustered to identify transcripts that are expressed in a species- and/or size-specific manner. We have successfully used this method to compare five species of Darwin's finches (*G. magnirostris*, *G. conirostris*, *G. fortis*, *G. scandens*, and *G. difficilis*) that differed in beak morphology (Abzhanov et al. 2006). The Sharp-beaked Finch (*G. difficilis*), the most basal species, served as the reference.

MATERIALS

The recipes for items marked with <R> begin on page 498.

CAUTION: See the Cautions Appendix for appropriate handling of materials marked with <!>.

Reagents

2xYT medium
Agarose gels (1% and 2%)
BioPrime Array CGH Genomic Labeling System (Invitrogen)
Blocking DNAs (T7-dT primer/oligo[dA])
Blocking solution <R>
Cot-1 DNA
Cy5- and Cy3-dCTP (GE Life Sciences)
DMSO (30%) <!>
dNTPs (20 mM)
Ethanol (75%) <!>
Finch tissue, preserved in RNA*later*® (see Protocol 1)
Hybridization buffer for microarray analysis <R>
M13 forward and reverse primers (20 µM each)
Mung bean nuclease (Promega)
pBluescript II XR cDNA Library Construction Kit (Stratagene)
PBS (1x; sterile)
PCR buffer (10x) with $MgCl_2$ (Roche)
Proteinase K (1 mg/ml) <!>
QIAprep 96 Turbo Miniprep Kit (QIAGEN)
QIAquick PCR Purification Kit (QIAGEN)
SMART PCR cDNA Synthesis Kit (Clontech)
0.2x SSC <R> (with and without 0.1% SDS <!>)
Taq DNA polymerase (Roche)
TRIzol <!>
   A column-based RNA extraction kit (e.g., RNeasy, QIAGEN) can be used instead of TRIzol.
UltraPure Phenol:Chloroform:Isoamyl Alcohol (25:24:1; Invitrogen) <!>

Equipment

96-deep-well plates
96-pin disposable replicators
384-well microarray plates with V-shaped bottom (Genetix)
Agilent 2100 BioAnalyzer System (Agilent Technologies)
Amicon-30 spin columns (GE Life Sciences)
BioRobotics MicroGrid II TAS with 16 MicroSpot 10K quill pin (Genomic Solutions)
Centrifuge (benchtop)
Disposable plastic replicators (Genetix)
Gel box and power sources for electrophoresis
GenePix Scanner with software (Molecular Devices)
GS Gene Linker UV Chamber (Bio-Rad) <!>
Heat block (preset to 95°C)
Humidity chamber

> To construct the humidity chamber, place moist paper towels on the bottom of 12-inch square plastic culture dishes. Tape two 10-ml serological pipettes into each culture dish. (The pipettes support the slides.) Finally, wrap the chambers in aluminum foil.

Hybridization chamber (preset to 42°C; Corning)
Ice
MATLAB (The MathWorks, Inc.)
PCR plates (96 well)
Shaking incubator (preset to 37°C)
Spectrophotometer for reading 96-well plates (e.g., SpectraMax Plus, Molecular Devices)
Thermal cycler
UltraGAPS slides (Corning)

> The choice of slides for printing the microarrays is very important. We have tested a wide variety of platforms, and the best results were obtained using UltraGAPS slides.

Water baths, incubators, and/or high-temperature ovens (set at 65°C and 100°C; see Step 28)

METHOD

Preparation of a cDNA Microarray

1. Synthesize a cDNA library from RNA*later*®-preserved tissue of the reference species using the pBluescript II XR cDNA Library Construction Kit. Follow manufacturer instructions.

2. Size-fractionate the cDNA inserts to about 1 kb in length. Use the manufacturer's protocol from the pBluescript II XR cDNA Library Construction Kit.

   The cDNAs spotted onto the microarray must be about 1 kb in length to ensure that they can be used for comparisons across multiple related species.

3. Using disposable plastic replicators, inoculate 96-deep-well plates with *E. coli* containing cDNA inserts. Each well should contain 1.2 ml of 2xYT medium.

4. Grow cultures for 20 hours at 37°C under constant shaking at 200 rpm and then pellet the cells via centrifugation (100–150 ml for 10 minutes at room temperature). Purify the plasmids using the QIAprep 96 Turbo Miniprep Kit. Adjust the concentration of plasmid DNA to 100 µg/ml.

5. Using 96-pin disposable replicators, transfer 10 µg of each plasmid template to a PCR plate containing 50 µl of reaction mix in each well, for a total volume of 60 µl per reaction. The reaction mix (per well) is as follows:

Reaction	Quantity
dNTPs (20 mM)	1 µl
M13 forward primer (20 µM)	3 µl
M13 reverse primer (20 µM)	3 µl
PCR buffer (10x) with $MgCl_2$	6 µl
Taq DNA polymerase	2 units
H_2O	to 50 µl

6. Place the plate on a thermal cycler and perform 32 cycles of the following amplification profile:

Step	Time	Temperature (°C)
Incubation	45 seconds	95
Annealing	45 seconds	53
Extension	2 minutes 45 seconds	72

7. Purify the polymerase chain reaction (PCR) products using the QIAquick PCR Purification Kit. Elute the products in 100 µl of H_2O.

8. Quality-score and size the PCR products by running them on a 2% agarose gel.
   Products should range in size from 700 to 2000 bp.

9. Measure the total PCR yield for each well on a spectrophotometer. For wells containing less than 1.5 µg of product, reamplify and combine until the total mass exceeds this value.

10. Dry the plate for 10 minutes at room temperature. Resuspend each product in 50 µl of H_2O. Transfer the products to a 384-well V-bottom spotting plate.

11. Dry the spotting plate for 5 minutes at room temperature. Resuspend each product in 7.5 µl of 30% DMSO.

12. Use a BioRobotics MicroGrid II TAS with 16 MicroSpot 10K quill pins to array the products at a pitch of 260 µm onto UltraGAPS slides under a relative humidity of 45% ± 2%.

13. Rehydrate each array in a humidity chamber for 30 seconds.

14. Denature the arrays on a heat block for 30 seconds at 95°C.

15. Crosslink at 125 mJ for 25 seconds in a GS Gene Linker UV Chamber.

16. Block the slides in blocking solution for 1 hour at room temperature.

Preparation of Target

17. Rehydrate the RNA*later*®-preserved tissue samples in sterile 1x PBS for 1 to 2 hours (depending on the size of the sample).

18. Place 1–2 µg of tissue in 800 µl of TRIzol.
    The samples can be stored in TRIzol at –70°C until the RNA extraction procedure is performed.

19. Extract total RNA from the tissue samples using TRIzol. Follow manufacturer instructions.

20. For quality control, run each sample of total RNA through a 1% agarose gel and use an Agilent 2100 BioAnalyzer to check for possible contamination and degradation.

21. Normalize all of the RNA samples to 1 µg/µl. Aliquot 10 µg for each experiment.

22. Use each RNA sample to produce cDNA targets using reverse transcription–PCR (RT-PCR). Reverse-transcribe the RNA samples and amplify the resulting cDNA for 10–16 cycles using the SMART PCR cDNA Synthesis Kit.
    The number of cycles necessary to generate the required amount of cDNA must be determined empirically. The quantity of the amplification product should be sufficient for five to ten target-labeling reactions/hybridizations.

23. Following PCR amplification, remove unincorporated primers by adding 1 µl of mung bean nuclease and incubate for 10 minutes at room temperature.

24. Treat the cDNA with 1 mg/ml proteinase K for 10 minutes (use 300 µl for each slide or 20 ml for a 24-slide chamber) and then extract the cDNA with UltraPure Phenol:Chloroform: Isoamyl Alcohol (25:24:1) using the manufacturer's protocol.

25. Concentrate the purified cDNA with Amicon-30 spin columns.

26. Label 10 µg of cDNA by incorporating either Cy5- or Cy3-dCTP during random-hexamer-primed primer extension in the presence of Klenow DNA polymerase using the BioPrime Array CGH Genomic Labeling System.

Hybridization

27. For each hybridization, mix the labeled target pairs together in a single tube and add 10 µg of Cot-1 DNA and 20 µg T7-dT primer/oligo(dA) as blocking DNAs. Precipitate the mixture.

28. Wash the hybridization mix in 75% ethanol, air-dry the pellet for 10 minutes, and resuspend the target in 10 µl of hybridization buffer. To ensure that the targets are completely in solution, incubate the mix for 15 minutes at 65°C, heat-denature for 5 minutes at 100°C, and snap-cool on ice.

29. Apply the target mix from Step 28 directly to the preblocked array (from Step 16) and place the array in a hybridization chamber overnight at 42°C.

30. During the next day, wash and dry the array at room temperature as follows:

    i. Wash in 0.2x SSC/0.1% SDS for 2 minutes.

    ii. Wash twice in 0.2x SSC for 2 minutes each.

    iii. Dry the array in a benchtop centrifuge at low speed.
    Store the array in the dark at room temperature until scanned.

Microarray Analysis

31. Scan the array using the software on a GenePix scanner. Use the software to generate raw GPR (GenePix Results) files.
    See Troubleshooting.

32. Study the expression of candidate molecules by cluster analysis. Use MATLAB to compute and visualize the clustering analysis. Agglomerative hierarchical clustering should be performed using the Euclidean distance measure: The average linkage and ward heuristics should be used to connect the gene clusters. For *k*-means clustering, use the *k*-means algorithm to partition the genes into *k* discrete clusters based on their expression.

TROUBLESHOOTING

Problem (Step 31): Hybridization signal is variable both within and between slides.
Solution: Make sure that the hybridization solution is evenly spread over the entire slide. In addition, ensure that the hybridization chambers are properly assembled to allow the uniform spread of the targets via the capillary effect.

Problem (Step 31): Scans are inconsistent and of poor quality.
Solution: Make sure that slides are washed sufficiently in Step 30.

Recipes

CAUTION: See the Cautions Appendix for appropriate handling of materials marked with <!>.

ACETIC ANHYDRIDE–TEA SOLUTION

Reagent	Quantity (for 875 ml)	Final concentration
Acetic anhydride <!>	625 ml	71% (v/v)
Triethanolamine (TEA; 0.1 M) <!>	250 ml	29 mM

BCIP-NBT SOLUTION

Reagent	Quantity (for 20 ml)	Final concentration
5-Bromo-4-chloro-3-indolyl-phosphate (BCIP; 50 mg/ml) <!>	70 µl	175 µg/ml
4-Nitro blue tetrazolium chloride (NBT; 100 mg/ml) <!>	67.5 µl	338 µg/ml
NTMT (pH 9.5)	to 20 ml	

BLOCKING SOLUTION

Reagent	Quantity (for 10 ml)	Final concentration
BSA (10%)	100 µl	0.1%
Formamide <!>	5 ml	50%
SDS (10%) <!>	100 µl	0.1%
20x SSC	2.5 ml	5x
H$_2$O	to 10 ml	

HYBRIDIZATION BUFFER FOR MICROARRAY ANALYSIS

Reagent	Quantity (for 10 ml)	Final concentration
50x Denhardt's reagent	1 ml	5x
Formamide, deionized <!>	5 ml	50%
Potassium phosphate (0.1 M)	0.5 ml	5 mM
SDS (10%) <!>	0.5 ml	0.5%
20x SSC	2.5 ml	5x
H$_2$O	0.5 ml	

HYBRIDIZATION SOLUTION FOR IN SITU HYBRIDIZATION ANALYSIS

Reagent	Quantity (for 100 ml)	Final concentration
50x Denhardt's reagent	2 ml	1x
Dextran sulfate	10 g	10%
EDTA (0.5 M) <!>	0.2 ml	1 mM
Formamide <!>	50 ml	50%
NaCl (5 M)	12 ml	600 mM
SDS (20%) <!>	1.25 ml	0.25%
Tris-Cl (1 M, pH 7.5)	1 ml	10 mM
tRNA from yeast (Gibco)	20 mg	200 µg/ml
H$_2$O	to 100 ml	

Store at –20°C.

MABT

Reagent	Quantity (for 100 ml)	Final concentration
Maleic acid <!>	1.2 g	100 mM
NaCl (5 M)	3 ml	150 mM
Tween 20	100 µl	0.1%
H$_2$O	to 100 ml	

Adjust the pH to 7.5 with NaOH <!>.

NTMT (pH 9.5)

Reagent	Quantity (for 100 ml)	Final concentration
MgCl$_2$ (1 M) <!>	5 ml	50 mM
NaCl (5 M)	2 ml	100 mM
Tris-Cl (1 M, pH 9.5)	10 ml	100 mM
Tween 20	100 µl	0.1%
H$_2$O	to 100 ml	

20x SSC

Reagent	Quantity (for 1 liter)	Final concentration
NaCl	175.3 g	3.0 M
Sodium citrate <!>	88.2 g	0.3 M

Dissolve the reagents in 800 ml of H$_2$O. Adjust the pH to 7.0 with a few drops of 14 N HCl <!>. Adjust the volume to 1 liter with H$_2$O. Sterilize by autoclaving. Prepare working concentrations (5, 2, 1, and 0.2x) by diluting the 20x stock solution with H$_2$O. Store at room temperature.

TNE SOLUTION

Reagent	Quantity (for 100 ml)	Final concentration
EDTA (0.5 M) <!>	0.2 ml	1 mM
NaCl (5 M)	10 ml	500 mM
Tris-Cl (1 M, pH 7.5)	1 ml	10 mM
H$_2$O	to 100 ml	

REFERENCES

Abzhanov, A., Protas, M., Grant, B.R., Grant, P.R., and Tabin, C.J. 2004. Bmp4 and morphological variation of beaks in Darwin's finches. *Science* **305**: 1462–1465.

Abzhanov, A., Kuo, W.P., Hartmann, C., Grant, B.R., Grant, P.R., and Tabin, C.J. 2006. The calmodulin pathway and the evolution of elongated beak morphology in Darwin's finches. *Nature* **442**: 563–567.

Bowman, R.I. 1961. Morphological differentiation and adaptation in the Galápagos finches. *Univ. Calif. Publ. Zool.* **58**: 1.

Brent, A.E., Schweitzer, R., and Tabin, C.J. 2003. A somitic compartment of tendon progenitors. *Cell* **113**: 235–248.

Burns, K.J. and Skutch, A.F. 2003. Tanagers and tanager-finches. In *The firefly encyclopedia of birds* (ed. C Perrins), pp. 629–663. Firefly Books, Buffalo, New York.

Burns, K.J., Hackett, S.J., and Klein, N.K. 2002. Phylogenetic relationships and morphological diversity in Darwin's finches and their relatives. *Evolution* **56**: 1240–1256.

Darwin C. 1839. *The voyage of the beagle* (1962 edition). Natural History Library, Anchor Press, Norwell, Massachusetts.

Francis-West, P.H., Tatla, T., and Brickell, P.M. 1994. Expression patterns of the bone morphogenetic protein genes *Bmp-4* and *Bmp-2* in the developing chick face suggest a role in outgrowth of the primordia. *Dev. Dyn.* **201**: 168–178.

Francis-West, P., Ladher, R., Barlow, A., and Graveson, A. 1998. Signalling interactions during facial development. *Mech. Dev.* **75**: 3–28.

Francis-West, P.H., Robson, L., and Evans, D.J. 2003. Craniofacial development: The tissue and molecular interactions that control development of the head. *Adv. Anat. Embryol. Cell Biol.* **169**: 1–138.

Futuyma, D. 2005. *Evolution*. Sinauer, Sunderland, Massachusetts.

Grant, P.R. 1986. *Ecology and evolution of Darwin's finches*. Princeton University Press, Princeton, New Jersey.

Grant, P.R. 1999. *Ecology and evolution of Darwin's finches*. Princeton University Press, Princeton, New Jersey.

Grant, P.R. and Grant, B.R. 2002. Adaptive radiation of Darwin's finches. *Am. Sci.* **90**: 130–139.

Grant, B.R. and Grant, P.R. 2003. What Darwin's finches can teach us about the evolutionary origins and regulation of biodiversity. *BioScience* **53**: 965–975.

Grant, P.R. and Grant, B.R. 2008. *How and why species multiply. The radiation of Darwin's finches*. Princeton University Press, Princeton, New Jersey and Oxford, United Kingdom.

Grant, P.R., Grant, B.R., and Abzhanov, A. 2006. A developing paradigm for the development of bird beaks. *Biol. J. Linn. Soc.* **88**: 17–22.

Hamburger, V. and Hamilton, H.L. 1951. A series of normal stages in the development of the chick embryo. *J. Morphol.* **88**: 49–92.

Hanken, J. and Hall, B.K. 1993. *The skull: Patterns in structural and systematic diversity*. The University of Chicago Press, Chicago, Illinois.

Helms, J.A., Cordero, D., and Tapadia, M.D. 2005. New insights into craniofacial morphogenesis. *Development* **132**: 851–861.

Herrel, A., Podos, J., Huber, S.K., and Hendry, A.P. 2005. Evolution of bite force in Darwin's finches: A key role for head width. *J. Evol. Biol.* **18**: 669–675.

Huber, S.K., De León, L.F., Hendry, A.P., Bermingham, E., and Podos, J. 2007. Reproductive isolation of sympatric morphs in a population of Darwin's finches. *Proc. Biol. Sci.* **274**: 1709–1714.

Lack, D. 1940. Evolution of the Galápagos finches. *Nature* **146**: 324–327.

Lack, D. 1945. The Galápagos finches (Geospizinae): A study in variation. *Occas. Pap. Calif. Acad. Sci.* **21**: 1–152.

Landsborough Thomson, A., ed. 1964. *A new dictionary of birds*. McGraw-Hill, New York.

Monroe, B.L. and Sibley, C.G. 1993. *A world checklist of birds*. Yale University Press, New Haven, Connecticut.

Petren, K., Grant, B.R., and Grant, P.R. 2000. A phylogeny of Darwin's finches based on microsatellite DNA length variation. *Proc. Roy. Soc. Lond. Biol. Sci.* **266**: 321–329.

Petren, K., Grant, P.R., Grant, B.R., and Keller, L.F. 2005. Comparative landscape genetics and the adaptive radiation of Darwin's finches: The role of peripheral isolation. *Mol. Ecol.* **14**: 2943–2957.

Podos, J. 2001. Correlated evolution of morphology and vocal signal structure in Darwin's finches. *Nature* **409**: 185–188.

Price, T.D. and Grant, P.R. 1985. The evolution of ontogeny in Darwin's finches: A quantitative genetics approach. *Am. Nat.* **125**: 169–188.

Sato, A., O'hUigin, C., Figueroa, F., Grant, P.R., Grant, B.R., Tichy, H., and Klein, J. 1999. Phylogeny of Darwin's finches as revealed by mtDNA sequences. *Proc. Natl. Acad. Sci.* **96**: 5101–5106.

Sato, A., O'hUigin, C., Tichy, H., Grant, P.R., Grant, B.R., and Klein, J. 2001. On the origin of Darwin's finches. *Mol. Biol. Evol.* **18**: 299–311.

Schneider, R.A., Hu, D., Rubenstein, J.L., Maden, M., and Helms, J.A. 2001. Local retinoid signaling coordinates forebrain and facial morphogenesis by maintaining FGF8 and SHH. *Development* **128**: 2755–2767.

Sibley, C.G. and Ahlquist, J.E. 1990. *Phylogeny and classification of birds*. Yale University Press, New Haven, Connecticut.

Sulloway, F.J. 1982. The *Beagle* collections of Darwin's finches (Geospizinae). *Bull. Br. Mus. (Nat. Hist.) Hist. Ser.* **43**: 49–94.

Weiner, J. 1995. *The beak of the finch: A story of evolution in our time*. Vintage Books, Random House, New York.

21 Japanese Quail
An Efficient Animal Model for the Production of Transgenic Avians

Greg Poynter, David Huss, and Rusty Lansford

California Institute of Technology, Division of Biology and the Biological Imaging Center, Beckman Institute, Pasadena, California 91125

ABSTRACT

The ability to generate transgenic mice has been a powerful tool in studying functional genomics, and much of our knowledge about developmental biology has come from the study of chicken embryology. Unfortunately, the availability of molecular genetic techniques, such as transgenics and knockouts, has been limited for developmental biologists using avian animal models. Efforts to develop a system for the rapid production of transgenic chickens have met with many obstacles including high animal husbandry costs and long generational times. Recently, the Japanese quail has proven to be an excellent model organism for the production of transgenic avians using lentiviral vectors. The relatively small size of the adults, short time to sexual maturity, and prodigious egg production of the Japanese quail make development of transgenic lines less labor- and space-intensive compared to chickens. The high degree of homology between chicken and quail genomes allows researchers to design highly specific DNA constructs for the production of transgenic birds. In addition, transgenic quail offer all the advantages of the classic avian developmental model system, such as the ability to readily produce quail:chick transplant chimeras. Finally, Japanese quail are ideal for in ovo imaging of embryos expressing fluorescent reporters introduced from a transgene and/or electroporation. Here, we provide detailed methods for generating transgenic quail using high-titer lentivirus.

PROTOCOLS

1. Generation of High-titer Lentivirus for the Production of Transgenic Quail, 506
2. Injection of Lentivirus into Stage-X Blastoderm for the Production of Transgenic Quail, 510
3. Screening for Transgenic Japanese Quail Offspring, 514

BACKGROUND INFORMATION

The Japanese or Coturnix quail (*Coturnix coturnix japonica*) is a precocial ground-dwelling bird that belongs to the family Phasianidae, order Galliformes, and class Aves. Adults weigh between 150 and 250 g, and the sexual dimorphism in adult breast feather coloration makes identifying cocks and hens very easy (Fig. 1A,B). Eggs display a brown mottled appearance and typically weigh about 13 g (Fig. 1C). *C. coturnix japonica* is a domesticated subspecies of the common quail (*C. coturnix*) whose migratory path includes Europe, Africa, India, and Asia. The common quail was originally domesticated in China during the 11th century and arrived in Japan around the 12th

This chapter, with full-color images, can be found online at www.cshprotocols.org/emo.

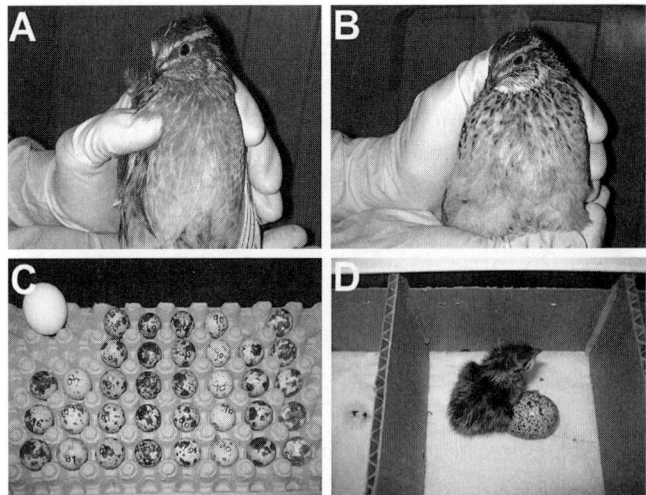

FIGURE 1. (*A*) Adult male Japanese quail. The breast feathers are cinnamon brown (color not shown). (*B*) Adult female Japanese quail. The breast feathers are speckled brown (color not shown). (*C*) Quail eggs, which are about one-fifth the size of chicken eggs. Japanese quail eggs are highly pigmented. Each column of quail eggs is from a different hen and displays distinct patterning. (*D*) Japanese quail hatchling with its empty eggshell 1 day after hatching. The hatching trays can be subdivided to isolate individual eggs and hatchlings for identification purposes.

century where it was initially raised for its song. By the later part of the 19th century, the Japanese quail was being raised and bred for its colorful plumage and as an important food source. By the mid-20th century, Japanese quail were introduced to surrounding countries including Korea, China, and various regions of Southeast Asia. All Japanese quail currently being used for research are descended from lines rederived in Japan after World War II (Roberts 1999).

The Japanese quail has been used as an animal model in laboratories all over the world for many years. It was first described as a research model in the United States in 1959 when Padgett and Ivey (1959) noted its practicality as a laboratory animal for avian developmental studies. In addition, they prepared a detailed developmental atlas (Padgett and Ivey 1960) based on the chicken staging system of Hamburger and Hamilton (1951). This atlas, along with that of Zacchei (1961), continues to serve as the gold standard for staging quail development. An exhaustive study of the anatomy and histology of the Japanese quail was published by Fitzgerald (1969), and many of the metrics associated with maintaining a laboratory breeding colony of Japanese quail were published by Woodard et al. (1973).

SOURCES AND HUSBANDRY

Fertilized quail eggs can be obtained from a number of sources. Our lab uses eggs from our aviary at the California Institute of Technology or from a local commercial vendor. Most researchers will be able to find a local producer of fertilized eggs because Japanese quail eggs are considered delicacies in many cuisines. Local game bird producers of pheasants, grouse, and bobwhite quail may be able to accommodate an additional colony of Japanese quail specifically for research purposes. Currently, colonies of Japanese quail with long historical breeding records are being managed at Auburn University, the University of British Columbia, and the University of Wisconsin, Madison. Quail lines that produce either wild-type or white feather color combined with wild-type or white eggshell pigmentation are available from Dr. Bernie Wentworth at the University of Wisconsin, Madison. Japanese quail with single-gene mutations, strains that have been selected for atherosclerosis and body weight, and random-bred wild-type populations are available from the Avian Genetic Resource Laboratory at the University of British Columbia.

Generally, the same conditions and equipment that are used for chicken egg incubation and hatching can be used with quail. Incubators, brooders, egg racks, and other equipment for raising quail are available from commercial poultry supply houses such as G.Q.F Manufacturing, Inc. (Savannah, Georgia). When incubated at 37.5°C with a relative humidity of 60%, Japanese quail hatch at about the 16th day of incubation (Fig. 1D). Hatchlings must be brooded with a supplemental heat source until their 4th week, and at 9–10 weeks of age, they reach sexual maturity. Single males can be bred to multiple females and can typically be housed in modified rodent cages. Egg production peaks at about five eggs per week per female; this continues until about 1.5 years of age before declining.

RELATED SPECIES

The blue-breasted quail (*Coturnix chinensis*), or button quail, is a close relative of the Japanese quail. Animals of this species grow to an adult body size of only about 50 g (compared to 150–250 g for the Japanese quail). This cost-saving trait makes the blue-breasted quail another potential animal model for generating transgenic avians (Ono et al. 2005).

USES OF THE JAPANESE QUAIL MODEL SYSTEM

A literature search quickly reveals that quail are being used in a wide array of laboratories. For example, in the field of neuroscience, the quail is useful in studies of animal learning and behavior using operant conditioning (Burns and Domjan 2001). The 16-day developmental period of quail makes it an excellent model for studying the effect of microgravity on embryonic development during short-duration spaceflights (Barrett et al. 2000). The ability of birds, but not mammals, to regenerate the mechanosensory hair cells in their auditory and vestibular sensory organs after damage has made quail a good model for many scientists in the field of otolaryngology (Ryals and Rubel 1988). These examples and those below are just a small number of ways in which Japanese quail are being used in the modern biomedical research laboratory.

Developmental Biology

The avian embryo, unlike that of a rodent, is easily accessible during its normal development. This allows the embryo to be manipulated and visualized as it grows and therefore makes avians highly valuable in the study of developmental biology. Our understanding of myogenesis, vasculogenesis, angiogenesis, skeletogenesis, virology, and teratology has progressed significantly as a result of studies on avian embryos (Mizutani 2002). The ability of scientists to culture both chicken and quail embryos ex ovo has greatly increased their usefulness as animal models (Perry 1988; Chapman et al. 2001; Ono et al. 2005).

The avian embryo is a warm-blooded vertebrate with early developmental patterns remarkably similar to those of humans. The developmental embryology of the quail closely mirrors that of the chicken. However, the condensed heterochromatin inside the quail cell nucleus clearly differentiates it from chicken cells. Developmental biologists have long used this difference to transplant tissue from one of these species into the other. The quail:chick chimera has proven to be a highly successful system for elucidating cell fates during development (Le Dourain and Kalcheim 1999; Lalloue et al. 2005). The ability to directly visualize the avian embryo during development in ovo has led to the use of computer-analyzed time-lapse video microscopy (Kulesa and Bronner-Fraser 2000). In addition, the small size of the quail egg has allowed researchers to produce an electronic developmental atlas using magnetic resonance imaging (MRI) technology (Ruffins et al. 2007).

Human Health and Disease

Japanese quail are used to study many diseases that affect human health. They have relatively long life spans and a physiology that is similar to that of humans, making them a useful model to examine age-related disease. Quail have been used in studies addressing senescence in immunology, endocrinology, and reproductive biology (Ottinger et al. 2004). Numerous Japanese quail strains have been established and derived from commercial birds in Japan. Some of these strains are specific for plumage color, eggshell color, blood type, and growth rate; other strains are models for human hereditary diseases, malformations, and abnormalities. For example, quail lines have been developed with myotonic dystrophy and acid maltase deficiency, also known as Pompe's disease (Kikuchi and Yang 1998; Mizutani 2002). The mechanisms underlying the development of vascular lesions and hypercholesterolemia have been studied using two separate strains of Japanese quail that are susceptible or resistant to atherosclerosis (Yuan et al. 1997).

GENETIC AND GENOMIC RESOURCES

The karyotype of the Japanese quail is similar to that of the chicken ($2n = 78$). It has 39 chromosomes: 10 distinct macrochromosomes (chromosomes 1–8 and sex chromosomes ZW) and 29 cytogenetically indistinguishable microchromosomes (Kikuchi et al. 2005). Until recently, Japanese quail genetics has been limited to studies based on blood protein markers and feather color (Ito et al. 1988a,b; Shibata and Abe 1996; Minvielle et al. 2000). In one of the first attempts to compare DNA homology between Japanese quail and chickens, comparative genomic analysis using cytogenetic banding revealed that chromosome homology between the two divergent species was highly conserved and showed very few chromosomal rearrangements (Shibusawa et al. 2001). Upon completion of the *Gallus gallus* (red jungle fowl) genome in 2004, it was shown that the Japanese quail and chicken share 99% sequence homology (Hillier et al. 2004; Wallis et al. 2004).

The completion of the chicken genome also enabled researchers to use various mapping techniques to begin direct comparisons of the chicken and Japanese quail genomes. A DNA-based genomic map using amplified-fragment-length polymorphisms (AFLPs) (Roussot et al. 2003) and a microsatellite-based linkage map (Kayang et al. 2004) were combined and analyzed with a new set of microsatellite markers and fluorescence in situ hybridization (FISH) (Kayang et al. 2006). This study found a high degree of homology between the chick and Japanese quail for the 8 macrochromosomes and 14 microchromosomes analyzed. Subsequent comparisons of orthologous genes on macrochromosomes in Japanese quail and chickens once again confirmed the high level of sequence homology between the two species (Sasazaki et al. 2006). Current work has focused on using the data generated from the linkage maps to begin identifying quantitative trait loci (QTL) in the Japanese quail. Chromosomal regions coding for traits such as growth rate, feed consumption, egg production, body temperature, fear response, and egg shape have been identified to date (Beaumont et al. 2005; Minvielle et al. 2005, 2006).

Until the genome of the Japanese quail is fully sequenced, chicken bacterial artificial chromosome (BAC) libraries appear to be a suitable alternative, given the high degree of sequence homology between the chicken and Japanese quail. The CHORI-261 Chicken BAC Library (Jungle Fowl UCD001, Inbred 256 [female]) was constructed by Michael Nefedov in Pieter de Jong's laboratory at BACPAC Resources (Children's Hospital Oakland Research Institute, Oakland, California). Clones from this library are available at a nominal cost to researchers and can be requested online at http://bacpac.chori.org/ordering_information.htm. DNA from the same bird was used in the chicken genome sequencing project at Washington University (St. Louis, Missouri).

TECHNICAL APPROACHES

We and other investigators have used lentivirus to create tissue-specific transgenic quail using promoters derived from human, rat, and mouse sequences (Scott and Lois 2005; Poynter and Lansford 2008). This procedure is described in full in Protocols 1–3. Care should be taken when selecting cross-species sequences to ensure proper spatial and temporal expression patterns in the quail. Characterization using immunohistochemistry or in situ hybridization should be of the utmost priority after the generation of all transgenic animals, especially those generated using nonavian DNA. If the lentiviral construct contains a fluorescent reporter, it may be possible to screen the hatchlings using whole-body fluorescent imaging. This type of screening method is quick and noninvasive. Ubiquitously expressed fluorescent proteins driven by a strong promoter, such as phosphoglycerol kinase (PGK), lend themselves well to this method of screening. For tissue-specific reporters, the exact localization and brightness (a function of both cellular abundance and promoter activity) will determine if this method can be used effectively. For example, a neuron-specific transgenic quail line expressing green fluorescent protein (GFP) can be screened by quickly illuminating the hatchlings under a dissecting microscope fitted with a UV lamp and GFP filter set. When the hair is removed, the emitted light from the GFP-labeled brain of positive hatchlings is easily bright enough to pass through the bone and skin (Fig. 2A,B). Likewise, a vascular-specific promoter may allow the hatchlings to be screened under a fluorescent stereomicroscope by observing the chorioallantoic membrane (CAM) blood vessels left inside the empty eggshell after hatching (Fig. 2C). Other areas with a high degree of vascularization, such as the shafts of blood feathers in juvenile birds, can be screened phenotypically by fluorescence (Fig. 2D). Hatchlings initially deemed positive by fluorescence screening can then be confirmed using more traditional methods such as polymerase chain reaction (PCR) (see Protocol 3) or Southern blotting.

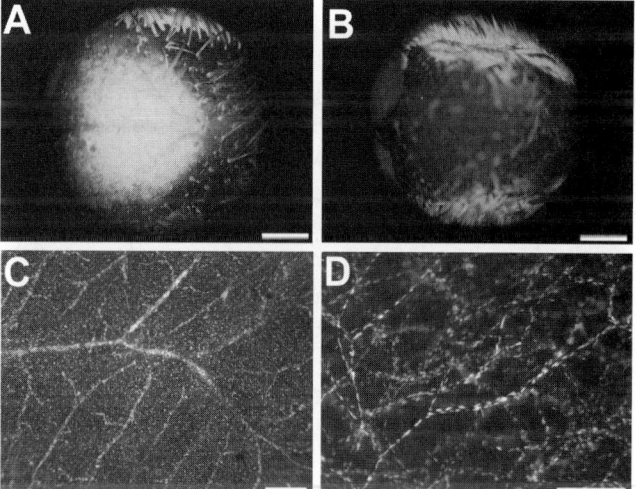

FIGURE 2. Phenotypic screening of quail hatchlings for a fluorescently tagged transgene. (*A*) Coronal view of the head of an alert 2-day-old quail hatchling that is positive for a neural-specific enhanced GFP transgene. The brightly labeled brain easily shines through the bone. Most of the autofluorescent feathers have been removed from the skin. Bar, 2 mm. (*B*) A wild-type 2-day-old hatchling, oriented as in *A*. The absence of strong brain fluorescence indicates that this bird lacks the transgene. (*C*) The chorioallantoic membrane from the inside of an eggshell of a hatchling that is positive for an endothelial-specific enhanced YFP (yellow fluorescent protein) transgene. The fluorescent protein was localized to the nucleus. The blood vessels in this highly vascularized tissue are brightly labeled. Bar, 50 μm. (*D*) Tissue from the base of a blood feather pulled from the breast of a juvenile quail that is positive for an endothelial-specific, nuclear-localized enhanced YFP transgene. Bar, 200 μm.

Protocol 1

Generation of High-titer Lentivirus for the Production of Transgenic Quail

This protocol describes how to generate high-titer lentivirus for the production of transgenic Japanese quail. The virus is pseudotyped with vesicular stomatitis virus with the envelope G glycoprotein (VSV-g), which gives a broad infectious range and allows concentration of viral supernatants by ultracentrifugation. Using this method, we typically produce titers greater than 1×10^8 TU/ml and recommend using a virus with a titer of at least this high for in vivo work.

MATERIALS

The recipes for items marked with <R> are on page 517.
CAUTION: See the Cautions Appendix for appropriate handling of materials marked with <!>.

Reagents

293FT cell line (Invitrogen)
Bleach <!>
Cells and medium for titering the lentivirus (see Steps 24–30)
Complete medium, with and without G418 <R> <!>
Dulbecco's modified Eagle's medium (DMEM), without serum or supplements
Ethanol (100%) <!>
Fetal bovine serum (FBS)
Gelatin (0.1% w/v), prepared in H_2O
1x Hank's balanced salt solution (HBSS)
Lentivirus expression construct
  Use a pLenti backbone (Invitrogen). The pLenti plasmid DNA must be of very high quality (i.e., prepared using QIAGEN Maxi Prep or similar) to ensure high transfection efficiency.
Lipofectamine 2000 Transfection Reagent (Invitrogen)
Opti-MEM I Reduced-serum Medium (Invitrogen)
  This medium does not contain serum.
Trypsin-EDTA solution (Mediatech) <!>
ViraPower Lentiviral Packaging Mix (Invitrogen)

Equipment

Aspirator
Centricon Plus-70 Centrifugal Filter Device, with Ultracel-30 membrane (30-kD MW cutoff; Millipore)
Centrifuge (tabletop or similar)
Centrifuge tubes (15 ml, conical)
Fluorescence microscope
Hemocytometer
Ice
Ice-water bath
Incubator, preset to 37°C and 5% CO_2

Microcentrifuge tubes (0.65 ml)
Syringe (30 ml) with 0.45-µm filter
Tissue culture dishes (6 well)
Tissue culture dishes (10 cm)
Tissue culture hood
Ultracentrifuge (e.g., Beckman TL-100 with TLS-55 swinging bucket rotor)
Ultracentrifuge tubes (e.g., 2.2-ml Beckman Ultraclear polycarbonate centrifuge tubes)

METHOD

When working with lentiviral vectors, which are replication-defective yet highly infectious human immunodeficiency virus (HIV), gloves should be worn at all times, and all virus work should be performed in a tissue culture hood. All materials exposed to lentivirus should be soaked or diluted with bleach for 30 minutes before disposing of them as biohazardous waste.

Preparing Cells for Transfection

1. Precoat each 10-cm tissue culture dish with gelatin as follows:
    i. Add 4 ml of 0.1% gelatin to the dish and allow it to sit for several minutes.
    ii. Remove the gelatin by aspirating.
    iii. Open the lid slightly and allow the plate to dry in the tissue culture hood for 10–15 minutes.
2. Using trypsin-EDTA, harvest a confluent plate of 293FT cells. Count the cells using a hemocytometer.
3. Plate 5×10^6 293FT cells per 10-cm gelatin-coated dish in 8 ml of complete medium. Place the dish in an incubator overnight at 37°C and 5% CO_2 to allow the cells to firmly adhere to the surface.

Transfection

4. In a sterile 15-ml conical tube, combine 1.5 ml of Opti-MEM I Reduced-serum Medium and 36 µl of Lipofectamine 2000 Transfection Reagent. Mix gently and incubate for 5 minutes at room temperature.
5. In a second sterile 15-ml conical tube, add 1.5 ml of Opti-MEM I Reduced-serum Medium, 9 µg of ViraPower Lentiviral Packaging Mix, and 3 µg of the lentivirus expression construct. Mix gently.
6. Combine the tubes from Steps 4 and 5 and mix gently. Incubate the mixture for 20 minutes at room temperature.
    This mixture now contains DNA:Lipofectamine 2000 complexes.
7. Mix 0.5 ml of FBS with 4.5 ml of Opti-MEM I Reduced-serum Medium and set aside.
8. After the 20-minute incubation in Step 6, aspirate the medium from the 293FT cells (from Step 3) and replace it with the medium that was prepared in Step 7.
9. Add the mixture from Step 6 to the 293FT cells in a dropwise fashion. Gently swirl the plate to ensure that the cells are completely covered.
10. Incubate the cells for 5–6 hours at 37°C and 5% CO_2.
11. Aspirate the medium and add 8 ml of complete medium without G418. Incubate the cells overnight at 37°C and 5% CO_2.

If the construct contains a fluorescent reporter, such as enhanced GFP, assess the transfection efficiency using a fluorescence microscope. Good transfection efficiencies are reached when more than 90% of the cells are positive by fluorescence. Lower transfection efficiencies will still produce virus, but at a much lower titer. If transfection efficiency is low, see Troubleshooting.

Collecting Viral Supernatants

12. Collect the 8 ml of medium from the 293FT cells (from Step 11) and place it in a 15-ml conical tube. Gently add 8 ml of fresh medium (complete medium without G418) to the cells and continue growing them at 37°C and 5% CO_2.

13. Centrifuge the medium at 400g for 5 minutes at room temperature to pellet any cell debris.

14. Transfer the cleared supernatant to a clean 15-ml conical tube. Place the tube at –80°C.

15. Repeat Steps 12–14 every 24 hours for a total of 3 days. Store all supernatants at –80°C.

Concentrating Lentivirus by Ultracentrifugation

16. Thaw all frozen viral supernatants in an ice-water bath.

17. Filter the supernatants through a 0.45-μm syringe filter to remove any remaining cell debris. Collect the supernatants in a clean tube or bottle and store them on ice.

18. Prewet a Centricon Plus-70 Centrifugal Filter Device by adding 70 ml of sterile 4°C 1x HBSS and centrifuge at 3000g for 5 minutes at 4°C. Discard the flow-through.

19. Add the supernatants to the filter system and centrifuge at 3000g for 30 minutes at 4°C. Discard the flow-through.

20. Invert the top portion of the filter apparatus into the collection cup and centrifuge at 1000g for 2 minutes at 4°C.

    The collection cup now contains concentrated lentivirus.

21. Presterilize an ultracentrifuge tube with 100% ethanol and rinse it twice with DMEM (without serum or supplements). Add the concentrated supernatants from Step 20 to the ultracentrifuge tube and centrifuge it at 50,000g for 2 hours at 4°C.

22. Aspirate the supernatant and add fresh DMEM (without serum or supplements) or 1x HBSS to the viral pellet.

    A good rule of thumb is to resuspend the pellet in approximately 25 μl of DMEM per 10-cm dish. For example, if the supernatants were collected from four 10-cm dishes for 3 days, the pellet should be resuspended in 100 μl of DMEM.

23. Pipette 10-μl aliquots of concentrated virus into 0.65-ml microcentrifuge tubes. Store the aliquots at –80°C.

Titering the Lentivirus

The following steps outline a method for titering a virus expressing off a ubiquitous promoter. If the construct contains a tissue-specific promoter, determine the appropriate cell line in which to visualize the expression of the transgene.

24. On the day before infection, seed either 293FT or NIH-3T3 cells into a 6-well dish at 2×10^4 cells/cm^2 in 1 ml of appropriate medium per well. Use one 6-well dish for each lentiviral prep. Incubate the cells overnight at 37°C and 5% CO_2.

25. Aspirate the medium from the well.

26. Add 1 μl of virus into 1 ml of medium in the first well.

    This will serve as the 1:1000 or 10^{-3} dilution.

27. Prepare four 1:10 serial dilutions of the virus from Step 26 and add them to the next four wells.

    The total dilution series will be 10^{-3} to 10^{-7}. The final well receives no virus.

28. Incubate the cells for 12 hours (or overnight) at 37°C and 5% CO_2 to allow the virus to infect the cells.

29. After 12 hours (or overnight), replace the medium with fresh medium and continue to grow for an additional 36 hours at 37°C and 5% CO_2.

30. At 48 hours postinfection, examine the cells under a fluorescent microscope. Count and record the number of positive cells for each dilution. Determine which dilution gives approximately 10–30 positive cells and calculate the viral titer using the following formula: viral titer (TU/ml) = (number of GFP$^+$ cells)(dilution)(1000). If more than ten positive cells are seen in the 10^{-7} dilution well, repeat the viral titer protocol with higher dilutions.

    We typically produce titers greater than 1×10^8 TU/ml using this method. If virus titer is low, see Troubleshooting.

TROUBLESHOOTING

Problem (Step 11): Transfection efficiency is low.
Solution: Ensure that the 293FT cells are at a low passage number before beginning the transfection.

Problem (Step 30): Low-titer lentivirus stocks are generated.
Solution: Maintain the 293FT cells in complete media containing 500 µg/ml G418 in order to retain the SV40 large T antigen in the cells, but remove this antibiotic before culturing the cells for virus production. In addition, maintaining the adherence of the 293FT cells to the gelatin-coated culture dish is important in producing high-titer stocks.

Protocol 2

Injection of Lentivirus into Stage-X Blastoderm for the Production of Transgenic Quail

This protocol describes how to generate transgenic quail by injecting a lentiviral vector into freshly laid eggs at stage X (Eyal-Giladi and Kochav 1975). The lentivirus infects primordial germ cells originating in the area pellucida.

MATERIALS

CAUTION: See the Cautions Appendix for appropriate handling of materials marked with <!>.

Reagents

Albumin (thin)

Thin albumin can be collected by cracking open a nonexperimental egg into a sterile Petri dish. Use a sterile transfer pipette to pull off the thin albumin. Avoid the thick albumin, which forms a dense layer around the yolk. Repeat with multiple eggs until about 10 ml has been collected. Store the thin albumin in a sterile tube at room temperature for use during the lentiviral injection procedure.

Eggs (freshly laid)

Fertile, freshly laid eggs can be obtained through local commercial vendors or through your own breeding colony. The most important factor to consider when choosing the source of the eggs is how the eggs were handled after laying and during shipping. Eggs can be stored for 3–7 days at 15–17°C with minimal consequences; longer storage or storage under suboptimal conditions tends to decrease hatchability. Eggs should always be stored in egg flats with the large end facing upward.

Ethanol (70% and 95%) <!>
1x Hank's balanced salt solution (HBSS), sterile
Lentivirus stock (concentrated; see Protocol 1) <!>

Equipment

Animal cages for adults

We use modified rodent cages from Lab Products that are 28,000 cm^3 in size. Cage racks specific for poultry are also commercially available.

Aspen shavings (Northeastern Products Corp.)
Brooder cages and heater (Lyon Technologies, Inc.)
Chukar egg trays (G.Q.F. Manufacturing Co.)
Dissecting scissors (curved)
Dissecting stereomicroscope
Dura Pads (Shepherd Specialty Paper)
Egg holder (for Step 4)

Foam pipe insulation that has been carved to cradle a quail egg works well as an egg holder.

Egg incubator (G.Q.F. Manufacturing Co.)

The incubator should be maintained at 37.5°C and 60% relative humidity.

Felt-tipped pens (blue or red)
Forceps (blunt dressing)

Forceps (fine)
Guinea pig huts (red plastic; Bio-Serv)
Hatching incubator with trays (Lyon Technologies, Inc.)
  The hatcher should be maintained at 37°C and 70% relative humidity.
Marbles
Mazuri Exotic Gamebird Breeder (PMI Nutrition International)
Mazuri Exotic Gamebird Starter (PMI Nutrition International)
Microloader pipette tips (Eppendorf)
Micromanipulator and microinjector (Harvard Apparatus PLI-90)
Needle puller (Sutter Instrument Company)
Paraffin wax (molten)
Pasteur pipettes (glass)
Plastic leg bands (sizes 2 and 4), numbered
Quartz glass needles (1 mm O.D., 0.7 mm I.D.) (Sutter Instrument Company)
Steri-Strip (3M Health Care), precut to about 1 cm^2
Ultrasonic humidifier (Holmes Products)
Water fount (plastic; G.Q.F. Manufacturing Co.)

METHOD

Preparation of Freshly Laid Eggs and Microinjection Apparatus

1. Remove freshly laid eggs from 15°C storage and set them horizontally in a chukar egg tray. Spray the eggs with 70% ethanol to disinfect the surface of the shell. Incubate the eggs for at least 1 hour at room temperature to allow the blastoderm to rotate to the area below the apex of the eggshell.

2. Mark the apex of the shell with a blue or red felt-tipped pen.
   This gives an appropriate reference mark for the theoretical location of the blastoderm.

3. Use a needle puller to pull the quartz glass needle to a smooth bevel. Fill the needle with about 10 µl of concentrated virus using a pipette fitted with a long Microloader pipette tip. Break the tip of the needle with a fine forceps to about 40 µm. Tighten the needle into the pressure injector and clamp it into the micromanipulator.

Injection of Stage-X Blastoderm

4. Place the egg on a foam egg holder under a dissecting microscope. Turn on the ultrasonic humidifier and wet the surface of the eggshell with 95% ethanol. Allow the shell to dry before proceeding to Step 5.

5. Use dissecting scissors to cut a 3–5-mm hole in the eggshell around the reference mark. Remove the piece of eggshell immediately and apply a drop of sterile 1x HBSS to the exposed blastoderm. Carefully remove any air bubbles with a small glass Pasteur pipette.

6. Add a small volume of 1x HBSS or thin albumin to float the embryo to the top of the egg opening.
   This raises the embryo to an appropriate level for injection and protects the embryo from drying out.

7. Angle the virus-filled quartz injection needle (from Step 3) so that it just touches the vitelline membrane. Slowly insert the needle through the vitelline membrane and epiblast and into the subgerminal cavity of the blastoderm.
   A slight light refraction will be seen when the needle comes in contact with the vitelline membrane.

8. Gently inject the virus at 0.1–0.5 psi. Slowly increase the pressure from the microinjector by manually adjusting the positive pressure, fill the entire subgerminal cavity with virus (1–2 µl), and then decrease the positive pressure.

   Injection at 0.1–0.5 psi is recommended because higher pressures can disturb fragile embryonic cells. During this step, a slight pink color should be seen building in the area pellucida if the lentivirus was stored in complete medium in Protocol 1; this is phenol red from the DMEM.

9. Withdraw the needle from the embryo. Adjust the level of the albumin in the egg, if necessary, so that the egg is filled to the rim of the opening. Remove all air bubbles from the albumin.

10. Carefully wipe the area around the opening with 95% ethanol. Do not allow ethanol to enter the egg because it will adversely affect its viability. Allow the ethanol to dry before proceeding to Step 11.

11. Use blunt forceps to seal the opening with a precut Steri-Strip. Ensure that the Steri-Strip forms a tight seal at all corners.

12. Apply a thin layer of molten paraffin to the top of the Steri-Strip.

    The paraffin prevents evaporation through the seal during the subsequent incubation period.

Incubation and Hatching

Be sure to include unmanipulated eggs as a control for correct incubation conditions. However, manipulated eggs appear to be 1–2 days behind, developmentally, when compared to unmanipulated eggs. Keep this in mind when transferring eggs to the hatcher; see Troubleshooting.

13. Place the eggs in chukar egg trays with their blunt ends up.

    This ensures proper orientation of the embryo during development.

14. Place the egg trays on the tilting shelves of an egg incubator. Incubate the eggs for the next 14 days at 37.5°C and approximately 60% relative humidity. Make sure that the incubator shelves tilt hourly through 90° to avoid detrimental effects on hatchability.

    See Troubleshooting.

15. On day 15, lay the eggs on their sides in an open hatching tray that is lined with Dura Pads. Load the tray into a hatching incubator set at 37°C and 70% relative humidity. To preserve this high level of humidity, do not open the door of the incubator during the hatch.

    Dura Pads are an absorbent soft paper that prevents leg splay in new hatchlings. The hatchlings' nutritional needs are met during the first 24 hours by the yolk sac that is adsorbed into their abdominal cavity.

Rearing and Breeding

16. One day after hatching, label each bird with a numbered size-2 plastic leg band. Transfer the chicks to a 37°C brooder for 1 week and provide the chicks food and water as follows:

    i. Using a blender, grind the Mazuri Exotic Gamebird Starter feed. Spread an unlimited supply of the feed on the floor of the brooder.

    ii. Provide water in a plastic water fount for quail. Add marbles to the water trough to prevent drowning of the hatchlings.

17. In the second week, decrease the brooder temperature to 32°C. Provide blended starter feed as described in Step 16.i.

18. In the third week, decrease the brooder temperature to 29°C. Offer a mix of blended and full-sized pellets of starter feed. Replace the initial leg band with a numbered size-4 plastic leg band.

19. In the fourth week, maintain the brooder at room temperature (22°C) and give the birds full-sized starter pellets only.

20. After 5 weeks, set up breeding pairs (one male and one female) in the adult cages. Pair potential founder birds, which may be mosaic for the transgene, with wild-type adults of the same age. If a founder is male, a second female may be added. Rear the breeding pairs under the following conditions:

    i. Spread a layer of aspen shavings on the bottom of the cage and place a red plastic guinea pig hut in the cage to give the females a safe haven during breeding. Because the breeding cycle is based on day length, provide 14 hours of light during each 24-hour period to maintain egg production.

    ii. Feed the birds a diet of Mazuri Exotic Gamebird Breeder ad libitum, which provides extra calcium for egg-laying females. Offer water ad libitum.

TROUBLESHOOTING

Problem (Step 14): Eggs that have been windowed typically show a high percentage of mortality during the first couple of days of incubation. Often, the surviving embryos are developmentally delayed. Furthermore, it can be lethal to move eggs into the hatcher, with its high humidity, before the embryo is ready to pip into the internal air space.

Solution: Candle the eggs to determine the approximate age of the embryo. Candling brown mottled eggs must be done with a very bright light in a dark room. The majority of the egg will appear to be dark red or brown where the embryo is growing. The air cell will appear to be bright white at the large end of the egg. The air cell will expand in size when the embryo is ready to pip into this interior space.

Protocol 3

Screening for Transgenic Japanese Quail Offspring

After mosaic founder breeding pairs start to produce fertile eggs, the hatchlings must be screened for germ-line transmission to the subsequent G_1 generation. This protocol describes how to isolate hatchling genomic DNA from the chorioallantoic membrane (CAM), which remains inside the egg after hatching. Collecting genomic DNA from the CAM decreases the hatchling's stress during handling and eliminates the need for a blood draw. By following this protocol, the CAM of a single egg will provide 50 µg or more of high-quality genomic DNA. The protocol also describes how to screen the genomic DNA samples for the transgene by PCR. PCR genotyping should be used for screening hatchlings with a nonfluorescent transgene or with a fluorescently labeled transgene that does not lend itself well to phenotypic screening.

MATERIALS

The recipes for items marked with <R> are on page 517.

CAUTION: See the Cautions Appendix for appropriate handling of materials marked with <!>.

Reagents

Agarose gel (1%) and standard electrophoresis buffer (see Step 12)
Ammonium acetate (7.5 M)
Chloroform <!>
Digestion buffer <R>
dNTPs
Ethanol (70% and 100%) <!>
H_2O (sterile)
Isoamyl alcohol <!>
Oligonucleotide primers (5′ and 3′), for the transgene and a housekeeping gene
Phenol (pH 8.0) <!>
Taq DNA polymerase and 10x buffer from supplier
TE buffer <R>

Equipment

Felt-tip pen
Gel electrophoresis equipment
Hatching equipment, as described in Protocol 2
Hatching trays with subdividers
  Subdividers can be fashioned out of cardboard or other materials (Fig. 1D). The dividers should be as tall as the hatching trays in order to keep the hatchlings from jumping into the adjacent cubicle.
Incubator, preset to 55°C
Microcentrifuge at room temperature and 4°C
Microcentrifuge tubes (1.5 ml, sterile)
PCR tubes
Plastic leg bands (size 2), numbered
Scoopula (with rounded end)
Spectrophotometer
Thermal cycler

METHOD

Hatching G₁ Quail

1. Collect eggs daily from founder cages (see Protocol 2).
   The eggs may be stored for up to 1 week at 15°C.

2. Incubate potential transgenic and known wild-type eggs as described in Protocol 2. At day 14 of incubation, place each egg on its side in a separate 8-cm² compartment within the hatching rack.
   This allows one to band the hatchling and number its egg for subsequent genotyping.

3. One day after hatching, mark each quail chick with a numbered size-2 plastic leg band. Number the corresponding empty eggshells with a felt-tip pen and allow the eggshells to dry for several hours at room temperature. Move the chicks to the brooder as described in Protocol 2.

DNA Extraction

4. Fill the eggshells from Step 3 with sterile H_2O and allow them to soak for 5 minutes. Use a scoopula with a rounded end to scrape the thin, transparent CAM containing the blood vessels from the inner shell membrane. Place it in a sterile 1.5-ml microcentrifuge tube.
   In older birds, sufficient genomic DNA can be obtained from the follicle portion of 5–10 feather shafts, especially those containing small amounts of tissue or blood.

5. Add 1 ml of digestion buffer to the chorioallantoic membrane (or feather shafts) and incubate it overnight at 55°C.

6. The next day, add 500 µl of the sample to a new 1.5-ml microcentrifuge tube. Add 500 µl of 25:24:1 phenol:chloroform:isoamyl alcohol and mix thoroughly by gently inverting the tube 100 times. Centrifuge the mixture in a microcentrifuge at 5000g for 10 minutes at room temperature and discard the lower layer.

7. To remove the phenol, repeat the extraction by adding 500 µl of 24:1 chloroform:isoamyl alcohol to the upper layer from Step 6. Centrifuge the sample in a microcentrifuge at 5000g for 10 minutes at room temperature and place the upper layer in a new tube.

8. To precipitate the genomic DNA, add 100 µl of 7.5 M ammonium acetate and 1 ml of 100% ethanol to the sample from Step 7. Centrifuge the mixture at 13,000g for 10 minutes at 4°C and discard the supernatant.

9. Wash the pellet twice in 1 ml of room-temperature 70% ethanol. Air-dry the pellet for about 15 minutes and resuspend it in 150 µl of TE buffer (pH 8).

10. Quantitate the genomic DNA by UV spectrophotometry.
    If the DNA yield from feather shaft samples is low, see Troubleshooting.

Genotypic Screening

11. In each PCR tube, set up a 25-µl reaction containing 100 ng of genomic DNA, 200 µM dNTPs, 0.5 µM 5′ and 3′ oligonucleotide primers, 1.5 units of Taq DNA polymerase, and 1x buffer. In addition to the primers to detect the fluorescent reporter gene delivered by the lentivirus, include primers to detect a cellular housekeeping gene, such as *GAPDH*. Amplify on a thermal cycler using standard procedures.

12. Run half of each resulting PCR product on a 1% agarose gel using standard procedures (Fig. 3).
    Individuals identified as transgenic by PCR can then be grown to adulthood and bred for homozygosity, if desired, or bred to the wild-type animals hatched in the same clutch to produce additional hemizygotes.

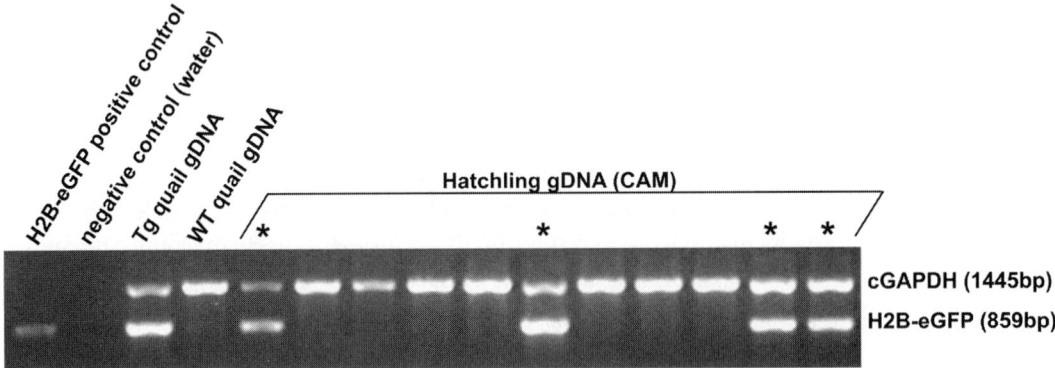

FIGURE 3. Screening G₁ hatchling genomic DNA by PCR. Japanese quail eggs from founder breeding pairs were set to hatch. After hatching, the chicks and their respective eggshells were numbered and genomic DNA was prepared using the chorioallantoic membrane (CAM) from the individual eggs. Multiplex PCR was performed on 100 ng of genomic DNA using oligos specific to chicken *GAPDH* (cGAPDH) and a histone 2B-eGFP (H2B-eGFP) transgene. Half of each PCR was separated on a 1% agarose gel and photographed under UV transillumination. The asterisks (*) denote G₁ offspring that were positive for the H2B-eGFP transgene. (Tg) Transgenic; (WT) wild type.

TROUBLESHOOTING

Problem (Step 10): Adult feather shafts yield low quantities of DNA.

Solution: Extracting genomic DNA from feather shafts can be difficult, given the paucity of tissue that adheres to the white follicles of mature feathers. Pluck only feathers with thick black follicles, which contain highly vascularized tissue. If these types of feathers are scarce, performing a blood draw from a leg vein will also provide material for genomic DNA isolation.

Recipes

CAUTION: See the Cautions Appendix for appropriate handling of materials marked with <!>.

COMPLETE MEDIUM

Reagent	Quantity (for ~1 liter)	Final concentration
DMEM (high glucose, with L-glutamine and phenol red, no sodium pyruvate) (Mediatech)	900 ml	90%
Fetal bovine serum (FBS), heat-inactivated (Omega Scientific)	100 ml	10%
G418 (50 mg/ml) (Sigma-Aldrich) <!>	10 ml	500 µg/ml
L-Glutamine (200 mM) (Mediatech)	10 ml	2 mM
HEPES (1 M) (Mediatech)	10 ml	10 mM
MEM nonessential amino acids (10 mM) (Mediatech)	10 ml	0.1 mM
Penicillin (10,000 IU/ml)/streptomycin <!> (10,000 µg/ml) (Mediatech)	10 ml	1%
Sodium pyruvate (100 mM) (Mediatech)	10 ml	1 mM

Do not include the G418 antibiotic when culturing cells for virus production.

DIGESTION BUFFER

Reagent	Quantity (for 100 ml)	Final concentration
EDTA (0.5 M, pH 8.0) <!>	5 ml	25 mM
NaCl (5 M)	1 ml	50 mM
Proteinase K (20 mg/ml) <!>	0.5 ml	0.1 mg/ml
SDS (10%) <!>	10 ml	1%
Tris-Cl (1 M, pH 8.0)	1 ml	10 mM
H$_2$O	to 100 ml	

TE BUFFER

Reagent	Quantity (for 100 ml)	Final concentration
EDTA (0.5 M, pH 8.0) <!>	0.2 ml	1 mM
Tris-Cl (1 M, pH 8.0)	1 ml	10 mM
H$_2$O	to 100 ml	

REFERENCES

Barrett, J.E., Wells, D.C., Paulsen, A.Q., and Conrad, G.W. 2000. Embryonic quail eye development in microgravity. *J. Appl. Physiol.* **88:** 1614–1622.

Beaumont, C., Roussot, O., Feve, K., Vignoles, F., Leroux, S., Pitel, F., Faure, J.M., Mills, A.D., Guemene, D., Sellier, N., et al. 2005. A genome scan with AFLP markers to detect fearfulness-related QTLs in Japanese quail. *Anim. Genet.* **36:** 401–407.

Burns, M. and Domjan, M. 2001. Topography of spatially directed conditioned responding: Effects of context and trial duration. *J. Exp. Psychol. Anim. Behav. Process.* **27:** 269–278.

Chapman, S.C., Collingnon, J., Schoenwolf, G.C., and Lumsden, A. 2001. Improved method for chick whole-embryo culture using a filter paper carrier. *Dev. Dyn.* **220:** 284–289.

Eyal-Giladi, H. and Kochav, S. 1975. From cleavage to primitive streak formation: A complementary normal table and a new look at the first stages of the development of the chick. I. General morphology. *Dev. Biol.* **49:** 321–337.

Fitzgerald, T.C. 1969. *The Coturnix quail, anatomy and histology.* Iowa State University Press, Ames, Iowa.

Hamburger, V. and Hamilton, H. 1951. A series of normal stages in the development of the chick embryo. *J. Morphol.* **88:** 49–92.

Hillier, L.W., Miller, W., Birney, E., Warren, W., Hardison, R.C., Ponting, C.P., Bork, P., Burt, D.W., Groenen, M.A., Delany, M.E., et al. 2004. Sequence and comparative analysis of the chicken genome provide unique perspectives on vertebrate evolution. *Nature* **432:** 695–716.

Ito, S., Kimura, M., and Isogai, I. 1988a. A sex difference in recombination values between extended brown and phosphoglucose isomerase loci in Japanese quail. *Jpn. J. Zootech. Sci.* **59:** 801–805.

Ito, S., Kimura, M., and Isogai, I. 1988b. Linkage between panda plumage and albumin loci in Japanese quail. *Jpn. J. Zootech. Sci.* **59:** 822–824.

Kayang, B.B., Vignal, A., Inoue-Murayama, M., Miwa, M., Monvoisin, J.L., Ito, S., and Minvielle, F. 2004. A first-generation microsatellite linkage map of Japanese quail. *Anim. Genet.* **35:** 195–200.

Kayang, B.B., Fillon, V., Inoue-Murayama, M., Miwa, M., Leroux, S., Feve, K., Monvoison, J.-L., Pitel, F., Vignoles, M., Mouilhayrat, C., et al. 2006. Integrated maps in quail (*Coturnix japonica*) confirm the high degree of synteny conservation with chicken (*Gallus gallus*) despite 35 million years of divergence. *BMC Genomics* **7:** 101–118.

Kikuchi, T. and Yang, H.W. 1998. Clinical and metabolic correction of pompe disease by enzyme therapy in acid maltase deficient quail. *J. Clin. Invest.* **101:** 827–833.

Kikuchi, S., Fujima, D., Sasazaki, S., Tsuji, S., Mizutani, M., Fujiwara, A., and Mannen, H. 2005. Construction of a genetic linkage map of Japanese quail (*Coturnix japonica*) based on AFLP and microsatellite markers. *Anim. Genet.* **36:** 227–231.

Kulesa, P. and Bronner-Fraser, M. 2000. In ovo time-lapse analysis after dorsal neural tube ablation shows rerouting of chick hindbrain neural crest. *Development* **127:** 2843–2852.

Lalloue, F.L. and Ayer-Le Lievre, C.S. 2005. Experimental study of early olfactory neuron differentiation and nerve formation using quail-chick chimeras. *Int. J. Dev. Biol.* **49:** 193–200.

Le Douarin, N. and Kalcheim, C. 1999. *The neural crest.* Cambridge University Press, Cambridge, United Kingdom.

Minvielle, F., Ito S., Inoue-Murayama, M., Mizutani, M., and Wakasugi, N. 2000. Genetic analyses of plumage color mutation on the Z chromosomes of Japanese quail. *J. Hered.* **91:** 499–501.

Minvielle, F., Kayang, B.B., Inoue-Murayama, M., Miwa, M., Vignal, A., Gourichon, D., Neau, A., Monvoisin, J.L., and Ito, S. 2005. Microsatellite mapping of QTL affecting growth, feed consumption, egg production, tonic immobility and body temperature of Japanese quail. *BMC Genomics* **6:** 87.

Minvielle, F., Kayang, B.B., Inoue-Murayama, M., Miwa, M., Vignal, A., Gourichon, D., Neau, A., Monvoisin, J.L., and Ito, S. 2006. Search for QTL affecting the shape of the egg laying curve of the Japanese quail. *BMC Genet.* **7:** 26.

Mizutani, M. 2002. Establishment of inbred strains of chicken and Japanese quail and their potential as animal models. *Exp. Anim.* **51:** 417–429.

Ono, T., Nakane, Y., Wadayama, T., Tsudzuki, M., Arisawa, K., Ninomiya, S., Suzuki, T., Mizutani, M., and Kagami, H. 2005. Culture system for embryos of blue-breasted quail from the blastoderm stage to hatching. *Exp. Anim.* **54:** 7–11.

Ottinger, M.A., Abdelnabi, M., Li, Q., Chen, K., Thompson, N., Harada, N., Viglietti-Panzica, C., and Panzica, G.C. 2004. The Japanese quail: A model for studying reproductive aging of hypothalamic systems. *Exp. Gerontol.* **39:** 1679–1693.

Padgett, C.S. and Ivey, W.D. 1959. *Coturnix* quail as a laboratory research animal. *Science* **129:** 267–268.

Padgett, C.S. and Ivey, W.D. 1960. The normal embryology of the *Coturnix* quail. *Anat. Rec.* **137:** 1–11.

Perry, M.M. 1988. A complete culture system for the chick embryo. *Nature* **331:** 70–72.

Poynter, G. and Lansford, R. 2008. Generating transgenic quail using lentiviruses. *Methods Cell Biol.* **87:** 281–293.

Roberts, M. 1999. *Quail, past and present (Gold Cockerel series).* Bartlett & Son, Exeter, St. Thomas, United Kingdom.

Roussot, O., Feve, K., Pilsson-Petit, F., Pitel, F., Faure, J.-M., Beaumont, C., and Vignal, A. 2003. AFLP linkage map of the Japanese quail *Coturnix japonica. Genet. Sel. Evol.* **35:** 559–572.

Ruffins, S.W., Martin, M., Keough, L., Truong, S., Fraser, S.E., Jacobs, R.E., and Lansford, R. 2007. Digital three-dimensional atlas of quail development using high-resolution MRI. *Sci. World J.* **7:** 592–604.

Ryals, B.W. and Rubel, E.W. 1988. Hair cell regeneration after acoustic trauma in adult *Coturnix* quail. *Science* **240:** 1774–1776.

Sasazaki, S., Hinenoya, T., Lin, B., Fujiwara, A., and Mannen, H. 2006. A comparative map of macrochromosomes between chicken and Japanese quail based on orthologous genes. *Anim. Genet.* **37:** 316–320.

Scott, B.B. and Lois, C. 2005. Generation of tissue-specific transgenic birds with lentiviral vectors. *Proc. Natl. Acad. Sci.* **102:** 16443–16447.

Shibata, T. and Abe, T. 1996. Linkage between the loci for serum albumin and vitamin D binding protein (GC) in the Japanese quail. *Anim. Genet.* **27:** 195–197.

Shibusawa, M., Minai, S., Nishida-Umehara, C., Suzuki, T., Mano, T., Yamada, K., Maikawa, T., and Matsuda, Y. 2001. A comparative cytogenetic study of chromosome homology between chicken and Japanese quail. *Cytogenet. Cell Genet.* **95:** 103–109.

Wallis, J.W., Aerts, J., Groenen, M.A., Crooijmans, R.P., Layman, D., Graves, T.A., Scheer, D.E., Kremitzki, C., Fedele, M.J., Mudd, N.K., et al. 2004. A physical map of the chicken genome. *Nature* **432:** 761–764.

Woodard, A.E., Abplanalp, H., Wilson, W.O., and Vohra, P. 1973. Japanese quail husbandry in the laboratory. Department of Avian Sciences, University of California, Davis, California.

Yuan, Y.V., Kitts, D.D., and Godin, D.V. 1997. Influence of dietary cholesterol and fat source on atherosclerosis in Japanese quail (*Coturnix japonica*). *Br. J. Nutr.* **78:** 993–1014.

Zacchei A.M. 1961. The embryonal development of the Japanese quail (*Coturnix coturnix japonica* T. and S.). *Arch. Ital. Anat. Embriol.* **66:** 36–62.

WWW RESOURCES

http://genome.wustl.edu/index.cgi The chicken (*Gallus gallus*) genome was sequenced at Washington University in St. Louis.

http://poultry.mph.msu.edu Chicken database of BAC clones associated with DNA markers from the U.S. Poultry Genome Project.

22 The Short-tailed Fruit Bat *Carollia perspicillata*
A Model for Studies in Reproduction and Development

John J. Rasweiler IV,[1] Chris J. Cretekos,[2] and Richard R. Behringer[3]

[1]Department of Obstetrics and Gynecology, State University of New York Downstate Medical Center, Brooklyn, New York 11203; [2]Department of Biological Sciences, Idaho State University, Pocatello, Idaho 83209; [3]Department of Molecular Genetics, University of Texas M.D. Anderson Cancer Center, Houston, Texas 77030

ABSTRACT

Carollia perspicillata has proven to be a valuable laboratory model for studies in reproduction and development. We present here an overview of the care and handling of *Carollia* in captivity and discuss some pertinent studies in reproductive biology. Finally, we describe various features of the genome and some of the genetic manipulations that are now possible.

PROTOCOLS
1 Feeding Short-tailed Fruit Bats, 525
2 Generating Timed Pregnancies, 528
3 Collection of Bats from the Wild, 531
4 Bat Embryo Collection, 533
5 Fixation and Storage of Bat Embryos, 535
6 Whole-mount In Situ Hybridization of Bat Embryos with RNA Probes, 537
7 Alcian Blue Staining of Cartilage, 544
8 Alcian Blue/Alizarin Red Staining of Cartilage and Bone, 546
9 Whole-mount Immunohistochemistry, 548

BACKGROUND INFORMATION

The short-tailed fruit bat *C. perspicillata* is a robust, moderate-sized bat (adult weight in the wild is approximately 14–24 g) that occurs from southern Mexico to Peru, Bolivia, Paraguay, most of Brazil, and northeastern Argentina, as well as on the islands of Trinidad and Tobago (Fig. 1). It is frequently found in abundance and may be the most common mammal inhabiting forested areas in the lowland tropics of the New World. Although predominantly frugivorous, *Carollia* also consumes some floral components (nectar and pollen) and insects. Although insects usually comprise only a small part of *Carollia*'s natural diet, they may provide important sources of essential amino acids as well as some vitamins and minerals (Rasweiler 1977; Rasweiler and Badwaik 2009b).

The genus *Carollia* contains six recognized species whose distribution frequently overlaps (Pine 1972; Fleming 1988; Eisenberg 1989; Simmons 2005). Although criteria have been developed for distinguishing them (Cloutier and Thomas 1992; see references in Simmons 2005), in practice, this can sometimes be challenging, even for expert bat systematists. It is also important to note that where these "species" overlap in distribution, it has not yet been established whether they reproduce in isolation from one another or are incapable of interbreeding.

This chapter, with full-color images, can be found online at www.cshprotocols.org/emo.

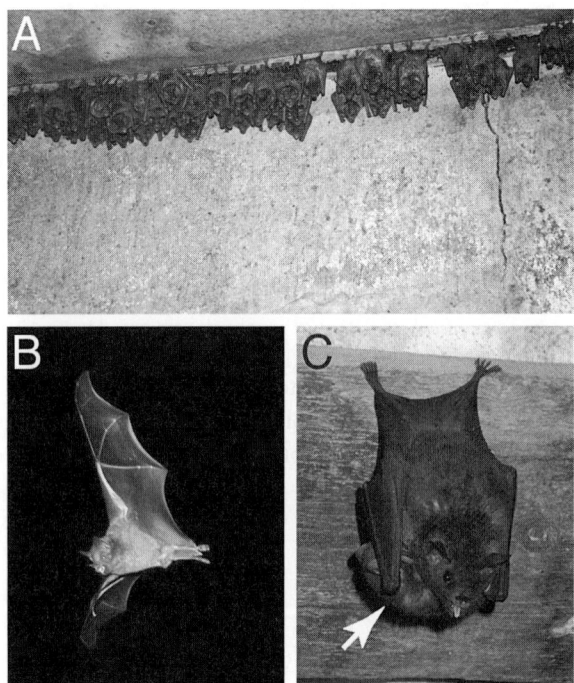

FIGURE 1. *Carollia perspicillata,* the short-tailed fruit bat. (A) Large group of *Carollia perspicillata* roosting at the edge of the ceiling in a room of an abandoned house in Trinidad. (B) *Carollia* in flight. (C) Mother carrying suckling pup (*arrow*). The mothers can fly with these sizable pups.

Fortunately, most information discussed here on the reproductive and developmental biology of *Carollia* has been derived from studies conducted with animals of Trinidadian origin or descent. As far as we know, only *C. perspicillata* occurs on that island. The earliest experimental efforts to examine these aspects of *Carollia's* biology in captive-maintained animals were conducted before Pine's (1972) taxonomic revision of the genus was published and apparently included *C. perspicillata* and *C. brevicauda* (Rasweiler and de Bonilla 1972, 1992; de Bonilla and Rasweiler 1974).

Carollia has been used as a laboratory animal model in several areas of reproductive and developmental biology. Like humans (in most cases), *Carollia* is monovular, has a simplex uterus, exhibits true menstruation, displays interstitial implantation of the blastocyst within a preferred region of the uterus, has highly invasive trophoblast, and forms a discoidal, hemochorial placenta (Badwaik and Rasweiler 2000; Rasweiler and Badwaik 2000). Alterations in several of these processes or characteristics are responsible for a number of human reproductive problems. For example, implantation is thought to be a stage at which human pregnancies frequently fail, and abnormal positioning of the implantation site is a common and serious complication of human pregnancy. Implantation is also a target in efforts to develop new methods of fertility regulation for humans. Despite the obvious need for relevant animal models to study these problems, those currently available are often too expensive (primates) or sometimes too different from humans (litter-bearing laboratory species) to be fully useful.

Carollia is also useful for a broader range of developmental studies. Although the utility of laboratory rodents (especially the mouse) and the rabbit as models in this area is generally appreciated, representatives of other major mammalian groups deserve attention in the interest of developing a more comprehensive and balanced view. Bats (order Chiroptera) are certainly one of these groups. According to a recent compilation, there exist at least 1116 species of bats (comprising >20% of all living mammalian species) (Simmons 2005; Wilson and Reeder 2005). Furthermore, bats are exceedingly abundant mammals, particularly in tropical areas of the world, and they exhibit remarkable diversity in habits, form, and function.

Among the interesting features of the early embryogenesis of *Carollia* are development to the blastocyst stage and shedding of the zona pellucida in the oviducts, formation of reticulated endoderm during implantation (as in humans), development of an unusually large inner cell mass (ICM), disposition of endoderm and Reichert's membrane around all or nearly all of the ICM, extensive apoptosis of epiblast cells during amniogenesis, and formation of a planar embryo at the primitive-streak stage (in contrast to the cup-shaped embryo of rodents) (Badwaik et al. 1997). *Carollia* is also unusual among mammals in that it sometimes takes embryos into prolonged periods of delay (diapause) after implantation, at the primitive-streak stage, which can last from weeks to months. This delay period occurs seasonally in the wild, but in response to stress in captivity (Rasweiler and Badwaik 1997), and is associated with major alterations in trophoblastic differentiation and placental development (Badwaik and Rasweiler 2001). The endocrine control of the delay remains to be elucidated. Finally, organogenesis in *Carollia* embryos is of interest, because they develop some morphological features (e.g., wings and facial specializations) that are quite different from more commonly studied mammals and even other bats (Cretekos et al. 2005). This has prompted us to use *Carollia* as a model to study gene expression during limb development (Chen et al. 2005; Sears et al. 2006; Weatherbee et al. 2006; Cretekos et al. 2007, 2008).

HUSBANDRY

Carollia has several great advantages as a captive bat model for reproductive and developmental studies. It is abundant in the wild, readily adapts to captivity, can be maintained conveniently and inexpensively (Rasweiler and Badwaik 1996, 2009a,b,c), is colonial (meaning that sizable numbers of animals can be housed together), remains in reproductive condition in captivity but abandons its usual reproductive seasonality of the wild, exhibits a limited period of estrus (thereby greatly facilitating the timing of pregnancies), and breeds very successfully. It is not difficult to maintain a captive colony containing hundreds of these animals.

In our research setting, *Carollia* are housed in custom-designed metal cages measuring about 170 cm wide by 81 cm high by 77 cm deep (Fig. 2A) (Rasweiler and Badwaik 1996, 2009c). The

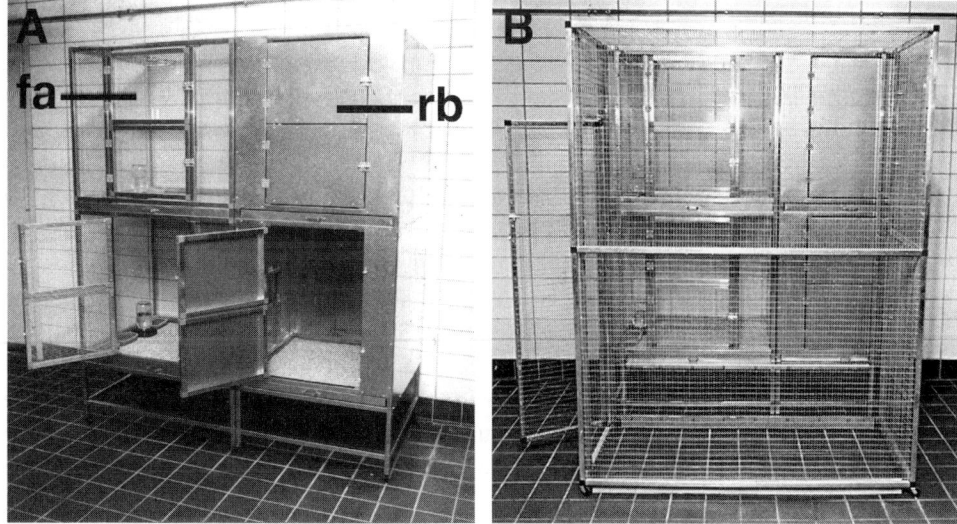

FIGURE 2. (A) Stacked pair of cages used to house small frugivorous and nectarivorous bats (e.g., *Carollia* and *Glossophaga*). (B) Light-weight, roll-up cage to enclose the animal handler positioned in front of a pair of bat cages, when the animals must be handled. This is open on the side abutting the animal cages, except at the bottom, so that all of the doors on each cage could be opened simultaneously. (Reproduced, with permission, from Rasweiler and Badwaik 1996.)

cages have an open feeding area (92 cm wide) connected to a darkened roosting box via an access hole (15 cm high × 30 cm wide). Both compartments are lined completely with one-quarter-inch galvanized mesh except for the floors, the interior of the doors to the roosting box, and a galvanized sheet metal shield (to preclude roosting by the bats) over the food and water dishes. The bats normally hang from the ceiling when not flying. Although the cages are large enough to permit the animals to fly, it would be beneficial to provide even more space for flight (assuming space is not at a premium). The roosting boxes should be kept relatively small and accessible, however, to facilitate repeated, noninjurious capture of the bats.

Each cage can accommodate up to about 20 bats. These groupings can include adult females and juveniles of both sexes, but only a single adult male. Once males mature, they are prone to fight with one another, which can result in crippling injuries that eventually prevent some of the males from flying.

When animals must be handled (e.g., on a daily basis to check for breeding activity), a lightweight roll-up cage that can accommodate the handler is positioned in front of the cage housing the *Carollia* (Fig. 2B). The access hole to the roosting box is blocked with a piece of Styrofoam wrapped in multiple layers of cloth, trapping the animals inside. They are then captured individually with a deep-pocketed aquarium net (Rasweiler and Badwaik 1996). The presence of the roll-up cage greatly restricts the flight of any animals that escape from the roosting box. The feeding area of the cage is darkened with a sheet of black plastic during handling periods and used as a temporary holding pen. This system of handling the animals is highly efficient and noninjurious to the bats.

The closely related long-tongued bat *Glossophaga soricina* is somewhat smaller than *Carollia* (i.e., 9–12 g), but it exhibits many biological similarities. It can also be successfully maintained and bred in captivity with the same husbandry methods (Rasweiler and Badwaik 2009a,b,c).

USES OF THE *C. PERSPICILLATA* SYSTEM: STUDIES IN REPRODUCTIVE BIOLOGY AND BREEDING

When *Carollia* are introduced into captivity from the wild, they are generally housed separately by sex, with the females grouped together and the males caged singly. The animals are usually allowed at least 8 months to adapt to their new environment. A single stud male with very prominent testes is then introduced into each cage of females for breeding purposes, and he generally has no problem servicing groups up to at least 18 females. In the wild, *Carollia* normally organize into small harems consisting of one male and two to ten females (Fleming 1988). In this situation, however, the largest harems may contain about the maximum number of females that the males can effectively defend.

When a good stud male is used, most females in a group will be inseminated within a month. This reflects the fact that adult female *Carollia* become polyestrous in captivity, are spontaneous ovulators, and have cycles about 25 days in length. Although females exhibit two relatively synchronized pregnancy periods in the wild on Trinidad, evidence of seasonality is lost in captivity. This means that embryos can be generated with similar ease at any time of the year.

Evidence of mating activity is obtained by aspirating a small volume of distilled water in the vagina of each female each morning with a microeyedropper (Fig. 3). The aspirate is then placed on a slide, dried, and examined microscopically for spermatozoa (Rasweiler and Badwaik 1996). The normal duration of estrus is not known, but sperm are usually seen in aspirates for 2–4 days. The first day of a sperm-positive aspirate is considered to be day 1 postcoitus (p.c.). Vaginal plugs are also occasionally seen, but they are not a reliable means of checking for mating activity. New plugs are sometimes seen on the second day of sperm-positive aspirates, indicating that estrus can extend for more than 24 hours.

About half of the females killed and checked on the morning of day 1 p.c have a large preovulatory follicle in one ovary, and the remainder have a newly ovulated oocyte in the oviduct. All

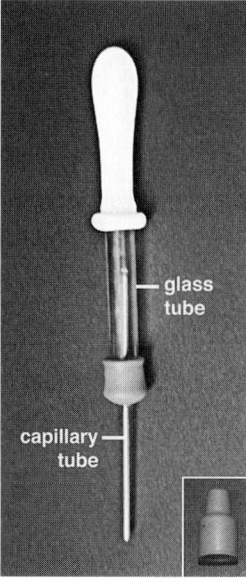

FIGURE 3. Microeyedropper used for taking vaginal aspirates from *Carollia* and *Glossophaga*. (*Inset*) Rubber stopper with thin sleeve projecting downward. To construct the microeyedropper, a hole is first made with a needle probe in the diaphragm of the stopper. The stopper is then placed in one end of the glass tube and its sleeve is everted over the outside of the tube. Finally, a capillary tube (lubricated with glycerol) is passed halfway through the stopper diaphragm.

females killed and checked on the evening of day 1 p.c. have ovulated (J.J. Rasweiler and N.K. Badwaik, unpubl.).

Much of embryonic development in *Carollia* has been characterized from studies of conceptuses obtained after timed matings in captive-bred animals (de Bonilla and Rasweiler 1974; Badwaik et al. 1997; Oliveira et al. 2000; Rasweiler et al. 2002; Cretekos et al. 2005).

GENETICS

Genome

The *C. perspicillata* genome is parceled into 20 chromosomes for females and 21 chromosomes for males (Baker 1979). The X chromosome has an autosomal translocation and the extra chromosome in males is the unpaired homolog that corresponds to the autosomal region of the X chromosome (Hsu et al. 1968). The genome of *C. perspicillata* has not been sequenced. However, a low (1.7x)-coverage genome assembly of another microchiropteran, the insectivorous little brown bat (*Myotis lucifugus*), has been generated by the Broad Institute (Boston). A high-coverage assembly for *Myotis* is scheduled for release in the summer of 2008. The current genome can be accessed at www.ensembl.org/Myotis_lucifugus/index.html. In addition, low-coverage (2x) sequencing of the genome of a megachiropteran, the large flying fox (*Pteropus vampyrus*), is in progress at the Baylor College of Medicine (Houston). The availability of the *Myotis* and *Pteropus* genomes will facilitate genome studies in all bat species.

Genome Resources

A bacteriophage λ genomic library has been generated for *C. perspicillata* (Cretekos et al. 2001). In addition, cDNA libraries have been generated from Stages 14–17 *Carollia* embryos and adult brain (Cretekos et al. 2007). Furthermore, a *Carollia* bacterial artificial chromosome (BAC) library with eightfold coverage has been generated and arrayed onto filters (Eric Green, National Human Genome Research Institute). *Carollia*-specific in situ hybridization probes have been established for *Hoxd13* (Chen et al. 2005), *Bmp4, Bmp7, Gremlin* (Weatherbee et al. 2006), *Fgf8* (Cretekos et al. 2007), and *Prx1* (Cretekos et al. 2008). Many mouse in situ hybridization probes also work well in *Carollia*, including *Bmp2, Fgf8, Msx2, Spry2* (Weatherbee et al. 2006), *Shh, Ptc1, Hand2, Lmx1b, Wnt5a*, and *Fgf10* (C.J. Cretekos, unpubl.).

Classical Genetics

The logistics of maintaining productive breeding colonies of captive bats precludes classical genetic studies for most bat species. However, *C. perspicillata* is easily maintained and bred in captivity, making simple genetic crosses possible.

TECHNICAL APPROACHES

The following series of protocols describe the collection, care, and handling of *C. perspicillata*, as well as the collection and use of embryos for various whole-mount studies, including situ hybridization and immunohistochemistry, for a variety of histological procedures.

Protocol 1

Feeding Short-tailed Fruit Bats

Carollia is predominantly frugivorous in the wild and will therefore accept a variety of noncitrus fruits in captivity (e.g., bananas, apples, melon, and peaches). To promote cage cleanliness, however, they are routinely fed a fruit-based liquid diet that cannot be carried away from the food dishes. The mixture includes canned apricot nectar and pureed peaches as a palatable base, with other components intended to enhance protein, vitamin, mineral, and essential fatty acid levels in the diet (Rasweiler and Badwaik 2009b). By intention, this diet is simple and inexpensive to prepare, and it can also be prepared in advance and refrigerated or frozen for later use.

MATERIALS

Reagents

Apricot or peach nectar (canned)
Calcium supplement (dibasic calcium phosphate; $CaHPO_4$)
Corn oil
High-protein Monkey Diet 5045 (for New World primates)

This monkey diet is normally sold in the form of pellets; however, the pellets are finely ground for the bat diet by the mill or feed supplier, so that the diet can pass through a #4 screen (4/64-inch mesh opening). This processing facilitates suspension of the monkey food in the final, liquid bat diet.

Peaches in pear juice (sliced and canned)
TestDiet® Monkey Mini MV Tablets without added iron (vitamin supplement)
Tween 80® (polyoxyethylenesorbitan monooleate; emulsifier)

Equipment

Can openers (for puncturing holes or removing lids)
Freezer dedicated to food storage
Graduated cylinders (glass or plastic; dedicated for preparing and measuring Tween 80 and corn oil)
Graduated cylinders (plastic, 1–2 liters; for measuring water to be added to diet mixture)
Kitchen blender
Kitchen spatula
Laboratory stand (heavy duty, rectangular) and clamp holder (for attaching to rods [up to 19 mm in diameter] at right angles) for mounting motorized jumbo stirrer
Ladle (for transferring peaches)
Plastic chick waterers

Each should consist of a plastic Mason jar and a base designed for jar slip on, not screw on.

Plastic containers (durable, multiple, for food storage in refrigerator or freezer; 3.5–4 liters)
Plastic containers (strong, cylindrical)

These containers should have covers for mixing and temporarily storing food (7.5, 12, or 18 liters, depending on the quantity of food to be prepared).

Plastic or glass vials (inexpensive)

The vials should have snap lids for weighing and storing daily aliquots of the calcium supplement.

526 / Chapter 22

>Plastic terra cotta saucers (durable, shallow, 6-inch diameter)
>>*These saucers, originally designed to be placed under flower pots, are used as food dishes for the animals.*
>
>Polypropylene beakers with handles (1000- and 5000-ml capacities)
>Refrigerator dedicated to food storage
>Top-loading balances (inexpensive and portable, 0–200- and 0–1000-g capacities)
>Variable speed jumbo stirrer (with 46-mm impeller shaft; Fisher Scientific)

METHOD

Preliminary Steps to Final Food Preparation

1. Preweigh multiple daily aliquots of the calcium supplement and store in vials for later use.
2. Prepare stock supply of corn oil containing Tween 80 emulsifier, sufficient for 1–2 months.
   The stock mix consists of 1000 ml of corn oil + 15 ml of Tween 80.
3. Assemble and precool all major dietary components except the calcium supplement and corn oil mix. The following diet formula provides enough food for approximately 32 animals.

   Diet Formula:

Apricot nectar	671 ml
Sliced peaches	215 ml
Calcium supplement	2.16 g
TestDiet® Monkey Mini MV Tablets without added iron	1.61 tablet
Corn oil/Tween 80 mixture	6.43 ml
Water	226 ml
High-protein Monkey Diet 5045	114 g

   Because the bats are relatively small and have a high surface-to-volume ratio, their daily food consumption is significantly affected by changes in environmental temperature.

Diet Preparation and Storage

4. Puree the sliced peaches, calcium phosphate, and Mini MV tablets together in a kitchen blender. Add the Mini MV tablets at the rate of 0.75 tablet per 100 ml of sliced peaches.
   The tablets are easily split by hand and generally added to the nearest 0.5 tablet required.
5. Prepare the final diet in a large plastic mixing container with agitation provided by a variable speed jumbo stirrer.
   A simple, kitchen blender is far too inefficient to prepare food for a sizable colony.
   For weekend and holiday feeding, quantities of the diet sufficient for each day are left refrigerated in plastic containers. Supplies for extended periods (e.g., weeks) can be prepared in advance, subdivided into quantities sufficient for each day, and frozen in plastic containers. To minimize microbial growth, these containers are transferred from freezer to refrigerator for thawing 3 days before use. They are never thawed at room temperature. Make sure to number the containers and provide the animal caretakers with a daily schedule for container thawing and use.

Food Serving

6. Feed the animals every day.
   Daily feeding is important because food deprivation will cause the animals to drop their body temperature and go into daily torpor. The effects of this reaction on other physiological processes are unknown. The following steps will help to minimize microbial development in food after serving.

i. Serve food in shallow dishes and provide only a modest excess each day to ensure that animals consume substantial amounts of anything in the diet that tends to sediment. As treats for the animals, the diet is occasionally supplemented with apple, banana, or melon slices.

ii. Always serve food cold, usually within 30 minutes and never more than 90 minutes before the start of the dark phase in the animal quarters. The dark phase (12 hours long) is set to begin at 3 pm. This timing is intended to minimize the possibility that food may be put out too early by caretakers on weekends or holidays.

iii. Provide water ad libitum in chick waterers. Always position food dishes and waterers beneath smooth (sheet metal) ceiling shields to prevent animals from roosting in that area and to minimize contamination of the food and water with excrement.

iv. Keep room temperature between 24°C and 26°C. Provide backup air conditioning to prevent the ambient temperature from exceeding 27°C.

Protocol 2

Generating Timed Pregnancies

When female *Carollia* are bred, they are first kept sexually segregated for some months to adapt to captivity (in the case of recently captured animals) and to ensure that none are already pregnant. In the case of captive-reared females that had previously been housed with males, 4 months should be sufficient. In the case of wild-caught females, 7 or 8 months may not be too long, because introduction into captivity can substantially prolong existing early pregnancies (i.e., at or earlier than the primitive-streak stage). Wild-caught females are also much more prone than captive-reared females to take pregnancies into delay after captive breeding. Captive-reared and -bred females will take pregnancies into delay if stressed (Rasweiler and Badwaik 1997).

It is important to note that females should not be housed in sexual isolation for prolonged periods, because this can eventually lead to the development of markedly hyperplastic uteri and reduced fertility. Although this is a very serious problem for zoos maintaining all-female colonies of *Carollia*, it is not the case with our captive colony because the adult females are regularly bred, both to replenish the colony and inhibit the development of hyperplastic uteri (J.J. Rasweiler, unpubl.).

For breeding purposes, a stud male with prominent testes is introduced into each cage of females. Vaginal aspirates are then checked each morning thereafter for the presence of spermatozoa.

MATERIALS

CAUTION: See the Cautions Appendix for appropriate handling of materials marked with <!>.

Reagents

Antisedan (atipamezole HCl) <!>
Bacitracin antibiotic ointment
Bacteriostatic (0.9% sodium chloride for injection; USP)
Ethanol (70%) <!>
Ketaset (ketamine HCl) <!>
Rompun (xylazine HCl) <!>

Equipment

Binocular magnifier (adjustable visor loop magnifier)
Forceps (one pair of 4-inch forceps with curved tips; one pair of 6-inch straight forceps)
Microeyedroppers (Fig. 3)
> *Each of these eyedroppers is assembled from a piece of glass tubing (50 mm long, 8 mm outer diameter), a 1-ml pipette nipple, and a serum sleeve rubber stopper (7 x 11 mm) through which is mounted a capillary tube (75 mm long, 1.2 mm inner diameter) with a fire-polished tip. About half of the capillary tube is positioned to either side of the rubber stopper, on which it pivots.*

Microscope
Microscope slides
Narrow-nosed pliers (3-mm width)

Plastic bird bands (size XCS, 2.5 mm inner diameter, colored and numbered; www.achughes.com).
Plastic transfer pipette (7 ml)
Small ventilated containers

> *For example, use 1-liter cardboard food containers lined with plastic mesh (aquaculture netting) to hold the bats during recovery from anesthetization (see Figure 1 in Rasweiler and Badwaik 2009a)*

Stainless steel bead chain (#3) and #3 connectors
Stiff, straight-tipped laboratory spatula (small)
Wire cutter (small)

METHOD

Fitting the Bead Chain Necklace for Identification

Before commencing the timed matings, each female must be fitted with a bead chain necklace carrying a numbered plastic bird band for identification purposes. For most adult females, a 14-bead segment of chain provides a proper fit.

1. Before placing the necklace on the bat, put a connector on one end of the bead chain and use a spatula to spread open the opposite end of the connector slightly.

2. Final placement of the necklace on the bat requires either the cooperative efforts of two people or anesthetization of the bat. The latter approach is easier and facilitates more careful closure of the bead chain connector.

    i. Anesthetize the bat lightly with ketamine HCl (about 11.25 mg/kg) and xylazine HCl (about 1.5 mg/kg) diluted in bacteriostatic saline. First, swab the injection site with 70% ethanol and then administer the saline solution as a single subcutaneous injection of 45 µl over the pectoral muscles.

        To avoid an extra handling step, assume that all captive animals have a mass of 20 g. They are generally a little larger than wild-caught females. Avoid injection into the richly vascularized pectoral muscles because rapid entry of too much anesthetic into the circulation can be lethal.

    ii. Place the necklace on the bat. Use a pliers to gently squeeze the connector closed.

        Oversqueezing must be avoided because it can expose irritating edges of the connector.

    iii. To accelerate arousal from the anesthesia, inject the bat with atipamezole HCl (a xylazine reversal agent; about 0.41 mg/kg) diluted in bacteriostatic saline, administered as a subcutaneous injection of 35 µl over the pectoral muscles.

        Again, assume a body mass of 20 g.

    iv. During arousal, keep the bat for about 1 hour in a small container that prevents potentially injurious efforts to fly.

        Infrequently, a necklace will irritate part of a bat's neck. Remove the necklace by snipping the bead chain and then apply Bacitracin ointment to the irritated area.

Collection of Temporal Data for Fertilization and Embryonic Development in *Carollia*

Females are checked for sperm during the same time period each morning to determine whether insemination has occurred.

3. Use soft, flexible goatskin work gloves to handle the animals.

    The flexibility facilitates gentle handling of the bats and proper operation of the microeyedroppers for obtaining vaginal aspirates. For appropriate handling procedures, discussions, and illustrations, see Rasweiler (1977).

4. Use a binocular magnifier, which is worn on the head, to facilitate reading numbers on the engraved bird bands (particularly when the paint wears off) and for the successful collection of vaginal aspirates. If necessary, rotate the bands with a forceps for proper reading.

   It may be useful to apply water with a plastic transfer pipette to free up the bands that have become stuck to their necklaces with caked-on food.

5. To collect a vaginal aspirate, draw distilled water into the capillary tube of the microeyedropper.

   Some practice is required to master operation of the microeyedropper with a gloved hand.

6. Insert the tip of the microeyedropper about 2–3 mm into the vagina of the bat and aspirate the distilled water into and out of the vagina. Repeat the insertion and aspiration procedure twice to ensure that an adequate sample has been obtained.

7. Expel the aspirate gently onto a slide and allow it to dry. With the microscope's iris diaphragm partially closed down, examine the aspirate microscopically for spermatozoa.

   A paucity of vaginal epithelial cells in the dried aspirate indicates that an inadequate sample was taken. To economize on slides, four or five successive daily aspirates from one animal can be placed next to one another on a single slide, and the slides can be washed easily for reuse.

   See Troubleshooting.

TROUBLESHOOTING

Problem (Step 7): Precautions are required to avoid cross-contamination of aspirates with sperm.
Solutions:

1. On any day, use a different microeyedropper for each bat.

2. Flush each microeyedropper with clean distilled water immediately after use and three times before reuse.

3. When expelling aspirates onto slides, make an effort to avoid the creation and popping of bubbles, which can spray sperm onto adjacent areas of a slide or even onto another slide.

Protocol 3

Collection of Bats from the Wild

Adult female *Carollia* exhibit a relatively high degree of reproductive synchronization in populations sampled carefully in Central America and on Trinidad (Fleming 1988; Rasweiler and Badwaik 1997; Badwaik and Rasweiler 2001; J.J. Rasweiler et al., unpubl.). On the basis of these studies and temporal data for pregnancies in captive-bred animals, one may be able to sample adult female *Carollia* from other populations and, during much of the year, predict when different embryonic stages might be prevalent.

Most adult female *Carollia* on Trinidad seem to carry two pregnancies each year, based on the observation that more than 90% are typically pregnant when sampled around the middle of each pregnancy period. For many females, the first pregnancy appears to be established between September and early November, includes a period of postimplantational developmental delay at the primitive-streak stage, and is completed in March or April. A peak in births has been observed around April 1 (Rasweiler and Badwaik 1997). Most parous females then conceive again at a postpartum estrus. In captive animals, this estrus usually occurs between 3 and 6 days after parturition, but sometimes, it is several days later (Rasweiler and Badwaik 1996). When parous females were collected by the authors in late May during 8 successive years, most carried conceptuses that had progressed to the somite stage or beyond. On the basis of studies of pregnancies in captive-bred *Carollia* (Rasweiler and Badwaik 1997; Badwaik and Rasweiler 2000; Cretekos et al. 2005), this result is what we would expect in normal (nondelayed) pregnancies if conception had occurred at a postpartum estrus during March or April in the wild.

Using the available temporal data on pregnancies in both captive and wild *Carollia*, it is possible to predict when embryos at particular stages of development are most likely to be carried by females in the wild population on Trinidad. How a similar approach might be used to collect embryos from another population is outlined below.

MATERIALS

Equipment

Cages for transporting and holding the bats (Fig. 4)
> *These are plywood boxes (60.9 cm long × 30.4 cm deep × 23.1 cm high) with galvanized one-quarter-inch mesh attached to the interior of its frame. The entire interior (except for the floor) is lined with mesh, and the bats normally hang from the ceiling. There are ventilation slots at the front. The roof is also hinged, so that it can be partially opened to provide additional ventilation and the bats can be easily removed. No more than about 25 bats should be held and transported in one cage.* Carollia *have a propensity to cluster tightly together when stressed. On a hot day, this can result in mortalities, possibly due to overheating. It should be noted that different cages and containers are used for shipping* Carollia *long distances by air (see Rasweiler and Badwaik 2009a).*

Nets for collecting bats in the wild
> *Hand (butterfly-type) or mist nets are suitable.*

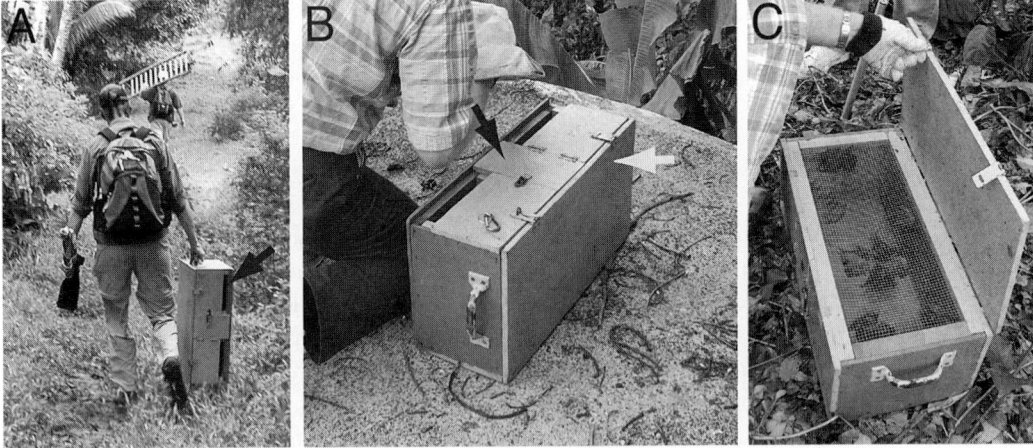

FIGURE 4. Field boxes/cages for transporting bats. (*A*) The field boxes are light weight and easy to transport in the field. They have openings at the bottom covered with wire mesh (*black arrow, left*) that serve as air vents. (*B*) The field box can be placed on its side with the front door (*black arrow*) on top to add bats. There is also a hinged top lid (*white arrow, right*) that can be opened to view the animals or placed ajar to provide more ventilation. (*C*) Field box containing bats hanging from wire mesh with the top lid open.

METHOD

1. Obtain a collecting license from the recognized wildlife management authority of the country in which the work is being conducted.

2. Collect the bats from their roosts in the daytime or by mist netting at night. To assess the population for reproductive synchronization, collect at least 10 adult females at the same location and on the same day.

   The animals should not be held for too long (<12 hours) in their collecting/transport cages and must be protected from exposure to heat. If collected in the first part of the night, it cannot be assumed that they have fed. Food (e.g., peeled, well-ripened bananas) and water can be offered, but there is no guarantee that any will be consumed.

3. Examine the bat's mammary glands and reproductive tracts to assess the reproductive state of each female.

   The condition of the mammary glands provides a useful initial indicator of possible or probable reproductive state. Lactating or recently lactational females will have enlarged teats surrounded by bare patches of skin, and milk will be evident in secretory portions of the mammary glands. Postlactational females will exhibit hair regrowth in the area around each of the teats that varies from slight to extensive. When fur regrowth in these areas is extensive, it often appears as a distinctive dark-gray coloration. The presence of any of these conditions raises the possibility that the female may have recently experienced a postpartum estrus and is pregnant again.

4. To export bat embryos and tissues (which are considered wildlife products) from the country of origin, obtain an export license from the recognized wildlife management authority.

   This license and possibly additional documentation must be presented for legal importation into other countries. For example, in the case of the United States, such documentation would have to be presented to U.S. Customs and Border Protection, as well as to the U.S. Fish and Wildlife Service, at the port of entry. It is also advisable to carry a letter on institutional stationery indicating how the specimens were fixed and that they are noninfectious. Potentially infectious specimens (e.g., nonfixed, frozen material) may require special import permits and packaging.

Protocol 4

Bat Embryo Collection

Bat embryos can be collected from females that are pregnant as a result of matings established in a laboratory colony or during the breeding season in the wild. The entire reproductive tract containing the single conceptus is dissected from the female and the uterus is measured (Fig. 5A). If the uteri contain embryos that can be dissected out, measured, and staged, this information can be compared to published data for normal pregnancies (Badwaik and Rasweiler 2001; Cretekos et al. 2005). The results can then be used to predict when conceptuses at other stages of development might be collected. Such data, by themselves, do not establish whether those pregnancies may have previously included a period of delay at the primitive-streak stage. The embryo is usually dissected from the reproductive tract free of decidual tissue and its extraembryonic membranes, including the yolk sac and amnion (Fig. 5B,C). Once isolated, the embryo is then processed for subsequent analysis (e.g., histology, in situ hybridization, skeleton preparations, and molecular biology, as described in the following protocols).

If females possess uteri between 2.2 and 3.3 mm, they could be either pregnant or not; therefore, the exact state should be determined histologically. This is done most efficiently by leaving the ovaries, oviducts, and uterus together and serial-sectioning them in a frontal plane. Pregnant tracts carrying primitive-streak-stage/trilaminar blastocysts in delay will also fall within this size range (J.J. Rasweiler and N.K. Badwaik, unpubl.). Abundant trophoblast present around maternal blood vessels in the myometrium, particularly in the proximal half of the uterus, would indicate that the pregnancy is indeed in delay (Badwaik and Rasweiler 2001). If females possess uteri of more than 4 mm in diameter, this would suggest that they (1) may have recently given birth (which should also be indicated by the condition of the mammary glands) or (2) are carrying embryos that have passed the primitive-streak stage. Postparturient involution of the uterus is rapid, however. Uterine diameters recorded for lactating females (three animals per group) at 5-day intervals following parturition are as follows: day-5 postpartum (4.3–5.2 mm), day 10 (3.1–4.5 mm), day 15 (2.7–3.5), and day 20 (3.0–3.2 mm) (Rasweiler and Badwaik, unpubl.). The first day that a baby is observed on the mother would be considered day 1. Two of the animals examined on day 10 had 3.1-mm uteri.

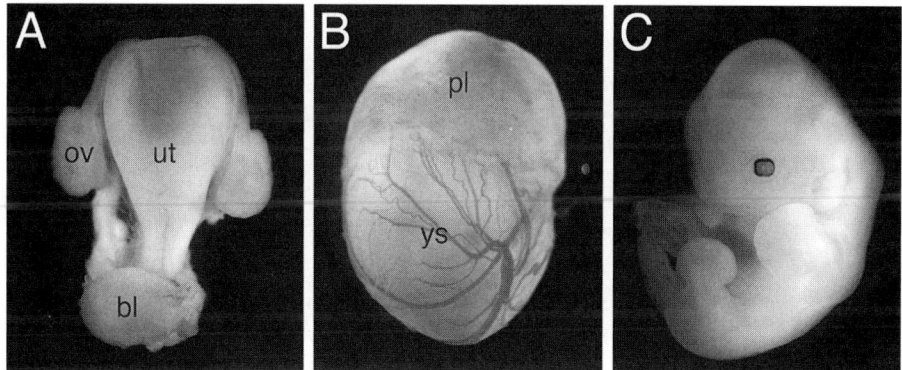

FIGURE 5. *Carollia* female reproductive tract and embryos. (*A*) Dissected female reproductive tract. (bl) Bladder; (ov) ovary; (ut) uterus. (*B*) Embryo dissected free of the uterus but still within the extraembryonic membranes. (pl) Discoidal placenta; (ys) yolk sac. (*C*) CS 16 embryo.

MATERIALS

The recipe for the item marked with <R> is on page 552.

CAUTION: See the Cautions Appendix for appropriate handling of materials marked with <!>.

Reagents

Carollia pregnant female(s), obtained from a laboratory colony or in the wild
Ethanol (70%) <!>
Phosphate-buffered saline (PBS) <R>

Equipment

Binocular magnifier (adjustable visor loupe magnifier)
Cardboard
Dissecting instruments (fine scissors, blunt and #5 watchmakers forceps)
Pins
Plastic dishes (100 mm)
Polyethylene transfer pipettes (for transferring smaller embryos into tubes; Samco 202-1S)
Spatula (for transferring larger embryos into tubes)
Stereodissecting microscope with overhead illumination

METHOD

1. Euthanize pregnant female bats according to local regulations.

2. Place the female ventral side up on cardboard (or similar material) and extend the wings and pin them to the cardboard to facilitate access to the abdomen. Wet the entire ventral surface of the abdomen and thorax with 70% ethanol.

3. Make T-shaped incisions through the skin on the ventral surface (i.e., from left to right across the lower abdomen, immediately cranial to the external genitalia, and midventrally from there to the throat). Pull up the strips of skin on either side of the midventral incision and in a cranial direction to create a hair-free field (these strips of skin should come off easily). Open the abdominal cavity broadly to expose the reproductive tract.

4. Dissect the entire reproductive tract, including uterus, oviducts, and the bladder, free of the female and place it in ice-cold PBS in a 100-mm plastic dish.

    Isolation of the tract is facilitated by holding the bladder with blunt forceps and lifting as the tract is dissected from the female. In addition, retaining the bladder with the tract during the dissection provides a landmark for distinguishing left from right that is useful for determining which ovary had ovulated.

5. Measure the greatest diameter or transverse dimension of the corpus of the uterus and compare with a table of uterine diameters from timed matings to give an approximation of the stage of embryo within (Badwaik and Rasweiler 2001; Cretekos et al. 2005).

6. Dissect the embryo/fetus from the uterus and free of the extraembryonic membranes. Finally, cut the umbilicus and rinse the embryo/fetus free of excess blood in PBS.

7. Determine the stage of the embryo by morphological comparison with the *Carollia* staging system established by Cretekos et al. (2005).

Protocol 5

Fixation and Storage of Bat Embryos

This protocol describes fixation of embryos and tissues of the bat *C. perspicillata*. The fixed embryos can be stored at –20°C for prolonged periods and can be used for whole-mount in situ hybridization (Protocol 6), whole-mount immunohistochemistry (Protocol 9), or for embedding and sectioning in paraffin or frozen blocks for a variety of histological procedures.

MATERIALS

The recipes for items marked with <R> begin on page 551.

CAUTION: See the Cautions Appendix for appropriate handling of materials marked with <!>.

Reagents

Carollia embryos or tissues obtained by dissection as described in Protocol 4
Diethyl pyrocarbonate (DEPC; e.g., Sigma-Aldrich) <!>
   DEPC is used to remove trace amounts of nuclease activity from water or buffer solutions.
DEPC-treated H_2O <!> <R>
Methanol (biotech grade; e.g., Sigma-Aldrich) <!>
Phosphate-buffered saline (PBS; RNase-free) <R>
   RNase-free PBS can be purchased as 1x solution or a 10x stock from a variety of sources, or it can be prepared according to the recipe and treated with DEPC to inactivate nucleases.
Paraformaldehyde (EM grade; Polysciences or equivalent) <!>
Paraformaldehyde (4%) <!> in 1x PBS (4% PFA) <R>
   Prepare fresh before use. Field alternative: In many cases, it is inconvenient to prepare fresh PFA. We have found that 16% paraformaldehyde solution, supplied under nitrogen gas in glass ampules by Electron Microscope Sciences, can be diluted in DEPC-H_2O and 10x PBS to a final concentration of 4% paraformaldehyde and 1x PBS and substituted for freshly prepared fix with excellent results.

Equipment

Disposable polypropylene tubes
   Various types and sizes of tubes can be used, depending on the size of the embryo/tissue to be fixed. Factors to consider when choosing a tube: Choose a type with a secure and reliable closure. Make sure that the diameter of the mouth of the tube is larger than the smallest diameter of the embryo/tissue to be fixed. Choose a tube whose volume is at least 10 times the volume of the embryo/tissue. A 2-ml screw-top Eppendorf tube works well for stages up to CS 17; 15-ml falcon-type tubes work well for larger specimens.

METHOD

1. After dissection, immerse embryos or dissected tissues in at least 10 volumes of cold 4% PFA and incubate overnight at 4°C.

2. Wash three times in 1x PBS for 15 minutes each time, with rocking.
3. Dehydrate the embryos or tissues through a graded series of washes, with rocking:

   25% methanol/PBS for 15 minutes
   50% methanol/PBS for 15 minutes
   75% methanol/PBS for 15 minutes
   100% methanol/PBS, twice, for 15 minutes each time

   Embryos can be stored at –20°C for several years without significant loss of signal.

Protocol 6

Whole-mount in Situ Hybridization of Bat Embryos with RNA Probes

Whole-mount in situ hybridization is a technique that allows direct visualization of the mRNA expression pattern of a gene of interest within embryos fixed at a wide range of developmental stages (Fig. 6). Due to limitations in probe penetration, the size of the specimen must be considered when using whole-mount procedures. This protocol is based on the methods of Harland (1991), Wilkinson (1992), and Braissant and Wahli (1998) and is presented in the following steps and sections.

Preparation of RNA Probes

The preparation of probe is an important step in whole-mount in situ hybridization. Two primary considerations for making a good probe are (1) a good template sequence and (2) an efficient labeling reaction. In our experience, the ideal template sequence for a whole-mount in situ probe should be 300–1000 bp in length. We have had little success using shorter probes, although they can be made to work in a case-by-case (and presumably sequence-dependent) manner. Typically, one begins with a cDNA or coding genomic fragment cloned into an in vitro transcription vector. The linearized vector is purified and used as the template for in vitro transcription. See the Troubleshooting section at the end of this protocol for suggestions on consistently obtaining efficient labeling reactions and the use of specificity controls.

In Situ Hybridization Procedure

The hybridization sequence is presented as a three-day schedule:

Day 1: Pretreatment and hybridization
Day 2: Posthybridization washes and antibody incubation
Day 3: Postantibody washes and color reaction

MATERIALS

The recipes for items marked with <R> begin on page 551.

CAUTION: See the Cautions Appendix for appropriate handling of materials marked with <!>.

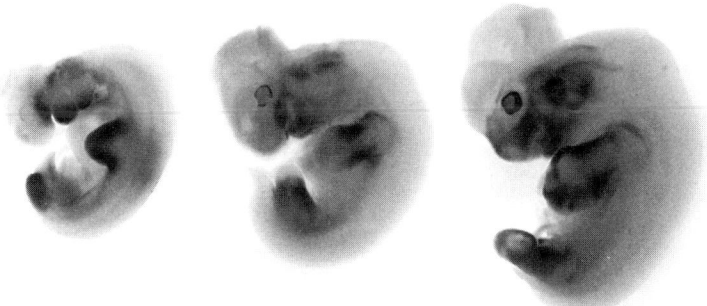

FIGURE 6. Expression of *Prx1* in *Carollia* embryos. Lateral views of embryos processed for whole-mount in situ hybridization using a bat *Prx1* probe. Stages are, from left to right, CS 14, CS 15, and CS 16.

Reagents

Use DEPC-H₂O (see below) for all solutions unless otherwise indicated. Store at room temperature unless otherwise indicated.

Acetic anhydride <!>
Antibodies
    Sheep serum, heat-treated
    Anti-digoxygenin (anti-DIG) alkaline-phosphatase (AP) antibody (Roche) *or* anti-fluorescein-AP antibody (Roche) *or* strepavidin-AP antibody (Roche)
    For detection, the choice of antibody depends on the label used in the rNTP-labeling mix in Step 1.
BABB (benzyl alcohol:benzyl benzoate [1:2, v/v]) <!>
    For BABB solutions, use glass containers.
BCIP (5-bromo-4-chloro-3-indolyl-phosphate; e.g., Roche) <!>
    Prepare a solution of 50 mg/ml in DMF <!> and store in aliquots at –20°C.
Blocking reagent (Roche)
BM purple (Roche)
5x Carbonate buffer <R>
Carollia embryos (fixed as described in Protocol 5)
DEPC-H₂O <!> <R>
Diethyl pyrocarbonate (DEPC; e.g., Sigma-Aldrich) <!>
    DEPC is used to remove trace amounts of nuclease activity from water or buffer solutions.
Dimethylformamide (DMF; e.g., Fisher) <!>
DTT (100 mM; dithiothreitol) <!>
Ethanol (80% and 100%) <!>
Glycine (2 mg/ml in PBT-H)
Hybridization buffer <R>
Methanol (biotech grade; e.g., Sigma-Aldrich) <!>
NBT (nitroblue tetrazolium salt; e.g., Roche) <!>
    Prepare a solution of 75 mg/ml in DMF <!> and store in aliquots at –20°C.
NTMT <R>
    Prepare fresh before each use.
Phosphate-buffered saline (PBS; RNase-free) <R>
PBT-H (PBS with 0.1% Tween 20) <R>
PBT-H/Hybridization buffer (1:1, v/v)
Paraformaldehyde (4%) <!> in 1x PBS (4% PFA) <R>
Proteinase K (10 µg/ml in PBT-H; e.g., Roche) <!>
RNA polymerase (T3, T7, SP6; e.g., Roche)
RNase A (100 µg/ml; e.g., Sigma-Aldrich)
RNase inhibitor (e.g., RNasin, Promega)
10x rNTP-labeling mix (Roche) <R>
Sodium acetate (3 M)
20x SSC <R>
TBST (Tris-buffered saline with Tween 20) <R>
TE (Tris EDTA) <R>
Template DNA (linearized; 1 mg/ml)
    The template is derived from a cDNA or coding genomic fragment cloned into an in vitro transcription vector (pBluescript, pGEMT, pT3T7, etc.). The vector is linearized by restriction enzyme digestion at the site that separates the template strand from the vector and the products are gel-purified.
TENS <R>
10x Transcription buffer (supplied with RNA polymerase)

Triethanolamine (TEA) <!>-HCl (100 mM) <!>
Tween 20 (molecular biology grade; e.g., Sigma-Aldrich)
Wash solution 1 <R>
Wash solution 2 <R>
Wash solution 3 <R>

Equipment

G50 Sephadex (DNA grade F; Amersham Bioscience)
G50 Sephadex spin column
   To prepare the spin column:

   i. Hydrate Sephadex G50 beads in DEPC-treated TE.
   ii. Remove the plunger and cap from a disposable 1-cc syringe (save the cap).
   iii. Plug the bottom of the syringe barrel with RNase-free (baked overnight) glass wool ("fluff" from the back of a sterile disposable pipette also works if glass wool is not available).
   iv. Back-fill syringe with G50 Sephadex using a Pasteur pipette.
   v. Hang the syringe in a disposable 14-ml round-bottom culture tube.
   vi. Centrifuge briefly at 600 rpm in a swinging bucket centrifuge to spin down the matrix.
      Use setting 2 on a standard benchtop blood centrifuge. If large buckets are available, use a 50-ml falcon tube as an improvised tube adapter.
   vii. Add approximately 400 µl of TE to the top of the syringe and centrifuge at 600 rpm for 4 minutes to pack the matrix.
   viii. Remove syringe and discard flowthrough liquid in the culture tube. Replace the cap on the syringe and put it back into the culture tube.

Tubes for in situ hybridization
   For Steps 11–18, disposable polypropylene tubes of various types and sizes can be used, depending on the size and number of embryos or tissue samples. Choose a tube whose volume is at least ten times the volume of the embryo/tissue. For Steps 19–32, it is critical that the tubes remain sealed for long periods at high temperature: 2-ml, screw-top Eppendorf tubes with silicone gaskets work well for stages up to CS 17; for larger or multiple specimens, 4- or 8-ml glass stab vials with Teflon rubber gaskets (Wheaton) work well; after Step 32, use only glass vials.

METHOD

Preparation of RNA Probes

1. Mix the following components in a clean, sterile Eppendorf tube on ice and incubate for 24 hours at 37°C.

Labeling reaction components	Typical reaction
10x Transcription buffer	5 µl
100 mM DTT	5 µl
10x rNTP-labeling mix	5 µl
1 mg/ml Linearized template DNA	2.5 µl
RNase inhibitor	0.5 µl
RNA polymerase (T3, T7, SP6)	90 units
DEPC-H$_2$O	to 50 µl

   See Troubleshooting.

2. Remove 1 µl of the reaction for gel electrophoresis with size and quantitative standards.
3. Add 1 µl of RNase-free DNase and incubate the reaction for 10 minutes at 37°C.
4. Add 50 µl of TENS and pass the reaction over a G50 Sephadex spin column to remove unincorporated nucleotides and degraded template. Add the probe mixture to the top of the matrix and centrifuge the column at 600 RPM for 4 minutes.
5. Remove the syringe from the tube and carefully remove the cap. Transfer the flowthrough to a clean, sterile Eppendorf tube; discard the syringe.
6. Add 10 µl of 3M sodium acetate to the flowthrough and mix. Add 300 µl of ethanol, mix, and incubate for 30 minutes at –20°C.
7. Recover the purified RNA probe by centrifugation at maximum speed in a microfuge for 10 minutes.
8. Rinse the pellet with 80% ethanol. Allow it to air dry for 10 minutes at room temperature.
9. *Optional:* If the probe is greater than 1000 bp in length, resuspend the pellet in 1x carbonate buffer and incubate for 10–60 minutes at 60°C until the average probe fragment is about 400 bp in length. Determine the incubation time empirically for each probe.
10. If the probe length is in the 300–1000-bp range, resuspend the pellet in hybridization buffer at 10 µg/ml.

    Probes are generally used at a concentration of 0.1–1 µg/ml; the final working concentration must be determined empirically for each probe. We generally test a new probe at 0.5 µg/ml and then increase the concentration if the signal is weak and the background is low or decrease the concentration if the signal is adequate but the background is high. This stock solution can be stored for years at –20°C.

In Situ Hybridization Procedure

Day 1: Pretreatment and Hybridization

11. If the embryos have been dehydrated and stored in 100% methanol, proceed to Step 12. Otherwise, dehydrate the embryos through a graded series of 15-minute washes, with rocking, in methanol/PBS to 100% methanol (25, 50, 75%, then twice in 100%).

    Embryos can be stored in methanol at –20°C for several years without significant loss of signal. Dehydration is important to improve permeability of tissue and should be performed even if no storage is necessary.

12. Rehydrate the embryos through 5–15-minute washes, with rocking, in 75, 50, and 25% methanol/PBS.
13. Wash the embryos twice in PBT-H.

    Optional: For older embryos that may have significant blood pooling or if performing double labeling with a HRP-conjugated antibody/DAB staining, bleach the embryo in 3–6% hydrogen peroxide:PBT-H for 1 hour at room temperature.

14. Permeabilize the embryo by treatment with proteinase K or with DEPC.

    There are several options for permeabilization, the choice of which depends on the embryo stages being examined and whether the target tissue is superficial or deep.

    Option 1: Treat the embryos with 10 µg/ml proteinase K in PBT-H at room temperature. See the table below for recommended digestion times.

    Option 2: For stages up to CS 14 and for superficial tissues up to CS 15, treat the embryos with 0.1% DEPC in PBT-H twice for 15 minutes each time at room temperature.

Stage	Deep tissues time (min)	Superficial tissues time (min)
Presomite	no treatment	no treatment
CS 10	2	no treatment
CS 11	4	no treatment
CS 12	8	2
CS 12L/13E	15	4
CS 13	20	6
CS 13L/14E	25	10
CS 14	30	15
CS 15	35	20
CS 15L	45	25
CS 16E	50	30
CS 16	65	35
CS 16L	NA	40
CS 17E	NA	45
CS 17	NA	50

For information on staging Carollia *embryos, see Cretekos et al. (2005). Note that whole-mount in situ is not recommended for deep tissues beyond CS 16 or for superficial tissues beyond CS 17.*

15. Remove the proteinase K (or DEPC) solution gently and carefully rinse the embryos with PBT-H.

    Care must be taken to treat the proteinase-K-treated embryos gently because they will be especially fragile and easily damaged until they are postfixed (Step 17 below).

16. Wash the embryos twice for 10 minutes each time in 2 mg/ml glycine in PBT-H, mixing gently.

    Optional: If nonspecific background is a problem with the probe of choice, replace Step 16 (glycine treatment) with an acetylation step:

    i. Wash the embryos twice for 15 minutes each time in 100 mM TEA-HCl in H_2O, mixing gently.

    ii. Add 2.5 μl/ml acetic anhydride. Mix gently and incubate for 15 minutes at room temperature.

    iii. Add another 2.5 μl/ml acetic anhydride. Mix gently, incubate for 15 minutes, and then rinse gently twice in PBT-H.

17. Postfix the embryos for 20 minutes in 4% PFA in PBT-H and then remove the fix and rinse with PBT-H.

18. Wash the embryos with PBT-H twice for 15 minutes each time, with rocking.

19. Wash with PBT-H/ hybridization buffer until embryos settle to the bottom of the tube.

20. Replace the PBT-H/hybridization buffer with hybridization buffer. Mix, allow the embryos to settle, and then prehybridize for 1–2 hours at the appropriate hybridization temperature (see Step 22 to select hybridization temperature).

21. Prepare hybridization buffer/probe mix (with 0.1–1 μg/ml labeled RNA probe) in sufficient quantity to cover the embryos and prewarm it to the selected hybridization temperature. Replace the hybridization buffer (in Step 20) with the prewarmed hybridization/probe mix.

22. Incubate the embryos overnight at the appropriate hybridization temperature.

    The optimal hybridization temperature must be determined empirically for each probe. Consider the following guidelines for selecting a useful range of temperatures.

    - If using an in-species probe (i.e., a probe transcribed from a cloned *Carollia* sequence on *Carollia* embryos), optimal results should be obtained in the 65–70°C range.
    - For cross-species probes (i.e., a probe transcribed from a mouse cDNA cross-species on a bat embryo), optimal results should be obtained in the 55–65°C range.

- In general, if the signal is weak and background is low, try temperatures at the lower end of the recommended range; if the signal is strong but background is high, try temperatures at the upper end of the range.

  See Troubleshooting.

Day 2: Posthybridization Washes and Antibody Incubation

23. Rinse the embryos twice with prewarmed (~65°C) Wash solution 1.
24. Wash twice for 30 minutes each time with Wash solution 1 at the temperature used for hybridization (Step 22).
25. OPTIONAL: Perform further washes to reduce the background signal:
    i. Wash the embryos twice with Wash solution 2 for 15 minutes each at room temperature.
    ii. Replace with Wash solution 2, including 100 µg/ml RNase A, and incubate for 20 minutes at room temperature.
    iii. Prepare a 1:1 mixture of Wash solution 2:Wash solution 3 and wash further in this mixture until the embryos settle.
    iv. Rinse with Wash solution 3 prewarmed to hybridization temperature.

    This step is designed to degrade any RNA probe that is not properly hybridized to target mRNA molecules and thereby reduce nonspecific background signal. In practice, we find that the stringency of the hybridization and washes is sufficient to remove any nonhybridizing probe and this step is unnecessary. Furthermore, when using nonbat species probes on bat embryos, this step should always be omitted.

26. Wash the embryos twice for 30 minutes with Wash solution 3 at the hybridization temperature (Step 22).
27. Rinse twice with TBST and then wash once with TBST for 15 minutes.
28. Prepare a preblocking solution of TBST with 10% heat-treated sheep serum and 1% Roche blocking reagent. Incubate the embryos in this solution for at least 1 hour.
29. While preblocking the embryos, dilute anti-DIG-AP antibody 1:2000 in preblocking solution (TBST with 1% heat-treated sheep serum and 1% Roche blocking reagent). Incubate at room temperature until ready to use.

    Note that the antibody used for detection depends on the label used in the rNTP-labeling mix in Step 1. Here, anti-digoxygenin (anti-DIG)-alkaline-phosphatase (AP) antibody is used to detect digoxigenin-labeled probe. Anti-fluorescein-AP antibody and strepavidin-AP antibody are used to detect fluorescein-labeled probe and biotin-labeled probe, respectively.

30. Replace the preblocking solution in Step 28 with diluted antibody solution prepared in Step 29 and incubate the embryos overnight at 4°C.

Day 3: Postantibody Washes and Color Reaction

31. Rinse the embryos twice with TBST and then wash at least five times for 1 hour each time in TBST, with rocking.

    Unless the probe is very clean, it is advisable to include additional washes or to add an extra wash overnight at 4°C.

32. Wash twice in NTMT for 15 minutes.
33. Incubate the embryos in BM Purple Substrate directly or in a solution of NBT/BCIP in NTMT (4.5 µl/ml NBT and 3.5 µl/ml BCIP in NTMT). Wrap the tubes in aluminum foil (BM Purple and NBT/BCIP are light-sensitive), rock for 15 minutes, and then incubate for 15 minutes to several days at room temperature. Observe the development of color during the course of incubation.

Incubation time depends on the expression level of the target gene and the quality of the probe. See Troubleshooting.

34. When the color has developed to the desired level, rinse and wash the embryros twice with PBT-H. Postfix them in 4% PFA for 2 hours at room temperature or overnight at 4°C.

35. Wash the embryos twice in PBT-H for 10 minutes each time.

    If the target gene is expressed primarily in superficial tissues, image them at this point or store them at 4°C in PBT-H with 0.1% sodium azide for later imaging.

36. *Optional:* For visualization of target gene expression in deep tissues, partially or fully clear the embryos before imaging.

    To partially clear the embryos, replace PBT-H with 50% glycerol in PBT-H and allow the embryos to settle before imaging.

    To fully clear the embryos, dehydrate the embryos through a graded series of washes for 15 minutes each, with rocking, in methanol/PBS to 100% methanol (25, 50, 75%, then twice in 100%) and then clear in BABB using glass vials. Because staining will fade over time in BABB, image the embryos immediately following clearing and then wash three times for 15 minutes in 100% methanol to remove BABB. After detection, store embryos in methanol at −20°C.

    Optional: For cellular resolution of target gene expression, we have had good results using the method for sectioning embryos after whole-mount in situ hybridization; see Protocol 9, Steps 12–18.

TROUBLESHOOTING

Problem (Step 1): It may be a challenge to achieve consistently efficient RNA probe synthesis.
Solution: We recommend the following considerations for preparing probe:

- After linearization, purify the template DNA using a mini spin-column method (e.g., QIAquick, QIAGEN).
- Use high-quality RNA polymerases within expiry date and at the indicated concentration.
- When checking the results of probe synthesis reaction by gel electrophoresis (see Protocol 6, Preparation of RNA Probes, Step 2), pay close attention to both the average length of product and its yield. Shorter than expected product length and unusually poor yield are indicators of inefficient probe synthesis. The most likely reason is carryover of gel contaminants during purification of template DNA (see above).

Problem (Step 22): It is important to minimize nonspecific background signals.
Solution: Many protocols recommend transcribing both sense and antisense strand probes from each template and promote the use of the former as a specificity control for the latter. Although this is generally accepted, we prefer to use a previously tested antisense probe against a gene known to be expressed during the stages in question as the specificity control for the following reasons: (1) Some genes have endogenous tissue-specific antisense transcripts that would be detected by a so-called sense strand control and (2) nonspecific background signal looks essentially the same for any given species and stage; researchers will quickly learn to recognize it.

Problem (Step 33): In some cases, particularly when NBT/BCIP is used for the color reaction, the embryos take on a general pinkish-purple cast.
Solution: This color can be washed out by dehydrating through a graded series of 15-minute washes in methanol/PBS to 100% methanol (25, 50, 75%, then twice in 100%), followed by washing 15 minutes to overnight in 100% methanol and then rehydrating back through a graded series to PBT-H.

Protocol 7

Alcian Blue Staining of Cartilage

This protocol (adapted, with modifications, from Nagy et al. 2003) is used to visualize the cartilaginous elements of the developing skeleton. It has been used with excellent results on *Carollia*, as well as several other bat species at stages from CS 14 through CS 22 (Fig. 7). Staining has been shown to last at least 2 years in BABB. However, staining has been observed to weaken during the course of months. It is therefore advisable to document results immediately.

MATERIALS

The recipe for the item marked with <R> begin on page 551.

CAUTION: See the Cautions Appendix for appropriate handling of materials marked with <!>.

Reagents

Acetic acid (5%) <!>
Acid alcohol (5% acetic acid <!> in 70% ethanol <!>)
Alcian blue (8GX; Fisher) stain solution 0.05% in acid alcohol <!>
  Best if made fresh or at least prepared periodically.
Ammonium hydroxide:ethanol solution <!> (NH_4OH <!>, 0.1% in 70% ethanol <!>)
BABB (benzyl alcohol:benzyl benzoate [1:2, v/v]) <!> <R>
  For BABB solutions, use glass containers.
Bat embryos (CS 14 through CS 22), collected as described in Protocol 4
  We typically fix bat embryos dissected in the field overnight in Bouin's solution and then wash and store for up to several years in 70% ethanol at room temperature.
Bouin's fixative (Polysciences)
Methanol <!>

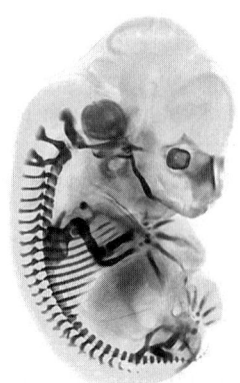

FIGURE 7. Cartilaginous skeleton of a CS 16 *Carollia* embryo. Lateral view stained with alcian blue for cartilage and cleared in BABB.

METHOD

Perform all steps at room temperature unless otherwise noted.

1. Wash the embryos in the ammonium hydroxide:ethanol solution for about 24 hours until the embryos appear white with no remaining yellow color. Six to eight changes may be required.

2. Equilibrate in acid alcohol (two changes, each for 1 hour).

3. Stain the embryos in 0.05% alcian blue solution for 2–6 hours, depending on the stage of the embryo.

   The appearance of deeper than background blue staining of cartilage in regions where skeleton is relatively superficial (e.g., ribs and neural arches of vertebrae) indicates that staining time is sufficient.

4. Wash the embryos twice with acid alcohol for 1 hour each.

5. Clear embryos by washing them in methanol (two times for 1 hour each) and then clear in BABB and store at room temperature.

   Use glass containers (e.g., scintillation vials) when preparing BABB.

Protocol 8

Alcian Blue/Alizarin Red Staining of Cartilage and Bone

This protocol (adapted, with modifications, from Nagy et al. 2003) is used to stain embryo skeleton at later stages of development, when there is significant replacement of the cartilaginous early skeleton with ossified bone. It has been used with good results on *Carollia* and other bat species from CS 20 through neonatal stages.

MATERIALS

CAUTION: See the Cautions Appendix for appropriate handling of materials marked with <!>.

Reagents

Acetic acid (5%) <!>
Acetone <!>
Acid alcohol (5% acetic acid <!> in 70% ethanol <!>)
Alcian blue (8GX; Fisher) stain solution 0.05% in acid alcohol <!>
  Best if made fresh or at least prepared periodically.
Alizarin red stain solution (50 mg alizarin red [Sigma-Aldrich] per liter of 1% potassium hydroxide; this is 0.015% [w/v] alizarin red in 1% potassium hydroxide)
Bat embryos (CS 20 through neonatal stages; collected as described in Protocol 4)
Clearing solution (1% potassium hydroxide in 20% glycerol)
Ethanol (70% and 95%) <!>
Glacial acetic acid (20 ml) <!>
Glycerol
Potassium hydroxide (1% w/v, in distilled water) <!>
Storage solution (glycerol:ethanol <!>; 1:1, v/v)

Equipment

Forceps and scissors (size depends on age of fetus)
Tubes (50-ml, screw-capped, clear plastic)

METHOD

Perform all steps at room temperature unless otherwise noted.

Preparation of the Embryos for Staining

1. Isolate the fetuses or euthanize neonates according to local regulations. Place the fetuses or neonates in tap water for 1–24 hours (*optional*).

2. Scald the fetuses or pups in hot tap water (65–70°C) for 20–30 seconds for easier maceration of the tissue later.

3. Use forceps to eviscerate the embryos, including the contents of the peritoneal and pleural cavities.

4. Fix the embryos in 95% ethanol overnight. Make sure to remove all of the bubbles from the body cavity.

   The fixed, field-collected specimens may be stored in 95% ethanol at 4°C for up to several months.

5. For embryos at stages 20 to 24, proceed to Step 6; for embryos at stages greater than CS 24 and for neonatal stages, remove as much of the skin as possible to facilitate penetration of alcian and alizarin stains.

Staining the Embryos

6. Rinse the fetuses or neonates briefly in deionized water. Stain them for cartilage by placing the embryo in enough alcian blue stain solution to cover the body completely. Allow the staining to continue for 6 hours to overnight, depending on the stage (with increasing incubation periods for older fetuses).

   A 50-ml, screw-capped plastic tube is convenient, or for newborn mice or younger fetuses, a 6-well tissue culture plate is useful. Make sure to remove all of the bubbles from the body cavity.

7. Wash the embryos in 70% ethanol for 6–8 hours. A few changes of the ethanol should be sufficient.

8. Clear the sample by placing it in 1% KOH for about 2 hours.

9. Counterstain bone in alizarin red stain solution overnight.

   Younger fetuses may require less time. The tissue will continue clear while it stains.

10. Clear the samples by placing them in clearing solution overnight. Replace the clearing solution with several changes of storage solution during several days until the tissue is completely clear.

Preparing Samples for Visualization and Documentation

11. If desired, remove the limbs from the skeleton preparation, which is easily performed with forceps.

    The removal of limbs facilitates visualization of the rest of the body or documentation of the limbs in isolation.

12. Dissect the rib cage away from the skeleton preparation by cutting the ribs close to the vertebrae. Place a glass slide on top of the isolated rib cage to create a flat mount for photodocumentation.

Protocol 9

Whole-mount Immunohistochemistry

This protocol (adapted, with modifications, from Nagy et al. 2003) is used for direct visualization of the expression pattern of a protein of interest within embryos fixed at a wide range of developmental stages and has been used with good results on *Carollia* embryos at stages from CS 11 through CS 15.

MATERIALS

The recipes for items marked with <R> begin on page 551.

CAUTION: See the Cautions Appendix for appropriate handling of materials marked with <!>.

Reagents

Antibodies
   Primary antibody (e.g., an affinity-purified, polyclonal rabbit antiserum), directed against antigen or protein of interest
   Secondary antibody (e.g., horseradish peroxidase [HRP]-coupled goat antirabbit IgG [Boehringer Mannheim or Jackson ImmunoResearch Laboratories])
Aqua Poly/Mount (Polysciences)
BABB (benzyl alcohol:benzyl benzoate, 1:2) <!> <R>
   BABB is used to clear the embryos after staining and provides a mountant to observe and photograph the stained embryo.
Bovine serum albumin (BSA, 0.2%; Sigma-Aldrich)
Carollia embryos of the desired stage
   Embryos should be prepared for fixation and stored in 100% methanol, as described in Protocol 4.
DAB <!> (3,3′-diaminobenzidine tetrahydrochloride; Sigma-Aldrich)
   Store desiccated at –20°C. Warm to room temperature before weighing.
DAB/NiCl$_2$ <!> <R>
   Use within 1 hour of preparing and protect from exposure to light.
Hydrogen peroxide (H$_2$O$_2$) <!>
   This is generally supplied as a 30% solution. Store at 4°C. The solution lasts about 1 month at this temperature.
Methanol (50, 80, and 100%) <!>
Methanol/dimethylsulfoxide (DMSO, 4:1; prepare fresh) <!>
Methanol/DMSO/H$_2$O$_2$ (4:1:1; prepare fresh) <!>
2-Methylbutane <!>
OCT/30% sucrose mixture: OCT <!> (Tissue-Tek) and 30% sucrose/PBS (1:1, v/v)
Paraformaldehyde (4%) <!> in 1x PBS (4% PFA) <R>
PBSMT <R>
PBT-I <R>
   Make fresh before use. Note that PBT-I is different from the PBT used in whole-mount in situ hybridization (PBT-H).
Phosphate-buffered saline (PBS) <R>
Sucrose (30% in 1x PBS [pH 7.4])

Equipment

Cryotome
Dissecting microscope
Glass slides
Petri dishes (35-mm glass)
Rocking platform
Tubes (15-ml, screw-capped)

METHOD

Preparation of Embryos for Antibody Treatment

1. Rehydrate the fixed embryos at room temperature in microfuge tubes in the following series; perform all steps with gentle rocking.

   Rocking is important to facilitate penetration of the antibody, so make sure that the solution is well mixed. Note, however, that the embryos are fragile once they are fixed and so should not be subjected to vigorous rocking.

   50% methanol (1 ml) for 30 minutes. Remove solution with a Pasteur pipette or micropipetter
   PBS (1 ml) for 30 minutes. Siliconize the tubes if the embryos stick to the sides
   PBSMT (1 ml), twice for 1 hour each

2. Incubate the embryos with rocking in microfuge tubes with 1 ml of primary antibody diluted in PBSMT overnight at 4°C.

   The correct dilution must be determined empirically, but 1:200 is a typical dilution. This and all subsequent procedures can be performed at room temperature if the antibody used is stable. If the antibody is unstable, perform procedures at 4°C.

3. Wash the embryos in PBSMT five times with rocking:

   Once with 1 ml for 1 hour at 4°C. Transfer to 15-ml tubes
   Once with 10 ml for 1 hour at 4°C
   Three times in 10 ml for 1 hour at room temperature

4. Transfer the embryos to microfuge tubes. Incubate them in 1 ml of secondary antibody diluted 1:500 in PBSMT with rocking overnight at 4°C.

5. Wash the embryos again as in Step 3.

6. Rinse the embryos in 5 ml of PBT-I. Transfer them into microfuge tubes and then wash them in 1 ml of PBT-I with rocking for 10 minutes.

Staining Embryos for HRP Activity

7. Incubate the embryos in microfuge tubes with 1 ml of DAB/$NiCl_2$ for 30 minutes at room temperature.

   This step allows full penetration of the substrate into the embryo.

8. Add hydrogen peroxide to a final concentration of 0.03% and rock until the color intensity in the embryos looks appropriate (i.e., to the point at which specific regions of staining are obvious but before background staining comes up; usually after 2–10 minutes).

   If necessary, check the color in the dissecting microscope. If the color reaction occurs too quickly, use a lower concentration of hydrogen peroxide.

9. Postfix the embryos in 4% PFA.

   Without postfixation, the color of the stained embryos will fade under strong light, particularly in light used for photography. Although embryos can be stored in BABB in the dark, it is advisable to obtain a photographic record as soon as possible. Use a tungsten color film without filters or a daylight color film with a blue filter. For photography, place the embryo on a depression slide with a coverslip on top.

An alternative procedure is to use AP-coupled secondary antibody and a substrate that is insoluble in water and alcohol (e.g., Naphtol-AS-MX-phosphate/fast red TR_f).

10. Rinse embryos in the following series with rocking:

    PBT-I (1 ml) quick rinse
    PBT-I (1 ml) for 30 minutes at room temperature
    50% methanol (1 ml) for 30 minutes at room temperature
    80% methanol (1 ml) for 30 minutes at room temperature
    100% methanol (1 ml) for 30 minutes at room temperature

11. Replace the rinse with BABB (500 µl) and incubate for 10 minutes. Visualize and image or photograph the stained embryos in a glass petri dish (do **not** use polystyrene dishes).

Sectioning Embryos after Whole-mount Staining

If desired, the embryos may be sectioned after whole-mount staining to study the cellular resolution of protein expression.

12. To prepare whole-mount embryos for sectioning, remove BABB by washing the embryos three times in 100% methanol for 30 minutes each.

13. Rehydrate the embryos through the following series with rocking:

    75% methanol/PBS for 5–15 minutes
    50% methanol/PBS for 5–15 minutes
    25% methanol/PBS for 5–15 minutes
    PBS twice for 5–15 minutes

14. Transfer the embryos into 30% sucrose in 1x PBS and incubate with gentle shaking until the embryos sink to the bottom of the tube.

    This may take up to 2 days, depending on the size of the embryos.

15. Transfer the embryos to OCT/30% sucrose and shake gently for 2 hours at room temperature.

    Embryos can be stored for up to 1 week at 4°C in 30% sucrose or the OCT/30% sucrose mixture.

16. Embed the embryos in OCT under a dissecting microscope to ensure the desired orientation of the specimens and then snap-freeze by immersion in a solution of 2-methylbutane cooled on dry ice.

17. Store the embedded blocks, wrapped in aluminum foil, at –80°C.

18. Section the blocks on a cryotome onto glass slides and mount the sections in Aqua Poly/Mount.

Recipes

CAUTION: See the Cautions Appendix for appropriate handling of materials marked with <!>.

BABB (BENZYL ALCOHOL:BENZYL BENZOATE [1:2, V/V]) <!>

Prepare a 1:2 solution (v/v) of benzyl alcohol in benzyl benzoate. Use glass containers only.

5x CARBONATE BUFFER

Reagent	Quantity (for 10 ml)	Final concentration
Sodium bicarbonate (1 M)	2.0 ml	200 mM
Sodium carbonate (1 M) <!>	3.33 ml	300 mM

Adjust the final volume to 10 ml with distilled or deionized H_2O.

DAB/NiCL$_2$ <!>

Mix 0.03 g of DAB and 0.03 g of $NiCl_2$ in 50 ml of PBT-I <R>. Use within 1 hour of preparing and protect from exposure to light. The nickel enhances the sensitivity of the color reaction and produces a slate-gray to purple precipitate. If necessary, vary the amount of nickel to alter the intensity of the color. Cobalt can be substituted for nickel.

DEPC <!>-TREATED H_2O

Add 1 ml of DEPC to 1 liter of distilled or deionized H_2O and shake vigorously to mix. Autoclave for 20 minutes on liquid cycle to sterilize and deactivate DEPC.

HYBRIDIZATION BUFFER

Reagent	Quantity (for 100 ml)	Final concentration
20x SSC (pH 5.0)	10 ml	2x
SDS (20%, pH 5.4) <!>	5 ml	1%
Heparin (50 mg/ml) <!>	0.1 ml	50 µg/ml
Yeast (torula) RNA (Sigma-Aldrich) (20 mg/ml)	0.25 ml	50 µg/ml
Formamide <!>	50 ml	50%

Adjust the final volume to 100 ml with DEPC-treated H_2O and store at $-20°C$.

NTMT

Reagent	Quantity (for 100 ml)	Final concentration
NaCl (5 M)	2 ml	100 mM
Tris-Cl (1 M, pH 9.5)	10 ml	100 mM
$MgCl_2$ (1 M) <!>	5 ml	50 mM
Tween 20	0.1 ml	0.1%
Levamisole <!>	48 mg	2 mM

Adjust the final volume to 100 ml with DEPC-treated H_2O. Prepare fresh before each use.

PARAFORMALDEHYDE (4%) <!> IN 1x PBS (4% PFA)

Reagent	Quantity	Final concentration
Paraformadehyde powder <!>	4 g	4%
DEPC-H_2O <!>	80 ml	
NaOH (10 M) <!>	2 drops	
10x PBS	10 ml	1x

Incubate the paraformaldehyde in H_2O, treated with 2 drops of 10 M NaOH, at 65°C, mixing occasionally until the powder is completely dissolved. Chill the solution on ice and then add 10x PBS.

10x PHOSPHATE-BUFFERED SALINE (PBS)

Reagent	Quantity (1 liter)	Final concentration
NaCl	80 g	1.37 M
KCl <!>	2 g	27 mM
KH_2PO_4 <!>	2.4 g	20 mM
Na_2HPO_4	14.4 g	100 mM

Dissolve the component in 800 ml of distilled or deionized H_2O and adjust the pH to 7.4 with HCl <!>. Bring the final volume to 1 liter with H_2O, dispense into aliquots and sterilize by autoclaving for 20 minutes on liquid cycle. Store at room temperature.

PHOSPHATE-BUFFERED SALINE (PBS), RNASE-FREE

RNase-free PBS can be purchased as a 1x solution or a 10x stock from a variety of sources or it can be prepared according to the preceding recipe and treated with DEPC to inactivate nucleases. Before autoclaving, add 1 ml of DEPC and shake vigorously and then autoclave as directed to sterilize the solution and inactivate the DEPC.

PBT-H

Prepare PBS with 0.1% Tween 20

For whole-mount in situ hybridization.

PBT-I

Prepare PBS with 0.2% bovine serum albumin (BSA; Sigma-Aldrich) and 0.5% Triton X-100.

Make fresh before use. Note that PBT-I for immunohistochemistry is different from PBT used in whole-mount in situ hybridization (PBT-H).

PBSMT

Prepare PBS with 2% nonfat instant skim milk (to block all nonspecific protein-binding sites in the tissues) and 0.5% Triton X-100 (to facilitate permeability of the tissues).

> Prepare fresh before use. Note that the brand of nonfat skim milk is important. Carnation gives consistently good results whereas other brands do not.

10x rNTP-LABELING MIX

Reagent	Quantity (for 500 µl)	Final concentration
rATP (not dATP) (100 mM)	50 µl	10 mM
rCTP (not dCTP) (100 mM)	50 µl	10 mM
rGTP (not dGTP) (100 mM)	50 µl	10 mM
rUTP (65 mM)	50 µl	6.5 mM
Digoxigenin-11-UTP <!> (or fluorescein-12-UTP or biotin-16-UTP) (35 mM)	50 µl	3.5 mM

The labeling mix may be purchased as a 10x digoxigenin-rUTP/rNTP mix (Roche) or it can be prepared according to the recipe. For detection, use anti-digoxigenin-alkaline-phosphatase (AP) antibody (Roche), anti-fluorescein-AP antibody (Roche), or strepavidin-AP (Roche), respectively. Store in aliquots at –20°C.

20x SSC

Reagent	Quantity (g)	Final concentration
NaCl	175.3	3 M
Sodium citrate	88.2	3 M

Dissolve the components in 800 ml of distilled or deionized H_2O and adjust the pH to 5.4 with HCl <!>. Add water to a final volume of 1 liter, dispense into aliquots, and sterilize by autoclaving. Store at room temperature.

TBST (Tris-buffered saline with Tween-20)

Reagent	Quantity (100 ml)	Final concentration
Tris-Cl (1 M, pH 7.5)	2.5 ml	25 mM
NaCl (5 M)	2.8 ml	140 mM
KCl (1 M) <!>	270 µl	2.7 mM
Tween 20	100 µl	0.1

Adjust the volume to 100 ml with H_2O. Dispense into aliquots and sterilize the solution by autoclaving for 20 minutes on liquid cycle. Store at room temperature.

TE

Reagent	Quantity (for 100 ml)	Final concentration
Tris-Cl (1 M, pH 7.5)	1 ml	10 mM
EDTA (0.5 M, pH 8) <!>	2 ml	10 mM

Adjust the volume to 100 ml with H_2O. Sterilize the solution by autoclaving for 20 minutes on liquid cycle. Store at room temperature.

TENS

Reagent	Quantity (for 1 liter)	Final concentration
NaCl (5 M)	20 ml	100 mM
Tris-Cl (1 M, pH 7.5)	20 ml	20 mM
EDTA (0.5 M, pH 8) <!>	20 ml	20 mM
SDS (20%) <!>	20 ml	1%

10x TRANSCRIPTION BUFFER

Reagent	Quantity (for 10 ml)	Final concentration
Tris-Cl (1 M, pH 7.5)	4 ml	0.4 M
NaCl (5 M)	1 ml	0.5 M
MgCl$_2$ (1 M) <!>	800 ml	80 mM
Spermidine (1 M) <!>	200 ml	20 mM

Adjust the volume to 10 ml with nuclease-free H$_2$O. Store in aliquots at –20°C.

WASH SOLUTION 1

Reagent	Quantity (for 100 ml)	Final concentration
20x SSC	25 ml	5x
20% SDS <!>	5 ml	1%
Formamide <!>	50 ml	(50%)

WASH SOLUTION 2

Reagent	Quantity (for 100 ml)	Final concentration
NaCl (5 M)	10 ml	0.5 M
Tris-Cl (1 M, pH 7.5)	1 ml	10 mM
Tween 20	100 µl	0.1%

WASH SOLUTION 3

Reagent	Quantity (for 100 ml)	Final concentration
20x SSC	10 ml	2x
Formamide <!>	50 ml	(50%)
Tween 20	100 µl	0.1%

REFERENCES

Badwaik, N.K. and Rasweiler IV, J.J. 2000. Pregnancy. In *Reproductive biology of bats* (ed. E.G. Crichton and P.H. Krutzsch), pp. 221–294. Academic, San Diego.

Badwaik, N.K. and Rasweiler IV, J.J. 2001. Altered trophoblastic differentiation and increased trophoblastic invasiveness during delayed development in the short-tailed fruit bat, *Carollia perspicillata*. *Placenta* **22**: 124–144.

Badwaik, N.K., Rasweiler IV, J.J., and Oliveira, S.F. 1997. Formation of reticulated endoderm, Reichert's membrane and amniogenesis in blastocysts of captive-bred, short-tailed fruit bats, *Carollia perspicillata. Anat. Rec.* **247:** 85–101.

Baker, R.J. 1979. Karyology. In *Biology of bats of the new world family Phyllostomidae*, (ed. R.J. Baker et al.), vol. 3, pp. 107–155. Texas Tech University (Museum special publication no. 16). Texas Tech University, Lubbock.

Braissant, O. and Wahli, W. 1998. A simplified in situ protocol using non-radioactive probes to detect abundant and rare mRNAs. *Biochemica* **1:** 10–16.

Chen, C.-H., Cretekos, C.J., Rasweiler IV, J.J., and Behringer, R.R. 2005. *Hoxd13* expression in the developing limbs of the short-tailed fruit bat, *Carollia perspicillata. Evol. Dev.* **7:** 130–141.

Cloutier, D. and Thomas, D.W. 1992. *Carollia perspicillata. Mamm. Spec.* **417:** 1–9.

Cretekos, C.J., Rasweiler, J.J., and Behringer, R.R. 2001. Comparative studies on limb morphogenesis in mice and bats, a functional genetic approach towards a molecular understanding of diversity in organ formation. *Reprod. Fertil. Dev.* **13:** 691–695.

Cretekos, C.J., Weatherbee, S.D., Chen, C.-H., Badwaik, N.K., Niswander, L., Behringer, R.R., and Rasweiler IV, JJ. 2005. Embryonic staging system for the short-tailed fruit bat, *Carollia perspicillata*, a model organism for the mammalian order Chiroptera, based upon timed pregnancies in captive-bred animals. *Dev. Dyn.* **233:** 721–738.

Cretekos, C.J., Deng, J.M., Green, E.D., NISC Comparative Sequencing Program, Rasweiler IV, J.J., and Behringer, R.R. 2007. Isolation, genomic structure and developmental expression of Fgf8 in the short-tailed fruit bat, *Carollia perspicillata. Intl. J. Dev. Biol.* **51:** 333–338.

Cretekos, C.J., Wang, Y., Green, E.D., NISC Comparative Sequencing Program, Martin, J.E., Rasweiler IV, J.J., and Behringer, R.R. 2008. Regulatory divergence modifies limb length between mammals. *Genes Dev.* **2:** 141–151.

de Bonilla, H. and Rasweiler IV, J.J. 1974. Breeding activity, preimplantation development, and oviduct histology of the short-tailed fruit bat, *Carollia*, in captivity. *Anat. Rec.* **179:** 385–404.

Eisenberg, J.F. 1989. *Mammals of the neotropics. I. The northern neotropics: Panama, Colombia, Venezuela, Guyana, Suriname, French Guiana.* The University of Chicago Press, Chicago.

Fleming, T.H. 1988. *The short-tailed fruit bat.* The University of Chicago Press, Chicago.

Harland, R.M. 1991. In situ hybridization: An improved whole-mount method for *Xenopus* embryos. *Methods Cell Biol.* **36:** 685–695.

Hsu, T.C., Baker, R.J., and Utakoji, T. 1968. The multiple sex chromosome system of American leaf-nosed bats (Chiroptera–Phyllostomidae). *Cytogenetics* **7:** 27–38.

Nagy, A., Gertsenstein, M., Vintersten, K., and Behringer, R. 2003. *Manipulating the mouse embryo: A laboratory manual*, 3rd ed. Cold Spring Harbor Laboratory Press, Cold Spring Harbor, New York.

Oliveira, S.F., Rasweiler IV, J.J. and Badwaik, N.K. 2000. Advanced oviductal development, transport to the preferred implantation site, and attachment of the blastocyst in captive-bred, short-tailed fruit bats, *Carollia perspicillata. Anat. Embryol.* **201:** 357–381.

Pine, R.H. 1972. Bats of the genus *Carollia*. Agricultural Experimental Station, technical monograph 8. Texas A & M University Press, College Station.

Rasweiler IV, J.J. 1977. The care and management of bats as laboratory animals. In *Biology of bats* (ed. W.A. Wimsatt), vol. 3, pp. 519–617. Academic, New York.

Rasweiler IV, J.J. and Badwaik, N.K. 1996. Improved procedures for maintaining and breeding the short-tailed fruit bat (*Carollia perspicillata*) in a laboratory setting. *Lab. Anim.* **30:** 171–181.

Rasweiler IV, J.J. and Badwaik, N.K. 1997. Delayed development in the short-tailed fruit bat, *Carollia perspicillata. J. Reprod. Fertil.* **109:** 7–20.

Rasweiler IV, J.J., and Badwaik, N.K. 2000. Anatomy and physiology of the female reproductive tract. In *Reproductive biology of bats* (ed. E.G. Crichton and P.H. Krutzsch), pp. 157–220. Academic Press, San Diego.

Rasweiler IV, J.J. and Badwaik, N.K. 2009a. Special considerations for the capture, handling and transport of *Glossophaga soricina* and *Carollia perspicillata*. In *Bats in captivity* (ed. S. Barnard), vol. 2. Krieger, Melbourne, Florida. (In press.)

Rasweiler IV, J.J. and Badwaik, N.K. 2009b. Additional comments on the nutrition and feeding of some phyllostomid bats in the laboratory. In *Bats in captivity* (ed. S. Barnard), vol. 3. Krieger, Melbourne, Florida. (In press.)

Rasweiler IV, J.J. and Badwaik, N.K. 2009c. The laboratory environment for maintaining and breeding some bats in the family Phyllostomidae. In *Bats in captivity* (ed. S. Barnard), vol. 3. Krieger, Melbourne, Florida. (In press.)

Rasweiler IV, J.J. and de Bonilla, H. 1972. Laboratory maintenance methods for some nectarivorous and frugivorous phyllostomatid bats. *Lab. Anim. Sci.* **22:** 658–663.

Rasweiler IV, J.J. and de Bonilla, H. 1992. Menstruation in short-tailed fruit bats (*Carollia* spp.). *J. Reprod. Fertil.* **95:** 231–248.

Rasweiler IV, J.J., Oliveira, S.F., and Badwaik, N.K. 2002. An ultrastructural study of interstitial implantation in captive-bred, short-tailed fruit bats, *Carollia perspicillata*: Trophoblastic adhesion and penetration of the uterine epithelium. *Anat. Embryol.* **205:** 371–391.

Sears, K.E., Behringer, R.R., Rasweiler IV, J.J., and Niswander, L.A. 2006. Development of bat flight: Morphologic and molecular evolution of bat wing digits. *Proc. Natl. Acad. Sci.* **103:** 6581–6586.

Simmons, N.B. 2005. Order Chiroptera. In *Mammalian species of the world. A taxonomic and geographic reference*, 3rd ed. (ed. D.E. Wilson and D.A. Reeder), vol. 1, pp. 312–529. John Hopkins University Press, Baltimore.

Weatherbee, S.D., Behringer, R.R., Rasweiler IV, J.J., and Niswander, L.A. 2006. Interdigital webbing retention in bat wings illustrates genetic changes underlying amniote limb diversification. *Proc. Natl. Acad. Sci.* **103:** 15103–15107.

Wilkinson, D.G. 1992. *In situ hybridization: A practical approach.* Oxford University Press, New York.

Wilson, D.W. and Reeder, DA. 2005. Introduction. *Mammalian species of the world. A taxonomic and geographic reference*, 3rd ed. (ed. D.E. Wilson and D.A. Reeder), vol. 1, pp. xxv–xxix. John Hopkins University Press, Baltimore.

WWW RESOURCES

http://embryology.med.unsw.edu.au/OtherEmb/bat.htm This site describes the bat (Chiroptera) order, its distribution, and various developmental studies.

http://www.ensembl.org/Myotis_lucifugus/index.html Presented here is the first release of the low-coverage 1.7x assembly of the genome of the microbat or little brown bat (*Myotis lucifugus*). *Myotis lucifugus* is one of 16 mammals that will be sequenced as part of the Mammalian Genome Project.

23 Opossum (*Monodelphis domestica*)
A Marsupial Developmental Model

Anna L. Keyte and Kathleen K. Smith

Department of Biology, Duke University, Durham, North Carolina 27708

ABSTRACT

Monodelphis domestica is the most commonly used laboratory marsupial. In addition to the many factors that make it a convenient laboratory animal (small size, ease of care, nonseasonal breeding), it is the first marsupial whose genome has been sequenced. In this review, we present an overview of aspects of its biology and use as a model organism and discuss basic care, breeding, embryo manipulation, and modifications of common techniques for the study of the development of this species.

> **PROTOCOLS**
> 1. Basic Maintenance and Breeding of *Monodelphis*, 561
> 2. Harvesting *Monodelphis* Embryos, 564
> 3. *Monodelphis* Whole-embryo Culture, 566
> 4. Whole-Mount In Situ Hybridization in *Monodelphis* Embyros, 568

BACKGROUND INFORMATION

Monodelphis domestica (Didelphidae, Marsupialia), the gray, short-tailed, or laboratory opossum, is an increasingly popular model organism, and colonies are found in laboratories in numerous locations worldwide. Its establishment as a laboratory model was largely through the efforts of the Southwest Foundation for Biomedical Research, which maintains a fully pedigreed colony of approximately 2400 animals (VandeBerg and Robinson 1997; VandeBerg 1999). Its use in biomedical research is likely to increase as a result of the recently available genome sequence (Mikkelsen et al. 2007).

Monodelphis is a small (~100 gram) member of the marsupial family Didelphidae (the term "didelphid," meaning "two uteri," was given to this group as a whole because the pouch was originally thought to be a "second uterus"; it does not refer to the divided or double uterus possessed by all marsupials). The species is native to Brazil, northern Argentina, and adjoining regions. Individuals are omnivorous, eating insect prey, fruit, and small vertebrates in the wild and are often found in or near human dwellings (hence the specific descriptor "*domestica*"). The name *Monodelphis* ("single uterus") refers to the fact that, like many small marsupials, it is pouchless (Fig. 1).

Like all marsupials, *Monodelphis* has a distinctive reproductive strategy. It is born after an extremely short gestation period and in an embryonic state. At birth, many organ systems are no more developed than those of a 10- to 12-day-old embryonic mouse; development is completed during a prolonged period of lactation. But unlike most marsupials, *Monodelphis* breed year-round and reach sexual maturity relatively rapidly, at approximately 5 to 7 months. The young are weaned at about 60 days, and in the lab, animals generally live for 2.5 to 3 years (Macrini 2004). Females may raise two to three litters each year.

This chapter, with full-color images, can be found online at www.cshprotocols.org/emo.

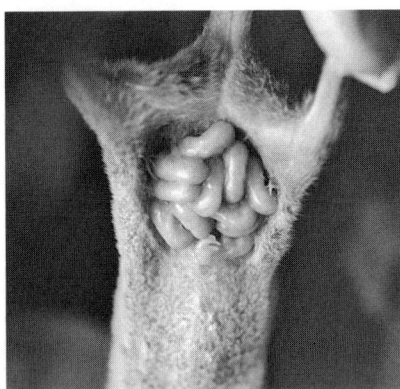
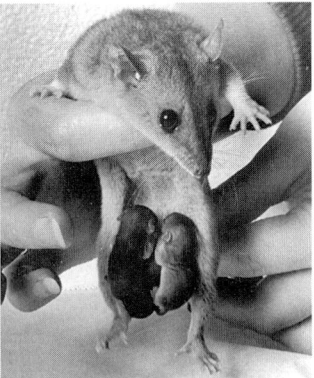

FIGURE 1. *Monodelphis domestica* mother with a litter of 4-day postnatal embryos (*left*) and young, approximately 30-day-old postnatal pups (*right*).

SOURCES AND HUSBANDRY

Monodelphis are relatively docile and easy to maintain and breed in the laboratory (for general husbandry and breeding guidelines, see Protocol 1). They generally become quite tame and, if regularly handled, rarely bite, struggle, or attempt to escape. However, because *Monodelphis* are solitary in the wild, adults will fight if kept together. Therefore, when females and males are placed together for breeding, special care must be taken to minimize injury. We have found high rates of breeding success (75–90%), with minimal injury due to fighting, after following the few simple steps described in Protocol 1.

At the time of publication, *Monodelphis domestica* individuals do not appear to be available for purchase for research, although they are increasingly common in the pet trade. The Southwest Foundation for Biomedical Research (www.sfbr.org), which maintains a very large colony, will in some cases collaborate with investigators and provide material (J.L. VandeBerg, pers. comm.). To do extensive embryonic work, a reasonably large breeding colony must be maintained. We currently maintain approximately 100 animals (~3:1 female:male ratio), which allows for sacrifice of up to 12 pregnant females per month for experimental purposes, as well as for replenishment of the colony.

RELATED SPECIES

Monodelphis is one of the few widely available marsupial species used in the lab. Some labs have also attempted to work with *Didelphis virginiana*; however, these animals are much larger and more difficult to keep. They are seasonal breeders, breeding once a year, and it takes more than 120 days to raise a litter, from birth to weaning. Several Australian investigators maintain colonies of *Macropus eugenii* (tammar wallaby), *Sminthopsis macroura* (stripe-faced dunnart), and *Trichosurus vulpecula* (brush possum) but these animals are of limited availability, even in Australia.

USES OF THE *MONODELPHIS* MODEL SYSTEM

Monodelphis has proven to be useful as a model organism for many types of studies. Quite a few important events in the formation of the nervous system occur postnatally (Saunders et al. 1989). For example, during the first week of life, the spinal cord is capable of repair following severe damage, so these animals are used as a model for spinal cord injury (Mollgard et al. 1994; Saunders et al. 1995). Because *Monodelphis* is one of the few species in which UV radiation has been shown to induce melanoma, they are used as a convenient model for the development of UV-induced melanoma (Ley et al. 2000). *Monodelphis* is also used as a model for comparative studies of the immune system (Miller and Belov 2000) and as a model for genetic influences on cholesterol (Rainwater et al. 2001). In addition, *Monodelphis* has been used to understand the basic functions of the olfactory system and, in particular, the role of various olfactory chemicals on social and reproductive behavior (Zuri et al. 2003).

Recently, *Monodelphis* has been used by many labs to understand fundamental aspects of marsupial development, anatomy, evolution, and, in particular, evolutionary consequences of the derived marsupial mode of development and reproduction (Selwood and Johnson 2006; Freyer et al. 2007). For example, Smith and colleagues have detailed differences in the relative developmental timing of craniofacial and neural structures that result from the unique requirements of its reproductive strategy (Smith 1994, 2001, 2006; Vaglia and Smith 2003). These studies have demonstrated that marsupial development, when compared to other amniotes and especially eutherian mammals, is characterized by a complex pattern of heterochrony (shifts in the rate and timing of development of various elements). In the head, for example, structures important in suckling develop early, relative to eutherian mammals, whereas neural elements are delayed. Within the nervous system, forebrain structures are particularly delayed. Further work (Smith 2006) has shown that this heterochrony extends to the first differentiation of the neural crest cells, which differentiate and delaminate from the neural plate early, relative to events of neural or mesodermal development. These heterochronies in the differentiation of neural cells and neural crest cells can now be traced back to shifts in gene expression patterns (K.K. Smith, unpub.).

In addition to events in the cranial region, marsupials exhibit a steep anterior-posterior gradient in the development of the body axis, with advanced development of the cervical region and forelimbs and delayed development of posterior somites and hind limbs. Work on the developmental basis of observed heterochronies in the limbs and body axis is underway (Fig. 2; Keyte et al. 2006).

GENETICS, GENOMICS, AND ASSOCIATED RESOURCES

Monodelphis reaches sexual maturity by about 6 months, but some labs are able to effectively use classical quantitative genetics using long-term pedigreed colonies (Rainwater et al. 2001). The Southwest Foundation for Biomedical Research has partially inbred strains that exhibit certain traits (e.g., differing levels of plasma cholesterol after being fed a high-cholesterol diet); however, at this time, there are no developmental mutant strains.

The genomic resources for *Monodelphis* were recently reviewed (Samollow 2006). A database of single-nucleotide polymorphisms (SNPs), bacterial artificial chromosome (BAC) libraries (Duke et al. 2007), a genetic linkage map (Samollow et al. 2007), and the genomic sequence (Mikkelsen et al. 2007) are currently available. An expressed sequence tag (EST) database and microchip expression arrays are in preparation (Samollow 2006). To date, there are no established guidelines for naming *Monodelphis* genes.

TECHNICAL APPROACHES

RNA and DNA extraction, immunohistochemistry, and in situ hybridization techniques for *Monodelphis* embryos are all standard protocols similar to those used in any vertebrate. Specifics for immunohistochemistry on *Monodelphis* neonates have been described (Smith 1994). See Protocol 2 for embryo harvesting and Protocol 3 for in vitro whole-embryo culture. See Protocol 4 for whole-mount in situ hybridization (WMISH) techniques currently used in our lab.

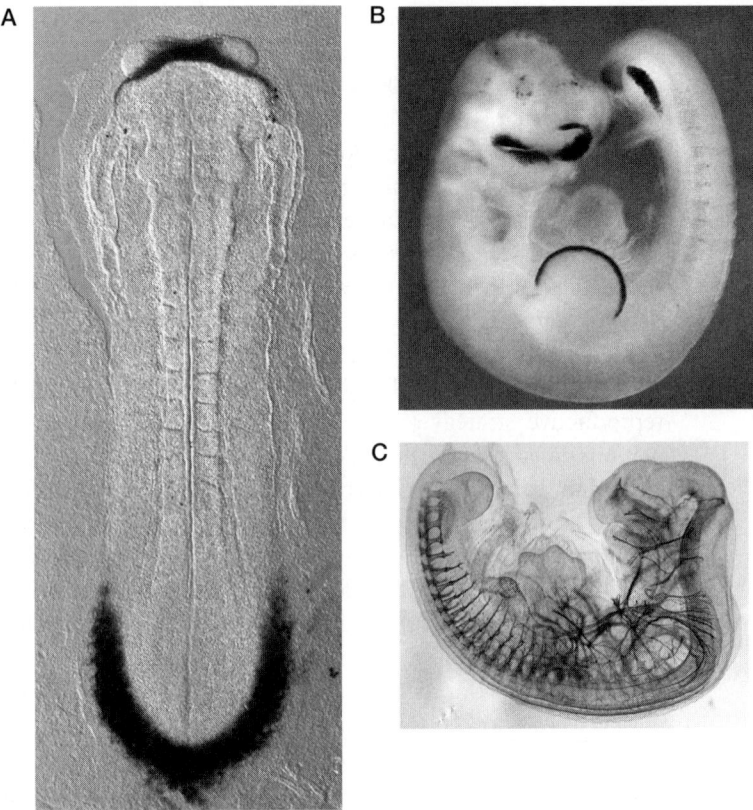

FIGURE 2. *Monodelphis* embryos: stage 24 hybridized with in situ probe to *Pitx1* (A), stage 29 hybridized with in situ probe to *Fgf8* (B), and stage 31 stained with antibody to neurofilament (C).

Although in vitro embryo electroporation has not been attempted as far as we know, it is likely to work in a manner very similar to that used with mouse or chick. Injection into the embryonic vesicle could facilitate DNA uptake. Germ-line transformation has not yet been accomplished and gene knockdown techniques have not been tried thus far (R. Behringer, pers. comm.).

Protocol 1

Basic Maintenance and Breeding of *Monodelphis*

Monodelphis are easily maintained and bred in the lab. However, because adults will fight and often kill one another if kept in the same cage for prolonged periods, we developed a special protocol for breeding. We have found high rates of breeding success (75–90%), with minimal injury due to fighting, using this procedure. Here, we outline this breeding strategy and describe how to successfully maintain a colony of *Monodelphis* in a laboratory setting.

MATERIALS

Reagents

Complete Reproduction Fox Food Pellets (National Fur Foods)
Mealworms or small pieces of fruit

Equipment

Bedding, aspen
Cages for breeding

Breeding cages must be large (e.g., 20 × 16 × 8.5 inch); Econo-Cages (Maryland Plastics) are suitable (cage bottom, VWR 10718-302; wire top, VWR 10707-708).

Cages for housing

Standard polycarbonate rat cages (10.5 × 19 × 8 inch) may be used and are available from Allentown Caging Equipment Co. (cage bottom, PC10198HT; filter top, MBT1019HT; wire lid, WBL1019RMB).

Food bowls, stainless steel (12 cm diameter, 5 cm deep)
Houses

Cut a 1–2-inch hole in the side of a 15-cm-diameter rubber bowl (e.g., Rubbermaid food storage bowl); when inverted, this serves as a "house."

Newsprint, shredded
Video equipment (optional; see Steps 5–7)

A low-light video camera and time-lapse video recorder may be obtained from any audiovisual supplier. Part numbers and components change frequently, but equipment that is sold for routine surveillance is appropriate and reasonably inexpensive. Red light bulbs (25 watt) may be used at "night" (see Step 5).

Water bottles

Water bottles and accessories available from Allentown Caging Equipment Co. (16-oz. water bottle, PC16BHT; water bottle stopper, S1011N; 3.5-in. sipper, OT1001).

METHOD

Basic Colony Maintenance and Animal Husbandry

Monodelphis adults are solitary animals and are kept singly in standard polycarbonate rat cages (10.5 × 19 × 8 inch) with wire covers, standard water bottles, and a layer of aspen bedding on the floor. Animals, particularly subadults, are quite adept at escaping, so we cover each cage with a plastic filter cover (with the filter removed) as a second line of defense. We provide animals with a small "house" (an inverted 15-cm-diameter plastic bowl with a 1–2-inch-diameter hole in the side), shredded newsprint for nesting material, and ad libitum dry food in a stainless steel bowl. We supplement dry food with mealworms or small pieces of fruit once or twice a week. Cages are cleaned weekly and animals are provided with a complete change of food, water, bedding, and nesting material. Animals that are handled regularly become quite tame and can easily be picked up by their prehensile tail when changing cages. Animals are kept on a 14-hour-light/10-hour-dark cycle at a temperature of approximately 75°C and a humidity level of 60–70%. High humidity is particularly important because animals develop health problems when kept at lower humidity levels for prolonged periods (see also VandeBerg 1999).

Breeding

Males and females reach sexual maturity at 5–6 months. We have found that animals rarely mate the first time that they are paired. For males, it appears to be a matter of experience and general size: Older, more experienced males are more successful at breeding. Female *Monodelphis* generally undergo no regular reproductive cycle, but are instead brought into estrus and ovulate in response to the presence of male pheromones (Fadem et al. 1982; Trupin and Fadem 1982; Fadem 1985). They appear to require at least one cycle of exposure to male hormones before successfully breeding. When females reach 5 months of age, we generally place them in the male room for 1–2 weeks to be exposed to male pheromones and then breed them for the first time at approximately 7 months of age.

After mating, gestation lasts for 14.5 days. Litter sizes average about 8 pups, but can number up to 13 pups. Females rarely keep litters of less than 4. Like many marsupials, *Monodelphis* naturally superovulate and produce more young than are ever raised. We have counted up to 24 intrauterine embryos.

Because *Monodelphis* are solitary in the wild, adults will fight if kept together for prolonged periods. Therefore, we developed the following protocol for breeding.

Mating Adult Monodelphis

1. Maintain males and females in different rooms or chambers during nonbreeding periods.
   This ensures that male pheromones do not reach the females except during breeding.

2. Three to 6 days before breeding, place the females in the male room and in close proximity to their future mate.
   Olfaction is critically important in Monodelphis *social behavior and in inducing estrus and ovulation in the females. This introductory period minimizes aggression and reduces the time during which the animals must be maintained in the same cage. It also appears to increase breeding success.*

3. After 2 to 3 days, switch cages so that the female is in the male's cage and the male is in the female's cage. Leave them in these cages for the remaining olfactory contact period.

4. After the above period of olfactory contact, place one male and one female together in a large breeding cage with water bottles and food bowls, but without a "house." Keep the animals in the same cage for 5–7 days.
   With a prior 6-day period of olfactory contact, they generally mate 4–5 days after introduction into a common cage. At this point, one of two procedures may be followed: To obtain embryos of a specific age, follow Steps 5–7; for postnatal pups, follow Steps 8–10.

Obtaining Stage-specific Embryos

5. If embryonic specimens of a specific age are desired, film the males and females with a low-light video camera and time-lapse video recorder to document mating.

    By documenting mating, embryos of a specific age can be obtained. During the "night" period, light may be supplemented with 25-watt red light bulbs.

6. Review the tapes to determine whether mating has occurred and record the specific time of mating.

    Well over 95% of documented mating events produce pregnancy in the female.

7. At the appropriate stage in development, harvest the embryos from the female (see Protocol 2).

    Age-stage systems of intrauterine development are available in Mate et al. (1994) and Selwood et al. (1997).

Obtaining Postnatal Pups

8. If postnatal individuals are desired for experiments or colony replenishment, do not film the mating. Simply check for pups on a daily basis starting at 14 days after first placing the females and males together.

    Young are generally born 14.5 days after mating. They are permanently attached to the female for the first 10–12 days of life and begin to move independently (although they still feed frequently) after about 35 days. Females may be kept with their young until weaning.

9. As the animals mature, clean the cages two to three times per week, depending on the litter size. The young become quite independent after about 35 days, so take great care to prevent escapes after this time.

10. At 55–60 days after mating, wean the young by separating them from their mother. Keep two or three same-sex siblings in the same cage for an additional 1–2 months after weaning.

    Eventually, they will begin to fight and must be moved to individual cages.

Protocol 2

Harvesting *Monodelphis* Embryos

Monodelphis embryos are easily harvested. Depending on how the embryos will be used in downstream procedures, there may be slight differences in euthanasia procedure, fixation, and embryo treatment. Most commonly, specimens will be used for anatomical or molecular (in situ hybridization) techniques, in which case they will be fixed in standard fixatives appropriate for the particular protocol.

MATERIALS

CAUTION: See the Cautions Appendix for appropriate handling of materials marked with <!>.

Reagents

Females, pregnant (see Protocol 1)
Fixative or culture medium as appropriate for further procedure (see Step 8)
PBS (pH 7.4)
   PBS should be prewarmed to 32.6°C if embryos will be cultured and at 4°C if embryos will be fixed. It can be prepared with DEPC-treated H_2O if embryos will be used in RNA-sensitive applications.

Equipment

CO_2 <!> chamber, for standard euthanization of small rodents
Dissection tools (fine forceps, scissors, scalpels)
Embryo spoon (e.g., Moria spoon)
Petri dishes, plastic

METHOD

1. Euthanize the female in CO_2 with secondary exsanguination.

   Monodelphis may be euthanized following the same procedures applied to rodents. Guidelines for proper CO_2 euthanasia of rodents have been established by the National Institutes of Health. Most research institutions have specific procedures and require training in the use of CO_2. Investigators should consult with their institutions to ensure that proper procedures are followed.

2. After euthanasia, make a midline incision from the vaginal opening to the lower part of the ribcage.

   In a pregnant female, the bilateral uteri will be swollen and easily apparent.

3. Remove the uteri by incision of the midline vagina and the lateral uterine ligaments.

4. Place the uteri in a Petri dish containing PBS and wash off as much blood as possible.

5. Separate the two uteri and place them in fresh PBS.

6. Gently open each uterus by carefully cutting through the uterine wall.

   To prevent damage to embryos within the uterus, take care not to cut too deep into the tissue.

7. Remove the embryos.

    i. If the embryos are less than 11 days old and they have not yet implanted, the free-floating embryos (vesicles) may simply be rolled out of the uterus.

    ii. Embryos between days 11 and 12 are beginning the process of vesicle breakdown and implantation; care must be taken when removing embryos.

    iii. Embryos more than 12 days old have established placental contact and must be dissected free.

    Illustrations and descriptions of embryos at various stages may be found in Mate et al. (1994) and Selwood et al. (1997).

8. Using the embryo spoon, move the embryos into either culture media or fixative as appropriate for the procedure of interest.

    Protocol 3 describes how to culture Monodelphis *embryos in vitro. As discussed above, most standard techniques used to study vertebrate development may be used for* Monodelphis *embryos and neonates; fixation protocols specific to those techniques should be followed. See Protocol 4 for whole-mount in situ hybridization of* Monodelphis *embryos.*

Protocol 3

Monodelphis Whole-embryo Culture

The embryos of *Monodelphis*, like those of other marsupials, may be cultured in vitro. The length of embryo viability depends in part on the stage at which culture begins, but embryos of different species of marsupials have been cultured for 18 to almost 72 hours by various investigators. Good culture results for *Monodelphis* have been obtained by the following method, which was modified slightly from that of Selwood and colleagues (Baggott and Moore 1990; Selwood and Vandeberg 1992; Moore and Taggart 1993; Yousef and Selwood 1993; Mollgard et al. 1994; Gardner et al. 1996; Selwood et al. 1997; Cruz et al. 2000). Other investigators have found that for some marsupials, longer culture periods may be achieved with 95% oxygen and 5% carbon dioxide. Several different culture media have been tried with different marsupial species, but for *Monodelphis*, high-glucose DMEM supplemented with 10% heat-inactivated fetal calf serum works well. We have not found significant differences with higher levels of calf serum or with the addition of serum derived from opossum rather than from calf. Embryos may be manipulated (e.g., labeled with Di-I) and then placed in the incubator. We have applied this technique most commonly to embryos at stages 23–25; they have retained viability and normal development though stage 26 when embryos would begin to implant in vivo. We have not had experience culturing postimplantation embryos.

MATERIALS

The recipe for the item marked with <R> is on page 572.

CAUTION: See the Cautions Appendix for appropriate handling of materials marked with <!>.

Note: All reagents and tools must be sterile and free of toxic materials such as fixatives or organic solvents.

Reagents

Calcium- and magnesium-free PBS (e.g., Dulbecco's PBS), equilibrated in incubator
Culture medium <R>, freshly prepared and equilibrated in incubator
Female, pregnant (see Protocol 1)
High-glucose DMEM, equilibrated in incubator
Isofluorane <!>

Equipment

Culture plate (24 well)
Dissection tools (forceps, scissors, scalpels)
Embryo spoon (e.g., Moria spoon)
Incubator, humidified (saturated) and equilibrated to 32.6°C and 5% CO_2
   Standard CO_2 incubators may be used; follow manufacturer directions.
Petri dishes (50 x 15 mm), plastic
Rocker, placed inside incubator

METHOD

1. Euthanize the female with an anaesthetic overdose of isofluorane, with secondary exsanguination.

    Euthanization procedures are generally regulated by individual institutions following federal guidelines; investigators should contact their institution for accepted procedures. Generally, procedures appropriate for larger rodents are appropriate for Monodelphis.

2. Remove the uteri from the females (see Protocol 2) and wash them in prewarmed (32.6°C) Dulbecco's PBS.

3. Place one uterus in a sterile 50 × 15 mm Petri dish containing prewarmed DMEM and then place it in the incubator. Dissect the second uterus by continuing to Steps 4–6.

4. Rinse the uterus several times in fresh culture medium and stroke the blood vessels with forceps to remove as much blood as possible. Place the uterus in clean culture medium.

5. Gently open the uterus by making an incision along its long axis. Gently roll the preimplantation embryos into the culture medium.

6. Gently pick up the embryos with the embryo spoon and place them singly into each well of a 24-well culture dish (each well should contain 1 ml of prewarmed culture medium). If the vesicle breaks, discard the embryo.

7. Repeat Steps 4–6 with the second uterus until as many embryos as desired are in the culture dish. Place the dish on a rocker in an incubator.

8. To change the medium, gently remove 0.5 ml of fluid and replace it with 1 ml of fresh, prewarmed medium.

    The medium usually needs to be replaced after about 12 hours. If using indicator DMEM, change the medium after the color has changed.

Protocol 4

Whole-mount In Situ Hybridization in *Monodelphis* Embryos

This protocol details whole-mount in situ hybridization of *Monodelphis* embryos, but it is broadly applicable to any marsupial. It was adapted from a protocol provided by O. Pourquié that is routinely used on mouse and chick (Dequeant et al. 2006). Special conditions have been included throughout the protocol for various stages of marsupial embryos. Whole, preterm embryonic stages (approximately stage 33 to birth) have proven to be difficult to work with because formation of the cuticle prevents probe and antibody penetration. It is not absolutely necessary to use DEPC-treated H_2O to make all of the solutions used before and during hybridization, but you may want to if you are having trouble with RNA degradation. See Wilkinson (1999) for tips on troubleshooting RNA degradation and high background.

MATERIALS

The recipes for items marked with <R> begin on page 572.

CAUTION: See the Cautions Appendix for appropriate handling of materials marked with <!>.

Reagents

Anti-digoxigenin (DIG) AP (Roche 11093274910) <!>
Blocking solution <R>
Female, pregnant (see Protocol 1)
Fixative solution <R>
Glycerol (optional; see Step 29)
Hybridization mix <R>
Methanol <!>
NTMT <R>
NTMT-NBT-BCIP <!> <R>
Paraformaldehyde (4%) <!> or formaldehyde (4%) <!> (see Step 27)
PBS (pH 7.4)
PBS containing 0.01% sodium azide <!> (optional; see Step 28)
PBS-EGTA <R>
PBT <R>
PBT-proteinase K <R>
Postfixative solution <R>
RNA probe, DIG labeled <!>

Before starting this protocol, synthesize the probe according to the specifications in the Roche DIG RNA Labeling Kit (11175025910) or similar protocol. In general, probes made to other marsupial species do have high enough specificity for Monodelphis *embryos, but mouse probes do not.*

TBST <R>

Equipment

Coverslips, glass (optional; see Step 29)
Culture plates (12 well) (optional; see Steps 5–10)
Culture tubes (5 ml)
Digital camera, mounted on a dissecting microscope
Dissection tools (forceps, scissors, scalpels, pin holder)
Embryo spoons (e.g., Moria spoon) or sterile transfer pipettes
> *The pipettes must be large enough for the embryos with which you are working.*

Incubator, preset to 70°C
Microcentrifuge tubes (1.5 ml and 2 ml), RNase-free
Netwell inserts (15 mm; membrane size: 74 µm), mesh (optional; see Steps 5–10)
> *Clean Netwell inserts with RNAse Away (Molecular BioProducts) or a similar RNase decontaminant before use.*

Parafilm (optional; see Step 3.iv)
Petri dishes (100 × 15 and 50 × 15 mm)
Pins, small (very fine insect pins are suitable)
Rotator
Transfer pipettes, sterile

METHOD

Perform all steps at room temperature unless otherwise indicated. The embryos are fragile; use a transfer pipette or embryo spoon to move them.

Embryo Collection, Fixation, and Dehydration

1. Remove the uteri from the female and rinse in PBS in a 10 × 15 mm Petri dish (see Protocol 2). Dissect the embryos from both uteri in PBS-EGTA as follows:

    i. For embryos younger than stage 28, remove the outer eggshell coat and open the embryonic vesicle opposite the embryo.
    > *It is best to leave most of the vesicle around the embryo in younger stages (stage 26 and younger). This makes it easier to manipulate the embryos when recording images.*

    ii. For embryos stage 28 or older, remove the amnion and other extraembryonic membranes.

    iii. If the neural tube has closed anteriorly, puncture it once or twice with a small pin.
    > *All of these procedures help to minimize dye-trapping.*

2. Place the embryos in fixative for 3 hours at room temperature or overnight at 4°C.
    > *We fix all the embryos from one female in one 50 × 15 mm Petri dish.*

3. Wash and dehydrate the embryos as follows:

    i. Wash the embryos twice in PBT for 5 minutes each.

    ii. Wash the embryos in PBT containing 50% methanol for 5 minutes.

    iii. Wash the embryos twice in 100% methanol for 5 minutes each.

    iv. Replace the 100% methanol one final time if the embryos will be stored before continuing protocol.
    > *At this stage, embryos can be kept at –20°C for several weeks. Embryos are conveniently stored in 5-ml culture tubes with Parafilm around the cap to prevent evaporation.*

Rehydration, Proteinase Digestion, and Postfixation

The following steps can be performed in either 5-ml culture tubes or 12-well culture plates containing Netwell inserts.

5. Rehydrate the embryos as follows:

    i. Wash the embryos in PBT containing 50% methanol for 5 minutes.

    ii. Wash the embryos twice in PBT for 5 minutes each.

6. Incubate the embryos in PBT-proteinase K on a rotator. Use the following chart to determine the appropriate incubation time:

Stage of embryo	Incubation time (min)
22	6–8
23–24	10
25	15
26–27	20
28–29	25
30–31	30

7. Wash the embryos in PBT for 5 minutes.

8. Fix the embryos in postfixative solution for 20 minutes on a rotator.

9. Rinse the embryos twice with PBT, then once in PBT/hybridization mix (50:50).

10. Transfer the embryos into hybridization mix.

    Depending on the number and size of embryos, 1.5-, 2.0-, or 5-ml tubes can be used. The embryos can be kept at –20°C for weeks.

Hybridization and Washes

11. Incubate the embryos at 70°C for at least 1 hour in hybridization mix. At the same time and at the same temperature, incubate the hybridization mix for the hot washes.

12. Decant the hybridization mix, leaving the embryos at the bottom of the tube. Refill the tube with hot hybridization mix and add 10–20 µl of probe per 2-ml volume.

    It is not necessary to denature the probe before adding it to the hybridization mix.

13. Incubate at 70°C overnight.

    Gentle rotation or agitation will improve probe penetration into older embryos.

14. Rinse the embryos twice with hot hybridization mix.

    At this point, embryos can be transferred back into 5-ml tubes or 12-well culture plates.

15. Wash the embryos twice (for 30 minutes each) at 70°C with hot hybridization mix.

16. Wash the embryos with a hot hybridization mix/TBST solution (50:50) for 15 minutes at 70°C.

Immunochemistry, Washes, and Detection of Probe

17. Rinse the embryos twice with TBST at room temperature.

18. Wash the embryos once in TBST for 15 minutes on a rotator.

19. Incubate the embryos in 4–5 ml of blocking solution for at least 1 hour.

20. Incubate the embryos overnight at 4°C in 2 ml of blocking solution with 1 µl of anti-DIG antibody (1:2000 dilution).

The 2-ml tubes work well for this step. The antibody solution can be recovered and reused multiple times. This step greatly reduces background in younger embryos (stage 21 and younger).

21. Rinse the embryos three times with TBST.
22. Wash at least three times in TBST on a rotator for 1 hour each.

    The number of washes depends on the size of the embryos. An overnight wash at 4°C can also be included to further reduce background.

23. Wash twice in freshly made NTMT on a rotator for 10 minutes each.
24. Incubate at 37°C in NTMT-NBT-BCIP solution for as long as needed.

    This step could take anywhere from 30 minutes to many days depending on the probe and the quality of the embryos. For longer incubations, maintain embryos at 4°C or room temperature.

25. When the signal is sufficient, rinse the embryos with PBT.
26. (Optional) Dehydrate and rehydrate the embryos in a methanol series.

    This turns the signal color from indigo to dark blue and tends to help to distinguish the signal from the background.

27. Fix the embryos in 4% paraformaldehyde or 4% formaldehyde for 20 minutes.
28. Rinse the embryos in PBS.

    The embryos can be stored long-term in PBS containing 0.01% sodim azide.

29. Take pictures of the embryos in PBS or glycerol using a digital camera mounted on a dissecting microscope and a combination of incident and transmitted light to obtain the best results.

    Embryos stage 26 or younger can be laid flat under a piece of glass coverslip to include more of the specimen in the plane of focus.

Recipes

The recipes for items marked with <R> are also listed here.

CAUTION: See the Cautions Appendix for appropriate handling of materials marked with <!>.

BLOCKING REAGENT STOCK SOLUTION

Reagent	Quantity (for 500 ml)	Final concentration
Blocking Reagent (Roche 11096176001)	50 g	10%
MABT <R>	to 500 ml	

Autoclave and store in 10-ml aliquots at –20°C.

BLOCKING SOLUTION

Reagent	Quantity (for 50 ml)	Final concentration (%)
Blocking reagent stock solution <R>	10 ml	2
Goat serum, decomplemented for 30 minutes at 56°C	10 ml	20
TBST <R>	30 ml	

CULTURE MEDIUM

Reagent	Quantity (for ~10 ml)	Final concentration
Fetal bovine serum, heat-inactivated	1 ml	10%
GlutaMAX™ (L-glutamine), filter-sterilized (GIBCO 35050-061)	10 µl	0.2 mM
Penicillin/streptomycin <!>, filter-sterilized (GIBCO 15140-148)	100 µl	100 units/ml penicillin 100 µg/ml streptomycin
Dulbecco's modified Eagle's medium (DMEM), high glucose	9 ml	

Equilibrate the DMEM in an incubator (32.6°C and 5% CO_2) for at least 1 hour before use (or place it in the incubator the night before). The culture medium must be prepared shortly before use.

FIXATIVE SOLUTION

Reagent	Quantity (for ~50 ml)	Final concentration
EGTA (0.5 M)	200 µl	2 mM
Formaldehyde (37%) <!>	5 ml	3.7%
NaOH (1.0 M) <!>	150 µl	3 mM
PBS (pH 7.4)	45 ml	

HYBRIDIZATION MIX

Reagent	Quantity (for 1 liter)	Final concentration
EDTA (0.5 M) <!>	10 ml	5 mM
Formamide <!>	500 ml	50%
Heparin (50 mg/ml) <!>	2.0 ml	100 μg/ml
SDS (20%) <!>	5 ml	0.1%
20x SSC (pH 5.0) <R>	65 ml	1.3x
tRNA (20 mg/ml)	2.5 ml	50 μg/ml
Tween 20	2.0 ml	0.2%
H_2O	to 1 liter	

MABT (pH 7.5)

Reagent	Quantity (for 1 liter)	Final concentration
Maleic acid <!>	11.6 g	100 mM
NaCl	8.7 g	150 mM
Tween 20	1 ml	0.1%

It can be difficult to mix the MABT solution and reach the correct pH. Start by adding NaOH pellets <!> until the pH is close to 7.5 and then add dilute NaOH until the pH is exactly 7.5. Adjust the final volume to 1 liter with H_2O.

NTMT

Reagent	Quantity (for 50 ml)	Final concentration
$MgCl_2$ (1 M) <!>	2.5 ml	50 mM
NaCl (5 M)	1 ml	100 mM
Tris-Cl (2 M, pH 9.5)	2.5 ml	100 mM
Tween 20	0.5 ml	1%
H_2O	to 50 ml	

NTMT-NBT-BCIP

Reagent	Quantity (for ~20 ml)	Final concentration
5-Bromo-4-chloro-3-indolyl-phosphate (BCIP) (50 mg/ml) <!>	70 μl	175 μg/ml
4-Nitro blue tetrazolium chloride (NBT) (100 mg/ml) <!>	67.5 μl	338 μg/ml
NTMT <R>	20 ml	

PBS-EGTA

Reagent	Quantity (for ~500 ml)	Final concentration
EGTA (0.5 M)	2 ml	2 mM
PBS (pH 7.4)	500 ml	

PBT

Reagent	Quantity (for ~500 ml)	Final concentration
Tween 20	500 µl	0.1%
PBS (pH 7.4)	500 ml	

PBT-PROTEINASE K

Reagent	Quantity (for ~50 ml)	Final concentration
Proteinase K (10 mg/ml) <!>	50 µl	0.01 mg/ml
PBT <R>	50 ml	

POSTFIXATIVE SOLUTION

Reagent	Quantity (for ~50 ml)	Final concentration
Formaldehyde (37%) <!>	5 ml	3.7%
Glutaraldehyde (25%) <!>	200 µl	0.1%
PBT <R>	45 ml	

20x SSC (pH 5.0)

Reagent	Quantity (for 1 liter)	Final concentration (20x)
NaCl	175.3 g	3.0 M
Sodium citrate <!>	88.2 g	0.3 M

Dissolve the ingredients in 800 ml of H_2O. Adjust the pH to 5.0 with 14 N HCl <!> and then adjust the final volume to 1 liter with H_2O. Dispense into aliquots. Sterilize by autoclaving.

TBST

Reagent	Quantity (for 1 liter)	Final concentration
KCl <!>	0.02 g	268 µM
NaCl	0.8 g	13 mM
Tris-Cl (1 M, pH 7.8) <!>	2.5 ml	2.5 mM
Tween 20	1 ml	0.1%
H_2O	to 1 liter	

Alternatively, prepare a stock solution of 10x TBS, dilute it, and add Tween 20.

ACKNOWLEDGMENTS

The authors wish to thank Lynne Selwood and Paul Trainor for suggestions regarding embryo culture technique, Olivier Pourquié and his lab for the in situ hybridization protocol, and Richard Behringer for answering questions regarding germ-line transformation in *Monodelphis*. Many members of the Smith lab during the past 20 years have helped to refine the procedures for *Monodelphis domestica* maintenance, breeding, and laboratory technique.

REFERENCES

Baggott, L.M. and Moore, H.D.M. 1990. Early embryonic-development of the gray short-tailed opossum, *monodelphis-domestica*, in vivo and in vitro. *J. Zool.* **222:** 623–639.

Cruz, Y.P., Hickford, D., and Selwood, L. 2000. A staging scheme for assessing development *in vitro* of organogenesis stage embryos of the stripe-faced dunnart, *Sminthopsis macroura* (Marsupialia: Dasyuridae). *J. Reprod. Fertil.* **120:** 99–108.

Dequeant, M.L., Glynn, E., Gaudenz, K., Wahl, M., Chen, J., Mushegian, A., and Pourquié, O. 2006. A complex oscillating network of signaling genes underlies the mouse segmentation clock. *Science* **314:** 1595–1598.

Duke, S.E., Samollow, P.B., Mauceli, E., Lindblad-Toh, K., and Breen, M. 2007. Integrated cytogenetic BAC map of the genome of the gray, short-tailed opossum, *Monodelphis domestica*. *Chromosome Res.* **15:** 361–370.

Fadem, B.H. 1985. Evidence for the activation of female reproduction by males in a marsupial, the gray short-tailed opossum (*Monodelphis domestica*). *Biol. Reprod.* **33:** 112–116.

Fadem, B.H., Trupin, G.L., Maliniak, E., VandeBerg, J.L., and Hayssen, V. 1982. Care and breeding of the gray, short-tailed opossum (*Monodelphis domestica*). *Lab. Anim. Sci.* **32:** 405–409.

Freyer, C., Zeller, U., and Renfree, M.B. 2007. Placental function in two distantly related marsupials. *Placenta* **28:** 249–257.

Gardner, D.K., Selwood, L., and Lane, M. 1996. Nutrient uptake and culture of *Sminthopsis macroura* (stripe-faced dunnart) embryos. *Reprod. Fertil. Dev.* **8:** 685–690.

Keyte, A.L., Imam, T., and Smith, K.K. 2006. Limb heterochrony in a marsupial, *M. domestica*. *Dev. Biol.* **295:** 249.

Ley, R.D., Reeve, V.E., and Kusewitt, D.F. 2000. Photobiology of *Monodelphis domestica*. *Dev. Comp. Immunol.* **24:** 503–516.

Macrini, T.E. 2004. *Monodelphis domestica*. *Mamm. Species* **760:** 1–8.

Mate, K.E., Robinson, E.S., Vandeberg, J.L., and Pedersen, R.A. 1994. Timetable of in-vivo embryonic-development in the grey short-tailed opossum (*Monodelphis-domestica*). *Mol. Reprod. Dev.* **39:** 365–374.

Mikkelsen, T.S., Wakefield, M.J., Aken, B., Amemiya, C.T., Chang, J.L., Duke, S., Garber, M., Gentles, A.J., Goodstadt, L., Heger, A., et al. 2007. Genome of the marsupial *Monodelphis domestica* reveals innovation in non-coding sequences. *Nature* **447:** 167–177.

Miller, R.D. and Belov, K. 2000. Immunoglobulin genetics of marsupials. *Dev. Comp. Immunol.* **24:** 485–490.

Mollgard, K., Balslev, Y., Janas, M.S., Treherne, J.M., Saunders, N.R., and Nichols, J.G. 1994. Development of spinal cord in the isolated CNS of a neonatal mammal (the opossum *Monodelphis domestica*) maintained in longterm culture. *J. Neurocytol.* **23:** 151–165.

Moore, H.D. and Taggart, D.A. 1993. In vitro fertilization and embryo culture in the grey short-tailed opossum, *Monodelphis domestica*. *J. Reprod. Fertil.* **98:** 267–274.

Rainwater, D.L., Kammerer, C.M., Singh, A.T.K., Moore, P.H., Poushesh, M., Shelledy, W.R., VandeBerg, J.F., Robinson, E.S., and VandeBerg, J.L. 2001. Genetic control of lipoprotein phenotypes in the laboratory opossum, *Monodelphis domestica*. *GeneScreen* **1:** 117–124.

Samollow, P.B. 2006. Status and applications of genomic resources for the gray, short-tailed opossum, *Monodelphis domestica*, an American marsupial model for comparative biology. *Aust. J. Zool.* **54:** 173–196.

Samollow, P.B., Gouin, N., Miethke, P., Mahaney, S.M., Kenney, M., VandeBerg, J.L., Graves, J.A.M., and Kammerer, C.M. 2007. A microsatellite-based, physically anchored linkage map for the gray, short-tailed opossum (*Monodelphis domestica*). *Chromosome Res.* **15:** 269–281.

Saunders, N.R., Adam, E., Reader, M., and Mollgard, K. 1989. *Monodelphis domestica* (grey short-tailed opossum): An accessible model for studies of early neocortical development. *Anat. Embryol.* **180:** 227–236.

Saunders, N.R., Deal, A., Knott, G.W., Varga, Z.M., and Nicholls, J.G. 1995. Repair and recovery following spinal cord injury in a neonatal marsupial (*Monodelphis domestica*). *Clin. Exp. Pharmacol. Physiol.* **22:** 518–526.

Selwood, L. and Johnson, M.H. 2006. Trophoblast and hypoblast in the monotreme, marsupial and eutherian mammal: Evolution and origins. *BioEssays* **28:** 128–145.

Selwood, L. and VandeBerg, J.L. 1992. The influence of incubation temperature on oocyte maturation, parthenogenetic and embryonic development in vitro of the marsupial *Monodelphis domestica*. *Anim. Reprod. Sci.* **29:** 99–116.

Selwood, L., Robinson, E.S., Pedersen, R.A., and VandeBerg, J.L. 1997. Development in vitro of marsupials: A comparative review of species and a timetable of cleavage and early blastocyst stages of development in *Monodelphis domestica*. *Int. J. Dev. Biol.* **41:** 397–410.

Smith, K.K. 1994. Development of craniofacial musculature in *Monodelphis domestica* (Marsupialia, Didelphidae). *J. Morphol.* **222:** 149–173.

Smith, K.K. 2001. Early development of the neural plate, neural crest and facial region of marsupials. *J. Anat.* **199:** 121–131.

Smith, K.K. 2006. Craniofacial development in marsupial mammals: Developmental origins of evolutionary change. *Dev. Dyn.* **235:** 1181–1193.

Trupin, G.L. and Fadem, B.H. 1982. Sexual behavior of the grey short-tailed opossum *Monodelphis domestica*. *J. Mammal.* **63:** 409–414.

Vaglia, J.L. and Smith, K.K. 2003. Early differentiation and migration of cranial neural crest in the opossum, *Monodelphis domestica*. *Evol. Dev.* **5:** 121–135.

VandeBerg, J.L. 1999. The laboratory opossum (*Monodelphis domestica*). In *UFAW handbook on the management of laboratory animals: Terrestrial vertebrates* (ed. T. Poole and P. English), vol. 1, pp. 193–209. Blackwell, Oxford.

VandeBerg, J.L. and Robinson, E.S. 1997. The laboratory opossum (*Monodelphis domestica*) in laboratory research. *ILAR J.* **38:** 4–12.

Wilkinson, D.G., ed. 1999. *In situ hybridization: A practical approach.* Oxford University Press, Oxford.

Yousef, A. and Selwood, L. 1993. Embryonic development in culture of the marsupials *Antechinus stuartii* (Macleay) and *Sminthopsis macroura* (Spencer) during preimplantation stages. *Reprod. Fertil. Dev.* **5:** 445–458.

Zuri, I., Su, W., and Halpern, M. 2003. Conspecific odor investigation by gray short-tailed opossums (*Monodelphis domestica*). *Physiol. Behav.* **80:** 225–232.

General Cautions

Please note that the Cautions Appendix in this manual is not exhaustive. Readers should always consult individual manufacturers and other resources for current and specific product information. Chemicals and other materials discussed in text sections are not identified by the icon (<!>) used to indicate hazardous materials in the protocols. However, without special handling, these materials may be hazardous to the user. Please consult your local safety office or the manufacturer's safety guidelines for further information.

The following general cautions should always be observed.

- **Before beginning the procedure,** become completely familiar with the properties of substances to be used.
- **The absence of a warning** does not necessarily mean that the material is safe, because information may not always be complete or available.
- **If exposed** to toxic substances, contact your local safety office immediately for instructions.
- **Use proper disposal procedures** for all chemical, biological, and radioactive waste.
- **For specific guidelines on appropriate gloves to use,** consult your local safety office.
- **Handle concentrated acids and bases** with great care. Wear goggles and appropriate gloves. A face shield should be worn when handling large quantities.

  Do not mix strong acids with organic solvents because they may react. Sulfuric acid and nitric acid especially may react highly exothermically and cause fires and explosions.

  Do not mix strong bases with halogenated solvent because they may form reactive carbenes that can lead to explosions.

- **Handle and store pressurized gas containers** with caution because they may contain flammable, toxic, or corrosive gases; asphyxiants; or oxidizers. For proper procedures, consult the Material Safety Data Sheet that must be provided by your vendor.
- **Never pipette** solutions using mouth suction. This method is not sterile and can be dangerous. Always use a pipette aid or bulb.
- **Keep halogenated and nonhalogenated** solvents separately (e.g., mixing chloroform and acetone can cause unexpected reactions in the presence of bases). Halogenated solvents are organic solvents, such as chloroform, dichloromethane, trichlorotrifluoroethane, and dichloroethane. Nonhalogenated solvents include pentane, heptane, ethanol, methanol, benzene, toluene, *N,N*-dimethylformamide (DMF), dimethylsulfoxide (DMSO), and acetonitrile.
- **Laser radiation,** visible or invisible, can cause severe damage to the eyes and skin. Take proper precautions to prevent exposure to direct and reflected beams. Always follow the manufac-

turer's safety guidelines and consult your local safety office. See caution below for more detailed information.

- **Flash lamps**, due to their light intensity, can be harmful to the eyes. They also may explode on occasion. Wear appropriate eye protection and follow the manufacturer's guidelines.
- **Photographic fixatives, developers, and photoresists** also contain chemicals that can be harmful. Handle them with care and follow the manufacturer's directions.
- **Power supplies and electrophoresis equipment** pose serious fire hazards and electrical shock hazards if not used properly.
- **Microwave ovens and autoclaves** in the lab require certain precautions. Accidents have occurred involving their use (e.g., when melting agar or bacto-agar stored in bottles or when sterilizing). If the screw top is not completely removed and there is inadequate space for the steam to vent, the bottles can explode and cause severe injury when the containers are removed from the microwave or autoclave. Always completely remove bottle caps before microwaving or autoclaving. An alternative method for routine agarose gels that do not require sterile agar is to weigh out the agar and place the solution in a flask.
- **Ultrasonicators** use high-frequency sound waves (16–100 kHz) for cell disruption and other purposes. This "ultrasound," conducted through air, does not pose a direct hazard to humans, but the associated high volumes of audible sound can cause a variety of effects, including headache, nausea, and tinnitus. Direct contact of the body with high-intensity ultrasound (not medical imaging equipment) should be avoided. Use appropriate ear protection and display signs on the door(s) of laboratories in which the units are used.
- **Use extreme caution when handling cutting devices,** such as microtome blades, scalpels, razor blades, or needles. Microtome blades are extremely sharp! Use care when sectioning. If unfamiliar with their use, have an experienced user demonstrate proper procedures. For proper disposal, use the "sharps" disposal container in your lab. Discard used needles *unshielded*, with the syringe still attached. This prevents injuries (and possible infections) when manipulating used needles because many accidents occur while trying to replace the needle shield. Injuries may also be caused by broken Pasteur pipettes, coverslips, or slides.
- **Procedures for the humane treatment of animals** must be observed at all times. Consult your local animal facility for guidelines. Animals, such as rats, are known to induce allergies that can increase in intensity with repeated exposure. Always wear a lab coat and gloves when handling these animals. If allergies to dander or saliva are known, wear a mask.

GENERAL PROPERTIES OF COMMON CHEMICALS

The hazardous materials list can be summarized in the following categories.

- Inorganic acids, such as hydrochloric, sulfuric, nitric, or phosphoric, are colorless liquids with stinging vapors. Avoid spills on skin or clothing. Spills should be diluted with large amounts of water. The concentrated forms of these acids can destroy paper, textiles, and skin and cause serious injury to the eyes.
- Inorganic bases, such as sodium hydroxide, are white solids that dissolve in water and under heat development. Concentrated solutions will slowly dissolve skin and even fingernails.
- Salts of heavy metals are usually colored, powdered solids that dissolve in water. Many of these are potent enzyme inhibitors and therefore toxic to humans and the environment (e.g., fish and algae).
- Most organic solvents are flammable volatile liquids. Avoid breathing the vapors, which can cause nausea or dizziness. Also avoid skin contact.

- Other organic compounds including organosulphur compounds, such as β-mercaptoethanol or organic amines, can have very unpleasant odors. Others are highly reactive and should be handled with appropriate care.
- If improperly handled, dyes and their solutions can stain not only the sample, but also skin and clothing. Some are also mutagenic (e.g., ethidium bromide), carcinogenic, and toxic.
- Nearly all names ending with "ase" (e.g., catalase, β-glucuronidase, or zymolyase) refer to enzymes. There are also other enzymes with nonsystematic names such as pepsin. Many of them are provided by manufacturers in preparations containing buffering substances, etc. Be aware of the individual properties of materials contained in these substances.
- Toxic compounds are often used to manipulate cells. They can be dangerous and should be handled appropriately.
- Be aware that several of the compounds listed have not been thoroughly studied with respect to their toxicological properties. Handle each chemical with appropriate respect. Although the toxic effects of a compound can be quantified (e.g., LD_{50} values), this is not possible for carcinogens or mutagens where one single exposure can have an effect. Also realize that dangers related to a given compound may also depend on its physical state (fine powder vs. large crystals/diethylether vs. glycerol/dry ice vs. carbon dioxide under pressure in a gas bomb). Anticipate under which circumstances during an experiment exposure is most likely to occur and how best to protect yourself and your environment.

HAZARDOUS MATERIALS

Acetic acid (concentrated) must be handled with great care. It may be harmful by inhalation, ingestion, or skin absorption. Wear appropriate gloves and goggles. Use in a chemical fume hood.

Acetic acid (glacial) is highly corrosive and must be handled with great care. It may be a carcinogen. Liquid and mist cause severe burns to all body tissues. It may be harmful by inhalation, ingestion, or skin absorption. Wear appropriate gloves and goggles and use in a chemical fume hood. Keep away from heat, sparks, and open flame.

Acetic anhydride is extremely destructive to the skin, eyes, mucous membranes, and upper respiratory tract. It may be harmful by inhalation, ingestion, or skin absorption. Wear appropriate gloves and safety glasses and use in a chemical fume hood.

Acetone causes eye and skin irritation and is irritating to mucous membranes and the upper respiratory tract. Do not breathe the vapors. It is also extremely flammable. Wear appropriate gloves and safety glasses. Keep away from heat, sparks, and open flame.

Acetosyringone is an irritant and may be harmful by inhalation, ingestion, or skin absorption. Wear appropriate gloves and safety goggles. Do not breathe the dust.

$AgNO_3$, *see* **Silver nitrate**

$Al_2(SO_4)_3$, *see* **Aluminum sulfate**

Aluminum sulfate, $Al_2(SO_4)_3$, may be harmful by inhalation, ingestion, or skin absorption. Wear appropriate gloves and safety glasses.

Aminobenzoic acid may be harmful by inhalation, ingestion, or skin absorption. Wear appropriate gloves and safety glasses.

Ammonium chloride, NH_4Cl, may be harmful by inhalation, ingestion, or skin absorption. Wear appropriate gloves and safety glasses and use in a chemical fume hood.

Ammonium hydroxide, NH_4OH, is a solution of ammonia in water. It is caustic and should be handled with great care. When ammonia vapors escape from the solution, they are corrosive and toxic and can be explosive. Use only with a mechanical exhaust. Wear appropriate gloves and use only in a chemical fume hood.

Ammonium molybdate, $(NH_4)_6Mo_7O_{24} \cdot 4H_2O$, or its **tetrahydrate,** may be harmful by inhalation, ingestion, or skin absorption. Wear appropriate gloves and safety glasses and use in a chemical fume hood.

Ampicillin may be harmful by inhalation, ingestion, or skin absorption. Wear appropriate gloves and safety glasses and use in a chemical fume hood.

Antisedan, *see* **Atipamezole hydrochloride**

Note: In general, proprietary materials are not listed here. Kits and other commercial items as well as most anesthetics, dyes, fixatives, and stains are also not included. Anesthetics and antibiotics also require special care. Follow the manufacturer's safety guidelines that accompany these products.

Atipamezole hydrochloride (Antisedan) may be harmful by inhalation, ingestion, or skin absorption. Wear appropriate gloves and safety glasses.

BCIP, see **5-Bromo-4-chloro-3-indolyl-phosphate**

BCP, see **1-Bromo-3-chloropropane**

Benzoic acid is an irritant and may be harmful by inhalation, ingestion, or skin absorption. Wear appropriate gloves and safety glasses. Do not breathe the dust.

Benzyl alcohol is an irritant and may be harmful by inhalation, ingestion, or skin absorption. Wear appropriate gloves and safety glasses. Keep away from heat, sparks, and open flame.

Benzyl benzoate is an irritant and may be harmful by inhalation, ingestion, or skin absorption. Avoid contact with the eyes. Wear appropriate gloves and safety glasses.

Blasticidin may be fatal if swallowed. It is harmful by inhalation, ingestion, or skin absorption. Wear appropriate gloves and safety glasses. Do not breathe the dust.

Bleach (Sodium hypochlorite), NaOCl, is poisonous, can be explosive, and may react with organic solvents. It may be fatal by inhalation and is harmful by ingestion and destructive to the skin. Wear appropriate gloves and safety glasses and use in a chemical fume hood to minimize exposure and odor.

Bleomycin sulfate is toxic, a possible carcinogen, and may cause heritable genetic damage. It may be harmful by inhalation, ingestion, or skin absorption. Wear appropriate gloves and safety goggles and use in a chemical fume hood. Do not breathe the dust.

Boric acid, H_3BO_3, may be harmful by inhalation, ingestion, or skin absorption. Wear appropriate gloves and goggles.

BrdU, see **5-Bromo-2′-deoxyuridine**

5-Bromo-4-chloro-3-indolyl-phosphate (BCIP) is an irritant and may be harmful by inhalation, ingestion, or skin absorption. Wear appropriate gloves and safety glasses.

1-Bromo-3-chloropropane (BCP) has a narcotic effect and may be harmful by inhalation, ingestion, or skin absorption. Wear appropriate gloves and safety glasses. Do not breathe the vapor.

5-Bromo-2′-deoxyuridine (BrdU) is a mutagen. It may be harmful by inhalation, ingestion, or skin absorption. It may cause irritation. Avoid breathing the dust. Wear appropriate gloves and safety glasses and always use in a chemical fume hood.

Bromophenol blue may be harmful by inhalation, ingestion, or skin absorption. Wear appropriate gloves and safety glasses and use in a chemical fume hood.

n-Butanol is irritating to the mucous membranes, upper respiratory tract, skin, and especially the eyes. Avoid breathing the vapors. Wear appropriate gloves and safety glasses and use in a chemical fume hood. n-Butanol is also highly flammable. Keep away from heat, sparks, and open flame.

$CaCl_2$, see **Calcium chloride**

Cacodylate contains arsenic, is highly toxic, and may be fatal if inhaled, ingested, or absorbed through the skin. Wear appropriate gloves and safety glasses and use in a chemical fume hood. See also **Potassium cacodylate** and **Sodium cacodylate**

Calcium chloride, $CaCl_2$, is hygroscopic and may cause cardiac disturbances. It may be harmful by inhalation, ingestion, or skin absorption. Do not breathe the dust. Wear appropriate gloves and safety goggles.

Calcium nitrate, $Ca(NO_3)_2$, is a strong oxidizer and reacts violently upon contact with many organic substances. Handle with great care. It may be harmful by inhalation, ingestion, or skin absorption. Wear appropriate gloves and safety glasses. Keep away from heat, sparks, and open flame.

$Ca(NO_3)_2$, see **Calcium nitrate**

Carbon dioxide, CO_2, in all forms may be fatal by inhalation, ingestion, or skin absorption. In high concentrations, it can paralyze the respiratory center and cause suffocation. Use only in well-ventilated areas. In the form of dry ice, contact with carbon dioxide can also cause frostbite. Do not place large quantities of dry ice in enclosed areas such as cold rooms. Wear appropriate gloves and safety goggles.

Cefotaxime may cause allergic reactions. It may be harmful by inhalation, ingestion, or skin absorption Wear appropriate gloves and safety goggles and always use in a chemical fume hood. Do not breathe the dust.

Cesium chloride, CsCl, may be harmful by inhalation, ingestion, or skin absorption. Wear appropriate gloves and safety glasses.

CHAPS, see **3-[(3-Cholamidopropyl)dimethyl-ammonio]-1-propanesulfonate**

$CHCl_3$, see **Chloroform**

Chloroform, $CHCl_3$, is irritating to the skin, eyes, mucous membranes, and respiratory tract. It is a carcinogen and may damage the liver and kidneys. It is also volatile. Avoid breathing the vapors. Wear appropriate gloves and safety glasses and always use in a chemical fume hood.

3-[(3-Cholamidopropyl)dimethyl-ammonio]-1-propane-sulfonate (CHAPS) is an irritant and may be harmful by inhalation, ingestion, or skin absorption. Wear appropriate gloves and safety glasses.

Citric acid is an irritant and may be harmful by inhalation, ingestion, or skin absorption. It poses a risk of serious damage to the eyes. Wear appropriate gloves and safety goggles. Do not breathe the dust.

CO_2, see **Carbon dioxide**

Cobalt chloride, $CoCl_2$, is toxic, a possible carcinogen, and is dangerous to the environment. It can cause burns and may be harmful by inhalation, ingestion, or skin absorption. Wear appropriate gloves and safety glasses.

$CoCl_2$, see **Cobalt chloride**

Copper sulfate, $CuSO_4$, is toxic and a danger to the environment. It may be harmful by inhalation, ingestion, or skin absorption. Wear appropriate gloves and safety glasses and use in a chemical fume hood.

CsCl, *see* **Cesium chloride**

CuSO$_4$, *see* **Copper sulfate**

Cysteine is an irritant to the eyes, skin, and respiratory tract. It may be harmful by inhalation, ingestion, or skin absorption. Wear appropriate gloves and safety glasses. Do not breathe the dust.

DAB, *see* **3,3′-Diaminobenzidine**

DAPI, *see* **4,6-Diamidine-2-phenylindole dihydrochloride**

DCM, *see* **Dichloromethane**

DEPC, *see* **Diethyl pyrocarbonate**

4,6-Diamidine-2-phenylindole dihydrochloride (DAPI) is a possible carcinogen. It may be harmful by inhalation, ingestion, or skin absorption. It may also cause irritation. Avoid breathing the dust and vapors. Wear appropriate gloves and safety glasses and use in a chemical fume hood.

3,3′-Diaminobenzidine (DAB) is a carcinogen. Handle with extreme care. Avoid breathing the vapors. Wear appropriate gloves and safety glasses and use in a chemical fume hood.

Dichloromethane (DCM), CH$_2$Cl$_2$ (also known as **Methylene chloride**), is toxic if inhaled, ingested, or absorbed through the skin. It is also an irritant and is suspected to be a carcinogen. Wear appropriate gloves and safety goggles and use in a chemical fume hood. Do not breathe the vapors.

Diethyl pyrocarbonate (DEPC) is a potent protein denaturant and is a suspected carcinogen. Aim bottle away from you when opening it; internal pressure can lead to splattering. Wear appropriate gloves, safety goggles, and lab coat and use in a chemical fume hood.

DIG, *see* **Digoxigenin**

Digoxigenin (DIG) may be fatal if inhaled, ingested, or absorbed through the skin. Wear appropriate gloves and safety glasses and use in a chemical fume hood. Do not breathe the dust.

Dimethyl benzyl ammonium chloride may be harmful by inhalation, ingestion, or skin absorption. Wear appropriate gloves and safety glasses and use in a chemical fume hood.

N,N-Dimethylformamide (DMF), HCON(CH$_3$)$_2$, is a possible carcinogen and is irritating to the eyes, skin, and mucous membranes. It can exert its toxic effects through inhalation, ingestion, or skin absorption. Chronic inhalation can cause liver and kidney damage. Wear appropriate gloves and safety glasses and use in a chemical fume hood.

Dimethylpimelimidate (DMP) is irritating to the eyes, skin, mucous membranes, and upper respiratory tract. It can exert harmful effects by inhalation, ingestion, or skin absorption. Avoid breathing the vapors. Wear appropriate gloves, face mask, and safety glasses.

Dimethylsulfoxide (DMSO) may be harmful by inhalation or skin absorption. It easily penetrates the skin and anything dissolved or mixed with it will be absorbed. Wear appropriate gloves and safety glasses and use in a chemical fume hood. DMSO is also combustible. Store in a tightly closed container. Keep away from heat, sparks, and open flame.

Dithiothreitol (DTT) is a strong reducing agent that emits a foul odor. It may be harmful by inhalation, ingestion, or skin absorption. When working with the solid form or highly concentrated stocks, wear appropriate gloves and safety glasses and use in a chemical fume hood.

DMF, *see* **N,N-Dimethylformamide**

DMP, *see* **Dimethylpimelimidate**

DMSO, *see* **Dimethylsulfoxide**

Dry ice, *see* **Carbon dioxide**

DTT, *see* **Dithiothreitol**

EMS, *see* **Ethylmethanesulfonate**

Epon resin, *see* **Resins**

ETDA, *see* **(E)-11-tetradecenyl acetate**

(E)-11-tetradecenyl acetate (ETDA) is combustible and may be harmful by inhalation, ingestion, or skin absorption. Wear appropriate gloves and safety glasses. Keep away from heat, sparks, and open flame. Do not breathe the vapor or mist.

Ethanol (EtOH), CH$_3$CH$_2$OH, is highly flammable and may be harmful by inhalation, ingestion, or skin absorption. Wear appropriate gloves and safety glasses. Keep away from heat, sparks, and open flame.

Ethylmethanesulfonate (EMS) is a volatile organic solvent, mutagen, and carcinogen. It is harmful if inhaled, ingested, or absorbed through the skin. Discard supernatants and washes containing EMS in a beaker containing 50% sodium thiosulfate. Decontaminate all material that has come in contact with EMS by treatment in a large volume of 10% (w/v) sodium thiosulfate. Use extreme caution when handling. When using undiluted EMS, wear protective appropriate gloves and use in a chemical fume hood. Store EMS in the cold. DO NOT mouth-pipette EMS. Pipettes used with undiluted EMS should not be too warm; chill them in the refrigerator before use to minimize the volatility of EMS. All glassware coming in contact with EMS should be immersed in a large beaker of 1 N NaOH or laboratory bleach before recycling or disposal.

EtOH, *see* **Ethanol**

Eukitt mounting medium is highly flammable and may cause serious injury to the eyes. It may be harmful by inhalation, ingestion, or skin absorption. Wear appropriate gloves and safety goggles and work in a well-ventilated area. Keep away from heat, sparks, and open flame. Do not breathe the vapors.

Fast Green is a carcinogen and may be harmful by inhalation, ingestion, or skin absorption. Wear appropriate gloves and safety glasses and always use in a chemical fume hood.

Fast Red may cause methemoglobinemia through overexposure. It may be harmful by inhalation, ingestion, or skin absorption. Wear appropriate gloves and safety glasses.

FeCl$_3$, *see* **Ferric chloride**

Ferric chloride, FeCl$_3$, may be harmful by inhalation, ingestion, or skin absorption. Wear appropriate gloves and safety glasses and use only in a chemical fume hood.

Ferrous sulfate, FeSO$_4$ · 7H$_2$O, may be harmful by inhalation, ingestion, or skin absorption. Wear appropriate gloves and safety glasses.

FeSO$_4$ · 7H$_2$O, *see* **Ferrous sulfate**

FITC, *see* **Fluorescein isothiocyanate**

Fluorescein isothiocyanate (FITC), may be harmful by inhalation, ingestion, or skin absorption. Wear appropriate gloves and safety glasses.

Formaldehyde, HCHO, is highly toxic and volatile. It is also a possible carcinogen. It is readily absorbed through the skin and is irritating or destructive to the skin, eyes, mucous membranes, and upper respiratory tract. Avoid breathing the vapors. Wear appropriate gloves and safety glasses and always use in a chemical fume hood. Keep away from heat, sparks, and open flame.

Formalin is a solution of formaldehyde in water. *See* **Formaldehyde**

Formamide is teratogenic. The vapor is irritating to the eyes, skin, mucous membranes, and upper respiratory tract. It may be harmful by inhalation, ingestion, or skin absorption. Wear appropriate gloves and safety glasses and always use a chemical fume hood when working with concentrated solutions of formamide. Keep working solutions covered as much as possible.

G418 (an aminoglycosidic antibiotic) is toxic and may cause harm to an unborn child. It may be harmful by inhalation, ingestion, or skin absorption. Wear appropriate gloves and safety goggles and use in a chemical fume hood. Do not breathe the dust.

Glacial acetic acid, *see* **Acetic acid (glacial)**

Glutaraldehyde is toxic. It is readily absorbed through the skin and is irritating or destructive to the skin, eyes, mucous membranes, and upper respiratory tract. Wear appropriate gloves and safety glasses and always use in a chemical fume hood.

Guanidine thiocyanate may be harmful by inhalation, ingestion, or skin absorption. Wear appropriate gloves and safety glasses.

H$_3$BO$_3$, *see* **Boric acid**

HCHO, *see* **Formaldehyde**

HCl, *see* **Hydrochloric acid**

H$_3$COH, *see* **Methanol**

Heparin is an irritant and may act as anticoagulant subcutaneously or intravenously. It may be harmful by inhalation, ingestion, or skin absorption. Wear appropriate gloves and safety glasses.

Heptane may be harmful by inhalation, ingestion, or skin absorption. Wear appropriate gloves and safety glasses. It is extremely flammable. Keep away from heat, sparks, and open flame.

Hexamethylenetetramine is flammable and may be harmful by inhalation, ingestion, or skin absorption. Wear appropriate gloves and safety glasses. Avoid breathing the dust. Keep away from heat, sparks, and open flame.

Hexane is extremely flammable and may be harmful by inhalation, ingestion, or skin absorption. Wear appropriate gloves and safety glasses and use only in a chemical fume hood. Keep away from heat, sparks, and open flame.

H$_2$O$_2$, *see* **Hydrogen peroxide**

HOCH$_2$CH$_2$SH, *see* **β-Mercaptoethanol**

H$_3$PO$_4$, *see* **Phosphoric acid**

Hydrochloric acid, HCl, is volatile and may be fatal if inhaled, ingested, or absorbed through the skin. It is extremely destructive to mucous membranes, the upper respiratory tract, eyes, and skin. Wear appropriate gloves and safety glasses and use with great care in a chemical fume hood. Wear goggles when handling large quantities.

Hydrogen peroxide, H$_2$O$_2$, is corrosive, toxic, and extremely damaging to the skin. It may be harmful by inhalation, ingestion, and skin absorption. Wear appropriate gloves and safety glasses and use only in a chemical fume hood.

Hygromycin B is highly toxic and may be fatal if inhaled, ingested, or absorbed through the skin. Wear appropriate gloves and safety goggles and use only in a chemical fume hood. Do not breathe the dust.

Isoamyl alcohol (IAA) may be harmful by inhalation, ingestion, or skin absorption and presents a risk of serious damage to the eyes. Wear appropriate gloves and safety goggles. Keep away from heat, sparks, and open flame.

Isofluorane (Isoflurane) is an irritant and may be harmful by inhalation, ingestion, or skin absorption. Chronic exposure may be harmful. Wear appropriate gloves and safety glasses.

Isopentane (2-Methylbutane) is extremely flammable. Keep away from heat, sparks, and open flame. It may be harmful by inhalation, ingestion, or skin absorption. Wear appropriate gloves and safety glasses.

Isopropanol is flammable and irritating. It may be harmful by inhalation, ingestion, or skin absorption. Wear appropriate gloves and safety glasses. Do not breathe the vapor. Keep away from heat, sparks, and open flame.

Isopropyl alcohol, *see* **Isopropanol**

KBr, *see* **Potassium bromide**

KCl, *see* **Potassium chloride**

KI, *see* **Potassium iodide**

KNO$_3$, *see* **Potassium nitrate**

KOH, *see* **Potassium hydroxide**

Lentiviral vectors are based on human immunodeficiency virus (HIV), the virus responsible for the development of acquired immunodeficiency syndrome (AIDS). Lentiviruses are a subclass of retroviruses that are able to infect both proliferating and nonproliferating cells. Lentivirus may be transmitted by penetration of the skin and/or mucous membrane exposure of the eyes, nose, and mouth. Wear gloves and safety goggles. All virus work should be performed in a tissue culture hood. All materials exposed to lentivirus should be soaked or diluted with bleach for 30 minutes before being disposed of as biohazardous waste. Consult your local safety office for further information and follow their guidelines for handling and disposal.

Levamisole is toxic if ingested. It may be harmful by inhalation or skin absorption. Wear appropriate gloves and safety glasses.

LiCl, *see* **Lithium chloride**

Liquid nitrogen (LN$_2$) can cause severe damage due to extreme temperature. Handle frozen samples with extreme caution. Do not breathe the vapors. Seepage of liquid nitrogen into frozen vials can result in an exploding tube upon removal from liquid nitrogen. When possible, use vials with O-rings. Wear cryomitts and a face mask. Do not allow liquid nitrogen to spill on your clothes. Do not breathe the vapors.

Lithium chloride, LiCl, is an irritant to the eyes, skin, mucous membranes, and upper respiratory tract. It may be harmful by inhalation, ingestion, or skin absorption. Wear appropriate gloves and safety goggles and use in a chemical fume hood. Do not breathe the dust.

LN$_2$, *see* **Liquid nitrogen**

Magnesium chloride, MgCl$_2$, may be harmful by inhalation, ingestion, or skin absorption. Wear appropriate gloves and safety glasses and use in a chemical fume hood.

Magnesium sulfate, MgSO$_4$, presents chronic health hazards and affects the central nervous system and gastrointestinal tract. It may be harmful by inhalation, ingestion, or skin absorption. Wear appropriate gloves and safety glasses and use in a chemical fume hood.

Maleic acid is toxic and harmful by inhalation, ingestion, or skin absorption. Reaction with water or moist air can release toxic, corrosive, or flammable gases. Do not breathe the vapors or dust. Wear appropriate gloves and safety glasses.

Manganese chloride, MnCl$_2$, may be harmful by inhalation, ingestion, or skin absorption. Wear appropriate gloves and safety glasses and use in a chemical fume hood.

MeOH or H$_3$COH, *see* **Methanol**

β-Mercaptoethanol (2-Mercaptoethanol), HOCH$_2$CH$_2$SH, may be fatal if inhaled or absorbed through the skin and is harmful if ingested. High concentrations are extremely destructive to the mucous membranes, upper respiratory tract, skin, and eyes. β-Mercaptoethanol has a very foul odor. Wear appropriate gloves and safety glasses and always use in a chemical fume hood.

MES, *see* **2-(*N*-Morpholino)ethanesulfonic acid**

Methanol, MeOH or H$_3$COH, is toxic and can cause blindness. It may be harmful by inhalation, ingestion, or skin absorption. Adequate ventilation is necessary to limit exposure to vapors. Avoid inhaling these vapors. Wear appropriate gloves and safety goggles and use only in a chemical fume hood.

2-Methylbutane, *see* **Isopentane**

Methylnadicanhydride (MNA) is corrosive and causes burns. It may be harmful by inhalation, ingestion, or skin absorption. Wear appropriate gloves and safety goggles.

***N*-Methyl-*N*′-nitro-*N*-nitrosoguanidine (MNNG)** is a mutagen and carcinogen. It may be harmful by inhalation, ingestion, or skin absorption. Consult your local safety office for specific handling and disposal procedures. Use extreme caution and avoid breathing the vapors. Perhaps the most dangerous part of handling MNNG is the point at which the bottle of MNNG is opened. When the vapor pressure is released, the solid MNNG crystals and powder can be dispersed and inhaled if precautionary measures are not taken. Therefore, always open a bottle of solid MNNG with protective appropriate gloves under a chemical fume hood. One recommended method of weighing out MNNG to prepare a stock solution is to use sterile glass scintillation vials with plastic screw caps or glass vials used for slants. First, weigh the vial on a Mettler or similar precision balance; typical vials are 16 g or less. Then, under the hood, place a small amount of MNNG in the vial, cover, and weigh again. From the added weight, calculate the weight of MNNG. In this manner, the bottle of MNNG is never opened away from the hood and the solid MNNG is always in a covered container. Do not weigh on paper in a Mettler! To prepare a stock solution, ideally about 35–50 mg of MNNG should be placed in a vial. All materials coming in contact with MNNG should be immersed in a large beaker of 1 N NaOH or laboratory bleach before recycling or disposal.

MgCl$_2$, *see* **Magnesium chloride**

MgSO$_4$, *see* **Magnesium sulfate**

MNA, *see* **Methylnadicanhydride**

MnCl$_2$, *see* **Manganese chloride**

MNNG, *see* ***N*-Methyl-*N*′-nitro-*N*-nitrosoguanidine**

MOPS, *see* **3-(*N*-Morpholino)-propanesulfonic acid**

2-[*N*-Morpholino]ethanesulfonic acid (MES) may be harmful by inhalation, ingestion, or skin absorption. Wear appropriate gloves and safety glasses.

3-(*N*-Morpholino)propanesulfonic acid (MOPS) may be harmful by inhalation, ingestion, or skin absorption. It is irritating to mucous membranes and the upper respiratory tract. Wear appropriate gloves and safety glasses and use in a chemical fume hood.

NaBH$_4$, *see* **Sodium borohydride**

Na$_2$CO$_3$, *see* **Sodium carbonate**

Nadicmethylanhydride, *see* **Methylnadicanhydride**

NaN$_3$, *see* **Sodium azide**

NaOCl, *see* **Bleach**

NaOH, *see* **Sodium hydroxide**

NBT, *see* **4-Nitro blue tetrazolium chloride**

NH$_4$Cl, see **Ammonium chloride**

(NH$_4$)$_6$Mo$_7$O$_{24}$ · 4H$_2$O, see **Ammonium molybdate**

NH$_4$OH, see **Ammonium hydroxide**

7-NI, see **7-Nitroindazole**

Nickel chloride, NiCl$_2$, is toxic and may be harmful by inhalation, ingestion, or skin absorption. Do not breathe the dust. Wear appropriate gloves and safety glasses.

Nickel sulfate, NiSO$_4$, is a possible carcinogen and may cause heritable genetic damage. It is a skin irritant and may be harmful by inhalation, ingestion, or skin absorption. Wear appropriate gloves and safety glasses and use in a chemical fume hood. Do not breathe the dust.

NiCl$_2$, see **Nickel chloride**

NiSO$_4$, see **Nickel sulfate**

4-Nitro blue tetrazolium chloride (NBT) may be harmful by inhalation, ingestion, or skin absorption. Wear appropriate gloves and safety glasses.

Nitrogen (gaseous or liquid) may be harmful by inhalation, ingestion, or skin absorption. Wear appropriate gloves and safety goggles. Contact your local safety office for proper precautions.

7-Nitroindazole (7-NI) is toxic and may impair fertility. It is a possible mutagen and causes severe irritation. It may be harmful by inhalation, ingestion, or skin absorption. Wear appropriate gloves and safety goggles and use only in a chemical fume hood. Do not breathe the dust.

NPG (n-Propyl gallate), see **Benzoic acid**

Optimal cutting temperature (OCT) is composed of polyvinyl alcohol, polyethylene glycol, and dimethyl benzyl ammonium chloride. Follow the manufacturer's guidelines for handling OCT.

Orthophosphoric acid, see **Phosphoric acid**

Paraformaldehyde is highly toxic and may be fatal. It may be a carcinogen. It is readily absorbed through the skin and is extremely destructive to the skin, eyes, mucous membranes, and upper respiratory tract. Avoid breathing the dust or vapor. Wear appropriate gloves and safety glasses and use in a chemical fume hood. Keep away from heat, sparks, and open flame.

Phalloidin is extremely toxic and may be fatal by inhalation, ingestion, or skin absorption. Great care should be taken when using phalloidin. Wear appropriate gloves and safety glasses and use in a chemical fume hood. Do not breathe the dust. Do not use if skin is cut or scratched.

Phenol is extremely toxic and highly corrosive and can cause severe burns. It may be harmful by inhalation, ingestion, or skin absorption. Wear appropriate gloves, goggles, and protective clothing and always use in a chemical fume hood. Rinse any areas of skin that come in contact with phenol with a large volume of water and wash with soap and water; do not use ethanol!

Phenol red may be harmful by inhalation, ingestion, or skin absorption. Wear appropriate gloves and safety glasses and use in a chemical fume hood.

Phenylmethylsulfonyl fluoride (PMSF), C$_7$H$_7$FO$_2$S, is a highly toxic cholinesterase inhibitor. It is extremely destructive to the mucous membranes of the respiratory tract, eyes, and skin. It may be fatal by inhalation, ingestion, or skin absorption. Wear appropriate gloves and safety glasses and always use in a chemical fume hood. In case of contact, immediately flush eyes or skin with copious amounts of water and discard contaminated clothing.

Phleomycin may cause DNA damage and is harmful by inhalation, ingestion, or skin absorption. Wear appropriate gloves and safety goggles and use only in a chemical fume hood. Do not breathe the dust.

Phosphoric acid, H$_3$PO$_4$, is highly corrosive and extremely destructive to the tissue of the mucous membranes and upper respiratory tract, eyes, and skin. It is harmful by inhalation, ingestion, or skin absorption. Wear appropriate gloves and safety glasses. Do not breathe the vapors.

PMSF, see **Phenylmethylsulfonyl fluoride**

Potassium bromide, KBr, may be harmful by inhalation, ingestion, or skin absorption. Wear appropriate gloves and safety glasses and use in a chemical fume hood.

Potassium cacodylate, see **Cacodylate**

Potassium chloride, KCl, may be harmful by inhalation, ingestion, or skin absorption. Wear appropriate gloves and safety glasses.

Potassium hydroxide, KOH, is highly toxic and may be fatal if swallowed. It may be harmful by inhalation, ingestion, or skin absorption and should be handled with great care. Solutions are corrosive and can cause severe burns. Wear appropriate gloves and safety goggles.

Potassium iodide, KI, may be harmful by inhalation, ingestion, or skin absorption. Wear appropriate gloves and safety glasses and use in a chemical fume hood.

Potassium nitrate, KNO$_3$, is a strong oxidizer; contact with other material may cause fire. It is also corrosive and may be harmful by inhalation, ingestion, or skin absorption. Wear appropriate gloves and safety goggles.

2-Propanol, see **Isopropanol**

Propylene oxide is highly flammable and toxic and may be carcinogenic. High concentrations are extremely destructive to the mucous membranes and upper respiratory tract. It may be harmful by inhalation, ingestion, or skin absorption. Wear appropriate gloves and safety glasses and use only in a chemical fume hood. Keep away from heat, sparks, and open flame.

n-Propyl gallate (NPG), see **Benzoic acid**

Proteinase K is an irritant and may be harmful by inhalation, ingestion, or skin absorption. Wear appropriate gloves and safety glasses.

Resins are flammable and are suspected carcinogens. The unpolymerized components and dusts may cause toxic reactions, including contact allergies with long-term exposure. Avoid breathing the vapors and dust. Wear appropriate gloves and safety goggles and always use in a chemical fume hood. Sensitivity to these chemicals may develop with repeated contact. Keep away from heat, sparks, and open flame.

Rhodamine phalloidin, *see* **Phalloidin**

Sarkosyl, *see* **Sodium *N*-lauroylsarcosinate**

SDS, *see* **Sodium dodecyl sulfate**

Serotonin may be harmful by inhalation, ingestion, or skin absorption. Wear appropriate gloves and safety glasses. Do not breathe the dust. Overexposure may cause reproductive disorders.

Silica is an irritant and may be harmful by inhalation, ingestion, or skin absorption. Wear appropriate gloves and safety glasses. Do not breathe the dust.

Silver methenamine is corrosive and causes burns. It is harmful by inhalation, ingestion, or skin absorption. Wear appropriate gloves and safety goggles and use in a chemical fume hood. Do not breathe the dust.

Silver nitrate, AgNO$_3$, is a strong oxidizing agent and should be handled with care. It may be harmful by inhalation, ingestion, or skin absorption. Avoid contact with skin. Wear appropriate gloves and safety glasses. It can cause explosions upon contact with other materials.

SnCl$_2$, *see* **Stannous chloride**

Sodium azide, NaN$_3$, is highly poisonous. It blocks the cytochrome electron transport system. Solutions containing sodium azide should be clearly marked. It may be harmful by inhalation, ingestion, or skin absorption. Wear appropriate gloves and safety goggles and handle with great care. Sodium azide is an oxidizing agent and should not be stored near flammable chemicals.

Sodium borohydride, NaBH$_4$, is corrosive and causes burns. It may be harmful by inhalation, ingestion, or skin absorption. Wear appropriate gloves and safety goggles and use in a chemical fume hood.

Sodium cacodylate may be carcinogenic and contains arsenic. It is highly toxic and may be fatal by inhalation, ingestion, or skin absorption. It also may cause harm to an unborn child. Effects of contact or inhalation may be delayed. Do not breathe the dust. Wear appropriate gloves and safety goggles and use only in a chemical fume hood. *See also* **Cacodylate**.

Sodium carbonate, Na$_2$CO$_3$, may be harmful by inhalation, ingestion, or skin absorption. Wear appropriate gloves and safety glasses and use in a chemical fume hood.

Sodium citrate, *see* **Citric acid**

Sodium dodecyl sulfate (SDS) is toxic, an irritant, and poses a risk of severe damage to the eyes. It may be harmful by inhalation, ingestion, or skin absorption. Wear appropriate gloves and safety goggles. Do not breathe the dust.

Sodium hydroxide, NaOH, is highly toxic and caustic and should be handled with great care. Wear appropriate gloves and a face mask. All other concentrated bases should be handled in a similar manner.

Sodium hypochlorite, NaOCl, *see* **Bleach**

Sodium *N*-lauroylsarcosinate may be harmful by inhalation, ingestion, or skin absorption. Wear appropriate gloves and safety glasses. Do not breathe the dust.

Spermidine may be corrosive and cause severe eye and skin burns. It may be harmful by inhalation, ingestion, or skin absorption. Effects may be delayed. Wear appropriate gloves and safety goggles and use in a chemical fume hood.

Spermine may be corrosive and cause severe eye and skin burns. It may be harmful by inhalation, ingestion, or skin absorption. Effects may be delayed. Wear appropriate gloves and safety goggles and use in a chemical fume hood.

Stannous chloride, SnCl$_2$, is corrosive and causes burns. It is destructive to the mucus membranes and upper respiratory tract and may be harmful by inhalation, ingestion, or skin absorption. Wear appropriate gloves and safety goggles. Do not breathe the dust.

Streptomycin is toxic and a suspected carcinogen and mutagen. It may cause allergic reactions. It may be harmful by inhalation, ingestion, or skin absorption. Wear appropriate gloves and safety glasses.

Sulfadiazine is an irritant and may be harmful by inhalation, ingestion, or skin absorption. Wear appropriate gloves and safety goggles. Do not breathe the dust.

TCA, *see* **Trichloroacetic acid**

TEA, *see* **Triethanolamine**

Tetracycline may be harmful by inhalation, ingestion, or skin absorption. Wear appropriate gloves and safety glasses and use in a chemical fume hood.

11-Tetradecenyl acetate is combustible and may be harmful by inhalation, ingestion, or skin absorption. Wear appropriate gloves and safety goggles.

Tetramethylrhodamine isothiocyanate (TRITC) may be harmful by inhalation, ingestion, or skin absorption. Wear appropriate gloves and safety glasses.

Thiourea may be carcinogenic and harmful by inhalation, ingestion, or skin absorption. Wear appropriate gloves and safety glasses and use in a chemical fume hood.

Tricaine methane sulfonate is an irritant and may be harmful by inhalation, ingestion, or skin absorption. Wear appropriate gloves and safety glasses.

Trichloroacetic acid (TCA) is highly caustic. Wear appropriate gloves and safety goggles.

1,1,1-Trichloro-2-methyl-2-propanol may be harmful by inhalation, ingestion, or skin absorption. Wear appropriate gloves and safety goggles. Do not breathe the dust.

Triethanolamine (TEA) may be harmful by inhalation, ingestion, or skin absorption. Wear appropriate gloves and safety glasses and use only in a chemical fume hood.

Triethylamine is highly toxic and flammable. It is extremely corrosive to the mucous membranes, upper respiratory tract, eyes, and skin. It may be harmful by inhalation, ingestion, or skin absorption. Wear appropriate gloves and safety glasses and use in a chemical fume hood. Keep away from heat, sparks, and open flame.

Tris is an irritant and may be harmful by inhalation, ingestion, or skin absorption. Wear appropriate gloves and safety glasses.

Tris-(hydroxymethyl)aminomethane, *see* **Tris**

TRITC, *see* **Tetramethylrhodamine isothiocyanate**

Triton X-100 causes severe eye irritation and burns. It may be harmful by inhalation, ingestion, or skin absorption. Wear appropriate gloves and safety goggles. Do not breathe the vapor.

TRIzol may be fatal if absorbed through the skin, inhaled, or swallowed. It can also cause severe burns. Wear appropriate gloves, safety goggles, and protective clothing and always use in a chemical fume hood. Rinse any areas of skin that come in contact with TRIzol with a large volume of water and wash with soap and water; do not use ethanol!

Trypsin may cause an allergic respiratory reaction. It may be harmful by inhalation, ingestion, or skin absorption. Do not breathe the dust. Wear appropriate gloves and safety goggles. Use with adequate ventilation.

Tryptophan may be harmful by inhalation, ingestion, or skin absorption. Wear appropriate gloves and safety glasses.

Urea may be harmful by inhalation, ingestion, or skin absorption. Wear appropriate gloves and safety glasses.

UV light and/or **UV radiation** is dangerous and can damage the retina. Never look at an unshielded UV light source with naked eyes. Examples of UV light sources that are common in the laboratory include handheld lamps and transilluminators. View only through a filter or safety glasses that absorb harmful wavelengths. UV radiation is also mutagenic and carcinogenic. To minimize exposure, make sure that the UV light source is adequately shielded. Wear protective appropriate gloves when holding materials under the UV light source.

Valine may be harmful by inhalation, ingestion, or skin absorption. Wear appropriate gloves and safety glasses.

Vancomycin may cause allergic reactions. It is an irritant and may be harmful by inhalation, ingestion, or skin absorption. Wear appropriate gloves and safety glasses. Do not breathe the dust.

Xylazine may be harmful by inhalation, ingestion, or skin absorption. Wear appropriate gloves and safety glasses.

Zinc sulfate, $ZnSO_4$, may be harmful by inhalation, ingestion, or skin absorption. Wear appropriate gloves and safety glasses.

$ZnSO_4$, *see* **Zinc sulfate**

Index

A

Acanthamoeba, 33. *See also Dictyostelium discoideum*
Acanthoecidae, 2. *See also* choanoflagellate
Achaearanea tepidoriorum (spider), 349. *See also Cupiennius salei*
Actinopterygii (ray-finned fishes). *See Polypterus* (bichir)
Acytostelium, 31. *See also Dictyostelium discoideum*
African butterfly. *See Bicyclus anynana*
Alectrion obsoleta. *See Ilyanassa obsoleta* (snail)
American Wandering Spider. *See Cupiennius salei*
Amoebazoa, 31. *See also Dictyostelium discoideum*
Amphimedon queenslandica (sponge)
    background information, 140–141
    genetics, genomics, associated resources, 143
    model system uses, 142–143
    protocols
        cell movement analysis, 155–157
        genotyping, 158–160
        isolation of developmental material, 144–147
        recipes, 161–164
        whole-mount in situ hybridization, 148–154
    related species, 141–142
    sources and husbandry, 141
Antirrhinum majus (snapdragon)
    background information, 105–106
    genetics, genomics, associated resources, 109–110
    model system uses
        biochemistry and pollination, 36
        development studies, 108–109
        ecology and population genetics, 109
    protocols
        cultivation, 111–113
        propagation, 114–116
        recipes, 117
    related species, 106–108
    sources and husbandry, 106
Antirrhinum pseudomajus, 107. *See also Antirrhinum majus*

Arabidopsis, 108. *See also Antirrhinum majus*
Arthropod development study. *See Parhyale hawaiensis* (crustacean)
Astyanax mexicanus (blind Mexican cave fish)
    background information, 469–470
    genetics and genomics resources, 473
    model system uses, 472–473
    protocols
        breeding technique, 476
        handling of eggs and fry, 479
        recipes, 480
        sex determination, 474–475
        in vitro fertilization, 477–478
    sources and husbandry, 470–472

B

Bat. *See Carollia perspicillata*
Beroe ovata, 169, 170
Bichir. *See Polypterus*
Bicyclus anynana (African butterfly)
    background information, 292–294
    behavior and life-history evolution studies, 296–297
    evolution and development studies, 294–296
    genetic resources, genomics and transgenesis, 298
    genetic resources, populations, 297–298
    protocols
        adult pheromones analysis, 316–317
        body composition analysis, 318–319
        CO_2 measurement, 320–321
        culture and propagation of populations, 299–300
        dissection of wings, 307–308
        fixation and dissection of embryos, 305–306
        hemolymph extraction, 322–323
        immunohistochemistry staining of embryos, 312–313
        immunohistochemistry staining of wing discs, 314–315

Bicyclus anynana (African butterfly) (*continued*)
    injection of chemicals into pupae, 324
    recipes, 325–328
    in situ hybridization of embryos and wings, 309–311
    surgical manipulations on wings, damage and cauteries, 301–302
    surgical manipulations on wings, grafts, 303–304
  related species, 294
  sources and husbandry, 294
Blind Mexican cave fish. *See Astyanax mexicanus*
Butterfly. *See Bicyclus anynana*

C

Caenorhabditis elegans, 275. *See also Pristionchus pacificus* (nematode)
Callorynchus milii (chimera), 434. *See also Scyliorhinus canicula* (spotted dogfish)
Camarhynchus. *See* Darwin's finches
Carollia perspicillata (short-tailed fruit bat)
  background information, 519–520
  genetics, genomics, associated resources, 523–524
  husbandry, 521–522
  protocols
    cartilage and bone staining with alcian blue/alizarin red, 546–547
    cartilage staining with alcian blue, 544–545
    collection from the wild, 531–532
    embryo collection, 533–534
    embryo fixation and storage, 535–536
    feeding, 525–527
    recipes, 551–554
    timed pregnancies generation, 528–530
    whole-mount immunohistochemistry, 548–550
    whole-mount in situ hybridization, 537–543
  reproductive biology and breeding studies, 522–523
Ceratodon purpureus, 71–72. *See also Physcomitrella patens*
Certhidea. *See* Darwin's finches
Choanoflagellate, 1–2. *See also Monosiga breviocollis*
Chondrichthyans. *See Scyliorhinus canicula*
Codosigidae, 2. *See also* choanoflagellate
Comb jellies. *See Mnemiopsis leidyi*
Crustaceans. *See Parhyale hawaiensis*
Ctenophora. *See Mnemiopsis leidyi*
Cupiennius salei (American Wandering Spider)
  background information, 347–349
  development and behavior studies, 349
  genetics, 350
  protocols
    cell-death detection using TUNEL, 356–358
    cell proliferation detection, 362–365
    collection and fixation of embryos, 351–352
    embryo dissection, 366–369
    gene silencing using embryonic RNAi, 359–361
    recipes, 370–372
    whole-mount in situ hybridization, 353–355
  related species, 349
  sources and husbandry, 349

D

Darwin's finches
  background information, 481–483
  genetics, genomics, associated resources, 486
  model system uses, 483–486
  protocols
    embryo collection, 487–489
    microarray analysis of beak mRNA, 494–497
    recipes, 498–499
    in situ hybridization analysis of beak tissue, 490–493
  sources and husbandry, 483
Dictyostelia, 31. *See also Dictyostelium discoideum*
Dictyostelium discoideum
  gene nomenclature guidelines, 37
  genetics, 36–37
  genomics resources, 38
  history of study, 31–32
  model system uses
    cytoskeleton function, 34–35
    development and pattern formation, 34
    intraspecies and interspecies interactions, 36
    molecular basis of human disease, 35–36
  natural history and distribution, 29–30
  organism description, 30
  protocols
    DNA extraction, 56–58
    electroporation, 52–53
    growth and maintenance of cells, 39–42
    multicellular development, 43–45
    permanent stocks creation, 46–48
    plasmid DNA used to transform, 49–51
    recipes, 62–66
    RNA extraction, 59–61
    transformants selection, 54–55
  related species, 33–34
  sources and husbandry, 32–33
  strain nomenclature guidelines, 37
  taxonomic information, 30–31
Diplogastridae, 277. *See also Pristionchus pacificu* (nematode)
Dogfish. *See Scyliorhinus canicula* (spotted dogfish)
Drosophila melanogaster, 331. *See also Gryllus bimaculatus* (two-spotted cricket)
Dugesia japonica (planarians), 196, 197. *See also Schmidtea mediterranea* (planarian)
Dugesiidae, 196. *See also Schmidtea mediterranea* (planarian)

E

Edwardsiella lineata, 175
Embryonic development studies. *See Ilyanassa obsoleta* (snail)
Ephydatia, 142

F

Finches. *See* Darwin's finches
Fruit bat. *See Carollia perspicillata*

G

Geodia, 142
Glossophaga soricina, 522. *See also Carollia perspicillata* (short-tailed fruit bat)
Glycosylation disorder, 35
Gryllus bimaculatus (two-spotted cricket)
    background information, 331–332
    genetics and genomics resources, 343
    model system uses
        body size regulation, 339
        calling song analysis, 341–342
        circadian rhythms, 342
        embryogenesis, 333–335
        learning and memory, 343
        leg regeneration, 339–341
        morphological change, 336–339
        segment patterning, 335–336
    related species, 332–333
    RNAi analysis, 344
    sources and husbandry, 332
    transgenesis, 344
    whole-mount in situ hybridization, 343

H

Helobdella (leech)
    background information, 245–246
    genetics and genomics resources, 248
    model system uses
        neurogenesis, 248
        stem cell biology and regeneration, 247
    protocols
        devitellinization of living embryos, 256–257
        embryo handling, 249–250
        immunostaining, 260–262
        microinjection, 251–255
        recipes, 270–272
        silver staining, 258–259
        in situ hybridization, 263–267
        whole-mount preparation for microscopy, 268–269
    related species, 247
    sources and husbandry, 246–247

Heteropsis iboina, 294. *See also Bicyclus anynana*
Heterostelids, 31. *See also Dictyostelium discoideum*

I

Ilyanassa obsoleta (snail)
    background information, 219–220
    genetics and genomics resources, 222
    model system uses, 221
    protocols
        embryos and larvae fixation, 233–234
        genomic DNA isolation, 235–236
        larval metamorphosis induction, 227–229
        obtaining embryos, 223–226
        pressure injection of embryos, 230–232
        protein isolation from embryos, 237–238
        recipes, 239–241
    related species, 220–221
    sources and husbandry, 220
    technical approaches overview, 222

J

Japanese quail
    background information, 501–502
    genetics and genomics resources, 504
    model system uses
        developmental biology, 503
        human health and disease, 504
    protocols
        generation of high-titer lentivirus, 506–509
        injection of lentivirus into eggs, 510–513
        recipes, 517
        screening for germ-line transmission, 514–516
    related species, 503
    sources and husbandry, 502–503
    technical approaches overview, 505

K

Kickxiella, 107. *See also Antirrhinum majus*

L

Leeches. *See Helobdella*
Leucetta, 142
Lobata. *See Mnemiopsis leidyi*
Locusta migratoria (grasshopper), 332. *See also Gryllus bimaculatus*
Lophotrochozoa. *See Helobdella* (leech); *Ilyanassa obsoleta* (snail)
Lycopersicon (tomato). *See Solanum lycopersicum*

M

Mexican tetra. *See Astyanax mexicanus*
Misopates, 107. *See also Antirrhinum majus*
Mnemiopsis leidyi (comb jellies)
    background information, 167–168
    genetics, genomics, associated resources, 175
    model system uses
        feeding, 172–173
        morphology, 172
        muscles, 173–174
        nerves and sensory structures, 174–175
        population growth, 175
        reproduction and development, 169–171
    protocols
        recipes, 191–193
        spawning and embryo collection, 177–178
        tissue preparation and DNA extraction, 186–188
        tissue preparation and RNA extraction, 189–190
        whole-mount antibody staining, 179–181
        whole-mount in situ hybridization, 182–185
    related species, 169
    sources and husbandry, 168–169
Monodelphis domestica (opossum)
    background information, 557
    genetics, genomics, associated resources, 559
    model system uses, 558–559
    protocols
        harvesting embryos, 564–565
        maintenance and breeding, 561–563
        recipes, 572–574
        whole-embryo culture, 566–567
        whole-mount in situ hybridization, 568–571
    related species, 558
    sources and husbandry, 558
    technical approaches overview, 559–560
Monosiga brevicollis (choanoflagellate)
    background information, 1–2
    genomics resources, 4
    model system uses, 4
    protocols
        clonal cultures establishment, 5–7
        culture start and maintenance, 8–9
        genomic DNA preparation, high-molecular weight, 18–20
        genomic DNA preparation, rapid, 16–17
        genomic DNA separation from bacterial prey DNA, 21–24
        recipes, 25–27
        RNA preparation, 14–15
        single-cells isolation, 5–7
        storage of cultures, long-term frozen, 10
        subcellular localization of actin, β-tubulin, DNA, 11–13
    related species, 3
    sources and husbandry, 3
Monosiga ovata (choanoflagellate), 4
Moss. *See Physcomitrella patens*
Mycetozoa, 31. *See also Dictyostelium discoideum*

N

Nassa obsoleta. *See Ilyanassa obsoleta* (snail)
Nassarius obsoletus. *See Ilyanassa obsoleta* (snail)
Nassarius reticulatus. *See Ilyanassa obsoleta* (snail)
Nematode. *See Pristionchus pacificus*

O

Oncopeltus fasciatus (milk-weed bug), 332. *See also Gryllus bimaculatus*
Oopsacas, 142
Opossum. *See Monodelphis domestica*
Oscarella, 142
Osteichthyes. *See Polypterus* (bichir)

P

Parhyale hawaiensis (crustacean)
    background information, 374–375
    genetics, genomics, associated resources, 382–383
    model system uses
        body plan, 376–378
        developmental staging, 378–382
    protocols
        antibody staining of embryos, 393–395
        blastomere injection with labeled tracers, 389–392
        embryo fixation and dissection, 384–388
        recipes, 400–403
        in situ hybridization of labeled RNA probes, 396–399
    related species, 376
    sources and husbandry, 375–376
Parvisporids, 31. *See also Dictyostelium discoideum*
Petromyzon marinus (sea lampreys)
    background information, 405–406
    genetics, genomics, associated resources, 409
    model system uses, 407–408
    protocols
        DiI cell labeling, 415–416
        embryo culturing, 410–412
        immunostaining of embryos, 421–424
        injection of RNA and morpholino oligos into embryos, 413–414
        recipes, 425–428
        whole-mount in situ hybridization, 417–420
    related species, 406–407
    sources and husbandry, 406
Physarum, 33. *See also Dictyostelium discoideum*

Physcomitrella patens (moss)
    background information, 70–71
    genetics, genomics, and resources, 73
    model system uses, 72
    protocols
        culturing methods, 75–79
        isolation and regeneration of protoplasts, 80–81
        isolation of DNA, RNA, and proteins, 93–96
        mutagenesis of spores and protonemal tissue, 84–86
        PEG-induced protoplast fusion hybridization, 82–83
        recipes, 97–102
        somatic hybridization, 82–83
        transformation of gametophytes, 91–92
        transformation using direct DNA uptake, 87–88
        transformation using T-DNA mutagenesis, 89–90
    related species, 71–72
    sources and husbandry, 71
    technical approaches overview, 73–74
Pinaroloxias. *See* Darwin's finches
Planarians. *See Schmidtea mediterranea*
Platyctenes, 169
Platyspiza. *See* Darwin's finches
Pleurobrachia pileus, 169
Polypterus (bichir)
    background information, 447–450
    genetics and genomics resources, 457
    model system uses
        early cleavage stages, 452–453
        gastrulation, 454–457
    protocols
        microinjection and animal cap assay, 458–459
        recipes, 463–465
        whole-mount in situ hybridization, 460–462
    related species, 452
    sources and husbandry, 450–452
Polypterus endlicheri. *See Polypterus* (bichir)
Polypterus senegalus. *See Polypterus* (bichir)
Polysphondylium, 31. *See also Dictyostelium discoideum*
Pristionchus pacificus (nematode)
    background information, 275–277
    genetics and genomics resources, 282–283
    model system uses
        bacterial interactions, 281–282
        beetle interactions and biogeography, 280–281
        behavior and chemoattraction, 282
        vulva development, 277–280
    protocols
        isolation from beetles, 284–285
        olfactory response assessment, 286–287
        recipes, 288
    related species, 277
    sources and husbandry, 277
Proterospongia sp. (choanoflagellate), 4

Q

Quail. *See* Japanese quail

R

Raja erinaceai (little skate), 434. *See also Scyliorhinus canicula* (spotted dogfish)
Ray-finned fishes (Actinopterygii). *See Polypterus* (bichir)
Rhizostelids, 31. *See also Dictyostelium discoideum*

S

Sairocarpus, 107. *See also Antirrhinum majus*
Salpingoecidae. *See also* choanoflagellate
Schistocerca sp. (grasshopper), 332. *See also Gryllus bimaculatus*
Schmidtea mediterranea (planarian)
    background information, 195–197
    genetics and genomics resources, 199–200
    model system uses
        memory, learning, and behavior, 199
        stem cells and germ cells, 198–199
        tissue regeneration, cellular turnover, and aging, 197–198
    protocols
        establishing and maintaining colonies, 201–205
        gene knockdown using RNAi, 206–209
        live imaging, 210–214
        recipes, 215
    related species, 197
    sources and husbandry, 197
    technical approaches overview, 200
Scyliorhinidae. *See Scyliorhinus canicula* (spotted dogfish)
Scyliorhinus canicula (spotted dogfish)
    background information, 432–433
    genetics, genomics, associated resources, 440–443
    model system uses
        body plan, 435–439
        physiological systems, origin and evolution of, 439–440
    related species, 434
    sources and husbandry, 433–434
    technical approaches, 443–444
Sea lampreys. *See Petromyzon marinus*
Short-tailed fruit bat. *See Carollia perspicillata*
Snail. *See Ilyanassa obsoleta*
Snapdragon. *See Antirrhinum majus*
Social amebas. *See Dictyostelium discoideum*
Solanum lycopersicum (tomato)
    background information, 119–121
    genetics and genomics resources, 125–126
    model system uses
        compound leaf development, 124–125
        floral system and plant architecture, 124

Solanum lycopersicum (tomato) (*continued*)
    fruit characteristics, 122–124
    protocols
        crossing, 130
        grafting plants, 131
        growing, 128–129
        recipes, 135–136
        transformation, 132–134
    related species, 121–122
    sources and husbandry, 121
Spider. *See Cupiennius salei*
Sponges. *See Amphimedon queenslandica*
Spongilla, 142
Spotted dogfish. *See Scyliorhinus canicula*

Squalus acanthias (spiny dogfish shark), 434. *See also Scyliorhinus canicula*
Streptosepalum, 108. *See also Antirrhinum majus*
Suberites, 142
Sycon, 142

T

Tegenaria atrica (spider), 349. *See also Cupiennius salei*
Tomato. *See Solanum lycopersicum*
Tortula ruralis, 71–72. *See also Physcomitrella patens*
Tribolium castaneum (beetle), 332. *See also Gryllus bimaculatus*
Two-spotted cricket. *See Gryllus bimaculatus*